Handbook of Financial Stress Testing

Stress tests are the most innovative regulatory tool to prevent and fight financial crises. Their use has fundamentally changed the modeling of financial systems, financial risk management in the public and private sector, and the policies designed to prevent and mitigate financial crises. When financial crises hit, stress tests take center stage. Despite their centrality to public policy, the optimal design and use of stress tests remains highly contested.

Written by an international team of leading thinkers from academia, the public sector, and the private sector, this handbook comprehensively surveys and evaluates the state of play and charts the innovations that will determine the path ahead. It is a comprehensive and interdisciplinary resource that bridges theory and practice and places financial stress testing in its wider context. This guide is essential reading for researchers, practitioners, and policymakers working on financial risk management and financial regulation.

J. DOYNE FARMER is Director of Complexity Economics at the Institute for New Economic Thinking at the Oxford Martin School, and is the Baillie Gifford Professor at Mathematical Institute at the University of Oxford, as well as an External Professor at the Santa Fe Institute. His current research is in economics, including financial stability, sustainability, technological change, and economic simulation.

ALISSA M. KLEINNIJENHUIS is a Research Scholar at the Stanford Institute for Economic Policy Research (SIEPR), at Stanford University. She is also a Senior Research Fellow at the Institute for New Economic Thinking at the Oxford Martin School, at the University of Oxford. Her focal areas of research are financial crises and climate finance, linked by their emphasis on addressing externalities emerging from too-big-to-fail (or too-many-to-fail) financial institutions and climate change. Subtopics of special relevance in her studies are financial regulation, models of contagion and systemic risk, financial stress testing, financial intermediation, monetary policy, asset pricing, and climate financial risks and opportunities.

TIL SCHUERMANN is a partner at Oliver Wyman where he advises private and public sector clients on stress testing, capital planning, enterprise-wide risk management, model risk management, climate risk, and governance including board effectiveness.

THOM WETZER is Associate Professor of Law and Finance at the University of Oxford, Founding Director of the Oxford Sustainable Law Programme, and Senior Research Fellow at the Institute for New Economic Thinking at the Oxford Martin School. His research examines how law and finance can generate value and advance the public good, with a particular focus on financial regulation, corporate governance, financial risk management, and climate risk.

"This well-documented compendium on financial stress testing could not arrive at a more timely moment. As the world embarks on a daunting mission to reign in global warming, stress testing promises to be a key tool for helping central banks and supervisors assess climate-related risks, not only on their own balance sheets, but in the economy as a whole, as well as the books of the banks it supervises."

Christine Lagarde, President of the European Central Bank

"Stress tests have grown from their beginnings as a simple, practical tool for communicating risk in a particular portfolio into a much broader framework. This Handbook is a thought-provoking package of 30 essays by leading academics, regulators, and practitioners. Topics range widely, from fundamental scenario design to transparency considerations, feedback effects, micro versus macro perspectives, as well as the implications for different types of financial institutions. The Handbook has something for anyone interested in the state of the art, including risk professionals, regulators, policymakers, and academics."

Wilson Ervin, Former Chief Risk Officer

"The 2009 bank stress tests were one of the turning points of the global financial crisis, and they are now a basic part of the supervisory toolkit. This volume provides a comprehensive overview of what we have learned about stress testing and what we still need to know to make it even more effective."

Ben Bernanke, Former Chair of the United States Federal Reserve

"This excellent compilation of contributions on stress testing covers a vast spectrum ranging from the economic history of evolution of stress tests as a centerpiece in prudential regulation of the financial sector to the challenges going forward. The book covers both micro- and macro-prudential stress tests, provides a conceptual foundation for the use of both, touches upon ongoing issues such as stress tests for central counterparties, and is a must-read for practitioners, policymakers, and academics interested in creating a robust financial sector."

Viral Acharya, New York University Stern School of Business

"This is by far the most comprehensive available reference work on financial stress testing. One need only review the list of contributors to appreciate its definitive quality—many of those who invented stress testing following the [Global] Financial Crisis are authors. The *Handbook of Financial Stress Testing* covers the subject in all key dimensions, and from philosophy to execution. I strongly recommend that this magnificent compendium be read by anyone concerned with the risk of future financial crises."

Darrell Duffie, Stanford University

Handbook of Financial Stress Testing

EDITED BY

J. Doyne Farmer
University of Oxford

Alissa M. Kleinnijenhuis
Stanford University

Til Schuermann
Oliver Wyman

Thom Wetzer
University of Oxford

Foreword by
Timothy F. Geithner

CAMBRIDGE
UNIVERSITY PRESS

CAMBRIDGE
UNIVERSITY PRESS

University Printing House, Cambridge CB2 8BS, United Kingdom

One Liberty Plaza, 20th Floor, New York, NY 10006, USA

477 Williamstown Road, Port Melbourne, VIC 3207, Australia

314–321, 3rd Floor, Plot 3, Splendor Forum, Jasola District Centre, New Delhi – 110025, India

103 Penang Road, #05–06/07, Visioncrest Commercial, Singapore 238467

Cambridge University Press is part of the University of Cambridge.

It furthers the University's mission by disseminating knowledge in the pursuit of
education, learning, and research at the highest international levels of excellence.

www.cambridge.org
Information on this title: www.cambridge.org/9781108830737
DOI: 10.1017/9781108903011

First published 2022

Printed in the United Kingdom by TJ Books Limited, Padstow Cornwall

A catalogue record for this publication is available from the British Library.

Library of Congress Cataloging-in-Publication Data
Names: Farmer, J. Doyne, editor
Title: Handbook of financial stress testing / [edited by] J. Doyne Farmer,
University of Oxford, Alissa M. Kleinnijenhuis, University of Oxford,
Til Schuermann, Oliver Wyman Limited, Thom Wetzer, University of Oxford.
Description: Cambridge, United Kingdom ; New York, NY : Cambridge
University Press, 2022. | Includes bibliographical references and index.
Identifiers: LCCN 2021041871 (print) | LCCN 2021041872 (ebook) |
ISBN 9781108830737 (hardback) | ISBN 9781108903011 (epub)
Subjects: LCSH: Financial risk management. |
BISAC: MATHEMATICS / Applied Classification: LCC HG173 .H36 2022 (print) |
LCC HG173 (ebook) | DDC 658.15/5–dc23
LC record available at https://lccn.loc.gov/2021041871
LC ebook record available at https://lccn.loc.gov/2021041872

ISBN 978-1-108-83073-7 Hardback

For Rafa

Contents

Contributors

Tobias Adrian
International Monetary Fund

Nicola Anderson
Bank of England

William F. Bassett
Federal Reserve Board

Nathanaël Benjamin
Bank of England

Richard B. Berner
New York University

Anthony Bousquet
European Central Bank

Alex Brazier
Bank of England

Eduardo Canabarro
Barclays

Stephen G. Cecchetti
Brandeis University

Tim P. Clark
Federal Reserve Board (formerly)

Daniel Cope
Oliver Wyman

Christine M. Cumming
Columbia University

Udaibir Das
International Monetary Fund

Kieran Dent
Bank of England

Gonzalo Fernandez Dionis
George Washington University

J. Doyne Farmer
University of Oxford

Robert Engle
New York University

William B. English
Yale University

Mark D. Flood
University of Maryland

John Geanakoplos
Yale University

Timothy F. Geithner
Warburg Pincus; Yale School of Management

Itay Goldstein
University of Pennsylvania

Charles A. E. Goodhart
London School of Economics and Political Science

Marco Gross
International Monetary Fund

Abigail Haddow
Bank of England

Grzegorz Hałaj
Bank of Canada

Andrew Haldane
Bank of England

Jérôme Henry
European Central Bank

Richard J. Herring
Wharton

Greg Hopper
Goldman Sachs

Carey Hsu
Oliver Wyman

David Jacobs
Bank of England

Jonathan Jones
Office of the Comptroller of the Currency

Kathryn Judge
Columbia University

Alissa M. Kleinnijenhuis
Stanford University

Yaron Leitner
Washington University in St. Louis

Iman van Lelyveld
De Nederlandsche Bank

Clinton Lively
Oliver Wyman

Jacques Longerstaey
Nuveen

James Morgan
Oliver Wyman

James Morsink
International Monetary Fund

Patricia C. Mosser
Columbia University

Paul Nahai-Williamson
Bank of England

Brian Peters
American International Group

Matthew Pritsker
Federal Reserve Bank of Boston

Amar Radia
Bank of England

Elena Rancoita
European Central Bank

David E. Rappoport
Federal Reserve Board

Natasha Sarin
University of Pennsylvania

Kermit L. Schoenholtz
New York University

Til Schuermann
Oliver Wyman

Liliana Schumacher
International Monetary Fund

Miguel Segoviano
International Monetary Fund

Evan Sekeris
PNC

Akhtar Siddique
Office of the Comptroller of the Currency

Lawrence H. Summers
Harvard University

Stefan Thurner
Medical University of Vienna

Virginie Traclet
Bank of Canada

Cees Ullersma
De Nederlandsche Bank

Thom Wetzer
University of Oxford

Dawid Żochowski
European Central Bank

Foreword

My first exposure to the power and limits of stress testing came too late, just a year or so into my tenure at the Federal Reserve Bank of New York (FRBNY). I had asked Gerry Corrigan in the fall of 2004 to reprise his post–Long-Term Capital Management (LTCM) Counter party Risk Management Policy Group work with a fresh assessment of the state of risk management in the major financial institutions. His assessment was damning. The vast majority of the major banks and investment banks could not capture, in anything close to real time, the full exposure of the institution to individual counterparties or particular types of risks. And not many of these institutions made more than a cursory attempt to examine the losses they might face in a terrible recession or major financial crisis. We undertook a coordinated effort with the other supervisors, US and non-US, of the major financial institutions to try to remedy those failings, but those efforts came late in the boom and were insufficiently forceful.

Around the same time, within the FRBNY, I started to ask why we were confident that the "8 percent" minimum capital against risk-weighted assets—a ratio that had prevailed as the standard for almost two decades—was "enough" capital and how much would be left if we faced a severe recession. Those discussions led to an early attempt at a system-wide stress test, which produced the observation that with losses two or three times peak losses in the last few recessions, the major banks would be left with thinner but still comfortable levels of capital.

As this history suggests, we approached the end of one of the largest and longest credit booms of the prior half century with a limited appreciation of the limits of the prevailing regulatory capital regime. That regime, as we discovered too late, had many failures, and those failures were magnified by the structure of the US financial system, with a vast and diverse array of financial institutions operating outside the constraints of the bank capital regime and without access to the safety net of deposit assistance and access to the Fed's standing lender-of-last-resort facility for banks.

The regulatory capital regime was limited in scope, applying only to banks, and on the eve of the crisis, banks were responsible for less than half of intermediation in the US financial system. The capital regime that did apply to banks was exacting enough to encourage the migration of risk outside of banks, but it was too weak overall in terms of the level and quality of capital required against risk. As the crisis intensified and the recession deepened, expected losses approached levels that would have consumed the total amount of common equity in the major banks, which worked to accelerate the run on the system. The capital regime was not just too thin to allow banks to withstand the effects of the near collapse of much of the rest of the financial system but also too thin to allow banks to expand to compensate for the reduction in lending capacity as a diverse mix of nonbank institutions came under acute pressure.

The bank capital regime provided little warning of the approaching abyss. It was not forward-looking in capturing the increasing probability of large expected losses. And in part because of this, the authority available to supervisors was not used effectively to preserve or build capital in the quarters leading up to the acute phases of the crisis. It is worth noting that no market-based or other indicators of bank capital were any more effective in foreshadowing the approaching fear of systemic insolvency of the financial system.

How significant was the regulatory failure in the design of the capital regime? It was obviously substantial. Actual gross losses during the crisis totaled about $800 billion relative to about $2.4 trillion in capital at banks and broker-dealers at year-end 2007. For the largest banks and investment banks, the actual erosion in Tier 1 common capital ratios due to losses net of income averaged a bit over 300 basis points, but with a wide range from over 1,000 basis points for the weakest bank to almost 0 for the strongest. Market-based expectations of losses at the peak of the crisis were substantially higher than actual losses, reflecting the collapse in liquidity and the fear of a much-worse recession. If you applied the tougher risk weights of the postcrisis capital regime to the precrisis balance sheets of the major banks, their reported regulatory capital ratios would have been around 200 basis points weaker on average. Total public capital injections plus the required capital raises for the major banks in the crisis totaled $350 billion, which was about 500 basis points relative to risk-weighted assets in the middle of 2008.

These estimates, although large, understate the scale of potential losses that would have occurred without the full power of the other policy actions deployed by the United States in the crisis. Without the aggressive fiscal and monetary policy response, the dramatic expansion of explicit and implicit guarantees of liabilities, the broader purchase and funding facilities that put a floor under asset prices, and the government-sponsored enterprise (GSE) and housing market interventions, the recession would have been much more severe, there would have been an even larger number of failures across the financial system, and losses for the core of the financial system would have been much higher.

The "stress test" adopted by the United States in early 2009 played an important role, although not the decisive role, in the resolution of the crisis. It proved, beyond our expectations, an effective way to complete the recapitalization of the US financial system with a very modest additional use of public resources. Critical to its effectiveness were several aspects of the design. The requirement was set in terms of common equity, rather than capital ratios, to limit the risk that banks would cut assets further, which would have worsened the recession. The loss estimates were conservative relative to market estimates. The results were disclosed to the public with a fair amount of detail by institution. We made it clear that we would provide public capital, if the firms were unable to raise private capital, and set the price for that capital, which helped put a floor on bank equity prices. The capital backstop helped reinforce our commitment to not allow another run on the system to jeopardize the immediate survival of any of the major banks. The capital backstop allowed the creditors of banks to be more confident in funding them, and it allowed equity investors to be more confident that they could purchase bank equity again and be compensated for the risks in doing so.

The fact that we were able to recapitalize the US financial system so quickly with so little public money—and earned a substantial financial return on those public equity investments—was possible because of the forcefulness of the US macroeconomic policy response, with the Fed's asset purchase program and the fiscal stimulus; the Federal Deposit Insurance Corporation's guarantee of the liabilities of the banking system; the Treasury's guarantee of money market funds; and the array of funding backstops to the commercial paper, asset-backed securities, and other markets provided by the Fed, in some case with support from the Treasury.

Observers have spent some time since the crisis debating why the stress test "worked" despite the minimal amounts of incremental capital needs identified in May 2009. It worked because it was credible and well designed, but perhaps more importantly, it worked because of the force of these other policy measures in limiting the depth of the recession and containing the collapse of the broader financial system. Together, the US strategy was to recapitalize the financial system quickly and definitively and by doing so, to limit the risk of a deeper recession, and in parallel, to use the full Keynesian arsenal of the United States to limit the severity of a recession, which in turn improved the effectiveness of the other financial programs in preventing the collapse of the financial system.

The most important lessons of the financial crisis are not just about capital and stress testing. The resilience of the financial system depends not just on the level and distribution of common equity in the entire financial system but on the strength of the Keynesian arsenal and the power of the funding protections provided by the central banks and the government to the financial system as a whole. Capital was necessary but not sufficient.

No capital regime, even with the more systematic integration of stress testing, can fully compensate for the weakness of these other policy tools. But a capital regime built on a foundation of tough, independently designed stress tests can provide a greater margin of safety in future crises. By forcing the financial system to hold more capital against the severe recession, we should be able to reduce the probability and severity of a future crisis. And when we next face a major crisis, stress testing can be used as part of a strategy to prevent the collapse of the financial system and restore it to health more quickly.

We live in an uncertain and challenging world, always vulnerable and with a limited capacity to predict and preempt crisis. This makes it critically important to maintain a strong Keynesian macroeconomic policy arsenal, a full range of emergency financial authorities and tools, and strong capital defenses. Stress testing should remain a critical part of the defenses of the major financial systems.

Timothy F. Geithner
November 15, 2019

1

Introduction and Overview

J. Doyne Farmer, Alissa M. Kleinnijenhuis, Til Schuermann, and Thom Wetzer

As we looked back on the 10th anniversary of the financial crisis, and in particular the 10-year anniversary of the 2009 US bank stress test, we thought it a good time to assemble both a retrospective and current state of financial stress testing.

Stress testing, although hardly new, was applied like never before during the Global Financial Crisis (GFC), which is roughly dated 2007–2009, although it stretched out in Europe. At a time of substantial market turmoil and significant uncertainty about the health of individual banks and thus the banking system, assessing banks' resilience against a specific and clear macroeconomic and market scenario turned out to be a very useful mechanism for providing transparency into which banks needed capital (and how much) and which did not. Sufficiently rich disclosure was needed to allow the market to assess the credibility of the results.

With the success of crisis or wartime stress testing, this approach also became the tool of choice in peacetime by regulators and bank risk managers alike. Of course, we did not expect when we started this project that 2020 would bring a new but different crisis, namely, the COVID-19 pandemic. The chapters were all in their final draft form when the pandemic hit, but the tools, methods, ideas, and approaches described in this handbook are being used widely to help manage the current crisis. The banking system is undoubtedly better prepared than it would otherwise have been, and the stress tests conducted during 2020 bore this out. Future innovations in stress test design might help regulators to have an even better grasp of the system-wide effects of such monumental crises.

The aim of the *Handbook of Financial Stress Testing* is to provide policymakers, practitioners, scholars, and students with a comprehensive resource, a common point of reference into cutting-edge work on stress testing. We asked experts and leading scholars from a variety of jurisdictions, professional backgrounds, and disciplines to contribute a critical reflection on stress testing from their particular vantage point. It has been a real challenge to put together a volume of this ambition. We are proud to say that this has resulted in a rich collection of insights about the past, present, and possible future of stress testing, with direct practical benefit for those working in this field.

This handbook would not have been possible without the generosity of the International Monetary Fund (IMF), which organized a conference in October 2019, "Rethinking Financial Stability: The FSAP at 20."[1] The FSAP is arguably the grandfather of stress testing programs, so it was only fitting for the first of this 2-day conference to be devoted to

---

[1] FSAP: Financial Sector Assessment Program. Conference information can be found at www.imf.org/en/News/Seminars/Conferences/2019/09/20/rethinking-financial-stability-the-fsap-at-20.

contributions from the *Handbook of Financial Stress Testing*. The conference provided a valuable venue for the authors of this handbook to engage in exchange and dialogue, which helped create a joint effort with a coherent trajectory. Adrian, Morsink, and Schumacher also contributed a chapter on the latest thinking on stress testing at the IMF—so we are doubly thankful.

The *Handbook of Financial Stress Testing* is organized into five parts:

  I. History and Objectives
 II. Inputs and Outputs
III. Microprudential Stress Testing
 IV. A Macroprudential Perspective on the Financial System
  V. Macroprudential Stress Testing

The first part, History and Objectives, starts with a discussion of the objectives and challenges of stress testing by Herring and Schuermann. They expand on the crisis response motivation and present some of the basic ideas that are explored in more depth throughout the handbook, including the distinction of the discipline into microprudential (resilience of banks) and macroprudential (resilience of banking and financial systems), disclosure regimes, and challenges confronted (e.g., the focus on capital rather than liquidity). They present six fundamental choices in structuring a stress testing exercise: (1) the design of stress scenarios; (2) the risk exposures to be stressed; (3) the range of institutions to be tested, the length of the scenario, and the intervals over which shocks are measured; (4) the development of models to map shocks into outcomes and impacts on individual bank financials and on the banking system; (5) the choice of criteria to determine whether banks pass or fail the stress test; and (6) the decision about what to disclose to the public.

Das, Dent, and Segoviano discuss the historical evolution of stress testing and identify areas where more integrated approaches are needed. They highlight the need to better capture all enterprise risk, as well as the amplification of shocks through the banking and nonbanking systems. Improving data and information technology (IT) infrastructure would support increased automation of stress test development and analysis and thereby reduce the resource burden for stress-tested institutions. Finally, Anderson, Brazier, Haldane, Nahai-Williamson, and Radia from the Bank of England provide a retrospective on why banks failed the stress test and how the discipline has evolved over the subsequent decade, as well as how we build on these foundations to complete the agenda set out a decade ago. Further development of system-wide stress tests for macroprudential policymaking is needed, the authors argue, but such stress tests would need to involve a deeper study of the "empirical and theoretical foundations for modeling of institutions' behavior" and incorporation of a broader set of markets. They further emphasize that such stress tests should start taking the evolving and global nature of financial markets into account, as well as their feedbacks with the real economy. To guard against complacency, exploratory scenarios could be used to capture unexplored tail risks in stress tests that do not have any historical precedent.

Part II of this handbook, Inputs and Outputs, expands on some of the important ingredients and outputs of stress testing. The first two chapters provide insight into the scenario design challenges from a system perspective that a regulator would need to address, whether with a micro- or a macroprudential objective. Gross, Henry, and Rancoita present a summary of recent developments in the field of macrofinancial scenario design. Various areas are in need of refinement, including properly designing countercyclical scenarios, capturing state-dependent nonlinearities, calibrating the appropriate horizon for market-risk shocks, and avoiding double-counting of shocks in a system-wide stress test. In particular, scenarios should not double-count the initial and higher-order shocks. Flood, Jones, Pritsker, and Siddique explore the challenges that come with a heterogeneous banking system: the less homogeneous, the more complex the risk surface, the more complex the scenario design problem. They propose variants of reverse stress testing and apply their approach to data from 28 large bank holding companies in the US, which reveals significant heterogeneity.

Hopper provides the scenario design perspective for a financial institution, highlighting that a well-designed stress scenario depends on a well-developed risk identification process: if you don't know your risks, it's hard to know what to stress. Hopper points out that more complex financial institutions will likely face complex interrelationships across business activities and thus risks. This makes capturing the full risk profile with just one scenario quite challenging, calling for the need for multiple scenarios.

Stress tests use a mixture of public and private data. Banks' internal stress tests and most supervisory or regulatory stress tests make use of private or proprietary data not available to market participants. However, market data can be very informative, and Engle lays out what we have learned by using information, especially from equity markets, to provide insights into bank and banking system resilience to shocks. Engle makes use of the SRISK measure, based on dynamically estimated market betas, to showcase the advantages of these market-based approaches: they are easy to perform (so it is easy to try many scenarios), they are flexible in their severity, and it is straightforward to conduct sensitivity analysis around the stress estimates. Sarin and Summers express the importance of developing stress tests that integrate both market and book capital. They note that it is "puzzling that [current] bank stress tests entirely eschew market information about financial stability because the regulatory capital ratios that they rely on are known to be a static and easily arbitraged measure of the bank's true capital position." Yet, stress tests that rely solely on market measures of financial sector health are imperfect because they can be driven by noise and are procyclical. Sarin and Summers therefore argue that "stress tests should take into account both market and regulatory indicia of bank health, rather than being mechanically tied to market performance alone."

Ullersma and Van Lelyveld survey the opportunities that the increasingly ubiquitous granular data of the financial system offer for stress testing. Highly granular (transaction-level) data allow a regulator to tailor the data aggregation to the needs of each of its competence domains. System-wide stress tests, for instance, typically require disaggregated data, whereas macro analyses can often be conducted with aggregate statistics. They also discuss the logistical challenges associated with using more granular data, such as adopting solid data-governance processes, having the right skills and tools within the central bank to work with and interpret the data, and ensuring data quality and consistency across the financial system.

A key feature of stress testing is information production; the stress test should tell us something new about the risk profile of a bank or a banking system. But what should one disclose, how much should be disclosed, and who should disclose? Goldstein and Leitner provide a rich discussion of these issues and point out that in addition to the obvious benefits of a rich disclosure regime, there are also costs. The benefits are clear: more information to the market to allow investors and other agents to arrive at a better understanding of the risk profile of banks subjected to stress testing—in short, it promotes market discipline. The costs are less obvious, but they include the possible crowding out of other sources of information (regulators have access to private information about banks, so their disclosure must be especially valuable), and it could invite gaming by banks. Judge wraps up Part II with a discussion of the benefits of stress testing during times of war. She argues that after a financial crisis has taken hold, stress tests can offer useful guidance about the location of weaknesses impeding market functioning, enabling more tailored government interventions. Stress tests can provide market participants with credible information that underlying problems have been or will be addressed, thereby facilitating market functioning. However, given that regulators will rationally be hesitant to produce, much less disclose, information that could exacerbate the very crisis regulators are seeking to contain, crisis-time stress testing is only viable if regulators also have the tools needed to address any bad news the testing may reveal.

Part III launches into a discussion of microprudential stress testing for a variety of financial institutions. Cope, Hsu, Lively, Morgan, Schuermann, and Sekeris present a detailed map of stress testing for commercial, investment, and custody banks, covering the range of businesses, products, and services that are provided by banks. Because "modern" stress testing was developed for and by banks, the methodology has influenced the approach to stress testing in other parts of financial services. Longerstaey provides a discussion of stress testing for asset managers, bearing in mind that they manage assets on behalf of their clients. Put differently, asset management is a balance-sheet-light business model, and as such, the financial risk is borne by the clients, which presents its own challenges. Peters delves into the stress testing challenges and approaches for insurers, which typically face a broader set of risks than banks but have a less fragile funding structure. All of these approaches make clear that models are central to any and all stress testing, and with the proliferation of and dependence on models comes model risk. Canabarro explains how formal model risk management has become an important feature of post-GFC risk management broadly and stress testing in particular. After all, it is hard enough to forecast performance in ordinary market conditions, let alone under rare stress conditions.

Next, Clark, who oversaw the development and implementation of the stress testing program at the Federal Reserve, provides the supervisory perspective on stress testing, noting that "[t]he use of stress testing in supervision is a valuable tool in a business where good tools are scarce, and its prominence has created large benefits for supervisors, banks, and the stability of the financial system." Powerful and popular as stress testing has become, Cumming rounds out Part III with a rich discussion of the strengths but also the weaknesses of this now widely used tool. To name just two of those remaining challenges: joint capital and liquidity stress testing (this volume focuses almost exclusively on capital stress testing, a reflection of the state of practice) and the more comprehensive inclusion of nonfinancial risks.

Part IV moves away from stress tests of individual institutions and pivots to the systemwide concerns that stress tests seek—or could seek—to address. Fernandez Dionis and Mosser provide an overview of the structure of the financial system, stressing that banks are just one of the many agents operating in that system. Their discussion of the intense interconnectedness of the various elements of the financial system (banks, asset managers, exchanges, central counterparties, etc.) provides a rationale for the development and use of stress tests that are more systemic—or more macroprudential—in orientation. Building on such thinking, Goodhart discusses how stress tests can aid regulators in developing "holistic bank regulation." He writes that "such tests, both at the micro and systemic level, should, in theory, be able to take account of relative capital, liquidity, and profitability concerns, all together in a single package." Currently, however, "regulators tend to have tunnel vision when introducing, or amending, financial regulation. They consider each proposal in isolation, without attempting to see how it might fit, holistically, into the broader canvas of the whole set of financial regulations affecting banks."

Geanakoplos points out that leverage is a fundamental driver of financial instability. Any understanding of financial stability, he argues, needs to explore the system's sensitivity to changes (usually increases) in leverage, and stress tests can help do that. English presents a discussion of monetary policy and financial stability. He highlights how the two interact and how stress tests can—for example, through the choice of stress scenarios—help inform (and should be informed by) these two policy perspectives. Finally, Berner, Cecchetti, and Schoenholtz focus on stress test practices from a network perspective, highlighting central counterparties (CCPs) as an example of a critical network hub. Because these networks are highly interconnected, shocks propagate quickly and widely and can even be amplified along the way. Such networks tend to be opaque, and thus designing and executing a cogent stress test is challenging. The authors suggest using market-based risk indicators such as CoVaR and SRISK as measures of systemic (network) vulnerabilities.

Part V covers macroprudential stress tests, which serve to assess systemic risk in interconnected financial systems and to design and evaluate macroprudential policies to mitigate this risk. Bassett and Rappoport present a way in which microprudential stress tests, such as those run by the Federal Reserve Board, can be extended to become more macroprudential. They show how contagious spillovers, specifically funding spillovers, can be modeled as an add-on to such microprudential stress tests, using a reduced-form relation between banks' funding cost, bank capital, and economic activity. Hałaj and Traclet discuss how contagion mechanisms are incorporated in bank stress test models of the Bank of Canada. Their chapter introduces the Macro-Financial Risk Assessment Framework (MFRAF) and the Bank Dynamic Balance Sheet (BDBS) model of the Bank of Canada. MFRAF captures three second-round effects (i.e., the interaction of liquidity and solvency risk, overlapping portfolio contagion, and exposure loss contagion) but does not capture how banks adjust their balance sheets in response to shocks and regulatory constraints under a profit-oriented optimization objective. The BDBS model does just that. It uses optimal portfolio choice theory to model how banks might adjust the balance sheet (i.e., their assets, funding, and equity) in response to shocks to maximize the risk-adjusted return on capital while adhering to regulatory constraints (i.e., risk-weighted capital ratio, leverage ratio, and liquidity coverage ratio). The chapter insightfully puts their work in a broader discussion of the challenges ahead to improve systemic stress tests.

Adrian, Morsink, and Schumacher provide a comprehensive overview of stress testing at the IMF. They not only present an overview of the historical evolution of stress testing at the IMF but also discuss how the IMF's stress tests are currently conducted and what the remaining challenges are to improve the value of stress tests for policymaking in addressing emerging risks. So far, both micro- and macroprudential policies have been informed mainly by microprudential stress tests, but increasingly, macroprudential stress tests are used to calibrate macroprudential policies. Challenges remain, such as the need to better capture behavioral responses driving contagious spillovers, incorporating physical and transition risks from climate change for the financial system, and measuring systemic risk emerging from novel financial technologies (e.g., fintech) and cybersecurity breaches. The authors conclude that the degree to which these challenges can be met in part depends on the availability of suitable data.

Bousquet, Henry, and Żochowski set out how a top-down macroprudential stress test can be developed for the euro area, building on microprudential stress tests that have already been developed for various subsets of the financial system. They call for a top-down approach with macroprudential policymaking to be able to capture system-wide interactions within the banking sector as well as cross-feedbacks with other parts of the financial system, such as insurers, other financial institutions (e.g., money market funds), and central clearing parties. The authors argue that agent-based models are well suited to model interactions among financial intermediaries in a systemic stress test, as long as they strike the right balance between policy interpretability and real-world complexity. Whether system-wide modeling is possible, they emphasize, hinges on obtaining more granular firm-specific data, especially on nonbanks. Next, Thurner offers a complex systems perspective on macroprudential regulation. The science of complex systems—the science of self-organized, networked dynamical systems—can contribute to the understanding of the origins of systemic risk and how it extends current stress testing technology by systematically focusing on the interconnections of financial contracts and their implications for systemic risk. To illustrate that approach, he argues that a systemic risk tax levied on new financial contracts that would increase systemic risk reduces this risk in a self-organized manner without efficiency losses or additional economic burdens to the financial system.

Finally, Benjamin, Haddow, and Jacobs offer a thought-provoking piece on stress testing a central bank's own balance sheet. They set out how a central bank can use forward-looking stress tests as a tool for its own risk management. A central bank's risk exposure stems from its contingent balance sheet expansions associated with its role as a backstop of the financial system. Put differently, central banks, by design, have to expose themselves to wrong-way risk. Central bank stress tests specify the severely adverse scenarios for which the central bank has to stand ready to deliver its policy goals of monetary and financial stability. The authors stress that the severity of adverse shocks is endogenous because these are affected by mitigatory interventions of the central bank. Therefore, it will be critical to incorporate "the feedback effects of central bank actions through the broader system of banks, nonbanks, and markets that compose the financial system." Macroprudential stress tests that include central banks, the authors conclude, should therefore be developed.

The handbook concludes with a forward-looking discussion by Farmer, Kleinnijenhuis, and Wetzer. Finance is characterized by rapid change and exploding complexity, creating serious supervisory and regulatory challenges. Technological advances, including in model development and big data, enable the development of truly system-wide stress tests that can help supervisors and regulators to meet those challenges. The authors sketch what it would take to develop and implement a stress test of the full financial macrocosm, what such stress tests might look like, and how such tests can help enable more dynamic financial regulatory policy.

Part I

History and Objectives

2

Objectives and Challenges of Stress Testing[*]

Richard J. Herring and Til Schuermann

This chapter lays out the objectives and challenges of stress testing. Stress testing originated in the field of engineering as a way of testing the degree of stress a material or structure could withstand, but it has been adapted and applied to many other disciplines, from medicine to risk management in financial institutions. Its traditional use in financial institutions was largely limited to examining the impact of a specific shock on individual positions or portfolios. This use of stress tests was recognized by the Basel Committee on Banking Supervision (BCBS) in the Market Risk Amendment (1996), which required banks that adopted the internal-models approach to conduct stress tests on market positions, and in Basel II (2004), which required banks adopting the internal-ratings-based approaches to conduct stress tests of their credit-risk models. The International Monetary Fund (IMF) and World Bank developed broad financial-sector stress analyses as part of their Financial Sector Assessment Programs,[1] but during the Global Financial Crisis (GFC), a qualitatively different kind of stress test was introduced to evaluate the capital adequacy of the major banks (and the banking system more broadly) under a plausible adverse scenario. This served two objectives: (1) to identify and remediate banks that lacked sufficient capital to comply with minimum required capital ratios over the entire stress scenario and (2) to restore confidence in the core of the banking system by requiring that banks eliminate any capital shortfall promptly either by raising capital in private markets or from retained earnings or, if necessary, from a government backstop fund.[2]

The stress test embedded in the 2009 Supervisory Capital Assessment Program (SCAP) in the US was widely regarded as the turning point in restoring confidence in the US financial system, which had been seriously undermined by the failure of regulatory capital ratios to reflect the deteriorating condition of major banks or to differentiate between relatively strong and weak banks. This success led to parallel efforts to use stress tests to restore confidence in other banking systems and the adoption of stress testing as a principal tool of banking supervision under normal financial conditions.

To ensure the credibility of stress tests (and to enhance the ability of banks to measure and manage their own risks), supervisors have placed increased demands on banks to improve

---

[*] We would like to thank Tim Clark, Thom Wetzer, and an anonymous referee for helpful comments and Siddharth Chandrasekhar for excellent research assistance. Any remaining errors are, of course, our responsibility.
[1] See Chapter 3 in this handbook.
[2] For a historical overview, see Schuermann (2014).

their governance of risk and risk management processes. This has required tremendous investments in information technology (IT), data collection, and modeling. An evaluation of these qualitative aspects of stress tests accompanied the report of quantitative results. To pass the stress test in the US regime, banks were required not only to meet these qualitative standards but also to meet the required capital ratios throughout the stress scenario. European supervisors use the qualitative evaluation as a guide for possible additional capital through Pillar 2, ostensibly to motivate the bank to improve those processes. The stress test also provided a more transparent way to conduct bank supervision on the safety and soundness of individual banks, the primary microprudential objective. Moreover, relative to the traditional emphasis on point-in-time regulatory ratios, this approach was more forward-looking and provided a more plausible way to evaluate the safety and soundness of individual banks.

By subjecting the major banks to the same stress scenarios at the same time, the authorities were better able to evaluate the resilience of the financial system, the primary macroprudential objective.[3] Thus, stress testing provides a way for the authorities to evaluate not only the resilience of individual banks but also the resilience of the entire banking system and has become an important tool in prudential supervision.

Although stress testing practices have advanced markedly over the past decade, many challenges remain, including expansion of the scope of stress tests to encompass risks in other sectors of the financial system to develop a more robust view of national financial resilience; extension of the perimeter of stress tests across national borders to provide a better understanding of the resilience of the global financial system; incorporation of second-round effects of shocks, which are often more important than the first-round impacts measured in most current stress tests; and expansion of the range of shocks in stress scenarios beyond the macroeconomic and market-rate shocks embedded in most current scenarios.

Depending on institutional mandates, the authority overseeing financial stability (usually the central bank) takes primary responsibility for designing and implementing scenarios with a macroprudential emphasis, and the bank supervisory authority (if different) takes the primary role with respect to scenarios with a microprudential emphasis. Of course, these authorities (if different) must cooperate in stress tests designed to accomplish both macro- and microprudential objectives. The objectives of a stress test will determine the other fundamental choices in structuring the exercise: (1) the design of stress scenarios; (2) the risk exposures to be stressed; (3) the range of institutions to be tested, the length of the scenario, and the intervals over which shocks are measured; (4) the development of models to map shocks into outcomes and trace their impact on individual bank financials and on the banking system; (5) the choice of criteria to determine whether banks pass or fail the stress test; and (6) the decision about what to disclose to the public. A brief review of the SCAP stress test illustrates the fundamental choices that were made in this pioneering case.

1 The Example of SCAP (2009)

SCAP was arguably the first large-scale public application of stress testing designed to assess the capital adequacy of individual banks and the health of the banking system.[4] The basic

---

[3] See Chapters 24 and 26 in this handbook.
[4] The Bank of England conducted a private stress test in late 2008/early 2009, referred to in Haldane (2009).

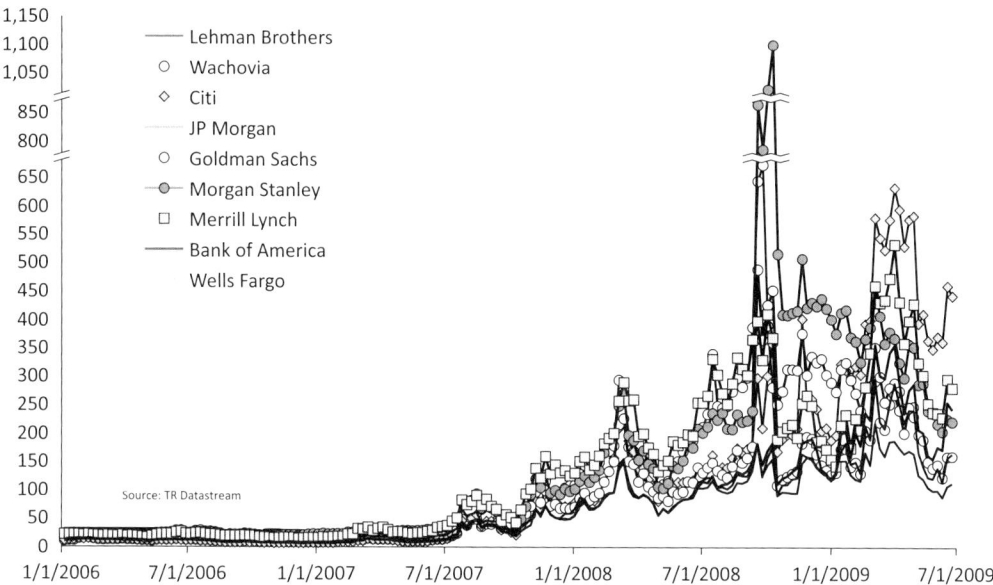

Figure 2.1 Weekly 5-Year CDS spreads (basis points), large US institutions, January 2006 to June 2009. Source: Refinitiv Datastream.

framework shaped the development of subsequent stress tests. SCAP was designed with the objective of restoring a loss of confidence in the banking system and bank supervisors. Banks had incurred massive losses that were expected to rise. Markets took an increasingly pessimistic view of the solvency of banks, as reflected in 5-year credit default swap (CDS) spreads, which showed the cost of insuring against the default of individual banks (see Figure 2.1).

The principal regulatory measure of bank capital adequacy, the Tier 1 capital ratio, had failed to reflect the obvious deterioration in the condition of banks or distinguish between strong and weak banks, which led markets to question whether the regulatory authorities could be trusted to halt the deterioration in financial conditions. Haldane (2011), using analysis conducted by the Bank of England, compares an index of capital ratios for banks that required intervention during the crisis with banks that did not. The two indices are nearly indistinguishable; if anything, "crisis" banks looked a little stronger than "noncrisis" banks. Moreover, the ratios failed to detect the decline in the health of the banking system before and into the start of the worst banking crisis since the Great Depression. The standard regulatory metric of Tier 1 capital did not signal the decline in the capital adequacy of the banking system, nor did it help identify, ex ante, banks that turned out to be so weak as to need public support.

SCAP was designed and implemented by the Federal Reserve System (Fed), which has both macroprudential and shared microprudential responsibilities (Hirtle et al., 2009). Banks were reporting increasingly large credit losses, and so the stress scenario focused on exposures to credit risk. The Fed provided two 2-year scenarios: a baseline scenario representing consensus forecasts of the macroeconomic variables and a hypothetical adverse scenario that could be regarded as a severe but plausible worst-case outlook.[5] In addition, five banks with significant trading operations were required to execute an instantaneous shock to the trading book to assess market and counterparty credit-risk impact. The Fed applied the test to the 19 bank holding companies that had assets greater than $100 billion, which accounted for roughly two-thirds of bank assets and could be considered the core of the banking system. The modeling strategy combined a top-down approach that relied on Fed data and models with bottom-up modeling based on each bank's own internal data and models. The models traced the impact of the shock scenario on each bank's balance sheet and income statement to test whether the bank had sufficient capital to meet capital requirements in each period of the severe stress scenario. Banks that were able to meet capital requirements each period passed the test. If a bank failed to pass the test, the policy consequence was that the bank was required to raise additional capital to eliminate any shortfall or seek a capital infusion from the government. Although it was not a central objective of the stress test, the exercise increased the understanding of the regulators about the health of the banking system and provided some insights into the risk management practices and data gaps at individual banks.

In a sharp break with the long-standing supervisory tradition of maintaining secrecy about most supervisory actions and hiding virtually all data that might undermine confidence in individual banks, the Fed decided to disclose the results of the stress test and the methodology underlying the results. The central objective of the stress test was to restore confidence in the financial system and the regulatory process, and so it was essential that the public believe the results of the stress test. The revelation that 10 of the 19 banks failed the severe stress test helped persuade the public that the test was rigorous. And the accompanying requirement that those 10 banks raise sufficient capital within 6 months to fill the shortfall forestalled most of what might otherwise have been damaging consequences for the banks that failed SCAP. Only one of the 10, GMAC, found it necessary to draw on the government backstop and was, in fact, nationalized.

The success of SCAP and this unprecedented disclosure of information about the financial condition of individual banks in restoring confidence in the banking system encouraged authorities in other jurisdictions to design stress tests and make similar disclosures. The failure of the European stress testing exercises in 2010 and 2011 to restore confidence highlighted the importance of credibly rigorous stress tests accompanied by the ability and will to take swift remedial action with regard to banks that do not have sufficient ability to absorb stressed losses. The European authorities understood that it was essential to disclose the results of the stress testing, but they lacked the resources to remediate shortfalls in loss-absorbing capital. And so, in order to avoid reporting negative results that might exacerbate the condition of weak banks, the authorities appear to have moderated the degree of stress in their scenario and its translation into the impact on bank capital ratios so that almost all banks passed the test (Schuermann, 2014). This gamble might have worked, except that soon after the results were released, major banking crises erupted in Europe. Subsequent country-level stress tests in Europe were more successful because scenarios stressed bank

---

[5] Ironically, by the time the results were released, the unemployment rate had risen to an even higher level than that specified in the adverse scenario.

capital levels more credibly, and a variety of capital-support measures were made available to banks that failed the tests.[6]

The following sections discuss how tests have evolved since SCAP. We examine choices to be made regarding the following: (1) the design of stress scenarios, (2) the inclusion of risk factors; (3) design features to reduce the procyclicality of stress tests, (4) criteria for passing stress tests, (5) the scope and length of stress scenarios and the intervals over which results are measured, (6) modeling strategies, and (7) communication challenges. The chapter concludes with a discussion of the limitations of stress testing and highlights some of the most important remaining challenges.

2 The Design of Stress Scenarios

The institutions that take primary responsibility for the design of stress scenarios vary across jurisdictions. Some jurisdictions lodge banking supervision and oversight, the financial stability mandate, deposit insurance responsibilities for deposit insurance, and lender-of-last-resort (LLR) responsibilities in several agencies or authorities. The United States, for instance, combines banking supervision, the financial stability mandate, and LLR under the Federal Reserve, although banking supervision is shared with other agencies (e.g., the Office of the Comptroller of the Currency [OCC], an independent regulatory unit of the Department of the Treasury), including the deposit insurer. In Canada, the Bank of Canada has LLR and financial stability responsibility but not bank-supervision responsibility, which is overseen by the Office of the Superintendent of Financial Institutions, whereas the Canada Deposit Insurance Corporation lacks a bank-supervisory mandate. The UK had carved out banking supervision from the Bank of England in 1997, only to fold it back in following the financial crisis in 2013. And the European Central Bank (ECB) assumed supervisory responsibility through the Single Supervisory Mechanism in 2014 following an EU-wide comprehensive assessment that combined an asset-quality review (AQR) and stress test.[7] In a crisis, there may be additional nondomestic institutions, such as the IMF or the European Commission and European Stability Mechanism (for EU countries), with important roles, including the ability to provide emergency funding and thus impose conditions and constraints on the stress testing program. With a range of agencies and official institutions, often with overlapping responsibilities, the allocation of roles and coordination make governance a significant challenge.

Borio et al. (2012) have observed that it is much more straightforward to design an appropriate stress scenario in a crisis (wartime) than during apparently tranquil financial conditions (peacetime).[8] In wartime, the risks to the banking system are all too obvious, and so the choice of exposures to stress and the shocks to consider is clear. The key objective is to restore confidence in the system rather than to probe for a range of additional weaknesses

---

[6] Two examples are Ireland (2011, www.centralbank.ie/regulation/industry-sectors/credit-institutions/ Documents/The%20Financial%20Measures%20Programme%20Report.pdf) and Spain (2012; www.bde.es/f/webbde/SSICOM/20120928/informe_ow280912e.pdf).

[7] See the results document at www.bankingsupervision.europa.eu/ecb/pub/pdf/ aggregatereportonthecomprehensiveassessment201410.en.pdf.

[8] See Schuermann (2016) for an extended discussion of differences in the use of stress tests in wartime and peacetime.

that might appear in the future. During peacetime, the exposures that should be stressed and the relevant shocks to consider are much less clear. The peacetime objective is to ensure that banks and the banking system will be resilient against a wide range of "severe but plausible" shocks.

This is a difficult challenge because a failure to select the right shocks or the right degree of severity may engender a sense of complacency, which can contribute to financial fragility. And during extended intervals of relative tranquility in the financial sector, disaster myopia is likely to develop. The disaster-myopia hypothesis is a way of reconciling two different observations about bank lending: (1) "The worst of loans are made in the best of times"; (2) "Loan officers never intend to make bad loans." Both observations seem plausible but appear to be contradictory. They can be reconciled if lenders' subjective probabilities of a shock tend to decline over a sustained period in which outcomes have been favorable so that they are willing to take larger exposures without consciously intending to take larger exposures to loss. Borio et al. (2012) generalize this point as the paradox of financial instability: "the system looks strongest precisely when it is most vulnerable" because credit growth is strong; asset prices are high; leverage, as measured at market prices, appears to be low; bank profits and asset quality seem robust; and risk premiums and volatilities in asset prices are low. Technological and financial innovations often flourish in periods of extended growth, so it is possible to argue that historical experience is no longer relevant.

Shocks that occur frequently generate a statistical record that enables analysts to identify tail events with confidence and generally do not pose a threat to financial stability. But shocks that occur infrequently and at irregular intervals do not provide a comparable statistical record. Moreover, we generally lack a sufficient understanding of the underlying forces that generate such shocks to enable us to make predictions with any degree of confidence. Under such circumstances, in the absence of statistical evidence or a clear understanding of the mechanisms that produce shocks, decision-makers generally rely on heuristics to formulate probabilities that such shocks will occur. These heuristics—the availability heuristic and the threshold heuristic—lead to a bias in assessing these potential threats to financial stability that cause decision-makers to underestimate or even disregard such shocks.[9]

Although these biases have been shown to affect most decision-makers, bankers may have a variety of additional institutional incentives to ignore such shocks (Guttentag and Herring, 1986). Although regulators are not subject to these additional incentives that exacerbate disaster myopia, they face a serious practical problem: it is very difficult for supervisory authorities to influence the behavior of banks that appear to be profitable and in good condition but are vulnerable to shocks of unknown probability. Stress tests provide an opportunity to counteract disaster myopia by forcing banks to consider the consequences of hypothetical shocks.

The stress scenario should reflect the risk aversion (or, in the parlance of risk management, the *risk appetite*) of the authorities. Because the stress scenario reflects the authorities' view of the severity of shocks that banks should be able to sustain on their own resources,

---

[9] The concept of disaster myopia is based on the work of Tversky and Kahneman (1982) and Simon (1978). See Guttentag and Herring (1984) for the first applications of the hypothesis to the evolution of financial fragility, and see Haldane (2009) for an application to stress testing.

it implicitly identifies circumstances when official resources are likely to be made available, although no authority makes such a commitment.

The design and severity of the stress scenario will be an inherent point of contention between the supervisory authorities and banks, which are well aware that the consequence of a severe stress test, or the inclusion of an unfamiliar shock, may be a requirement to increase capital or liquidity positions or some other more intrusive supervisory action. Banks will tend to complain that scenarios include implausible shocks or shocks that are implausibly stressful. The criterion of plausibility is crucial for sustaining confidence in stress tests, but it poses serious limits on the ability of the authorities to stress tail events or introduce new shocks. The general approach to designing stress tests is to rely on actual historical episodes of banking crises or on historical evidence about the distribution of risk factors. The obvious danger is that history seldom repeats itself. For example, the historical record did not enable the authorities in any country to anticipate the GFC. Shocks were more severe than historical precedents. They affected risk exposures that had not been carefully monitored or adequately understood. Moreover, innovations in financial instruments and markets reduced the relevance of historical data and introduced new, damaging channels of contagion.

Because it is expensive to build the infrastructure to conduct stress tests, there is an understandable reluctance to introduce different kinds of shocks that will require additional data and new models. This dampens enthusiasm for innovative new stress scenarios and causes stress scenarios to be more similar over time than they should be to truly test the resilience of the system. Another factor that tends to discourage innovations in stress scenarios is the concern expressed by one leading regulator (Quarles, 2019) that to better facilitate banks' capital planning, stress tests should not exacerbate volatility in capital requirements.

As an alternative to history-bound stress tests, the authorities could rely more heavily on expert judgment in the construction of hypothetical scenarios, although they are even more likely to be contested on the grounds of plausibility. These may be informed by exploratory scenarios that investigate new kinds of risks to inform the design of future stress scenarios but do not involve immediate sanctions or public disclosures.[10] This seems an especially useful way to deal with changes in market structure or new instruments.

It is difficult and potentially misleading to summarize the severity of a stress test in terms of any one risk factor because many risk factors are involved and bank positions differ. One way to calibrate the severity of stress scenarios is to employ reverse stress tests that attempt to identify the kind and size of shock that would seriously jeopardize the solvency of a bank. This is effectively what is done by banks during internal stress testing, whether it is the internal stress scenario developed for Comprehensive Capital Analysis and Review (CCAR) in the United States or for the Internal Capital Adequacy Assessment Process (ICAAP) in other jurisdictions. It is carried to its logical extreme for resolution planning, where banks contemplate a shock so severe that it pushes the bank past the point of nonviability.

---

[10] The biennial stress tests of the Bank of England serve this function.

3 The Choice of Risk Factors

Most stress tests have focused on macro and financial shocks to banks, especially credit shocks. This emphasis reflects the fact that credit problems have been the dominant source of losses in most banking crises. But banks are exposed to a wider range of risks, many of which are not directly related to macro shocks or changes in market prices, and so attempts to stress such exposures are more ad hoc. Shocks to operational risks have only the most tenuous relation to macro or financial variables, and although market risks are probed for banks with substantial trading portfolios, these stress tests are not well integrated with broader stress scenarios.[11]

Borio et al. (2012) raise an additional concern about typical scenarios based on assumed macro shocks, such as the unemployment rate or gross domestic product (GDP) growth. Unfortunately, these variables tend to be lagging indicators of financial crisis and thus are unlikely to uncover vulnerabilities in the financial system in peacetime. Borio et al. suggest peacetime scenarios should place much greater emphasis on financial-cycle indicators, such as the joint behavior of credit and property prices.

Some regulators have also been exploring the inclusion of nonfinancial shocks, such as fines for conduct violations or other legal exposures. In the case of insurers, nonfinancial risk factors are often more important than macroeconomic and financial shocks (see Chapter 15 in this volume). For example, in 2019, the Bank of England constructed a stress scenario for insurers in its jurisdiction that incorporates severe natural disasters, such as a hurricane in the United States, earthquakes in California and Japan (where, for good measure, Tokyo is also hit by the concomitant tsunami), and a severe windstorm and flood in the UK.[12]

The choice of risk factors for the stress scenario reflects a trade-off between the goal of spanning the full range of risk factors to which banks are exposed and the objective of minimizing operational complexity and compliance costs. This problem is particularly acute with regard to global banks, which are exposed to risk factors in multiple countries or regions and exchange rates. The Fed specifies 28 risk factors, of which 16 are domestic and 12 are foreign: 3 factors each for the euro area, the UK, developing Asia, and Japan. The Bank of England specifies 48 variables, and the European Banking Authority (EBA) uses the scenario designed by the European Systemic Risk Board, which includes risk factors for 28 countries in the EU plus risk factors for nine non-EU countries, along with relevant exchange rates and several other regional or country-specific variables, adding up to almost 500 risk factors. All scenarios have the typical macroeconomic metrics, such as GDP, prices (or inflation), unemployment, consumption, and government rates. In addition, equity and volatility indices, credit spreads, real estate price indices, and foreign exchange rates are usually included. Stress tests also typically include a trading shock for banks with significant capital-market activity, which reflects the impact of a severe, instantaneous shock to the trading portfolio, including the impact on counterparty credit risk through derivative positions. The specification of this scenario is an order of magnitude more complex, reaching more than 20,000 parameters in the case of the Federal Reserve's Global Market Shock.

---

[11] That said, there is some evidence that operational losses are actually procyclical (Curti et al., 2019).

[12] See www.bankofengland.co.uk/-/media/boe/files/prudential-regulation/letter/2019/general-insurance-stress-test-2019-scenario-specification-guidelines-and-instructions.pdf.

4 Options to Modify Stress Scenarios to Reduce the Procyclicality of Bank Behavior

Risk-sensitive capital requirements are procyclical by design because capital requirements increase as measured risks rise, and capital requirements fall as measured risks decline. This heightens incentives for banks to reduce lending in a contraction and increase lending in an expansion, thus amplifying the business cycle. Stress scenarios are countercyclical to the extent that they cause banks to take account of potential future shocks in meeting capital requirements, offsetting to some extent the incentive to increase lending in an expansion. Moreover, the requirement that banks have sufficient capital throughout the severely adverse stress scenario so that they will be able to meet regulatory minimums at all times is intended to ensure that banks will not be obliged to reduce lending to meet regulatory minimums. The Fed has chosen to strengthen the countercyclical properties of stress tests so that the severity of stress scenarios rises automatically as the unemployment rate falls. It constructs the severely adverse stress scenario so that whenever the unemployment rate falls below 10 percent, the increment in the unemployment rate in the severely adverse stress scenario must be large enough to raise it to 10 percent. Thus, the lower the initial unemployment rate, the larger the increase in the unemployment rate in the severely adverse stress scenario. Moreover, the Fed uses the large-scale Federal Reserve Board (FRB)/US macroeconomic model to specify the other macroeconomic variables so that they are consistent with the assumed path of the unemployment rate and are likely to take on more extreme values as well. In the most recent amendment to its design framework (Fed, 2019), the Fed has stated its intention to reduce the severity of the severely adverse scenario if the unemployment is elevated at the start of the scenario.[13]

The Fed supplements the fundamental recession-based stress scenario with shocks to banks' risk exposures to worrisome trends that the Fed deems salient. These salient risks have included oil-price shocks, a severe recession in the euro area, a hard landing in the Chinese economy, and excessive leverage in the corporate sector.[14] In Chapter 5 of this volume, Anderson et al. note that the Bank of England makes similar adjustments to reflect judgments about the extent to which key risk factors depart from long-term trends, augmenting the basic countercyclical approach in its Annual Cyclical Scenario (ACS).[15]

Ironically, despite these efforts to ensure that stress scenarios increase in severity as the unemployment rate falls below 10 percent, Kohn and Liang (2019) find that the most important countercyclical aspect of stress tests over a period in which the unemployment rate has declined to record lows has had less to do with the increasing severity of the stress scenario than the unrelated requirement in the US regime that banks' planned capital distributions be included in the severely adverse stress scenario. Because banks tend to

---

[13] The former rule specified that the increase in the unemployment rate typically must be at least 4 percent above its initial level or a sufficient increment in the unemployment rate to reach 10 percent in the stress scenario if it is larger. The new rule (Fed, 2019) clarified that when the unemployment rate is elevated at the start of the scenario, the Fed may choose to reduce the increment to 3 percent rather than the presumed 4 percent increment.

[14] In Chapter 24 of this volume, Bassett and Rappoport point to funding dynamics, especially short-term wholesale funding, as another channel for procyclicality that can be addressed with stress testing.

[15] See Bank of England (2015).

increase payouts in good times, this increases the amount of capital they must have to pass the stress test. Kohn and Liang (2019) note that this increase in financial flexibility may be enhanced by the informal limit that dividends be no more than 30 percent of planned capital distributions. Because of perceived adverse signaling effects, banks have been much more reluctant to cut dividends in a downturn than to reduce share buybacks. Their conjecture is that requiring capital distributions beyond 30 percent of projected baseline earnings to take the form of share buybacks will make it easier for banks to reduce capital distributions in a downturn.

The Financial Policy Committee of the Bank of England attempts to ensure that stress scenarios have a countercyclical impact by monitoring imbalances in the growth of credit, asset prices, and distortions in corporate and household balance sheets and adjusting the severity of stress tests accordingly. Criteria set for passing the severely adverse stress test may also be an important way to enhance the countercyclical impact of stress tests.

5 Criteria for Passing Stress Tests

The specification of the rules banks must follow in projecting their capital over the given stress scenario to pass the stress scenario are crucial. In the absence of such restrictions, banks may choose to achieve their Tier 1 capital requirements by reducing lending to shrink risk-weighted assets, which is precisely the sort of procyclical behavior that the authorities aim to discourage. Thus, banks are usually not permitted to pass the stress test by reducing extensions of credit. Indeed, the Fed assumes that balances sheets will grow slightly to place additional emphasis on the policy objective of maintaining the credit capacity of the banking system under stressful conditions, and other authorities (e.g., Bank of England, EBA, ECB) require the balance sheet to stay the same size.

In addition to imposing constraints on balance-sheet size over the stress scenario, the authorities impose a range of additional restrictions on banks' portfolio choices to curb reliance on implausible and macroprudentially undesirable behavior. For example, during the stress scenario, banks may be tempted to substitute retail deposits for increasingly expensive wholesale funding, but this strategy would not be feasible if many were to pursue it simultaneously. On the asset side of the balance sheet, banks may be motivated to substitute new lower-risk borrowers for existing higher-risk borrowers or borrowers experiencing financial distress, thus adding to recessionary pressures.

Some observers make unduly optimistic interpretations about the extent to which stress tests can mitigate procyclical bank behavior. The most the authorities can accomplish through such restrictions on bank responses to the stress test is to ensure that banks have sufficient capital throughout the stress scenario so that they will not be obliged to reduce their risk-weighted assets to meet minimum regulatory requirements. But binding capital requirements are not the only or, perhaps, the most important reason banks tend to reduce lending in a downturn. Even without pressure to meet regulatory capital requirements, banks may choose to reduce their risk exposures because they believe it is the prudent way to maximize returns to shareholders.

The authorities may specify both quantitative and qualitative criteria for passing stress tests. Quantitative criteria are defined in the form of capital thresholds commensurate with the jurisdiction's regulatory requirements. They are typically anchored in the Basel minimums and nearly always use the risk-based ratios, such as Tier 1 and CET1. The Fed and the EBA have taken contrasting approaches. In the United States, banks must clear five different capital hurdles to pass the stress test. Although the EBA set a CET1 hurdle rate of 5.5 percent in its 2014 stress test, it abandoned pass/fail thresholds in the 2016 stress test. Instead of defining a pass/fail threshold, it will now use the results as "an input for the Pillar 2 assessment of banks." This may have the same impact on changes in bank capital requirements as the Fed's approach, but the change in capital requirements is much less transparent, which makes it more difficult for the public to monitor banks and hold regulators accountable for taking appropriate responses to the stress test results.

The Bank of England sets the hurdle rate according to an individual bank's capital requirements, including buffers for systemically important banks. The results are used to set an individual bank's capital buffers under Pillar 2 and to inform macroprudential policy and determination of the countercyclical buffer (Dent et al., 2016).[16]

6 The Scope and Length of Stress Scenarios and Test Intervals

Most stress tests have focused on the banking system in part because banks have been regarded as the core of the systemic risk problem and, undoubtedly, also because the authorities charged with oversight of financial stability generally have unambiguous authority over banks and access to data, which does not necessarily extend to other financial institutions.

Stress testing programs, whether in wartime or peacetime, cover some notion of the "core" of the banking system. Coverage may expand or shrink over time in line with the objectives of the stress test. For example, the United States began, in 2009, with a size cutoff of $100 billion, which included 19 institutions. The Dodd–Frank Act in 2010 reduced the size cutoff to $50 billion, and the number of institutions included in the stress program (along with large US subsidiaries of foreign banks in 2017) peaked in 2018 with 38 institutions. The number of institutions included in the 2019 stress test declined to 18, in keeping with the Economic Growth, Regulatory Relief, and Consumer Protection Act (2018). The stress test conducted by the EBA in 2014 covered 123 institutions across 22 countries, with size cutoffs varying by country to account for differences in the relative importance of banking systems. This shrank to 51 institutions in 2016 and 48 in 2018 across 15 countries. Differences in the scope of the stress tests across countries may reflect differences in market concentration; the importance of foreign banks; the role of government banks; and the importance of special business models, such as capital markets and trading activities, custody services, and specialty financing, such as for example, ship finance in Greece.

As the share of bank assets in the financial system declines, the question of whether the scope of stress tests should expand becomes more salient. Can the authorities discharge their responsibility to oversee the resilience of the financial system without understanding how stress scenarios will affect other major participants in the financial system? Some nonbank regulators require their regulatees to conduct stress tests. In some jurisdictions, these firms

---

[16] The hurdle rate was once expressed as a percentage of risk-weighted assets (RWAs), but the new approach will use different scaling factors for different risks. For example, credit risk will be scaled by credit RWAs, market risk and credit counterparty risk by market RWAs, operational risk by total assets, and interest rate risk in the banking book by total banking-book assets.

include insurance companies, securities firms, mutual funds, and central counterparties, but these stress tests are generally conducted for microprudential purposes and are not coordinated with each other or the stress tests applied to banks. The ability to assess overall financial resilience would surely be advanced if these stress tests were coordinated to occur at the same time and employed at least some common risk factors[17] because interactions among various sectors of the financial system may have important implications for financial stability.

Because many shocks take considerable time to manifest themselves in bank balance sheets and income statements, stress scenarios generally extend over a number of years. An obvious exception is shocks to mark-to-market portfolios, which can have an instantaneous impact, which is one reason they are modeled separately. But over what period should stress tests extend? The typical horizon is 2 to 3 years. The EBA specifies stress scenarios over 3 years, whereas the Fed specifies nine-quarter scenarios (2 years plus the quarter it takes to run the stress test). In contrast, the Bank of England designs stress scenarios that extend over a 5-year horizon.

Although longer periods provide more time to assess the damage to bank balance sheets and income statements, clear trade-offs are involved. Modeling the evolution of bank balance sheets and income statements is more uncertain the longer the period over which projections must be made. And the longer the period over which a stress scenario extends, the more important it is to model the behavioral responses of banks and market participants and interactions with the real economy. But this has proven very difficult; see Anderson et al. (Chapter 5 in this volume) and Hałaj and Traclet (Chapter 25 in this volume).

Over what intervals should the impact of shocks be measured and evaluated? All stress testing regimes except the US regime define scenarios in 1-year increments. The Federal Reserve defines the scenario variables in quarterly intervals and expects banks to produce corresponding financial estimates on a quarterly basis. The principal rationale is that reporting, whether to regulators or to the market, occurs at a quarterly frequency. But depending on loss and revenue dynamics, a bank could fall below required thresholds during one or more quarters during a year, yet recover by year-end. Measured at quarterly intervals, the bank would be recorded as having failed the stress test, but if it were measured at annual intervals, it would pass.

7 Modeling Choices

One fundamental modeling decision is the extent to which the authorities rely on banks' own estimates of the impact of the scenario on their balance sheets and capital versus the regulator's independent assessment. For quality assurance and to evaluate the reasonableness of the banks' own projections, all regulators have some analytical and modeling machinery in place to arrive at their own estimates of the scenario impact. But regulatory models differ in complexity and granularity, with the Federal Reserve likely at the most extreme end.[18]

---

[17] Of course, core macro and financial-risk stress scenarios would need to be augmented with additional stress factors designed to probe the special vulnerabilities of each nonbank sector.

[18] See the Federal Reserve's "Dodd–Frank Act Stress Test 2019: Supervisory Stress Test Methodology," available at www.federalreserve.gov/publications/files/2019-march-supervisory-stress-test-methodology.pdf.

Most regimes allow the banks' estimates to be the primary (but certainly not the only) determinant of the final loss under stress, but the Fed relies solely on its own models to determine whether a bank passes or fails the quantitative hurdles in the stress tests.[19]

The authorities typically rely on their own models to ensure that shocks embedded in the stress scenario are mutually consistent. Once the stress scenario is specified, the shocks must be transformed into impacts on the bank's balance sheet, RWAs, net income, and resulting capital levels and regulatory capital ratios.

Although authorities concerned mainly with macroprudential oversight may rely only on highly aggregated, top-down models, authorities with microprudential responsibilities must rely on much more granular models that show how various elements of the bank's income statement and balance sheet will respond to the shock scenario. The general approach is to model various components of a bank's income statement and balance sheet and aggregate them to determine the change in the bank's capital position. This approach is tightly linked to accounting statements, but it may not fully capture the sharp changes in bank value that are associated with financial crises.

Calomiris and Nissim (2014) argue that balance-sheet-oriented models fail to capture the most important sources of loss for the largest banks. They show that the recognition of credit losses during the GFC was much less important for the largest banks than the decline in the value of their intangible assets, such as the market value of loan and deposit relationships and the market value of noninterest sources of income, such as mortgage servicing fees and trading activities. In some cases, these sources of loss were due to endogenous changes in macroeconomic policy that are not typically factored into stress scenarios. For example, more expansive monetary policy tends to drive down interest rates, reducing the value of core deposits, and flatten the term structure of interest rates, which reduces the value of a carry trade. Calomiris (2017) asserts that models based on managerial accounting—cash flows associated with lines of business—would yield more accurate information about how the value of banks would be likely to change in stress scenarios.

Other approaches to assessing financial vulnerability rely more heavily on market data and often convey a very different message about the fragility of banks than the accounting data-based models used in stress scenarios. Sarin and Summers (2016) analyze a variety of market-valuation measures for the largest US banks and reach a much less optimistic conclusion about the health of US banks than that conveyed by the stress test results. Similarly, Acharya et al. (2014) have developed a stress test methodology based entirely on publicly available market data, which they use to estimate a bank's capital shortfall in response to a common shock.[20] Their results also convey a strikingly different message than conventional stress tests with regard to which banks had the largest capital shortfall and the overall condition of the banking system. In addition, they show that hurdle rates based on RWAs, the usual standard, have a much lower correlation with market-based measures of risk than ratios of Tier 1 capital to total assets.

---

[19] Although the Fed inevitably has less insight into the idiosyncratic risks facing individual banks, it argues that independent estimates are important to the credibility of stress tests and for producing comparable results across banks based on the same methodology, which will give it better insights into the overall stability of the banking system.

[20] See also the contributions from Sarin and Summers (Chapter 9) and Engle (Chapter 8) in this handbook.

The authorities have consistently resisted the introduction of market data regarding the valuation of banks in stress tests. In part, this is based on concerns that the signal-to-noise ratio in share prices would inject uninformative volatility in the results of stress tests and exacerbate their procyclical impact. Yet numerous studies have shown that accounting data tend to overstate the value of assets in banking books, particularly when macroeconomic conditions deteriorate.[21] This means that the starting point for a stress scenario, the accounting value of bank capital, may overstate a bank's ability to absorb loss. The stress test provides information about incremental losses over the length of the scenario, but if the starting point overstates the loss-absorbing capacity of capital, the stress test will also yield misleading estimates of a bank's capital adequacy. Flannery (2019) suggests a relatively simple way to moderate this bias by taking a weighted average of the market value and the accounting value of bank capital to use as the starting point for the stress test. A 90-day moving average of the market value of bank capital would remove a substantial amount of the noise in day-to-day share-price fluctuations and provide a useful indication of the extent to which market values differ from accounting values. At a minimum, a large discrepancy between a smoothed measure of the market value and the accounting value of bank capital should prompt a much closer look at reasons for the gap, which may have important implications for the validity of the stress test.

Although stress tests are explicitly forward-looking, they do require that the current conditions of the balance sheet are accurately reflected in the accounting. Importantly, loans and other assets that are not performing should be classified as such and appropriately reflected in provisions. AQRs serve this purpose. For example, in the 2014 AQR conducted by the ECB, portfolios of banks were sampled and carefully examined to ensure that they were appropriately classified or valued. The ECB AQR was part of the transfer of oversight of many European banks from national authorities to the Single Supervisory Mechanism (SSM). This AQR was combined with a subsequent stress test in a nearly year-long effort to form a comprehensive assessment of the major European banks.[22] Most country-level stress tests in Europe preceding the ECB exercise also combined an AQR with a stress test.

8 Communications Challenges

The authorities face a number of important decisions about whether and what to communicate about the results of stress tests.[23] Because the central object of stress tests is to either restore or maintain confidence in the banking system and bank supervision, there is a presumption that a considerable amount of disclosure is necessary. Candor about the state of the banking system is the cornerstone of a successful communications strategy. In wartime, this could be understandably uncomfortable for the authorities under whose watch the crisis arose, but that is when the potential value of transparency is greatest.

---

[21] For example, Calomiris and Herring (2013) have shown that a mark-to-market approach to measuring capital adequacy based on share prices did a much better job than accounting or regulatory measures in distinguishing strong banks from weak banks and identifying which banks would require official intervention during the GFC in both the United States and Europe.

[22] See www.bankingsupervision.europa.eu/banking/tasks/comprehensive_assessment/html/index.en.html.

[23] For a fuller discussion of the challenges of stress testing disclosure, please see the chapter by Goldstein and Leitner (Chapter 11) in this handbook.

The first decision concerns disclosure regarding the stress testing process and construction of scenarios. To be convincing, the stress scenario should be explained and shown to be sufficiently adverse to produce losses that will test the ability of banks to absorb loss. Additionally, the procedure for projecting losses at individual banks should be credible. The original methodology document published before the release of the SCAP test results emphasized the conservative nature of the test by disclosing the assumed loss rates for each major asset class. If the public doubts the severity of the stress scenario, the results can do little to bolster confidence.

Second is whether to disclose individual bank results. Banks and the authorities are justifiably concerned that disclosure of capital shortfalls could exacerbate strains on these institutions and further weaken the financial system. For that reason, it is crucial that the revelation of capital shortfalls be accompanied by plausible plans to fill the gap in a relatively brief period of time.

Disclosure of individual bank results has several advantages. More information about individual banks is likely to bolster confidence in the system by reducing some uncertainty about the condition of banks.[24] Moreover, this kind of information will facilitate market discipline on individual banks. Goldstein and Sapra (2013) have argued that this advantage may be offset to some extent by a reduction in the incentive for the private sector to produce information regarding the condition of banks. Flannery et al. (2017) have provided evidence that this offset has been inconsequential so far by showing that the number of security analysts covering each bank subjected to stress tests has increased after the publication of stress test results.

Equally, and perhaps more importantly, such disclosures will increase the accountability of regulators and reduce the scope for forbearance, which has frequently contributed to the buildup of losses in the banking system, exacerbated financial crises, and greatly complicated the resolution of insolvent institutions. Disclosure about the links between stress test results and prudential policy may also strengthen public confidence in the supervisory system.

The SCAP exercise introduced the term *buffer* to describe the additional capital banks needed to pass the stress tests above the regulatory minimum. This concept was taken up in the Basel III reforms with the introduction of various buffers above the minimum capital requirement. The Bank of England has taken measures to link stress results to the buffers for individual banks and also to the countercyclical buffer for the banking system, and the ECB has taken a similar approach.

The credibility of the stress testing process can be enhanced by a description of how the stress scenario is translated into losses at individual banks. Some transparency regarding the stress test models is necessary for regulatory oversight, but many banks would like complete transparency about the models regulators employ. Banks argue that full transparency about regulatory models would enable external experts to evaluate the models, which could lead to substantial improvements.

---

[24] Unfortunately, the information disclosed is a point estimate rather than a distribution. Because the results are the outcome of numerous models, each of which is estimated with error, stating the outcome as a single number conveys an unwarranted sense of precision.

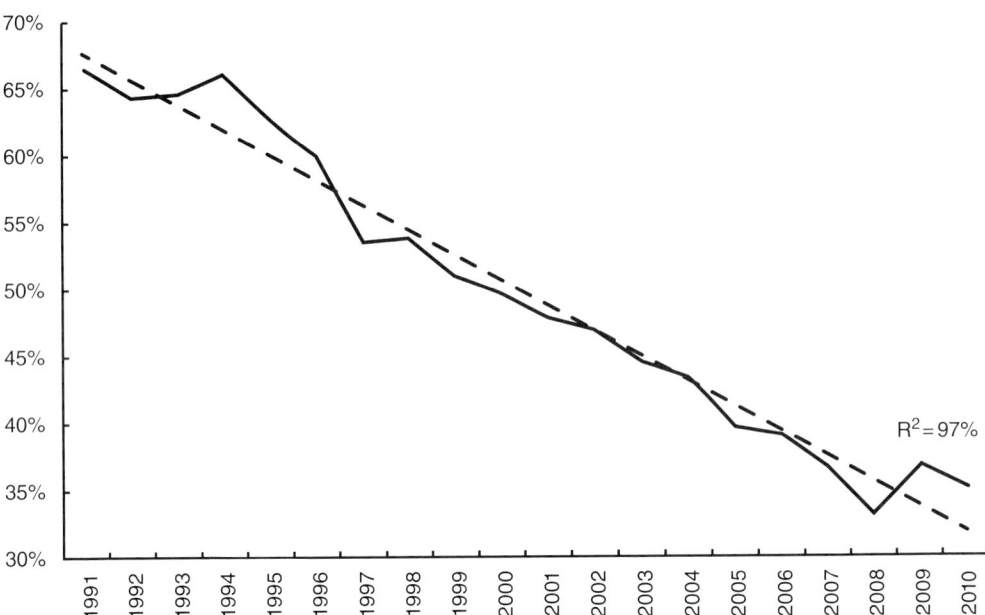

Figure 2.2 Historical development of the ratio of RWAs to total assets for systemically important banks. The banks in sample are as follows: UBS, Barclays, BNP Paribas, Citigroup, HSBC, Credit Agricole, Royal Bank of Scotland, Deutsche Bank, Bank of America, ABN AMRO, Societe Generale, ING Bank, Banco Santander, UniCredit, and Credit Suisse. Source: From Slovik (2011), Fig. 1.

In addition, banks note that the impact of shocks on bank financials can be modeled in a variety of equally valid ways, but uncertainty about the regulatory models obliges them to accumulate "excess" capital to compensate for regulatory model risk.[25] Flannery (2019) observes that even though banks may incur private costs if they feel obliged to rely more heavily on equity finance than they would prefer, the social costs are much less clear. The social costs depend on the value of a reduction in the costs of bank failure and the extent to which other institutions not subject to the stress test can substitute for any reduced lending by the banks included in the stress test program.

But even if the benefits of reducing "excess" capital were large, they would be offset by two consequences for full transparency. Full disclosure would open opportunities for regulatory arbitrage in which banks adjust their portfolios to reduce the regulatory measure of stressed losses without actually reducing their exposures to risk. Banks have a long history of precisely this sort of behavior. One clear example is the substantial decline in the ratio of RWAs to total assets in the years leading up to the GFC. Although when judged by the decline in RWAs to total assets, banks appeared to have reduced the riskiness of their portfolios, the events of 2007–2009 showed that the truth was more likely the reverse (see Figure 2.2).

---

[25] For a fuller discussion of model risk, see the chapter by Canabarro in this handbook (Chapter 16).

Even if banks do not attempt to game the regulatory models, individual banks' models will tend to converge to the regulatory model to minimize uncertainty about the outcome of stress tests. This model "monoculture" will tend to reduce diversification across banks and increase systemic risk (Schuermann, 2013; Hirtle and Lehnert, 2014). Moreover, regulatory models have often proven to be wrong.

In addition to the disclosure of the quantitative results of stress tests, the Fed has also disclosed the results of qualitative assessments of each bank's risk management and capital-planning processes. Failure to meet these qualitative standards has been disclosed alongside the outcomes of the quantitative tests.

Many observers have concluded that the qualitative assessments have produced substantial improvements in each bank's risk management processes and capabilities, as well as the quantity and quality of data available to evaluate the safety and soundness of banks (Kohn and Liang, 2019). Moreover, the approach has ensured that senior management and boards of directors will be engaged in monitoring the institution's risk exposures relative to its capital resources and has disciplined the capital-planning process.

It seems likely that public disclosure of the qualitative assessment has led banks to place greater emphasis on improving risk management capabilities than might otherwise have occurred (Kohn and Liang, 2019). The qualitative assessments have been quite rigorous, with more institutions failing them than the quantitative tests (in the Federal Reserve's CCAR program). This has led to complaints that the qualitative standards are opaque and may convey misleading information to the public. But in the last round of US stress tests, virtually all of the banks passed the qualitative assessment (or came very close to doing so). The Fed has concluded that most banks have made sufficient progress in upgrading their risk management and capital-planning procedures so that, starting with the 2019 stress testing cycle, oversight of qualitative standards will be shifted from the stress test program to the normal Pillar 2 supervisory oversight process under which information about individual banks is disclosed only under the unusual circumstances in which enforcement actions are taken. Although removal of the possibility of public disclosure of failure to meet the qualitative standards may reduce the incentives for some banks to continue to improve their risk management processes, this policy change better aligns Fed practices with those in other countries, where qualitative standards are generally monitored as part of the Pillar 2 process.

9 Limitations of Stress Testing and Crucial Remaining Challenges

Stress tests have become an important part of the toolkit for both microprudential and macroprudential regulatory authorities, yet they cannot be, nor are they intended to be, the only tool. Stress scenarios can portray only one or a small number of a vastly larger set of possible future shocks, and the stress scenarios the authorities select may have little relation to the shocks that actually occur. Moreover, even if the shocks are properly identified, the severity of the actual shocks may be substantially greater. This problem is especially acute during peacetime, when there is considerable uncertainty about the shocks that may threaten the resilience of the financial system. Moreover, stress tests cannot address a number of issues, such as money laundering, the verification of bank asset values, and compliance with laws and regulations, that concern the supervisory authorities.

Progress over the last decade in designing and implementing concurrent stress tests for banks has been remarkable in terms of the availability of more and higher-quality data; advances in designing coherent scenarios; and in modeling how the stress scenario affects various elements of the banks' income statements, balance sheets, and capital positions. These advances have required substantial investments in IT and human capital by both banks and the regulatory authorities. They have enhanced market discipline and the transparency of bank regulation and supervision. It is hoped that these advances will increase the resilience of the financial system and reduce the probability and severity of future banking crises. Nonetheless, a number of challenges remain, many of which have been mentioned in the preceding discussion of the key decisions in designing stress tests.

First, with regard to the design of stress scenarios, it is crucial that the authorities resist disaster myopia and continue to explore a wide range of potential weaknesses in the banking system. This becomes increasingly challenging with longer time intervals since the last major financial crisis. Banks will be inclined to argue that any new or more extreme shock is not plausible. In addition, the costs of changing models and collecting new data will dampen enthusiasm for introducing new shocks. Yet, applying qualitatively similar stress scenarios year after year will tend to lead to a convergence in bank models, result in a reduction in the diversification of assets in the banking system, and produce a false sense of confidence that may inadvertently increase the fragility of the financial system.

It is reassuring that large banks in the United States appear to be well prepared to sustain a substantial recession, but the next major shock to the system may not be related to the business cycle. It may involve a widespread IT failure, cyberattacks, or a major collapse in nonbank financial institutions. Stress tests can be employed to investigate the consequences of such events, but this will require substantial additional investments in data collection and modeling. The introduction of exploratory, biennial stress scenarios by the Bank of England is a promising effort to facilitate the exploration of a broader range of shocks that could be included in future scenarios that determine capital buffers. Moreover, with more and better-quality data and sophisticated modeling techniques, it may be possible for the authorities to run multiple stress scenarios without imposing additional compliance costs on banks.

In addition, stress scenarios need to be expanded to incorporate subsequent-round effects. Current approaches produce, at best, partial equilibrium results with a focus on first-round impacts on bank financial statements. We know, however, that this seriously understates the severity and contagious nature of financial crises. For example, Brunnermeier (2009) has provided dramatic evidence from the GFC broadly quantifying the importance of subsequent-round effects. He explains how $300 to $400 billion of losses in subprime mortgages led to trillions of dollars of potential write-downs in the financial system and the worst global financial crisis since the Great Depression. The transmission and amplification of these losses greatly exceeded the direct contractual losses that are the basis for loss calculations in most stress tests. Rather, the high leverage and heavy dependence on short-term liabilities in combination with the opacity of bank balance sheets and interconnections among institutions led to runs on institutions suspected of being in a weak condition, pressures to deleverage, demands for more collateral in trading markets, and the forced sale of assets at fire-sale prices, which generated losses at other institutions, thus exacerbating downward pressure on asset prices. Funding available to institutions suspected of being near insolvency evaporated. In response, institutions began to hoard liquidity, and some previously liquid markets seized up. The essence of a financial crisis is that solvency shocks can cause liquidity shocks and vice versa.[26]

Although there have been attempts to introduce liquidity factors in macro stress scenarios, most such efforts have been ad hoc and do not attempt to model the full range of mechanisms through which suspected solvency problems can ignite liquidity problems. Other authorities have been experimenting with the introduction of channels through which liquidity shocks amplify macro shocks to banks. But more systematic efforts to incorporate feedback effects will require modeling the behavior of key market participants, which is quite challenging. Meanwhile, the authorities have made extensive efforts to develop liquidity-shock scenarios and stress tests. Although solvency shocks and liquidity shocks often interact, efforts to integrate the two kinds of stress scenarios are difficult, at least in part because of the time dimension: liquidity shocks play out rapidly, well within the span of a single observation in a macro-shock scenario. In that respect, liquidity shocks are more like market shocks, which have an immediate impact rather than the kinds of recession scenarios central to most macro stress tests.

Although the differences in objectives between wartime and peacetime scenarios are widely recognized, less attention has been placed on the transition between peacetime scenarios and wartime scenarios. Kohn and Liang (2019) have produced an interesting example in which a moderate recession would deplete the capital resources of large banks to the extent that they would be challenged to pass the next stress test without reducing lending, which is, of course, precisely the behavior that stress tests seek to limit.

More broadly, this raises a question about the extent to which the authorities can implement their intentions to reduce the severity of stress tests in wartime. Under conditions of uncertainty, market participants may be skeptical about reductions in the severity of stress scenarios, particularly because the advent of wartime conditions means that peacetime stress tests have not been effective. Kohn and Liang's (2019) example points to the serious challenges faced by supervisors in transitioning from peacetime to possible wartime (or at least skirmish) stress testing.

Looking ahead, the stress testing framework should be broadened to explore the wide range of nonfinancial threats to financial stability. To date, most of the emphasis has been placed on examining the consequences of macroeconomic, financial shocks. Yet, nonfinancial shocks may be an even greater threat to financial stability. The universe of such shocks is larger, more varied, and more difficult to identify and model than macroeconomic shocks. Even though threats such as cyber-risk and climate change are widely recognized, they have yet to be incorporated in broad stress scenarios. The Bank of England has pioneered the use of exploratory scenarios to explore some longer-term trends that may jeopardize financial resiliency, such as increasing competition from the nonbank sector; the decline of industries that have been important bank customers, such as the auto or oil industry; a sustained fall in the volume of international trade; an extended period of very low interest rates with a

---

[26] Even without careful attention to the linkages between solvency and liquidity shocks, conventional macro stress tests may reduce the probability of a liquidity shock by reducing the leverage of banks.

flattened term structure; or climate change.[27] Such an exercise is necessarily highly specu-lative but may yield substantial value by prodding banks and their supervisors to consider longer-term strategic challenges. Moreover, it should generate insights for the supervisory authorities about emerging risks that should be incorporated in future stress tests.

Second, the scope of stress tests should be expanded. Because the banks' share of total credit is declining in many countries (e.g., banks' share is much less than half in the United States), it is increasingly important to understand how stress scenarios will affect nonbank financial institutions and financial markets. Although many nonbank financial institutions do conduct stress tests, they are not concurrent with bank stress tests and do not include the same stress factors. This lack of coordination sharply limits the value of these tests in producing insights about how major financial institutions will interact under stress and about the resilience of the financial system more generally. Countries that have an integrated finan-cial supervisor like the UK and Canada are likely to make more progress in this sphere than jurisdictions that have multiple market and institutional supervisors, in which coordination costs are much higher.

The scope of stress tests should be extended across national borders as well. The GFC demonstrated clearly that financial crises have become global. Better oversight of global financial stability requires a global understanding of international transmission mechanisms and emerging risks. Oftentimes the most serious shocks to financial stability, especially in small open economies, originate outside national borders. Progress in this domain will involve unprecedented international coordination and sharing of sensitive bank-specific information, but the Financial Stability Board is taking steps that may ultimately facilitate a more systematic stress testing of global financial stability (see Anderson et al., Chapter 5 in this volume).

Third, the tendency to develop model monoculture needs to be resisted. A drift toward uniformity in modeling pervades many parts of the stress testing process. The first is in the risk identification process and scenario design. For a number of reasons, both supervisory authorities and banks appear to prefer to stick with familiar risks and thus familiar stress scenarios. The Federal Reserve's CCAR scenario paths for most key variables, such as GDP, unemployment, house prices, and equity prices, have followed a similar path every year: a sharp early decline followed by a relatively quick recovery. The Bank of England's use of the BES may help break this pattern.

The second and perhaps more obvious source of evidence of convergence in model-ing may be found in the models that translate the scenarios into outcomes such as losses, balance-sheet dynamics, and net revenues. In an effort to capture recent downturns, modelers

---

[27] The first Bank of England Biennial Exploratory Scenario (BES) (Bank of England, 2017) focused on the strategic responses of banks to adverse trends in the operating environment. The emphasis was not on capital adequacy but rather on how banks would meet regulatory requirements and change business models in adverse circumstances. Thus, unlike with the ACS, banks were permitted to adjust levels of activity in response to the scenario. And because long-term trends were highlighted, banks were required to project results over a much longer (10-year) horizon. The Bank of England made clear that the results of the BES would not be factored directly into additional capital or liquidity requirements for banks. This may undercut, to some extent, the usual opposition from banks to considering nontraditional shocks of unknown probability and thereby facilitated acceptance of the approach. See Anderson et al., in this volume (Chapter 4) for additional discussion of the BES.

are at risk of sacrificing robustness by overfitting their models to the GFC. The trade-off between robustness and accuracy is central to statistical modeling. One reason why the root-mean-squared error is targeted in regressions is that it optimally trades off bias and variance in a quadratic loss world. When aiming for a near-term forecast, however, where the recent sample is more relevant, it makes sense to overweight the minimization of bias. In contrast, for longer-term forecasts and/or forecasting away from the mean, it may be preferable to overweight the minimization of variance. That might mean accepting lower R^2 and other goodness-of-fit measures for simpler but perhaps more robust models, especially ones that are robust to unknown structural breaks (Pesaran et al., 2009).[28]

In part, this can be accomplished by varying and expanding the range of shocks in stress scenarios to reduce incentives to calibrate models to the last crisis in an attempt to increase "accuracy." The substantial investments the banks and authorities have made in existing models are likely to lead to an innate resistance to change, but because the financial system is inherently dynamic, such changes are necessary if the exercise is to continue to be useful in sustaining financial stability.

One other technical detail deserves attention. Most of the models that convert shocks embedded in the stress scenario into losses at banks are linear and include very little feedback from the behavior of other market participants. But we know that as crisis conditions approach, many responses are nonlinear. Although linear approximations are standard practice, one unfortunate consequence is that the shocks in stress scenarios in peacetime need to be increasingly severe to generate substantial losses in banks. But as the severity of shocks increases, critics will argue that the stress scenarios are implausibly severe, thus undermining support for meaningful stress tests.

Finally, stress tests have found a role in a wider range of applications than just capital adequacy. Banks make use of results and insights to help set the internal risk appetite and limits and to support internal capital allocation and performance measurement (return on stressed capital). As the stress scenarios become more extreme, they feed into recovery planning at banks where more drastic actions, such as, for instance, the sale of a business, may be needed to allow the bank to recover. At the limit, a stress may be so severe that it pushes the bank past the point of nonviability, which is how resolution planning is approached at many banks. This stress continuum, from relatively mild stresses for day-to-day risk management all the way to orderly resolution, implies a consistent set of internal limits, escalation protocols, and corresponding countermeasures. It is best to work backward: having established the point of nonviability for resolution planning, what, then, is the point where management may need to break the glass and implement a recovery plan? And typically, such drastic actions should not be required to survive a standard supervisory stress test such as the CCAR. Stress testing and, in particular, the corresponding scenarios are the common thread across these different prudential regulatory exercises.

These are difficult challenges, but in most cases, substantial efforts are being made to conquer them. The progress that has been made over the last decade provides grounds for optimism that stress tests will become an even more valuable tool in achieving micro- and macroeconomic stability.

---

[28] See also the excellent discussion on model challenges in stress testing by Anderson et al. (Chapter 5 in this volume).

References

Acharya, V., R. Engle, and D. Pierret (2014), "Testing macroprudential stress tests: The risk of regulatory risk weights," *Carnegie-Rochester Public Policy Conference Volume of the Journal of Monetary Economics*, 65, 36–53.

Bank of England (2015), "The Bank of England's approach to stress testing the UK banking system," available at www.bankofengland.co.uk/financialstability/Documents/stresstesting/2015/approach.pdf.

Bank of England (2017), "Stress testing the UK banking system: Key elements of the 2017 stress test," available at www.bankofengland.co.uk/-/media/boe/files/stress-testing/2017/stress-testing-the-uk-banking-system-key-elements-of-the-2017-stress-test.

Basel Committee on Banking Supervision (1996), "Amendment to the capital accord to incorporate market risks," January 4.

Basel Committee on Banking Supervision (2004), "International convergence of capital measurement and capital standards: A revised framework," June 9.

Borio, C. E. V., M. Drehmann, and K. Tsatsoronis (2012), "Stress-testing macro stress testing: Does it live up to expectations?" *Journal of Financial Stability*, 12(1), 3–15.

Brunnermeier, M. K. (2009), "Deciphering the liquidity and credit crunch 2007–2008," *Journal of Economic Perspectives*, 23(1), 77–100.

Calomiris, C. W. (2017), *Reforming financial regulation after Dodd-Frank*, Manhattan Institute.

Calomiris, C. W., and R. J. Herring (2013), "How to design a contingent convertible debt requirement that helps solve our too-big-to-fail problem," *Journal of Applied Corporate Finance*, 25, 66–89.

Calomiris, C. W., and D. Nissim (2014), "Crisis-related shifts in the market valuation of banking activities," *Journal of Financial Intermediation*, 23, 400–435.

Curti, Filippo, Marco Migueis, and Robert T. Stewart (2019), "Benchmarking operational risk stress testing models," FEDS Working Paper No. 2019–038, available http://dx.doi.org/10.17016/FEDS.2019.038.

Dent, Kieran, Ben Westwood, and Miguel Segoviano (2016), "Stress testing of banks: An introduction," *Bank of England Quarterly Bulletin*, 56(3), 130–143.

Federal Reserve System (2019), "Amendments to policy statement on the scenario design framework for stress testing, a rule by the Federal Reserve System," *Federal Register*, February 28.

Flannery, Mark (2019), "Transparency and model evolution in stress testing," Fed Conference, Stress Testing: A Discussion and Review Conference, available at www.bostonfed.org/-/media/Documents/events/2019/stress-testing/dynamism-and-transparency-in-stress-testing-paper.pdf?la$=$en.

Flannery, Mark, Beverly Hirtle, and Anna Kovner (2017), "Evaluating the information in the federal reserve stress tests," *Journal of Financial Intermediation*, 29, 1–18.

Goldstein, I., and H. Sapra (2013), "Should banks' stress test results be disclosed? An analysis of the costs and benefits," *Foundations and Trends in Finance*, 8, 1–54.

Guttentag, J., and R. Herring (1984), "Credit rationing and financial disorder," *Journal of Finance*, 39(5) pp. 1359–1382

Guttentag, J., and R. Herring (1986), "Disaster myopia in international banking," Essays in International Finance, 164, 1-40.

Haldane, Andrew G. (2009), "Why banks failed the stress test," Speech at the Marcus Evans Conference on Stress Testing, available at www.bankofengland.co.uk/speech/2009/why-banks-failed-the-stress-test

Haldane, Andrew G. (2011), "Capital discipline," Speech at the American Economic Association Conference, available at www.bankofengland.co.uk/speech/2011/capital-discipline-speech-by-andy-haldane

Hirtle, Beverly J., and Andreas Lehnert (2014), "Supervisory stress tests," Federal Reserve Bank of New York Staff Report No. 696, available at www.newyorkfed.org/medialibrary/media/research/staff_reports/sr696.pdf.

Hirtle, B. J., T. Schuermann, and K. Stiroh (2009), "Macroprudential supervision of financial institutions: Lessons from the SCAP," Federal Reserve Bank of New York Staff Report No. 409.

Kohn and Liang (2019), "Understanding the effects of the U.S. stress tests," Fed Conference, Stress Testing: A Discussion and Review available at www.bostonfed.org/-/media/Documents/events/2019/stress-testing/effects-of-stress-test-paper.pdf?la$=$en.

Pesaran, M. Hashem, Til Schuermann, and L. Vanessa Smith (2009), "Forecasting economic and financial variables with global VARs," *International Journal of Forecasting*, 25(4), 642–675, 703–715.

Quarles, Randall (2019), "A decade of continuity and change," Fed Conference, Stress Testing: A Discussion and Review, available at www.federalreserve.gov/newsevents/speech/quarles20190709a .htm.

Sarin, N., and L. Summers (2016), "Understanding bank risk through market measures," *Brookings Papers on Economic Activity,* Fall, 57–127.

Schuermann, Til (2013), "The Fed's stress tests add risk to the financial system," editorial, *Wall Street Journal*, March 19, 2013.

Schuermann, Til. (2014), "Stress testing banks," *International Journal of Forecasting*, 30(3), 717–728.

Schuermann, Til (2016), "Stress testing in wartime and in peacetime," in Ronald W. Anderson (ed.), *Stress testing and macroprudential regulation: A trans-Atlantic assessment*, CEPR eBook.

Simon, H. A. (1978), "Rationality as process and as product of thoughts," *American Economic Review*, 68, 1–16.

Slovik, Patrick (2011), "Systemically important banks and capital regulation challenges," OECD Economics Department Working Paper No. 916, available at https://dx.doi.org/10.1787/5kg0ps8cq8q6-en.

Tversky, A., and D. Kahneman (1982), "Availability: A heuristic for judging frequency and probability under uncertainty: Heuristics and biases," in D. Kahneman, P. Slovic, and A. Tversky (eds.), *Judgment under uncertainty: Heuristics and biases*, Cambridge University Press.

3

Fit for Purpose? The Evolving Role of Stress Testing for Financial Systems

Udaibir Das, Kieran Dent, and Miguel Segoviano

1 Introduction

Stress tests are applied across a wide range of sectors, from medicine, aeronautics, and engineering to finance. Although the purpose and nature of the stress tests vary, the underlying concept is similar. In general, a stress test is used to assess the ability of an object, person, institution, or system to withstand extreme stress or an adverse event. In medicine, for instance, cardiologists use stress tests to assess heart and blood flow during physical exertion. In engineering, stress tests are used to establish the breaking point of components by subjecting them to extreme pressure. Stress tests used by actuaries evaluate several statistically defined possibilities to determine the most damaging combination of events and the likelihood of such an event.[1]

In banking, stress tests have traditionally been used to assess the impact of adverse scenarios on the resilience of balance sheets and sustainable profitability. The adverse scenarios can be systemic or idiosyncratic in nature. For example, they might be calibrated to reflect a severe economic downturn or financial stress that would affect many banks. Or they might be calibrated to reflect a significant increase in claims relating to the past miss-selling of financial products, which would affect banks that had sold these products.

Bank stress tests have their roots in the early 1990s where they first emerged as a discipline within banks' trading books. The use of bank stress tests increased steadily after the Asian financial crisis of 1997, and they have now become the sine qua non of supervision and systemic risk oversight. Several elements of the post–Global Financial Crisis (GFC) reforms are now striving to promote the implementation of stress testing on a continuing basis for internationally active banks. Stress tests have become a core part of both the microprudential toolkit to manage risk at individual banks and are now being incorporated in macroprudential policymaking to manage systemic banking risk.

The remainder of this chapter explores the evolution of stress tests—from their emergence in the early 1990s to the present day. Section 2 focuses on stress testing in financial institutions, tracing its development from small-scale trading-book exercises to sophisticated enterprise-wide activities. Section 3 addresses the use of bank stress tests by the official sector, which has undergone a step change in the wake of the financial crisis to become a key part of the regulatory toolkit. Finally, Section 4 considers some possible developments in stress tests over the coming years.

---

[1] See, for example, Heale (2013).

2 Stress Testing in Financial Institutions

Banks' desire to evaluate the sensitivity of their balance sheets and income statements predates the advent of formalized stress testing, with efforts being made to calculate risk-adjusted measures of return on capital in the mid-1970s. However, stress tests of the sort that would be recognized today, in which the implications of a clearly defined adverse scenario for bank balance sheets and profitability are analyzed, first emerged as a discipline within the trading books of large, internationally active banks in the early 1990s.

These early stress tests were used by trading-desk managers to evaluate their vulnerability to market risk, the risk of losses arising from adverse movements in market prices. The scenarios used could be historical or hypothetical. Under a historical scenario, the paths for economic and financial-market variables associated with past loss events were re-run to explore the impact of a repeat occurrence (see Table 3.1 for examples of such historical scenarios). In contrast under a hypothetical scenario, the variable paths specified were not associated with any historical event but rather were chosen to be consistent with an economic or financial-market shock.

Table 3.1. Historical scenarios in early stress tests

Scenario	Year	Description
Black Monday	1987	An international equity market crash, in which the value of major international stock markets fell by between 19% and 40% within a single month.
US interest rate shock	1994	A series of unanticipated increases in the federal funds rate, resulting in a significant reduction in the value of bond and other fixed income portfolios and an associated increase in yields. The resulting shock generated contagion into the US equity market.
Mexican peso crisis	1994	A sudden devaluation of the Mexican peso, followed shortly by the peso being allowed to float freely. Investors withdrew from exposures to Mexico and other emerging economies. The resulting contagion spread to financial markets throughout Latin America and Asia.
Asian financial crisis	1997	A series of currency devaluations following speculative attacks on Asian countries with fixed exchange rates. This spread throughout South East Asia, with affected countries experiencing significant currency devaluations, alongside collapse in the value of stock markets and other asset prices.
Russian financial crisis	1998	Economic turmoil leading to an erosion of investor confidence in Russia. This resulted in a severe shock to the value of Russian stock, bond, and currency markets. The Russian government responded by announcing a simultaneous devaluation of the rouble, a default on domestic debt obligations, and a moratorium on the repayment of foreign debt.

Source: Dent et al. (2016).

Udaibir Das, Kieran Dent, and Miguel Segoviano

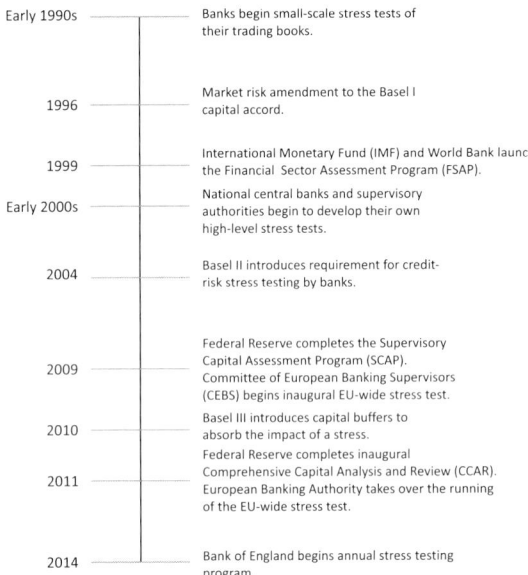

Early 1990s	Banks begin small-scale stress tests of their trading books.
1996	Market risk amendment to the Basel I capital accord.
1999	International Monetary Fund (IMF) and World Bank launch the Financial Sector Assessment Program (FSAP).
Early 2000s	National central banks and supervisory authorities begin to develop their own high-level stress tests.
2004	Basel II introduces requirement for credit-risk stress testing by banks.
2009	Federal Reserve completes the Supervisory Capital Assessment Program (SCAP). Committee of European Banking Supervisors (CEBS) begins inaugural EU-wide stress test.
2010	Basel III introduces capital buffers to absorb the impact of a stress.
2011	Federal Reserve completes inaugural Comprehensive Capital Analysis and Review (CCAR). European Banking Authority takes over the running of the EU-wide stress test.
2014	Bank of England begins annual stress testing program.

Figure 3.1 Timeline of key events in the development of stress testing.
Source: Dent et al. (2016).

The nature of the stress scenarios thus varied considerably across banks. Even where scenarios at different banks referenced the same historical event, the variable paths contained within varied widely across banks (Committee on the Global Financial System [CGFS], 2001). Similarly, there was significant variation in the way these early market-risk stress tests were used by banks. At most banks, stress tests were used to estimate maximum loss and set exposure limits on trades. Some banks went further, using stress tests to inform contingency plans for times of market stress or the allocation of capital across business lines (CGFS, 2000).

The active use of trading-book stress testing got a boost in the mid-1990s with an amendment to the prudential Basel I capital accord that incorporated market risks (the risk of losses on positions associated with changes in market prices) (see Figure 3.1) (Basel Committee on Banking Supervision [BCBS], 1996). This amendment made bank supervisors require banks to measure and apply capital charges for their market risks in addition to their credit risks. The market-risk amendment introduced two approaches for calculating capital charges for market risk. The first was a standardized approach under which banks' exposures would be bucketed according to their characteristics (e.g., instrument type and maturity) and a "standard" capital charge would be applied to all exposures that fall into that bucket. The second was an approach that allowed banks to use their own internal risk models to calculate market-risk capital charges. Banks seeking to use this internal-model approach were required to have a rigorous and comprehensive stress testing program in place.

The Basel accord amendment played a key role in making the supervisory requirement of stress testing for market risk a standard practice at large international banks. However, the development of stress testing for credit risk significantly lagged that of market risk. This was despite credit risk being by far the largest risk to which banks are normally exposed.

The 1999 BCBS report on credit-risk modeling at large international banks concluded that few institutions were performing stress tests for credit risk, and where they were, stress tests were only being performed sporadically. The report expressed concern that banks had not dedicated significant resources to devising appropriate credit-risk stress tests.

In 2000 and again in 2004, the CGFS of the Bank for International Settlements (BIS) carried out a survey of stress testing practices (CGFS, 2000, 2001, 2005). The findings confirmed that stress testing for market risk had become a standard risk management technique in select portfolios of internationally active banks. However, the surveys found that developments in credit-risk stress testing were lagging those of market risk by a large margin and also noted that no respondent had integrated market and credit risk into its trading-book stress tests. Difficulty in marking-to-market loan portfolios and insufficient data series were suggested as the factors impeding stress testing of credit risk.

In 2004, attempts were made at the international level to address the lag between the development of stress tests for market and credit risk. The BCBS Basel II accord introduced a more risk-sensitive approach to calculating capital requirements for credit risk, relative to the Basel I standard agreed upon in 1988 (BCBS, 1988, 2004). It effectively extended the principles of the market-risk amendment to credit risk, introducing a standardized approach, as well as an internal-ratings-based (IRB) approach that made it possible for banks to use their own internal risk models to calculate credit-risk capital charges. As with market risk, banks seeking to use the internal-model approach were required to have a program of credit-risk stress testing in place. In addition, Basel II made it a requirement for banks to use stress tests to assess the adequacy of the regulatory capital requirements calculated under the IRB approach. Banks were expected to operate with a level of capital equivalent to the higher of (1) the regulatory capital calculated under the IRB approach or (2) that necessary to absorb the impact of their stress tests. Basel II also suggested that the results of these stress tests be subject to review by national supervisors.

The GFC exposed significant shortcomings in the design and scope of the stress tests that were being run by the banks. A BCBS report concluded that precrisis stress tests relied too heavily on historical information and statistical relationships drawn from the period of benign economic operating conditions that preceded the crisis (BCBS, 2009). The scenarios used in such tests were of a duration and severity well below those of the crisis and typically resulted in loss estimates equivalent to a quarter's worth of earnings at most banks (well below most banks' crisis loss experience). Precrisis stress tests were also found to have significant blindsides, including for counterparty credit risk, funding liquidity risk, and the behavior of more complex financial instruments.

Ultimately, the introduction and proper implementation of credit-risk stress testing requirements came too late to have any meaningful impact on outcomes, leading to the 2008 GFC. Systemically important financial systems had not universally implemented the requirement. Some continued to use the Basel I standard, which carried no requirement for banks to perform credit-risk stress testing. In others, the decision to implement Basel II came late and had not been implemented in a timely manner, and the IRB permissions needed to carry out the credit-risk stress testing only started in 2006–2007. A BCBS investigation carried out in 2009 confirmed that although stress testing for market risk was well established on the eve of the GFC, stress testing for credit risk was still in the nascent stages (BCBS, 2009).

In general, prior to the financial crisis, stress testing was not carried out as an integrated enterprise-wide exercise but rather was conducted as separate exercises for risks, or portfolios by risk function, with little engagement from business areas or senior management (BCBS, 2009). To strengthen the bank stress test requirements, the BCBS established principles for sound stress testing practices and supervision in 2009. For banks, these principles sought to (1) improve the use of stress tests and their integration into risk management, (2) improve banks' stress test methodology and scenario selection, and (3) identify areas of focus for stress tests relating to risks highlighted during the financial crisis. In addition, supervisors were encouraged to assess banks' compliance with the principles for sound stress testing. The revised international capital standards agreed upon in the wake of the crisis, Basel III, also reiterated the requirements for IRB banks to perform stress tests for credit and market risk (BCBS, 2011).

Since the financial crisis, the sophistication of bank stress tests has improved significantly. This was confirmed by a fresh BCBS survey of stress testing approaches at 54 banks (including 20 globally systemically important banks) in 2016 (BCBS, 2017). The BCBS report found that stress testing for credit risk had become standard industry practice, with all banks surveyed including both credit and market risk in their stress tests. At the same time, enterprise-wide stress tests were becoming standard practice at large international banks, with two-thirds of survey respondents completing enterprise-wide stress tests at least annually.

However, challenges remain, and the BCBS (2017) report highlighted several issues that present barriers to the further development of stress testing at banks, including data quality and availability. At some banks, stress tests were still being viewed largely as a regulatory compliance exercise, with poor integration into business-as-usual risk monitoring and management. Given the significant amount of resources banks devote to stress testing, the need for better integration of stress tests into business-as-usual processes, and thus derive greater business benefits from them, was identified as a key priority for banks participating in the BCBS survey. To this end, the BCBS published a revised set of stress testing principles in 2018, including an expectation that stress tests should be used as a tool to inform risk appetite and business decisions around portfolio management and funding allocations at banks (BCBS, 2018).

3 Official-Sector Stress Testing

The development of banking stress tests in the official sector followed the introduction of stress testing at banks. Although central banks and banking supervisors had been exploring the impact of what-if analyses on their banks for some time, the use of stress testing as a policy tool for formalizing such analysis emerged only toward the end of the 1990s. In contrast to the stress tests performed by banks, the scope of these early official-sector stress tests was very broad. Whereas early bank stress tests tended to focus narrowly on risks within a business unit or portfolio, early official-sector stress tests tended to be very high level, attempting to explore risk across the entirety of the banking and financial systems. These exercises relied on highly partial and aggregated data, and as a result, their risk and transmission-channel coverage was severely limited, by both the lack of data and modeling limitations.[2]

---

[2] See, for example, Bank of England (2006), Box 6 in Section 3.

The idea that stress testing should become a frequently used tool by the official sector (other than bank supervisors for supervisory purposes) was spurred on by the International Monetary Fund (IMF) and World Bank (WB) Financial Sector Assessment Programme (FSAP) launched in response to the Asian crisis of 1997 (IMF, 2014). Stress testing was an essential component of the FSAP and meant to provide a quantitative measure of the vulnerability of the financial system (to the extent feasible) to different shocks and to complement the insights gathered from other components of the FSAP assessment. Unlike traditional bank supervisory stress tests, the FSAP stress test was aggregate in nature, using bank-by-bank data, and sought to aggregate, interpret, and communicate the results to avoid signaling out any one bank or a set of banks. Initially, the approach was to treat each risk separately, but the FSAP practices rapidly evolved to measuring and estimating different risks jointly within banks and other financial institutions, as discussed later in this chapter.

The early use of stress testing within the FSAP encouraged central banks and bank supervisors in many countries to perform their own banking stress tests. In some cases, these simply updated the stress test exercises carried out in previous FSAPs, whereas in other cases, central banks or supervisors sought to develop their own independent stress test exercises.[3] Eventually, these approaches would evolve into the concurrent stress testing frameworks widely used by central banks and supervisors today.

Until the GFC, the stress tests performed by central banks and bank supervisors tended to be on an ad hoc basis and were carried out at a relatively high level. Relatively few resources were allocated to official-sector stress tests precrisis, and the impact of the stress was often modeled by treating the banking or financial system as a single entity. This limited the tests' usefulness for supervisors of individual banks or financial institutions and made it difficult to capture the interactions between such entities. In addition, the results of these pre-GFC official-sector stress tests rarely had a direct impact on regulatory or financial policies beyond occasionally being published in central bank publications, such as financial stability reports, or being used for supervisory purposes. And even then, any disclosure was on an aggregated, system-wide basis, with no detail on the performance of individual banks.[4]

The severity of the GFC brought with it a step change in the sophistication and usage of stress tests by the official sector. This was also consistent with the availability of advanced stress testing methods. Official-sector stress tests went from being small-scale, isolated exercises to large-scale, comprehensive risk assessment programs, often leading directly to policy responses. Unsurprisingly, the substantial advancement in the official-sector stress tests has been made possible only by a significant increase in the resources devoted to stress testing in many jurisdictions.

Stress tests played a key role in the policy response to the financial crisis, demonstrating their usefulness as a crisis-management tool. A prominent example of this is the launch of the US Supervisory Capital Assessment Program (SCAP) conducted by the Federal Reserve (Fed) in 2009 (Bernanke, 2009, 2010; Geithner, 2014). This supervisory stress test was applied to 19 of the largest domestic US banks against a common stress scenario. The use of a common scenario facilitated more easy comparison and peer review of participants and offered insight into the possible banking-system-wide consequences of such a scenario.

---

[3] See, for example, Bunn et al. (2005) and Bank of England (2006, Box 6 in Section 3).
[4] See Hoggarth and Whitley (2003), Bunn et al. (2005), and Bank of England (2006).

The scenario was to unfold over a common time frame with a common balance-sheet starting point. Again, this facilitated easier comparison of banks' results and insight into the system-wide consequences. The results of the SCAP were published on an individual-bank basis. Those banks judged to need additional capital were given 6 months to raise it, with the US Treasury Department providing a backstop in the event that any bank was unable to do so in private markets.

Stress testing was also deployed as a crisis-management tool in Europe, with the Committee of European Banking Supervisors (CEBS) initiating a stress test of the EU banking sector in May 2009 (CEBS, 2009). The CEBS stress test included 22 major European banking groups with cross-border operations. As with the SCAP, participating banks were subjected to a common stress scenario designed to be significantly more severe than the outlook for EU economies at that time. The Tier 1 minimum capital standard in the CEBS test was the same as that of the SCAP, 6 percent of risk-weighted assets (RWAs) (there was no CET1 hurdle rate in the CEBS test). In contrast with the SCAP, the results of the 2009 CEBS stress test were not published on a bank-by-bank basis, the test had no funding backstop, and the CEBS was unable to compel banks to raise capital in response to the test (this point is explored further in Chapter 12). The CEBS exercise was repeated in 2010, and this time, the results were published on a bank-by-bank basis (CEBS, 2010).

The use of concurrent stress testing as a crisis-management tool in the wake of the GFC has been followed by a proliferation of less ad hoc, public, concurrent stress testing frameworks across the official sector. The first of these was the Fed's Comprehensive Capital Analysis and Review (CCAR) in the United States (Bernanke, 2013; Hirtle and Lehnert, 2014). Launched in 2011, the CCAR involves both an annual concurrent stress test of major US banks, in which banks are subjected to a common stress scenario over a common time frame with a common starting point, and a review of their capital-planning processes. The use of the latter provides a qualitative assessment of the adequacy of banks' risk management processes to complement the quantitative assessment of their capital adequacy provided by the stress test. The CCAR stress test includes US bank holding companies with assets over $50 billion, with results published on a bank-by-bank basis. The stress test incorporates common baseline, adverse, and severely adverse scenarios and requires participating banks to submit results from their own bespoke stress scenarios. Results from these bespoke scenarios provide insight into banks' own views on the risks to which they are most exposed and may address risks relevant to particular banks that are not covered in the common scenarios. Ultimately, the results of the severely adverse scenario generally determine whether or not participating banks are required to amend their forward-looking capital plans.

The EU-wide stress test was repeated for the third time in 2011, this time under the auspices of the European Banking Authority (EBA). There was then a 3-year gap before the EBA launched what has now become a biennial concurrent stress test of major European banks, with the results published in October 2014. The EBA stress test does not have a single inclusion threshold because it seeks to capture the largest banks in each EU member state, but banks must have total assets of at least €30 billion to be considered for inclusion. The EBA test incorporates common base and stress scenarios.

The EBA test is run on a static-balance-sheet basis, with the scenario shocks applied directly to banks' starting balance sheets. The use of a static balance sheet effectively prevents firms from proposing to mitigate the impact of the stress by deleveraging, which could exacerbate the effects of the initial stress on the real economy by constraining the supply of credit to households and firms. Stress impacts are projected by banks themselves, with projections reviewed, and adjustments made, by a combination of national supervisors and EBA staff. There is no minimum capital requirement in the EBA test, with the requirement for banks to take remedial actions at the discretion of their national supervisors. Results are published on a bank-by-bank basis, and as with the CCAR, the focus of the EBA test is predominantly supervisory. But the test results are taken into account by the European Systemic Risk Board, the body responsible for the management of systemic risk in the European Union.

Shortly after the launch of the EBA test, the Bank of England introduced its own public, concurrent stress test with the results of its inaugural test published in December 2014. The Bank of England's concurrent stress testing framework currently captures all UK banks with retail deposits of more than $50 billion. The stress test utilizes a dual-scenario approach, with an annual macroeconomic and financial-market stress scenario and a biennial exploratory scenario used to explore new and emerging threats to financial stability. The Bank of England test is run on a dynamic-balance-sheet basis, with banks expected to produce annual projections for 5 years after the onset of the stress scenario. The use of a dynamic balance sheet enables the Bank of England to consider the adequacy of participating firms' forward-looking capital plans and evaluate management actions proposed by firms to mitigate the impact of the stress scenario as it unfolds.

The approach of the Bank of England is closer to that of the EBA than the Federal Reserve, with banks' own stress projections forming the basis of the stress test results and Bank of England models used to challenge and inform adjustments to these starting projections. Banks participating in the Bank of England test are expected to continue to lend through the stress while maintaining sufficient capital resources to exceed the sum of the Pillar 1 minimum CET1 capital and Tier 1 leverage requirements and any buffers for global and domestic systemic importance.[5] Banks that fall below the hurdle rate are likely to be expected to submit a new capital plan setting out how any capital shortfall will be addressed. The results of the Bank of England test are published on an individual-bank basis and used to inform both macro- and microprudential policy. For example, the setting of both the Pillar 2 supervisory buffer and the UK countercyclical capital buffer rate is informed by the results of the concurrent stress test.

The FSAP stress testing framework has also been quick to respond to the lessons of the GFC. For a detailed exposition of the evolution of the IMF FSAP stress test, see Chapter 26. The focus is shifting from a supervisory orientation to a macroprudential one to help identify macrofinancial risk channels and contagion. The attempt is to focus not on the size of the shock and the initial buffers of individual institutions alone but on responses to the shock and how they interact with each other and with other financial agents. Capturing these responses and interactions is important because they can serve to amplify the effects of an initial shock well beyond the first-round effects captured by existing stress tests. Thus, FSAPs are incorporating assessments of systemic risk and interconnectedness across firms

---

[5] The Bank of England's hurdle-rate framework is currently under review and may change following the implementation of International Financial Reporting Standard (IFRS) 9.

in financial markets. For more discussion of systemic risk and interconnectedness in stress tests, see Chapters 20 and 23.

4 The Future of Stress Testing

The broad contours of how modern-day bank stress testing practices have evolved reveal that the process has been a gradual one. One of the key lessons of the GFC has been the need for ensuring resilience and sound bank balance sheets. The public accountability resulting from having the ability to identify entity-level as well as systemic risks has gone up. The significance of stress testing has thus risen and continues to develop in terms of technical rigor, application, and policy use. The future of bank stress testing is most likely going to be conditioned by the six considerations discussed in the following subsections.

4.1 A Broader Risk Perimeter

Incorporation of a broader range of risks for enterprise-wide stress tests. Considerable progress has been made on this front since the GFC, with most banks incorporating both credit and market risk in their solvency stress tests. However, full integration of risks within enterprise-wide stress testing remains some way off, with most banks continuing to run solvency and liquidity stress tests as separate standalone exercises. The official sector is also capturing as wide a perimeter as possible for its macroprudentially oriented stress testing.

4.2 Incorporating Nonbanking Channels

Official stress tests should continue extending their reach beyond the core banking sector to the wider financial system. The interconnections in a system can serve to transmit a shock originating in one sector of the financial system (e.g., asset managers) to other sectors in the system (e.g., banks). These transmissions can go through multiple iterations, resulting in a system-wide impact that far exceeds the impact on the initially afflicted sector (Cortes et al., 2018). To enhance the ability of stress tests to support macroprudential policy, it is important that stress testing frameworks begin to investigate the interconnections between the banking sector and the broader financial system (Alla et al., 2018). These interconnections include both direct links (through financial transactions between institutions in different market segments) and indirect links (through the impact of the behavior of institutions in one market segment on institutions in another).

A direct link is created when institutions in various parts of the financial system enter into a financial transaction, for example, a repurchase agreement (or "repo"). In times of crisis, the haircut, or discount applied to the collateral in repo transactions, demanded by repo purchasers tends to increase (as was the case in the GFC; Gorton and Metrick, 2012). This reduces the amount of funding a financial institution can obtain for a given asset via a repo, and this can exacerbate funding pressures faced by repo sellers, who may already be constrained by a lack of market liquidity.

Even in the absence of direct links between financial institutions, the behavior of institutions in one part of the financial system can still affect institutions operating in another part. One example of this is the behavior of asset managers who have investor mandates that set out a minimum credit quality for the assets that the fund can invest in. During the recent financial crisis, widespread rating downgrades reduced the credit quality of a sizeable number of financial assets below investment grade. This led to a large-scale liquidation of affected assets, significantly reducing the prices of these assets. The price reductions, in turn, reduced the balance-sheet value of marked-to-market assets and therefore imposed losses on other financial institutions' asset holdings (Deb et al., 2011).

4.3 Capturing Systemic Effects—Design

Capturing the impact of systemic risk amplification and linkages between macroeconomic and financial stability. This would improve the ability of stress tests to support macropru-dential policy. As noted earlier, the results of concurrent stress tests already have a direct effect on macroprudential policy. However, their use in informing macroprudential policy has been more limited where concurrent stress testing frameworks have been introduced. In part, this is likely to reflect the fact that system-wide results have typically been produced by aggregating the results of individual banks. Although such an approach may be useful in understanding the impact of a stress on individual banks in isolation, it is less useful for understanding systemic risk. This is because these approaches fail to account for the feedback and amplification channels that can exacerbate the impact of an initial shock (Aymanns et al., 2018). Recently, authorities have prioritized the development of tools that attempt to quantify losses from systemic effects. Major central banks, including, among others, the European Central Bank, the Bank of England, the Bank of Canada, and the IMF, are investing significant resources in this objective. For more details on system-wide stress testing, see Chapters 28 and 30.

Systemic effects can endogenously amplify losses through macrofinancial feedback mechanisms and have the potential to magnify moderate exogenous shocks into substantial negative financial outcomes with large welfare effects. Systemic risk-amplification mechanisms are complex and change across time. These mechanisms are fueled by intricate macrofinancial loops and interconnectedness (direct and indirect) across financial entities and markets, which pave the road for loss contagion. Therefore, when modeling systemic risk, it is essential to identify such structures and how they might change at different stages of the financial cycle. Interconnectedness might not be apparent in normal times but can become significant in periods of financial distress, possibly giving rise to the nonlinear increases in magnitude and the speed of loss propagation observed during financial crises. For example, during the GFC, a substantial deterioration in the macroeconomic outlook coupled with significant losses that eroded banks' capital resources brought into question banks' ability to continue to operate as going concerns. This led to widespread market uncertainty over banks' long-term viability, resulting in a virtual shutdown of bank funding markets as solvency concerns crystallized in a liquidity crisis. In such circumstances, the behavioral responses of banks can become another important amplification channel. If all banks respond in the same manner, for example, by looking to substitute retail deposits for wholesale funding, increased competition for retail deposits is likely to drive up deposit rates. This increase in funding costs is, in turn, likely to have an adverse impact on bank profitability, worsening the impact of the initial stress on bank balance sheets and income statements.

4.4 Capturing Systemic Effects—Implementation

Conducting comprehensive system-wide stress test, developing adequate data, and building technical capability. Data constraints, the understanding of the intricacies of amplification mechanisms, how best to model those mechanisms, and how they might interact in complex financial systems impose significant impediments to both researchers and policymakers trying to implement stress-testing frameworks that can adequately quantify systemic effects.

The applied literature has taken two main approaches to quantifying systemic risk (Aymanns et al., 2018). These are (1) the development of simulated models (e.g., network and agent-based models [ABMs]) that attempt to capture the network structures and the behavioral responses of financial entities that ultimately underpin the interconnectedness structures among such entities and (2) the estimation of empirical models that attempt to extract the interconnectedness structures that can lead to systemic risk from market or balance-sheet data.

Simulated models have made important contributions to the analysis of contagion and have highlighted specific systemic-risk channels. These frameworks, however, require highly granular data that are not available in many countries (for more details, see Chapter 10). They frequently assess limited sets of amplification mechanisms and often become highly complex when such sets are expanded. This may have limited their role in policy analysis (Elsinger et al., 2013). Alternatively, empirical models derive metrics of interconnectedness and systemic risk from either balance-sheet or market data. However, such metrics are "reduced form," meaning that although they capture the effects of amplification mechanisms, they do not provide information on the specific channels of contagion that can lead to the materialization of systemic risk. Moreover, metrics derived from empirical models might be subjected to issues related to market data, especially when markets are illiquid. Finally, such metrics are usually not comparable to metrics obtained from standard stress tests; thus, empirical models to quantify systemic risk have not been embedded into macroprudential stress test frameworks.

Given the complexity of modeling and implementing stress tests that capture systemic effects, going forward, we believe that a promising approach for further progression of macroprudential stress testing involves the development of "encompassing frameworks" (as proposed by Alla et al. [2018]), that is, frameworks that aim to integrate diverse types of data and approaches as a way to maximize the information content of heterogeneous data sources and minimize potential model error. Encompassing frameworks can be seen as pragmatic approaches to the development of robust and implementable macroprudential stress tests. As an example, encompassing frameworks can be developed to combine the use of microprudential stress tests with reduced-form approaches to estimate systemic-effect losses. Such approaches could provide important benefits, as follows:

- They could permit the quantification of systemic-effect losses, even in cases when highly granular data to model amplification mechanisms are not available. Similarly, reduced-form approaches do not require modelers to explicitly model agents' behavioral reactions, which are difficult to incorporate in a comprehensive way and complex to properly calibrate. Approaches of this type would also be able to utilize microprudential stress test models and expertise already existing in many countries.

- They can be developed in multivariate dimensions by characterizing financial systems as portfolios of entities or sectors. This facilitates the integration of nonbank financial intermediaries into the analysis of systemic risk; thus, interactions between banks and nonbanks can be considered when quantifying systemic-risk-amplification losses.
- They can be developed as stochastic frameworks, allowing the estimation of the firms' asset values conditional on different states of nature, including specific valuations of other firms in the financial system. The stochastic feature of these models allows the modeler to quantify the probabilities of these events happening and the intensity of losses under these events. This feature can be extended to estimate distributions of macroeconomic scenarios and estimate their probabilities of occurance (vs. the discrete scenarios that are usually employed in stress tests, for which probabilities of occurrence cannot usually be estimated, thus describing such scenarios as "extreme but plausible").

However, reduced-form approaches to estimate systemic-effect losses imply costs in terms of not having detailed information on the channels that amplify systemic risk. Therefore, rather than an end per se, reduced-form approaches to estimate systemic-effect losses can also be seen as useful tools to support the further development of complementary methods, including ABMs and general-equilibrium frameworks. As publicly available data continue to improve and behavioral and theoretical models continue to develop, alternative encompassing frameworks could be developed by combining behavioral or theoretical models with calibrations based on reduced-form empirical approaches (akin to what has been done, e.g., in macroeconomics, where empirical moments are used to calibrate dynamic stochastic general equilibrium [DSGE] models).

For example, reduced-form approaches could be very useful to calibrate the nonlinear effects (e.g., decrease in prices, increase in probabilities of distress) and changes in behavioral assumptions, especially in times of distress, that can lead to systemic risk materializing. Espinoza et al. (2020) propose a general-equilibrium framework that incorporates macroeconomic and systemic-risk interactions. The authors show how empirical measurements of systemic-effect losses estimated with a reduced-form approach can be used to calibrate the parameters of the theoretical model that incorporates various systemic risk-amplification channels, including interbank lending, common asset exposures, and a "Minsky effect." The combination of such frameworks has the potential to provide policymakers with improved information, including the benefits of reduced-form approaches with the better insights of behavioral or theoretical models.

Pursuing system-wide stress testing is likely to deliver several important benefits for macroprudential policy. For example, a system-wide stress test could support a better understanding of how shocks propagate through the financial system and contribute to the buildup of systemic risk. This could inform policy efforts aimed at ensuring that banks and other regulated entities are insulated from contagion risk emanating from unregulated sectors of the financial system. System-wide stress testing might also help to guard against any perverse incentives created by the significant increase in banking regulation that has followed the financial crisis. Increased regulation creates incentives for more risky activities to be moved outside of the core banking sector and into the "shadow" banking sector (for more details see Chapter 20). A system-wide stress test would have the potential to shine a light on such practices and thus inform the appropriateness of the regulatory perimeter.

4.5 Greater Business Benefits

As noted earlier, for banks, one priority is likely to be deriving greater business benefits from stress testing, particularly given the substantial resources devoted to it. Most banks are already using stress tests for capital planning and to inform their risk appetites, but for some banks, stress tests continue to be seen as a regulatory compliance exercise (BCBS, 2017a). Given the significant resources devoted to stress testing, improving the efficiency of stress testing processes and increasing the extent to which their results inform everyday business and strategic decisions are likely to be areas of focus for banks going forward.

4.6 Impact of Technology

Efforts at reducing the resource burden of stress testing on a continuing basis. Further improvements in data and information technology (IT) infrastructure would support increased automation of stress test development and analysis. A constraint to greater automation at banks is the often lack of predictability in the stress testing requirements from the regulators and the variation in the requirements imposed across different jurisdictions. For example, changes in data requirements from one year to the next can make it difficult for banks to automate the collection of the information they must submit in support of their regulatory stress test results. Although definitional differences across jurisdictions limit banks' ability to utilize a common stress testing infrastructure, such an infrastructure is becoming more important when testing cross-border activities. This would call for greater predictability and coordination in official-sector stress tests.

5 Concluding Thoughts

Since first emerging in the early 1990s, stress tests have grown from isolated exercises conducted within banks' trading books to become a comprehensive and sophisticated risk management tool used extensively by both banks themselves and the official sector. Within banks, stress testing programs have moved beyond market risk to encompass a broad range of risks, including credit and liquidity risk. At the same time, their use has extended from simply quantifying risk exposure to informing the setting of risk appetites and capital planning. Within the official sector, concurrent stress testing programs that stress banks against a common scenario at the same time are becoming commonplace. The use of concurrent stress tests to inform microprudential and, to some extent, macroprudential policy is also becoming standard practice. Disclosure has also increased significantly since the crisis, with many authorities now regularly publishing the results of their stress tests on both an aggregated and individual-firm basis (for more detailed exposition on the issue of disclosure within stress testing, see Chapters 11 and 12).

Although noteworthy progress has been made in stress testing within both the private and public sectors, there are areas where more integrated approaches are needed. Within banks, there is more to be done to ensure that stress testing is viewed not merely as a regulatory compliance exercise but as a tool that can yield significant business benefits. At the same time, there are still some significant blindsides in the risk types explored in bank stress tests, and there is still more to do to ensure that enterprise-wide stress tests address all the risks to which banks are exposed. Within the official sector, progress toward using stress tests to inform macroprudential policy has been more limited than for microprudential policy. To this end, work is needed to ensure that official-sector stress tests capture the amplification and feedback channels that exacerbate the effects of an initial shock and thus contribute to systemic risk. There is also a need to extend stress testing beyond the core banking sector to ensure that the benefits stress testing can deliver for macroprudential policy are not limited to banks.

Many of the areas in which stress tests are likely to develop further over the coming years are explored in more detail in other chapters in this book.

References

Alla, Zineddine, Raphael A. Espinoza, Qiaoluan H. Li, and Miguel A. Segoviano (2018), "Macroprudential Stress Tests: A Reduced-Form Approach to Quantifying Systemic Risk Losses," IMF Working Paper No. 18/49.

Anderson, Ron, Jon Danielsson, Chikako Baba, Udaibir S. Das, Heedon Kang, and Miguel Segoviano (2018), "Macroprudential stress tests and policies: Searching for robust and implementable frameworks," IMF Working Paper No. 18/197.

Aymanns, C., J. Doyne Farmer, A. Kleinnijenhuis, and T. Wetzer, (2018), "Models of financial stability and their application in stress tests," in C. Hommes and B. LeBaron (eds.), *Handbook of computational economics*, Vol. 4, Elsevier.

Bank of England (2006), "Financial stability report, July," available at www.bankofengland.co.uk/publications/Documents/fsr/2006/fsrfull0606.pdf.

Bank of England (2014), "Stress testing the UK banking system: 2014 results," available at www.bankofengland.co.uk/-/media/boe/files/stress-testing/2014/stress-testing-the-uk-banking-system-2014-results.pdf.

Basel Committee on Banking Supervision (1988), "International convergence of capital measurement and capital standards," July.

Basel Committee on Banking Supervision (1996), "Amendment to the capital accord to incorporate market risks," January.

Basel Committee on Banking Supervision (1999), "Credit risk modelling: Current practices and applications," April.

Basel Committee on Banking Supervision (2004), "Basel II: International convergence of capital measurement and capital standards: A revised framework," June.

Basel Committee on Banking Supervision (2009), "Principles for sound stress testing practices and supervision," May.

Basel Committee on Banking Supervision (2011), "Basel III: A global regulatory framework for more resilient banks and banking systems—Revised version June 2011," June.

Basel Committee on Banking Supervision (2017a), "Basel III: Finalising post-crisis reforms," December.

Basel Committee on Banking Supervision (2017b), "Supervisory and bank stress testing: Range of practices," December.

Basel Committee on Banking Supervision, (2018), "Stress testing principles," October.

Bernanke, B. (2009), "The Supervisory Capital Assessment Program," available at www.federalreserve.gov/newsevents/speech/bernanke20090511a.htm.

Bernanke, B. (2010), "The Supervisory Capital Assessment Program—One year later," available at www.federalreserve.gov/newsevents/speech/bernanke20100506a.htm.

Bernanke, B. (2013), "Stress testing banks—what have we learned?" available at www.federalreserve.gov/newsevents/speech/bernanke20130408a.htm

Bunn, P., A. Cunningham, and M. Drehman (2005, June), "Stress testing as a tool for assessing systemic risk," *Bank of England Financial Stability Review*.

Committee of European Banking Supervisors (2009), "CEBS's press release on the results of the EU-wide stress testing exercise," available at https://eba.europa.eu/cebs-press-release-on-the-results-of-the-eu-wide-stress-testing-exercise.

Committee of European Banking Supervisors (2010), "Aggregate outcome of the 2010 EU-wide stress test exercise coordinated by CEBS in cooperation with the ECB," available at https://eba.europa.eu/sites/default/documents/files/documents/10180/15938/95030af2-7b52-4530-afe1-f067a895d163/Summaryreport.pdf.

Committee on the Global Financial System (2000), "Stress testing by large financial institutions: Current practice and aggregation issues," CGFS Paper No. 14.

Committee on the Global Financial System (2001), "A survey of stress tests and current practice at major financial institutions," April 2001.

Committee on the Global Financial System (2005), "Stress testing at major financial institutions: Survey results and practice," CGFS Paper No. 24.

Cortes, Fabio, Peter Lindner, Sheheryar Malik, and Miguel Angel Segoviano (2018), "A comprehensive multi-sector tool for analysis of systemic risk and interconnectedness (SyRIN)," IMF Working Paper *No.* 18/14.

Deb, P., M. Manning, G. Murphy, A. Penalver, and A. Toth (2011), "Whither the credit ratings industry?" Bank of England Financial Stability Paper No. 9, available at www.bankofengland.co.uk/financialstability/Documents/fpc/fspapers/fs_paper09.pdf.

Dent, K., M. Segoviano, and B. Westwood (2016), "Stress testing of banks: An introduction," *Bank of England Quarterly Bulletin*, 56(3), 130-143, available at https://www.bankofengland.co.uk/-/media/boe/files/quarterly-bulletin/2016/stress-testing-of-banks-an-introduction.pdf?la=en&hash=3C57129C772A42925EDABF0145129001AE7B245F.

Elsinger, H., A. Lehar, and M. Summer. "Network models and systemic risk assessment," in by J.-P. Fouque and J. A. Langsam (eds.), Handbook on systemic risk, Cambridge University Press.

Espinoza R., M. Segoviano, and J. Yan (2020), "Systemic risk modeling: How theory can meet statistics," IMF Working Paper No. 20/54.

Geithner, T. (2014), *Stress test: Reflections on financial crises*, Broadway Books.

Gorton, G., and Metrick, A. (2012), "Securitized banking and the run on the repo," *Journal of Financial Economics*, 104(3), 425-451.

Heale, B. (2013), "Stress and scenario testing: How insures compare with banks" available at www.moodysanalytics.com/risk-perspectives-magazine/stress-testing-europe/principles-and-practices/stress-and-scenario-testing-how-insurers-compare-with-banks.

Hirtle, B., and A. Lehnert (2014). "Supervisory stress tests," *Federal Reserve Bank of New York Staff Reports*, No. 696.

Hoggarth, G., and J. Whitley (2003), "Assessing the strength of UK banks through macroeconomic stress tests," *Bank of England Financial Stability Review Report*, June.

International Monetary Fund (2014), *Review of the Financial Sector Assessment Program: Further adaptation to the post crisis era*.

4

Why Banks Failed the Stress Test: A Progress Report on Stress Testing 10 Years On

Nicola Anderson, Alex Brazier, Andrew Haldane,
Paul Nahai-Williamson, and Amar Radia*

1 A Recap: Why Banks Failed the Stress Test—and Why Stress Testing Failed the Economy

In June 2004, the Basel Committee on Banking Supervision (BCBS) published a long-awaited revision to its framework around capital measurement and standards: Basel II.[1] Notable in this were the frequent mentions of stress testing, including under the first principle of supervisory review:

> Rigorous, forward-looking stress testing that identifies possible events or changes in market conditions that could adversely impact the bank should be performed.

In other words, stress testing intended not to consider what *will* happen but what *could* happen. A worthy aim, but when the crisis unfolded just 3 years later, banks across the globe were found wanting. Plainly, they had failed to run the very tests that could have warned them of the massive losses they went on to incur.[2]

There were several reasons for this failure. First, the phenomenon of "disaster myopia." This is the propensity to underestimate the probability of outcomes that are not quite unprecedented but are rare—those that lie in the tail of the distribution. As Haldane (2009) describes, disaster myopia is a well-established phenomenon in psychology, describing a situation where the distant past is forgotten—by risk takers, risk managers, and the authorities alike—along with the risk of the bad outcomes seen to occur in the past. The result is that risks can be grossly underestimated.

* The views expressed in this chapter are those of the authors and not necessarily those of the Bank of England or its committees. We are grateful to Sandro Chen, Cian O'Neill, Gerardo Martinez, Tord Krogh, Ben King, Pavel Chichkanov, Nick Vause, and Stephen Burgess for their support in producing this chapter. Special thanks to Mathieu Chavaz for his invaluable insights on the debate around stress test disclosure. Any remaining errors are our own.

[1] See "Basel II: International Convergence of Capital Measurement and Capital Standards: A Revised Framework".

[2] As Schuermann (2013) puts it, "corporate or enterprise-wide stress testing, whereby all businesses were subjected to a common set of stress scenarios, was in developmental phase at best." And where banks did engage in stress testing, the severity of the scenarios considered was generally mild (BCBS, 2009).

In the decade or so before the 2008 crisis, this tendency was reinforced by the increasing reliance on models for business-as-usual risk management purposes inside banks. Developed in the 1990s, value-at-risk (VaR) models, for example, had been used to manage traded risks, reinforced by their inclusion in Basel's market-risk amendment in 1996.[3] And although models of credit risk were somewhat slower to develop, these were specifically promoted by the introduction under Basel II of the internal-ratings-based (IRB) approach for measuring regulatory capital requirements—itself based on a standardized model, using the bank's own model outputs as inputs.

These models aim to indicate the losses potentially incurred on a portfolio of financial assets at some specified point in the tail of the distribution. Key determinants of this calculation are the volatilities of asset prices and the correlations between them. Intuitively, although such a model calibrated to a period of low market volatility and/or correlations may successfully measure risk as long as that continues, it is much less likely to capture the losses that might be incurred during a sudden period of heightened uncertainty and/or flight to quality—that is, at those extreme points of the distribution against which risks should be managed.

The benign economic conditions in the decade before the crisis—a period that, with hindsight, appears historically anomalous (see, e.g., Haldane, 2009)—resulted in risk models reinforcing the very human tendencies of risk takers, risk managers, and authorities alike to believe that "this time is different" and that extremely bad outcomes were not possible. Market signals also supported this narrative: systemic risk indicators constructed with market data were low and falling in the run-up to the crisis, suggesting that the risks of such an event were diminishing even as storm clouds gathered (Danielsson et al., 2016).

The belief that severe shocks were not a possibility led to a lack of imagination in considering the consequences were disaster to strike—what *could* happen if the "impossible" became reality. A second key reason for the failure of stress testing before the crisis was that understanding the risks facing the financial system—let alone hoping to model the consequences of those risks materializing—had become an ever more complex challenge. The increasingly interconnected nature of the financial system—and a lack of data on those interconnections—meant that attempting to understand and think through the ways in which shocks would propagate and amplify in a severe stress posed an enormous challenge. Similarly, the role that the shadow banking system would play in hugely amplifying losses in the crisis was only understood after the event.

Third, not only were such challenges potentially beyond the ability of individual banks to address—given the fact that any one bank would only have a very limited view of the financial system—but also they did not have the incentives to try. Buoyed by the belief that risks were low, there was a motive for banks to engage in excessive risk taking—and make excessive returns. In such situations, the concerns of any risk managers who valiantly resisted disaster myopia and exercised their imaginations were likely to fall on deaf ears within firms.[4] As recorded by Haldane (2009), a candid member of staff at one firm admitted to a further reason for having little or no incentive to stress test against severe shocks: "First, because if there were such a severe shock, they would very likely lose their

---

[3] See Basel Committee on Banking supervision (1996).
[4] See Haldane (2009) and references therein.

bonus and possibly their jobs. Second, because in that event the authorities would have to step in anyway to save a bank and others suffering a similar plight." And the incentives problem had a more systemic element: individual firms did not want to lose out by being prudent when others were not. As Chuck Prince put it before the crisis took hold, "as long as the music is playing, you've got to get up and dance" (*Financial Times*, 2007).

Authorities, meanwhile, had the incentive to conduct robust stress tests, but a focus on bank solvency missed growing funding and market-liquidity risks: stress tests by authorities in Iceland, for example, found that the three largest banks had sufficient capital to weather "large, concurrent shocks to financial markets and loan quality" (Ong and Čihák, 2010), but—in common with international best practice at the time—they did not account for the liquidity risks that would prompt the government to take over the banks.

So stress testing failed in large part due to the combined factors of disaster myopia, financial system complexity, and misaligned incentives. But the fundamental principle was sound. Focusing on what could happen, and how bad things could get when the economy takes a turn for the worse, means that the financial system can be prepared, its resilience can be reinforced, and disasters can be contained to downturns.

Drawing from the weaknesses in precrisis stress testing practices identified by authorities in the aftermath of the crisis, and the solutions prescribed to address them, Section 2 of this chapter outlines some of the necessary ingredients (as set out by Haldane [2009]) to establish a strong foundation for stress testing, from a macroprudential perspective in particular. In doing so, it records the progress that has been made since the financial crisis, alongside outstanding challenges.

Importantly, however, this is not the end of the story. Stress testing will need to continually evolve if it is to keep pace with a changing financial system and evolving risks. Section 3 of this chapter outlines some recent innovations and emerging issues that authorities will need to consider in order to continually improve stress testing and build for the future.

2 Learning Lessons: Establishing the Foundations

A decade on, how have we fared in addressing the shortcomings in stress testing laid bare by the financial crisis?

The crisis prompted widespread soul-searching on the part of authorities to understand what had gone wrong and how to establish firmer foundations for banking-system stress testing regimes.

Based on the lessons from the crisis, authorities identified a number of necessary ingredients for building a strong foundation for stress testing the banking system—for example, in Haldane (2009), BCBS (2009), and the International Monetary Fund (2012). We group these together into a set of five broad ingredients (as in Haldane [2009]):

  i. Setting comprehensive stress scenarios that represent tail events
 ii. Regular evaluation of the impact of common stress scenarios
iii. Assessing the impact of second-round effects: contagion and spillover effects
iv. Integrating stress testing into firms' risk management frameworks and their capital and liquidity planning
 v. Disclosure of stress testing frameworks, results, and their link to regulatory actions

We use these ingredients as an organizing framework for a progress report, evaluating developments over the past decade and highlighting where gaps remain.[5]

In short, our view is that progress has been steady and widespread. Stress testing today is a core part—and in many cases, the centerpiece—of macro- and microprudential toolkits worldwide. And as a tool, it is more effective than ever—barely recognizable from what existed before the crisis. But it remains incomplete, with more to do on a number of fronts.

2.1 Setting Comprehensive Stress Scenarios That Represent Tail Events

Authorities identified several areas for improvement in the design of stress scenarios. Chief among these was that authorities themselves should lead on scenario design and that scenarios should be sufficiently extreme to constitute a tail event.

Before the crisis, the prevailing approach was for banks to be firmly in the lead when it came to setting the stress scenario—"setting their own exam questions," so to speak. This led to scenarios that varied widely in severity but tended to reflect mild shocks—and "typically resulted in estimates of losses that were no more than a quarter's worth of earnings (and typically much less)" (BCBS, 2009).

Now authorities almost universally lead on scenario design. That is—in theory—the optimal arrangement. Although authorities may still be at risk of disaster myopia, their incentives are more aligned with taking a more cautious, prudent approach to evaluating tail risks, and so they are better placed institutionally to mitigate that risk. And it should be down to those who have statutory responsibility for financial stability to decide what state of the world banks should be able to withstand—to set the toughness of the test or articulate their *risk tolerance*.

In practice, has the new generation of authority-designed scenarios been sufficiently extreme to avoid the charge of failing to imagine what could go wrong? At the same time, have they avoided scenarios so Armageddon-like that demanding resilience against them would deliver only the stability of the graveyard (Cunliffe, 2014)—a state in which banks must build so much capital and be so cautious in their lending that financial conditions tighten and leave the economy moribund?

Figure 4.1 shows start-to-trough falls in domestic real gross domestic product (GDP) from recent stress scenarios published by several authorities. It shows them in a global and historical context, comparing them to the distribution of annual changes in real GDP for a panel of 21 advanced-economy countries between 1960 and 2015.

Figure 4.1 shows that these scenarios reside in the tail of the historical distribution of GDP outturns[6]—a distribution of events taking place over a sufficiently wide set of geographies and sufficiently long sweep of history to avoid falling into the trap of claiming disasters do not occur.

---

[5] We focus in this chapter on in-depth stress tests conducted by authorities. Market-based stress tests have been developed in parallel as a complementary approach to assessing potential bank vulnerabilities, as discussed in detail in Chapters 8 and 9. The relevance of market-based indicators to the ongoing assessment of bank vulnerabilities is recognized by the Bank of England—for example, via the inclusion of major listed UK banks' average price-to-book ratio in its core indicators (see, e.g., Bank of England, 2019a).

[6] Similarly, if we consider the deviation of GDP from the trend in these stress scenarios in a historical context, we find that it lies firmly in the tail of the historical distribution.

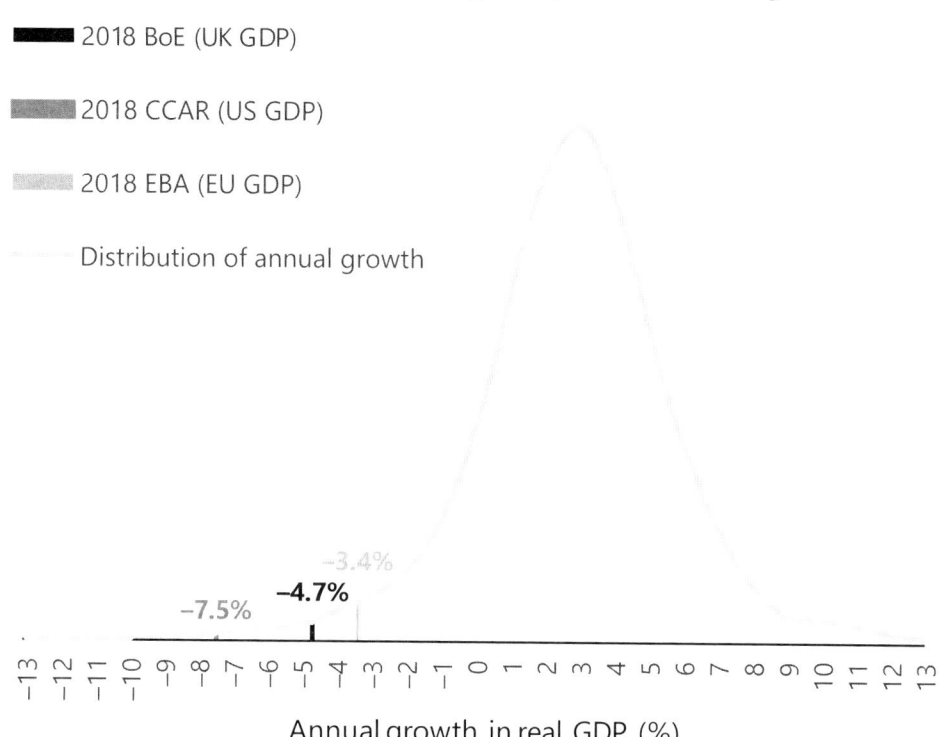

Figure 4.1 Stress scenario GDP start-to-trough falls in the context of the distribution of annual GDP changes in international advanced economies between 1960 and 2015.[7]

Recent stress scenarios generally lie between modest recessions and financial crises. For example, the severely adverse scenario in the 2018 Comprehensive Capital Analysis and Review (CCAR) exercise includes a start-to-trough fall in US real GDP of 7.5 percent, significantly more severe than the drop of a little over 4 percent in the most recent financial crisis.

Stress tests can be more severe than suggested by GDP alone. In the Bank of England's 2018 Annual Cyclical Scenario (ACS), for example, whereas UK GDP falls 4.7 percent—less than the fall of a little over 6 percent in UK GDP observed during the financial crisis—house prices plummet by 33 percent, and unemployment jumps from 4.3 percent to 9.5 percent, both substantially more severe than outturns in the financial crisis. Consistent with such severity, the scenario features severe market shocks, with accompanying increases in liquidity-risk premia across a number of financial markets and significant increases in

---

[7] Annual GDP changes across 21 advanced economies, since 1960 (where data are available). The CCAR US GDP fall is taken from the 2018 severely adverse scenario. The European Banking Authority (EBA) EU GDP fall and the Bank of England UK GDP fall are taken from their respective 2018 stress scenarios. The EBA stress scenario reports calendar-year GDP falls, whereas the CCAR and Bank of England scenarios include quarterly paths. Thus, the figure may understate the full implied start-to-trough fall in the EBA stress scenario if the trough were to occur intra-year. Historical data are sourced from the Organisation for Economic Co-operation and Development (OECD) Economic Outlook and national sources.

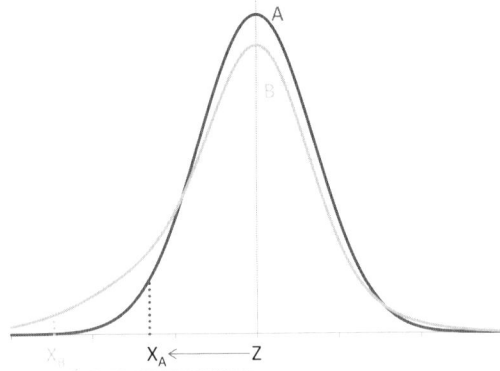

Figure 4.2 Stylised conditional distributions of expected GDP outturns relative to trend in different risk environments.

wholesale funding spreads. This in turn means that stress tests are more comprehensive now—incorporating credit risk, liquidity risk, counterparty credit-risk and market risk, for example.[8]

These sorts of comparisons help underscore that today's stress scenarios are sufficiently extreme to constitute a genuine tail risk when compared to historical experience.

But how do we know that they are appropriate given the risks faced *today*? And even if they are, that they will continue to be *tomorrow* when the risk environment changes?

What precisely constitutes a "sufficiently extreme" scenario—an appropriate "tail event"—will vary over time, most notably as the state of the financial cycle varies (Bank of England, 2015). When markets are exuberant near the top of the cycle, vulnerabilities are higher—and there is farther to fall. Thus, the stress scenario should become more severe as cyclical risks build.

Figure 4.2 illustrates this point. When UK GDP growth is around its trend (Z),[9] potential outturns in UK GDP relative to trend may be relatively symmetrically distributed, as shown in distribution A. In these circumstances, a tail-risk shock—at a low percentile of the distribution—would be a negative shock to GDP of X_A. In a period of exuberance—for example, when credit and asset prices are growing rapidly and risk premia are compressed— the left-hand-side tail of the distribution becomes "fatter" compared to a more "standard" risk environment, distribution B in Figure 4.2 (Adrian et al., 2017). Now, the tail-risk shock to GDP, at the same probability, has moved to X_B—a more severe outturn, putting a greater amount of GDP "at risk."

To what extent have the stress scenarios that authorities have been setting since the crisis taken into account the time-varying nature of risk in order to ensure the appropriate degree of severity?

---

[8] BCBS (2009) and IMF (2012) stress the importance of including all material risks in stress tests. In the Bank of England's stress test, for example, banks project the impact of the scenario on their liquidity positions—with regard to both asset values and projected outflows and inflows—and should identify management actions relating to the usage of central-bank facilities (Bank of England, 2019b).

[9] The relevant period of growth might be more than 1 year, reflecting the time taken for lower GDP to feed through to losses for banks and the potential length of recessions.

The answer appears to be "to some extent"—but increasingly so. For example, in the US CCAR, some degree of time-varying severity is incorporated in the severely adverse scenario by specifying a shock to the unemployment rate that will be higher when cyclical risks are high and lower if cyclical systemic risks are low (Federal Reserve, 2019).[10]

Others, such as the Bank of England and Bank of Japan, go further still. Their scenarios are *explicitly* and *systematically* countercyclical. Both the Bank of Japan's "tail scenario" and the Bank of England's ACS are intended to operationalize this framework.

At the Bank of England, the Financial Policy Committee (FPC) has a statutory duty to identify and monitor risks—and a toolkit (the stress test) to assess the impact on banks' resilience should these risks crystallize. To gauge whether risks may be elevated—and thus what the distribution of risks may look like—the FPC assesses indicators of *potential* imbalances in credit, asset prices and household and corporate balance sheets. Where indicators suggest a potential imbalance, the severity of the scenario is adjusted.

To illustrate how this works, in 2016, the Bank of England's ACS included a start-to-trough fall in Chinese real GDP of 0.5 percent. By 2017, policymakers had assessed that strong property price growth and an associated increase in household indebtedness, alongside a widening in credit-gap measures, meant that vulnerabilities in China had increased. Thus, the 2017 scenario featured a more severe fall in GDP of 1.2 percent.[11]

More recently, the IMF has further begun to include assessments of "GDP at risk"—a similar concept—in its *Global Financial Stability Report* (IMF, 2017).

2.2 Regular Evaluation of the Impact of Common Stress Scenarios

Regular, concurrent stress testing, whereby banks are simultaneously subjected to common scenarios, is now normal practice in a number of jurisdictions.

Table 4.1 shows the number of common regulatory stress test exercises run each year since 2007 for a selection of authorities—and illustrates that as stress testing frameworks have matured, a number of jurisdictions now regularly evaluate two distinct common stress scenarios.

Many authorities also run internal stress testing exercises using common scenarios and "top-down" internal models to stress test their banking systems against a range of scenarios. Several of these—the Bank of Japan, Norges Bank, and the Monetary Authority of Singapore, for example—have been running such exercises since before the crisis. And many more have adopted the practice since.

Across both concurrent stress test exercises and internal top-down types of exercises, the evaluation of the impact of common stress scenarios on banks has spread far and wide since

---

[10] CCAR has a minimum of three scenarios: a baseline scenario, an adverse scenario, and a severely adverse scenario. The severely adverse scenario reflects conditions that characterize postwar US recessions and may be augmented with particularly salient risks. The adverse scenario consists of a set of economic and financial conditions that are more adverse than those in the baseline scenario but less severe than those in the severely adverse scenario. The adverse scenario is intended to be flexible, for example, to enable supervisors to explore risks that, if included in the severely adverse scenario, would perhaps render it implausibly severe. This could, for instance, include large shifts in the yield curve that are atypical in a recession (Federal Reserve, 2019).

[11] We note that this approach provides a framework for incorporating shocks in the stress scenario that are plausible but outside of historical precedent.

Table 4.1. Number of different concurrent stress scenarios set each year by a selection of authorities since 2007

Year	BoE	US Fed (CCAR)	EBA/SSM	RBNZ[a,b]	APRA[b]
2007					
2008					
2009		1	1	1	1
2010			1		
2011		1	1	1	1
2012		1			
2013		1			
2014	1	2	1	2	2
2015	1	2		1	
2016	1	2	1		
2017	2	2		2	2
2018	1	2	1		

APRA, Australian Prudential Regulation Authority; BoE, Bank of England; EBA, European Banking Authority; RBNZ, Reserve Bank of New Zealand; SSM, Single Supervisory Mechanism. Data sourced from authorities' publications and speeches.
Where authorities are reported as running two stress scenarios in a given year, these generally refer to two different scenarios, for example, the severely adverse and adverse scenario in the Federal Reserve's CCAR exercise or the ACS and Biennial Exploratory Scenario in the UK.
[a] The RBNZ also ran a concurrent reverse stress testing exercise in 2016—which didn't feature a common stress scenario across banks but did produce insights that informed the design of future scenarios (RBNZ, 2018).
[b] In 2015, the RBNZ and Reserve Bank of Australia (RBA) also ran an exercise in which firms used a common regulatory stress scenario as part of their internal capital adequacy assessment processes.

the crisis. This adoption of concurrent stress testing is a striking development, representing one of the defining features of postcrisis regulatory reform.

Concurrent exercises bring a number of benefits. One such benefit is that they facilitate peer review and benchmarking. Table 4.2 shows the criteria for inclusion in concurrent stress tests for the same authorities as in Table 4.1 and the number of banks meeting those criteria in the recent exercises.

The table shows some variation between the approaches taken by different authorities—with some setting explicit thresholds for inclusion, some exercising judgment, and others simply explicitly including their largest banks—but the common thread is that the criteria are set such that the stress testing exercises include the major banks in each jurisdiction. This is consistent with the proposal by the IMF (2012) that macroprudential stress testing should include systemically important institutions.

We note that many authorities that run internal stress testing exercises take similar approaches to inclusion—with the Bank of Japan being a notable exception, with its stress testing framework covering over 350 banks.

Another important benefit of concurrent exercises is that they support assessment of the resilience of the system as a whole. They can reveal common vulnerabilities among banks and the buildup of risks in particular economic and financial sectors—and draw on insights from sectoral analysis to challenge the collective projections of banks.

Table 4.2. Criteria for inclusion of banks in a sample of recent international stress testing exercises

	Inclusion criteria	Number of banks	Share of banking sector covered[b]
BoE	$50bn retail deposits	7	Around 80%
U.S. Fed[a]	$50bn total assets	35	Around 80%
EBA/SSM	€30bn total assets	48	Around 70%
RBNZ	Four largest banks	4	Around 90%
APRA	Thirteen largest banks	13	Nearly 90%

Data sourced from authorities' publications.

[a] These criteria were in place for the 2018 stress test. In May 2018, US Congress enacted the Economic Growth, Regulatory Relief, and Consumer Protection Act (EGRRCPA), which amended section 165 of the Dodd–Frank Act to increase the threshold for application of enhanced prudential standards to bank holding companies. Eighteen months after the date of EGRRCPA's enactment, bank holding companies with total consolidated assets of less than $250 billion will no longer be subject to Section 165, provided that the Federal Reserve Board may apply any enhanced prudential standard to a bank holding company with between $100 billion and $250 billion in total consolidated assets if the Board determines that application of the prudential standard is appropriate to prevent or mitigate risks to the financial stability of the United States or to promote safety and soundness. For the 2019 stress test, the Federal Reserve Board announced that it would not take action to require bank holding companies with total consolidated assets greater than or equal to $50 billion but less than $100 billion to comply with the Board's stress test rules. The 2019 CCAR exercise consequently covered 18 banks.

[b] For the UK, this figure relates to the share of the outstanding stock of Prudential Regulation Authority (PRA)–regulated banks' lending to UK households and businesses at the time of the publication of the 2018 stress test results. For other jurisdictions, the shares of total banking-sector assets covered by the banks included in their stress tests are reported.

Concurrent stress testing also allows authorities to account for potential amplification mechanisms that could arise as a result of the banking system's response to an adverse shock (Bank of England, 2013)—and thus support a fuller exploration of macroprudential risks.

2.3 Assessing Second-Round Effects: Contagion and Spillover Effects

The role of contagion and spillover effects in the financial crisis has been well documented (see, e.g., Brunnermeier, 2009). The combined impact of second-round effects was to amplify roughly $300 billion of losses related to subprime mortgages into well over $2.5 trillion of potential write-downs in the global banking sector within a year (Brazier, 2017).

This amplification took place through both direct and indirect contagion mechanisms. Losses on subprime mortgages led to a loss of confidence in asset-backed commercial paper, causing funding problems for a number of institutions. Market losses were amplified by correlated selling by quantitative hedge funds. Increases in concerns around the solvency and liquidity risks of banks led to a freeze in the interbank market. Crowded trades and subsequent price dislocations led to margin calls that pushed many institutions into failure and led to the bailout of American International Group (AIG). And the problems at Lehman

Brothers led to losses for Reserve Primary Fund—a large money market mutual fund – which "broke the buck." A more general run on money market funds followed as investors feared that they, too, would face similar losses (see Brunnermeier [2009] for a summary of the crisis).

It is no surprise, then, that there is a broad consensus across the regulatory and academic communities that modeling these effects is a critical component of stress testing (see, e.g., Borio et al., 2012; BCBS, 2015; Demekas, 2015). But how to incorporate them?

Haldane (2009) proposed that stress tests should be dynamic, with the results of the common stress tests being "the starting point, not the end point" of an evaluation of systemic risk. This recognizes the fact that firms' first-round results and management actions—such as asset sales or liquidity hoarding—have feedback effects—or externalities—that influence the eventual size and impact of the stress.

Authorities around the world have made a significant amount of progress in modeling network externalities, supported by intensive academic effort.[12]

A first set of efforts has focused on modeling **direct contagion through contractual obligations**. Historically, models of potential spillovers focused on potential *domino effects following an outright default* as counterparties fail to meet their payment obligations (e.g., Eisenberg and Noe, 2001; Furfine, 2003).

These typically find that losses from direct spillovers of this kind are not generally sufficient to trigger crises. But this approach ignores the propagation of shocks before institutions default, which can amplify shocks and turn tremors into earthquakes.

Solvency shocks can reduce the value of interbank claims even without a default event. Barucca et al. (2016) proposed a general framework to model this type of *distress contagion* using a structural credit-risk model to revalue exposures. Bardoscia et al. (2019) set out an adaptation of this framework for application in banking-sector stress testing.

And they can be self-reinforcing. Shocks to solvency can raise banks' funding costs and thereby place additional pressure on solvency conditions. This amplification grows with the initial shock, with several studies documenting a nonlinearity between solvency and funding costs (Aymanns et al., 2016; Korsgaard, 2017). Models of this relationship find a natural home in bank stress testing (see the model of Dent et al. [2017], used in the Bank of England's [2017a] concurrent stress test).

Liquidity stress can amplify shocks still further. In the face of funding pressures and uncertainty, banks' active hoarding of liquidity can propagate liquidity stress in interbank networks (Kapadia et al., 2012). And the potential liquidity risks associated with margining of derivatives transactions have sparked further modeling efforts (Paddrik et al., 2016; Bardoscia et al., 2018). But direct connections are only a part of the story. **Indirect contagion**, from fire sales and information asymmetry, are to be found at the epicenter of systemic crises (see Clerc et al., 2016).

When two firms hold the same assets, the **fire sale** of assets by one can affect the other if such sales affect market prices and balance sheets are marked to market. Sales could be caused by factors such as short-term liquidity needs to meet margin calls (Brunnermeier and Pedersen, 2009; Shleifer and Vishny, 2011), investor redemptions

---

[12] For more comprehensive reviews of the literature and outstanding challenges for stress testing, see, for example, Glasserman and Young (2016), Battiston and Martinez-Jaramillo (2018), and Aymanns et al. (2018).

(Baranova et al., 2017), or pressures to deleverage to meet regulatory constraints (Greenwood et al., 2015; Cont and Schaanning, 2017). A key challenge in modeling contagion risk via this channel is quantifying the impact of sales on market prices. Some researchers have modeled the price-formation mechanism directly—Baranova et al., for example, model how a wave of corporate bond sales affects dealers' costs and how this in turn affects market prices. But for bank stress testing, a simpler price–response function is often used, based on regulatory expertise or a reduced-form approach relating price impact to a measure of market depth (see, e.g., Fique, 2017; European Central Bank [ECB], 2017).

Turning to **information asymmetries**, adverse news about one bank can lead to increases in the borrowing costs of its peers (Acharya and Yorulmazer, 2008). And in crises, negative revelations about one institution can trigger runs on both that institution and on connected institutions.[13]

Although modeling creditor behavior poses a significant challenge for stress testing, progress has been made here too. For example, the ECB has developed a systemic liquidity-risk framework that incorporates how informational spillovers can lead to funding-cost increases for banks with business models similar to those of the directly affected bank (ECB, 2017). Ahnert et al. (2016), meanwhile, model the interaction of a bank with its unsecured creditors as a coordination game with strategic complementarities to identify the solvency threshold below which creditors run in sufficient numbers to cause illiquidity.

These contagion models have been used to answer pressing financial stability questions (Banco Central do Brasil, 2017), in macro stress testing models that examine the resilience of the system (Bank of Japan, 2016), and to move toward the vision of adding second-round effects to common regulatory stress tests.

The Bank of Canada, for example, takes stress test results from the Office of the Superintendent of Financial Institutions as an input to its framework for modeling second-round effects (Fique, 2017). And the Bank of England has applied models of individual amplification mechanisms as part of its concurrent stress testing to assess whether second-round effects are accounted for in the results (see, e.g., Bank of England, 2017a; Churm and Nahai-Williamson, 2019). In a similar vein, the ECB has used its comprehensive toolkit to quantify potential second-round effects as a macroprudential extension to the EBA stress test (ECB, 2017).

These approaches represent the forefront of efforts to incorporate severe liquidity and solvency shocks into the same scenario. Although existing concurrent stress testing exercises generally feature a disruption to funding markets and a degree of liquidity stress, incorporating the risks of additional severe idiosyncratic or broader-based funding outflows into solvency-focused stress tests has remained the preserve of authorities' internal modeling.

This should be unsurprising. In part, it reflects the reduction in contagion risks associated with postcrisis regulatory reforms. Banks' reliance on short-term unsecured funding markets—and the risks that bank failures could spread and amplify distress via these markets—have reduced to the point of being negligible (Bardoscia et al., 2018). The com-

---

[13] The run on money market funds that followed the run on Reserve Primary Fund illustrates this point. Iyer and Peydro (2011) found that a similar story can play out in depositor runs: following the failure of a large bank in India, those banks with larger exposures to the failed bank saw larger depositor outflows.

plex web of derivatives, meanwhile, has been simplified through central clearing, reducing one of the key channels of uncertainty that operated in the crisis.

Postcrisis regulatory reforms have also significantly increased banks' resilience to severe liquidity shocks. Banks have more stable funding profiles, with less reliance on unsecured short-term funding, and material usable buffers of liquid assets, calibrated to be sufficient to meet funding outflows in a severe idiosyncratic liquidity stress. And they have recourse to developed central-bank liquidity facilities.

So when considering the risks around severe liquidity shocks, the emphasis increasingly lies less on whether banks *can* withstand such a shock and more on *what impact their responses to it may have* on the wider system and their own profitability—something that must be modeled centrally.[14]

The fact that severe funding shocks are not generally included in concurrent stress testing scenarios also reflects the fact that the crystallization of such risks is contingent on the resilience of banks to the solvency shock resulting from those scenarios. A bank that breezes through the stress test would not be expected to face a severe funding squeeze, and building one into the scenario a priori would risk rendering it implausible.

Indeed, a key macroprudential purpose of solvency stress testing is to ensure that banks can remain resilient to severe solvency shocks and continue lending to the real economy—thus reducing both the risks of financial-system contagion due to bank distress and the risks of spillovers to the real economy due to banks cutting lending.

It is for this reason, too, that modeling potential feedback effects between banks' lending decisions and the real economy has primarily been tackled outside of a regulatory stress testing context. Concurrent stress testing exercises generally do not allow for significant real economy deleveraging, for example, by stipulating that banks are not permitted to dynamically change their balance sheets (EBA, 2017), or are required to meet credit demand over the course of the stress period (Bank of England, 2017b). Although the ECB has, for example, applied its stress testing apparatus to consider the impacts of banks deleveraging as an overlay to regulatory stress test results (see, e.g., ECB, 2017), modeling of the feedback between bank lending and economic outturns in stress scenarios has generally come in the form of macroprudential stress test models designed and run internally by authorities, such as the ECB's semistructural model (Angelini et al., 2019) or the IMF's Macro-Financial System Simulator (Gross et al., 2019).

Thus, the modeling of second-round effects is likely to remain the preserve of authorities' internal models—both in the context of concurrent stress testing and beyond. We still face significant challenges in modeling these risks. The quest to understand what drives market (il)liquidity in different markets in a stress—and the implication this has for the potential impacts of forced asset sales—continues to provide fertile ground for researchers to make an impact of their own. So, too, does the quest to address the challenge of modeling the interaction between solvency and liquidity risks. And more work is needed to understand how institutions will respond to shocks in the new regulatory paradigm of risk-based capital

---

[14] These questions are the focus of the Bank of England's 2019 Biennial Exploratory Scenario (see Section 3).

requirements, leverage-ratio requirements, the liquidity coverage ratio, and the net stable funding ratio. Although much progress has been made, there is still more to do.

2.4 Integrating Stress Testing into Firms' Risk Management Frameworks and Their Capital and Liquidity Planning

Moving from the authorities to the banks, what has changed? Prior to the crisis, as already discussed, stress testing suffered from an "internal incentive problem" (Haldane, 2009). Inside firms, the incentives were all geared toward making decisions in favor of risk takers and against risk managers.

As the Basel Committee's 2009 "Principles for Sound Stress Testing Practices and Supervision" concluded: "Prior to the crisis, stress testing at some banks was performed mainly as an isolated exercise by the risk function with little interaction with business areas. This meant that, amongst other things, business areas often believed that the analysis was not credible. Moreover, at some banks, the stress testing programme was a mechanical exercise. Prior to the crisis, many banks did not have an overarching stress testing programme in place but ran separate stress tests for particular risks or portfolios with limited firm-level integration."

We can contrast that assessment with the recent "vision" for stress testing laid out by the Group Board of HSBC (Paisley, 2017): "Stress testing is a core part of robust capital planning and risk management. It helps the authorities assess the resilience of the bank and the system." The very fact that the Group Board has developed its own vision for stress testing reflects the step change in attitudes toward stress testing at the biggest banks.

For the firms that are subject to the Bank of England's stress test, we are able to see how stress tests now receive extensive board-level engagement. And when firms come in to present their stress test results, these presentations are fronted by senior executives, and managers from the stress testing and risk management functions tend to be joined by those from all of the relevant business areas.

All of this means that stress testing—both in concurrent stress testing exercises and as part of firms' own internal capital adequacy assessment processes—can be a driver of firms' liquidity and capital planning. And some authorities have also taken steps to facilitate that. For example, the systematic nature of the Bank of England's ACS means that banks can broadly anticipate the shape and severity of the scenario. And so, in turn, they can take steps to ensure that they have sufficient capital to withstand it, reducing the probability that they fall short of their hurdle rate and are subject to an adverse market reaction.

Overall—on the surface, at least—progress here appears to be in the right direction. The link to capital planning and the clear and public expectations on firms to conduct robust stress testing have strengthened the role of risk managers, helping to redress the balance between risk taking and risk managing.

But the incentive problem is intrinsically hard to observe and overcome, and so authorities have devoted time and effort to examining the robustness and quality of banks' stress testing processes. Indeed, one of the stated goals of regulatory stress testing is to drive improvements in banks' own risk management practices (Bank of England, 2013). Authorities have supported this goal through qualitative evaluations of banks' processes for producing stress test projections. In the United States, for example, banks can fail the stress test on qualitative grounds as well as quantitative grounds.[15] Indeed, in recent years, more banks have failed due to weaknesses in their capital planning processes than for reasons of quantitatively falling short of the bar. But the number of failures has been falling. In 2017, for the first time, all banks passed the test, testifying to the improvements in banks' frameworks (although in 2018, one bank did fail on qualitative grounds).

In the UK, in 2017 the Bank of England highlighted that although banks were on the right trajectory, some weaknesses remained in their stress testing analysis (Bank of England, 2017a), and in 2018, it published a supervisory statement on model risk management principles for stress testing to support improvements (Bank of England, 2018). Although policymakers recognized that the demands of producing stress test results limit the capacity of firms to make giant leaps in this area, they have made it clear that they expect further progress (Bank of England, 2017a).

So banks' processes and capabilities should—and must—continue to improve over time. With improved processes and beefed-up toolkits, running a larger set of different stress scenarios should become more feasible and less costly for banks. With the new generation of regulatory stress testing, authorities are aiming to avoid disaster myopia and reflect emerging risks at the macro level, and banks' own internal stress testing can complement this by identifying emerging risks in their business models.

Taken together, stress tests run by regulators and individual banks should be a potent combination for informing capital and liquidity planning for banks—as well as informing policy and supervisory actions to support resilience against a range of potential crises.

2.5 Disclosure of Stress Testing Frameworks, Results, and Their Link to Regulatory Actions

Much has changed in stress testing in the last decade—but how much of this change has been visible to the public, and which previously dark corners have been exposed to the light?

Transparency over the stress testing process has a number of potential benefits. Giving market participants more information on banks' tail-risk exposures via credible stress tests can bolster confidence in the banking system[16] and support market discipline (Goldstein and Sapra, 2013).[17] Stress tests provide a means to use forward-looking assessments of banks' resilience to inform capital requirements—and in doing so, increase the accountability of regulators to political authorities.

For these benefits to be realized, an appropriate degree of transparency is needed—which maximises the "signal" and minimizes the noise.

---

[15] A failure here refers to the US Federal Reserve objecting to a firm's capital plans – meaning that the firm may not make any capital distribution unless expressly permitted by the Federal Reserve.

[16] The 2009 US Supervisory Capital Assessment Program (SCAP), for example, restored confidence to shattered markets. See Chapter 12 for an in-depth discussion of the risks and benefits of disclosure in times of distress.

[17] See Chapter 11 for a comprehensive discussion of stress testing disclosure.

Table 4.3. Disclosure metrics for recent stress testing results published in the UK, United States, and EU[a]

	BoE 2019	**CCAR 2018**	**EBA 2018**
Participants	7	35	48
Pages (excluding appendices for firm-specific disclosures)[b]	16	25	39
Tables	20	80	7
Charts	19	1	46
Firm-specific data points (per firm)[c]	105	30	~18000

Data sourced from authorities' publications.

[a] We have used some discretion in assessing how many tables, charts, and so forth are included—for example, in counting separately charts and tables that are grouped together in the report.

[b] Page counts omit title pages, tables of contents, lists of figures, and similar content.

[c] Firm-specific data points identify unique data points, including data on banks' current balance sheets as well as stress projections. Data on firm-specific hurdle rates are also included.

There has been a major shift in transparency over stress testing in the past decade—with a greater degree of disclosure around the stress testing process becoming the norm.

Several authorities now publish detailed stress scenarios and the hurdle rates—the standards to which banks will be held—that apply in the test.

Stress tests have come to play a crucial role in capital buffer setting—and indeed have become a key determinant of banks' capital requirements in some jurisdictions. In the United States, this role has even been enshrined in legislation.[18]

This role has been supported by transparency around stress test results, with several authorities publishing stress test results for individual banks and information about the policy actions that those results have informed. They also talk to the press and to equity analysts about these results to avoid market participants and the public misunderstanding the results or "chasing shadows."

But there are some interesting differences in the approach to communications around stress test results. Table 4.3 illustrates the different emphases of the EBA, Bank of England, and US Federal Reserve in their results publications. The EBA, for example, publishes a huge amount of data on banks' balance sheets, giving the market a window into their balance sheets. The Bank of England, meanwhile, is liberal with words and charts to explain the key findings and metrics used for policy decisions and rather less focused on publishing underlying data points. The US Federal Reserve is somewhat less focused on the narrative of the stress test results than the Bank of England in publishing the results of CCAR but similarly limits data to the main results.[19]

There are also risks to turning up the lights all the way. Being fully transparent about the models and analytical tools used to produce stress test results can facilitate engagement and improvement by academics and market participants. But banks can, in principle, adjust their

---

[18] Via the Dodd–Frank Act (2010), implemented by federal regulations 12 CFR 252 and 21 CFR 225.8.

[19] Additional firm-specific projection data are included in the Dodd–Frank Act Stress Test (DFAST) results, including data on projected losses, revenue, and net income items.

portfolios to look less risky according to the model—"game the test"—with no guarantee that they would be less risky in reality (Goldstein and Sapra, 2013).

Disclosure of models could also add to noise and uncertainty in some cases. Where stress test projections are produced using a combination of banks' and authorities' models and analysis, supervisory insights and peer review—as at the Bank of England—disclosing the details of the authority's models may risk distracting from the key judgments and uncertainties in the stress test results.

The Bank of England has recognized the importance of communicating clearly about these judgments and uncertainties in the stress test results, for example, by including in its most recent stress test results a discussion of the key judgments about what would happen in the stress scenario and the rationale for them and the sensitivity of the stress test results to those judgments (Bank of England, 2019d).

There are other risks too. As Goldstein and Sapra (2013) point out, there is a potential for destabilizing "noise" to slip into publications on stress test results, with adverse effects on stability. And greater information provision by regulatory authorities could reduce market participants' own incentives to gather information independently.

To contain such risks, precision in communications is important—transparency should improve the information set of market participants. Where bad news is forthcoming, "it is imperative that supervisors disclose the corrective actions to remedy the bad news and such corrective actions should be credible" (Goldstein and Sapra, 2013). It is no coincidence that authorities have gone to great efforts to explain stress test results and to accompany their release with clear communications on the policy actions taken in response.

Stress tests are now an accepted part of the regulatory framework. The publication of stress scenarios and stress test results is closely watched by journalists and the market alike. And there is some evidence to suggest that the publication of results is becoming predictable or "boring"—frameworks are increasingly well understood, and there are indications that analysts are becoming increasingly adept at anticipating these results given the scenario (Glasserman and Tangirala, 2015; Chavaz et al., 2015). Transparency has been at the heart of this evolution.

As stress testing becomes run-of-the-mill and authorities' modeling and analytical toolkits mature, perhaps transparency will increase further. More information on the machinery behind the curtain could be revealed. Authorities could layer and target disclosures to maximize signal and minimize noise: communicating distillations of the key insights, results, and policy actions to the public while providing a tome of data and technical detail for analysts and market participants to pore over.

The increasing layering of publications—for example, the edifice of executive summaries, chapters, and annexes in the Bank of England's recent stress testing results publication or the EBA's publication of thousands of data points online—may provide a sign of things to come.

3 Building for the Future

Although there is more to do in some areas to complete the agenda set out a decade ago, the core foundations of banking-sector stress testing are firmly established.

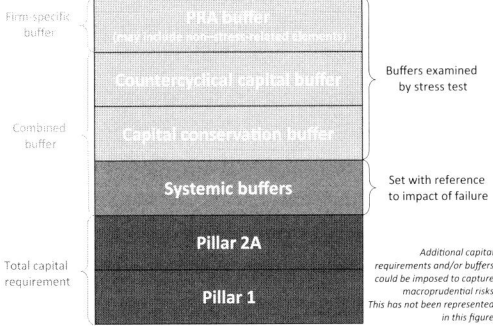

Figure 4.3 Capital buffers calibrated in response to stress test results.

This has provided a launchpad for policymakers to be more ambitious in their uses of banking-sector stress testing. In recent years, stress testing has moved beyond the narrow aim of ensuring the resilience of individual banks to ensuring the ability of the banking system to keep serving the real economy in a severe shock—and to go further still, in setting stress scenarios that go beyond "typical" macrofinancial tail events to explore a more diverse set of risks.

New horizons beckon, including expanding the field of vision to encompass nonbank financial institutions and working toward an international assessment of systemic risks that reflects the global nature of our financial system.

In this section, we set out some recent innovations in stress testing through the lens of the UK framework and explore some key challenges as regulatory stress testing enters its second decade.

3.1 Stress Testing for Macroprudential Policymaking

Concurrent stress tests provide an ideal setting to inform both micro- and macroprudential policy.

They can quantify idiosyncratic risks in the face of tail-risk scenarios, informing the calibration of individual banks' capital buffers; and assess the collective resilience of the banking system, informing the calibration of system-wide capital buffers.

In the UK, this marrying of micro- and macroprudential capital buffer setting was introduced as an explicit feature of the stress testing framework in 2015 (Bank of England, 2015). It was done in a manner consistent with the capital framework, with stress testing results used to inform time-varying components of the capital stack (Figure 4.3).

As we discussed earlier, banks subject to the ACS in the UK are required to maintain lending in the face of a severe shock scenario, calibrated to reflect the current risk environment.

Stress test projections are then used to assess the aggregate "UK impact" of the scenario on stress test participants,[20] which in turn implies a countercyclical capital buffer (CCyB) that banks would require to ensure their ability to maintain lending to the real economy.

---

[20] To do so, the Bank of England receives data from participating firms on the UK element of CCyB-relevant items, including, for example, firms' estimates for UK income and expenses, UK impairment charges, and UK credit-risk risk-weighted assets (RWAs).

If this buffer is larger than the capital conservation buffer (CCoB), it implies that a positive CCyB rate may be required. This information acts as one input to the FPC decision on the level of the UK CCyB rate.

The stress may have a larger impact on some banks than that captured by the combination of the CCoB and the CCyB. In such cases, the Prudential Regulation Committee (PRC) can increase regulatory capital buffers to address those bank-specific risks.

Macroprudential policymakers stand ready to cut the CCyB and release capital in the face of shocks. Coordination between the FPC and PRC in using stress test results to inform capital buffers means that if the FPC cuts the CCyB, banks' individual capital requirements will fall in lockstep, ensuring that banks can absorb shocks and maintain lending.

3.2 Beyond Traditional Macroeconomic Stress Scenarios

Banking systems must be able to survive severe macrofinancial shocks. Judicious use of stress testing can protect against disaster myopia and misaligned incentives, against a misplaced optimism—or hope—that "this time is different."

But authorities should also be cognizant of the fact that *different* does not necessarily mean *better*. The global financial crisis *was* different—it was genuinely global for the first time, and much worse for it (Haldane, 2015).

The apparatus of stress testing can be harnessed to explore resilience against a wider range of risks than those embodied by more "typical" macroeconomic crises—those risks that may just be emerging, that lurk in the darker corners of the economy and financial system, or that pose more subtle threats to the resilience of the banking system. Imaginative scenarios can complement more systematic ones.

The RBNZ blazed a trail here, using its apparatus to stress test New Zealand's major dairy lenders against a scenario featuring sustained low milk prices and sharp falls in dairy land values (RBNZ, 2016). Although this sector-specific stress generated significant increases in loss rates, the banking system remained resilient. And the exercise highlighted uncertainties around the time taken to resolve stressed dairy exposures. The RBNZ has also harnessed exploratory stress tests to run a reverse stress testing exercise in 2016, the findings of which motivated an exploratory scenario focusing on mortgage conduct risk as part of its 2017 stress test (RBNZ, 2018).

The Bank of England has introduced the Biennial Exploratory Scenario (BES) to its concurrent stress testing framework to increase the scope and value of stress testing. In 2017, the first BES was designed to examine banks' strategic responses to a structurally more challenging operating environment (Bank of England, 2017b). Whereas the ACS tested banks' ability to weather a (relatively) short, sharp shock, the BES focused on the resilience and adaptability of banks under chronic strain. It posed the question of whether banks could adapt their business models to be sustainable over the course of a 10-year scenario featuring a sustained period of stagnating global trade, falling cross-border activity, low interest rates, and increasing competition supported by advances in financial technology.

Just as more traditional stress testing has driven improvements in banks' risk management, so has the BES encouraged banks to grapple with difficult longer-term strategic questions—and revealed areas for improvement. Their projections suggested that, in aggregate, banks expect to be able to build sustainable business models in such a challenging environment.

But the BES revealed several areas where banks may be underestimating risks. First, banks may be underestimating the potential disruption to their business models caused by emerging financial technology. Second, banks' ability to generate acceptable returns on equity for their investors rested on their ability to achieve significant cost reductions and limit spending on cyber-risk. In both areas, the Bank of England identified clear risks around banks' assumptions. Third, the sustainability of banks' business models rested on the assumption that their cost of equity would fall over the course of the scenario. Overall, the exercise generated a number of insights to guide subsequent engagement between supervisors and banks about banks' strategic resilience.

In 2019/2020, the BES is being used to explore the implications of a severe and broad-based liquidity stress that simultaneously affects the major UK banks (Bank of England, 2019a)—and how the reactions of banks and authorities would shape its impact on the broader financial system and UK economy. To support that effort, the BES features two rounds, with the second round giving the Bank of England the option to update the scenario, given banks' first-round responses.

The exploratory nature of the BES presents both opportunities and challenges. Policy proposals arising from BES exercises are tailored to each exercise, and individual banks' results are not tied directly to policy actions. The 2019/2020 BES, for example, will *not* be used to set new liquidity standards—the Bank of England already monitors banks' resilience to liquidity risks closely, and banks are expected to use their regulatory liquidity buffers in a stress.

Rather, it will support the FPC and PRC in understanding whether there are barriers to banks using their liquid-asset buffers in stress. It will help to raise awareness of how the Bank of England's liquidity facilities—which include the ability to lend to banks in all major currencies—operate in a liquidity stress and how they interact with the PRA's regulatory framework. And it will enhance the Bank of England's understanding of how banks will use its liquidity insurance facilities in stress and any risks that the Bank of England itself may be exposed to through providing that liquidity (Bank of England, 2019a)—and so inform the design of those facilities.

So as a tool for enhancing strategic thinking, this means that the BES is less vulnerable to the incentive problems that pose a risk to capital-setting stress testing exercises.

It does, however, pose novel data and modeling challenges, both for banks and for the Bank of England. Each BES may have its own data requirements, and may require the development of new analytical frameworks and models. Indeed, one aim of the BES is to prompt both banks and the Bank of England to think about problems in a new way.

Future exploratory scenarios will continue to explore very different risks, with the hope of yielding similarly important insights. The discipline of a regular exploratory scenario should provide an additional spur to policymakers to scan the horizon for new and emerging risks.

For example, in 2021, the BES will focus on the financial system's resilience to the financial risks of climate change (see Bank of England, 2019b). It will test the resilience of the current business models of the largest banks, insurers, and the financial system to both the physical and the transitional risks posed by climate change. The goal of the exercise is to provide a comprehensive assessment of the scale of adjustment needed in coming decades, rather than set capital requirements. To that end, it will include three different scenarios and use a 30-year modeling horizon. The first two scenarios will capture the risks associated with earlier and later policy action to reach the Paris Agreement target.[21] The third will capture the physical risks associated with "no additional policy action" and a failure to meet the Paris Agreement target.

The principles motivating the introduction of the BES are alive, too, in the FPC's plans to test critical financial companies' ability to restore vital services following a cyberattack (Bank of England, 2018).

Information technology (IT) failures and cyber-incidents can impair processes and data supporting the services—such as payment processing, the insuring against and dispersing of risk (via derivatives contracts), and channeling savings into investment via debt and equity instruments—that the financial institutions provides to the real economy. So measures to support the financial system's resilience to cyber-risk are as important as those to support financial resilience. To guide firms in their planning around cyber-resilience, "the FPC is establishing its tolerance of the length of any period of disruption to the delivery of vital services the financial system provides to the economy" (Bank of England, 2018).

The Bank of England will work with others, especially the UK National Cyber Security Centre, to test that firms will be able to meet the FPC's standards for recovering services. As with financial stress testing, it is not expected that firms should be resilient in the most extreme conceivable circumstances—stress testing scenarios should be severe but plausible and used as a tool to identify where firms require remedial action plans to improve their resilience.

The expansion in the application of the philosophy and tools of stress testing to assessing and improving the resilience of the financial system to new and different threats is testament to their ongoing potential.

3.3 Beyond Banks: System-Wide Stress Simulations

The resilience of the banking system has transformed since the crisis. Stress tests have demonstrated that the UK banking system can now absorb within its capital buffers shocks of such severity that they would have wiped out the entire capital base of the system in 2008 (Baranova et al., 2017).

But as the resilience of the banking system has grown, so, too, has the importance of nonbank financial institutions. And with them, a new set of risks has emerged. Nonbank financial institutions now account for a significant share of financial system assets—around half in the UK (Baranova et al., 2017). And these institutions are connected to banks both indirectly via their activities in common asset markets and directly via derivatives, funding, and securities lending exposures. The potential for the actions of institutions such as money

---

[21] The Paris Agreement falls within the United Nations Framework Convention on Climate Change. It was agreed in 2015 at the 21st Conference of the Parties. Over 190 countries committed to put measures in place to limit the global temperature rise to "well below 2°C." Each signatory must make financial flows consistent with a path to low greenhouse gas-emissions.

market funds, hedge funds, and insurers to have a direct impact on the banking system—and vice versa—was starkly illustrated during the financial crisis.

This expanding financial ecosystem has become increasingly important for the real economy too: since the crisis, nearly all net finance raised by private companies in the UK has been through the issuance of equity and bonds (Brazier, 2018).

Thus, authorities have recently turned their attention to assessing the resilience of the wider financial system (e.g., see Bank of Korea, 2018) and, in particular, the potential for the collective actions of institutions such as hedge funds and asset managers to amplify financial market shocks (Bookstaber, 2014; Baranova et al., 2017).[22] This represents a natural expansion of the work already done to model second-round effects within the banking system.

Understanding how stress can manifest and propagate in the financial system as a whole is core to the remit of macroprudential authorities to ensure that the financial system can serve the real economy in good times and bad. Authorities could do this, for example, by identifying growing risks and taking action to mitigate these—and by recommending that sectors that increasingly pose a risk are brought inside the regulatory perimeter. The focus is not on the resilience of individual institutions but on the functioning of the system as a whole.

Expanding the focus to the wider financial system, however, need not—and should not—be a totally separate endeavor from banking-sector stress testing. On the contrary, understanding wider financial-system dynamics, and the role of the banking system within these, has the potential to enhance the robustness of banking-sector stress testing in two ways.

First, some authorities design regulatory stress scenarios to account for cyclical risk, informed by historical distributions of outturns in macroeconomic and financial variables. But the growth in the importance of nonbank financial institutions—alongside postcrisis changes in bank regulation—may have had an impact on *structural* risks in some markets. Building capacity to model the impacts of such structural changes on market resilience could inform judgments on the appropriate severity of regulatory stress scenarios for banks. And it can reduce the universe of "unknown unknowns" facing authorities.

Second, macroprudential authorities seek to account for potential second-round impacts of banks' actions in stress testing. Currently, the focus is predominantly on interactions between banks. But banks' responses to stress could spread distress to the wider financial system, for example, if they reduce their market-making activities or willingness to fund counterparties that are nonbank financial institutions. Authorities should be aware of these possibilities in stress test projections—and use those projections to inform understanding of how banks' actions could affect wider financial-system resilience in turn.

For all of these reasons, the Bank of England has been working to develop a system-wide stress simulation to help understand how the financial system as a whole is likely to respond to shocks.

---

[22] We note that a focus on the resilience of nonbank financial institutions themselves is not an entirely new development—the Monetary Authority of Singapore, for example, has included insurance companies and capital market intermediaries in its annual industry-wide stress test since 2003.

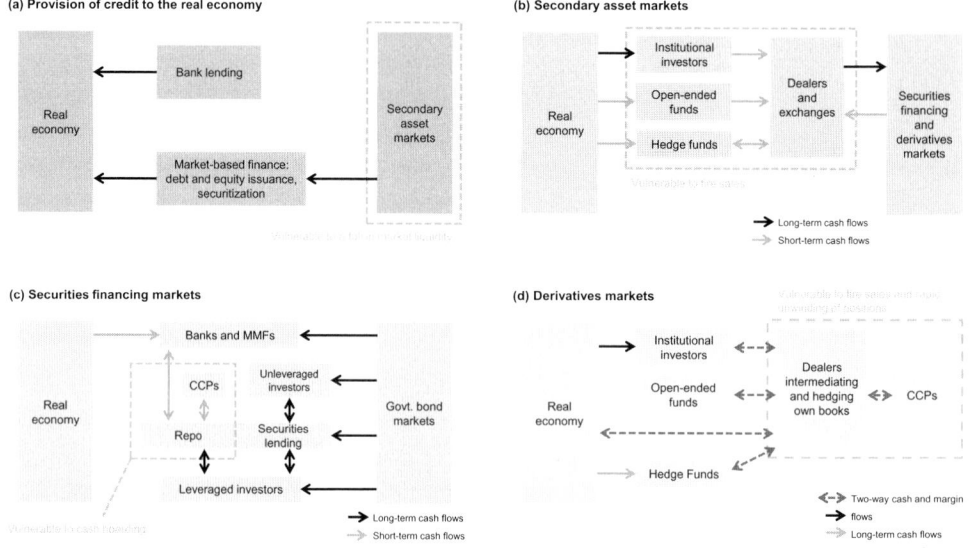

Figure 4.4 Stylized illustration of the functioning of the market-based finance system. The real economy sources credit from the banking system; and via market-based finance (panel [a]). The latter relies on secondary asset markets (panel [b]), in which a variety of investors trade securities and determine market prices. The functioning of secondary asset markets is in turn supported by securities financing markets (Panel [c]) and derivatives markets (Panel [d]). Securities financing markets are separated into repo markets, where cash is exchanged for securities, and securities lending markets, where securities are exchanged for other securities. Dealers intermediate both markets, recycling cash from cash providers—money market funds, CCPs, and commercial banks—to leveraged investors in the repo market and matching securities lenders and securities borrowers in the securities lending market. Both markets rely on the government bond market as the principal source of collateral. Derivatives markets are used to hedge exposures in secondary asset markets, and to speculate and gain leverage. They are dealer-intermediated, with CCPs sitting in the majority of trades. CCP, central counterparty clearing house; MMF, money market fund. Source: Aikman et al. (2019).

The Bank of England's first step in this direction focused on the potential risks posed by the growth of open-ended investment funds to market liquidity (Baranova et al., 2017). The authors modeled the impact of fund sales in response to investor redemptions on prices in the corporate bond market, with prices modeled as a function of the costs to dealer banks of continuing to make markets. The authors found that weekly redemptions from bond funds of 1 percent of total net assets—similar to the level observed in October 2008—would increase the liquidity component of corporate bond spreads by around 40 basis points. A further 30 percent increase in the level of redemptions could exhaust dealers' ability to absorb sales, leading to a market freeze.

But that work captured only a relatively small part of the market-based finance system. Figure 4.4 illustrates the layers of markets and institutions that compose the system and can act as a roadmap for the development of a broader model of system-wide dynamics.

The Bank of England is developing a model of the financial system that aims to capture, in a high-level way, a much larger share of this system (Aikman et al., 2019). This model aims to assess the resilience of the UK financial system and the ways in which the behaviors of different types of financial institutions can amplify initial shocks to financial asset prices in a general-equilibrium setting.

To do so, some trade-offs must be made. Seeing the big picture means foregoing some of the detail. The model features seven representative agents, each representative of a UK domiciled financial sector—heterogeneity of sectors rather than heterogeneity within sectors. These include two long-term investors (a pension fund and an insurance company), one short-term investor that arbitrages between the price of securities and economic fundamentals (a hedge fund), two cash providers (a money market fund and a commercial bank), one investment agent (an asset manager that invests on behalf of investors in an open-ended fund), and an intermediary (broker-dealer).

These agents interact via markets for UK-issued equities, corporate bonds, and government bonds and are linked by secured funding and derivatives markets.

In the face of shocks, sectors respond as a function of their behavioral objectives and the constraints they face—including their strategic asset allocations, internal risk targets, regulatory constraints, the behavior of their investors, and contractual obligations. Following an initial shock to, for example, the expected returns on corporate bonds, investors will adjust their portfolios, selling corporate bonds and depressing the market price. This fall in prices may push the dealer below its regulatory minimum-leverage ratio, forcing it to deleverage by reducing the repo funding it provides, in turn forcing its borrowers to sell assets. These forced sales depress asset prices below their fundamental value—amplifying the real economy shock.

Overall, the model provides a means to explore how stress can propagate via asset sales, disruption to repo markets, and liquidity stress due to collateral calls. And it does so in a general-equilibrium setting: market prices and quantities adjust such that aggregate demand matches supply in each market. The model therefore provides a means to assess how asset prices and funding availability are affected endogenously following an initial shock and the extent to which shocks to asset prices deviate from fundamentals.

Just as taking a big-picture view of the system foregoes the details of within-sector heterogeneity, the simplifying assumption of homogeneous assets within an asset class means that the model abstracts from potential heterogeneity in cross-holdings within an asset class (e.g., of high-yield vs. investment-grade corporate bonds). This has the benefit, however, of making the general-equilibrium approach more feasible.

So this type of exploratory model has the potential to shine a light on how and when each sector may amplify—or dampen—shocks and the relative importance of different asset and funding markets to system-wide resilience. Authorities can then dig deeper into these sectors and markets to develop more detailed satellite models that could inform policymaking—filling the details back in. And these sorts of models can track how risks evolve over time, as sectors become more or less active in certain markets and regulations change.

The challenges facing regulators, policymakers, and academics in further developing the capacity to model stress in the wider financial system are significant. Key challenges include the following:

- Striking the right balance between zooming out enough to capture the system while retaining sufficient detail in those places that really matter—reintroducing heterogeneity where it matters

- Extending modeling to a broader set of markets and behaviors—in particular to incorporate securities lending markets and short-selling behavior, as well as credit default swap (CDS) and foreign exchange (FX) derivatives markets
- Modeling the dynamic evolution of systemic stress—the timescales over which it evolves and the nonequilibrium stages it passes through as the system responds to the shock
- Further developing the empirical and theoretical foundations for modeling of institutions' behavior—both how (and when) institutions are likely to act as they approach and meet constraints and how institutions behave when they are unconstrained and so have the potential to support market functioning
- Continuing efforts to fill data gaps—the availability of data on the balance sheets of different types of financial institutions has not kept pace with their increasing importance. Without the raw materials to underpin them, models of stress in the wider financial system will be built on shaky foundations.
- Accounting for the role of nonresident financial sectors that also participate in domestic markets
- Making the link between stress in the financial system and how it affects—and is affected by—the real economy

Making inroads in these areas can continue to increase the power and benefits of stress testing. Recent work to develop a framework for simulating stress dynamics in a system of heterogeneous agents interacting via multiple financial networks points to one promising avenue for taking on some of these challenges (Kleinnijenhuis et al., 2020).

But for both banking-sector and system-wide stress testing, a further obstacle remains. Financial markets are global. Both within the banking system and in the wider financial system, restrictions on the sharing of confidential data across borders (and even between regulatory agencies in the same jurisdiction) have meant that authorities are limited to focusing on parts of their domestic financial ecosystem and do not have the information available to model global systemic resilience. The solution to this problem requires unprecedented levels of international data sharing and cooperation.

3.4 The Importance of International Coordination

Happily, the foundations of such cooperation started being laid around the same time that the new generation of regulatory stress testing made its big entrance in 2009. In November 2009, G20 finance ministers and central bank governors laid down the first pillar when they called on the Financial Stability Board (FSB) to improve data collection and sharing on the risks posed by global systemically important financial institutions (IMF and FSB, 2009).

In the last 5 years, the FSB has developed common data templates that support the collection and sharing of granular data on G-SIBs' bilateral exposures, funding dependencies, and balance sheets—supporting mapping of the global network of systemically important banks and opening up the possibility of stress testing the global banking network.

The FSB's workplan goes beyond the banking system—collecting data to support monitoring of nonbank financial intermediation, supporting the development of securities holdings databases in G-20 economies, and making progress in the technical and governance work for the collection of global data on securities financing transactions (IMF and FSB, 2017).

Continuing progress in these areas will be core to taking the next, most important step in stress testing: taking our models global.

The need for cooperation goes beyond data sharing. International organizations can bring together and focus the attention of researchers and policymakers from around the globe to accelerate progress and improve the regulatory toolkit. The IMF, FSB, and Bank for International Settlements (BIS) have consistently supported and encouraged progress in stress testing, documenting and signposting the journey that authorities have been on in the last decade (see, e.g., Anderson et al. [2018] for a recent account of progress in macroprudential stress testing).

But they can go—and are going—further by providing the crucible in which new toolkits can be forged and tested. The FSB, for example, has been coordinating ongoing international work to simulate stress in the global market-based finance system—a hint at the possibilities for stress testing over the next decade (FSB, 2017).

3.5 Challenges into the Next Decade and Beyond

The crisis held many lessons for authorities regarding what would be needed for stress testing to fulfill its promise as a tool to support micro- and macroprudential goals. In this chapter, we argue that these lessons have in general been learned and acted upon.

Stress testing has been reformed and enhanced, becoming a central plank of the regulatory architecture. Stress tests are routinely used by authorities to provide a forward-looking assessment of banks' resilience—and where required, to take actions to increase that resilience. The transparency around them has both increased authorities' public accountability and supported public confidence, even as the tests have demonstrated the need for banks to strengthen their capital positions and supported regulators in requiring them to do so. And stress tests have driven improvements in both banks' own risk management and the ability of authorities to model and understand risks. As such, stress tests have played an important role in the recovery of the financial sector postcrisis.

The last crisis supercharged the development of effective stress testing but must not constrain it. We must continue to look forward. At the core of stress testing is the simple idea that we should challenge ourselves always to consider "what could happen": to imagine future extreme events, whatever may cause them and whenever they may occur.

As stress tests have become more routine, authorities have started to expand their horizons—challenging themselves to remain agile and imaginative—and in so doing, to harness the apparatus and philosophy of stress testing to explore new risks.

It is vitally important that this continues. The financial system is dynamic and adaptive, so any regulatory regime must also be adaptive if it is to contain risk (Haldane et al., 2017). As the global economy continues to evolve, so too must stress testing. Authorities—and financial institutions—must develop different types of scenarios to explore new risks. They must continue to develop toolkits and data sets to understand and model existing and developing financial networks and contagion risks. And stress tests must evolve and expand to keep pace with the changing nature of the financial system.

Although huge strides have been made since 2009, there is—and may always be—more to do if stress testing is to continue to support macroeconomic and financial resilience in the coming decade and beyond.

References

Acharya, V. V., and T. Yorulmazer (2008), "Information contagion and bank herding," *Journal of Money, Credit and Banking*, 40, 215–231.

Adrian, T., N. Boyarchenko, and D. Giannone (2017), "Vulnerable growth," Federal Reserve Bank of New York Staff Report No. 794.

Ahnert, T., K. Anand, P. Gai, and J. Chapman (2016), "Asset encumbrance, bank funding and financial fragility," Bank of Canada Staff Working Paper 2016–16.

Aikman, D., P. Chichkanov, G. Douglas, Y. Georgiev, J. Howat, and B. King (2019), "System-wide stress simulation," Bank of England Staff Working Paper No. 809.

Anderson, R., J. Baba, C. Danielsson, U. S. Das, H. Kang, and M. A. Segoviano (2018), "Macroprudential stress tests and policies: Searching for robust and implementable frameworks," IMF and LSE Systemic Risk Centre Report.

Angelini, E., N. Bokan, K. Christoffel, M. Ciccarelli, and S. Zimic (2019), "Introducing ECB-BASE: The blueprint of the new ECB semi-structural model for the euro area," ECB Working Paper Series No. 2315, available at www.ecb.europa.eu/pub/pdf/scpwps/ecb.wp2315~73e5b1c3cd.en.pdf.

Aymanns, C., C. Caceres, C. Daniel, and L. Schumacher (2016), "Bank solvency and funding cost," IMF Working Papers 16/64.

Aymanns, C., J. D. Farmer, A. M. Kleinnijenhuis, and T. Wetzer (2018), "Models of financial stability and their application in stress tests," in C. Hommes and B. LeBaron (eds.), *Handbook of computational economics*, vol. 4, Elsevier.

Banco Central do Brasil (2017), *Financial Stability Report, Volume 16*.

Bank of England (2013), "A framework for stress testing the UK banking system—discussion paper."

Bank of England (2015), "The Bank of England's approach to stress testing the UK banking system."

Bank of England (2017a), "Stress testing the UK banking system: 2017 results."

Bank of England (2017b), "Stress testing the UK banking system: Key elements of the 2017 stress test."

Bank of England (2018, June), *Financial Stability Report, Issue No. 43*.

Bank of England (2019a, July), *Financial Stability Report, Issue No. 45*.

Bank of England (2019b), "Stress testing the UK banking system: 2019 guidance for participating banks and building societies," available at www.bankofengland.co.uk/-/media/BoE/Files/stress-testing/2019/stress-testing-the-uk-banking-system-2019-guidance.

Bank of England (2019c, December), "The 2021 biennial exploratory scenario on the financial risks from climate change," Bank of England Discussion Paper.

Bank of England (2019d, December), "The results of the 2019 stress test of UK banks," in *Financial Stability Report, Issue No. 46*.

Bank of Japan (2016), "Macro stress testing in the financial system report (October 2016)," available at www.boj.or.jp/en/research/brp/fsr/fsrb161026.htm/.

Bank of Korea (2018), "Results of non-bank financial institution stress test model development," *June 2018 Financial Stability Report*.

Baranova, Y., J. Coen, R. Lowe, J. Noss, and L. Silvestri (2017), "Simulating stress across the financial system: resilience of corporate bond markets and the role of investment funds," Bank of England Financial Stability Paper No. 42.

Bardoscia, M., P. Barucca, A. Brinley Codd, and J. Hill (2019), "Forward-looking solvency contagion," *Journal of Economic Dynamics and Control*, 108, 103755.

Bardoscia, M., G. Bianconi, and G. Ferrara (2018), "Multiplex network analysis of the UK OTC derivatives market," Bank of England Staff Working Paper No. 726.

Barucca, P., M. Bardoscia, F. Caccioli, M. D'Errico, G. Visentin, G. Caldarelli, and S. Battiston (2016), "Network valuation in financial systems." available at https://papers.ssrn.com/sol3/papers.cfm?abstract_id=2795583.

Basel Committee on Banking Supervision (1996), "Amendment to the capital accord to incorporate market risks."

Basel Committee on Banking Supervision (2009), "Principles for sound stress testing practices and supervision."

Basel Committee on Banking Supervision (2015), "Making supervisory stress tests more macroprudential: Considering liquidity and solvency interactions and systemic risk," Basel Committee for Banking Supervision. *Working Paper 29*

Basel Committee on Banking Supervision (2017), "Supervisory and bank stress testing: Range of practices."

Battiston, S., and S. Martinez-Jaramillo (2018), "Financial networks and stress testing: Challenges and new research avenues for systemic risk analysis and financial stability implications," *Journal of Financial Stability*, 35, 6–16.

Bookstaber, R., M. Paddrik, and B. Tivnan, "An agent-based model for financial vulnerability," Office of Financial Research Working Paper 14–05.

Borio, C. E. V., M. Drehmann, and K. Tsatsaronis (2012), "Stress-testing macro stress testing: Does it live up to expectations?" BIS Working Paper No. 369.

Brazier, A. (2017), "How to: MACROPRU. 5 principles for macroprudential policy," Speech at the London School of Economics Financial Regulation Seminar.

Brazier, A. (2018), "Market finance and financial stability: Will the stretch cause a strain?" Speech at the Brevan Howard Centre for Financial Analysis, Imperial College Business School.

Brunnermeier, M. K. (2009), "Deciphering the liquidity and credit crunch 2007–2008," *Journal of Economic Perspectives*, 23(1), 77–100.

Brunnermeier, M. K. and L. H. Pedersen (2009), "Market liquidity and funding liquidity," *The Review of Financial Studies*, 22, 2201–2238.

Chavaz, M., J. Chiu, and E. Stoya (2015, November 9), "The 'question' or the 'answer'? Market reaction to UK stress tests," Bank Underground blog post.

Churm, R., and P. Nahai-Williamson (2019), "Four years of concurrent stress testing at the bank of england: Developing the macroprudential perspective," in Akhtar Siddique, Iftekhar Hasan, and David Lynch (eds.), Stress testing, 2nd ed., Risk Quantum.

Clerc, L., A. Giovannini, S. Langfield, T. Peltonen, R. Portes, and M. Scheicher (2016), "Indirect contagion: The policy problem," ESRB Occasional Paper Series No. 9.

Cont, R., and E. Schaanning (2017), "Fire sales, indirect contagion and systemic stress testing," Norges Bank Working Paper 02/2017.

Cunliffe, J. (2014), "Regulatory reform and returns in banking," speech at Bank of England.

Danielsson, J., R., Macrae, D. Tsomocos, and J.-P. Zigrand (2016, December 15), "Why macropru can end up being procyclical," VOXEu, available at http://voxeu.org/article/why-macropru-can-end-being-procyclical

Demekas (2015), "Designing effective macroprudential stress tests: Progress so far and the way forward," IMF Working Paper 15/146.

Dent, K., S. Hacioglu Hoke, and A. Panagiotopoulos (2017), "Solvency and wholesale funding cost interactions at UK banks," Bank of England Staff Working Paper No. 681.

Dodd–Frank "Wall Street reform and Consumer Protection Act," Pub. Law 111-203, H.R. 4173, 111th Cong. (2010).

Eisenberg, L., and T. H. Noe (2001), "Systemic risk in financial systems," *Management Science*, 47, 236–249.

European Banking Authority (2017), "2018 EU-wide stress test methodological note," www.eba.europa.eu/sites/default/documents/files/documents/10180/2106649/2ff7a78b-ffdc-4fb6-9244-8ec994928c96/2018%20EU-wide%20stress%20test%20-%20Methodological%20Note.pdf?retry=1.

European Central Bank (2017), "STAMP€: Stress-test analytics for macroprudential purposes in the euro area," available at www.ecb.europa.eu/pub/conferences/shared/pdf/20170511_2nd_mp_policy/DeesHenryMartin-Stampe-Stress-Test_Analytics_for_Macroprudential_Purposes_in_the_euro_area.en.pdf.

Federal Reserve (2019), "Policy statement on the scenario design framework for stress testing," 12 *CFR* Part 252.

Financial Services Authority (2009), "Stress and scenario testing," Policy Statement 09/20.

Financial Stability Board (2017, February 28), "FSB assesses implementation progress and effects of reforms," available at: www.fsb.org/2017/02/fsb-assesses-implementation-progress-and-effects-of-reforms/.

Financial Times (2007, July 9), "Citigroup chief stays bullish on buy-outs."

Fique (2017), "The MacroFinancial Risk Assessment Framework (MFRAF), version 2.0," Bank of Canada Technical Report No. 111.

Furfine (2003), "Interbank exposures: Quantifying the risk of contagion," *Journal of Money, Credit and Banking,* 35, 111–128.

Glasserman, P., and G. Tangirala (2015), "Are the Federal Reserve's stress test results predictable?" Office of Financial Research Working Paper 15–02.

Glasserman, P., and H. P. Young (2016), "Contagion in financial networks," *Journal of Economic Literature*, 54(3), 779–831.

Goldstein, I., and H. Sapra (2013), "Should banks' stress test results be disclosed? An analysis of the costs and benefits," available at https://papers.ssrn.com/sol3/papers.cfm?abstract_id=2367536.

Greenwood, R., A. Landier, and D. Thesmar (2015), "Vulnerable banks," *Journal of Financial Economics*, 115, 471–485.

Gross, M. D., M. Leika and L. Valderrama (2019), "The macro-financial System simulator (MASS)" (forthcoming).

Haldane, A. G. (2009), "Why banks failed the stress test," Speech at the Marcus Evans Conference on Stress Testing.

Haldane, A. G. (2015), "On microscopes and telescopes," Speech at the Lorentz Centre Workshop on Socio-economic Complexity, Leiden, Netherlands.

Haldane, A. G., D. Aikman, S. Kapadia, and M. Hinterschweiger (2017), "Rethinking financial stability," Speech at the Peterson Institute for International Economics.

International Monetary Fund (2012, August), *"Macrofinancial stress testing—principles and practices,"* prepared by the Monetary and Capital Markets Department of the IMF.

International Monetary Fund (2017), "Financial conditions and growth at risk," in *Global Financial Stability Report, October*.

International Monetary Fund and Financial Stability Board (2009), "The financial crisis and information gaps," Report to the G-20 Finance Ministers and Central Bank Governors.

International Monetary Fund and Financial Stability Board (2017), "The financial crisis and information gaps: Second phase of the G-20 Data Gaps Initiative (DGI-2) second progress report."

Iyer, Rajkamal, and Jose-Luis Peydro (2011), "Interbank contagion at work: Evidence from a natural experiment," *Review of Financial Studies*, 24(4), 13371377.

Kapadia, S., M. Drehmann, J. Elliott, and G. Sterne (2012), "Liquidity risk, cash-flow constraints and systemic feedbacks," Bank of England Working Paper 456.

Kleinnijenhuis, Nahai-Williamson, Wetzer and Farmer (2020), "Foundations of system-wide financial stress testing with heterogeneous institutions," Bank of England Staff Working Paper No. 861

Korsgaard, S. (2017), "Incorporating funding costs *to* top down stress tests," *Danmarks* NationalBank Working Papers 110.

Ong, L., and M. Čihák (2010), "Of runes and sagas: Perspective on liquidity stress testing using an iceland example," IMF Working Paper No. 10/156.

Paddrik, M., S. Rajan, and H. P. Young (2016), "Contagion in the CDS market," Office of Financial Research Working Paper 16–12.

Paisley, J. (2017), "Stress testing: Where next?" *Journal of Risk Management in Financial Institutions*, 10, 224–237.

Royal Bank of New Zealand (2016), "Summary of the dairy portfolio stress testing exercise," *Reserve Bank of New Zealand Bulletin*, vol. 79, no. 5.

Royal Bank of New Zealand (2018), "The Reserve Bank's philosophy and approach to stress testing," *Reserve Bank of New Zealand Bulletin*, vol. 81, no. 8.

Shleifer, A., and R. Vishny (2011), "Fire sales in finance and macroeconomics," *Journal of Economic Perspectives*, 25, 29–48.

Schuermann, T. (2013), "Stress testing banks," *International Journal of Forecasting*, 30, 717–728.

Part II

Inputs and Outputs

5

Macrofinancial Stress Test Scenario Design—for Banks and Beyond

Marco Gross, Jérôme Henry, and Elena Rancoita[*]

1 Introduction

Macrofinancial scenario design constitutes an essential component of bottom-up and top-down stress test analyses carried out by financial firms and oversight authorities, respectively. Stress tests are generally conducted to inform policy measures to be taken at the firm or system level, which renders their results critical for the agents involved. The analysis can therefore have far-reaching implications, either at the financial firm level (e.g., influencing the way risk management or business-line assessment are conducted) or at the institutional level (e.g., informing policy decisions on regulatory capital or liquidity requirements). Because stress tests are by nature conditional forecast exercises, a great amount of attention is devoted to scenario design by many private and public institutions as a first step in the overall process. Of course, other elements, such as the models and methodologies adopted for the impact assessment, also play a crucial role in defining the stress test results and therefore are also key drivers of the policy actions hinging on the stress test results.[1]

The methodologies to design adverse scenarios have evolved significantly in recent years, in line with the expansion of stress testing activity. In the immediate aftermath of the Global Financial Crisis (GFC) of 2007–2009, stress test exercises have been employed as a crisis-management tool in order to evaluate the ability of credit institutions to withstand extreme tail events (e.g., similar in terms of severity to the GFC) and to calibrate firm-level recapitalization needs that would render them and the whole system more resilient to large shocks. In recent years, the scope of stress tests has widened in various dimensions. First, sector-wide stress tests are currently run not only for the banking sector but also for nonbank financial institutions, such as insurance companies, pension funds, central counterparties, money market funds, investment funds, and so forth. Second, the notion of feedback effects has gained prominence, which links to the need felt to account for the mutual dependence between financial developments (e.g., in terms of credit) and macroeconomic outcomes.

---

[*] The views expressed in this chapter reflect those of the authors and not that of the European Central Bank (ECB) or the International Monetary Fund (IMF). We benefited from comments from the editors and a referee as well as from participants in the IMF FSAP@20 conference and in seminars at Bank of Greece and Central Bank of Luxembourg.
[1] Consider, for example, caps and floors or other constraints embedded in the methodologies of some supervisory stress tests that affect the way macrofinancial scenarios translate into stress test results (see Niepmann and Stebunovs, 2018).

Third, a number of institutions have started using their stress test bottom-up and top-down analytical toolkits for the calibration of macroprudential policies (e.g., the Bank of England [BoE] calibrates the countercyclical capital buffer based on the Annual Cyclical Scenario [ACS] analysis; see BoE, 2016). This evolution has motivated the development of specific methodologies to ensure that the scenario design is aligned with the ultimate goal of a stress test.[2]

This chapter provides an overview of various modeling elements that are instrumental for the design and calibration of different types of scenarios, with a particular focus on a narrative-based approach for system-wide stress testing. The content of the chapter is largely based on the experience gained in scenario design at the European Central Bank (ECB), although not exclusively so. Since 2009, ECB staff has provided analytical support to the design of scenarios for EU-wide banking-system stress tests, first with the Committee of European Banking Supervisors (CEBS) and thereafter with the European Banking Authority (EBA) since 2011, under the aegis of the European Systemic Risk Board (ESRB). Over time, this task extended to designing scenarios for the stress tests run by the three European supervisory authorities (ESAs): the EBA, the European Securities and Markets Authority (ESMA), and the European Insurance and Occupational Pensions Authority (EIOPA).[3] In addition, ECB staff regularly produces macrofinancial scenarios for internal and external financial stability assessment studies. The latter is published in the ECB's biannual *Financial Stability Review* (FSR), where the impact of main identified risks on banks and insurers are reported.[4]

The chapter is structured as follows: In Section 2, an introduction to narrative-based scenario design is provided, together with an overview of the approach to scenario design at various policy institutions. In Section 3, several specific model methodologies are introduced, including, inter alia, an overview of calibration methodologies for financial and macroeconomic shocks and methodologies for quantifying scenario probabilities, designing reverse stress test scenarios, and generating scenarios. In Section 4, practical applications of scenario design for different types of financial institutions are presented (banks, insurance companies, pension funds, central counterparties). Section 5 concludes and provides an overview of possible future conceptual and methodological developments.

2 Narrative-Based Approach to Macrofinancial Scenario Design

Among a number of international institutions in charge of macrofinancial oversight, a common approach to scenario design is *narrative-based* design, meaning that the scenario design is linked to a risk assessment (e.g., some scoreboard or set of financial risk indicators) and a concrete storyline for an unfolding stress scenario. The scenarios of the EBA system-wide stress test exercises and of the IMF Financial Sector Assessment Programs (FSAPs) represent two notable examples in this respect. In practice, there are various ways to connect the risk assessment with the scenario: identified risk factors can be mapped into some exogenous

---

[2] See Baudino et al. (2018) for the more generic need for all stress test exercises to have features aligned with their respective objectives.

[3] The resulting documentation of all scenario inputs from the ECB/ESRB side is collected at the following link: www.esrb.europa.eu/mppa/stress/html/index.en.html.

[4] See Chapter 27 in this book for such illustrative scenario impact analyses and models employed to that end.

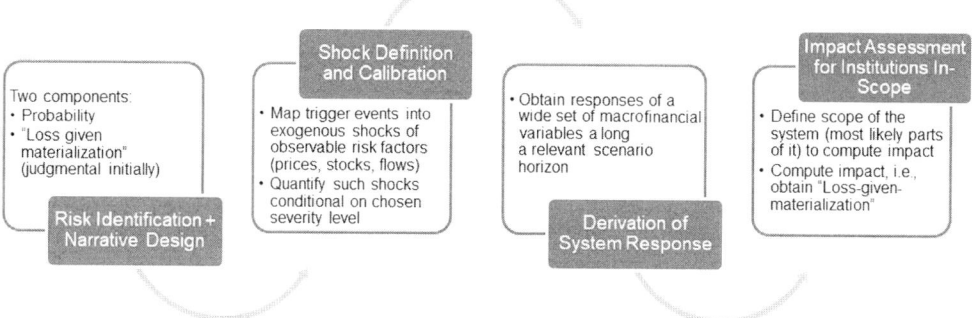

Figure 5.1 Scenario design process.

shocks to the model used for the scenario calibration; alternatively, some features or level of stress can be imposed directly on the path for variables reported in the scenario (including for those that are endogenous in econometric models). Narrative-based scenarios are usually accompanied by a description linking the risk assessment to the events materializing in the scenario as a result of a specific set of adverse shocks.

The design of narrative-based scenarios normally entails four steps, as depicted in Figure 5.1. The first step consists of the risk identification and definition of the economic events materializing in the scenario. *Risk* may be defined as the product of a probability of an event to materialize (which is meant to be small for an adverse scenario to be severe) and a "loss given materialization" (which is usually sizeable) in macroeconomic terms.[5] The latter can be difficult to assess; hence, in most cases, its evaluation is based on up-front expert judgment, possibly informed by prior model simulations. For example, global risks may cause potentially more material losses for open economies than local risks. Most importantly, the probability of events that may trigger a major shock is often not easy or even impossible to quantify. For instance, the probability that trade tensions escalate or that geopolitical risks in certain regions of the world materialize is hardly quantifiable, in which case a pure judgmental stance on the perceived probability of materialization has to be formed.

The second and third steps entail the derivation of the entire set of variables included in the scenario. In the second step, based on the risk assessment, risks are mapped into key events occurring in the scenario, and the desired severity level is defined. Then, models are used to derive consistently the path of a larger set of variables comprised in the scenario. Most economic models identify the response of endogenous variables to some exogenous shocks—for example, real gross domestic product (GDP) usually responds to shocks such as a repricing of risk premia and demand confidence. In such a case, as depicted in Figure 5.2, the second step can be performed by mapping the risk assessment into the set of exogenous

---

[5] If one were to consider adverse events that happen frequently, then, by nature of being frequent, they would not come along with material impacts. Events qualifying as "systemic risks" are generally characterized by "low probability and high (adverse) impact."

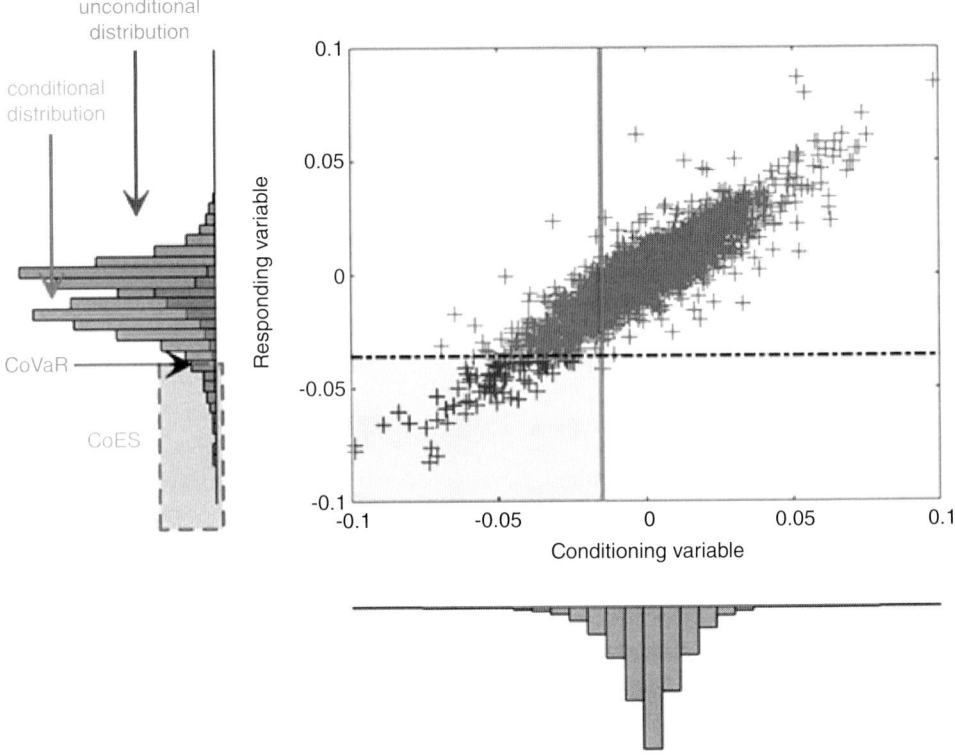

Figure 5.2 Graphical illustration of the multivariate copula methodology.

shocks "available in the model" (e.g., a repricing of risk premia and a decline of demand confidence). Another way of performing the second step would be to map directly the risk assessment into the path of some endogenous variables of the models (e.g., real GDP drops by a given amount) and then use statistical tail correlations or the covariance structure of the exogenous shocks of the model to derive the remaining endogenous variables (e.g., house prices). The severity can be assessed in both cases by focusing, respectively, on that of the (exogenous) shock or on that of the (endogenous) outcome.

Examples of both approaches can be found in various institutions. The first approach is adopted in particular to design EBA scenarios, that is, by mapping risks into exogenous shocks.[6] To that end, ECB staff employed a number of models, described in Sections 3.1 and 3.2.[7] Instead, institutions such as the Federal Reserve (Fed), IMF, BoE, and Bank of Japan (BoJ) start the scenario calibration by defining the scenario severity in terms of a path for one or more specific "headline" economic variables. For instance, the Fed starts the calibration of its "severely adverse" scenario by defining the unemployment rate and house price paths in this scenario, whereas the IMF, the BoE, and the BoJ usually start the scenario calibration from the definition of a path for real GDP. The remaining variables are

---

[6] See, for example, ESRB (2018).
[7] See also Henry (2015) in relation to the EBA exercise process.

inferred by means of models to derive their trajectories consistent with the given scenario severity for specific variables and in line with the scenario narrative.

The fourth and last step entails the use of models for the impact assessment on the set of entities under scrutiny in the envisaged stress test. This step relates back to the narrative design and can be either judgmental or quantitative. Although the judgmental assessment of the "loss given materialization" can inform the scenario from the outset, one could also adjust a narrative based on a preliminary quantitative impact assessment that shows relevant blind spots. Such interplay is also close to the notion of reverse stress testing and worst-case search methodologies (see Section 4). Such methodologies fundamentally rest on the notion of "loss given materialization," their key metric of interest, and thereby bridge the gap between the first and fourth stage shown in Figure 5.1.

The scenario calibration process (i.e., the first two components in Figure 5.1) can be accomplished by different means. We can distinguish three main approaches:

1. Historical scenarios: The replication of crisis episodes as they were observed historically, including a whole set of macrofinancial indicators.
2. Synthetic scenarios: Scenario profiles that result from models using historical data, reflecting past relationships but involving specific shock assumptions that generate scenario patterns that were not observed in history.
3. Reverse scenarios: Involving macrofinancial models while targeting a shock size and scenario profile that achieves a certain loss amount (e.g., the maximum loss) for a pre-defined set of financial institutions or other economic agents. These could still be bound by historical dependencies among macroeconomic variables, captured by models (see Section 2.4 for follow-up discussion).

Most of the institutions that made their scenario design methodologies public employ one or a mixture of these approaches.[8] Table 5.1 provides a summary of statistical and econometric modeling methods in relation to scenario design, which will shape the discussion in the remainder of this chapter. The information in the table is not meant to be exhaustive in various respects (e.g., the references to model types) but instead to offer a high-level overview perspective.

In Table 5.1, we distinguish between four model structures, starting in the first column with the "purest statistical" ones, in general not capturing any time-series dependencies. From the second column onward, the model types instead capture multivariate time-series dependencies. Satellite models that link bank-level risk metrics to macrofinancial conditions fall into this category. Taken together, such risk-parameter equation systems imply large-scale appended equation systems to models of a semistructural large-scale structure (macrofinancial scenario design purposes) in the third and fourth column of the table.

2.1 High-Frequency Asset Price Scenario Design Methodologies

The calibration of shocks to financial asset prices represents a particularly relevant element of scenarios for stress testing exercises for financial institutions, both because these

---

[8] See also Baudino et al. (2018) and Bassanin et al. (2021) for a survey on the methodologies adopted across institutions.

Table 5.1. Overview of model structures from a scenario design perspective

	Pure distributional models	Parametric time-series models	Semistructural macrofinancial models	Structural macro-financial models
Dimensionality/ Scope	Univariate or multivariate, small to medium to large scale		Multivariate, medium to large scale	
Frequency of data, typically involved (spec. in scenario design applications)	High frequency (e.g., daily)	Lower frequency (e.g., quarterly)	Lower frequency (e.g., quarterly)	
Notion of causality	Absent	Granger causality, shock identification methods (Cholesky, sign restrictions)	Causal structures built into the systems' equation structures from the start	
Accounting for nonlinear dependencies	Frequently done in distribution terms whenever deviating from Gaussian distributions (e.g., t-distribution/copula, extreme value copulas, nonparametric distribution/copulas, etc.)	Familiar but not too frequently considered yet for scenario design (consider for example regime switching models)	Not generally considered	DSGE models, usually linearized around a steady state, i.e., nonlinearities effectively not considered
Overall use specifically for macro-financial scenario design purposes	Multivariate distributions used for shock-conditional high-frequency market-risk scenario design; serial dependence hardly ever captured	Impulse responses (IRs) in four categories: non-factorized, generalized, Cholesky, sign-restricted; conditional-on-observables and conditional-on-shocks forecasting; in addition: vector error correction models for estimation of equilibrium (and deviations from equilibrium) with respect to specific price and volume measures (e.g., credit, house prices)	Not used in general for macrofinancial scenario design for the purpose of assessing financial sector resilience; focus instead on counterfactual monetary policy scenario analysis. At times employed for assessing second-round macro effects of capital shortfalls of banks. Semistructural models often involve error-correcting model (ECM) structures.	
Example models/ keywords	Univariate: VAR, ES; multivariate: CoVaR, CoES; copulas	ARMA, ARCH, GARCH, (S-)VARs, (S-)GVARs, (S-)BVARs, (S-)MS-VARs, (V)ECMs	Large-scale semistructural models at central banks (e.g., FRB/US, ECB's NAWM, EC's QUEST, BoC's GEM), Stress Test Elasticities (STEs) in use at the ECB	DSGE models

ARCH, autoregressive conditional heteroskedasticity; ARMA, autoregressive moving average; BoC, Bank of Canada; CoES, conditional expected shortfall; CoVar, covariance; DSGE, dynamic stochastic general equilibrium; EC, European Commission; ES, expected shortfall; FRB, Federal Reserve Board; GARCH, generalized autoregressive conditional heteroskedasticity; GEM, Global Economy Model; NAWM, New Area-Wide Model; (S-)BVAR, (structural) Bayesian vector autoregression; (S-)GVAR, (structural) global vector autoregression; (S-)MS-VAR, (structural) Markov-switching vector autoregression; (S-)VAR, (structural) vector autoregression; VAR, vector autoregression; (V)ECM, (vector) error-correction model.

institutions usually have large direct exposures to financial securities; and also because they constitute a starting-point narrative for ensuing downturns for the real economy. For example, shocks to interest rates affect the net interest margins, and all shocks to financial risk factors determine the market-risk losses. Stress test models and methodologies require very granular information on shocks to different financial security prices (e.g., in terms of a whole array of interest rates across a maturity spectrum, corporate bond yields, volatility metrics, etc.).

Designing financial asset price stress test scenarios is challenging because it generally involves a large number of risk factors. Table 5.2 provides a comparison of the number and type of risk factors included in scenarios used in different stress test exercises. For example, the EBA market-risk scenario for 2018 included about 900 risk factors. A similar number of risk factors was included in the EIOPA/ESMA scenarios for stress testing pension funds, insurance firms, and central counterparties. Examples of risk factors include equity prices; currency swap rates; government, financial, and nonfinancial corporate bond yields and spreads; foreign exchange (FX) rates; implied term structures of FX volatilities; interbank market rates, and credit default swap (CDS) spreads, among others. Many of these factors need—in the EU in particular—to be considered for numerous countries, regions, and subsectors of the economies, hence easily implying a total set of several thousands of risk factors.

Asset return modeling methods from the field of portfolio analysis, multivariate distribution, and dependence modeling are instrumental for scenario design based on high-frequency data (see first column in Table 5.1). One way of modeling joint distributions is *copulas* (Latin for "link, tie, or bond"), a statistical concept that, in conjunction with a description of individual factor-specific distributions, characterizes the full multivariate distributions of multiple factors. The development of this methodology stems from the fact that almost all financial data exhibit nonnormal features, including excess kurtosis, skewness, asymmetric tail dependence, and volatility clustering.[9]

Copula models can be divided into two groups: *parametric* and *nonparametric*. Parametric copulas approximate the joint and marginal distribution of the risk factors with some known distributions, which can be categorized to fall into three categories: related to the family of elliptical distributions (e.g., multivariate Gaussian and Student's *t*), Archimedean copulas (e.g., Gumbel, Joe, Kimeldorf–Sampson), and extreme-value copulas (e.g., Gumbel, Galambos, Huesler–Reiss). Nonparametric copula models are based on the empirical distribution without imposing ex ante any distributional shape.[10] For the design of adverse scenarios, the nonparametric copula method is particularly suitable because it allows the modeler to capture the nonnormal, nonlinear dependencies of financial asset returns. Parametric copulas, too, represent a useful tool for the design of financial shock scenarios,

---

[9] *Volatility clustering* refers to the empirical regulatory trend that "large changes tend to be followed by large changes, of either sign, and small changes tend to be followed by small changes" (Mandelbrot, 1963), which has been further empirically documented by Granger and Ding (1995), Cont (2007), and numerous others.

[10] Some useful entry points to the wide field of copula modeling include Deheuvels (1979; nonparametric/empirical copulas), Genest and MacKay (1986; Archimedean copulas), and Joe (1997; extreme-value copulas). A comprehensive overview of copulas and related concepts and methodologies can be found in Nelsen (1999). A specific focus on financial copula applications can be found in Cherubini et al. (2004).

Table 5.2. Comparison of the number and type of risk factors included in scenario, used in different supervisory stress test exercises in Europe

| | | No. of risk factors | Swap | | | Government bonds | Corporate bonds | Equities | | | | Macro-economic | Real estate | Volatilities | Liquidity |
			LIBOR	OIS	Inflation			Generic indices	Commodities	FX	CDS				
EBA	Macroeconomic	300–350	X			X		X	X	X	X	X	X		
	Market risk	900–1000	X			X	X	X	X	X	X			X	
EIOPA	Insurance	900–1,000	X		X	X	X	X	X		X		X		
	IORP	900–1,000	X		X	X	X	X	X		X		X		
ESMA	MMF	900–1,000	X			X	X	X	X	X	X				
	CCP	700	X			X	X	X	X	X	X			X	X

CCP, central counterparty; IORP, institutions for occupational retirement provision (IORP); LIBOR, London Interbank Offered Rate; MMF, money market fund; OIS, overnight index swap.

although they are, as of yet, mainly used by ECB staff for benchmarking the nonparametric copula results and for coping with scarce data (because they approximate the distribution with a precise distributional form).

The use of copulas and joint distribution modeling can be combined with other models, such as parametric (whether linear or nonlinear) models that can capture serial dependence or variance clustering in the data (i.e., which is not yet independent and identically distributed [i.i.d.]), to then apply the copula on the residuals of that model, which would be i.i.d.[11] There is a great variety of ways to combine marginals with copula functions in empirical applications, and hence some set of criteria is needed for the related choices to be made. One criterion can be the empirical, historical fit, with varying focus on the full distributions or the tails. A second, practical criterion relates to the purpose and use of the method. From a scenario design perspective, less interest may lie in obtaining estimates of parametric copulas to characterize tail dependencies than in obtaining unconditional or conditional scenario simulation outcomes. This argument, coupled with the desire to not risk mis-specifying tail dependencies if some suboptimal parametric marginal or copula function were employed, makes the use of nonparametric marginals and copulas for scenario design purposes appealing. Moreover, the high dimensionality of the data sets that are involved for financial factor scenario design at policy institutions often implies that most parametric distributions cannot be estimated (because of computational limits).

The simulation method used for scenario design by ECB staff is mainly based on a nonparametric copula methodology.[12] In this framework, about 3,000 high-frequency risk factors are jointly simulated forward in time up to a self-defined horizon (e.g., 60 business days) by bootstrapping from historical data. First, either a single factor (e.g., FTSE100) or a group of risk factors (e.g., all European stock prices) is assumed to be in the tail of the joint distribution; these are referred to as *conditioning factors*. Then the conditional expected shortfall (i.e., the tail of the conditional distribution) of all other risk factors is calculated (Figure 5.2). Overall, the approach is related to the one promoted by Adrian and Brunnermeier (2016).[13, 14]

Dependency breaks can be a relevant means to generate scenarios that do not merely reflect historical marginal distributions and joint dependence. Parametric copula methods

---

[11] For example, a generalized autoregressive conditional heteroskedasticity (GARCH) model filter can be applied to the data, with the resulting residuals being the input to the copula.

[12] This approach has been developed and operationalized by the authors of this chapter. See also Ojea Ferreiro and Rancoita (2019) for more details.

[13] A distinction can be made between a "plain" and "smooth" historical bootstrap methodology at this point. "Plain" bootstrapping means a direct draw of raw historical data vectors. "Smooth" bootstrapping means that a kernel estimate of the multivariate distribution is involved; the drawing from the kernel-based estimate requires the use of an accept-reject algorithm, which is a standard methodology that can be found in most textbooks on statistics (e.g., Robert and Casella, 2004). Kernel estimates in more than three dimensions are usually not feasible; when the focus is on scenario (shock)-conditional responses, one can employ many bivariate, pair-wise kernel estimates of all risk factors vis-à-vis the one (or the weighted aggregate distribution of a set of factors) that is shocked, to operationalize (i.e., involve) the kernel. The purpose of the kernel is to avoid replicating fine, possibly spurious details of the data in the tails of their distributions.

[14] One counterargument whose weight in the choice between parametric and nonparametric models should be nonzero is the loss of efficiency (precision) of the nonparametric estimates, conditional on a certain parametric functional form being correct (by assumption).

can be useful in that context because they contain, by their nature, *parameters*, which can be changed "manually." Correlation and covariance matrices in Gaussian copulas, for instance, can be made state dependent (e.g., crisis dependent), in which case they would still be historical (sub-)episode dependent. They can also be modified to contain completely hypothetical values for strengthening selected relationships between markets or even reverting the signs of dependencies. Likewise, semi-structural or structural macrofinancial models (third and fourth columns in Table 5.1) can be used to consider targeted dependency breaks.

Whether employing parametric or nonparametric statistical and econometric methods, the following "scenario simulation parameters" are important:

- **The scenario horizon.** The scenario horizon is usually defined as part of the stress test methodology and is mainly based on the type of risks that are relevant for a stress testing exercise. For example, if the relevant metric is the impact of changes of market prices on the valuation of securities, then a short time horizon is usually chosen (e.g., for market risk in banking-sector stress testing exercises).
- **The probability defining the tail of the distribution of the "conditioning factors."** This determines the severity of the scenario: a smaller probability implies a more severe scenario.
- **The historical sample period used for the calibration**. All else equal, the choice of a sample window influences the severity of a resulting adverse scenario, to the extent that historical recession/crisis events are included in the sample (i.e., it can be used to "go closer" to a certain type of crisis).

2.2 Lower-Frequency Scenarios and Macrofinancial Path Derivations for Longer Horizons

There are different methodologies for designing macrofinancial scenario paths for a longer horizon (e.g., up to 3 to 5 years). Some of the most common methodologies are based on estimated empirical regression models (structural value-at-risk [VaR] types of models) or on general-equilibrium models (see e.g., Vitek, 2015). In this case, the scenario calibration starts with the calibration of the exogenous shocks for these models.[15] These shocks usually include impulses to equity prices, short- and long-term interest rates, and other selected asset prices (e.g., for residential real estate). In addition, shocks to external demand and even to endogenous domestic variables can be considered in this step. Then, macroeconomic structural or semistructural models are used to derive a consistent macrofinancial forward path for all relevant macrofinancial variables and countries.

Many of the models regularly applied for this purpose have been developed mainly to support monetary policy. For ECB staff, the workhorse model suite used for scenario design purposes is called *stress Test Elasticities* (STEs). The STE model is based on a set of elasticities that are estimated by national central banks. The elasticities capture the response of a large set (50–60) of macroeconomic variables to exogenous shocks (to real economic

---

[15] One should note that an exogenous shock to a variable in a model could be an endogenous response in another model. For example, a variation in house prices might be due to some exogenous shock in a model, whereas in another model, could also be given by the response to the movement of other endogenous variables.

variables or selected financial asset prices) for all 28 EU countries. Another important feature is that the STEs also incorporate intra-EU trade spillovers.

Next to the STEs, the National Institute Global Econometric Model (NiGEM) is usually employed for the projections of the paths for non-EU countries. This model is a large-scale estimated multicountry/multiregional macroeconomic model. To calibrate international spillover effects (of, e.g., stock price or bond yield shocks), NiGEM can be complemented with a global VaR (GVAR) model. A typical output of NiGEM when employed for stress testing purposes is the impact on EU external demand from some imposed shock to the global environment. These EU external demand shocks are then in turn an input to the STEs to derive the resulting real economic implications across the EU countries.

2.3 Quantifying Scenario Probabilities

Scenario design practitioners are interested, at times, in quantifying a scenario probability, which is, to some extent, challenging and requires a number of assumptions and modeling choices that can influence the outcome. The probability of a realization of a set of continuous variables to attain a specific hypothesized value or path of values is zero, strictly speaking. This means that a scenario probability can be computed only from the perspective of all realizations beyond a certain predefined path. For evaluating this range, a model is required, which implies that scenario probability estimates are subject to *model dependence*. Challenging in this respect is, moreover, the fact that such model-based probability estimates are subject to what can be called *dimensional dependence*.

Concerning model dependence, one can choose to employ the macrofinancial model that was used to design the scenario to compute the scenario probability. Ideally, the way the predictive scenario conditional densities are produced, which the probabilities rely on, would reflect coefficient and residual uncertainty. The probability estimates would still be conditional on the model at hand. One can also choose to neglect the scenario-generating model and evaluate the probability of selected or a set of scenario variables based on historical distributions only—that is, the reverse of the conventional way of operating with VaR metrics (define a probability to determine the VaR, a tail-risk realization) by determining the probability from the tail-risk path.[16] Macrofinancial model-based scenario probabilities have been computed internally at the ECB on an explorative basis, using models such as large-scale multicountry vector autoregressions (VARs) and global VARs (GVARs).

A dimensional dependence of the probability estimates implies that a probability estimate corresponding to one and the same scenario package would be different if computed based on a selected single risk factor or subsets/groups of risk factors. The dimensional dependence, although not problematic if understood and handled and communicated properly, is an aspect that also links to "plausibility measurement" in the context of systematic, reverse stress testing and worst-case search methodologies (Section 4).

---

[16] The scenario design process at the IMF, for instance, frequently entails the use of expressing GDP growth deviations from baseline in multiples of standard deviations (which can easily be mapped into a probability estimate), with the latter being a measure that is based on data, not on model residuals nor in any other way involving a macrofinancial model for the purpose of making the statement about multiples of standard deviations and, hence, implicitly a likelihood.

The dimensional dependence does not only concern the *types* of macrofinancial variables (e.g., GDP, unemployment, interest rates, etc.) but also applies to the *inclusion of multiple countries* in a scenario. The scenarios designed by ECB staff, for instance, naturally include the cross-section of EU (or at least euro area) countries and a list of about 20 core macro-financial variables per country. A choice as to whether the commonly used notion of "severe yet plausible," and hence an appropriate severity criterion, is to apply to the conjunction of all countries or the countries individually has to be made. In the latter case, the joint probability of materialization across all EU countries would, if judged against historical dependencies (whether based on model-based or model-free distributions), virtually be 0 percent.

Moreover, when some hypothetical elements are embedded in a scenario, the meaning of a quantification of a scenario probability may largely vanish. One can still use the model (reflecting the dependency breaks therein) to compute a scenario probability. At the same time, based on a model-free, distribution-based method, the scenario probability would certainly be 0 percent if the hypothetical scenario features are dominant enough—that is, let the scenario trajectories deviate substantially enough from what historical dependencies would imply to be "yet plausible." Hence, if anything, scenario designers may choose to make a model-conditional statement, saying that conditional on the concrete dependency breaks reflected in the model, the scenario is expected to materialize with a small non-negative probability. Whether the scenario is then deemed "plausible" would all hinge on how meaningful (plausible) the structural-break assumptions embedded in the scenario are, which would always be a pure judgmental step. Even the purely hypothetical elements ought to be informed by knowledge of ongoing structural changes in financial markets and the economy at large (consider, e.g., the appearance of FinTech and related structural change, knowledge about which accumulates slowly as the system develops and will require significant time until it may be reflected in quantitative models for scenario design purposes).

2.4 Reverse Stress Test Scenario Design and Worst-Case Search Methodologies

Systematic scenario simulations aim to explore the impact of several scenarios in order to counteract the risk of overlooking plausible scenarios that might result in more material losses than a specific scenario at hand. Institutions in charge of macrofinancial scenario analysis usually employ a small number of alternative scenarios (see Table 5.3 in Section 3.2), whereas integrated worst-case scenario search (WCS) methodologies that automatize the search for "severe yet plausible" scenarios are not widely used yet. Work of the latter type has been appearing, for example, in Breuer and Krenn (2000) and Breuer et al. (2009); see also Chapter 6 in this handbook.[17] Although these early methodological developments were coupled with examples focusing on market risk, larger-scale macrofinancial applications have not been developed or published yet. One exception can be found in Breuer et al. (2009), where a GVAR model is employed to capture the joint dependencies

---

[17] The authors propose a concrete measure of plausibility that is not subject to the problem of dimensional dependence, which earlier approaches (using a probability mass concept), such as in Studer (1999) and Breuer and Krenn (2000), did yet imply.

Table 5.3. Comparison of some features of scenarios for the banking system stress tests of major institutions

	EBA	Fed	BoE	IMF (FSAPs)	BoJ
Narrative	Extended	Brief	Brief	Extended	Brief
No. of adverse scenarios	1	2 (adverse + severely adverse)	2 (ACS + BES)	1 to 2	2 (tail event + tailored)
No. of domestic countries	28	1	1	1	1
No. of foreign countries	16	4	9	8	5
Horizon	3y	3y	5y	5y	3y
Modeling techniques	VAR/BVAR + copula	Statistical targets + simple equations	Statistical targets + simple equations	Statistical targets + DSGE/panels	Statistical targets + system of VARs
Severity target	no	yes	yes	Yes	yes

BES, Biennial Exploratory Scenario; BVAR, Bayesian vector autoregression; DSGE, dynamic stochastic general equilibrium.

of GDP, domestic and foreign interest rates, and an exchange rate to then conduct a credit-risk-oriented WCS for a sample of Austrian banks.

For what concerns market risk, the WCS methodology can be helpful because it is often not easy to assess and rank the impact of market-risk scenario profiles ex ante, neither in quantitative nor even in directional terms. Some financial firms may even benefit from certain shocks, conditional on their trading positions. Beyond that, for larger-scale macrofinancial stress test assessments, the WCS methodology can be useful to fine-tune the initial narrative-based, handpicked scenarios, either in quantitative terms (calibration-wise) or even in qualitative terms (narrative-wise), if the deviation between the initial and the worst-case scenario would be substantial enough.[18] This could be part of the iterative process when designing the scenario, similar to the use made of impact assessments.

The reverse stress testing philosophy (as underlying the WCS methodology) can also be found in the solvency assessment of central counterparties (CCPs) (ESMA, 2018, Chapter 4.5). The objective there is to move beyond an initial consistent market-risk scenario, by increasing the number of defaulting CCP members and factoring up the

---

[18] IMF staff currently develops a large-scale integrated macro-financial scenario and impact-assessment framework to obtain full distributions of the projected solvency position of banks (Gross et al., 2019). The framework accounts for macrofinancial feedback using a nonlinear structural VAR (SVAR) core (state dependence) in both directions and brings some elements of a conditional-on-shocks methodology into the model, based on structural time-series model approaches, as exemplified in Antolin-Diaz et al. (2018). It is natural to embed the WCS methodology in such a framework, the focus of which lies in the stochastic scenario-generation principle.

severity of the market-price shocks, in order to obtain an assessment as to how vulnerable the CCPs are to a slowly increasing severity level.

3 Applications of Scenario Design: ESRB, ECB, and ESA Stress Tests

3.1 Banking-Sector Stress Test

A well-known application of the scenario methodology presented in the previous sections consists of the scenario developed for the EBA system-wide stress test exercises (see, e.g., ESRB, 2014, 2016, 2018). A similar approach has also been used for the calibration of scenarios for the comprehensive assessment of banks joining the Single Supervisory Mechanism (SSM), for the stress tests of less significant institutions (LSIs), and for top-down financial stability analysis of the banking sector, as presented in the ECB's *Financial Stability Review* (see, e.g., the 2018 review). These scenarios entail both a macrofinancial scenario with a 3-year horizon and a market-risk scenario with a set of one-off immediate shocks for a large number of financial risk factors.

In these applications, the scenario design starts from the definition of a narrative that is based on the ESRB risk assessment and reflects EU-wide systemic financial stability risks. Financial stability risks are ranked on the basis of the probability of materialization and their expected impact, as explained in Section 2. For example, an abrupt repricing of risk premia at the global level was deemed the most relevant financial stability risk in the risk assessment underlying the EBA scenario in 2014, 2016, and 2018. The main reason is that, even if the probability of materialization of this risk might have diminished with the improvement of the economic cycle, its materialization could lead to that of other risks and could have larger effects as a result of its global dimension. Then the various elements of the narrative are mapped into possible exogenous shocks, for example, shocks to asset prices and to demand components (originating in confidence losses).[19]

The calibration of stress test scenarios for stress testing the EU banking sector is a demanding exercise because it involves the calibration of the evolution of several variables for 28 domestic countries and several non-EU countries (i.e., all countries where the EU banks individually have relevant exposures). Table 5.3 compares some features of the scenarios of different institutions. The EBA scenario designed for the whole EU implies, in particular, the calibration of a larger number of countries with respect to scenarios designed by the Fed, BoE, IMF, and BoJ. Taking into consideration the number of countries and variables reported, the consistent calibration (taking into account spillover effects and consistency of the relative size of the shocks) of all these parameters is a complex task.

The set of models used for the calibration of the EBA scenario is illustrated in Figure 5.3. The scenario calibration starts with the calibration of the shock, exogenous in the available model, based on the narrative and risk assessment. The copula model described in Section 3.1 is used for the calibration of shocks to financial variables, such as swap rate curves, sovereign credit spreads, and equity prices. Shocks to consumption and investment confidence are calibrated using a Bayesian VAR (BVAR) model. Shocks to financial

---

[19] The EBA banking-sector system-wide stress test scenario narrative, the corresponding list of risks and implied exogenous shocks, and the resulting path of macroeconomic and financial variables are presented in ESRB (2018).

variables together with other shocks related to the international sector and linked to the narrative (e.g., shocks to tariffs) are then used as inputs to a medium-scale macroeconomic model (e.g., NiGEM or Oxford Economics) to define the path of macroeconomic variables in the nondomestic countries and the effect on demand for European goods from non-EU countries (foreign demand) and on commodities prices. Finally, shocks to financial variables, to foreign demand, and to domestic demand are then used as inputs to the STEs (Section 3.2) to define the paths of all other relevant domestic macroeconomic variables (e.g., real GDP, consumer prices, residential property prices, etc.). Shocks to these input variables can be calibrated using an ex ante estimation of their probabilities—solely based on partial information, that is, their own historical distribution or across financial prices. As already mentioned, this would not, however, mechanically deliver a probability measure for the overall impact as reflected in the published scenario.

Importantly, the macrofinancial scenario design process could also involve a method to account for some initial economic state dependency across countries and adjust shock probabilities accordingly. Depending on structural and cyclical factors that are country specific, shock probabilities may require adjustment. This could be warranted in a framework in which models (NiGEM, BVAR, STEs; see Figure 5.3) are mostly linear and hence cannot as such capture economic state dependence. As an example, certain European economies were facing recession conditions at the outset of the scenario horizon in 2014;

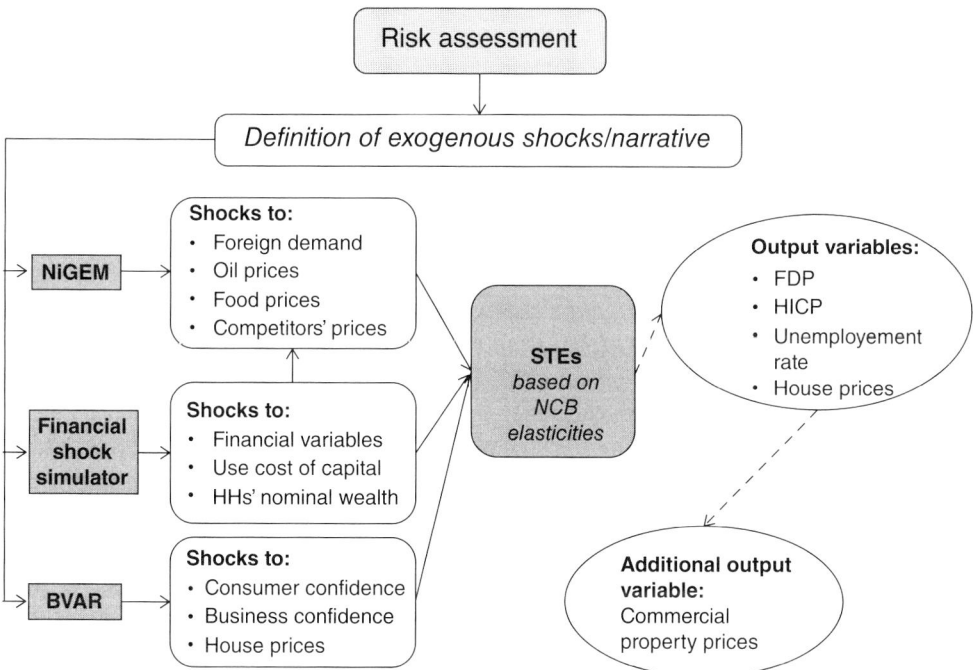

Figure 5.3 Calibration process of the EBA scenario for 2014, 2016, and 2018. NCB, national central banks; HICP, Harmonized Indices of Consumer Prices. Note: The calibration of the 2020 scenario adopted a very similar structure, although the NiGEM and the BVAR models were replaced by other ones (see ESRB, 2020). HH, household; NCB; HICP.

because linear models do not account for that, they would have suggested very strong adverse scenario trajectories, up to a point that would be deemed implausible.

3.2 Stress Tests for the Insurance Sector, Pension Funds, and Money Market Funds

The adverse scenarios for the stress tests for the insurance sector, institutions for occupational retirement provision (IORP), and money market funds (MMFs) are used to assess the impact of changes in market prices on the valuation of the assets and liabilities of these institutions. In Europe, EIOPA carries out the insurance-sector and IORP stress tests, whereas ESMA carries out the one for MMFs. Other assumptions that should belong to the scenario are developed by the competent supervisory authorities, such as the assumptions for long-term risk premia, which are needed for pension fund stress tests, and assumptions on redemptions, which are needed for the MMF stress tests.

Scenarios for nonbanks reflect specific vulnerabilities due to both the composition of their balance sheets and their economic and financial role, in a context where the solvency assessment is also performed using specific metrics. This all calls for narratives and sets of shocks that differ quite substantially from those developed for banks, with the latter warranting a deep and rich use of real economy-related variables (see Table 5.2 for a comparative overview of the respective risk factors at play).

Scenarios for nonbanks are characterized by one-off shocks that are applied to the value of assets and liabilities at a cutoff date. Thus, contrary to the scenario for the banking sector, these scenarios include only short-lived shocks to asset prices (e.g., equities, bonds, options, property prices) and to key parameters affecting the short-term liquidity of nonbank financial institutions but do not include projections for real GDP, unemployment rates, and so forth. The main reason is that these institutions are less affected by credit-risk losses, which usually materialize with some lag. In addition, these stress tests are relatively more recent and still in a developing phase. Shocks to financial asset prices are defined following a similar process to the one for the EBA scenario: on the basis of a risk assessment, exogenous shocks to a set of preselected shock-originating markets are calibrated to then derive the responses for a wider set of risk factors (>900) using the nonparametric copula methodology.

Each of these sectors is vulnerable to particular financial stability risks, and the narrative of the scenarios is therefore tailored to the specific vulnerabilities of each sector. Pension funds and the insurance sector are particularly vulnerable to movements in the interest rate curve, and the calibration of the shocks to interest rates is normally a core element of the scenario. An inversion of the yield curve generally increases the net present value of liabilities because the discount rate of the present value of future benefits is discounted at a lower rate while it decreases the net present value of the assets (which usually have shorter maturity).[20] Examples of particularly adverse scenarios are a "double-hit scenario" (e.g., those of the IORP stress tests in 2015 and 2017; see also ESRB, 2015, 2017) and a curve-inversion scenario. In a double-hit scenario, both stock prices and interest rates decline,

---

[20] In the banking sector, an inversion of the yield curve is also detrimental but as a result of a different mechanism because the balance-sheet structure is different. On average, a decline in long-term interest rates reduces the income and increases the expenses because assets have a longer maturity than liabilities. See Chapter 27 by Bousquet et al. for a comparison between the origin and impact of stress on banks' versus insurers' balance sheets.

which is an event that was rarely seen historically, and the calibration of these scenarios was then achieved using a specific calibration sample. A scenario where the interest rate curve inverts, will determine a larger increase of the values of liabilities than the one of assets (see ESRB, 2019).

For the first time in 2019, a financial-shock scenario for an MMF stress test was designed. This occasion was also used to design a single scenario for two different sectors. Namely, a unique scenario was designed that was used both for the EU-wide pension fund stress test and for the MMF stress test in 2019. Designing a single scenario increases the coherence between the stress testing exercises, allowing an appropriate response from the institutions and companies included in the scope of the tests. This was the first attempt to align stress tests for different sectors and supervisory authorities (i.e., ESMA and EIOPA). The challenge in designing a unique scenario is that the relevant risk factors differ between the stress testing exercises. For example, the MMF stress testing guidelines focus on short maturities (between 1 month and 2 years), whereas pension funds are more sensitive to stress on long maturities (10 years). In addition, the MMF stress test exercise also considers asset liquidity as a relevant source of risk, whereas the pension funds exercise also considers equity prices, residential real estate prices, and inflation swap rates.

3.3 CCP Stress Test Scenario Design

The CCP stress test scenario design also largely relies on the nonparametric copula methodology described previously. Contrary to the banking-sector stress test scenario design, the application of the methodology is still under development for the CCPs stress test, for which less experience has comparably been accumulated.

For example, the aforementioned concept of dependency breaks becomes very relevant for CCPs. One way of operating with the concept of dependency breaks in conjunction with the nonparametric copula concept specifically has been developed by EBC staff to design the scenario for the 2017 CCPs stress test.[21] Upon having defined some narrative-dependent shock origin, the nonparametric conditional expected shortfall estimates for a large number of factors were derived based on data over time-varying windows of 3 years, rolled over a 20-year period. The maximal (most adverse) responses of different risk-factor groupings were then computed (e.g., sovereign bond market factors, corporate bond markets, FX, etc.), based on historical windows identified for each group of factors, which were chosen to maximize the responses in such different groups. This approach entails a hypothetical scenario component in the sense of assuming that the strong tail dependencies of different factors that materialized at different points in history were assumed to *prevail simultaneously*. For a concrete application of this nonparametric modeling philosophy, including the "sample reshuffling" method, see ESMA (2018; see also Section 3.2).

The reason for considering the dependency breaks specifically for CCP scenario design lies in the fact that CCPs are subject to very stringent adverse stress testing requirements on a daily basis, which is implied by regulation (European Market Infrastructure Regulation [EMIR]). The rules imply that very severe shocks (historical maxima/minima) are to be applied on a factor-by-factor basis, thereby effectively ignoring historical dependencies

---

[21] See ESRB (2016).

altogether. In order for a (partially, hybrid) consistent scenario based on historical data to be of avail, methodological variants (e.g., "sample" reshuffling) were deemed to be useful.

4 Conclusions and Way Forward

The purpose of this chapter was to present a summary of recent developments in the field of scenario design, with an emphasis on the methodologies developed and employed, in particular, by ECB staff over the post-GFC years and explaining how the toolkit was adapted and tailored for the design of different types of scenarios and for different institution types. The following high-level conclusions result from the chapter.

First, a relevant aspect that deserves emphasis is to what extent scenario severity should be time varying (albeit, in probabilistic terms constant after accounting for economic state dependence) in order to avoid procyclical economic effects. This topic is particularly relevant when macrofinancial oversight authorities conducting stress tests are facing a protracted period of economic expansion and risks are building up (because the use of stress testing has become more widespread in the aftermath of the GFC). Stress tests in conjunction with scenario design principles that take explicit account of an economy's position in the business and financial cycle can be instrumental for informing the time-varying calibration of capital buffers. A macroprudential stress test designed in this way can help counteract the inherent procyclicality stemming from many sources, including accounting, regulation, and others.[22] Countercyclical scenario design should make sure that "shifts" (shocks) are more sizable during booms and less sizable during recessions, although, as concluded earlier, the scenario severity measured in probabilistic terms would in fact suggest that such time-varying shock profiles would be of rather constant likelihood (which requires nonlinear models).

Second, and related to the procyclicality aspect just highlighted, we conclude that statistical methodologies based on high-frequency data to calibrate market-risk shocks (which also often form the starting point of longer-horizon scenario paths for assessing the credit risk, interest rate risk, and so forth of financial institutions) ought to be used wisely in the sense of their key parameters being set in a way (in a time-varying manner) to not induce procyclicality. This—as does the previous point—links to the notion that "variance does not measure ex ante risk" (Minsky, 1982/1986; Danielsson et al., 2012/2016).[23] How to operationalize such time-varying settings, beyond pure judgmental variation, is a rather open question and will deserve more conceptual work and research.[24, 25] The same conclusion carries over

---

[22] For what concerns regulation, see Blum and Hellwig (1995), Borio et al. (2001), Kashyap and Stein (2004), and Repullo and Suarez (2013).

[23] The notion that "variance does not measure risk" describes the fact that variance metrics, such as realized variance measures based on stock returns or many other asset prices, tend to fall during economic expansions, at times when risk actually rises, generally as a result of rising leverage. Closer to the turning points of an economic cycle, such variance measures are clearly not forward-looking—that is, they are not ex ante measuring the risk that is building up and soon to materialize. Once a risk materializes, the variance measure signals that materialization ex post.

[24] Moreover, more technical, methodological work with respect to the use of very high-dimensional distributions and copulas and the related computational challenges shall be warranted.

[25] The MacroFin copula approach presented in Bassanin et al. (2021) is a step in this direction. It allows generating a scenario whose severity is linked to a financial-conditions indicator and applies the copula approach to both macroeconomic and financial data. More conceptual work with large-scale structural stress test frameworks, embedded in nonlinear macro models, is forthcoming in Gross et al. (2021).

to the models used to derive longer-horizon macrofinancial scenario paths, which, as of yet, most often do not feature the relevant state dependencies (i.e., nonlinearities) to make them suitable for designing scenarios that do not risk inducing procyclicality. For the EU, this is partially due to limitations in the available data, which usually include one crisis at most, making it difficult to capture nonlinear dependencies. Until the state-dependent nonlinearities are not on board, judgmental overlays can be considered with respect to shock probabilities, or the use of model-free approaches, such as the one pursued by the US Fed, can be considered.[26]

Third, an aspect that should be explored in more detail concerns the best horizon for the market-risk shock calibration. The choice should relate in principle to the time to market for the particular trading positions of financial institutions while taking proper account of hedges and derivatives as well. The horizon may need to be set differently for one and the same scenario depending on different time-to-market characteristics in the portfolio of the banks or entities subject to the stress testing. It is a topic that has become particularly relevant in the scenario design and stress testing of the insurance and pension funds sectors and, more recently, the CCPs (in 2017 and 2019) and the MMFs (in 2019).

Fourth, the notion of "second-round effects" ought to be explored and refined (defined) more rigorously. The first attempts of the institutions involved in macrofinancial, macro-prudential stress test modeling have come along with some inconsistencies—for example, in that the initial scenarios usually are dynamic balance-sheet scenarios that, in princi-ple, reflect financial-accelerator effects and the presence of financial institutions and their dynamic state-dependent reactions to economic capital shortages during recessions. A sub-sequent second-round impact assessment stemming from the assumed shortfalls of the banks would imply a "double counting," to an extent. Integrated model frameworks that bundle the scenario generation and impact assessment in one framework are required to that end.[27]

Fifth, several institutions are thinking about developing system-wide stress testing frame-works. Such system-wide frameworks would incorporate nonbank financial institutions along with banks, such as insurance companies, pension funds, other nonbank investment bank funds, and so forth. From a scenario design perspective, such system-wide frameworks imply various challenges because they would require the alignment of the risk factors, the time horizon, and other technical requirements that are generally specified in a different way for each individual sector's stress test exercise. In 2019, for example, for the first time, the ESRB published a scenario that was employed at the same time for the ESMA MMFs and the EIOPA IORP stress tests.[28] In addition, it is also more difficult to understand which combination of current risks might generate some amplification effects across different

---

[26] The growth-at-risk methodology (Adrian et al., 2019) is useful in this context because it is a nonlinear, state-dependent model framework, but it is not sufficiently broad in scope in terms of the variables that are needed for a full-fledged macrofinancial scenario (it focuses on only GDP instead). A larger-scale nonlinear model framework with a similar underlying philosophy is currently under development by Gross et al. (2021).

[27] See, for example, Dees et al. (2017) for a framework that has been developed by ECB staff for second-round impact assessments. See also Gross et al. (2021).

[28] For further details see "Adverse scenario for the European Insurance and Occupational Pensions Authority's EU-Wide Pension Fund Stress Test and for the European Securities and Markets Authority's Money Market Fund Stress-Testing Guidelines in 2019" at www.esrb.europa.eu/mppa/stress/shared/pdf/esrb .stress_test190402_EIOPA_insurance~c5c17193da.en.pdf?172d96eff093ab8ed90c18efd3cf979f.

sectors of the financial system at differing horizons and frequencies. This might point in the direction of multiple scenarios or reverse stress testing.

Sixth and finally, models used for scenario design and impact assessment may be used to explicitly reveal the uncertainty surrounding the scenarios and the resulting impact estimates, which stem from model uncertainty, coefficient and residual (exogenous shock) uncertainty, and scenario uncertainty. The latter is itself a reflection of model, coefficient, and residual uncertainty from the models that are used to design the scenarios. Meanwhile, a positive, reassuring remark to keep in mind is that the likely failure to imagine all ex ante conceivable events that may materialize ex post and imply sizable losses may not be overly problematic because a sufficient capitalization may shield not only against the hypothesized scenarios that informed and implied their build, but also against many nonconceived events in fact, too.

References

Adrian, T., N. Boyarchenko, and D. Giannone (2019), "Vulnerable growth," *American Economic Review*, 109(4), 1705–1741.

Adrian, T., and M. K. Brunnermeier (2016), "CoVaR," *American Economic Review*, 106(7), 1705–1741.

Antolin-Diaz, J., I. Petrella, and J. F. Rubio-Ramirez (2018), "Structural scenario analysis with SVARs," CEPR Discussion Paper Series.

Bank of England (2016), *The Financial Policy Committee's approach to setting the countercyclical capital buffer.*

Bassanin, M., J. Ojea Ferreiro, and E. Rancoita (2021), "The MacroFin copula: A Probabilistic approach for countercyclical scenario calibration," unpublished manuscript.

Baudino, P., R. Goetschmann, J. Henry, K. Taniguchi, and W. Zhu (2018)," Stress-testing banks–a comparative analysis," FSI Papers No. 12.

Blum, J., and M. Hellwig (1995), "The macroeconomic implications of capital adequacy requirements for banks," *European Economic Review*, 39(3), 739–749.

Borio, C., C. Furfine, and P. Lowe (2001), "Procyclicality of the financial system and financial stability: Issues and policy options," in *Marrying the macro- and micro-prudential dimensions of financial stability*, BIS.

Breuer, T., M. Jandacka, K. Rheinberger, and M. Summer (2009), "How to find plausible, severe and useful stress scenarios," *International Journal of Central Banking*, 5(3), 205–224.

Breuer, T., and G. Krenn (2000), "Identifying stress test scenarios," available at https://citeseerx.ist.psu.edu/viewdoc/download?doi=10.1.1.27.118&rep=rep1&type=pdf.

Cherubini, U., E. Luciano, and W. Vecchiato (2004), *Copula methods in finance*, Wiley Finance Series, John Wiley & Sons.

Cont, R. (2007), "Volatility clustering in financial markets: Empirical facts and agent-based models," in G. Teyssière and A. Kirman (eds.), *Long memory in economics*, Springer.

Danielsson, J., H. S. Shin, and J.-P. Zigrand (2012), "Procyclical leverage and endogenous risk," mimeo.

Danielsson, J., M. Valenzuela, and I. Zer (2016), "Learning from history: Volatility and financial crises," LSE Research Online Documents on Economics 66046.

Dees, S., J. Henry, and R. Martin (eds.) (2017), "STAMP€: Stress-Test Analytics for Macroprudential Purposes in the euro area," available at www.ecb.europa.eu/pub/pdf/other/stampe201702.en.pdf.

Deheuvels, P. (1979), "La fonction de dépendance empirique et ses propriétés—un test non paramétrique d'indépendance," *Académie Royale de Belgique—Bulletin de la Classe des Sciences*, 65, 274–292.

European Securities and Markets Authority (2018), "EU-wide CCP stress test 2017," ESMA Report.

European Systemic Risk Board (2014), "EBA/SSM stress test: The macroeconomic adverse scenario," available at https://eba.europa.eu/documents/10180/669262/2014-04-29_ESRB_Adverse_macroeconomic_scenario_-_specification_and_results_finall_version.pdf.

European Systemic Risk Board (2015), "Scenarios for the European Insurance and Occupational Pensions Authority's EU-wide pension fund stress test in 2015," available at www.esrb.europa.eu/mppa/stress/shared/pdf/2015-03-20_GB_21_EIOPA_pension_fund_ST_after_ESRB_GB.pdf?5766cc43d3819803bed6fad0d1c62112.

European Systemic Risk Board (2016), "Adverse macro-financial scenario for the EBA 2016 EU-wide bank stress testing exercise. Link: https://eba.europa.eu/documents/10180/1383302/2016$+$EU-wide$+$stress$+$test-Adverse$+$macro-financial$+$scenario.pdf.

European Systemic Risk Board (2017), "Adverse scenario for the European Insurance and Occupational Pensions Authority's EU-wide pension fund stress test in 2017," available at www.esrb.europa.eu/mppa/stress/shared/pdf/20170518_EIOPA_stress_test_scenario_pension_funds.en.pdf?6540d4e22efe90bb06489c96189b35ca.

European Systemic Risk Board (2018), "Adverse macro-financial scenario for the 2018 EU-wide banking sector stress test," available at www.esrb.europa.eu/mppa/stress/shared/pdf/esrb.20180131_EBA_stress_test_scenario__macrofinancial.en.pdf?cc581649420ff67ea29cd19f0dbb96ba.

European Systemic Risk Board (2019), "Adverse scenario for the European Insurance and Occupational Pensions Authority's EU-wide pension fund stress test and for the European Securities and Markets Authority's money market fund stress-testing guidelines in 2019," available at www.esrb.europa.eu/mppa/stress/shared/pdf/esrb.stress_test190402_EIOPA_insurance_c5c17193da.en.pdf?172d96eff093ab8ed90c18efd3cf979f.

European Systemic Risk Board (2020), "Adverse macro-financial scenario for the 2020 EU-wide banking sector stress test," available at https://eba.europa.eu/sites/default/documents/files/document_library//2020%20EU-wide%20stress%20test%20-%20Macroeconomic%20scenario.pdf.

Genest, C., and J. MacKay (1986), "The joy of copulas: Bivariate distributions with uniform marginals," *American Statistician*, 40, 280–283.

Granger, C. W. J., and Z. Ding (1995), "Some properties of absolute return: An alternative measure of risk," *Annales d'Economie et de Statistique*, 40, 67–91.

Gross, M., M. Leika, and L. Valderrama (2021), "The macro-financial system simulator (MASS)," IMF Working Paper, forthcoming.

Henry, J. (2015), "Macrofinancial scenarios for system-wide stress tests: Process and challenges," in M. Quagliariello (ed.), *Europe's new supervisory toolkit: Data, benchmarking and stress testing for banks and their regulators*, Risk Books.

Joe, H. (1997), "Multivariate models and dependence concepts," in *Monographs on statistics and applied probability*, vol. 73, Chapman & Hall.

Kashyap, A. K., and J. C. Stein (2004), "Cyclical implications of the Basel II capital standards," available at https://scholar.harvard.edu/files/stein/files/cyclical.pdf.

Mandelbrot, B. B. (1963), "The variation of certain speculative prices," *Journal of Business*, 36(4), 394–419.

Minsky, H. P. (1982), *Can "it" happen again? Essays on instability and finance*, M. E. Sharpe.

Minsky, H. P. (1986), *Stabilizing an unstable economy*, Yale University Press.

Nelsen, R. B. (1999), *An introduction to copulas*, Springer.

Niepmann, F., and V. Stebunovs (2018), "Modeling your stress away," CEPR Discussion Paper 12624.

Ojea Ferreiro, J., and E. Rancoita (2019), "Technical note on the financial shock simulator (FSS)," available at www.esrb.europa.eu/mppa/stress/shared/pdf/esrb.stress_test190402_technical_note_EIOPA_insurance_dcd7f1ed08.en.pdf?11ee83b5db7eb1079465d06cc78bb42c.

Repullo, R., and J. Suarez (2013), "The procyclical effects of bank capital regulation," *Review of Financial Studies*, 26(2), 452–490.

Robert, C. P., and G. Casella (2004), *Monte Carlo statistical methods*, 2nd ed. Springer.

Studer, G. (1999), "Market risk computation for nonlinear portfolios," *Journal of Risk*, 1(4), pp. 33–53.

Vitek, F. (2015), "Macrofinancial analysis in the world economy; A panel dynamic stochastic general equilibrium approach," IMF Working Paper, 15/227.

6

The Role of Heterogeneity in Scenario Design for Financial Stability Stress Testing[*]

Mark D. Flood, Jonathan Jones, Matthew Pritsker, and Akhtar Siddique

1 Introduction

Scenario-based stress testing has emerged over the past 30 years as an expedient response to unforeseen financial crises. Figure 6.1 depicts the waves of stress testing programs that followed three significant crisis episodes (shaded areas). In the United States, authorities' deployment of stress testing in the modern era dates to the regulatory response to the interest rate and credit risks revealed in the savings and loan (S&L) crisis of the 1980s (Cornyn and Jones, 1999; Frame et al., 2015). The International Monetary Fund (IMF) and World Bank implemented their Financial Stability Assessment Program (FSAP) after the Asian financial crisis of 1997. More recently, the United States implemented stress testing for banking institutions (bank holding companies, national banks, and federal thrifts) to help restore confidence in the US banking system in response to the 2007–2009 global financial crisis (Bernanke, 2013; Tarullo, 2014). Formal stress testing has since become a central element of the financial supervisory process in the United States and elsewhere.[1]

These frameworks did not emerge fully formed, and there is still no definitive recipe for the "right" way to implement financial stability stress tests. Instead, stress testing tools continue to evolve with the benefit of experience and as the financial system itself evolves. This chapter focuses on one aspect of the stress testing process that deserves closer attention

[*] The authors thank Til Schuermann (co-editor) and an anonymous referee for numerous helpful comments on this chapter. An earlier version of this chapter was presented on the first day of the October 14–15, 2019, International Monetary Fund (IMF) Rethinking Financial Stability: The FSAP at 20 Conference. A significant part of the chapter is based on work that was presented at the Second Annual ECB Macroprudential Policy and Research Conference, held at the European Central Bank (ECB) in May 2017. We also thank Nellie Liang, who was the discussant at the ECB conference, for her many helpful comments on the work that was presented. The authors are solely responsible for any remaining errors.

The views and opinions expressed are those of the authors and do not necessarily represent official positions or policy of the Office of the Comptroller of the Currency, the Federal Reserve Bank of Boston or others in the Federal Reserve System, or the US Department of the Treasury.

[1] The US stress tests in 2009 were known as the Supervisory Capital Assessment Program (SCAP); the current US supervisory stress test programs are the Comprehensive Capital Assessment Review (CCAR) and the Dodd–Frank Act Stress Tests (DFAST). Stress testing programs elsewhere include the Risk Assessment Model for Systemic Institutions (RAMSI) developed by the Bank of England (Alessandri et al., 2009), the Systemic Risk Monitor (SRM) developed by the Austrian Central Bank (Boss et al., 2006; Summer, 2007), the FSAP of the IMF and World Bank (Cihák, 2007; Schmieder et al., 2011), and the macro stress tests developed by the European Central Bank (ECB) (Henry and Kok, 2013; Dees et al., 2017).

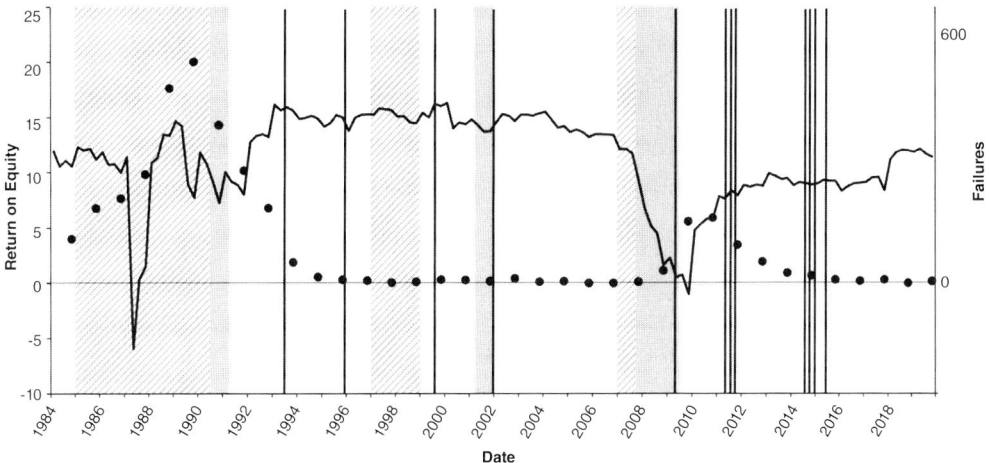

Initiated	Program	Notes
1980s Savings and Loan Crisis		
1993	Office of Thrift Supervision	Thrifts: quarterly, 1993–2010
1996	Basel Market Risk Amendment	Banks: annual
2002	OFHEO Risk-Based Capital	GSEs: annual, 2002–2008
1997 Asian Financial Crisis		
1999	IMF FSAPs	Nations: ongoing
2007–2009 Global Financial Crisis		
2009	Federal Reserve SCAP	Banks: one time
2009	CEBS EU-wide stress test	Banks: annual, 2009–2010
2011	Federal Reserve CCAR	Banks: annual
2011	EBA EU-wide stress test	Banks: biannual since 2014
2011	EIOPA stress test	Insurance: annual since 2014
2014	OCC DFAST program	Banks: annual
2014	ECB supervisory stress test	Banks: annual
2014	Bank of England stress test	Banks: annual
2015	EIOPA-IORP stress-test	Pensions: biannual since 2015

Figure 6.1 Stress testing programs as responses to three financial crises.
Crisis episodes are highlighted in light gray; official US recessions are highlighted in gray. The line is the return on average equity for US banks (left axis); points indicate the count of annual failures of US banks (right axis). Vertical black bars indicate the introduction of new stress testing programs motivated by the crisis episodes. CCAR, Comprehensive Capital Analysis and Review; CEBS, Committee of European Banking Supervisors; DFAST, Dodd-Frank Act Stress Tests; EBA, European Banking Authority; ECB, European Central Bank; EIOPA, European Insurance and Occupational Pensions Authority; IORPS, Institutions for Occupational Retirement Provision; FSAP, Financial Sector Assessment Program; GSE, government-sponsored enterprise; IMF, International Monetary Fund; OCC, Office of the Comptroller of the Currency; OFHEO, Office of Federal Housing Enterprise Oversight; SCAP, Supervisory Capital Assessment Program.

in this evolution, namely, the interaction between the fundamental heterogeneity of the financial system and the design of stress testing scenarios to address that diversity. We sketch some of the general challenges that heterogeneity implies for financial stability stress testing, and we outline a methodological framework for addressing these challenges. We also present a preliminary empirical illustration of two related ways that this framework might work: systemically chosen scenarios and applied reverse stress testing.

Broadly, there are two main approaches to scenario-based stress testing. Microprudential approaches focus on the resilience of individual financial firms under stress. Macroprudential approaches focus on the resilience of the financial system as a whole in a macroeconomic stress scenario. The fact that some very large financial institutions can individually be systemically important often blurs the distinction between the two approaches. Stress tests of these very large firms can simultaneously help explore the stability of the financial system as a whole. Additionally, a stress test that considers multiple firms in isolation but applies a common adverse scenario to them all may also reveal aspects of the stability of the overall financial system.

Constâncio (2016) emphasizes that macroprudential stress tests can have microprudential implications, and vice versa. Rather than trying to distinguish whether a particular stress testing exercise is microprudential or macroprudential, this chapter considers a financial stability stress test as one that generally aims to ensure the resilience of the financial system as a whole. We refer to such exercises interchangeably as *financial stability* or *systemic risk* stress tests. Financial stability stress tests may need to target very large financial institutions because these firms can be systemically important in their own right. Financial stability stress tests may also emphasize shocks with broad impact—those that can affect many financial institutions—because these scenarios can also have systemic consequences.

If systemic resilience is a primary objective, then supervisory authorities should naturally deploy financial stability stress testing system-wide (see Aymanns et al. [2018] for an overview of the state of the art). In particular, stress testing methods should apply not just to banks but also to other financial institutions, such as insurance companies, central counterparties (CCPs), and government-sponsored entities. The diversity of business models across these types of financial firms can be a fundamental source of heterogeneity. With this in mind, we generally refer to financial institutions (FIs) rather than banks in this chapter. If FIs are sufficiently homogeneous, then a common, severe recession may be the most important source of risk. On the other hand, if FIs' risk exposures are significantly heterogeneous, the design of scenarios to assess the resilience of the financial system as a whole becomes more challenging.

In practice, most financial stability stress tests have a macroprudential focus, assessing the resilience of the financial system in a small number of adverse stress scenarios. These scenarios typically countenance a severe macroeconomic deterioration, on the rationale that a severe recession would impose large losses at many FIs. For example, the IMF FSAP stress tests envision an adverse scenario with real gross domestic product (GDP) as the anchor variable; see Adrian et al. (2020). In 2020, the severely adverse scenario for the US CCAR involved a deep global recession, sharp declines in asset prices, and shocks to corporate credit spreads. The shock of the COVID-19 pandemic introduced a significant dose of reality into CCAR stress testing, however, with the Federal Reserve Board (FRB) adding sensitivity tests based on pandemic scenarios when it produced the results for CCAR 2020; see FRB (2020). European regulators, including the Bank of England, canceled or postponed their scheduled 2020 stress tests in light of the pandemic.

Conversely, as exemplified by the events of 2007–2009, a financial crisis can generate severe macroeconomic deterioration. Alfaro and Drehmann (2009) study 43 banking crises in 30 countries from 1974 to 2008, finding that roughly half occurred before the economy contracted, suggesting that macroeconomic downturns can sometimes be the outcome of financial crises rather than their source. Additionally, if FIs are heterogeneous in their risk exposures, a broad-based recession may not be the best stress scenario for systemic-risk objectives. For example, in the S&L crisis of the 1980s, thrifts faced significant interest rate risk, whereas commercial banks had much less exposure in this dimension. It is questionable whether stressing thrifts (and the full system) against general macroeconomic weakness, such as negative GDP growth, would have identified these vulnerabilities.

The goal of financial stability policy is to control systemic risk. Roughly, this is the hazard that the system will fail to provide an economically significant portion of the financial services, such as credit, that the rest of the economy needs; Bisias et al. (2012) survey the related issues. Financial stability stress testing addresses three facets of this objective. First, during crises, financial stability stress testing can reduce uncertainty to help identify which FIs are well capitalized. For example, describing the 2009 SCAP stress test, Schuermann (2014, p. 718) notes that "the cascading of defaults, and the resulting deep skepticism of stated capital adequacy by the market, forced regulators to turn to other tools for assessing, in a credible way, the capital adequacy of banks. That tool turned out to be stress testing."[2] Second, outside of crisis episodes, financial stability stress testing can help identify systemic vulnerabilities, such as significant clusters of FIs that would likely become undercapitalized or liquidity constrained in a stress episode. Addressing these vulnerabilities can reduce the likelihood that financial crises ultimately materialize. The stress test can also point toward steps, such as increasing capital buffers, that would remediate the vulnerabilities. Third, financial stability stress testing, by simulating crisis-like conditions, may help FIs and policy authorities better prepare for crises.

The importance of a particular hazard may also depend on the system's response to it. The ultimate impact may depend on transmission channels and feedback effects such as cascading defaults, asset fire sales, and precautionary hoarding of liquidity. To address these dynamics, financial stability stress tests sometimes model these higher-order effects as part of the scenario. Stress testing should include a consideration of the possibility of heterogeneity in the propagation channels of a shock. Analysis of transmission dynamics can also help prioritize among different sources of stress, given that a minor initial shock might amplify through feedback effects to become very important.

This chapter sets out a conceptual framework for thinking critically about the role of heterogeneity in the design of scenarios for financial stability stress tests. We consider two related reverse stress testing approaches for addressing the challenges of heterogeneity: the systemically chosen scenarios approach and the applied reverse stress testing approach. We also explore empirically how these approaches might work in practice, with an illustrative application to banking data. The chapter proceeds in three sections. Section 2 provides an overview of financial stability stress testing and how it is affected by heterogeneity. Section 3 discusses our methodological approaches for addressing heterogeneity. Section 4 concludes.

---

[2] There is a terminological question of whether the SCAP was "macroprudential," given that the format of the SCAP involved a system-wide application of microprudential stress tests using a common shock. The ensuing UK-wide and EU-wide stress tests were similarly conducted by microprudential supervisors. Administrators of the SCAP clearly considered their exercise to be macroprudential (Hirtle et al., 2009).

2 Challenges of Heterogeneity for Stress Testing

A microprudential stress test examines the viability of a given FI (in isolation) along a number of dimensions, such as capital or liquidity. The exercise typically imposes several general stress scenarios, but it may also include additional scenarios aimed at the FI's particular vulnerabilities. Plausibly, if microprudential stress testing ensures that each FI in the system can (independently) withstand a worst-case stress, then financial stress testing with only a microprudential focus should also ensure the viability of the financial system as a whole. This reasoning, although plausible, suffers from a fallacy of composition. Stressing each FI in isolation ignores interactions that might allow the stress to amplify as it propagates through the financial system.

To avoid this fallacy, a financial stability stress test emphasizes overall systemic resilience. This requires ensuring the viability of enough FIs, with high probability, to provide continued availability of credit extension, risk sharing, and payment services. To achieve this objective, a financial stability stress test should account for all relevant risk sources, their interconnections, and possible endogenous amplification mechanisms, which can account for much of the realized magnitude of a financial crisis.[3] The heterogeneity of the financial system and the hazards it faces is a critical factor for these financial stability goals.

2.1 Dimensions of Heterogeneity

We discuss three main types of heterogeneity. First, the state space of possible stress scenarios has many dimensions. Scenarios can differ in the risk factors that are required to be modeled, their geographic incidence, their time path, and the time path of FIs' exposures to the variables. For example, the risk factors might include GDP growth, stock prices, and unemployment in a variety of countries, and the time path might specify the sequence in which these shocks occur. Second, there is variety in FIs' business models, which manifests itself in differing risk-exposure profiles. For example, because of their specialized business models, life insurers and commercial banks have distinct risk-exposure postures. The third source of heterogeneity lies in the transmission channels through which shocks propagate and may be amplified. For example, an initial default might trigger a causal chain of subsequent defaults, or it might propagate instead (or in addition) through fire-sale impacts on the prices of other assets as surviving FIs rush to stay solvent.

We can summarize these types of heterogeneity in a response function:

$$\mathcal{L}_i = f_i(\mathbf{s}, \mathbf{x}_i). \tag{6.1}$$

The loss (or decrease in income), $\mathcal{L}_i$, experienced by entity i (FI or financial system) is a function of its portfolio holdings, $\mathbf{x}_i \in \mathbb{R}^K$ (i.e., business model[s]), their time path, and on a vector-valued stress scenario, $\mathbf{s} \in \mathbb{R}^d$. The vector-valued nature of both the business model ($\mathbf{x}_i$) and stress scenario ($\mathbf{s}$) incorporates possible diversity—inputs that can differ by other than simple differences in scale. Heterogeneity affects scenario design because scenarios need sufficient diversity to probe the key business-model vulnerabilities and transmission

---

[3] The literature on amplification mechanisms and connections to the real economy is extensive; see, for example, Daníelsson (2002), Daníelsson and Shin (2003), Daníelsson et al. (2013), Demekas (2015), Bookstaber et al. (2014), Borio et al. (2014), and Anderson et al. (2018).

pathways in the system and to identify which of them are of particular concern. These considerations are especially important when (as is typical) political and/or operational forces constrain the number of possible scenarios. Good scenario design alone is inadequate; the quantitative models that assess the scenarios' impact must also be up to the task.

In a systemic context, where i refers to a system of multiple FIs, heterogeneity in transmission channels appears in diverse choices for the response function, f_i. Farmer et al. (2020), for example, similarly focus on the diversity of FIs' business models and the central role of transmission channels in system-wide stress propagation. They implement a simulation framework that elaborates the interconnections that FIs form through contracting and the markets that establish values for these contracts. In some configurations, these interconnections can generate destabilizing amplifications; at the same time, capital buffers can help dampen shocks. In a complex financial system, it can be difficult to predict how these countermanding forces will play out.

Figure 6.2 illustrates the importance of the transmission mechanism in the context of a bank run. An initial exogenous shock, depicted on the left, triggers the process. Panel (a) depicts an unfolding scenario with two transmission channels. The initial shock (0) generates both a direct financial impact (1) on borrowers' solvency and a separate news release (3) that alters depositors' information sets. Banks are then hit from both sides: borrowers' defaults (2) reduce banks' asset values and capital, and nervous depositors' withdrawals

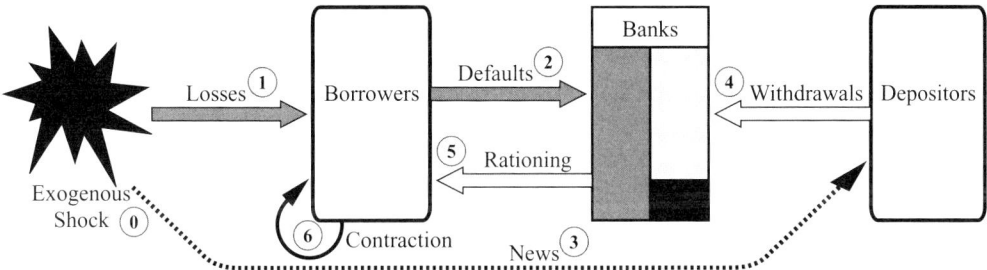

(a) Run driven by underlying fundamental losses in the economy: (0) an initial shock triggers (1) losses to borrowers and (2) defaults on bank loans; (3) news of the shock spreads, and (4) depositors withdraw, fearing a bank failure, (5) constraining the banks' ability to lend, ultimately (6) generating a general economic contraction.

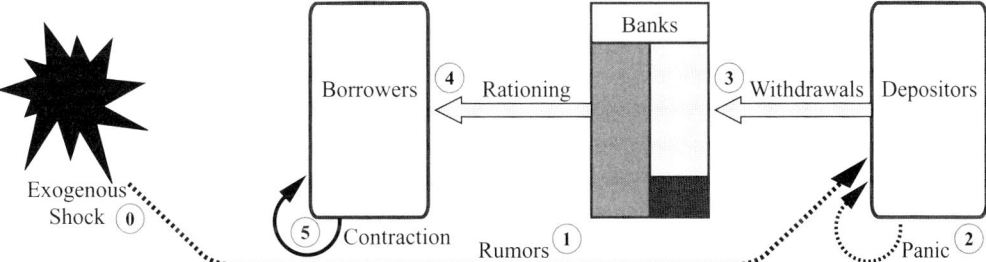

(b) Run driven by depositors' information (or perhaps misinformation): (0) an initial shock generates (1) rumors (which may be accurate) that reduce depositors' trust in bank solvency; initial withdrawals reinforce the general mistrust, leading to (2) panic feedback and (3) widespread depositor withdrawals, (4) constraining the banks' ability to lend, ultimately (5) generating a general economic contraction.

Figure 6.2 Diverse transmission channels for bank runs.

(4) shrink banks' liabilities. The contraction in the banking sector feeds back as a second-round effect (6) into the real economy through a reduction (5) in aggregate credit availability. Goldstein et al. (2020) emphasize that the strategic complementarity among depositors in a bank run—the incentives for an individual depositor to run depend on her perception of the likelihood her fellow depositors will run—are amplified at the system level by a strategic complementarity among FIs. Due to fire-sale externalities, each FI's incentive to liquidate long-term assets to meet withdrawals depends on the FI's perception of the liquidation plans of its peers.

Panel (b) of Figure 6.2 depicts a similar dynamic to that in panel (a), but where the initial shock (0) has no direct financial impact on borrowers in the real sector. Instead, the stress propagates initially only through an information channel (1). The new information (or mis-information) triggers a self-reinforcing panic (2) among the depositors, whose simultaneous withdrawals (3) provoke a contraction in the banking sector and reduction (4) in aggregate credit available to borrowers. The end result (5) is very similar to the contraction in panel (a) of Figure 6.2, but the transmission mechanism is fundamentally different. Incorporating such information-driven dynamics in a stress test would require modeling, as part of the stress scenario, the connection between the initial shock and market participants' expectations.

In terms of equation (6.1), the stress-test problem operates in a space with overall dimensionality of at least $d \times K$, the product of the scenario space and the business model space. Brief reflection confirms that this is a very large number. For example, the business model of securities broker-dealers may involve taking on and hedging exposures to equity, foreign exchange, and interest rate risks; see Pritsker (2017). Moreover, the dimensionality $d \times K$ is a lower bound. K counts the number of distinct types of financial obligations on the firm's books, but a fully specified business model should also encompass strategic plans, deals still in negotiation, and the firm's planned response to prospective stressful conditions. Even within a financial subsector, FIs can exhibit heterogeneity in business models. For the banking sector, for example, Roengpitya et al. (2017) and Ayadi et al. (2016) find that banks tend to cluster around canonical bank business models, such as "retail funded," "universal," and so forth.

The heterogeneity of FIs' business models affects (and is affected by) their exposures to risk. For example, every long position in a derivatives market is matched by a short position elsewhere. Thus, the portfolio choices that are a first-order manifestation of a financial institution's business model create heterogeneity in the cross-section, by construction. Empirical research shows how these different bank business models affect FIs' overall levels of risk, as well as their risk vulnerabilities; see Clark et al. (2007), Demirgüç-Kunt and Huizinga (2010), Altunbas et al. (2017), and Lucas et al. (2019). From a theoretical perspective, an FI's mix of activities (its business model) determines the overall risk profile of the firm. At the level of the financial system, the migration of an activity from one FI where it supplies major diversification benefits to another FI where the activity has few diversification benefits can make both FIs and the system more vulnerable. Empirical research confirms the role of business models in the risk-transfer effects of activity migration among FIs; see Köhler (2015) and Stiroh and Rumble (2006). Even for a given mix of business models, risk exposures evolve because the economic environment changes, including the range of available investment opportunities. The proliferation of subprime mortgages preceding the 2007–2009 crisis is an example.

Lastly, it matters how financial vulnerabilities emerge, creating pathways for the transmission and amplification of shocks. For example, if a single FI is small, and those connected to it are financially resilient, shocks to the FI are unlikely to spread. However, commonalities in the economic environment, such as prices or regulations, can incentivize FIs to take similar risks, leading to correlated vulnerabilities. For example, in the US S&L crisis of the 1980s, many thrifts shared a positive duration-gap exposure to rising interest rates. In the 2007–2009 crisis, many banks (and nonbanks) shared exposure to residential housing credit (see, e.g., Altunbas et al., 2017; Nijskens and Wagner, 2011). By shedding idiosyncratic risks, FIs may inadvertently converge on a common business model, thus forming a risk monoculture across the industry.

To some extent, it is possible to incorporate shock-transmission channels in the stress scenario itself. Default propagation is a relatively simple example because it transmits distress largely mechanically through the financial system. Simple defaults, of course, are not the only transmission channel in a crisis; the policy responses of regulators and FIs themselves can significantly affect the propagation of distress. Ideally, FIs' and regulators' reactions will be stabilizing; for example, the Federal Reserve's introduction of the SCAP in 2009 is widely credited with restoring a basic level of confidence in the US banking sector and providing a foundation for its recovery. Less discretionary rules and policies, such as capital and liquidity regulations, and authorities' transparency policies can also affect the evolution of a stress episode. Some macroprudential stress testing programs, such as the ECB's recent STAMP€ model (Dees et al., 2017) or the Bank of England's RAMSI model (Burrows et al., 2012), explicitly incorporate certain intertemporal and systemic propagation and amplification channels. Some crisis responses, however, are too idiosyncratic to build into a formal stress testing protocol. Citigroup's decision to reacquire $17 billion worth of fully divested structured investment vehicles (SIVs) during the turmoil of 2008 is an example; see Financial Crisis Inquiry Commission (2011, pp. 379–380).

In sum, financial stability stress testing should account for FIs' mixes of businesses and activities, and stress scenarios should evolve as the economic environment changes. Aspirationally, a financial stability stress testing framework would encompass all of this: a large number of FIs with highly heterogeneous business models, a diverse set of possible initial stress events, and a wide range of transmission channels, which depend on both the predefined financial obligations of participants in the system and their behavioral choices as they unfold over time in the wake of a shock. The operative question is how we might move closer to this ideal in ways that are realistically possible.

2.2 A Motivating Example

This subsection presents a stylized example to motivate some of the implementation challenges that heterogeneity implies. For purposes of exposition, we focus on banks (rather than FIs generally). The illustration centers on three benchmarks for the structure of risks in the banking system:

(H) High heterogeneity across banks. We assume banks' vulnerabilities are independently distributed, so the risks affecting each bank differ.
(L) Low heterogeneity across banks. We assume all banks are vulnerable solely to a single common factor, with the same directional exposure to that factor.

(M) Moderate heterogeneity across banks. We assume banks have different directional exposures to a single common factor, or they face multiple common factors, which determine their joint asset returns and/or propensity for joint funding problems.

The overall financial stability objective is to ensure, with high probability, the viability of enough FIs to provide continuity of financial services to the economy.

In the high-heterogeneity benchmark (H), banks' individual marginal probabilities of failure completely determine their joint probabilities of failure. In this case, explicitly micro-prudential policies are sufficient to control financial stability risks. In benchmark (L), when banks are instead highly homogeneous, their joint and marginal probabilities of default are solely determined by the marginal distribution of the single risk factor confronting all the banks. Under benchmark (L), explicitly microprudential regulations are again sufficient to achieve financial stability objectives, and vice versa. For example, if a downward movement in the single risk factor generates bank losses, then a single stress test requiring banks to be well capitalized against first-percentile changes in the risk factor would ensure with 99 percent probability that all banks will remain well capitalized, thereby achieving both macroprudential and microprudential objectives.

The moderate heterogeneity benchmark (M) is the interesting case. In these circumstances, there is a reasonable likelihood that enough banks will experience distress together to impair overall financial intermediation, but there are diverse pathways leading to this undesirable outcome. A simple prudential capital standard based only on banks' individual risk of failure will not achieve the financial stability objective. Nor will a single stress scenario reveal the necessary capitalization levels across the system.

A simple example illustrates the interaction between the benchmark structures of banking system risks and the financial stability objective. Again, the objective is to contain the probability that too many banks fail simultaneously. Specifically, we want each bank to have enough capital so that the undesirable financial stability outcome—that more than 4 percent of the banks in the system fail together—occurs with a probability of under 2 percent. The example assumes $N = 200$ banks of equal size and considers two stylized cases:

I: Banks' failures are independently distributed. This corresponds to the high-heterogeneity benchmark (H). Capitalizing each bank so that its marginal probability of failure is approximately 2 percent would then just achieve the systemic risk objective.[4]

II: Groups of banks have the same business models. This corresponds to the moderate-heterogeneity benchmark (M). More specifically, 100 of the banks make loans only to oil companies (petroleum sellers), and 100 of the banks make loans only to airlines (petroleum buyers); the banks are otherwise identical. For simplicity, we assume that oil-price shocks alone drive bank failures (via loan defaults).

Figure 6.3 illustrates the basic structure of case II. When oil prices fall too low, the representative bank lending to oil companies will fail; by extension, all its petroleum-lending peers

---

[4] If each bank's marginal probability of failure is $p = .01997$, the probability that more than 4 percent of banks default is 2.0 percent, from the complement of the binomial cumulative distribution function (CDF): $\sum_{j=9}^{N} \binom{N}{j} p^j (1-p)^{N-j} = .02$, where a 4 percent bank failure rate corresponds to $j = 8 = .04 \times 200$.

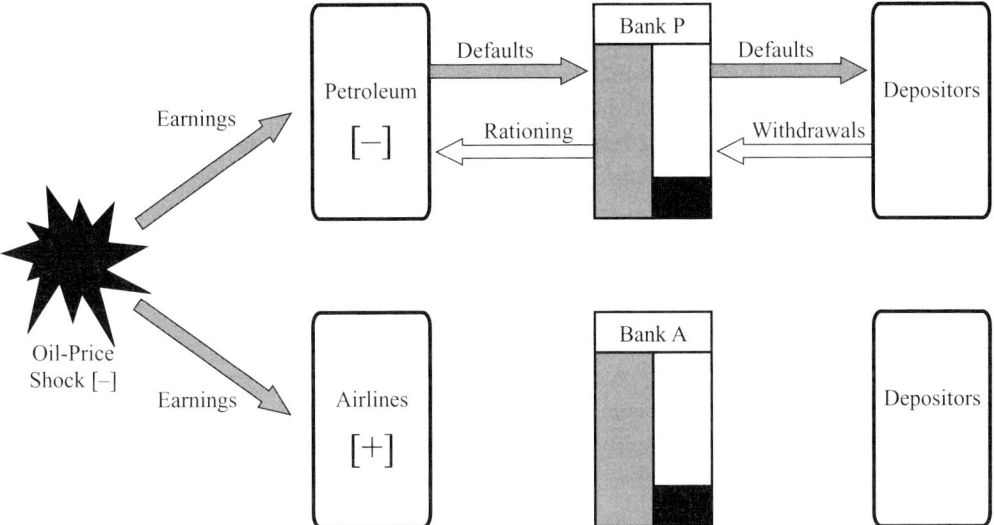

Figure 6.3 Negative oil-price shocks in a heterogeneous real economy and heterogeneous banking system. A common underlying shock can produce diverse outcomes across economic sectors and specialized banks. Bank P denotes a bank that makes loans to oil companies, and Bank A denotes a bank that makes loans to airlines.

fail, too. In the opposite outcome (not shown), when oil prices rise too high, all banks that lend to airlines will fail.

As a thought experiment, suppose each bank in case II sets capital so that its marginal probability of failure is 2 percent—the level of capitalization that achieves the financial stability objective in case I. In case II, this same amount of capitalization fails to achieve the financial stability objective. If oil prices are high enough to bankrupt a representative airline-lending bank, then all 100 banks in the airline-lending peer group fail together. In other words, if each bank sets its capital for a marginal probability of failure of 2 percent, the high-oil-price event, with probability 2 percent, will bankrupt 100 banks en masse. Symmetrically, the low-oil-price event will bankrupt 100 (different) banks together, again with probability 2 percent. Taken together, these two stress events imply a 4 percent probability that 100 banks will fail. The probability of widespread bank failure is twice the level considered acceptable under the financial stability objective.

Requiring banks to be well capitalized against a large increase in oil prices would reduce to 2 percent the probability that 4 percent or more of the industry fails en masse. Unfortunately, this capital policy still falls short of the systemic risk objective because there is a discontinuity in the incidence of failure; it implies a probability of 2 percent that 100 banks will fail. Instead, because of the heterogeneity of their risk exposures, the banks must be well capitalized against both the low- and high-oil-price stress scenarios to achieve the financial stability objective. For example, suppose $P_{oil} = G^{-1}(.009)$ and $P_{oil} = G^{-1}(.991)$ represent the high- and low-oil-price scenarios, respectively, where $G(x)$ is the CDF of oil prices. Then, if banks are well capitalized against both scenarios, the probability that more than 4 percent of banks fail is $0.9 + 0.9 = 1.8$ percent, which is consistent with the financial stability objective.

Extending case II, suppose that the two sets of banks have hedged their oil exposure by purchasing oil options. Specifically, banks lending to airlines purchase options that pay off when oil prices are high, and banks lending to oil companies purchase options that pay off when oil prices are low. Given these hedges, the oil-price risk confronting the banks becomes counterparty risk in the options market. The issue now becomes whether the banks' counterparties in the options market can meet their obligations in the event of extreme market moves. Onward hedging by those options counterparties would, in turn, further expand the network of possible transmission channels more broadly through the system. This expansion of scope is likely to encompass firms that are exposed to factors other than oil prices (e.g., interest rates). If possible, stress scenario design should incorporate the possibility of these indirect exposures.

In sum, this stylized example illustrates three key points. First, in the relatively realistic case of moderate heterogeneity, purely microprudential capital standards do not adequately address risk concentration and default correlation, potentially leaving systemic-risk objectives unsatisfied. Second, if there is heterogeneity in banks' business models, the scenario event space, or shock transmission channels, then system-level financial stability objectives may necessitate multiple stress scenarios. Third, as the scope of the stress test expands to consider various propagation channels through the system, interactions among multiple risk factors (e.g., oil prices and interest rates) will increasingly matter for the design of stress scenarios.

3 A Framework for Financial Stability Stress Testing under Heterogeneity

This section presents two complementary approaches for designing stress scenarios that take heterogeneity into account. The first approach defines a financial stability objective function, which then guides the design of stress scenarios that encourage FIs to act collectively to achieve that objective. The second approach heuristically (without a specific objective function) searches the scenario space for "hot spots"—regions of the scenario space that may be of financial stability concern—targeting those particular outcomes for closer evaluation. Each approach has its advantages. The first approach, by being explicit about the objective, helps ensure that the stress test will achieve financial stability policy goals, even in the presence of heterogeneity. The second approach addresses the challenge that a particular systemic-risk objective function will tend to focus attention on certain scenarios; by remaining agnostic about the objective, the second approach may find worrisome scenarios that a more focused search would overlook.

Both approaches envision regulatory stress tests with a "top-down" structure, which might motivate subsequent "bottom-up" stress tests to probe the identified hot spots more deeply. In a top-down stress test (as defined here), financial stability authorities use supervisory models and information sets to evaluate FIs' losses in one or more scenarios, thus highlighting hot spots in the scenario space. (The Capital and Loss Assessment under Stress Scenarios (CLASS) model at the Federal Reserve Bank of New York is one example of a top-down model; see Hirtle et al., 2015.) Top-down models often gloss over idiosyncratic differences among individual institutions; bottom-up analyses, based on FIs' own models and more detailed data, can supplement the top-down results. We are especially interested in the lessons the stress tests provide to help ensure capital adequacy for systemic-risk purposes.

Section 3.1 illustrates a particular systemic risk objective function that supports stress tests that account for heterogeneity. Section 3.2 introduces two top-down approaches to finding stress scenarios. One approach uses a simplified version of the objective function from Section 3.1; the other is a more heuristic approach. Section 3.3 illustrates the use of the two approaches in a stylized empirical example.

3.1 Establishing Financial Stability Objectives

This section presents methodological scaffolding for finding stress tests that account for heterogeneity to achieve particular financial stability objectives. The first step in the methodology is to state the financial stability goals in the form of an objective function that measures the achievement of those goals. The objective function may also help identify further steps toward achieving the goals. To maintain tractability, the objective function should focus on the most important aspects of heterogeneity. For example, if the goal centers on capital adequacy, then the objective function should capture the primary risk drivers for those business lines and transmission channels most closely related to fluctuations in capital. For example, if borrowers' income is an important determinant of loan defaults, the variables should include key drivers of borrowers' income, such as gross domestic product (GDP) and the unemployment rate. Similarly, if asset fire sales are an important channel of shock transmission, the objective function should capture variables that help track price impacts, such as the capitalization of arbitrageurs. Finally, the objective function should also build in the policy tools that achieve financial stability goals. For example, if supervisory authorities will require capital infusions to address shortfalls, then the objective function should incorporate possible infusions.

Following Pritsker (2014), suppose the financial stability goal is to ensure, with a high probability, the availability of financial intermediation services to the real economy and that FIs' ability to provide these services depends on their capitalization. We translate this goal into an objective function that measures the probability that financial distress sidelines too large a share of financial intermediation services simultaneously. Specifically, the objective is to keep this probability below some suitably low threshold; we can model this probability by simulating random scenarios.

Following Pritsker (2017), begin by randomly drawing N scenarios, indexed by $n \in \{1, \ldots, N\}$. Each FI, indexed by i, has an *assets in distress* function (AD), given by $AD_i(n) = A_i \times D_i(n, Z_n, X_i, C(0)_i + CI_i)$. $AD_i(n)$ represents the loss of financial intermediation capacity for institution i in scenario n. Z_n denotes the vector of variables that define the scenario. For example, Z_n might be realizations of the unemployment rate, GDP growth, and so forth for stress scenarios for the banking book, or Z_n might capture market-risk variables, such as changes in yield curves, stock prices, and so forth, for the trading book. A_i represents i's assets at date 0, which we assume to be proportional to i's intermediation capacity. Very simply, a large institution (many assets) has more capacity for intermediation than a small institution. The total intermediation capacity of the financial system is $\alpha \sum_i A_i$, where α is a constant of proportionality. $D_i(.)$, i's distress function, represents the proportion of i's intermediation capacity lost in scenario n, given its risk exposures to the scenario variables, X_i, and its capital. Capital consists of initial capital $C(0)_i$, plus any capital infusion, CI_i, required as a result of the stress testing exercise. We assume $D_i(.)$ is increasing in each i's losses but decreasing in its capital. All else equal, well-capitalized firms are less likely to default, and they can therefore intermediate more without experiencing financial distress.

With this notation, the economy's loss of intermediation capacity (the numerator in equation [6.2]) as a fraction of maximal total lending capacity (the denominator in equation [6.2]) in scenario n is given by the *system assets in distress* (SAD):

$$\text{SAD}(n, Z_n, X, C(0) + CI) = \frac{\sum_{i=1}^{N} \alpha A_i D_i(n, Z_n, X_i, C(0)_i + CI_i)}{\sum_{i=1}^{N} \alpha A_i}$$

$$= \frac{\sum_{i=1}^{N} A_i D_i(n, Z_n, X_i, C(0)_i + CI_i)}{\sum_{i=1}^{N} A_i}$$

$$= \sum_{i=1}^{N} w_i D_i(n, Z_n, X_i, C(0)_i + CI_i) \tag{6.2}$$

where X, $C(0)$, and CI are vectorized versions, indexed by i, of firms' exposures, initial capital, and injected capital, respectively, and $w_i = \frac{A_i}{\sum A_i}$. ($X$ might also include information about counterparty exposures and the value of those exposures given default risk; we do not explore this extension here.) In equation (6.2), the constant of proportionality (α) cancels out, and SAD(.) reduces to a weighted average of the proportional loss of each institution's intermediation capacity. The weights are the share of i's assets in the total, corresponding to i's share of the maximal intermediation capacity. A reasonable parameterization interprets $D_i(.)$ as i's probability of default over the next year, conditional on scenario n and the other arguments of the distress function. We will use a parameterization similar to this in some of our examples.

Let ζ denote a systemic-loss threshold, stated as a fraction of maximal intermediation capacity. When losses to intermediation capacity exceed ζ, systemic problems ensue. Looking across all scenarios, Prob(SAD(.) $\geq \zeta$) is a measure of systemic risk. We can control this risk through capital injections (CI) or through reductions in FIs' risk-exposure profiles. To measure systemic risk, we need to estimate Prob(SAD(.) $\geq \zeta$). A simple method to do this is to simulate identical and independently distributed (i.i.d.) draws of scenarios from their distribution. Let $\widehat{\Psi}$ denote the empirical fraction of scenarios that exceed the systemic-risk threshold, ζ:

$$\widehat{\Psi}(X, C(0) + CI) = \frac{1}{N} \sum_{n=1}^{N} 1_{\{\text{SAD}(n, Z_n, X, C(0)+CI) \geq \zeta\}}$$

$$\xrightarrow{p} \text{Prob}(\text{SAD}(.) \geq \zeta). \tag{6.3}$$

In other words, $\widehat{\Psi}$ is a consistent estimate of systemic risk.

The expressions for SAD (equation [6.2]) and systemic risk (equation [6.3]) account for heterogeneity both explicitly and implicitly. Heterogeneity is explicitly accounted for by modeling differences in institutions' risk exposures X_i, capital $C_i(0)$, and capital injections CI_i for each i. The details of the distress functions implicitly capture further aspects of heterogeneity. For example, if interbank loans or other counterparty relationships connect institutions, then one firm's distress can affect others. The distress functions can incorporate such linkages to account for shock transmission. Alternatively, if a scenario posits firms' reactions to distress, such as asset fire sales, the distress functions could incorporate the propagation to counterparties' responses; agent-based simulation is one approach to such detailed dynamics (see, e.g., Farmer et al., 2020).

Given the systemic-risk objective function, we consider two use cases. The first detects capital shortfalls in achieving systemic-risk objectives and identifies the specific banks in need of more capital. The second identifies potential systemic-risk hot spots—scenarios in which many banks suffer distress together. One way to identify hot spots is to find the set of scenarios where SAD exceeds the systemic risk threshold, ζ. More generally, one can use reverse stress tests, as in Flood et al. (2017), to identify hot spots by finding the worst stress scenarios for each institution and their intersections across institutions. An alternative approach to reverse stress testing uses dimensionality reduction to identify linear combinations of variables that drive systemic risk. Pritsker (2017), for example, uses sliced inverse regression factor analysis to create factors that explain variation in SAD, identifying factor movements of sufficient size to cause significant financial distress. The factors also help identify stress scenarios to achieve financial stability objectives. We return to reverse stress testing in Section 3.2.

We wish to determine if the financial system is undercapitalized from a systemic risk perspective. To do this, we apply equation (6.3) at current capital levels and then assess whether the level of systemic risk exceeds the regulatory threshold. For example, if we interpret any occurrence of excessive SAD—too much intermediation capacity is lost—as a systemic crisis, then we could state the systemic-risk threshold in terms of the probability of a systemic crisis over the next year. Denote this systemic-risk threshold as T_{sys}. A systemic-risk threshold of 1 percent implies a regulatory goal of keeping the probability of a systemic crisis over the next year at 1 percent or less, roughly a 1-in-100-year event. Given a level of T_{sys}, we can assess whether the financial system as a whole and/or individual institutions are undercapitalized by evaluating whether, at current capital levels ($C(0)$): $\Psi(X, C(0)) > T_{sys}$. Furthermore, if systemic risk is too high, then the gradient of $\Psi(X, C(0))$ with respect to each FI's initial capital identifies the marginal benefit for systemic risk of increasing that firm's capital.

Finally, following Pritsker (2014), if systemic risk exceeds the threshold, we can use the systemic-risk measure, Ψ, to find capital injections CI that drive systemic risk below the threshold. There may be many such choices of CI that achieve the financial stability objective. One approach is to solve for the choices of CI that minimize the total capital injected while satisfying the financial stability objective. If we interpret the capital injections as a cost, then we can optimize that cost by solving the minimization problem:

$$\min \sum_{i=1}^{I} CI_i \quad \text{such that } \widehat{\Psi}(X, C(0) + CI) \le \bar{T}_{sys}. \tag{6.4}$$

An advantage of this approach is that it solves a relatively low-dimensional optimization problem to search for least-cost capital injections that satisfy the financial stability objective. This is easier than specifying a small set of particular (potentially high-dimensional) scenarios for the stress testing exercise.

3.2 Two Approaches to Reverse Stress Testing

Forward and reverse stress testing represent two distinct approaches to designing scenarios. In a forward stress test, the set of stress scenarios is the starting point, and the stress test analysis returns a computation of solvency under each scenario. In the notation of equation (6.1), when stress testing a single FI, McNeil and Smith (2012) parameterize a forward stress test as a search for the least solvent among a set of "likely" events, $s \in \mathbf{s}$:

$$s_f^* = \arg\max_{s \in \mathbf{s}} \mathcal{L}_i(s). \tag{6.5}$$

In contrast, a reverse stress test starts with the set of undesirable outcomes. McNeil and Smith (2012) refer to hot spots as the set of "ruin" events, R; these might be the events representing insolvency of an individual FI or a crisis in a financial system as a whole. Reverse stress testing then works backward to find the single most likely stress scenario that will produce an outcome among the ruin events:

$$s_r^* = \arg\max_{s \in R} \lambda(s), \tag{6.6}$$

where $\lambda(s)$ is a measure of the likelihood of s occurring. This focus on a unique scenario is the extreme case, however. More generally, reverse stress testing can help identify hot spots in the scenario space that imply an unacceptable likelihood of ruin.

The financial crisis of 2007–2009 piqued regulators' interest in reverse stress testing. The Counterparty Risk Management Policy Group (CRMPG, 2008, p. 26) "recommends that firms think creatively about how stress tests can be conducted to maximize their value to the firm including the idea of a reverse stress test where the emphasis is on the contagion that could cause a significant stress event to the firm." The Basel Committee on Banking Supervision (BCBS, 2009, p. 14) recommends that a "stress testing programme should also determine what scenarios could challenge the viability of the bank (reverse stress tests) and thereby uncover hidden risks and interactions among risks." For the IMF's FSAP stress tests, Ong et al. (2010) propose a "breaking-point method," which is essentially a reverse stress test to assess the level of nonperforming loans that would cause a bank to become insolvent. For individual financial institutions, reverse stress testing is now a regulatory requirement in the UK and the EU; see Grigat and Caccioli (2017).

By defining the set of undesirable outcomes (the ruin events) directly, reverse stress testing immediately addresses many of the key challenges posed by heterogeneity. However, this comes at a cost. Reverse stress testing implicitly involves a search over the scenario space, which is typically very large. Furthermore, a reverse stress testing exercise that focuses on the single most likely ruin event, using this event to ensure the financial system is well capitalized, may fail to achieve the financial stability objective. In the oil-price example in Section 2.2, imagine that a large oil-price increase is slightly more likely than a large oil-price decline. A reverse stress test that ensures the financial system is well capitalized only against the oil-price increase would fail to achieve the financial stability objective. Reverse stress testing methods should seek broader "dangerous regions" rather than isolated hot spots.

Various approaches to dimensionality reduction are available to help find the relevant reverse stress scenarios. For example, Flood and Korenko (2015) propose a parsimonious method to generate an arbitrary number of scenarios spread approximately evenly over the scenario space. Alternatively, revisiting the oil-price example of Section 2.2, the worst-stress scenarios in the applied reverse stress testing approach (see Flood et al., 2017) would identify two clusters of unfavorable scenarios, one in which oil prices are high and another in which they are low, and then identify the drivers behind those scenarios. In contrast, the dimension-reduction techniques in Pritsker (2017) would identify oil prices as the important linear combination of factors that cause banks' joint distress, then identify large upward and downward movements as scenarios that cause too much joint distress.

3.2.1 Systemically Chosen Scenario Approach

One possibility for scenario design is the *systemically chosen scenario* (SCS) approach (or "SCSA" in Pritsker [2017]) because the selection criterion for scenario factors is their ability to explain systemic risk. The SCS approach builds on three ideas. First, even though FIs' risks depend on many variables, such as the performance of individual loans or points along the yield curve, we assume these variables are determined by a smaller underlying set of (possibly latent) economic factors, denoted F. We approximate F with linear combinations of the variables affecting FIs' profits and losses. Second, FIs manage risks by hedging against a subset of F, denoted F_B, while remaining exposed to others, denoted F_A. Third, we assume that joint financial distress through FIs' unhedged risk exposures is an important source of systemic risk. One way to use stress tests to control systemic risk is to choose scenarios based on the factors driving systemic risk and then ensure that FIs are sufficiently well capitalized against those scenarios. Under some circumstances, this approach can approximately satisfy the systemic-risk objective in equation (6.3). The key steps are identifying the FIs' unhedged factors, F_A, and then defining scenarios based on those factors.

To implement the SCS approach, we assume that a top-down stress test has identified the variables, Z, that could affect the FIs, and we know how FIs' distress functions respond. To illustrate the approach, suppose Z affects FIs' net worth, which then affects financial distress. To recover the factors, we use the following steps:

1. Generate N i.i.d. draws (scenarios) of Z as the matrix $Z = [Z_1, Z_2, ... Z_N]$.
2. Compute FIs' profit and loss (P&L) in each scenario.
3. Compute $SAD_n(.)$ as the $N \times 1$ vector $Y = (SAD_1(.), ... SAD_N(.))$.
4. Estimate $E(Z|Y)$; and $\Sigma_{E(Z|Y)}$ and Σ_Z, the covariance matrices of $E(Z|Y)$ and of Z, respectively.
5. Recover the factors using information from step 4.

In step 1, Z_n is a function of the full set of factors, F_A and F_B. In step 2, P&L depends only on the unhedged factors, represented by F_A. In step 3, because Z only affects SAD through P&L, SAD and (hence) the vector Y depend only on the unhedged factors, F_A. Step 4 projects each of the Z variables onto Y. Because Y only depends on F_A, the projections of Z only depend on F_A. Finally, step 5 uses the matrices from step 4 to recover the space spanned by F_A. More specifically, step 5 recovers principal-component factors as linear combinations, denoted β_A, of the Z variables spanning the same space as the factors. (These linear combinations are the eigenvectors of the matrix $\Sigma_Z^{-1} \Sigma_{E(Z|Y)}$.)

We can rank the principal components by their ability to explain variation in SAD. The eigenvalues of $\Sigma_Z^{-1}\Sigma_{E(Z|Y)}$ provide this information, with larger eigenvalues corresponding to more explanatory power. Additionally, after identifying the linear combinations, the factors become observable as $F_A = Z\beta_A$, exposing which realizations of the F_A factors correspond to high levels of SAD. This process supports the design of stress scenarios to ensure FIs are adequately capitalized against systemic risk (see Pritsker [2017] for details). Finally, to construct the scenarios, we need the mapping between the factors and the Z variables. Pritsker (2017) simply approximates these via linear regression, with each Z variable being related to the factors as follows:

$$Z = \theta_0 + \theta_1 F_A + \epsilon_Z. \tag{6.7}$$

The variables Z in the stress scenario are then $Z = E(Z|F_A) = \theta_0 + \theta_1 F_A$.

The SCS approach derives conditions for a single scenario such that if enough banks are well capitalized against it, then systemic risk is low. These conditions will hold approximately if enough FIs have sufficiently directionally similar exposures to the identified factors. But this condition is not guaranteed. The oil-price example in Section 2.2 is a case in point. In that case, the financial stability objective requires two scenarios—both a high oil price and a low oil price.

In sum, SCS reduces dimensionality to identify the factors driving systemic risk. These factors can then support the design of stress scenarios that achieve a particular systemic-risk objective. Depending on FIs' risk exposures, it is possible that the objective will require more than one scenario.

3.2.2 Applied Reverse Stress Testing

Flood et al. (2017) develop an *applied reverse stress testing* (ARST) approach that can identify plausible, severe, and useful point-in-time scenarios for reverse stress testing FIs' portfolios. We present a modified version of that approach here. A true reverse stress test would work backward from a full search over an FI's portfolio-loss response as a function of the state variables. In contrast, ARST approximates this by considering a large set of forward scenarios in order to explore the state space broadly, tracking the impact of each scenario on one or more target variables that characterize the system. We focus on net charge-offs for each FI as the target variable in our empirical example in Section 3.3. Ultimately, we seek to identify a manageable *adversity set*—perhaps comprising multiple hot spots—of scenarios that are likely to expose system-level vulnerabilities. Implementation of the ARST approach involves the following steps:

1. Estimate univariate time-series models for a small set of macroeconomic and financial variables.
2. Combine the univariate models into a vector-valued process by modeling their dependence with a copula.
3. Estimate the target variable's sensitivity to the macroeconomic variables for each FI.
4. Generate a large set of scenarios and matching values for each FI's target variable.
5. Choose a handful of the most stressful scenarios as an adversity set for each FI.

In step 1, a manageable number of macroeconomic/financial time series is chosen to represent the high-dimensional scenario space. These variables should plausibly trigger risk responses in the financial institutions or system of interest and, as such, be related to the target variables of interest. In step 2, each series is modeled as a univariate time series, such as an autoregressive moving-average (ARMA) model; see, for example, Flood et al. (2017). Step 2 transforms the univariate models into a system of linked equations, following the approach in Rosenberg and Schuermann (2006), by applying a copula to model the dependence structure among the variables' point-in-time innovations. In step 3, there are many ways to estimate the target variables' sensitivity to the macroeconomic and financial variables; we use ordinary least squares (OLS) regression for the example in Section 3.3. Step 4 is a Monte Carlo exercise that generates a large number of scenarios and resulting P&L values for each FI. Step 5 chooses to focus on the outcomes for the target variables for each FI that are associated with the greatest stress; in most applications, these will be tail outcomes. We then further explore commonality in the adversity sets to cluster banks by their adverse scenarios. Techniques to explore commonality include the cosine similarity of adverse scenarios for different FIs and measuring the distance among FIs' scenarios with a metric such as Mahalanobis distance.

An important advantage of the ARST method is the richness of its empirical approach. ARST, for example, could generate multiperiod scenarios and then identify which of those scenarios correspond to adverse states for some function of the outcome variable, such as its discounted sum over the period of a stress test. In this version, ARST can identify which multiperiod scenarios cause many banks to experience financial stress together or spaced closely together through time. By contrast, it would be more difficult to adapt the SCS approach to a multiperiod framework.

The specification of the marginal distribution of the innovations to the macroeconomic/ financial variables can further enrich the scenarios in the ARST approach. ARST's flexibility can capture fat tails in the marginal distribution of the variables, as well as asymptotic tail dependence among the variables, as in Flood and Korenko (2015), Kole et al. (2007), and Hasan et al. (2015). Furthermore, if there are links across FIs so that one FI's target variable is affected by other FI's target variables, ARST might model that type of interaction, too. Although the ARST approach is in theory very rich in a multiperiod setting, for the purposes of parsimony and simplicity, we use a single-period framework in the example that follows.

3.3 An Empirical Example

This section walks through a simple empirical example of the SCS and ARST frameworks for systemic-risk reverse stress testing. Our goal is simply to illustrate the methodologies, not to present definitive results. The example considers a sample of N_b stylized banks, for which the only asset we examine is the loan book. We define asset losses, Y_i, at each bank i as total net charge-off rates:

$$Y_i = \frac{\text{total net charge offs}}{\text{total loans}}.$$

We model heterogeneity among the banks' loan books by allowing net charge-off rates to have different sensitivities, β_i, to a set of macroeconomic and financial variables, denoted X, and bank-specific determinants of charge offs, ϵ_i. For bank i at time t, we can write:

$$Y_{i,t} = \alpha_i + X_t \beta_i + \epsilon_{i,t}, \tag{6.8}$$

or, combining across i:

$$Y_t = \alpha + X_t \beta + \epsilon_t, \tag{6.9}$$

where Y_t, α, and ϵ_t are $1 \times N_b$, and β is $10 \times N_b$.

To add realism to the stylized banks, we estimated equation (6.9) for $N_b = 28$ banks and 10 macroeconomic variables. Our data cover the 28 largest US bank holding companies (BHCs; which we here generically call "banks") over the sample period 2000–2012. Using publicly available FR Y9-C data, we estimate the sensitivities of the 28 banks' total net charge-off rates to 10 macroeconomic variables in separate OLS regressions for each bank in the sample. The explanatory macroeconomic and financial variables, made public by the Federal Reserve, are quarterly observations of the following: real GDP growth rate, real disposable income growth rate, unemployment rate, Consumer Price Index (CPI) inflation rate, 3-month Treasury bill yield (quarterly change), 10-year Treasury bond yield (quarterly change), BBB corporate yield (quarterly change), Dow Jones Total Stock Market Index (DJMI) (quarterly percentage change), House Price Index (HPI; CoreLogic) (quarterly percentage change), and Volatility Index (VIX; quarterly change).

At the beginning of each quarter, net charge-off rates, Y, and the realizations of X and ϵ are unknown. By the end of the quarter, banks learn X and ϵ and use this information in determining their charge-offs. (In their quarterly financial reports, charge-offs based on date t information may not appear until date $t + 1$. For ease of exposition, all variables chosen based on information known by the end of date t have a date t subscript.)

Figure 6.4 depicts the heterogeneity in banks' risk exposures. Each row and column in Figure 6.4 corresponds to one of the 10 macroeconomic variables. The underlying data for each cell are the estimated regression sensitivities, β_i, for each bank and macroeconomic variable. Cells on the diagonal show the cross-sectional histogram (over 28 observations, smoothed with a kernel density estimator) of the 28 banks' sensitivities to the corresponding macroeconomic variable. Off-diagonal cells show bivariate scatter plots (28 observations each) of the estimated sensitivities to both macroeconomic variables (row and column). The estimated sensitivities, β_i, are scaled to fall on the unit circle ($\beta_i' \beta_i = 1$). Most cases exhibit little cross-sectional heterogeneity in sensitivities to merit attention. However, in some cases, the banks' sensitivities appear to cluster roughly into disparate groups.

We model the total net charge-off rate as resulting in a one-for-one depletion in banks' capital ratios. When the depletion rate for bank i erodes its capital ratio K_i below a buffer level B_i (i.e., when $K_i - Y_i < B_i$), the bank becomes financially impaired. Rearranging the equation, bank i becomes impaired if $Y_i - (K_i - B_i) > 0$. Using the impairment measures for each bank i, we create two systemic impairment measures. The first is aggregate impairment, Π_{agg}. When all banks are the same size, aggregate impairment in a scenario (S) is given by

$$\Pi_{\text{agg}}(S) = \sum_{i=1}^{28} Y_i(S) - (K_i - B_i). \tag{6.10}$$

Equation (6.10) describes a (very roughly) fixed-sum game. Suppose losses or gains of financial intermediation capacity for each bank are proportional to its impairment, $Y_i(S) - (K_i - B_i)$, with a constant of proportionality of 1 for simplicity. Further suppose intermediation is perfectly substitutable across banks, so some banks' loss of intermediation capacity in

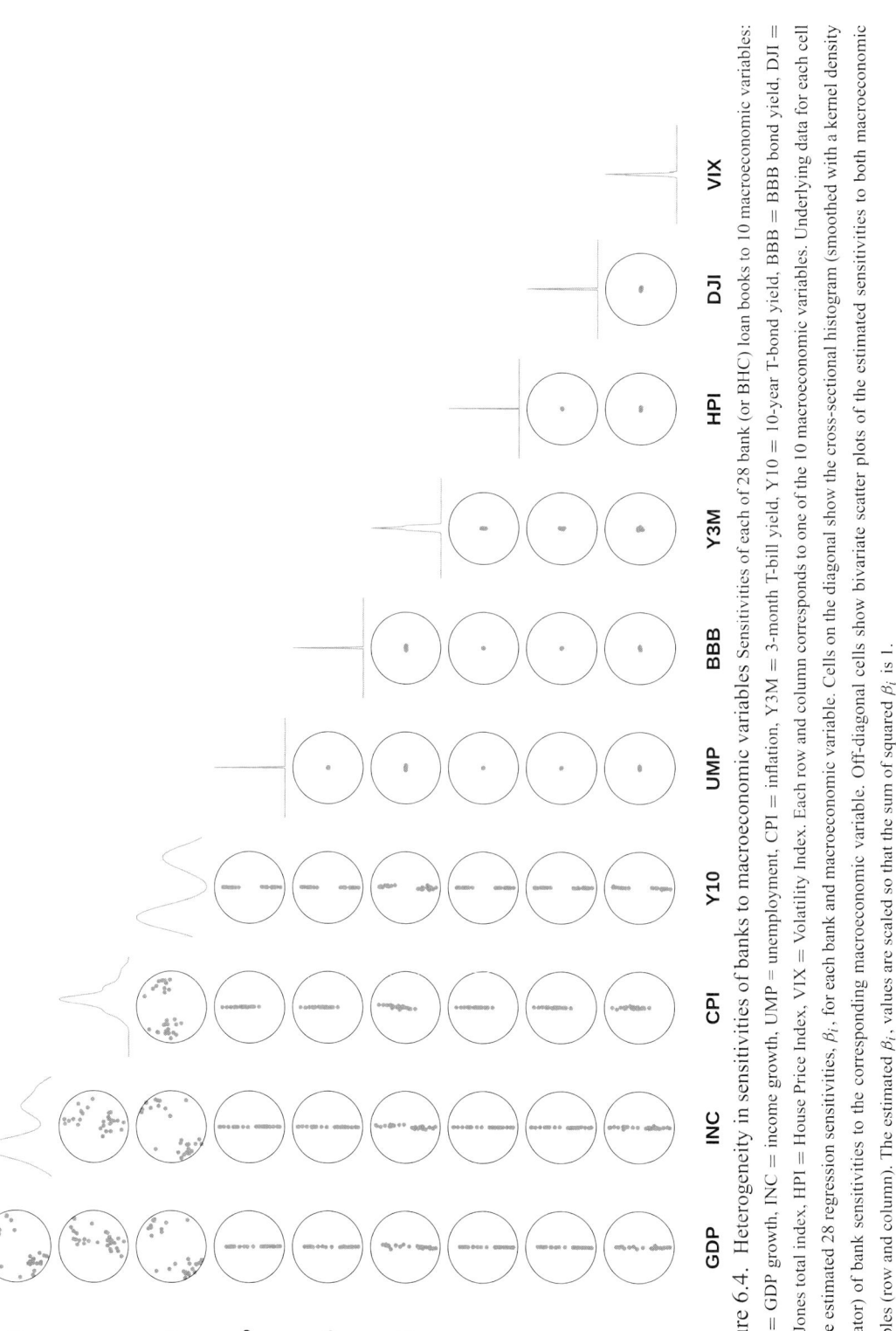

Figure 6.4. Heterogeneity in sensitivities of banks to macroeconomic variables Sensitivities of each of 28 bank (or BHC) loan books to 10 macroeconomic variables: GDP = GDP growth, INC = income growth, UMP = unemployment, CPI = inflation, Y3M = 3-month T-bill yield, Y10 = 10-year T-bond yield, BBB = BBB bond yield, DJI = Dow Jones total index, HPI = House Price Index, VIX = Volatility Index. Each row and column corresponds to one of the 10 macroeconomic variables. Underlying data for each cell are the estimated 28 regression sensitivities, β_i, for each bank and macroeconomic variable. Cells on the diagonal show the cross-sectional histogram (smoothed with a kernel density estimator) of bank sensitivities to the corresponding macroeconomic variable. Off-diagonal cells show bivariate scatter plots of the estimated sensitivities to both macroeconomic variables (row and column). The estimated β_i, values are scaled so that the sum of squared β_i is 1.

a scenario can be perfectly offset by others' gains. Then the aggregate impairment, $\Pi_{\text{agg}}(S)$, is a fixed sum to allocate somehow across all the banks. When $\Pi_{\text{agg}}(S) > 0$, it implies that the banking system as a whole has lost intermediation capacity in that scenario because loan losses, as reflected in net charge-offs, have caused net capital erosion. The analysis that follows abstracts from banks' size differences, treating all banks as having same size. (In a full analysis, if banks are of different sizes, we should account for this in computing individual and aggregate impairment. One way is to scale each bank i's impairment by its assets A_i: $\Pi_{\text{agg}} = \sum_{i=1}^{N} A_i (Y_i - (K_i - B_i))$.)

On the other hand, if intermediation is *not* substitutable across banks—the extreme case—so that one bank's losses cannot offset another's gains at all, then a better measure of impairment is the fraction of banks with positive impairment:

$$\Pi_{\text{frac}} = \frac{1}{N_b} \sum_{i=1}^{N_b} 1_{\{Y_i - (K_i - B_i) > 0\}}, \tag{6.11}$$

where $\mathcal{I}[.]$ is an indicator variable. Although we do not do so here, it is clear that one could create other measures to account for whether a bank has positive impairment, as well as the magnitude of that impairment, and to account for size differences across banks.

Based on these impairment measures, we create two systemic-risk measures. The first is the probability that aggregate impairment exceeds zero, $\text{Prob}(\Pi_{\text{agg}} > 0)$. The second measure is the probability that seven or more of the banks become financially impaired together, $\text{Prob}(\Pi_{\text{frac}} \geq .25)$. We view a quarter of the banks becoming impaired together as a meaningful notion of systemic impairment. For this example, we want to keep the probability that too many banks become impaired together low.

We consider two tools that regulatory authorities can use to control banks' risk and systemic risk. The first is minimum capital requirements. The second is stress tests. In our stylized analysis, we assume that minimum capital requirements are set so that each bank's marginal probability of impairment over the next period is 5 percent. (Impairment does not imply bank insolvency, and banks could have a 5 percent probability of impairment while simultaneously having a much lower probability of insolvency.)

To use the impairment function provided here in practice requires knowledge of the buffers, B_i, such that when capital ratios fall below the buffer, intermediation becomes impaired. For illustrative simplicity, we set the buffers for all banks to zero, which corresponds to financial intermediation becoming impaired only when banks become insolvent. We first examine how well both of these systemic-risk objectives are achieved on the basis of capital requirements alone. We simulate banks' net charge-offs in 10,000 random i.i.d. scenarios, compute Π_{agg} and Π_{frac} in each scenario, and then compute systemic risk.[5] In addition, to apply the SCS approach to generate stress scenarios, we estimate factors

---

[5] The time horizon for each scenario is one quarter. Each scenario is an i.i.d. draw from the distributions of X and ϵ. X and ϵ are independent. $X \sim \mathcal{N}(\mu_x, \Sigma_x)$, and $\epsilon \sim \sqrt{\frac{\nu-2}{\nu}} \times MVT(0, \sigma_\epsilon, \nu)$, where MVT is the multivariate Student's t distribution, with $\nu = 5$ degrees of freedom. We estimated μ_x, Σ_x, and Σ_ϵ from our data. For our simulations, we imposed a Gaussian distribution on X. To capture extreme tail dependence in our simulations, we assumed ϵ is multivariate Student's t, and chose $\nu = 5$ as a "reasonable" choice to capture fat tails. Multiplying the multivariate t vector by $\sqrt{\frac{\nu-2}{\nu}}$ guarantees that the variance-covariance matrix of ϵ is Σ_ϵ. The target variable Y is generated as in equation (6.9).

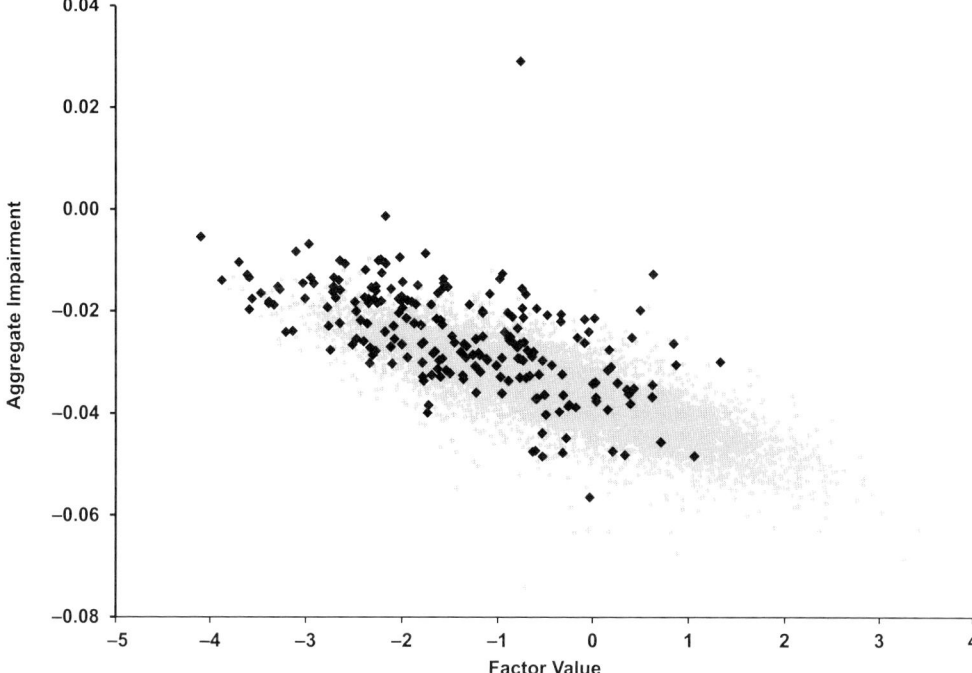

Figure 6.5 Aggregate and fractional impairment versus extracted systemic risk factor. A scatter plot of aggregate impairment—Π_{agg} in Section 3.3—of the financial sector in 10,000 simulated scenarios against an extracted risk factor, labeled "Risk Factor," chosen for its ability to explain aggregate impairment. The factor is highly correlated (-0.8) with aggregate impairment. Dark points represent scenarios in which the fraction of banks whose intermediation capacity is initially impaired in a scenario is at least 25 percent. Lower realizations of the factor are associated with a greater proportion of scenarios in which at least 25 percent of banks are initially impaired. Aggregate impairment represents the losses in intermediation capacity of the financial sector as a whole if banks' intermediation capacities are fixed but are perfect substitutes for each other. When aggregate impairment is less than or equal to zero, there is no aggregate loss in intermediation capacity. Fraction impaired—Π_{frac} in Section 3.3—measures the percentage of banks that initially lose intermediation capacity in a scenario. If banks cannot substitute intermediation capacity for each other, fraction impaired measures the fraction of banks that lose intermediation capacity in the scenario. For further details, see Section 3.3.

(linear combinations of the X variables) that are able to explain the impairment measures. For each of the impairment measures, one factor is much more important than the others. Because the factors for both measures are highly correlated, we only present results using the factor for Π_{agg}, which we denote F_A.

Figure 6.5 provides a scatter plot of aggregate impairment (Π_{agg}) against the factor (F_A) for 10,000 scenarios. In all but one scenario, aggregate impairment is negative, meaning when banks' intermediation capacities are fixed but are perfect substitutes, then the probability that the financial system will experience impairment is 1/10,000, a very low number. On the other hand, if banks' intermediation activities cannot substitute for each other, then we see there are many scenarios (dark points) where seven or more of the banks

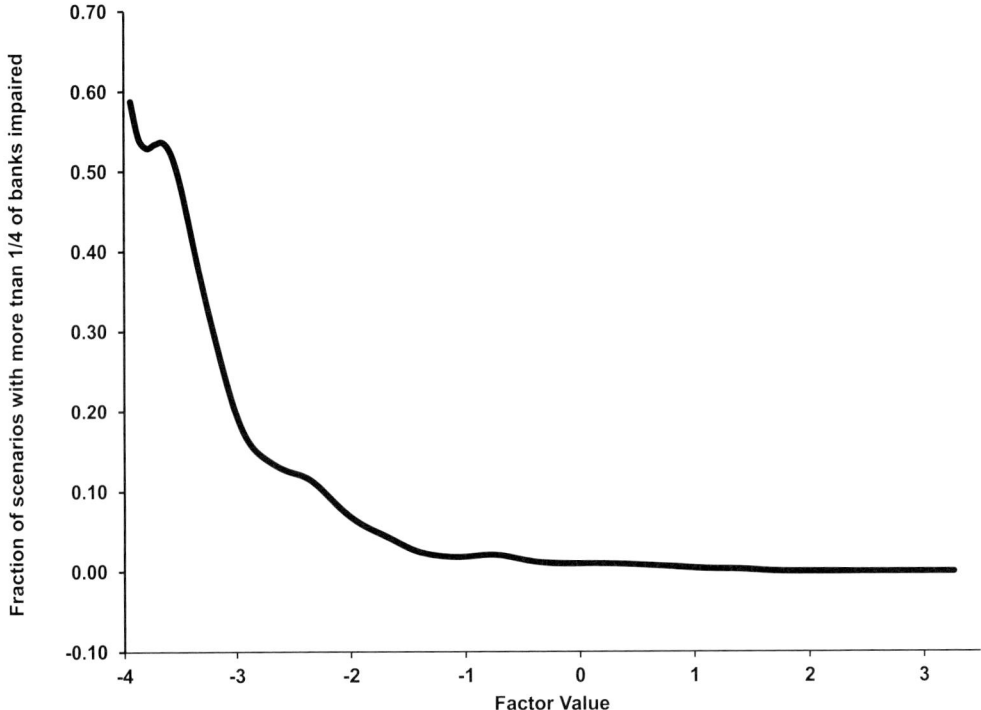

Figure 6.6 Fraction of scenarios with $\geq 1/4$ of banks impaired conditional on extracted systemic-risk factor. A nonparametric regression of whether at least 25 percent of banks that are financially impaired in a scenario against the systemic risk factor for that scenario for the random scenarios and systemic-risk factor in Figure 6.5. For lower values of the factor, the fraction of scenarios in which at least 25 percent of banks are initially impaired is higher. For further details, see Figure 6.5 and Section 3.3.

become impaired at the same time. The probability that 7 or more banks become impaired is approximately 2.2 percent—the fraction of scenarios in which 7 or more banks become impaired. If this fraction is viewed as too high, one possible policy response is to require banks that are more likely to become impaired together to have more capital to reduce the probability of joint impairment. Figure 6.5 shows that there are relatively more scenarios in which many banks become impaired, conditional on low values of the factor.

Figure 6.6 confirms this, presenting nonparametric estimates of the fraction of scenarios in which 25 percent or more of the banks become impaired, conditional on the factor. This fraction is increasing as the factor declines. These findings suggest that scenarios can be constructed on the basis of the factor itself, using the approach in equation (6.7). Banks with too little capital in the scenario need to recapitalize. Figure 6.7 shows that constructing stress scenarios in this way can reduce the probability that too many banks become jointly impaired, thus achieving (or coming closer to achieving) the systemic-risk objective.

In sum, this example shows how the SCS approach can combine capital requirements and stress tests to achieve systemic-risk objectives, even when banks are heterogeneous. It is important to note some caveats. First, for the method to work, factors must exist to

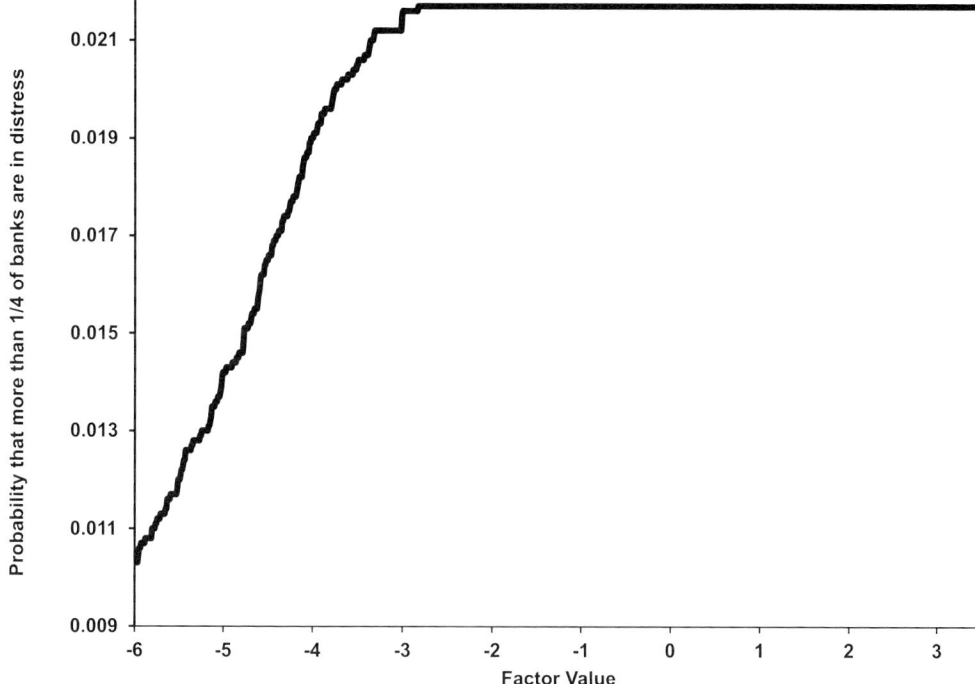

Figure 6.7 Probability $\geq 1/4$ of banks impaired, conditional on stress scenarios based on systemic factor. For the 10,000 random scenarios in Figure 6.5, the probability that more than 25 percent of banks become financially impaired (measured as the fraction of scenarios in which 25 percent of banks become impaired) is presented as a function of whether they are subject to a stress test based on the value of the systemic-risk factor ("Factor Value"). As the value of the systemic-risk factor becomes more negative, the probability that 25 percent or more of banks become impaired declines. For further details, see Section 3.3.

reduce the dimension of the stress testing problem; these may or may not exist in practice. Second, as the oil-price example in Section 2.2 shows, achieving systemic-risk objectives may require multiple scenarios.

ARST is a different approach to finding systemic-risk hot spots. ARST looks directly for commonality of the scenarios that are in the tails of the banks' P&L distribution, then uses this commonality to identify potential systemic hot spots—clusters of banks that can experience bad tail outcomes together. The identified hot spots can support both top-down and bottom-up stress scenarios. For this example, we rely on the same scenario-generating process described previously for the SCS example. (This process differs from the scenario-generation process of Flood et al. [2017].) We generate 100,000 scenarios using the approach described for SCS (see footnote 5), and for each bank, we focus on the scenario that generates each bank's 500th most adverse outcome among an adversity set of 500 scenarios. This scenario corresponds to the bank's 99.5th percentile of loss, which is the scenario that generates the median losses in the 99 percent tail of each bank's loss distribution.

We analyze banks' similarity in the upper tail by measuring the cosine similarity between the empirical scenarios that correspond to the 99.5th percentile of their loss distributions and

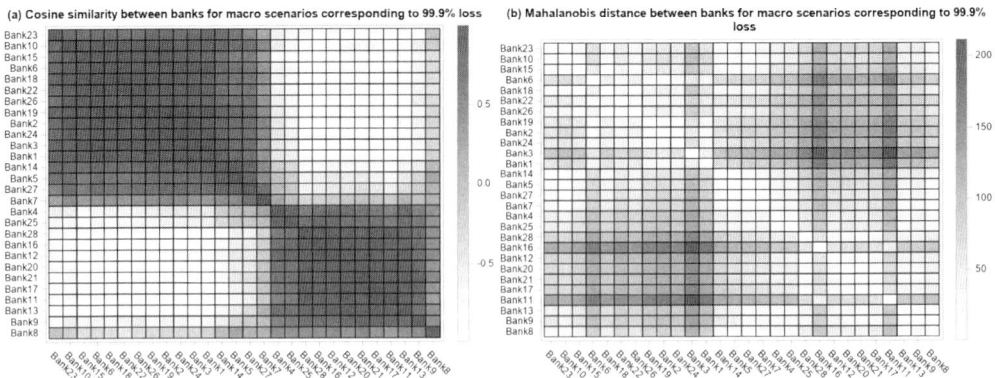

Figure 6.8 Heterogeneity in the most stressful macroeconomic scenarios across banks. The cosine similarity and Mahalanobis distances between the scenarios that correspond to the 99.9 percent loss (greatest change in net charge-off) for each of the 28 banks. The cosine similarity ranges from −1 to +1, with +1 (lighter shades) being most similar and −1 (darker shades) being most dissimilar. The Mahalanobis metric ranges from 0 to about 300, with 0 being most similar and 300 being most dissimilar.

by measuring the Mahalanobis distance between the scenarios. Panel (a) of Figure 6.8 shows the results for the cosine similarity exercise. Here, we have permuted the rows and columns of the cosine similarity matrix so that similar banks are grouped together. The matrix has a clear block-diagonal structure, with two clusters of banks. In each cluster, banks' 99.5 percentile stress scenarios have high cosine similarities, showing they are similar to each other. At the same time, the cosine similarities between banks that are in different clusters are highly negative, suggesting the tail scenarios in the two clusters are negatively related to each other. This situation is reminiscent of the oil-price example in Section 2.2, in which half the banks were stressed when oil prices rose enough, and the other half were stressed when oil prices fell enough. To further check this finding, we estimated the Mahalanobis distance between the tail scenarios when the banks are clustered according to the cosine similarity analysis. Panel (b) of Figure 6.8 shows the results. Although the Mahalanobis effects are not as striking, they do show that the distance between tail scenarios within each cluster is typically smaller than the distances between them. Together, the figures suggest there are two systemic-risk hot spots, implying that at least two stress scenarios may be needed to ensure banks are well capitalized.

Finally, for each of the banks, we select the 500 scenarios that are at 99.5 percent loss or higher. We then find the scenarios that are within this set for the largest number of FIs, that is, associated with a 99.5 percent or greater loss. We find the four scenarios that appear in common within the 500 scenarios for each bank. The most frequent scenario is associated with 16 banks, second most common with 14, third most common with 13, and the fourth most common with 12. Using all four scenarios, we cover all 28 banks. In other words, if each of the 28 banks is capitalized in at least one of the four scenarios in a stress test, then all 28 banks are capitalized at 99.5 percent or greater. With the top-two and top-three (of four) scenarios, three or four banks become undercapitalized, respectively. Figure 6.9 plots the cosine similarities between these four scenarios. These scenarios appear to be quite dissimilar.

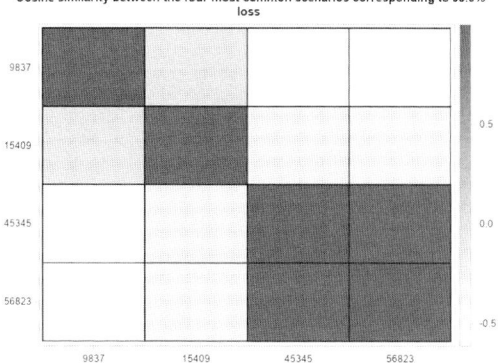

Figure 6.9 Heterogeneity in common macroeconomic scenarios. The cosine similarities between the four scenarios that correspond to the 99.9 percent (or greater) loss for the largest number of the 28 banks. The cosine similarity ranges from -1 to $+1$, with $+1$ being identical and -1 being most dissimilar.

Before concluding, it is important to contrast the SCS and ARST results. On its face, SCS suggests that for the banks considered, one stress scenario is sufficient to achieve systemic-risk objectives, whereas the ARST analysis suggests at least two may be needed. Why is there a difference? We believe the main difference driving the results is that SCS uses specific systemic-risk objective functions based on the probability that too many banks become undercapitalized together. It takes this probability into account in determining whether the systemic-risk objective is satisfied. It is possible to satisfy this criterion even if there are some scenarios in which many banks become undercapitalized together. ARST proceeds differently. There is not a systemic-risk objective function; instead, there is a search for tail commonality—the systemic-risk hot spots. ARST finds these tail hot spots, which can then serve as the basis for further analysis.

There is clearly value to both approaches. Both approaches take FIs' heterogeneity into account in finding stress scenarios. SCS helps find hot spots under the strong assumption that the systemic-risk objective function is known, identifying systemic risk on that basis. ARST is more agnostic about the form of the objective function or whether it is known. This reflects the idea that the objective function may not actually be fully known ex ante and may instead depend on what is learned from identification and further probing of the hot spots themselves.

4 Conclusions

Supervisory stress testing regimes have evolved largely in response to past financial crises. Stress tests began as a way to assess FIs' vulnerability to very large interest rate shocks following the disruptions of the US S&L crisis of the 1980s. More recently, the financial crisis of 2007–2009 has motivated sustained interest in stress testing for financial stability purposes. At the level of the financial system, heterogeneity can play an important role in shaping these stress tests. Heterogeneity appears in many guises, including in scenario design, FIs' business models, and shock-transmission channels.

This chapter sets out a conceptual framework for thinking critically about the role of heterogeneity in financial stability stress test scenario design. We begin by defining a financial stability objective function, system assets in distress (SAD), to frame the problem. A financial stability objective can also help guide the design of stress scenarios that encourage FIs to act collectively to achieve that objective.

We then present two approaches for scenario design in financial stability stress tests in the context of heterogeneity. Both approaches are forms of reverse stress testing, which begins by defining the undesirable outcomes, such as joint insolvency, and then searches the scenario space for those events most likely to produce those undesirable outcomes. This search of the scenario space is critical because heterogeneity implies that the worrisome scenarios may come from many directions.

The two approaches are complementary. The SCS approach is policy centered. The SCS approach bases the selection of scenario factors on their ability to explain systemic risk. By explicitly incorporating the policy objective, SCS helps ensure that the exercise will achieve financial stability goals. The ARST approach is more heuristic. The ARST approach explores the scenario space broadly to identify a manageable adversity set of scenarios that are likely to expose system-level vulnerabilities. By remaining agnostic about the policy objective function, the ARST approach is more likely to find worrisome scenarios that a more focused search would overlook.

We also explore the practical ramifications of these approaches, with an empirical implementation of each. The example is intended to be illustrative, not definitive. Using quarterly data on 28 BHCs and 10 macroeconomic factors, we implement simple versions of both the SCS and ARST approaches. Both approaches uncover evidence of heterogeneity in the data. We also see value in both approaches. SCS seeks (and identifies) systemic-risk hot spots under the strong assumption that the systemic-risk objective is known. ARST is a more data-driven approach, remaining agnostic about the form of the financial stability objective function.

We close with two topics for future research and discussion. First, the heterogeneity of risk models across FIs can have important implications for stress test results. We have focused here on top-down stress tests, in which supervisors use their own models and methods to identify hot spots. But FIs often deploy their own internal models and can analyze certain scenarios more deeply. In both cases, the models are inevitably imperfect, as all models are; model errors are thus themselves a source of risk. At one extreme, if the (heterogeneous) model errors are independent across FIs, it is likely that at least one FI will be blindsided, but it is also more likely that diversification across model errors will make the system as a whole more resilient. At the other extreme, a risk-modeling monoculture of homogeneous model errors may encourage correlated risk exposures at the system level. Second, it is important to consider the general-equilibrium effects of the policy responses to vulnerabilities identified through stress tests. For example, increasing liquidity requirements in response to stress test results can encourage FIs to hoard liquidity individually, in ways that decrease aggregate liquidity as a negative externality. The general-equilibrium implications of liquidity requirements—and the trade-offs relative to other policies, such as capital requirements—are a complex topic deserving of focused analysis when considering the effects of formulating policy responses to stress test results.

References

Adrian, T., J. Morsink, and L. Schumacher (2020), "Stress testing at the IMF," IMF Departmental Paper 20/04, available at https://tinyurl.com/y46xfq6n.

Alessandri, P., P. Gai, S. Kapadia, N. Mora, and C. Puhr (2009), "A framework for quantifying systemic stability," *International Journal of Central Banking*, 5(3), 47–81.

Alfaro, R., and M. Drehmann (2009, December), "Macro stress tests and crises: What can we learn?" *BIS Quarterly Review*, 29–41.

Altunbas, Y., S. Manganelli, and D. Marques-Ibanez (2017, October), "Realized bank risk during the great recession," *Journal of Financial Intermediation*, 32, 29–44.

Anderson, R., J. Danielsson, C. Baba, U. S. Das, H. Kang, and M. A. S. Basurto (2018, September), "Macroprudential stress tests and policies: Searching for robust and implementable frameworks." IMF Working Paper WP/18/197, available at www.systemicrisk.ac.uk/publications/stress-tests.

Ayadi, R., W. P. D. Groen, I. Sassi, W. Mathlouthi, H. Rey, and O. Aubry (2016), "Banking business models monitor 2015: Europe," Technical report, Alphonse and Dorimène Desjardins International Institute for Cooperatives and International Research Centre on Cooperative Finance, available at https://tinyurl.com/y454fcnh.

Aymanns, C., J. D. Farmer, A. M. Kleinnijenhuis, and T. Wetzer (2018), "Models of financial stability and their application in stress tests," in C. Hommes and B. LeBaron (eds.), *Handbook of computational economics* vol. 4, Elsevier.

Basel Committee on Banking Supervision (2009, May), "Principles for sound stress testing practices and supervision," BCBS Guideline 155, available at www.bis.org/publ/bcbs155.htm.

Bernanke, B. S. (2013), "Stress testing banks: What have we learned?" Technical Report, Board of Governors of the Federal Reserve, available at https://tinyurl.com/y2zv2ayg.

Bisias, D., M. Flood, A. W. Lo, and S. Valavanis (2012), "A survey of systemic risk analytics," *Annual Review of Financial Economics*, 4, 255–296.

Bookstaber, R., J. Cetina, G. Feldberg, M. Flood, and P. Glasserman (2014), "Stress tests to promote financial stability: Assessing progress and looking to the future," *Journal of Risk Management in Financial Institutions*, 7(1), 16–25.

Borio, C., M. Drehmann, and K. Tsatsaronis (2014, June), "Stress-testing macro stress testing: Does it live up to expectations?" *Journal of Financial Stability*, 12, 3–15.

Boss, M., G. Krenn, C. Puhr, and M. Summer (2006), "Systemic risk monitor: A model for systemic risk analysis and stress testing of banking systems," *Financial Stability Report (Oesterreichische Nationalbank)*, 11, 83–95.

Burrows, O., D. Learmonth, and J. McKeown (2012, September), "RAMSI: A top-down stress-testing model," Financial Stability Paper 17, Bank of England, available at https://tinyurl.com/y3x75qcr.

Cihák, M. (2007, March), "Introduction to applied stress testing," IMF Working Paper No. 07//59, available at http://imf.org/external/pubs/ft/wp/2007/wp0759.pdf.

Clark, T., A. Dick, B. Hirtle, K. J. Stiroh, and R. Williams (2007, December), "The role of retail banking in the U.S. banking industry: Risk, return, and industry structure," *FRBNY Economic Policy Review*, 13(3), 39–56.

Constâncio, V. (2016, October), "The role of stress testing in supervision and macroprudential policy," in R. Anderson (ed.), *Stress testing and Macroprudential regulation: A transatlantic assessment*, CEPR Press.

Cornyn, A., and J. Jones (1999), "An overview of the OTS net portfolio value model and recent trends in the thrift industry," in S. Nawalkha and D. Chambers (eds.) *Interest rate risk: Measurement and management*, Institutional Investor.

Counterparty Risk Management Policy Group (2008, August), "Containing systemic risk: The road to reform," Report of the CRMPG III, Counterparty Risk Management Group, available at www.crmpolicygroup.org/docs/CRMPG-III.pdf.

Daníelsson, J. (2002), "The emperor has no clothes: Limits to risk modelling," *Journal of Banking and Finance*, 26(7), 1273–1296.

Daníelsson, J., and H. S. Shin (2003), "Endogenous risk," in S. Jenkins (ed.), *Modern risk management: A history*, Risk Books.

Daníelsson, J., H. S. Shin, and J.-P. Zigrand (2013, January), "Endogenous and systemic risk," in J. Haubrich and A. Lo (eds.), *Quantifying systemic risk*, University of Chicago Press.

Dees, S., J. Henry, and R. Martin (2017, February), "STAMP€: Stress-test analytics for macroprudential purposes in the euro area," ECB Technical Report, available at https://tinyurl.com/y2lvqvu3.

Demekas, D. G. (2015, June), "Designing effective macroprudential stress tests: Progress so far and the way forward," IMF Working Paper WP/15/146, available at https://tinyurl.com/yxpbxwjp.

Demirgüç-Kunt, A., and H. Huizinga (2010, December), "Bank activity and funding strategies: The impact on risk and returns," *Journal of Financial Economics*, 98(3), 626–650.

Farmer, J. D., A. M. Kleinnijenhuis, P. Nahai-Williamson, and T. Wetzer (2020), "Foundations of system-wide financial stress testing with heterogeneous institutions," Staff Working Paper 861, Bank of England, available at https://tinyurl.com/yxftbmgj.

Federal Reserve Board (2015, March), "Assessment of bank capital during the recent coronavirus event," Technical Report, Board of Governors of the Federal Reserve, available at www.federalreserve.gov/publications/files/2020-sensitivity-analysis-20200625.pdf.

Financial Crisis Inquiry Commission, *The financial crisis inquiry report: Final Report of the National Commission on the Causes of the Financial and Economic Crisis in the United States, official government edition*, US Government Printing Office, January 2011.

Flood, M. D., J. Jones, G. Korenko, and A. Siddique (2017, October), "Applied reverse stress testing: Conditional forecasting for scenario formulation," Technical Report, Office of the Comptroller of the Currency.

Flood, M. D., and G. G. Korenko, "Systematic scenario selection: Stress testing and the nature of uncertainty," *Quantitative Finance*, 15(1), 43–59.

Frame, W. S., K. Gerardi, and P. S. Willen, (2015, March), "The failure of supervisory stress testing: Fannie Mae, Freddie Mac, and OFHEO," Working Paper 15–4, Federal Reserve Bank of Boston, available at https://tinyurl.com/yxrpg6lh.

Goldstein, I., A. Kopytov, L. Shen, and H. Xiang (2020, June), "Bank heterogeneity and financial stability," Technical Report, University of Pennsylvania, available at http://conference.nber.org/conf_papers/f140887.pdf.

Grigat, D., and F. Caccioli (2017), "Reverse stress testing interbank networks," *Scientific Reports*, 7(15616), 1–11.

Hasan, I., A. Siddique, and X. Sun (2015, January), "Monitoring the 'invisible' hand of market discipline: Capital adequacy revisited," *Journal of Banking and Finance*, 50, 475–492.

Henry, J., and C. Kok (2013, October), "A macro stress testing framework for assessing systemic risks in the banking sector," Occasional Paper Series 152, European Central Bank, available at www.ecb.europa.eu/pub/pdf/scpops/ecbocp152.pdf.

Hirtle, B., A. Kovner, J. Vickery, and M. Bhanot (2015, July), "Assessing financial stability: The Capital and Loss Assessment under Stress Scenarios (CLASS) model," Staff Report 663, Federal Reserve Bank of New York, available at https://tinyurl.com/y2qh9ox2.

Hirtle, B., T. Schuermann, and K. Stiroh (2009, November), "Macroprudential supervision of financial institutions: Lessons from the SCAP," Staff Report 409, Federal Reserve Bank of New York, available at www.newyorkfed.org/research/staff_reports/sr409.html.

Köhler, M. (2015, February), "Which banks are more risky? The impact of business models on bank stability," *Journal of Financial Stability*, 16, 195–212.

Kole, E., K. Koedijk, and M. Verbeek (2007, August), "Selecting copulas for risk management," *Journal of Banking and Finance*, 31(8), 2405–2423.

Lucas, A., J. Schaumburg, and B. Schwaab (2019, July), "Bank business models at zero interest rates," *Journal of Business and Economic Statistics*, 37(3), 542–555.

McNeil, A. J., and A. D. Smith (2012, May), "Multivariate stress scenarios and solvency," *Insurance: Mathematics and Economics*, 50(3), 299–308.

Nijskens, R., and W. Wagner (2011, June), "Credit risk transfer activities and systemic risk: How banks became less risky individually but posed greater risks to the financial system at the same time," *Journal of Banking and Finance*, 35(6), 1391–1398.

Ong, L. L., R. Maino, and N. Duma (2010, December), "Into the great unknown: Stress testing with weak data," IMF Working Paper No. 10/282, available at https://tinyurl.com/y669hx55.

Pritsker, M. (2014), "Enhanced stress testing and financial stability," Technical Report, Federal Reserve Bank of Boston, available at www.aeaweb.org/conference/2014/retrieve.php?pdfid=994.

Pritsker, M. (2017, September), "Choosing stress scenarios for systemic risk through dimension reduction," Risk and Policy Analysis Unit Paper RPA 17–4, Federal Reserve Bank of Boston, available at https://ssrn.com/abstract=3067266.

Roengpitya, R., N. A. Tarashev, K. Tsatsaronis, and A. Villegas (2017, December), "Bank business models: Popularity and performance," *BIS Quarterly Review*, 2014, 55–65.

Rosenberg, J. V., and T. Schuermann (2006, March), "A general approach to integrated risk management with skewed, fat-tailed risks," *Journal of Financial Economics*, 79(3), 569–614.

Schmieder, C., M. Hasan, and C. Puhr (2011, April), "Next generation balance sheet stress testing," IMF Working Paper 11/83, available at https://tinyurl.com/y5ren8ap.

Schuermann, T. (2019, July–September), "Stress testing banks," *International Journal of Forecasting*, 30 (3), 717–728.

Stiroh, K. J., and A. Rumble (2006, August), "The dark side of diversification: The case of US financial holding companies," *Journal of Banking and Finance*, 30(8), 2131–2161.

Summer, M. (2007, July), "Modeling instability of banking systems and the problem of macro stress testing," in O. Castren, G. Caviglia, and P. G. Teixeira (eds.), *Conference on stress testing and financial crisis simulation exercises*, European Central.

Tarullo, D. K. (2014, June), "Stress testing after five years," Technical Report, Board of Governors of the Federal Reserve, available at www.bis.org/review/r140627a.pdf.

7

Designing Coherent Scenarios: A Practitioner Perspective

Greg Hopper[*]

1 Introduction

Since the financial crisis, stress testing has continued to grow in importance as a key risk management tool at financial institutions. The increasing use of stress testing has been stimulated by the widespread recognition both in the regulatory and practitioner communities that the conventional risk management models, such as value at risk (VaR) and potential exposure (PE), although still very useful, are not fully capable of capturing all the risks that a financial institution may face. VaR models for market risk and PE models for credit risk are calibrated by either estimating their parameters from historical data or inferring their parameters from market expectations implied by asset prices. These models reflect recent history or recent market expectations, which can turn out to be badly wrong in an unusual or highly stressed market environment. Stress tests methodologies could be liberated from the dependence on history or market expectations that saddled traditional risk management models.[1]

However, once risk analysts began to develop stress test models seriously, they were confronted with a number of difficult challenges. If stress tests are not calibrated to history or market expectations, then how should they be calibrated? The most obvious answer, that stress tests are calibrated by an appeal to expert judgment, proved inadequate. How do we know when a stress test is plausible? VaR and PE models calculate their results at a prespecified confidence level, such as 99 percent, so their plausibility is clear. Unfortunately, there is no objective way to verify the plausibility of an expert judgment. Plausibility is an important feature of any risk management tool because the goal is to act on the results. Although stress tests must be sufficiently severe, they also need to be plausible if they are to be effective.

Estimated models also have the advantage that they can maintain coherency between financial variables, although that coherency is, of course, based on historically observed relationships. We want to break historical relationships in a stress test, but how do we know that the resulting broken relationships between variables are plausible or even possible? Traditional models also have the advantage that they simulate many possible outcomes and therefore calculate many possible losses. But a stress test is just one scenario, which may

[*] The views expressed in this chapter are those of the author and not necessarily the views of Goldman Sachs.

[1] See Aratan (2013) for a more detailed discussion of the evolution of banks' risk management models, from VaR models to stress testing.

fail to specify potential market environments that are important given a particular financial institution's risk profile.

Traditional risk management models, then, have many advantages, despite their short-comings. On the other hand, a chief advantage of stress tests is that they allow the measurement of different risks across the enterprise in a consistent fashion, which is difficult if not impossible to accomplish using the traditional risk models. This chapter suggests a stress test framework that can overcome the challenges of specifying plausible and coherent stress tests, resulting in a methodology that retains the advantages of traditional risk management models while allowing enterprise-wide risk measurement that is not tightly coupled to history or market expectations.[2] That framework consists of three essential components: (1) a thorough and robust firm-wide risk identification (ID) program that serves as an input into the scenario-design process; (2) a set of econometric stress testing models built by a specialized team with expertise in macroeconomics, financial economics, and time-series econometrics; and (3) an organizational structure and governance process that ensure that risk identification and scenario design are properly coordinated. In a nutshell, an effective stress test framework is an econometric model that combines historical observation, market expectations, and expert judgment to produce a stress test that reflects and stresses the particular material risks that a financial firm has identified while maintaining coherence and plausibility.

2 The Importance of Risk Identification in Designing a Stress Test

Much of the discussion around the perceived drawbacks of traditional risk models has centered around concerns about the distributional assumptions implicit in the conventional models. It is often pointed out that financial-asset returns have "fat tails," implying that large positive or negative returns occur more frequently and in greater magnitude than is implied by the distributional assumptions built into risk management models. Although in reality, well-specified models account for fat-tailed distributions, they can still become unrealistic under extreme or unusual market conditions. Stress tests are thought to be the remedy in these situations because asset shocks can be defined to be more extreme in a stress test than would have been observed historically.

The ability to include more extreme shocks in stress tests does not necessarily make them better than traditional models, however. Extreme shocks in a stress test may lower exposure in many circumstances, contrary to the intent of the stress test. For example, if the financial institution is short credit risk, more extreme shocks to credit spreads will create large gains, which may offset other losses and give an unrealistically favorable picture of the firm's true risk.

More generally, stress tests can fail if they do not stress the risk the financial institution actually faces. An equity-neutral hedge fund will not seem to have much risk no matter how large the specified equity shocks are in a stress test. For the stress test to be meaningful in such a case, it will have to include stressed basis risks between equities; that is, it will have to stress equity correlations. To take a different example, a financial institution that is long mezzanine tranches of collateralized debt obligations (CDOs) would misunderstand its

---

[2] See Dent et al. (2016) for a discussion of typical stress testing methodologies currently in use at banks.

risk if it stressed the default risk of the underlying names in the CDOs. Its true risk may be liquidity risk: a sufficiently large increase in spreads could require more margin to be posted than the firm's funding can support. Thus, before designing a stress test, it is vital for the financial institution to determine what its material risks are so that it can design the appropriate stress test.

3 Elements of an Effective Risk Identification Process

The risk ID process is an essential component of scenario design. To be effective, the process must thoroughly examine all risks that a firm faces, document them, and determine whether they are material. A risk ID process consists of the following elements:

- Risk taxonomy
- Risk ID process
- Risk reporting

4 Risk Taxonomy

A risk taxonomy is a classification scheme for all the possible risks a firm faces. The purpose of a risk taxonomy is to categorize all the risks so that no risks are missed in the risk ID process. When defining a risk taxonomy, it is important to consider the horizon over which risks are being measured. For some financial firms, such as hedge funds, the horizon may be very short, perhaps 10 days. Financial institutions may choose a horizon of 1–3 years, whereas insurance companies might define very long horizons. Once the horizon is specified, then all risks should be classified into levels of increasing granularity. Table 7.1 gives an example of a simplified risk taxonomy.

In this simplified taxonomy, we have defined the first two levels have been defined. Level 1 should be defined broadly enough to encompass all risks. The next levels should break those risks into more granular units of risk. In this case, we have divided market risk in level 2 into trading risk, which comprises all positions whose value varies with financial inputs, and credit valuation adjustment (CVA) risk, which captures the market risk of the

Table 7.1. Simplified risk taxonomy

Level 1	Level 2
Market risk	1. Trading
	2. CVA
Credit risk	1. Default risk
	2. Allowance risk
Operational risk	1. Intentional error
	2. Unintentional error
Liquidity risk	1. Inflow
	2. Outflow
Strategic risk	1. Business environment
	2. Business decisions

counterparty credit risk associated with a portfolio of derivative trades. Credit risk is divided into default risk, which would include defaults of all types of counterparties, and allowance risk, which measures the credit risk of loans under Financial Accounting Standards (FAS) 5 and FAS 114. Operational risk is divided into intentional errors (e.g., fraud, conduct risk) and unintentional errors, such as software bugs or trade-settlement errors. Liquidity risk is broadly divided into inflow and outflow, such as, for example, margin received or posted. And finally, strategic risk is divided into business events outside the scope of control of the financial institution, such as Brexit or changes in customer demand, and events within the business's control.

This sample taxonomy is not unique, of course. The same firm could have divided the risks in level 2 in a number of different ways. Moreover, different firms will have different risks. CVA risk may be relevant for a bank that trades derivatives but not necessarily for other types of financial institutions. An investment-management company may not have any default risk, or if it does, it may not have allowance risk. Each institution must define the risk-taxonomy categories consistent with its own risk profile and should divide the risks into more granular components in a way that is useful to it.

However a financial institution decides to specify its risk taxonomy, it must ensure that it is complete so that all risks the institution faces are included. The risk taxonomy serves as a convenient check to ensure that no risks have been missed in the risk ID process. It also establishes a common language a firm can use to discuss its risk. A common language is an important tool in the risk ID process because it enables people from across the financial institution to discuss potential risks in a consistent manner.

5 Risk Identification Process

The next step in establishing a risk ID process is to set up a risk ID team composed of experienced risk professionals with broad experience in risk. Because the results of the risk ID process will be used to define stress tests, the risk ID team should be integrated with the scenario-design team and may in fact have overlapping personnel. The risk ID team should be responsible for defining the risk taxonomy, developing a risk-surveillance process, and producing risk ID reports.

The risk-surveillance process must be set up so that it covers all risks that are categorized in the risk taxonomy. Because most financial firms will already have developed a robust process to identify and measure risks by risk silo, the risk ID team will not need to reinvent new risk processes in general but will be able to rely on the processes and reports that have already been produced. However, using the taxonomy, the risk ID team can determine whether there are any holes in the firm's risk process (i.e., risks that are either not included at all or not coordinated across risk silos). Many risks are multidimensional, and although each dimension may be adequately managed in traditional risk departments, the interplay between these risks may not be adequately captured.

Having made a determination on whether all risks are properly managed in the firm, the risk ID team must then establish a process that determines what risks the firm actually faces. Because the set of risks can be very large, the risk ID team must also establish an objective standard for materiality and develop a process by which materiality can be measured so that only the material risks are considered in designing a stress test.

Guided by the risk taxonomy, the risk ID team would then set up a series of risk workshops with the relevant areas that either produce or manage the risk in question. A risk workshop is a meeting managed by the risk ID team in which it leverages expertise around the firm in compiling a set of material risks. For example, to assess the market risks a firm may face, the risk ID team would meet with members of firm-wide market-risk management and review their risk management reports, as well as discuss developments in the markets. The risk ID team would also conduct risk workshops with traders to get their views on the risks they face. Whom the risk ID team invites to its risk workshops depends on the risk under consideration. Identification of legal risks would involve workshops with the legal staff, whereas strategic risks would involve workshops with business unit leaders or the firm's economics department and trading staff. In general, the risk ID team must develop a thorough understanding of the firm so that it can determine who the relevant people are to invite to a workshop.

Although the risk-workshop process can be a very effective tool to compile a holistic view of the firm's risk profile, it may need to be supplemented by an automated process when the firm's risks are very complicated. The more complex and numerous the risks, the greater the likelihood that a material risk will slip through the cracks. An automated process can help safeguard against overlooking a risk and has the further advantage that it can be used to measure materiality. For example, suppose that a financial institution has a large number of trades that have market risk. One way to automate the risk ID process for these trades is to leverage the pricing-model infrastructure of the firm by employing an automated process to make a list of all the inputs into each trade's pricing model. Each input could then be stressed according to a predefined standard, and the total loss associated with that input would then be tabulated and compared with a numerical materiality standard. Of course, any automated process will have its limitations and will not be able to establish materiality in all cases. Thus, it is also important for the risk ID process to develop qualitative criteria for the materiality of risks. These qualitative criteria would exploit the expert judgment of risk managers who specialize in particular risks.

6 Risk Reporting

The risk ID process should also have a reporting component, whose frequency would align with the risk ID exercise. The frequency of the risk ID process, and therefore the reporting, depends on how quickly the risk profile of the firm changes. There is no hard and fast rule for how frequently a risk identification should be conducted. Large banks may find that a quarterly or semiannual process is sufficient, whereas an asset-management institution may want to perform the process more frequently, depending on its strategy.

A key report that should result from risk identification would list all the material risks at a level of granularity useful for senior management to understand and also useful to input into the scenario-design process. The level of granularity itself will depend on the risk profile: if it is relatively simple, then the number of material risk factors reported would be relatively small. But if it is complex, with many basis risks and risks across the taxonomy, then the number could be relatively large, numbering in the hundreds. The report should tie back to the risk taxonomy so that the actual risks identified can be compared with all possible risks the firm faces.

A financial firm may also want to create other reports that are useful for scenario design. Those reports will be linked to the specifics of the scenario-design process. For example, the risk ID team could create a report that ranks the material risks in order of importance. Or it could design special reports around new and emerging risks that have been recently uncovered.

7 Using Risk Identification in Scenario Design

How should the list of all material risks be used to design a stress test? The reverse stress was one of the first methodologies that linked scenario design and risk identification. To design a reverse stress test, the risk analyst first states a target loss. Then, the stress test is structured so that the most important material risks are emphasized in order to produce a loss amount equal to the target. For example, if the most important risk factors are interest rates and equity prices, the risk analyst would keep increasing the shock size applied to interest rates and equity prices until the target loss is achieved.

Although this procedure does connect the firm's risk profile with the design of the stress test, risk analysts quickly discovered that reverse stress tests were not as useful as would have been expected. The target loss was generally arbitrary, and the necessary shocks to produce that loss were often highly implausible. As a result, reverse stress fell out of favor. But the reverse stress test did clarify the problem of plausibility in stress test design: When are shocks too big or too small? And how do we ensure that a set of stress test shocks is plausibly coherent?[3] What is needed is a stress test that lies somehow between the extremes of a reverse stress test that relies too much on the risk profile of the firm and a standard stress test that does not account for the risks of the firm at all.

To see how to specify such stress tests, it is useful to differentiate two types: strategy-specific stress tests (SSSTs) versus automatic scenario generation (ASG). An SSST a stress test that measures the risk profile that results from a firm following some specific strategy. SSSTs are usually very bespoke to build, although they do not generally require the specification of multiple scenarios. ASG is a model that generates many different scenarios. ASG is useful when the risk profile identified is very complex. ASG allows the risk profile of a firm to be explored with many different scenarios.[4]

8 Strategy-Specific Stress Tests

The need for an SSST generally arises in smaller firms that have a well-defined business model, such as asset managers or hedge funds. To understand how to develop an SSST, it is useful to consider a specific example.

Suppose that a financial firm has decided to specialize in selling mezzanine tranches of synthetic CDOs. Each CDO will consist of 125 names with credit spreads generally in

---

[3] See Schuermann (2013) for more discussion of the importance of incorporating plausible severity and coherence in good scenario design.

[4] See Hopper (2010, 2014a,b) for more thorough discussions of strategy-specific stress tests and automatic scenario generation. For an alternative view, see Rebonato (2010), who advocates a Bayesian methodology for developing coherent scenarios.

the neighborhood of 50 basis points (bps) or less. The firm expects that each CDO will have a $3.3 billion notional. The attachment and detachment points are 10 and 20 percent, respectively, which implies that the firm will have to pay any postrecovery default losses on the portfolio that are greater than $330 million but less than $660 million in notional losses. The firm expects to earn 30 bps in exchange for selling credit protection on each CDO. It will sell two 5-year $3.3 billion tranches each quarter over the next 2 years and then hold the trades to maturity. The firm has $100 million in capital.

In designing a stress test, it will be tempting for this firm to focus on the default risk of the underlying credits in the CDOs because it will have to make payments if a sufficient number of the names it is selling protection on end up deafulting. Market risk would seem to be zero because the firm is holding the trades to maturity. If this firm did design a stress test that stressed default risk, it would likely conclude that its risk is low except in a very extreme scenario.

To see this, we can construct a simple default scenario—a year-long recession in which the names in the portfolio simultaneously default. How can we specify the number of defaults in such a scenario in a plausible manner? Suppose we take the average spread to be the upper bound of 50 bps. Taking the 1-year spread to be, conservatively, the upper bound for the portfolio, 50 bps, and recovery to be 40 percent, we can estimate the 1-year risk-neutral default probability to be 0.83 percent. We know that the risk-neutral default probability is an overestimate of the true default probability because of the existence of a risk premium, which can easily double the true default probability. However, because we are interested in constructing a scenario in which there is a recession that lasts for 1 year, we assume that the default probability on average for each name in the synthetic CDOs is 0.83 percent.

To estimate the probability that portfolio losses are between $330 million and $660 million, we use the simple Gaussian copula model. Once we specify the confidence level, CL, and ρ, the asset-level correlation, we can calculate a probability of default conditional on being in a recession and conditional on the confidence level. Then, we can calculate the probability that losses will be between $330 million and $660 million using a binomial distribution. We specify a correlation of 40 percent for various confidence levels in Table 7.2.

Table 7.2 suggests that the chance that the firm will experience losses between 10 and 20 percent of the notional value of the portfolio is an extreme tail event. The probability is only 11 percent given a 98th percentile credit event during a 1-year recession and zero for lower confidence levels. Is the risk truly this low?

Table 7.2. Simple default scenario

Stressed correlation	40%
Stressed default probability	0.83%
Confidence level	Probability of loss
90%	0
95%	0
98%	11%
99%	60%

The problem with this analysis is that it focused on one very obvious risk–default risk. Had the firm gone through a thorough risk ID process, however, and made a list of all its risks, it would have discovered a far more important risk, liquidity risk, and would have realized that market risk cannot be ignored even though the trades are being held to maturity. Liquidity risk arises because of the requirement that the firm posts margin equal to the mark-to-market value of the trades. At the inception of the trades, margin requirements will generally be low or zero because the trades will have zero value at par, but as spreads increase, margin requirements will increase as well.

The relevant stress test will indeed be a recession, as in the default-risk stress test, but rather than focusing on default risk, the liquidity stress test will concentrate on spread increases consistent with a recession. Constructing the liquidity SSST will require building a stress test that reflects the strategy of the firm. Specifically, in the SSST, we will assume that the firm's strategy will be to sell two $3.3 billion tranches each quarter over the next year. At that point, we assume that the economic outlook will darken and spreads will double, a market event that is not atypical in a recession. How much margin will be required to be posted in such a scenario?

Building such a liquidity SSST is considerably more difficult than the default stress test because complex trades must now be priced in a hypothetical environment. Because the trades are complex, the risk ID process should have also reviewed model-risk questions. What kind of pricing model is appropriate? Will the industry-standard Gaussian copula model be adequate, or is the correlation smile important to include in the stress test? Should a different model altogether be used? Can the desk-pricing model be employed? Can it price aged trades under hypothetical scenarios, and is it fast enough to support experimentation with multiple scenarios?

Because a stress test does not usually require extreme estimation accuracy, we prioritize speed of execution over precision and choose to build the underlying pricing model to be used in the SSST rather than relying on the desk-pricing model. We keep the industry-standard Gaussian copula pricing model that employs a single correlation because the risk ID process will not have identified correlation basis risk as a significant risk factor. This choice might well be different, of course, from the model the firm typically uses to mark its trades. Rather than simulate the pricing model, we choose to use a semianalytic method that has reasonable computational speed and is broadly applicable across a variety of synthetic CDOs.[5]

Having constructed the pricing model, we specify a stress test in which the firm sells eight 5-year trades over the course of the year, two at the end of each quarter. Thus, at the end of the year, the firm will have sold two trades with a remaining maturity of 4.25 years, 2 trades with a remaining maturity of 4.5 years, two trades with a remaining maturity of 4.75 years, and two trades with a remaining maturity of 5 years. At the end of the first year, we assume that spreads double. We also assume that correlation increases by 20 percentage points in the stress test. Table 7.3 shows the results of the SSST.

As can be seen in Table 7.3, the margin required to be posted vastly exceeds the firm's capital. It should be noted that this scenario is entirely plausible: spreads can easily double

---

[5] For this example, we price the CDOs by approximating the conditional loss distributions by means of the characteristic function. Huang and Oosterlee (2011) offer another approach.

Table 7.3. Strategy-specific stress test results

Maturity	Margin required ($millions)
5 years	53
4.75 years	50
4.5 years	46
4.25 years	43
Total	192

before or during a recession. The implied correlation will rise in such a situation as well. In this case, it was not necessary to design a particularly stressful scenario, which is not uncharacteristic of SSSTs, nor was it necessary to introduce complications such as stochastic recovery. A good risk ID process will uncover the true risks a firm faces. Once these risks are understood, very plausible scenarios that stress those risks can often be very illuminating.

9 Automatic Scenario Generation

In most cases, the risk profile of a financial firm cannot be fully understood in terms of its business strategy. Of course, the risk ID process should, as much as possible, try to understand the risks created by a financial firm's business strategy. Generally in complex firms, however, the connection between business strategy and risk will involve risks only at a very coarse level: the firm may be long rate risk and short commodity risk, for example. But the risk ID process will not be able to determine the complex interrelationships between risks at different levels in the risk taxonomy.

Whenever the risk profile of the firm involves complex interrelationships, it is difficult to properly stress the firm's risks using a single scenario. If a firm is long rate risk, for example, how are those risks broken out by tenor and geography? If the firm is short commodity risk, are there basis risks between different oil prices, between oil and natural gas, or between commodities and power? How should the stress test be formulated so that it stresses the risk profile conservatively while maintaining plausibility? When constructing one scenario, the analyst must understand the interplay between all the risks in a complex risk profile a priori, an understanding that is difficult to obtain from the risk ID process. The advantage of running multiple scenarios, however, is that the alternative scenarios can complement the risk ID process by discovering how the risks relate to each other. Even if the firm ultimately runs a single scenario, that single scenario's specification can be informed by the risk ID process implicitly conducted by running multiple scenarios.

The need to run multiple scenarios in complex risk environments rules out a manual judgmental process for specifying a stress test. In practice, hundreds of alternative scenarios may be needed to understand the interplay between risks, a number that is not practical to perform manually. Thus, running multiple scenarios requires the development of a stress test model that is capable of automatically generating scenarios. Such a model is termed an *automatic scenario-generation model* (ASGM).

Building such a model is a significant effort. A financial institution with a complicated risk profile will need to have a model that can generate scenarios that broadly cover risks across the risk taxonomy. The model will need to generate scenarios over the short run for market risks and over the medium term, perhaps over a 1- to 3-year horizon, to capture credit, revenue, liquidity, and operational risks. These scenarios will need to be consistent across timescales, consistent across risks, and sufficiently conservative but not so conservative as to be implausible. To fully capture the risks, such a model will need to generate scenarios coherently that include hundreds if not thousands of financial variables, as well as macroeconomic variables such as real gross domestic product (GDP) and the unemployment rate, because many of the risks a firms faces will depend on the state of the economy.

To build such a stress test model, it is important to establish a scenario-design team that is well versed in macroeconomics, financial economics, and time-series econometrics. This skill set is not common in quantitative risk management, which has historically focused on derivative and other financial pricing models. To develop such a team, it is often helpful to make it interdisciplinary, staffing it with more conventional risk management quants along with economists. Economists can help guide the econometric modeling, but building a full stress test system will also require substantial skills in coding, quantitative risk management, and pricing.

To generate multiple scenarios, it is necessary to build an econometric model that can project both macroeconomic and financial variables. Generally speaking, optimizing models such as dynamic stochastic general equilibrium (DSGE) that are used for policy analysis are not necessary or even very useful for stress testing. Instead, effective stress models can be built using standard time-series econometric tools, such as vector autoregressions, cointegrated systems, factor models, generalized autoregressive conditional heteroskedasticity (GARCH), regime switching, and so forth. The generation of multiple scenarios can be accomplished in a number of ways. One method is to simulate the estimated residuals of the models, thereby creating alternative scenarios. Another method is to create a scenario-generation engine that takes combinations of stylized facts to alter the basic properties of the econometric system, such as the size of the GDP shock or the slope of the yield curve, and then allow the system to determine the other variables. The advantage of using these methods to generate alternative scenarios is that they allow the creation of scenarios that have no historical precedent. In effect, these mechanisms break the empirical correlations that have been observed between variables. But these correlations are broken in a nonarbitrary manner so that the generated scenarios can continue to be plausible.

To give an example of how correlations between financial variables might be broken, suppose a scenario designer needs to generate market-risk scenarios to stress a complex trading risk portfolio under severely stressed market conditions. An equity portfolio scenario simulation model might be specified by assuming that some market index, such as the Standard & Poor's (S&P) 500, follows an exponential GARCH (EGARCH)[6] process, while the other equity prices are estimated as a beta to the index. In this case, bootstrapping the residuals would not break the historical correlations but would, rather, reproduce them. Instead, we could add a liquidity model that would help to simulate how long it would take a financial institution to eliminate a position in a particular equity position, which would help to break empirically observed correlations in a coherent manner. If, for example, it would take on average 10 days to get out of a position of size X in stock A and 5 days to get out

---

[6] See Nelson (1991) for a description of the EGARCH model.

of position of size Y, then we would use the 10-day simulation for stock A and the 5-day simulation for stock B, both taken from the same set of simulations.

To create simulated scenarios, it is necessary to create a means to simulate the time to liquidate a position in each equity position in the portfolio. The simulation could be done as follows: either through expert judgment or empirical observation, determine how much of a position in stock A could be eliminated without incurring unusual transaction costs in a single day. Suppose the average amount per day for stock A is $100 million, and the size of the position in stock A is $1 billion. Then, on average, it would take 10 days to liquidate the position in stock A. To create scenarios, it is just necessary to randomize the liquidation process. For example, opportunities to liquidate a position of size $100 million on a particular day could be simulated from a normally distributed random variable, with a mean equal to $100 million and a variance to be judgmentally determined by the scenario analyst. The number of such events could be specified as a Poisson process with a mean equal to 10 days. Scenarios could be simulated by first simulating, for each equity position, the number of days it would take to fully liquidate the position. Suppose we have three stocks and a particular simulation gives 5, 7, and 12 days, respectively. Then, we use the 5-day return from the equity simulation model for stock A, the 7-day return for stock B, and the 12-day return for stock C. By repeating these simulations many times, we can trace out a high number of scenarios in which correlations are broken between equity returns but in a manner that makes intuitive sense: the changing correlations depend on the underlying liquidity of the individual stock positions, which in turn depends on the actual size of each position.

To break correlations in a reasonable way, it is important to combine econometrics with judgment. In the previous example, the judgmental component was the specification of how much of a particular equity position could be liquidated on a particular day. Detailed data on liquidation will not likely be available even for nonstressed periods and so will depend primarily on market judgment. By varying that market judgment, it was possible to create scenarios that are reasonable while having never been experienced historically.

In the equity example, the risks were built into the scenario-generation model by including the actual size of each equity position. More generally, the risk ID process should be united with scenario generation. Using the output of the risk ID process, the scenario-design team can decide how to build in the risks. For example, the list of market risks, including basis risks and concentration risks, will determine the granularity of the risk factors that need to be econometrically modeled. Credit risks will require the determination, for each generated scenario, of how credit spreads, credit ratings, and loss given default will evolve over time. If there is evidence that operational risk can be determined by scenario variables, those models will need to be incorporated.

For each scenario, we must be able to estimate the totality of the risk produced by that scenario over all risk types. In general, it will not be computationally feasible to estimate the effect of each scenario on all risks. Thus, reasonable approximations for the losses resulting from risks will need to be constructed inside the ASGM. For example, the estimation of the losses resulting from market risks could be suitably linearized or approximated in some other way. To capture business environment risk, time-series models for revenues could be estimated that are a function of the scenario variables, and revenue risk could then be estimated for each scenario. Guided by the risk ID process, the scenario-design team should build a model of the risk profile of the firm into the ASGM. Once that is done, every scenario that is automatically generated by the model will include an estimate of its effect on the whole of the risk profile. In effect, the ASGM is itself an extension of the risk ID process, exploring the risk profile of the firm under alternative scenarios.

The scenario-design team must decide how many scenarios to run and how to select a representative scenario from the set of scenarios that reflects the risk profile of the firm. This selection process should define a scenario that is either one of the scenarios produced by the model or a composite of the scenarios that were run. It should conservatively stress the firm's risk profile, breaking historical correlations, but at the same time, it should maintain plausibility. Once a composite scenario is specified, it can be run on the full set of risks without approximations. Depending on the specification of the scenario, the results could be used to verify the capital adequacy of a financial institution, to specify a strategic asset allocation for a pension fund, or to measure the sensitivity to geopolitical events such as Brexit. The advantage of investing in such a model is that it will allow the firm to measure all the risks across different timescales, giving a complete and integrated picture of the risks a financial firm faces.

10 How Much Should We Rely on History in Designing Scenarios?

One of the most difficult questions any scenario designer will have to answer is how much to rely on history in developing a scenario or set of scenarios. The answer depends crucially on both the outcome of the risk ID process and the uncertainty of empirical experience. A properly conducted risk ID process should give the scenario designer the most important risks to focus on in the scenario design. But the scenario designer must determine how well those risks can be understood from historical experience.

This question will arise even when the risks are very simple. For example, assume an asset-management fund with an investment horizon of several years has a predominantly equity-based asset allocation. The obvious risk is that a recession will strike, but how much can history tell us about equity declines over recessions? Figure 7.1 depicts the 1-year-forward S&P 500 return; the lower peaks before the shaded areas, which record National Bureau of Economic Research (NBER) recessions, indicate the worst 1-year returns experienced during each recessionary period. As can be seen in the figure, 1-year returns can vary a lot and are not necessarily determined by the severity of the downturn. The 2001 recession was milder than the 1990 recession, yet the equity declines were much larger during the 2001 period, for example.

Of course, equity declines during the 2001 recessionary period were much larger, despite the mildness of the recession, because stock prices had run up to very high levels, with many observers claiming that the stock market was a bubble at that time. Equity price returns depend not only on the state of the economy but also on whether they are perceived to be overvalued. Overvaluation cannot be determined from history, however, but must be decided based on the particular facts that exist at the time the scenario is being formulated. In defining even a simple 1-year-horizon equity scenario, the scenario analyst will need to analyze the degree to which the equity market may have been supported by monetary policy, the prices of other assets, and the more nebulous market sentiment. Historical returns will not be very useful, except perhaps to establish bounds on reasonable scenario specifications.

Greg Hopper

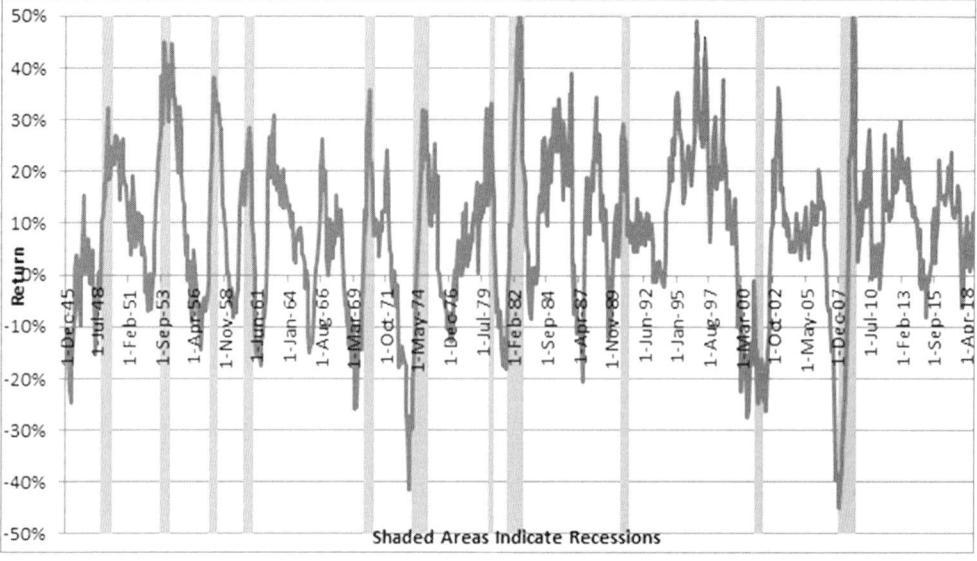

Figure 7.1 One-year-ahead S&P 500 return. Source: Yahoo Finance, NBER
(recession dates); author's calculations.

On the other hand, if the risk ID process had found that the risk was predominantly simple equity risk, was very liquid, and was also over a very short horizon, then the scenario designer could rely on history much more comfortably. In general, the more the risk ID process finds that the risks can be historically based, the less scenario analysis is needed: conventional risk models such as VaR or PE could be used in these cases. But the more the risk ID process identifies risks that are not well understood from a historical perspective, the more scenario analysis is needed.

Perhaps an extreme case in which history will be a poor guide to scenario specification is the nascent area of climate-change stress testing. If a financial institution is investing in, lending upon, or insuring properties that may be held for a long period of time, it may be necessary to perform climate-change physical-risk stress tests. Such stress tests do not depend much on history. To formulate such a physical stress test, the scenario designer could start with the output of a particular global circulation model (GCM), a model that projects physical quantities such as temperature, precipitation, and so forth at frequencies as high as daily over very long time periods, typically to 2100 AD. These projections depend on particular scenarios of how much carbon dioxide has been put into the atmosphere over time, which further depends on assumptions about economic growth, population growth, and climate-policy responses. The physical scenarios derived from the GCM would need to be transformed into metrics that can be used to assess economic consequences. For example, temperature scenarios could be used to project increases in energy costs. Sea-level rise combined with precipitation and hurricanes could be used to project potential damage to coastal properties, as well as changes in insurance premiums. Although there may be some reliance on history in making these economic assessments, the scenarios being projected are, by their nature, completely new, with no historical precedent.

11 Conclusion

The development of stress test models is the next frontier in risk management. Stress testing is a broad methodology that can address just about any question in risk management. Although stress testing has improved dramatically since the financial crisis, there are still many areas of development. For example, many firms are now focusing on measuring their climate-change risks, including the transition risks to an economy that relies less on carbon, as well as physical risks to property or investments resulting from the change in the environment. Stress testing is a natural tool because the models include a representation of the underlying real and financial components of the economy that have already been developed for other stress testing purposes. The measurement of systemic risk in the financial system is another area that could be clarified by an appropriate stress test methodology. Although this chapter has concentrated on applications of interest to financial firms, the methodologies discussed here could also be useful to firms in other industries. An ASGM founded upon an underlying economic model of the world economy could be used to stress the revenue and supply-chain risks of manufacturing firms. Looking forward, financial firms have an interest in developing methodologies to stress test cyber-risk. Those methodologies would obviously be useful to nonfinancial firms. However these new stress test methodologies are created and for what purpose, it will be necessary to marry risk identification and scenario design in order to create a truly informative and effective stress test.

References

Aratan, M. (2013), "The advancement of stress testing at banks," in J. Zhang (ed.), *CCAR and beyond: Capital assessment, stress testing, and applications*, Risk Books.

Dent, K., B. Westwood, and M. Segoviano (2016), "Stress testing of banks: An introduction," *Bank of England Quarterly Bulletin*, 56(3), 130–143.

Hopper, G. (2010), "Stress testing and scenario analysis: Some second generation approaches," in E. Canabarro (ed.), *Counterparty credit risk*, Risk Books.

Hopper, G. (2014a), "Dynamic stress testing of counterparty default risk," in E. Canabarro and M. Pykhtin (eds.), *Counterparty risk management: Measurement, pricing, and regulation*, Risk Books.

Hopper, G. (2014b), "The art and science of stress testing," in G. Hopper (ed.) Special Issue on Stress Testing, *Journal of Risk Management in Financial Institutions*, 7(1).

Huang, X., and C. Oosterlee (2011), "Saddlepoint approximations for expectations and an application to cdo pricing," *SIAM Journal of Financial Math*, 2(1), 692–714.

Nelson, D. (1991), "Conditional heteroskedasticity in asset returns: A new approach," *Econometrica*, 59(2), 347–370.

Rebonato, R. (2010), *Coherent stress testing: A Bayesian approach to the analysis of financial risk*, John Wiley & Sons.

Schuermann, T. (2013), "Stress testing banks," *International Journal of Forecasting*, 30(3), 717–728.

8

Stress Testing with Market Data[*]

Robert Engle

1 Introduction

Stress testing has become a key tool in the regulation of financial institutions. It provides a metric for both microprudential and macroprudential regulation. The existence of this handbook and its luminary authors is a testament to the importance of stress tests. A careful reading of this handbook, however, reveals that there are many issues that have yet to be solved to make regulatory stress tests sufficiently accurate that they can confidently ensure financial stability.

A stress test is a form of risk management that is focused on long-horizon risk. In contrast to value at risk (VaR) or conditional value at risk (CVaR), stress tests consider risks at much longer horizons, typically 1–5 years. Arguably, this is appropriate for assessing the risk of illiquid assets, such as those held by financial institutions. The key task is to assess the value of a firm or asset in the future under an adverse counterfactual scenario. Here we see the critical points of stress tests: the valuation model and the scenario.

This chapter mentions some of the difficulties in generating appropriate scenarios and valuing firms under these scenarios. In most cases, these difficulties can be solved if the regulator is better informed than the market. However, if this is not correct at all times and settings, then it is also sensible to carry out stress tests with market scenarios and market data. When these stress tests agree, the results gain added credibility. When they disagree, the parties can discuss whether the market has missed signals or whether the regulators' models are wrong or have been politically influenced.

The assumption that the regulator has superior information relative to the general market is not completely unreasonable. Banks are opaque, and the regulator has access to confidential data that are not available to the general market. The regulator also has access to the same information from the firm's competitors. Hence, the regulator may be able to better value a firm and possibly understand which weaknesses will be of most concern in the future.

The valuation model seeks to value all the assets of the firm from the bottom up. This is what analysts do to inform investors about firm values, and it is a fundamental part of the business of mergers and acquisitions. The job of the analyst, however, is easier because the

[*] This survey has been drawn in part from Engle (2018) and from Engle and Ruan (2019). The author is indebted to many colleagues for helpful comments and encouragement. Particular thanks go to Viral Acharya, Matt Richardson, Richard Berner, Kim Schoenholtz, Tianyue Ruan, Christian Brownlees, Rob Capellini, and participants at the Macro Financial Modeling (MFM) organized by Lars Hansen and Andy Lo. It has benefited immensely from suggestions from the editor, Til Schuermann, and reviewer, Mark Flood.

scenario is the current forecast, not a counterfactual adverse scenario. The valuations of the model can therefore only be compared with market valuations when a similar scenario is anticipated. Applying the value model under a benign scenario can provide a check on the valuation model. Alternatively, the market often provides valuations in adverse conditions, such as through options or response functions.

The challenges of the valuation model can be stated as follows:

1a. Its accuracy
2a. Its cost in time and money for both regulator and regulated
3a. Release of private information
4a. Interconnectedness. The valuation of one firm will depend on the response of other firms to the adverse scenario.

Point 1 includes not only the valuation model but the input data as well. Accounting data often have inherent biases, which may be incorporated into the results. Furthermore, accurate valuation of the firm should include assessing the liabilities under the adverse scenario because runs of liabilities may lead to insolvency. Point 2 is obvious but has an important corollary. If it is expensive to carry out a stress test, it will not be done frequently. Most stress tests are done at an annual or lower frequency. Yet for many decisions, especially including countercyclical capital regulation, a much higher frequency is needed.

The third point can be elaborated. Vast amounts of highly confidential information are submitted by the institution to the regulator with a guarantee of security. When the analysis is finished, some information is released back to the market, and this may reveal confidential information. Further, the regulator constructs an adverse scenario that is often considered to be private information. If this is released prior to the stress test, the financial institution may adjust its portfolio or business to circumvent the scenario. Not only does this invalidate the results, but it encourages the firm to manage its portfolio in a way different from its view of optimality.

The fourth point is fundamentally the difference between microprudential and macroprudential stress tests. In the latter, the interconnectedness should be considered.

The choice of scenario is also a challenge. Here again are four points:

1b. Which adverse scenario to use?
2b. How severe should it be?
3b. How can we compare one stress test with another?
4b. If multiple scenarios are used, how can the results be combined?

In various chapters of this handbook, numerous examples are provided of stress testing where the wrong scenario is used, the stress is too strong or too weak, or there is no comparability over time or across countries or regulators.

The use of multiple scenarios is potentially interesting, leading to reverse stress tests and worst-case scenarios. However, of course, if different firms find different scenarios to be worst cases, then the system as a whole may be more stable than expected. Ultimately, the insurance solution, which is to use thousands of equally likely scenarios, allows a full distribution of firm values for each firm. If the same scenarios are used by each firm, then covariances across firms can be computed. However, it is not easy to formulate many equally likely scenarios.

2 Market-Based Stress Tests

Many of these issues can be solved or mitigated by using market data. The market value of a firm is the market value of its assets, which is reflected in the market value of its equity. Of course, the accuracy of such measures depends on the liquidity and efficiency of the market. As asset values fluctuate, the equity values may change rapidly as a result of the leverage of the firm. If equity values fall to zero or to some low level relative to the liabilities, then the firm will be forced into bankruptcy. The distance to default in structural credit models is based on the ratio of market capitalization (market cap) to liabilities. A prudent financial firm will strive to keep this ratio above a minimal value to avoid the risk of failure. This value will be called "k^*."

An interpretation of k^* can be given in terms of capital requirements. Suppose the value of the firm is sufficiently well estimated by the sum of the accounting liabilities and the total market cap, which is sometimes called *quasi-assets*. Then today's capital adequacy ratio is the ratio of the market cap to quasi-assets. If there is a regulatory capital requirement, k, then

$$\frac{\text{market cap}}{\text{liabilities} + \text{market cap}} \geq k = k^*/(1 + k^*). \tag{8.1}$$

The capital shortfall is the difference between required capital and actual capital. It can be computed directly:

$$\text{Capital shortfall} = k \, (\text{liabilities} + \text{market cap}) - \text{market cap.} \tag{8.2}$$

To evaluate the capital shortfall under an adverse scenario, it is only necessary to estimate the impact on equity of the adverse scenario because this will imply a stressed market valuation of assets. This is commonly done by looking historically at the relation between firm equity values and market-risk factors to estimate the impact of the adverse scenario on the market cap. These impacts are the market betas on relevant factors.

We can perform a regression with several risk factors, each with its own beta, as

$$r_{i,t} = \mu + \sum_{j=1}^{J} \beta_{ij} f_{j,t} + e_{it}. \tag{8.3}$$

In this case, there are J risk factors, f, and it may be possible to transform a scenario into a hypothetical value for each of the factors. For example, there might be a real estate factor, an interest rate factor, an energy-cost factor, and a market factor. In this case, the expected change in market cap would be the same linear combination of the estimated betas for each firm. The betas will naturally differ across firms.

In reality, the betas will depend on the assets held by the firm at each point in time, so the betas should vary over time as well as across firms. If the betas are forced to be constant, the effect of the stress would not depend on the asset mix. If the stress is the same over time and the betas are constant, then the ratio of equity under stress to current equity will be constant.

A dynamic conditional beta (DCB) model, as in Engle (2016), recognizes that betas change. When there is just one risk factor, j, the DCB beta is simply

$$\beta_{i,j,t} = \rho_{i,j,t} \sqrt{\frac{\sigma_{i,t}}{\sigma_{j,t}}}, \tag{8.4}$$

where the volatilities and correlations evolve over time. More generally, this is given by a matrix inverse of the factor dynamic covariance matrix times the covariance vector of returns with factors.

With an estimated model such as equation (8.3), a stress test can be computed simply with a factor representation of the scenario. If there are more factors in the scenario than in the pricing model, then the other factors are not priced, which means they will not affect the stress test. If firms are very vulnerable to one factor and it does not get stressed in the scenario, then the stress test will miss the important weakness.

If, on the other hand, the pricing model has only one factor and the scenario has only one factor, the market factor, then it does not matter whether the market return is caused by a drop in real estate prices or a rise in energy prices. The fluctuation in beta in equation (8.4) will pick up the market concern about the factors that move the market. That is, if the fall in real estate prices is the key risk and the market knows it, then the market will give firms with big real estate portfolios big market betas. This assumes a great deal of information by the market; however, it is not implausible. Clearly, in the financial crisis, some firms had a much more severe collapse in market prices than others. This translates to bigger betas.

The logic of this also emphasizes how the market can price interconnectedness. If a firm has strong connections with a firm with a big real estate business, then this firm will also suffer a sell-off, and this will be translated into a bigger beta.

This argument suggests that it is not necessary for a stress scenario to foresee the major risks in the economy; it is sufficient for the market to foresee these risks. To be clear, this does not mean that the market can predict the outcome. It means that if some outcome is bad, the market knows which firms will be most affected. This is an important illustration of how market information can reduce the need for the regulator to foresee the cause of the next crisis. Every type of crisis will produce a drop in the broad equity market. When money managers understand these risks, they will adjust their firm-valuation models such that small changes in market values are accompanied by bigger changes in the equity prices of the most vulnerable firms.

3 The Role of Stress Tests in Macroprudential Analysis

It is widely believed that financial crises result from *excessive* credit growth. See, among others, Brunnermeier and Pedersen (2009) on liquidity spirals; Reinhart and Rogoff (2009), who claim that this time is *not* different; Borio (2014) and Drehman et al. (2012) on financial cycles; Adrian and Shin (2010, 2014) on leverage cycles; Schularick and Taylor (2012) on the predictive power of credit growth; and Mian and Sufi (2009) on household debt. However, the challenge is how to measure *excessive* credit growth. The argument is that credit growth is excessive if the financial sector does not have sufficient capital to cover market-value losses in a downturn. This is consistent with the notion that at the end of a credit cycle, increasingly risky credit will be issued, and the holders of this credit will be leveraged financial institutions with insufficient reserves to cover losses in a downturn. In this way, a "credit boom goes bust."

A small example motivates many measures of systemic risk. Consider the example of mortgages written by the financial sector. In the late stages of a credit boom, mortgages are likely to be issued to underqualified borrowers on overvalued properties. These mortgages will naturally have a market value that is less than their accounting value. If there is a downturn in the economy, the mortgages will fall further in value in response to declines in the collateral and repayment prospects. Financial institutions' market values will fall, reducing the market cap. If capital is insufficient, the institution will become insolvent and may fail or seek a bailout. The size of the capital shortfall under stress is a natural measure of excessive credit growth for this firm. When it is aggregated over financial firms, it becomes a natural measure of excessive credit growth for the economy. Two key features of this measure are that it should reflect the relation between the market and book values of the assets of the firm, including off-balance-sheet items, and it should account for the risk of the asset portfolio. Both of these vary over time, and estimation methods should take this into account. In the next section of this chapter, the SRISK measure is introduced and described in detail.

In a recent paper, Jordà et al. (2017) examine a long history of credit cycles and find that bank capital as measured by common Tier I capital/total assets has no predictability for financial crises, although it does ameliorate the associated social costs. They suppose this is because the risk on the asset side of the balance sheet is ignored in such a ratio. In other words, it is not just the capital available but also the capital under stress that implicitly adjusts for risk that should matter. Furthermore, their measure of capital is based on the accounting value of assets rather than the market value.

It is useful to focus conceptually on two features of an undercapitalized financial system:

1. An undercapitalized financial firm will be vulnerable to adverse external shocks.
2. An undercapitalized financial sector will naturally generate shocks that begin a financial crisis. The process of reducing risk and deleveraging is the catalyst that actually starts the crisis.

Although stress tests are designed to assess the vulnerability of banks to shocks, their real importance comes from the second point. An undercapitalized financial sector can cause the financial crisis. This process has been the focus of much macrofinancial research, where a variety of channels have been modeled, starting with Bernanke and Gertler (1989), Kiyotaki and Moore (1997), and Bernanke et al. (1999). Here we consider deleveraging or de-risking of the financial institutions. When capital is lacking, either the regulator or the risk manager of individual companies will recognize the risk and compel the firms to strengthen their balance sheets. This can be done in many ways, but the most commonly used approaches lead to sales of assets or equity. When such sales are in large volume, it is inevitable that there will be a price impact. This leads to a downward spiral of the financial sector and ultimately of the economy that has been written about extensively and is commonly called a *fire-sale externality*. See, for example, Cont and Schaanning (2017), Greenwood et al. (2015), and Pedersen (2009), among others. The initial conditions that make such a fire sale likely are precisely a large quantity of assets to be sold in a hurry and the lack of a large collection of willing buyers. The sale of mortgage-backed securities in the financial crisis of 2008–2009 is an illustration of this phenomenon.

4 SRISK

To measure the undercapitalization of a financial firm, we estimate the dollars that this firm would need to raise in order to function normally if we have another financial crisis. This is a widely used measure called *SRISK*, which stands for *systemic risk*. It is computed weekly and published on V-Lab[1] for more than 1,000 global financial firms. It measures capital shortfall under stress and can be interpreted as a market-based version of a regulatory stress test. It was initially introduced by Acharya et al. (2009, 2017) and then expanded by Brownlees and Engle (2017), Engle (2016), and Engle and Richardson (2015).

The exact calculation takes several steps:

1. Normal operation under stress conditions requires that the market value of equity divided by quasi-assets (book value of liabilities plus market value of equity) be above a minimum level. We typically use 8 percent, which is equivalent to a total leverage ratio of 12.5 to 1. This value corresponds to the typical leverage ratios of well managed financial firms in tranquil periods.[2,3] Users of the V-Lab website can change this value easily. The shortfall today is computed in equation (8.2) based simply on accounting and market-value numbers.

However, the power of a stress test is to estimate the firm condition under a counterfactual future stress condition. Hence, SRISK is defined as follows:

$$\text{SRISK}_t = \text{median (capital shortfall} \,|\, \text{crisis)} . \tag{8.5}$$

The use of the median is analytically convenient but also economically sensible. It simply says that half the time, the shortfall will be less than this, and half the time, it will be more. It is insensitive to low-probability negative-shortfall events, which sometimes arise when volatility is very high and firms may gamble on their survival.

The notional value of liabilities is assumed not to change with the stress even though short-term liabilities and some off-balance-sheet items, such as credit commitments, could change. More importantly, the value of equity will fall in a crisis. When the scenario happens, the equity will decline by a rate called the *long-run marginal expected shortfall* (LRMES). This is the important random variable that must be estimated with a statistical model:

$$\text{LRMES}_t = \text{median} \left(\frac{\text{equity}_t - \text{equity}_{t+T}}{\text{equity}_t} \,|\, \text{crisis} \right), \tag{8.6}$$

which implies:

$$\text{SRISK}_t = kD_t - (1-k)\,\text{equity}_t \,{}^*(1 - \text{LRMES}_t) . \tag{8.7}$$

2. To estimate LRMES, we use a standard financial approach, the market model:

$$r_t^j = \beta_t^j r_t^M + \varepsilon_t^j, \tag{8.8}$$

where r_t^j is the equity return on day t for firm j. Similarly r_t^M is the world equity market return, taken to be the return on the Morgan Stanley Capital International (MSCI) All-Country World Index (ACWI) exchange-traded fund (ETF).[4] Both returns are measured

---

[1] See vlab.stern.nyu.edu/welcome/risk.

[2] In June 2017, for example, the average of the six largest US bank leverage ratios was 9.1; in June 2014, it was 11.3; in June 2010, it was 15.5; and in June 2005, it was 9.5. The average of these is 11.3, or slightly less leverage than used in SRISK. Users of V-Lab have the option of setting the capital ratio at any desired level.

[3] Firms using International Financial Reporting Standards (IFRS) accounting rather than generally accepted accounting principles (GAAP) typically have a bigger balance sheet because there is less netting of derivatives. Consequently, percent is used for all European firms.

[4] The MSCI ACWI is an index of equity returns in 23 developed and 26 emerging markets.

as continuously compounded or log returns. This model could be estimated assuming that beta is constant and that the residual variance is constant; however, we will relax these assumptions.

This relation captures the market view of the impact of a market decline on firm asset values and consequent equity declines. It explicitly focuses on the comovement of this firm and the market, rather than just the volatility of the firm. Therefore, this is a macroprudential measure of risk rather than microprudential. A macro stress affects all financial firms, leading to a macroeconomic effect.

Letting p_t^j and p_t^M be the prices at the end of day t adjusted for corporate actions,

$$\frac{p_{t+T}^j}{p_t^j} = \exp\left(\sum_{i=1}^{T}\left(\beta_{t+i}^j r_{t+i}^M + \varepsilon_{t+i}^j\right)\right). \tag{8.9}$$

Assuming that the betas change little over the next 6 months, this becomes

$$\frac{p_{t+T}^j}{p_t^j} \approx \exp\left(\left(\beta_t^j\right)\left(\log\left(p_{t+T}^M/p_t^M\right)\right) + \sum_{j=1}^{T}\left(\varepsilon_{t+j}^j\right)\right). \tag{8.10}$$

The fractional return is a random variable, and we will consider the median, conditional on a stressed market return. Assuming the idiosyncratic errors have median zero, the median fractional change in equity prices in a crisis is LRMES, which is given by

$$\text{LRMES} \equiv \text{median}\left(\frac{p_t^j - p_{t+T}^j}{p_t^j}\left|\frac{p_t^M - p_{t+T}^M}{p_t^M} = \theta\right.\right) = 1 - \exp\left(\left(\beta_t^j\right)\log\left(1 - \theta\right)\right). \tag{8.11}$$

An alternative simulation approach to estimating LRMES has been implemented by Brownlees and Engle (2017, 2011) and in V-Lab's MESSIM. This is introduced and discussed later in this chapter.

For firms located in different time zones, part of the response will appear to be from the previous-day return. Hence,

$$r_t^j = \beta_t^j r_t^M + \gamma_t^j r_{t-1}^M + \varepsilon_t^j. \tag{8.12}$$

3. The parameters beta and gamma are estimated by DCB as described by Engle (2016). This approach recognizes that a regression coefficient is a product of a correlation between the firm return and the market return times the ratio of the standard deviation of the firm return and the market return. All three of these values are potentially time varying and can be estimated by GARCH and dynamic condition correlation (DCC). The DCB estimate of beta is therefore:

$$\hat{\beta}_t^j = \rho_t^j \sqrt{\frac{h_t^j}{h_t^M}}. \tag{8.13}$$

The same argument can be applied to gamma under the assumption of no autocorrelation in the market return.

It is natural to test the hypothesis that the betas are constant. Because this not a nested hypothesis, a useful strategy is to construct an artificial model that nests both the constant beta and the DCB. The nested model can be expressed as

$$r_t^j = \left(\phi_1 + \phi_2 \hat{\beta}_t^j\right) r_t^M + \left(\phi_3 + \phi_4 \hat{\gamma}_t^j\right) r_{t-1}^M + \varepsilon_t^j. \tag{8.14}$$

This model can be estimated assuming a Glosten–Jagannathan–Runkle GARCH (GJR-GARCH) error term to give estimates of the four coefficients. We would expect $\phi_2 = \phi_4 = 0$ for the beta constant or $\phi_1 = \phi_3 = 0$, $\phi_2 = \phi_4 = 1$ for the DCB. Typically, both hypotheses can be rejected, which means that some combination of constant and time-varying betas is favored by the data. Hence, we use these estimates with $\left(\hat{\phi}_1 + \hat{\phi}_2 \hat{\beta}_t^j\right)$ and $\left(\hat{\phi}_3 + \hat{\phi}_4 \hat{\gamma}_t^j\right)$ as the final estimates of beta and gamma.

In the formula for SRISK, we envision the stress lasting 6 months, so both beta and gamma are equally important and just sum together. Letting

$$\tilde{\beta}^j = \left(\hat{\phi}_1 + \hat{\phi}_2 \hat{\beta}_t^j\right) + \left(\hat{\phi}_3 + \hat{\phi}_4 \hat{\gamma}_t^j\right),$$

$$\text{SRISK}_t = kD_t - (1 - k)\, \text{equity}_t \exp\left(\tilde{\beta}^j \log(1 - \theta)\right). \tag{8.15}$$

4. The market variable used in SRISK is the global market MSCI ACWI ETF. Because this is traded in the United States, the closing price is the price at the New York close, and it reflects the information that agents know at that time. During the financial crisis, this measure declined approximately 40 percent over 6 months. Thus, we take this to be the stress for calculating SRISK; however, a V-Lab user can set the stress level wherever it is desired. By using a one-factor stress, we reduce the possibility that we will miss the cause of the next financial crisis. Whether the crisis is caused by a housing market collapse, sovereign debt collapse, commodity price collapse, exchange rate collapse, government shutdown, derivative market failure, or medical pandemic, the stock market is probably going to fall in anticipation of a decline in the real economy. It is almost inconceivable that a financial crisis could occur without a substantial fall in the stock market.

Interestingly, the firm betas reflect the characteristics of the crisis and their impact on each firm. For example, in the Global Financial Crisis (GFC), the betas of Bank of America and Citigroup rose to 3 and 4, whereas the betas of Goldman Sachs (GS) and BNP Paribas (BNPP) didn't move much at all. Neither GS nor BNPP were as involved in the subprime mortgage business or to the extent they were, they hedged their exposure. However, in the European sovereign debt crisis, Credit Agricole and BNPP's betas rose almost to 3 because these banks were very exposed to Greek debt and the debt of other peripheral countries.

5. Aggregate SRISK is the sum for a country of the capital shortfall of all the financial firms with positive values. This is the sum of capital needed by all the institutions that would have to raise capital in a crisis. We do not net the overcapitalized firms because the capital from these prudent firms is not immediately available to recapitalize the weak firms. Mergers do happen, but more often, the well-capitalized firms protect their reserves as bad events are approaching. Furthermore, some of the well-capitalized firms are not in exactly the same industry and therefore might not be capable of quickly merging or taking over business from a failing firm.

By examining the risk of a financial crisis based on aggregate SRISK, we implicitly incorporate a familiar externality emphasized by Acharya et al. (2009, 2017). The failure of a single medium-size company is unlikely to be systemic because its business is likely to be taken over by its competitors. However, if the competitors are also weak, then such mergers will not happen without raising capital. If one firm reduces its own SRISK, the whole industry benefits, and so does the economy, through reduced risk of a crisis. Thus, there is an externality that reduces the incentive to do costly deleveraging and asset sales. The simple domino analogy is incomplete because it suggests that any one company that fails would bring down the rest. It also suggests that firms more remote from the initial failure would have more time before the inevitable collapse. Investors in such a firm will not wait to sell out. A better analogy is a tsunami, where the whole financial sector is affected at the same time. Stress tests are designed to measure vulnerability to such a tsunami.

6. This estimation is carried out weekly for more than 1,000 financial firms around the world and published on V-Lab. At the end of each month, a final estimate is produced and is stored in V-Lab, where it can be viewed. This is a fixed value that will not be affected by future data or estimation. Plots of SRISK, in fact, are continuous using daily market data and quarterly or even annual accounting data as they become available. Thus, the SRISK estimates are all recursive and can be obtained in real time.

5 The Risk in SRISK

SRISK is an estimate of the capital shortfall of a firm conditional on a counterfactual scenario. Consequently, it is difficult to determine the accuracy with which it is estimated unless the scenario actually occurs. However, because financial crises are rare, the sample size in which the scenario occurs will always be small. Thus, the risk in such estimates can, at best, be studied econometrically within a particular model or empirically across several alternative model formulations. In this section, we examine the accuracy of SRISK estimation. We explore how the firm-specific and aggregate SRISK vary with stress and capital requirements. We develop bootstrap estimates of standard errors, and we compare estimates from different specifications, such as domestic SRISK (DMES) and the simulation estimates in MESSIM.

The formula for capital shortfall presented in equations (8.7) and (8.11) makes it clear that the estimation of LRMES is fundamental to the measurement of SRISK. Because debt and equity are known to the econometrician, these are not sources of uncertainty for SRISK. The conditional distribution of SRISK at any point in time depends only on LRMES. However, in examining SRISK over time, future observations on debt and equity are a major part of the forecast uncertainty. Furthermore, the policy parameters; k the stressed capital requirement; and the stress that is supposed to approximate the "crisis scenario" are all assumed to be given for this analysis, but their optimal values for any policy purpose can only be known with a great deal of uncertainty.

5.1 Sensitivity Analysis

We first examine the sensitivity of firm SRISK and aggregate SRISK for the United States as a function of the two policy parameters using data from the end of 2017. The default

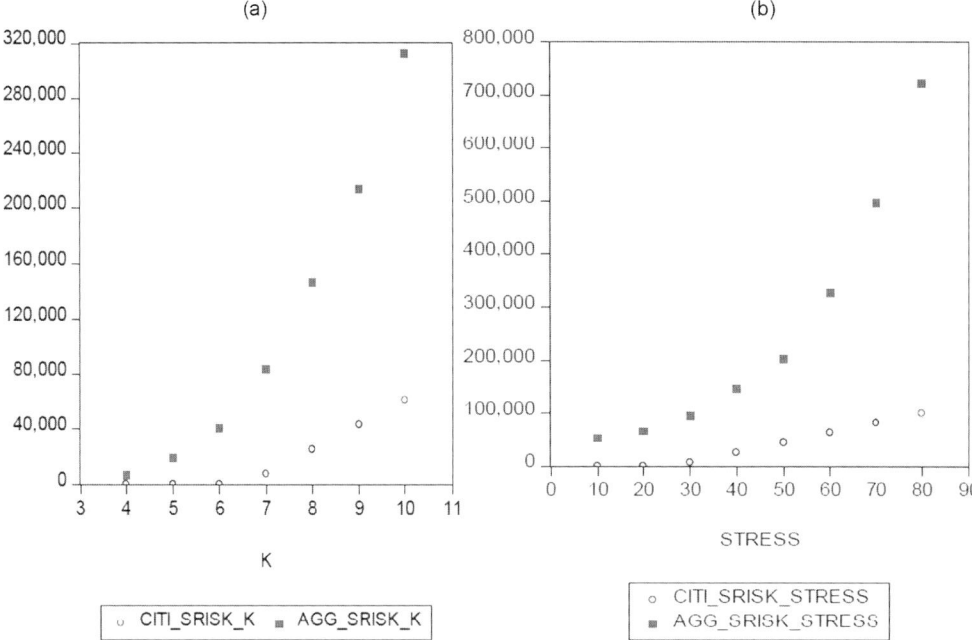

Figure 8.1 SRISK for Citi and United States as a whole as a function of capital requirements (panel [a]) and as a function of stress (panel [b]).
Source: V-Lab SRISK;[5] author's calculations.

value for the capital ratio under stress is 8 percent, but often policymakers would want to choose another value. If it is lower, then the risk of an individual firm will fall linearly until it hits zero, and the capital shortfall does not fall below this. Similarly, the aggregate SRISK will also fall, partly because the risk of each financial firm is reduced and partly because some firms hit zero and no longer contribute to aggregate risk. In panel (a) of Figure 8.1, the SRISK values at the end of 2017 due to Citi and for the United States as a whole are plotted against capital requirements for values from 4 to 10 percent.

Similarly, panel (b) of Figure 8.1 presents SRISK for Citi and the United States at the end of 2017 as a function of the stress while keeping the capital requirement at 8 percent. Rather than the standard assumption of a 40 percent decline in the global stock market, the values are shown for declines of 10 to 80 percent. Notice the nonlinear impact of higher stress and capital requirements in the aggregate figures.

5.2 Statistical Measures of SRISK Risk

This section presents a small bootstrap experiment to calibrate the confidence intervals expected for SRISK due to estimation errors. We examine the estimate of beta and LRMES for JP Morgan (JPM). The estimate of beta is done using either the straight DCB, as in equation (8.13), or the nested model of equation (8.14). Using data from 2000 to 2017,

---

[5] https://vlab.stern.nyu.edu/welcome/srisk.

we can estimate GJR-GARCH volatility models for the SPDR S&P 500 ETF Trust (SPY) and JPM and a DCC model on the residuals. From these estimates, we calculate a series of "idiosyncratic residuals" given by

$$e_t^i = \left(e_t^{JPM} - \rho_t * e_t^{SPY}\right) \Big/ \sqrt{1 - \rho_t^2}. \tag{8.16}$$

In this equation, $\left(e_t^{SPY}, e_t^{JPM}\right)$ are the standardized residuals from the SPY and JPM volatility models, respectively, and ρ is the DCC correlation between them. These idiosyncratic residuals are uncorrelated with SPY residuals and have unit variance.

A new set of disturbances is obtained by resampling the rows of $\left(e_t^{SPY}, e_t^i\right)$, which will preserve any contemporaneous dependence, such as tail dependence. Treating the parameters of the GARCH and DCC models as true, a new bootstrap data set (SPY_b, JPM_b) is constructed, and for this data set, we have a full time series of true volatilities, correlations, and betas. When we estimate this model, we will get different parameters and correspondingly different volatilities, correlations, and betas. The biases and variances of beta and LRMES are very similar because these are monotonic functions of each other at any point in time. Because LRMES is the input to SRISK, results on this are presented. Two metrics are computed: one is the bias and root-mean-squared error (RMSE) over time for a single-bootstrap simulation, which are then averaged over 1,000 simulations. The second is simply the error in the last or out-of-sample estimate of LRMES in each simulation from which bias and RMS errors over the 1,000 bootstrap simulations are computed.

A few of the DCC simulations fail to converge or converge to parameters that are outside the reasonable parameter space. In this case, a simple exponential smoothed estimate of the correlations is computed using a smoothing parameter of .96.

Table 8.1 shows that (unsurprisingly) the bias and RMSEs are quite similar whether they are computed just for the last observation or averaged over all observations in a simulation and then averaged over simulations. The magnitude of LRMES is exactly .4 if beta is 1 and will be greater or smaller depending on the beta at the time. The bias is therefore quite small at approximately 0.25 percent, or 1 part out of 400. The RMSE is approximately 3.5 percent for the DCB and a little greater for the nested DCB. Thus, if the estimated LRMES is .4, a confidence interval would be approximately $(.40 \pm .028)$. This is the uncertainty due to the measurement of beta. Notice that the standard error is slightly greater for the nested DCB. This is not surprising for correctly specified models because it is not the fully efficient estimator. However, it should give some protection for misspecified models.

Table 8.1. Bootstrap distribution of LRMES

Measure	DCB	Nested DCB
Bias	.0009	−.0067
Bias_last	.0010	−.0068
RMSE	.0144	.0164
RMSE_last	.0148	.0170

Author's calculations.

5.3 *Differences across Models*

V-Lab currently has three computations of SRISK for the United States and two for many other countries. These differ in the method to use in forecasting the beta and the precise stress considered. Because these sometimes give different answers, we want to see how different they really are and whether these differences are important for our understanding. These models differ only in the value of LRMES that is substituted into equation (8.7). Here is a quick summary:

1. Global marginal expected shortfall (GMES) uses a world ETF on the ACWI to stress and takes account of nonsynchronous returns. The default stress is exactly 40 percent.
2. Domestic marginal expected shortfall (DMES; marginal expected shortfall [DMES MES] for the United States) uses a domestic equity index for the stress and the beta. DSRISK is computed in local currency and does not need synchronization. The default stress is exactly 40 percent, but this is not equivalently stressful for all countries.
3. MESSIM is only available for the United States. It estimates the mean LRMES by simulation, as described by Brownlees and Engle (2017), using S&P as the market index. The crisis event includes paths with stress $\geq$ 40 percent.

Comparing MES and MESSIM, we see several differences. Most importantly, MES assumes today's beta will not change over the next 6 months. MESSIM allows it to change in response to negative returns, which naturally increase beta and SRISK, and mean reversion, which dampens extreme values of SRISK, which will lower SRISK. Furthermore, because the stress includes any scenario with loss greater than or equal to 40 percent, SRISK will be higher. However, the distribution of stressed returns is negatively skewed because it cannot exceed 100 because prices must be nonnegative. Hence, the mean LRMES is below the median, and in fact, when volatility is very high, the mean LRMES can be negative, reflecting the fact that even with an adverse scenario, there may be paths with sufficiently positive outcomes that less capital is expected to be needed under stress than without the stress. This counterintuitive outcome is prevented in V-Lab by truncating such negative paths, which introduces a further upward bias in MESSIM. The net difference is not clear theoretically.

Comparing MES and GMES, there are basically two differences. In the former, the market is S&P, whereas in the latter, it is the global equity market as measured by the ACWI. As a result, the betas are different and in some countries, the differences are important. This is the motivation for the model of Engle et al. (2015), who jointly estimate such betas for European markets. For banks that serve primarily domestic clients, it is natural that the beta would be bigger on the domestic market than on the global market. In addition, the GMES model allows for a response to yesterday's market return as well as today's market return. This is unnecessary for the United States but is needed for countries in time zones different from the United States.

The magnitude of these differences is fortunately quite small. Figure 8.2 shows the three SRISK measures for Bank of America (BAC) and GS. From an examination of these two plots, it is clear that the three measures move closely together. It is also clear that GMES is the lowest and MESSIM is the highest.

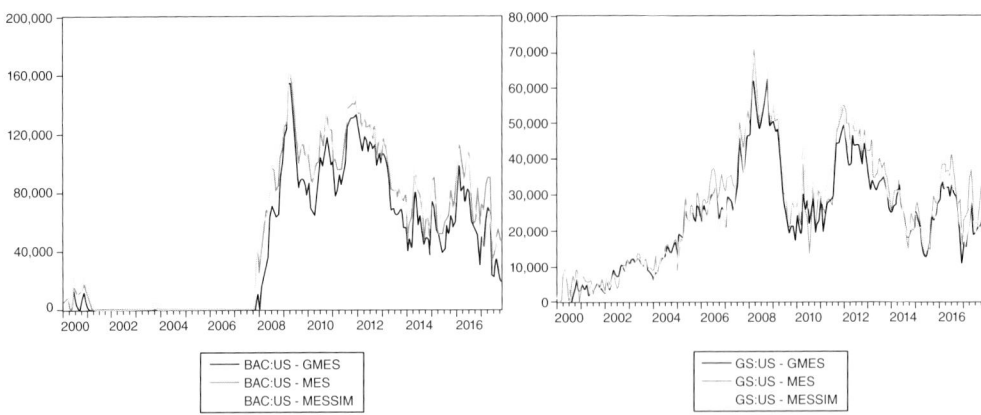

Figure 8.2 Three measures of SRISK for BAC and GS.
Source: V-Lab SRISK;[6]author's calculations.

For 10 large banks, these differences are systematic. Averaging over time and over banks, MES exceeds GMES by $3.1 billion, and MESSIM exceeds GMES by $6.9 billion. At the end of 2008, aggregate US SRISK reaches high levels in each of these measures. For MESSIM, it is $974 billion; for MES, it is $885 billion; and for GMES, it is $846 billion. The first two have the identical 82 firms, of which 47 and 46 are positive, respectively. In GMES, there are 155 firms, of which 68 are positive. The actual peaks occurred in the summer of 2008 before "Lehman weekend" for each of these measures, giving a prediction of things to come.

6 Comparison with SES, CoVar, and Regulatory Measures

Many alternative measures of systemic risk have been proposed and used to guide policy and regulation. This section compares SRISK with systemic expected shortfall (SES), covariance (CoVar), and regulatory measures. The SES measure is closely related to SRISK and was initially proposed by Acharya et al. (2009) and described in detail by Acharya et al. (2017). CoVaR was proposed by Adrian and Brunnermeier in 2008 and detailed by Adrian et al. (2016). Regulatory stress tests have been carried out by regulators in many countries with various mandates. This section compares the frameworks and the outcomes of these methods.

   Acharya et al. (2009) (hereafter APPR) develop a theoretical approach to systemic risk by postulating that the aggregate capital shortfall of the financial sector imposes a negative externality on the real economy. Whenever the capital shortfall exceeds some fraction of total assets, the externality becomes effective, and when it is large enough, there is a financial crisis. Thus, they must estimate the capital shortfall of the financial sector. The first step is to estimate the MES of a firm. This is defined as

$$\text{MES}_{j,t} = E\left(-r_t^j \,\middle|\, r_t^F < -c\right), \tag{8.17}$$

---

[6] https://vlab.stern.nyu.edu/welcome/srisk.

where r is the equity return on day t of firm j when the financial sector as a whole has return r_t^F that is more negative than a number c, which is set to the VaRisk. This is estimated in a simple and intuitive fashion. Suppose there are T days with the same joint distribution of returns. Then the 5 percent worst days for the sector as a whole will identify c, the VaR of the sector. On these days, the return of firm j is measured, and the average of these is an estimate of the MES. APPR (2009) choose T as a year and compute rolling estimates of MES for each of the firms in a financial sector.

When the joint distribution of returns is evolving, a bivariate volatility model can be a better approach:

$$\begin{pmatrix} r_t^j \\ r_t^F \end{pmatrix} | \mathrm{F}_{t-1} = D\left((0), H_t \right). \tag{8.18}$$

If the joint distribution D is normal or if the regression of the first variable on the second is linear, then we can write

$$r_t^j = \beta_t^j r_t^F + e_t, \quad \text{where } \beta_t^j = \frac{h_{12,t}}{h_{22,t}}, \quad \text{and}$$

$$\mathrm{MES}_t^j = \beta_t^j \mathrm{ES}_t^F, \quad \text{where } \mathrm{ES}_t^F = E_{t-1}\left(r_t^F \middle| r_t^F < -c \right), \quad \text{and } c = \mathrm{VaR}. \tag{8.19}$$

This is a natural alternative way to estimate MES. From a bivariate GARCH process of some kind, we can estimate the covariance and variance of system returns and also estimate the expected shortfall of the system, which is typically proportional to volatility. This would give an estimate of MES for each t.

A crisis may arise from shocks that are more extreme than the 5 percent point of the distribution, particularly when volatility is low. APPR (2009) propose to estimate the relation between the firm equity losses in a crisis and the MES. They define realized SES as the stock return of a company during the GFC, defined as July 2007 to December 2008. In a cross-sectional regression, this is regressed on precrisis measures of MES and leverage (LVG), as well as beta, expected shortfall (ES), volatility, log(assets), and industry dummies. The key variables are MES and LVG with a little evidence of size. The fitted value of this regression is their estimate of SES. This regression is repeated with longer lags of MES and LVG. The impact of lagged MES decays, but using the DCC, as in equation (8.19), is better because it is more persistent.

In this analysis, SES is an estimate of the same firm characteristic as LRMES described previously. However, it does not naturally have a stress other than the performance in the GFC. Interestingly, APPR (2009) argue that leverage is an input to the equity decline, whereas SRISK uses leverage to compute the capital shortfall but not to estimate LRMES. However, the leverage is implicitly incorporated in the beta estimated by the DCB because leverage changes slowly, and the estimate of beta will incorporate these changes.

As a validation exercise, APPR (2009) predict the outcome of the Supervisory Capital Assessment Program (SCAP) that was done in early 2009. This was an examination of the capital shortfall of 19 bank holding companies (BHCs) conducted by the Federal Reserve (Fed) has been widely praised as responsible for rebuilding confidence in the banking sector. SCAP estimated a capital shortfall of each BHC under an adverse stress scenario. APPR divide this capital shortfall by Tier 1 common capital to create a capital shortfall per dollar of market cap. They correlate this across firms, with MES and LVG measured from April 2008 to March 2009. MES has a 59 percent correlation, and LVG has a 31 percent correlation. In a regression, MES is significant, although LVG is not.

A related and very influential paper was written by Adrian and Shin (2010, 2014). They introduce a concept CoVaR that is a generalization of the familiar regulatory concept of VaR. The idea is simple and compelling. Let $r_{j,t}$ be the return on asset j, and let q be a such as 5 percent. VaR is defined implicitly by the following relation:

$$P\left(r_{j,t} < -\text{VaR}_{j,t}^q \mid M_{t-1}\right) = q, \tag{8.20}$$

where M is a set of state variables that are known the period before. Typically, M includes volatility forecasts such as GARCH or exponential smoothing or implied volatilities, but in principle, it could include other predictors.

CoVaR is defined for two measures of return, $(r_{j,t}, r_{m,t})$, where the first is the equity return on firm j, and the second is the return on a system, such as the entire equity market or the entire financial market. It satisfies the following equation:

$$P\left(r_{m,t} < -\text{CoVaR}_{j,t}^{p,q} \mid r_{j,t} = -\text{VaR}_{j,t}^q, M_{t-1}\right) = p. \tag{8.21}$$

This is a measure of the risk of a big decline in the market return when a particular firm has a big market decline that equals its VaR. It is the p quantile of the market return using the distribution conditional on the event that a particular firm return equals its VaR and the preceding state variables are M. This is a measure of how sensitive the overall market is to a decline in a particular financial company and is exactly the type of measure suggested in the definition of systemic risk. Adrian and Shin (2011, 2016) then define $\Delta\text{CoVaR}_{j,t}^q$ as

$$\Delta\text{CoVaR}_{j,t}^q = \text{CoVaR}_{j,t}^{q,q} - \text{CoVaR}_{j,t}^{q,0.5}. \tag{8.22}$$

This measures the increase in the risk of the market when a firm has extreme negative returns compared with when it has zero (median) returns. Finally, they define dollar delta CoVaR by multiplying by the firm market value:

$$\Delta\$\text{CoVaR}_{j,t}^q = W_{j,t}\Delta\text{CoVaR}_{j,t}^q. \tag{8.23}$$

This work not only defines the concept but introduces a novel estimation approach. They propose to estimate two equations:

$$r_{j,t} = \alpha_j + \beta_j M_{t-1} + \varepsilon_{j,t}, \tag{8.24}$$

$$r_{m,t} = \alpha_{m,j} + \beta_{m,j} M_{t-1} + \gamma_{m,j} r_{j,t} + \eta_{m,j,t}. \tag{8.25}$$

If these equations are estimated by least squares, the predictions will be the conditional mean of the dependent variable. But if they are estimated with quantile regression, as proposed by Basset and Koenker (1978) and applied by Engle and Simone (2004), among many others, they will predict the conditional quantile of the dependent variable. In this case, both equations are estimated for the q quantile.

Denoting the quantile estimates by a tilde, this yields

$$\text{VaR}^q_{j,t} = \tilde{\alpha}_j + \tilde{\beta}_j M_{t-1},$$

$$\text{CoVaR}^{q,q}_{j,t} = \tilde{\alpha}_{m,j} + \tilde{\beta}_{m,j} M_{t-1} + \tilde{\gamma}_{m,j} VaR^q_{j,t},$$

$$\Delta\text{CoVaR}^q_{j,t} = \tilde{\gamma}_{m,j} {}^* VaR^q_{j,t} = \tilde{\gamma}_{m,j} {}^* \left(\tilde{\alpha}_j + \tilde{\beta}_j M_{t-1}\right). \qquad (8.26)$$

Using this approach, they estimate time-series quantile regressions for 1,823 firms using weekly data from 1971 to 2013. The state variables include short-yield change, term-spread change, TED spread, credit-spread change, and market and real estate equity returns and equity volatility.

As pointed out by Acharya et al. (2012) and implemented by Adrian and Brunnermeier (2016), these measures can be estimated from a multivariate volatility model. If

$$\begin{pmatrix} r_{j,t} \\ r_{m,t} \end{pmatrix} |F_{t-1} \sim N\left((0), H_t\right) \qquad (8.27)$$

then, letting $\Phi(.)$ be the standard Gaussian cumulative distribution function,

$$\text{VaR}^q_{j,t} = \sqrt{h_{jj,t}}\Phi^{-1}(q),$$

$$\text{CoVaR}^{q,q}_{j,t} = \left(\sqrt{\left(1 - \rho^2_{j,m,t}\right) h_{m,t}} + \rho_{j,m,t}\sqrt{h_{m,t}}\right)\Phi^{-1}(q),$$

$$\Delta\text{CoVaR}^q_{j,t} = \rho_{j,m,t}\sqrt{h_{m,t}}\Phi^{-1}(q). \qquad (8.28)$$

Interestingly, the CoVaR and ΔCoVaR of a firm vary over time with the volatility of the market and the correlation between this firm and the market but not with the volatility of the firm. In the cross-section, only the correlation determines which firms are most systemic. Clearly, when this is multiplied by the firm market cap, as in equation (8.23), size also plays a role. Even without the assumption of bivariate normality, similar relations can be estimated with a nonparametrically estimated cumulative distribution function (CDF).

A key feature of the CoVaR definition is that it measures the effect of distress in one firm on the risk of the market as a whole. For this to be valid, it should hold other things constant—in particular, this would be the health of other firms. This is important because correlation between firms will lead to the inference that many firms are systemic even if only one is. It is also important because it ignores the externality that is the focus of many systemic-risk theories, including that by APPR (2009). The systemic risk of one firm will depend on the undercapitalization of other firms. The use of state variables to estimate CoVaR, as in equation (8.26), is a natural response. These variables are price variables rather than quantities as conventionally used to incorporate externalities, but these can potentially assess the health of the system. Although these play the role in CoVaR, they only enter ΔCoVaR through the VaR of a single firm. Thus, unless the externalities are adequately measured by the firm-specific VaR, they will be missed in this approach.

7 Comparison with Regulatory Assessments

Although the SRISK measure is grounded in economic theory and measured with reliable econometric methods, the outcomes that it predicts must be checked against reality and against the predictions by regulators, who generally have superior information (although they typically preclude themselves from using market data, so their data are not strictly a superset of public information).

Brownlees and Engle (2017) examine the capital injections carried out by the Fed during the crisis. They expect that the greater the capital shortfall, the greater the capital that the Fed would supply to a firm. They address this question using the Bloomberg Loan Crisis Data database, a data set containing details of such operations analyzed in a number of studies (e.g., Bayazitova and Shivdasani, 2012). The regression results show that SRISK is a significant predictor of the capital injections. The finding is robust to the inclusion of a number of controls, including firm size and alternative capital-shortfall indices.

Brownlees and Engle (2017) also investigate whether aggregate SRISK provides early warning signals of worsening macroeconomic conditions. Specifically, they use predictive regressions of the future growth rates of industrial production and the unemployment rate on the growth rate of aggregate SRISK (analogously to Allen et al. [2012]). The forecasting horizon of the regressions varies from 1 month to 12 months. The results show that an increase in SRISK predicts future declines in industrial production and increases in the unemployment rate and that the predictive ability of aggregate SRISK is stronger at longer horizons. The prediction results are robust to the inclusion of a large set of alternative control predictors that includes systematic risk (measured as the volatility of the market), the SRISK measure computed for nonfinancial firms, an aggregate capital-shortfall index computed from a structural Merton-type default-risk model, the default spread, the term spread, and an index measuring the degree of activity of the housing market.

Acharya et al. (2014) carry out a comparison between the capital-shortfall estimates of systemic-risk institutions provided by SRISK and regulatory stress tests (based on supervisory data). Their analysis compares the minimum Tier 1 leverage ratio with the V-Lab stressed total leverage ratio at the same time. They find the rank correlation for the 2012 Comprehensive Capital Adequacy Report (CCAR) is .846, and for the 2013 CCAR, it is .877. For the 2011 European Banking Association EBA) stress test, the rank correlation is only .570. The rank correlations between firm risk as measured by the ratio of risk-weighted assets to total assets, often called *risk density*, and by V-Lab risk show much lower correlations, and often these are negative. Firms with low risk-weighted capital may have the greatest risk. This is consistent with the observation that financial crises appear to arise from assets that have zero regulatory risk weights. Finally, because the 2011 EBA stress test was conducted just before the European sovereign debt crisis, it provides a good laboratory for comparing capital assessments. The correlation between stressed leverage ratios and subsequent stock returns was .354 for V-Lab and .046 for EBA. Overall, Acharya et al. show that regulatory capital shortfalls measured relative to total assets provide similar rankings to SRISK for US stress tests. On the contrary, rankings are substantially different when the regulatory capital shortfalls are measured relative to risk-weighted assets. Greater differences are observed in the European stress tests.

Brownlees et al. (2017) extend the analysis of US financial crises back to the mid 19th century. Using this novel and unique data set, which covers eight historical financial crises, they find that CoVaR and SRISK contain information that would allow regulators to identify systemically important financial institutions (SIFIs) in cross-sectional regressions. When the deposit flows were disproportionately withdrawn from banks that had high ex ante CoVaR or SRISK rankings, financial panics were likely to occur. In many of their analyses, SRISK appears to have a slight advantage over CoVaR. Nevertheless, CoVaR and SRISK provide fairly similar rankings of the most systemic institutions, and their rankings are correlated with rankings based on size or beta.

To compare SRISK measures across countries, we examine the Basel list of systemically important banks. In November 2017, the Financial Stability Board (FSB) released its most recent list of globally systemic important banks (GSIBs). This is a list of 30 global banks selected by a committee over the previous year blending five general criteria: size, interconnectedness, uniqueness, complexity, and cross-jurisdictional activity. Using data from December 2016, which would be input data used by the FSB for its report 9 months later, we can sort the global banks to find the 30 with the highest SRISK. The FSB and V-Lab lists are quite similar. There are 22 names in common. The eight that the FSB omits but are included via SRISK are Banco do Brasil, Bank of Communications, China CITIC, Industrial Bank Co., Bank of Nova Scotia, Bank of Montreal, Canadian Imperial Bank of Commerce, and Lloyds Banking Group Ltd. That is, three more Chinese banks, three Canadian Banks, one British, and one Brazilian should be added in place of four US, one Canadian, one Swedish, one Swiss, and one Dutch bank. Most of the FSB banks are either in or close to the top 30 on the SRISK measure, except for JPM and Wells Fargo (WF). These giant banks actually are measured to have a capital surplus as of December 2016, and because they are so large, this is a big surplus in dollars. This is a consequence of the leverage of these banks, which has fallen to 7 for WF and 8 for JPM. Because SRISK requires that banks have a leverage ratio below 12.5 under stress, these banks easily satisfy this measure of capital adequacy. It appears that the FSB has not effectively updated its data for the improved capital ratios of the US banks. Or it may be that the surcharge required by the FSB is sufficient to make these banks overcapitalized according to SRISK. In either case, this looks like a good outcome from the point of view of financial stability.

The list of 30 GSIBs for 2016 was announced in November 2016. We can match this with the SRISK list for Decemebr 31, 2015. In this year, the lists agree for 24 of these names. SRISK lists three more Chinese banks (Bank of Communication, China CITIC, and Industrial Bank), as well as one Brazilian, one German, and one Japanese bank. The list of 30 GSIBs for 2015 was announced in Novemeber 2015. Matching this with the SRISK list for Decemeber 31, 2015, again 24 names, including the first 20, are common. The FSB omits two Chinese banks, two Japanese banks, one Brazilian bank, and one German bank that SRISK identifies.

Overall, the evidence is clear that SRISK provides measures similar to regulatory systemic risk assessments within the United States, over a broad historical time frame and across countries. Some of the differences appear to be due to political objectives.

8 Conclusions

This survey has discussed the advantages of market-based stress tests relative to traditional regulatory stress tests. These are inexpensive and noninvasive and can be updated frequently, but they rely on the market to value complex firms and counterfactual outcomes. Thus, such stress tests can be a useful complement to conventional stress tests. When the measures agree, there is increased confidence in the results. When they disagree, there is an opportunity to dig deeper into the source of the disagreement and perhaps improve models in ways that resolve the differences.

A large and ongoing body of research on a particular implementation of a market-based stress test, SRISK, has been surveyed. The paper discusses details of the construction and uncertainties in the measure. It compares the outcomes of SRISK with regulatory systemic risk assessments and finds that the methods are complementary and typically similar. If deregulation again takes hold in the financial sector, measures like SRISK may have a unique and enduring role.

References

Acharya, V., R. Engle, and D. Pierret (2014), "Testing macroprudential stress tests: The risk of regulatory risk weights," *Journal of Monetary Economics*, 65, 36–53.

Acharya, Viral, Lasse Heje Pedersen, Thomas Philippon, and Matt Richardson (2009), "Regulating systemic risk," in Viral Acharya and Matt Richardson (eds.), *Restoring financial stability: How to repair a failed system*, Wiley.

Acharya, Viral, Lasse Heje Pedersen, Thomas Philippon, and Matt Richardson (2017), "Measuring systemic risk," *The Review of Financial Studies*, 30(8.1), 2–47.

Acharya, Viral V., Robert F. Engle, and Matthew Richardson (2012), "Capital shortfall: A new approach to ranking and regulating systemic risks," *American Economic Review: Papers and Proceedings*, 102, 59–64.

Adrian, Tobias, Nina Boyarchenko, and Domenico Giannone (2016), "Vulnerable growth," Federal Reserve Bank of New York Staff Report 794794.

Adrian, Tobias, and Markus Brunnermeier (2016), "CoVaR," American Economic Review, 106(7), 1705–1741.

Adrian, Tobias, and Hyun Song Shin (2010), "Liquidity and leverage," *Journal of Financial Intermediation*, 19, 418–437.

Adrian, Tobias, and Hyun Song Shin (2014), "Procyclical leverage and value-at-risk," *Review of Financial Studies*, 27, 373–403.

Allen, L., T. G. Bali, and Y. Tang (2012), "Does systemic risk in the financial sector predict future economic downturns?" *Review of Financial Studies*, 25, 3000–3036.

Bassett, G. Jr., and R. Koenker (1978), "Asymptotic theory of least absolute error regression," *Journal of the American Statistical Association*, 73(363), 618–622.

Bayazitova, D., and A. Shivdasani (2012), "Assessing TARP," *Review of Financial Studies*, 25, 377–407.

Bernanke, Ben, and Mark Gertler (1989), "Agency costs, new worth, and business fluctuations," *American Economic Review*, 79(1), 14–31.

Bernanke, Ben, Mark Gertler, and Simon Gilchrist (1999), "The financial accelerator in a quantitative business cycle framework," in J. Taylor and M. Woodford (eds.), *Handbook of macroeconomics*, Elsevier Science.

Borio, Claudio (2014), "The financial cycle and macroeconomics: What have we learnt?" *Journal of Banking and Finance*, 45, 182–198.

Brownlees, Christian, Ben Chabot, Eric Ghysels, and Christopher Kurz (2020), "Back to the future: Backtesting systemic risk measures during historical bank runs and the great depression," *Journal of Banking and Finance*, 113, 105736

Brownlees, Christian T., and Robert F. Engle (2011)," Volatility, correlation and tails for systemic risk measurement," Working Paper, available at https://faculty.washington.edu/ezivot/econ589/VolatilityBrownlees.pdf.

Brownlees, Christian T., and Robert F. Engle, (2017), "SRISK: A conditional capital shortfall index for systemic risk measurement," *Review of Financial Studies*, 30(1), 48–79.

Brunnermeier, Markus, and Lasse H. Pedersen (2009), "Market liquidity and funding liquidity," *Review of Financial Studies*, 22, 2201–2238.

Cont, Rama, and Eric Schaanning (2017), "Fire sales, indirect contagion and systemic stress-testing," Norges Bank Working Paper, available at https://papers.ssrn.com/sol3/papers.cfm?abstract_id=2541114.

Drehmann, Mathias, Claudio Borio, and Kostas Tsatsaronis (2012), "Characterising the financial cycle: Don't lose sight of the medium term!" BIS Working Paper 380.

Engle, R. (2018) "Systemic risk 10 years later," *Annual Review of Financial Economics*, 10, 13.1–13.28.

Engle, Robert, Eric Jondeau. and Michael Rockinger (2015), "Systemic risk in Europe," *Review of Finance*, 19(1), 145–190.

Engle, Robert and Matthew Richardson,(2015), "Systemic risk and the prospect for global financial stability," *Banking Perspective, the Quarterly Journal of the Clearing House*, 3(2).

Engle, Robert F., (2016), "Dynamic conditional beta," *Journal of Financial Econometrics*, 14, 643–667.

Engle, Robert F., and Simone Manganelli (2004), "CAViaR: Conditional autoregressive value at risk by regression quantiles," *Journal of Business and Economic Statistics*, 22(4), 367–381.

Engle, Robert F., and Tianyue Ruan (2019), "Measuring the probability of a financial crisis," *Proceedings of the National Academy of Science*, 116(37), 18341–18346.

Greenwood, Robin, Augustin Landier, and David Thesmar (2015), "Vulnerable banks," *Journal of Financial Economics*, 115, 471–485.

Jordà, O., B. Richter, M. Schularick, and A. M. Taylor (2017), "Bank capital redux: Solvency, liquidity, and crisis," NBER Working Paper 23287.

Kiyotaki, N., and J.H. Moore (1997), "Credit cycles," *Journal of Political Economy*, 105, 211–248.

Mian, Atif, and Amir Sufi (2009), "The consequences of mortgage credit expansion: Evidence from the U.S. mortgage default crisis," *Quarterly Journal of Economics*, 24(1), 1449–1496.

Pedersen, Lasse Heje (2009), "When everyone runs for the exit," *The International Journal of Central Banking*, 5, 177–199.

Reinhart, Carmen, and Kenneth Rogoff (2009), *This time is different: Eight centuries of financial folly*, Princeton University Press.

Schularick, Moritz, and Alan M. Taylor (2012), "Credit booms gone bust: Monetary policy, leverage cycles and financial crises," *American Economic Review*, 102, 1029–1061.

9

On Market-Based Approaches to the Valuation of Capital

Natasha Sarin and Lawrence H. Summers

1 Introduction

Since the Great Recession, large banks have been subject to annual stress testing, a reform heralded by bank regulators as "perhaps the most successful supervisory innovation of the post-crisis era" (Powell, 2019). These annual exercises build on the success of the original crisis stress testing, which is given much of the credit for the successful recapitalization of the financial system in 2009.

The idea behind the crisis stress tests, or Supervisory Capital Adequacy Program (SCAP), was to provide the market credible information about losses banks had already suffered and those that they would likely continue to accumulate as the crisis raged on. Policymakers believed that resolution of uncertainty with respect to bank stability would help these institutions successfully raise capital from private markets. The crisis exercise was a resounding success: large banks addressed their deficiencies by raising capital in private markets over just a few months after the results were made public. As such, the crisis stress tests laid the groundwork for annual "normal-times" stress testing. Rather than wait for a downturn to take hold to measure banks' capital adequacy, regulators now subject banks to hypothetical stresses each year and force them to address capital weaknesses proactively, in the hopes that they will be well equipped to weather future downturns that arise.

For the last several years, banks' passage of the stress tests was celebrated as proof that large financial institutions are so well capitalized that they will be able to continue to intermediate as normal when the next crisis hits. We have previously questioned this conclusion and the regulatory complacency it begets (Sarin and Summers, 2017).

In this chapter, we build on our prior work to highlight four points. First, stress tests are dependent on regulatory measures of bank capital, which appear not to be a good proxy for economic capital measures that more accurately indicate the risk of insolvency. Second, stress tests have almost come to be seen as a panacea for resolving crises based on the successes of SCAP. We point out that although crisis stress tests were clearly valuable, it is hard to disentangle the role they played in stabilizing the financial sector from other factors, such as extraordinary fiscal interventions and good fortune. Third, banks perform far worse on a naive market-based stress test than they do on the annual regulatory exercise, which provides suggestive evidence that the regulatory tests are not painting a full picture of financial stability. Finally, we provide some early thoughts on the COVID-19 experience. The disparity between the optimistic view of bank health in recent stress tests and the reality today provides additional evidence that the current exercise is lacking.

2 Limitations of Regulatory Capital

Stress testing is meant to determine, on an annual basis, how much capital banks need to weather hypothetical downturns so that they will be in strong enough capital positions to continue to intermediate as normal when a real crisis hits. The methodology relies on regulatory capital ratios: like all financial regulatory frameworks, the stress tests place no attention on the price the market assigns to bank liabilities and equity in assessing the safety of financial firms. Instead, the premise of stress testing is that book capital levels are a good proxy for economic capital that protects banks from insolvency.

But regulatory capital is not quickly responsive to changes in banks' health. This is because bank accounting standards are largely based on historical cost-based accounting. Full fair-value accounting would require banks to measure and report assets and liabilities at the market price. Historical-cost accounting means assets and liabilities are reported at historical cost, and unrealized gains and losses due to subsequent changes in asset value are ignored until these are realized. In the aftermath of the savings and loan crisis, there was a shift toward more fair-value accounting in the financial industry (Furlong and Kwan, 2006; Li, 2017). The current approach is a mix of both accounting models, depending on the category of financial asset/liability in question: available-for-sale securities and loans, as well as trading securities, are to be reported at fair values on the bank balance sheet, whereas securities and loans that are held for investment are valued at historical costs. But the system is still weighted toward historical cost accounting, with only around 35 percent of large commercial banks' assets reported at close to fair value (Laux and Leuz, 2010).

To understand the differences between fair-value and historical-cost approaches, consider loans that are held for investment purposes and thus, at baseline, valued based on historical costs. Such loans are subject to impairment write-downs if it is probable that a creditor will be unable to collect all that is due, but in practice, these impairments are implemented with "large discretion, and oftentimes the impairment loss is recognized too late to alert investors or regulators about actual changes of bank performance" (Li, 2017). Given this discretion, banks that focus on traditional lending can essentially avoid fair-value accounting, and thus there is no tool by which to recognize when recapitalization is needed because capital ratios do not reflect changes in expected loan performance. The financial crisis provides clear evidence of this discretion and its consequences: past work compares loan losses implied by bank reporting with external estimates from both regulators and industry analyst reports. The lowest external estimate for each of the four largest US bank holding companies (BHCs) exceeded bank loss estimates by over 45 percent (Wells Fargo) and up to 76 percent (Bank of America). It is thus unsurprising that of the 31 BHCs that failed in the United States between 2007 and 2009, loans accounted for roughly three-quarters of their balance sheets, and mark-to-market trading securities played essentially no role (Laux and Leuz, 2010).

Figure 9.1 illustrates the divergence between regulatory capital measures and market-based capital measures for the largest financial institutions over the last two decades. Market measures of bank capital have been trending downward since the turn of the century, whereas book capital has moved in the opposite direction. In fact, although book capital ratios have risen dramatically since the Global Financial Crisis (GFC), as measured on a market basis, many large banks today are less well capitalized than they were in the precrisis period. It is thus concerning that regulators view the financial system as 7 to 10 times better capitalized

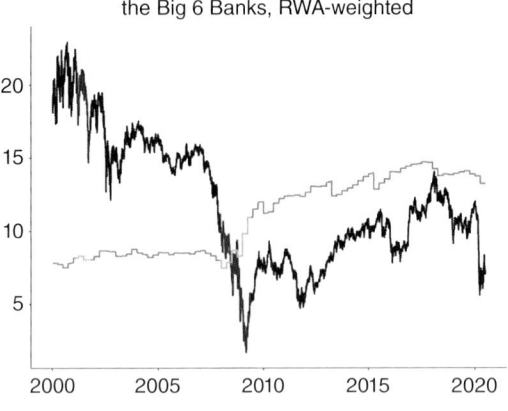

Figure 9.1 Comparing regulatory and market-based capital ratios. A risk-weighted-asset (RWA) weighted average of each measure is reported for the six largest BHCs in the United States. For the market-based capital metric, quarterly asset data are interpolated exponentially to daily. Source: Data from Bloomberg and FR Y-9Cs.

than prior to the crisis period (Carney, 2014) and have concluded that the era of capital building is in fact complete (Quarles, 2019) because banks are now safe enough to survive a downturn without substantial government intervention.

As we have previously highlighted (Sarin and Summers, 2017), the confidence—and thus complacency—of regulators with respect to financial-sector health is at odds with what market-risk measures suggest. Based on these measures, such as beta, historical volatility, implied volatility, and credit default swap (CDS) spreads, banks are riskier, not safer, today than in the precrisis period (Table 9.1). It is especially puzzling that bank stress tests entirely eschew market information about financial stability because the regulatory capital ratios that they rely on are known to be a static and easily arbitraged measure of a bank's true capital position. There are various examples of their deficiencies: Lehman was well capitalized right before its bankruptcy—despite market indicators revealing distress, its Tier 1 capital ratio was 11.6 percent, higher than the average of the other large banks at the time (averaging 8.4 percent). Figure 9.2 provides another vivid example, plotting the ratio of book equity levels to total assets between Q1 2006 and Q3 2008 for the five largest BHCs in the United States (Bank of America, JP Morgan Chase, Citibank, Wachovia, and Wells Fargo). In Q3 2008, Wells Fargo's book equity ratio was nearly 2 percentage points *lower* than Wachovia's; measured based on accrual accounting data, it had less capital than Wachovia did in September 2008. And yet a few weeks later, Wells Fargo announced plans to purchase Wachovia, whose early crisis loan losses has threatened its continued survival—a fact that the market well book equity ratio understood but these regulatory capital measures missed even up to the quarter preceding the firm's acquisition. More recently, Deutsche Bank had a Tier book equity ratio 1 capital ratio of over 11 percent in February 2016 when its share price dropped by nearly 10 percent in a single day. At the time, CEO John Cryan pointed to the firm's "strong capital and risk position" in attempts to assure bank employees and investors that it was "rock solid" despite market warnings of its instability (Cryan, 2016).

Table 9.1. *Comparing bank market measures before and after the Global Financial Crisis*

	Precrisis	Postcrisis	2019	Q1 2020	2020 Trough	June 15, 2020
Volatility	24.7	29.8	24.5	38.2	55.2	54.9
Ratio of bank volatility to market volatility	1.6	2.0	1.5	1.6	1.9	1.6
Implied volatility	23.2	27.9	24.2	53.3	180.8	65.5
Ratio of implied bank volatility to implied market volatility	1.6	1.7	1.6	1.6	3.1	1.9
Beta	1.2	1.5	1.1	1.2	1.4	1.4
CDS spread	30.7	102.5	56.2	73.5	187.8	71.5
Ratio of bank price-to-earnings (PE) ratio to market PE ratio	0.6	1.0	0.5	0.4	0.4	0.4
Preferred stock price	25.2	21.1	21.5	22.0	16.8	21.7
Book capital ratio (Tier 1)	8.3	14.0	14.7	14.0	13.8	13.8
Market capital ratio (MVE/A)	13.7	9.0	9.9	7.9	5.2	7.2

We use the three floating-rate preferred stock series issues before the crisis (BAC-E, GS-D, and MS-A). All measures are constructed as a simple average over the six largest US BHCs. Source: Data from Bloomberg and FR Y-9Cs.

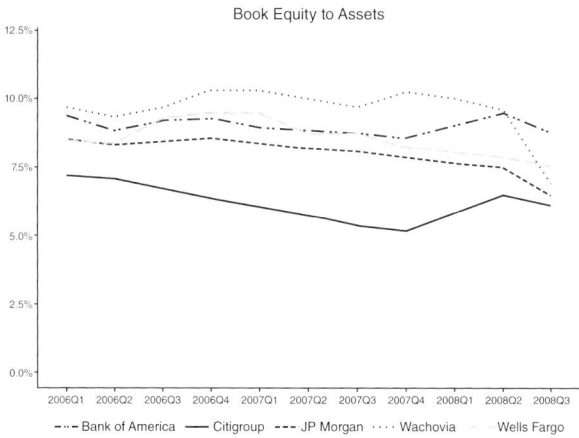

Figure 9.2 Regulatory capital ratios for largest BHCs precrisis. Source: Data from Compustat.

Unlike lagging and static regulatory capital ratios, market capital ratios paint a dynamic picture of the health of large financial institutions. Figure 9.3 compares the book and market capital ratios of the six largest financial institutions during the crisis. Although the market determination of banks' capital positions began to sour in 2007—contemporaneously with the onset of the downturn—regulatory capital ratios were largely unchanged. It is thus unsurprising that related work by Haldane (2011) finds that the banks that failed during the financial crisis had Tier 1 capital ratios that were indistinguishable from those of banks

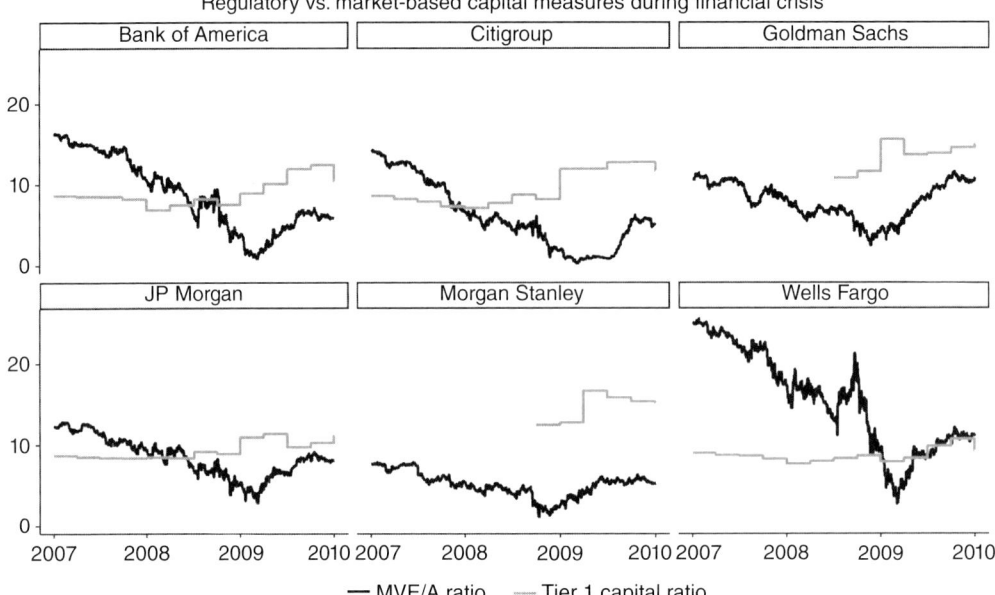

Figure 9.3 Regulatory capital ratios for largest BHCs precrisis. Tier 1 capital ratios for Goldman Sachs and Morgan Stanley unavailable until the second half of 2008, when the banks became BHCs. Source: Data from Compustat.

that successfully weathered the crisis, whereas market capital ratios were an accurate predictor of future distress. This suggests that measuring banks' equity positions based on market information can reveal important vulnerabilities about financial institutions that regulatory measures, for whatever reason—gaming incentives, difficulty in measurement, lack of dynamism—tend to miss.

Still, bank regulators tend to be reluctant to make too much out of market-based measures of bank health. One reason is that market indicators can be driven by noise as well as news, so there is a potential for false positives, which occur when market measures suggest cause for concern and yet banks are not impaired. Further, market measures of bank health are procyclical. This is why a regulatory regime focused on market measures alone is inadvisable.

3 Success of Crisis Stress Tests

In addition to the analytical arguments in favor of forward-looking regulation, one reason why stress tests have become so attractive is that they worked extraordinarily well in the aftermath of the GFC. When the Obama administration took office in early 2009, many industry observers and academics advocated nationalization of the largest financial institutions as the only way forward (e.g., Krugman, 2009). Instead of "preemptive nationalization," a decision was made to first determine how undercapitalized individual financial institutions were through a rigorous assessment of their losses. The hope was that rather than injecting hundreds of billions of dollars into the banking system without a full understanding of its exposures, the stress tests would provide enough credible information to the market so that capital deficiencies could be at least partially plugged by private markets, minimizing the need for an injection of additional taxpayer funds.

The stress tests were a resounding success. In retrospect, the decision to publicize bank-specific results, which was met with some resistance by inherently cautious Federal Reserve officials, made the exercise credible for the market. In fact, a contemporaneous report from Bridgewater announced "We Agree!" and confirmed that far from excessive optimism, the regulatory stress test estimates matched the firm's internal assessment almost identically. The tests found that the 19 firms subject to the stress tests needed to raise only $75 billion in new capital (a far cry from the hundreds of billions Krugman estimated would be necessary to nationalize Bank of America and Citigroup alone), and all but a few billion dollars was raised in private markets in the month following the announcement of the stress test results.

Then Federal Reserve chairman Ben Bernanke correctly identified the stress tests as one of the "critical turning points" in the financial crisis, restoring confidence in the system by providing investors credible information about prospective losses (Bernanke, 2013). The immediate market reaction makes this much clear: the two financial institutions closest to the brink, Bank of America and Citigroup, saw their stock prices rise by 63 and 35 percent, respectively, in the week following the announcement of the results, and the price of CDSs for the Big 6 banks dropped by a third (Geithner, 2014). The decision to be transparent about prospective losses helped to make the market "investable again," and the decision not to prematurely nationalize banks ended up saving taxpayers substantial sums.

Inevitably, the success of the crisis stress tests has led some to view stress testing as a panacea for preventing and resolving crises. But was the dramatic economic turnaround in 2009 ascribable to the performance of stress tests alone? It seems unlikely to be the case. Rather, there are at least two additional explanations: first, rather than a response to the performance of the stress tests, investors reacted positively to what came along with the tests: a government guarantee to secure the financial system. Second, regulators, in a sense, got lucky because the market was much more pessimistic than warranted, in part because the extraordinary fiscal stimulus provided much-needed support to the economy. We discuss these factors in turn.

3.1 Stress Tests Successful Because Provided Government Backstop

First, the original SCAP was performed in the context of an intense public debate on the inevitability of the nationalization of large public financial institutions. The prospect of direct government involvement was inextricably tied to the stress testing exercise: former Treasury secretary Tim Geithner presented the possibility of government intervention explicitly in *Stress Test*: "if a bank did turn out to be insolvent, we didn't intend to follow the Japan model of letting it limp along for years, too weak to lend, dragging down the economy." Part of the reaction of investors to the stress tests could be interpreted as a sign that banks were being substantially guaranteed, and the stress tests were a signal of the strength of public commitment to backstop bank liabilities and a procedure by which to quantify the support that was needed.

The market provides evidence that the provision of a government guarantee, rather than the credible information the stress tests provided, explains the recovery that followed. The stress tests were first introduced by Secretary Geithner on February 10, 2009, at a time when the plan for the tests and their consequences was not fully fleshed out. The lack of clarity was not well received by investors; indeed, the market began to turn downward in the midst

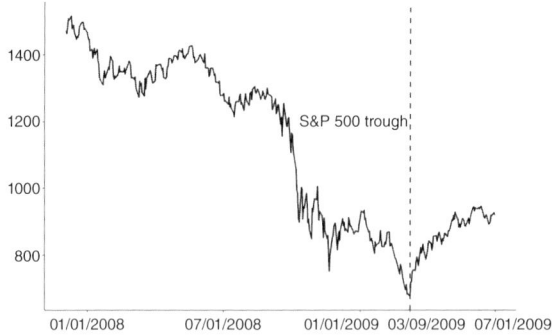

Figure 9.4 Standard & Poor's (S&P) 500 index throughout the GFC.
Source: Data from Bloomberg.

of the announcement, dropping by nearly 5 percent that day, with financials falling by more than twice that (Geithner, 2014).

A few weeks later, on February 25, 2009, the Treasury and Federal Reserve released the terms of the program: following the stress tests' assessments of banks' capital needs, they would first have the opportunity to turn to private sources of capital, and if this proved unavailable "in light of the current challenging market environment, the Treasury [would make] government capital available immediately." Financials reacted positively to this assertion that the government stood as a ready backstop of troubled financials: large bank stocks on average experienced a 12.1 percent abnormal return, and their CDS spreads (the cost of insurance against default) fell by 17 basis points (bps) on average (42 bps for the largest Big 6 banks). This is likely one reason why the market began to recover in March 2009 (Figure 9.4), after the stress tests were announced but well before the results were released in May.

In fact, as Table 9.2 makes clear, bank performance in the aftermath of the announcement of stress test results was less dramatic than in February when the terms of the capital provision were announced. Here, we follow the approach of Flannery et al. (2016) and trace out abnormal returns and absolute abnormal returns on 3 relevant days for the stress testing exercises: the announcement of SCAP by Secretary Geithner (February 10, 2009), clarity on the terms of the tests and government backstop provided by the Federal Reserve/Treasury a few weeks later (February 5, 2009), and the announcement of the stress test results (May 7, 2009). The early announcement of the tests was a negative event for bank financials, but the clarity provided just a few weeks later was a more significant event for banks (12.1 percent abnormal return for stress-tested banks) than the results (3.9 percent abnormal return).[1]

---

[1] Flannery et al. (2016) focus on abnormal returns following the announcement of stress test results, we extend this approach to the earlier days. They present both abnormal returns and absolute abnormal returns, to tease out the differences between the magnitude of and sign of equity movements in response to stress test news. Absolute abnormal returns are larger the more information (both positive and negative) is revealed on announcement days. What is interesting is that the absolute abnormal returns associated with the announcement of stress test results (11.3 percent) are much larger than the abnormal returns (3.9 percent), suggesting that this day featured both positive and negative news about bank financials.

Table 9.2. *Abnormal equity market returns on key SCAP news days*

	SCAP banks	Non-SCAP banks	Difference
	Cumulative abnormal returns		
Announcement[a]	−8.7%	−15.8%	7.1%
	(14.8%)	(12.3%)	(4.1%)
			9.1%
Clarity	12.1%	9.5%	2.7%
	(16.0%)	(13.3%)	(4.4%)
			55.2%
Results	3.9%	7.1%	−3.2%
	(14.0%)	(10.2%)	(3.8%)
			40.4%
	Absolute cumulative abnormal returns		
Announcement	12.6%	16.8%	−4.2%
	(11.4%)	(10.7%)	(3.2%)
			19.9%
Clarity	14.2%	11.7%	2.5%
	(14.1%)	(11.4%)	(3.9%)
			51.4%
Results	11.3%	10.1%	1.3%
	(8.8%)	(7.2%)	(2.4%)
			60.7%

Reported values are simple averages across the BHCs in our sample, along with standard errors and *p*-values on the *t*-test for the difference in means.
Source: Data are from the Center for Research in Security Prices and FR Y-9Cs.[b]

[a] Event dates are February 10, 2009, February 25, 2009, and May 8, 2009, for the periods marked announcement, clarity, and results, respectively.

[b] Specifically, we use the Q4 2008 and 2009 FR Y-9Cs to select BHCs with over $10 billion in assets. We match those BHCs using the Federal Reserve's RSSD-CRSP linking database. This generates a sample consisting of 17 BHCs subject to the 2009 stress test and 45 BHCs not subject to the 2009 stress test that have both RSSDs and PERMCOs. For each event date, we follow Flannery et al. (2016) in calculating a Fama–French three-factor model on the interval $[t - 130, t - 10]$ and reporting cumulated abnormal returns over the 3-day window $[t - 1, t + 1]$. The particular regression specification is $\log(p_{i,t}/p_{i,t-1}) = \alpha_i + \boldsymbol{\beta}_i^T \cdot (Rf_t, \text{HML}_t, \text{SMB}_t) + \varepsilon_{i,t}$, where abnormal returns are identified as $\varepsilon_{i,t}$ and absolute abnormal returns as $|\varepsilon_{i,t}|$.

It is also worth noting that the positive market response to the stress tests results being announced in May is statistically indistinguishable from the positive performance of non-stressed banks in their aftermath (Flannery et al., 2016). This is true more generally of the announcement days we consider (Table 9.2). This is, of course, at least partially explained by the interconnectedness of the financial system, but it also suggests that the dramatic success of stress testing is at least somewhat attributable to its role as a stabilizer of the financial sector, rather than the informational content of the results themselves.

Table 9.3. Comparing financial and nonfinancial equity index returns (March 2, 2009–October 1, 2009)

	S&P 500	Financials	Banks	Ex-Financials	Industrials
March 2, 2009 level	700.8	94.2	62.8	823.5	138.5
October 1, 2009 level	1,029.8	192.4	122.6	1,161.6	225.9
Return	46.9	104.1	95.1	41.1	63.1
Volatility (7 month)	21.4	52.9	68.4	18.6	27.0
σ-move	2.2	2.0	1.4	2.2	2.3
p-value (%)	1.4	2.4	8.2	1.4	1.0

Volatility is calculated using daily returns over the sample period. S&P Financials, Industrials, and Banks are indices consisting of the subset of S&P 500 companies classified by the Global Industry Classification Standard (GICS) as being in those respective industries, weighted by market capitalization; S&P Ex-Financials is an index consisting of all companies listed in the S&P 500, excluding firms classified by GICS as financial firms.
Source: Data from Bloomberg.

3.2 Stress Tests Successful Because Regulators Happened to Call the Bottom

It is also possible that in 2009, the market was excessively pessimistic on the state of the financial sector, and thus the stress tests genuinely brought out good news. If instead the stress tests had revealed capital shortfalls much greater than observers predicted—rather than much less severe—the exercise may have backfired. If many large banks were indeed very close to insolvency, making this information public at the bank level may well have realized some Federal Reserve officials' worst fears by panicking investors rather than reassuring them.

But changes in the broader market suggested that the economy was on the road to recovery. In fact, between March and October 2009, the S&P 500 as a whole increased in value by 47 percent, representing a 2.2-sigma move. Although bank financials increased by more (the S&P bank index increased by 95 percent), this represented a less extreme change (1.4-sigma shift). And non-financial firms—unaffected by bank stress tests—moved more significantly as well, increasing by 41 percent, which represented a 2.2-sigma move. At least one contributor to the recovery may have been successful fiscal measures independent of the stress tests; for example, the passage of the American Recovery and Reinvestment Act (Recovery Act) in February of 2009 successfully boosted the economy by providing direct checks to households and expanding unemployment benefits, tax credits, and funds for job preservation and job creation.

Given the many contributors to the success of the crisis stress tests, it seems unlikely that the simple act of stress testing in future crises will deliver such significant benefits. SCAP is perhaps best understood as regulators catching a falling knife—and if it were possible to time such remarkably good breaks in the market, investors would have a far easier time. One interpretation of the stress tests is that they rescued the financial system through a credible assessment of its weaknesses *and* the provision of a broad-based government guarantee. To the extent this is what took place, such a path forward will not be available in the future, either because there is already an expectation that government guarantees exist or because of a belief that the whole edifice of postcrisis stress testing is in service of not having to provide such guarantees.

These reflections are not meant to detract from the policy accomplishment of the successful resolution of the crisis, of which SCAP played a very significant part. But it is important to note that the original crisis stress tests are not, in fact, very probative on the role that stress tests will play as a preventative tool going forward. SCAP is also revelatory on the distinction between crisis and normal-times stress testing. The original tests essentially stressed banks against an existing downturn and were a mark-to-market accounting exercise aimed at quantifying the depletion of bank capital during the GFC. But had the tests been conducted just a few months prior, they would almost certainly have led to a misleading verdict. As evidence, consider that the capital positions of financial institutions were largely unchanged and did not reflect cause for concern even weeks prior to bankruptcy. So a stress test done to predict likely losses—before they accumulated—based on book capital measures was unlikely to signal a need for more capital-raising efforts. This is perhaps why normal-times stress tests are best understood as preserving infrastructure that can be quickly used to recapitalize financial institutions once downturns begin, rather than trusted to provide credible information about capital buffers. The failure of the COVID stress testing exercise casts doubt on whether this tool will be utilized as efficiently as it was in the GFC in downturns going forward.

4 The Case for Market-Based Stress Testing

In this section, we explain why we are skeptical of the clean bill of health banks receive each year in the regulatory stress tests and illustrate the deficiencies of the exercise by comparing it to a naive market-based approach.

It is worth noting that we are performing the analysis that follows based on the 2019 stress tests, so this is written from the pre-COVID perspective. It is possible that the COVID crisis represents an extreme tail event, and thus it is unreasonable to expect the stress tests to reflect such a possibility. But in light of the magnitude of the shocking events over recent history, we are skeptical that the severely adverse scenarios used by regulators correspond to scenarios that are severely adverse, as we discuss in Section 5.

Even leaving aside the issues with scenario design, we do not find it plausible that the banking system would withstand, without difficulty, shocks of the scope that are attached each year to the severely adverse scenario. In last year's stress tests, all large financial institutions passed, meaning the Federal Reserve determined that all large banks would be able to continue to intermediate as normal in the event of a downturn worse than the GFC, featuring unemployment rising to 10 percent, equity values falling by 50 percent, the Volatility Index (VIX) reaching a peak of 70 percent, and real estate values falling by 35 percent over nine quarters of severely adverse stress. Overall, we are sympathetic with the views espoused recently by some in the regulatory community, including former vice chairman for supervision Dan Tarullo, who has described the annual tests as nothing more than a "compliance exercise" (Tarullo, 2020).

4.1 Recent Stress Test Performance Overly Optimistic

In 2017, for the first time since annual stress tests for large banks began, all 34 of the nation's largest banks were deemed to have sufficient capital to weather a severely adverse shock.[2] Only one bank (Capital One Financial) was found to have any weakness in its capital position in the event of an adverse stress scenario, and even this did not precipitate a failing grade. Industry champions celebrated these stress test results and used this regulatory stamp of safety as ammunition to call for decreasing the intensity of the stress tests and of the financial regulatory regime more broadly. In response to the results, then governor and soon-to-be chair of the Federal Reserve Board Jerome Powell noted, "This year's results show that even during a severe recession, our large banks would remain well capitalized. This would allow them to lend throughout the economic cycle, and support households and businesses when times are tough" (Powell, 2017). Boosted by this performance, advocates of deregulation encouraged the adoption of higher asset thresholds for the stress test exercise to provide relief for smaller institutions. In May 2018, the Economic Growth, Regulatory Relief, and Consumer Protection Act passed and did exactly this: raising the asset threshold for annual supervisory stress tests from \$50 billion to \$250 billion.[3]

As a result, in 2019, only 18 banks were subject to annual stress testing, and all banks cleared the tests (Credit Suisse received a "conditional non-objection" to its capital plan). The 2019 exam at the time appeared to be a dire, severely adverse stress scenario, with unemployment more than doubling (rising to 10 percent in 2019), equity prices falling by 50 percent, a 70 percent increase in market volatility, and a 25 percent decline in home values (Board of Governors of the Federal Reserve System, 2019a,b). The results of the stress test suggest that, as a result of the severely adverse stress scenario, common Tier 1 capital ratios for large banks would decline from 12.3 percent in Q4 2018 to a minimum of 6.6 percent, and loan losses would be only 5.7 percent.

4.1.1 Stress Tests Suggest Some Banks Will Earn Profits, Not Suffer Losses, during Severely Adverse Stress

One data point indicative of the over-optimism of the stress tests is their estimates of bank income during the stress scenario. In Table 9.4, we show the Federal Reserve's estimates of banks' pretax net income, which includes the Federal Reserve's assumptions about banks' loan portfolios and securities losses. Remarkably, during a period of severely adverse stress—with unemployment reaching double digits and the stock market falling by 50 percent—the Fed's projections suggest that none of the largest banks will lose more than 2 percent of total assets during the crisis. And in fact, Wells Fargo is projected to *increase* its net income by approximately \$2.1 billion (0.11 percent of total assets) over nine quarters of stress.

---

[2] In this chapter, we refer to the stress test as a single exercise. Technically, it comprises two pieces: the Comprehensive Capital Analysis and Review (CCAR) and the Dodd–Frank Act Stress Testing (DFAST). They are coordinated jointly by the Federal Reserve (in hopes of reducing duplicate data collection) and aim to assess whether the largest BHCs in the United States (above \$10 billion in assets) have sufficient capital to continue operations in the event of a financial crisis. Importantly, we focus on the DFAST results, benchmarking these against a series of market-based approaches. DFAST does not take into account planned capital disbursements when it projects the impact of stress on bank equity positions. Instead, it uses a standardized set of capital action assumptions (see www.federalreserve.gov/supervisionreg/stress-tests-capital-planning.htm.

[3] The Board is still required to conduct "periodic" rather than annual supervisory tests for BHCs with between \$100 billion and \$250 billion in assets. See the Economic Growth, Regulatory Relief, and Consumer Protection Act at www.congress.gov/bill/115th-congress/senate-bill/2155.

Table 9.4. Projected losses in pretax income, 2019 severely adverse scenario

Bank	$B change	% assets change	% RWA change
Bank of America	(17.8)	(0.7)	(1.2)
Citigroup	(11.6)	(0.6)	(1.0)
Goldman Sachs	(18.0)	(1.8)	(3.3)
JP Morgan	(29.9)	(1.1)	(2.0)
Morgan Stanley	(16.9)	(1.9)	(4.6)
Wells Fargo	2.1	0.1	0.2

Source: Data from the 2019 DFAST results (Board of Governors of the Federal Reserve System, 2019b).

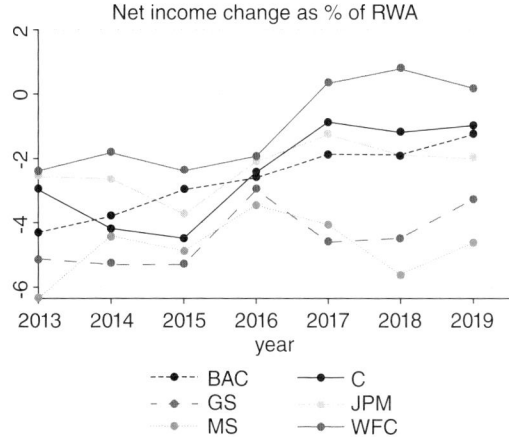

Figure 9.5 Projected change in net income before taxes over stress, scaled by RWAs. Source: Data from the 2013–2019 DFAST and CCAR results (Board of Governors of the Federal Reserve System, 2013a,b, 2018a,b, 2019a,b).

4.1.2 Stress Tests Have Grown More Optimistic over Time,
Potentially Because of Gaming by Banks

The Federal Reserve's stress tests have not always painted such an optimistic picture. The first annual postcrisis stress tests in 2013 predicted that the Big 6 banks would lose nearly $180 billion over nine quarters of severely adverse stress. Predicted losses have fallen by nearly 50 percent in the 7 years since, with the 2019 stress tests anticipating only $92 billion in losses. Figure 9.5 illustrates the changes over time for the Big 6 financial institutions: bank performance (scaled by RWA) over stress has improved over time, despite stress scenarios becoming more severe.

Relatedly, since the annual stress tests began in 2013, loan portfolio loss rates have fallen for the four large commercial banks (Figure 9.6). The same is not true for the investment banks: loss rates have risen for Goldman Sachs and are roughly unchanged for Morgan Stanley.

One interpretation of this finding may be that Bank of America, Citigroup, JP Morgan, and Wells Fargo are simply invested in less risky loan portfolios today than in the past, and so the decrease in portfolio losses is to be expected. But for all but Wells Fargo, the ratio of

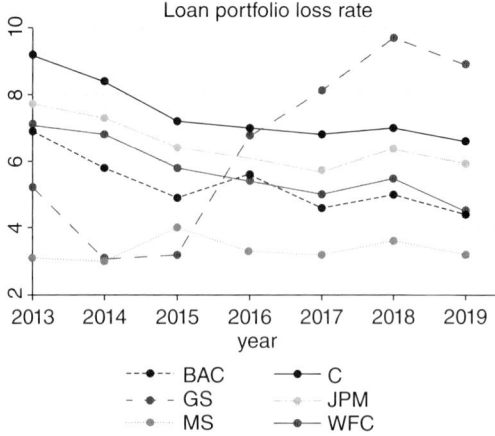

Figure 9.6 Stress test loan-loss rates. Source: Data from the 2013–2019 DFAST and CCAR results (Board of Governors of the Federal Reserve System, 2013a,b, 2018a,b, 2019a,b).

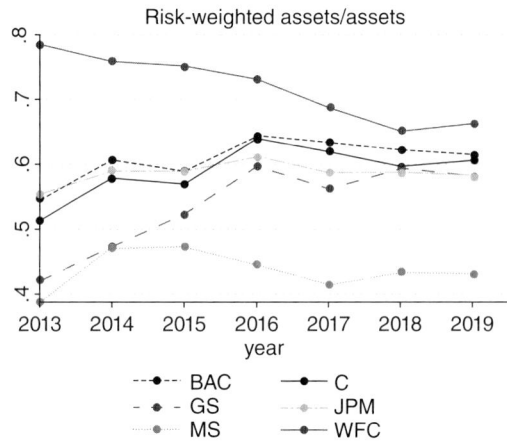

Figure 9.7 RWA share. Source: Data form FR Y-9Cs.

RWAs to assets has risen over time, suggesting that if anything, asset portfolios are riskier today (Figure 9.7).

It is interesting that the growth in stress test optimism is most pronounced for the commercial banks. A comparison with the initial SCAP is also informative: loan-loss rates have fallen from an average of 7.3 percent to just 3.6 percent, and securities losses from 1.7 to 1.2 percent, over a decade of testing (Table 9.5). This may be largely attributable to investment banks taking more risks after the crisis and regulators stressing against severe global shocks, whose impacts are borne most severely by firms with large trading books. It is also possible, though, that non–mark-to-market firms have become more clever about gaming the stress tests to minimize their exposure and maximize their capital disbursements. This latter interpretation is consistent with work by economists at the Federal Reserve who note that banks respond to a change that ties regulatory capital to the market value of

Table 9.5. Loan losses as percentage of average assets

	Loan losses % RWA	Securities losses[a] % RWA
2009 SCAP (market drops 50%, recession Q4 2007 to Q2 2009)		
Bank of America	7.9	2.5
Citigroup	8.0	2.5
JP Morgan	5.9	1.3
Wells Fargo	7.4	0.4
2013 Stress tests (market drops 50%, lasts 9 quarters)		
Bank of America	4.8	2.2
Citigroup	5.6	2.4
JP Morgan	4.2	2.0
Wells Fargo	5.0	1.2
2018 Stress tests (market drops 65%, lasts 9 quarters)		
Bank of America	3.5	1.5
Citigroup	4.3	2.0
JP Morgan	4.1	2.1
Wells Fargo	4.3	1.3
2019 Stress Tests (market drops 50%, lasts 9 quarters)		
Bank of America	3.0	0.8
Citigroup	4.0	1.6
JP Morgan	3.9	1.5
Wells Fargo	3.5	1.0

[a] Sum of realized losses, trading and counterparty losses, and other losses.
Source: Data from the 2009 SCAP and 2013, 2018, and 2019 DFAST and CCAR results (Board of Governors of the Federal Reserve System, 2009, 2013a,b, 2018a,b, 2019a,b).

their available-for-sale investment securities by reclassifying their portfolios to mitigate the impact of this change on stress test performance (Fuster and Vickery, 2018).

4.2 Market-Based Stress Test

We suspect that the excessive optimism of the annual stress tests is driven by their reliance on regulatory capital as a measure of bank health, which is known to be both static and unreliable as a predictor of financial stability. One way to make stress testing more informative would be to find ways to incorporate market information into the stress testing regime.

In Table 9.6, we attempt to illustrate the potential of this approach by providing a very rough estimate of what the market believes capital losses would look like in the event of the severe adverse stress scenario. We compute an average beta for each bank (based on the 5 years prior to the 2019 stress tests) and use these betas to estimate capital losses if a 50 percent decline in equity values were to occur. Two important caveats must be raised about these estimates. First, we ignore the other aspects of the Federal Reserve's severe adverse stress scenario (i.e., the implications of more than doubling current unemployment and a 25 percent decline in the housing market, among other elements). Second, we assume that betas are constant throughout the severely adverse stress scenario, ignoring important issues about the dynamic measurement of capital, for example, the fact that banks' assets become more volatile during downturns.

Table 9.6. Projected decline in common equity Tier 1 capital ratio under severely adverse stress scenario (50 percent decline in equity prices) versus imputed decline from bank betas

Bank	Beta	2018 TCE ratio	2018 RWA ($B)	2018 TCE ($B)	Severely adverse TCE ($B)	Market TCE ratio	Stress TCE ratio
Bank of America	1.4	11.6	1,437	167	65	4.9	9.7
Citigroup	1.4	11.9	1,174	139	54	5.0	9.5
Goldman Sachs	1.3	13.3	548	73	29	5.8	9.9
JP Morgan	1.2	12.0	1,529	183	81	5.7	8.2
Morgan Stanley	1.5	16.9	367	62	22	6.7	11.1
Wells Fargo	1.1	11.7	1,247	146	69	5.9	10.1

TCE, tangible common equity.
Source: Data from Bloomberg and the 2019 DFAST results (Board of Governors of the Federal Reserve System, 2019b).

We use historical bank betas and Q4 2018 tangible common equity reported by the six largest financial institutions in the United States to calculate what a 50 percent decline in the stock market would do to the capital position of these firms.

We compute the tangible common equity (TCE) that will remain after a 50 percent decline in the stock market as

$$0.5^\beta \times \text{TCE} = \text{TCE}_{s.\text{adverse}}.$$

Our market-based stressed TCE ratio is then

$$\text{TCE}_{s.\text{adverse}}/(\text{RWA} - \Delta\text{TCE}),$$

where

$$\Delta\text{TCE} = \text{TCE} - \text{TCE}_{s.\text{adverse}},$$

or how much TCE is lost in the severe adverse stress scenario.

This is an admittedly naive exercise. But the capital declines in the market-based approach cast aspersions on the regulatory community's claims that the largest financial institutions would continue to function as normal in the event of a recession-like shock (Office of the Comptroller of the Currency and Board of Governors of the Federal Reserve System, 2013). In Table 9.7, we naively attempt to account for the fact that bank betas will increase during times of distress. We adjust beta upwards so that after the first 25 percent decline in equity values, beta increases by 50 percent, after which banks are left with the following:

$$\text{TCE}_{s.\text{adverse}} = 0.5^{\frac{1}{2}(\beta_1 + 1.5\beta_1)} \times \text{TCE}_{2018}.$$

In this scenario, market-stressed TCE ratios fall below the 4.5 percent trigger for "prompt corrective action" for Bank of America and Citigroup and are substantially reduced relative to the Federal Reserve's approach for all of these large financial institutions.

It is helpful, given our focus on market-based measures, to compare the performance of mark-to-market firms like Goldman Sachs and Morgan Stanley under the current stress test regime to traditional commercial banks that do not mark the vast majority of their assets to market values. In Table 9.8, we do just this. Using the stress test results for the largest

Table 9.7. Projected decline in common equity Tier 1 capital ratio under severely adverse stress scenario (50 percent decline in equity prices) versus imputed decline from bank betas

Bank	Beta 1	Beta 2	2018 TCE ($B)	TCE first 25% ($B)	TCE second 25% ($B)	Market TCE ratio	Stress TCE ratio
Bank of America	1.4	2.1	167	104	51	3.9	9.7
Citigroup	1.4	2.0	139	87	43	4.0	9.5
Goldman Sachs	1.3	2.0	73	46	23	4.7	9.9
JP Morgan	1.2	1.8	183	122	66	4.7	8.2
Morgan Stanley	1.5	2.3	62	37	17	5.2	11.1
Wells Fargo	1.1	1.6	146	101	57	5.0	10.1

Source: Data from Bloomberg and the 2019 DFAST results (Board of Governors of the Federal Reserve System, 2019b).

Table 9.8. Common equity Tier 1, actual and projected at minimum and end of severely adverse stress scenario

Bank	Actual	Ending	Minimum	% Loss to ending	% Loss to minimum
Morgan Stanley	16.9	11.1	8.9	34.3%	47.3%
Capital One	11.2	6.0	6.0	46.4%	46.4%
Goldman Sachs	13.3	9.9	7.6	25.6%	42.9%
Deutsche Bank	22.9	15.0	14.8	34.5%	35.4%
HSBC	12.6	8.5	8.5	32.5%	32.5%
JP Morgan	12.0	8.2	8.1	31.7%	32.5%
Citigroup	11.9	9.5	8.2	20.2%	31.1%
Credit Suisse	25.8	22.3	18.4	13.6%	28.7%
UBS	21.7	16.8	16.0	22.6%	26.3%
TD	16.3	13.7	12.9	16.0%	20.9%
Barclays	14.5	12.4	11.6	14.5%	20%
Wells Fargo	11.7	10.1	9.5	13.7%	18.8%
Northern Trust	12.9	13.2	10.7	+2.3%	17.1%
Bank of America	11.6	9.7	9.7	16.4%	16.4%
PNC Financial Services	9.6	8.5	8.5	11.5%	11.5%
US Bancorp	9.1	8.1	8.1	11.0%	11.0%
State Street	11.7	11.8	10.9	+0.9%	6.8%
Bank of New York Mellon	11.7	13.1	11.3	+12.0%	3.4%

Source: Data from Bloomberg and the 2019 DFAST results (Board of Governors of the Federal Reserve System, 2019b).

banks (the "advanced-approach" firms: BHCs with assets greater than $250 billion or total foreign exposure of at least $10 billion), we compare projected declines in common equity Tier 1 capital ratios for the mark-to-market firms to their non–mark-to-market counterparts. Perhaps unsurprisingly, the mark-to-market firms perform worst in the severely adverse stress scenario: Goldman Sachs and Morgan Stanley experience a 47 and 43 percent decline in their capital position, respectively, significantly higher than the losses for any of the other advanced-approach firms, including Wells Fargo (19 percent), JP Morgan (33 percent), Bank of America (16 percent), and Citigroup (31 percent).

Combined, these two insights—(1) that a naive market-based approach results in capital losses that are nearly twice as severe as recent stress test results and (2) that mark-to-market firms perform worse on these regulatory stress tests—suggest that regulatory complacency is misplaced. The stress tests provide an overly optimistic view of how banks will perform in the next recession-like event, and an (admittedly naive) market-based approach raises significant cause for concern. The fact that the few trading firms that mark a larger share of their assets to market perform worse on the stress tests should caution that if market information were properly incorporated into the stress test exercise, non–mark-to-market commercial banks would appear significantly worse off in the adverse stress scenario. Bulow (2016) makes this point. He notes that because of "no move to mark to market accounting" and stress tests that "explicitly fail" to take into account market values, the current system does not force banks to respond quickly to signs of distress. Unlike commercial banks, trading firms that mark to market are forced to adjust capital requirements daily. This dynamic adjustment "[makes] their positions safer even with relatively smaller capital margins" (Bulow, 2016).

4.3 Some Concerns

There are at least two issues with a market-based approach to stress testing. The first is that market indicators can be driven by noise as well as news. This means that a rule that ties banks' capital requirements—and either limits disbursements or requires new capital to be raised—in response to market performance may overreact and mandate painful dilution of existing shareholders in response to blips that are not true crises. Comparing the last two recessions alone highlights this fact. The four largest commercial banks experienced decreases in their market-based capital ratios during the early 2000s that were as pronounced, if not more, than the decreases during the GFC (Figure 9.8). A market-based stress test that ties recapitalization to market performance would thus have demanded the same kind of equity raising in response to the dot-com bubble bursting as the GFC. Thus, transitioning to a such a regime could force banks to recapitalize even when such drastic action is not needed. This is one reason why the stress tests should take into account both market and regulatory indicia of bank health, rather than being mechanically tied to market performance alone. More generally, a dynamic regulatory approach that can successfully recapitalize the financial sector in moments of distress may at times overreact, but it will also allow banks to hold less capital in nonstress moments, reducing the distortions associated with holding excess capital in normal times.

Another issue for a market-based stress testing (and thus capital-raising) regime is its inherent procyclicality. Banks will perform poorly on these tests at the onset of a recession. During crises, the hope is for financial firms to continue to intermediate—lending to households and businesses—so that they can help bolster the economy. Imposing stricter capital regulation at these moments will have exactly the opposite result. Further, rules that tie capital raising to market indicia of financial stability will amplify death-spiral dynamics. Financial institutions that appear close to a stress test failure/market trigger requiring capital raising could be driven downward by market speculators. Finding ways to design a regulatory regime that incorporates market-based information—but is not solely reliant on

MVE/A Ratio

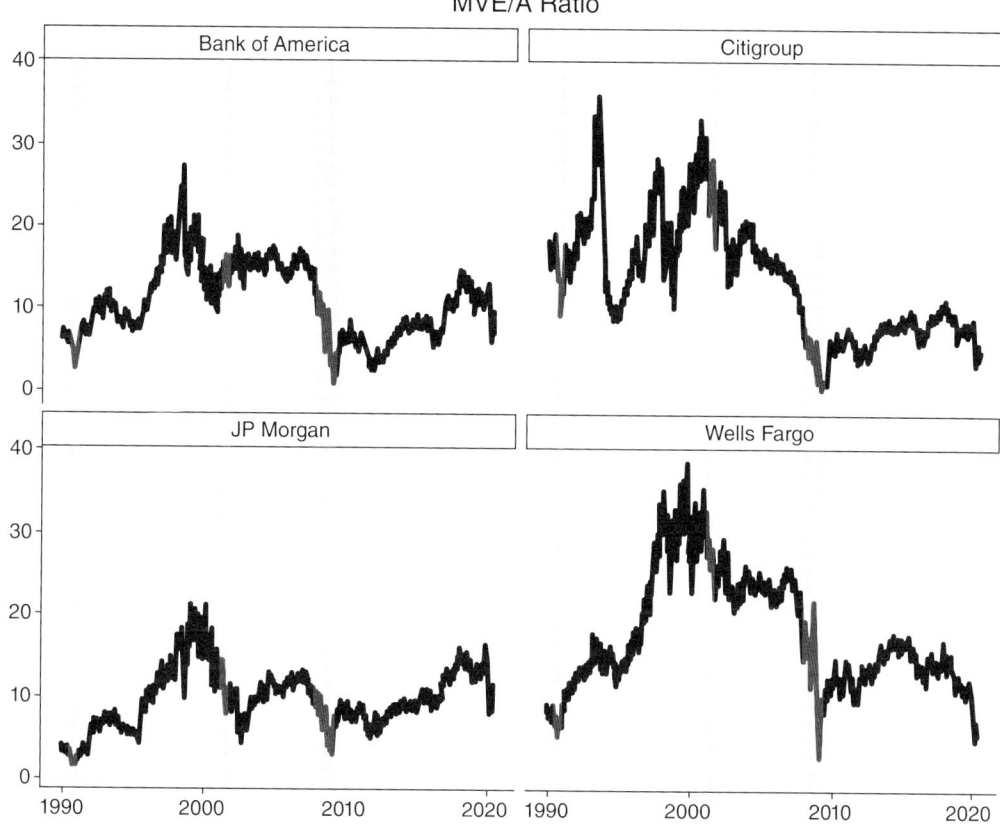

Figure 9.8 Market-based capital ratios over time. Quarterly asset data are interpolated exponentially to daily. Source: Data from Compustat and Bloomberg.

it—can mitigate these concerns as well as those associated with market noise. But more broadly, the procyclicality of the market-based stress tests relative to the regulatory tests is in itself an indictment of the current approach. The reason today's stress tests are not procyclical is because their conclusions do not reflect a dynamic and accurate assessment of financial stability. As bank losses mount and their capital buffers are depleted, forward-looking stress tests should require that banks bolster themselves against imminent distress.

5 Early Thoughts on the COVID Crisis

The recent experience of COVID-19 has laid bare the limitations of the regulatory regime with respect to adequate preparation for extreme tail events like the one we are experiencing.

Comparing changes in market and regulatory measures of financial stability in the recent months is revelatory: although regulatory capital ratios are essentially unchanged since the beginning of this year, market-based capital ratios have fallen to levels not seen since the financial crisis (Figure 9.9). Because stress tests rely on book capital measures alone as a predictor of banks' health, insofar as these ratios are not fully probative on economic capital, there is a limit to how useful this year's exercise could have been.

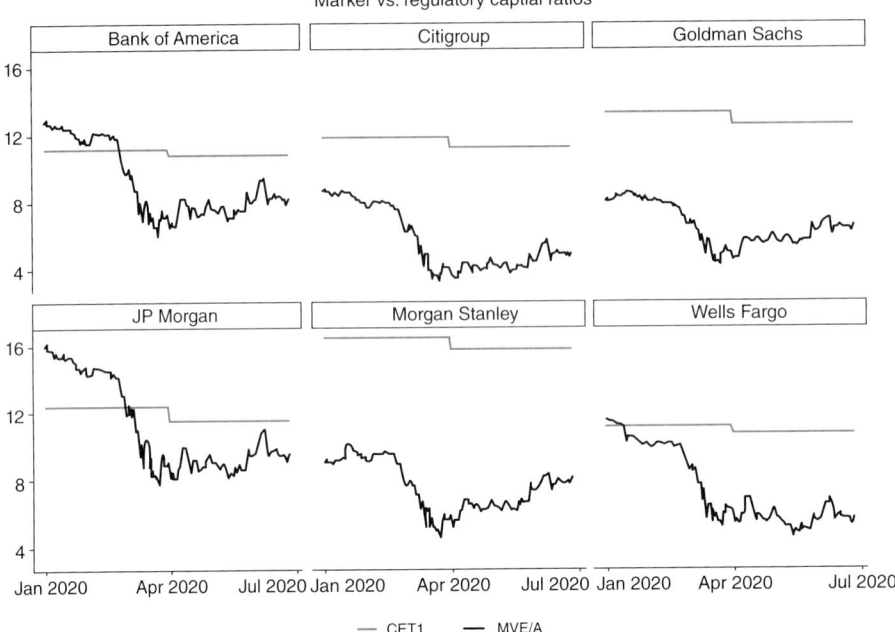

Figure 9.9 Comparing recent changes in market versus regulatory capital ratios since 2020. Quarterly assets are interpolated exponentially to daily values. Source: Data from Bloomberg, Compustat, and Bank FR Y-9Cs.

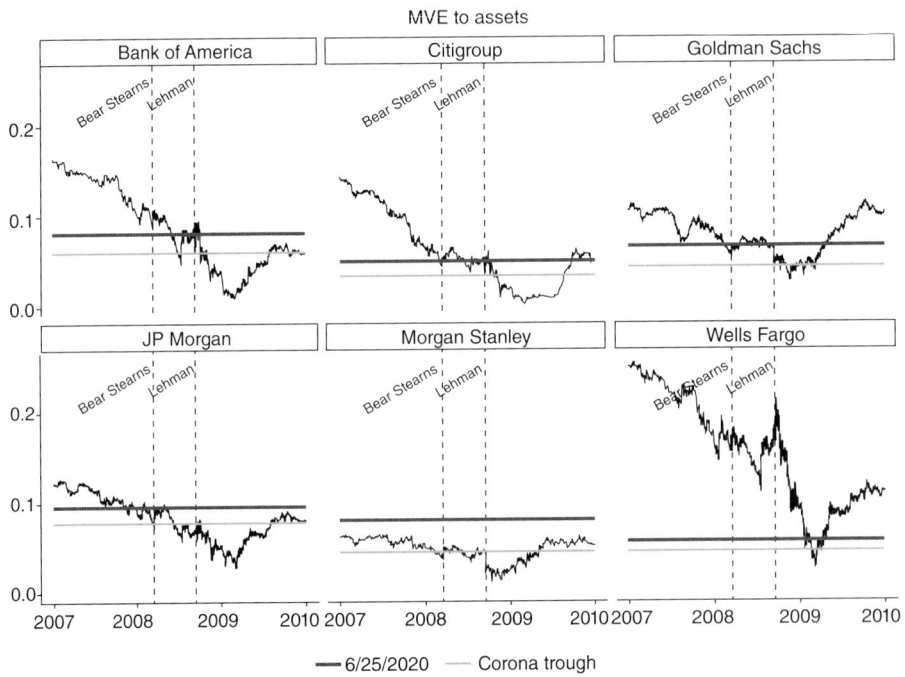

Figure 9.10 Comparing market-based capital ratios during GFC to coronavirus period. Quarterly asset data are interpolated exponentially to daily. Source: Data from Bloomberg.

Table 9.9. Comparing the 2020 DFAST severely adverse scenario to realized COVID-19 scenario

	2020 DFAST scenario (simulated)			Coronavirus shock (realized)		
	Q1 2020	Q2 2020	Trough	Q1 2020	Q2 2020	Trough
Unemployment	4.5	6.1	10.0	14.7	—	14.7
5Y Treasury	0.5	0.6	0.5	1.1	0.4	0.3
10Y Treasury	0.7	0.9	0.7	1.4	0.7	0.5
BBB corporate yield	5.2	6.1	6.6	3.3	3.4	5.5
Mortgage rate	3.9	4.2	4.4	3.5	3.2	3.7
Prime rate	3.4	3.4	3.4	4.4	3.3	4.8
Dow Jones Index	22,262	18,623	16,518	25,985	31,373	22,463
House Price Index	205	198	153	213	—	213
VIX	69.1	70.0	70.0	82.7	57.1	82.7

Realized Q2 2020 results refer to June 15, 2020.
Source: Data from Bloomberg, Federal Reserve Economic Data (FRED), and 2020 DFAST results (Board of Governors of the Federal Reserve System, 2020).

The baseline 2020 stress tests outlined a "severely adverse scenario" of unemployment reaching 10 percent by Q3 2021, the stock market falling by 50 percent, and real estate prices declining by around 30 percent. These scenarios, released in February 2020 before the crisis took hold, would in normal times be applied to bank balance sheets as of Q4 2019 to determine whether banks have sufficient capital to weather such distress.

Table 9.8 compares the severely adverse scenario to what occurred in the first half of 2020. The unemployment rate is significantly elevated with respect to what the severely adverse scenario anticipated, reaching nearly 15 percent in May. However, the market impact has been milder than the downturn anticipated by the severely adverse scenario: although the Dow Jones fell by 32 percent since January, it has since recovered and is down only 5 percent for the year to date.

It's hard to be sure what the recent experience tells us about the reasonableness of the stress test's assumptions. Stress tests are meant to determine the capital banks need in a hypothetical future downturn to require them to raise it preemptively, thus forestalling the need for significant government intervention. They do not assume the multi-trillion-dollar intervention to support the financial system and markets that we have seen in the last several months. Absent the Federal Reserve and Treasury's extraordinary interventions, the market deterioration would have been much more significant, very possibly exceeding the stresses that the severely adverse stress scenario involves. Perhaps the COVID crisis is such an extreme black swan event that it is unnecessary for banks to ex ante hold enough capital to be able to weather it independently of government intervention. But if this is the case, then it is important to understand that stress tests do not bolster banks enough to withstand the most left-tail events. This pushes against regulatory complacency, such as Vice Chairman Quarles's announcement that the "capital-building phase" of the postcrisis era is complete, and it also necessitates a regulatory regime that takes a dynamic view of capital and is quickly able to move toward a capital-hoarding phase in a severe downturn. The inexplicable failure to mandate that banks stop making capital disbursements in the last several months indicates that this dynamism is sorely lacking.

To be sure, Federal Reserve officials understood the need to adapt the stress tests to account for the current crisis, both because of an understanding that even the severely adverse scenario did not reflect the worst that may lie ahead and because bank balance sheets shifted so dramatically in the first half of 2020 as assets grew and lines of credit were drawn down that the baseline tests would not be informative. As Vice Chairman Quarles acknowledged, had the Federal Reserve not adapted the tests in light of the circumstances, "[we] wouldn't have been doing our job."

Thus, the Federal Reserve Board added a "COVID sensitivity analysis" to its baseline stress testing exercise. The goal of this portion of the analysis was, much like the original SCAP, to ascertain whether banks needed to hoard or even raise additional capital to manage the coming downturn. The fact that the economic capital of banks at the trough of these last few months has declined as much as it has certainly suggests that if the downward spiral continues for much longer—or exacerbates in the coming months—large banks will find themselves woefully undercapitalized, raising serious questions about their capacity to lend robustly and, of course, their solvency. Concerns about the weaknesses looming in the financial system are why regulators pushed for a foreclosure of disbursements, to help bolster banks against the coming storm (Kashkari, 2020; Brainard, 2020).

Although stress tests are a tool by which to quickly recapitalize the financial system, regulators missed out on this opportunity in the 2020 exercise. Despite a commitment to incorporating the unique risks posed by the coronavirus, the end result was "little more than a compliance exercise," as the former Federal Reserve vice chairman for supervision recently stated (Tarullo, 2020). Regulators made several missteps in conducting the sensitivity analysis. First, they failed to release bank-specific results on capital shortfalls, citing the "uncertainty" of their estimation (Quarles, 2020). But this ignores the lessons of SCAP: transparency provides credible information that investors can usefully rely on. Indeed, the success of SCAP relative to the EU crisis stress tests provides clear evidence that during a downturn, full disclosure of bank health is paramount (Schuermann, 2014).

The limited release of the sensitivity analysis makes clear that things may turn down very quickly in the financial system. At least 25 percent of large financial institutions "failed" the COVID stress tests, meaning that the Federal Reserve estimates that they will become undercapitalized should a severe double-dip COVID downturn arise. And yet inexplicably, the Federal Reserve essentially chose to let these banks continue to pay dividends until further notice, with the inevitable result that tens of billions of dollars that could be a buffer against COVID losses will instead be pocketed by shareholders in the next several months.[4] This is especially concerning given the substantial risks on the horizon; sovereign debt from emerging markets has already been downgraded to near-junk status, and growth forecasts are negative for the first time in six decades (Politi, 2020). Further, banks are also likely to experience large losses on their real estate portfolios: commercial real estate loan defaults are estimated to reach 2.7 percent in the COVID-19 downturn, nearing GFC levels (4.4 percent) and more than 25 times last year's loss rate (.1 percent) (Grant, 2020), and mortgage delinquencies are, by some estimates, expected to exceed their GCF peak (Passy, 2020).

---

[4] This is consistent with missteps made during the financial crisis, when large banks continued to pay dividends even after receiving infusions of government capital (Acharya et al., 2009). The objective of stress testing was to provide a process by which capital holes could be proactively identified and addressed. The recent COVID stress tests indicate that they are not living up to their promise.

6 Conclusion

The idea of doing explicit left-tail planning through bank stress testing is an excellent one, both in terms of assessing the health of institutions and in terms of bolstering them so that they maintain their capacity to lend in downturns. Unfortunately, judging the risk of financial institutions based on concepts of book capital alone is problematic. We believe that current claims by regulatory officials about the robustness of the financial system without extraordinary public measures are, at a minimum, substantially overstated. Recent experience in the midst of the COVID-19 pandemic illustrates the deficiencies of the current regime. Annual stress testing does not subject bank balance sheets to severe enough scenarios to mirror extreme left-tail outcomes, and even crisis stress tests adapted for this moment are too ineffectual to bolster the system against imminent losses.

It is imperative for regulators to find a way to incorporate market-based measures of bank health into their assessments of financial stability. As the foregoing analysis demonstrates, market-based capital measures update instantaneously and thus provide a dynamic view of bank health.

It is true that market measures of financial-sector health are imperfect, both because they can be driven by noise rather than updates on fundamental value and because they are procyclical, so weakness can be obscured by strong market performance. This is why we do not advocate for a purely market-based financial regulatory regime. The challenge will be to develop a framework for stress testing that integrates both market and book capital. It seems likely that the character of an appropriate regulatory regime is one that insists on satisfactory performance by both standards and is careful to understand and reconcile differences between the two measures. Practically, this means that the fundamental loss estimations that supervisors perform each year will still be crucial, but we will recognize the risks of their potential fallibility. It will also be important to reflect on the role that stress testing can play during a crisis to restore investor confidence and help banks recapitalize quickly. Although the original SCAP provided a model for successful crisis stress testing, the recent COVID experience hints that those lessons have been forgotten.

References

Acharya, V. V., H. S. Shin, and I. Gujral (2009), "Bank dividends in the crisis: A failure of governance," VOX CEPR Policy Portal, available at https://voxeu.org/article/amidst-crisis-banks-are-still-paying-dividends.

Bernanke, Ben (2013, April 8), "Stress testing banks: What have we learned?" Speech at Maintaining Financial Stability: Holding a Tiger by the Tail, Federal Reserve Bank of Atlanta, Stone Mountain, GA.

Board of Governors of the Federal Reserve System (2009, May), *The Supervisory Capital Assessment Program: Overview of results.*

Board of Governors of the Federal Reserve System (2013a, June), *Comprehensive Capital Analysis and Review 2013: Assessment framework and results.*

Board of Governors of the Federal Reserve System (2013b, June), *Dodd–Frank Act Stress Test 2013: Supervisory Stress Test Methodology and Results.*

Board of Governors of the Federal Reserve System (2018a, June), *Comprehensive Capital Analysis and Review 2018: Assessment framework and results.*

Board of Governors of the Federal Reserve System (2018b, June), *Dodd–Frank Act Stress Test 2018: Supervisory Stress Test methodology and results.*

Board of Governors of the Federal Reserve System (2019a, June), *Comprehensive Capital Analysis and Review 2019: Assessment framework and results.*

Board of Governors of the Federal Reserve System (2019b, June), *Dodd–Frank Act Stress Test 2019: Supervisory Stress Test methodology and results*.

Board of Governors of the Federal Reserve System (2020, June), *Dodd–Frank Act Stress Test 2020: Supervisory stress test methodology and results*.

Brainard, Lael (2020, June), "Statement by Governor Brainard," Press Release, available at www. federalreserve.gov/newsevents/pressreleases/brainard-statement-20200625c.htm.

Bulow, Jeremy (2016), "Discussion of understanding bank risk through market measures," *Brookings Papers on Economic Activity*.

Carney, Mark (2014, November 17), "The future of financial reform," Speech given at the Monetary Authority of Singapore Lecture, Monetary Authority of Singapore, Singapore.

Cryan, John (2016, February 9), "A message from John Cryan to Deutsche Bank employees," available at www.db.com/newsroom_news/2016/ghp/a-message-from-john-cryan-to-deutsche-bank-employees-0902-en-11392.htm.

Flannery, Mark, Beverly Hirtle, and Anna Kovner (2016, August), "Evaluating the information in the Federal Reserve stress tests," Federal Reserve Bank of New York Staff Report No. 744.

Furlong, Fred, and Simon Kwan (2006, August), "Safe and sound banking, 20 years later: What was proposed and what has been adopted." Federal Reserve Bank of San Francisco Working Paper Series No. 2006–27.

Fuster, Andreas, and James Vickery (2018, June), "Regulation and risk shuffling in bank securities portfolios," Federal Reserve Bank of New York Staff Reports No. 851.

Geithner, Timothy F. (2014), *Stress test: Reflections on financial crises*. Broadway Books.

Grant, Peter (2020, March 24), "Loss rate on banks' real-estate loans expected to soar." *Wall Street Journal*.

Haldane, Andrew (2011, January 9), "Capital discipline," Speech given at the American Economic Association, Denver, Colorado.

Kashkari, Neel (2020, April 16), "Big US banks should raise $200bn in capital now." *Financial Times*.

Krugman, Paul (2009, February 22), "Banking on the brink." *New York Times*.

Laux, Christian, and Christian Leuz (2010), "Did fair-value accounting contribute to the financial crisis?" *Journal of Economic Perspectives*, 24(1), 93–118.

Li, Jing (2017), "Accounting for banks, capital regulation, and risk-taking," *Journal of Banking & Finance*, 74, 102–121.

Office of the Comptroller of the Currency and Board of Governors of the Federal Reserve System (2013), "Regulatory capital rules. 12 CFR Parts 208, 217, and 225," available at www.occ.gov/news-issuances/news-releases/2013/2013-110a.pdf.

Passy, Jacob (2020, May 17), "Mortgage delinquencies caused by the coronavirus will exceed Great Recession levels, according to this forecast," *Market Watch*.

Politi, James (2020, June 8), "Emerging economies forecast to shrink for first time in 60 years." *Financial Times*.

Powell, Jerome (2017, June), "Federal Reserve Board releases results of supervisory bank Stress tests." Press Release, available at www.federalreserve.gov/newsevents/pressreleases/bcreg20170622a.htm.

Powell, Jerome (2019, July 9), "Welcome remarks," Speech given at Stress Testing: A Discussion and Review, Federal Reserve Bank of Boston, Boston, MA.

Quarles, Randal (2019), "Stress testing: A decade of continuity and change," Speech given at Stress Testing: A Discussion and Review, Federal Reserve Bank of Boston, Boston, MA.

Quarles, Randal (2020, June 19), "The adaptability of stress testing," Speech given at Women in Housing and Finance, Washington, DC.

Sarin, Natasha, and Lawrence H. Summers (2017), "Understanding bank risk through market measures," *Brookings Papers on Economic Activity*.

Scheurmann, Til (2014), "Stress testing banks," *International Journal of Forecasting*, 30(3), 717–728.

Tarullo, Daniel (2020, June), "Are we seeing the demise of stress testing?" Brookings Institution Upfront Blog, available at www.brookings.edu/blog/up-front/2020/06/25/stress-testing/.

10

Granular Data Offer New Opportunities for Stress Testing[*]

Cees Ullersma and Iman van Lelyveld

1 Introduction

Nowadays, more and more granular data are being collected. The 2008–2010 crisis has shown that authorities were missing crucial information for accurately identifying risks in the financial system. For example, due to the over-the-counter (OTC) nature of derivatives markets, there is no centralized overview of the market.[1] Participants only observe their own volumes and exposure concentrations. The major US investment banks therefore did not realize that *jointly*, they were massively exposed to a single entity, the lightly regulated insurer AIG. In setting their capital buffers and implementing other risk-mitigating procedures, they were thus ignoring an important yet unobserved concentration risk.

The realization that more information is needed to properly capture the risk in the financial system has led to a significant increase in the depth and scope of information being reported across the system.[2] At the same time, the cost of reporting has gone down because more and more economic interactions are recorded digitally. Also, as a side effect of internet-based activity (i.e., commercial and social networks), new aspects of economic activity have become measurable.[3] Furthermore, the cost of analysis has gone down, and (open-source) tools and methods (e.g., machine learning [ML] and artificial intelligence [AI]) are improving markedly. Not only are more data available for stress testing than ever before, but implementing more accurate models is within reach.

The spectacular growth in the granularity of data collected in the last century can be illustrated by comparing the level of detail available in Foster (1922) with that in Ehlers et al. (2018), shown in Figures 10.1 and 10.2, respectively. Arguably, Foster did not intend to depict each individual transaction contained in the flows, but the data were also simply not there. For the flows shown in Figure 10.2, the disaggregated information is, in principle, available. Combining this information is another matter, which we will discuss in detail later in the chapter.

---

[*] We are grateful to Roos van den Berg and Jiaqi Zheng for their assistance in constructing the figures and to three anonymous referees whose comments have improved the chapter significantly.

[1] See Abad et al. (2016) for a first effort to shed light on these "dark markets" and Levels et al. (2018) for further detail.

[2] See, for example, the second progress report on the implementation of Phase 2 of the G20 Data Gaps Initiative (Financial Stability Board [FSB] and International Monetary Fund [IMF], 2017).

[3] See, for example, Graff (2018), who shows Brexit-related worker migration in and out of the UK in a comprehensive and timely way using LinkedIn profiles.

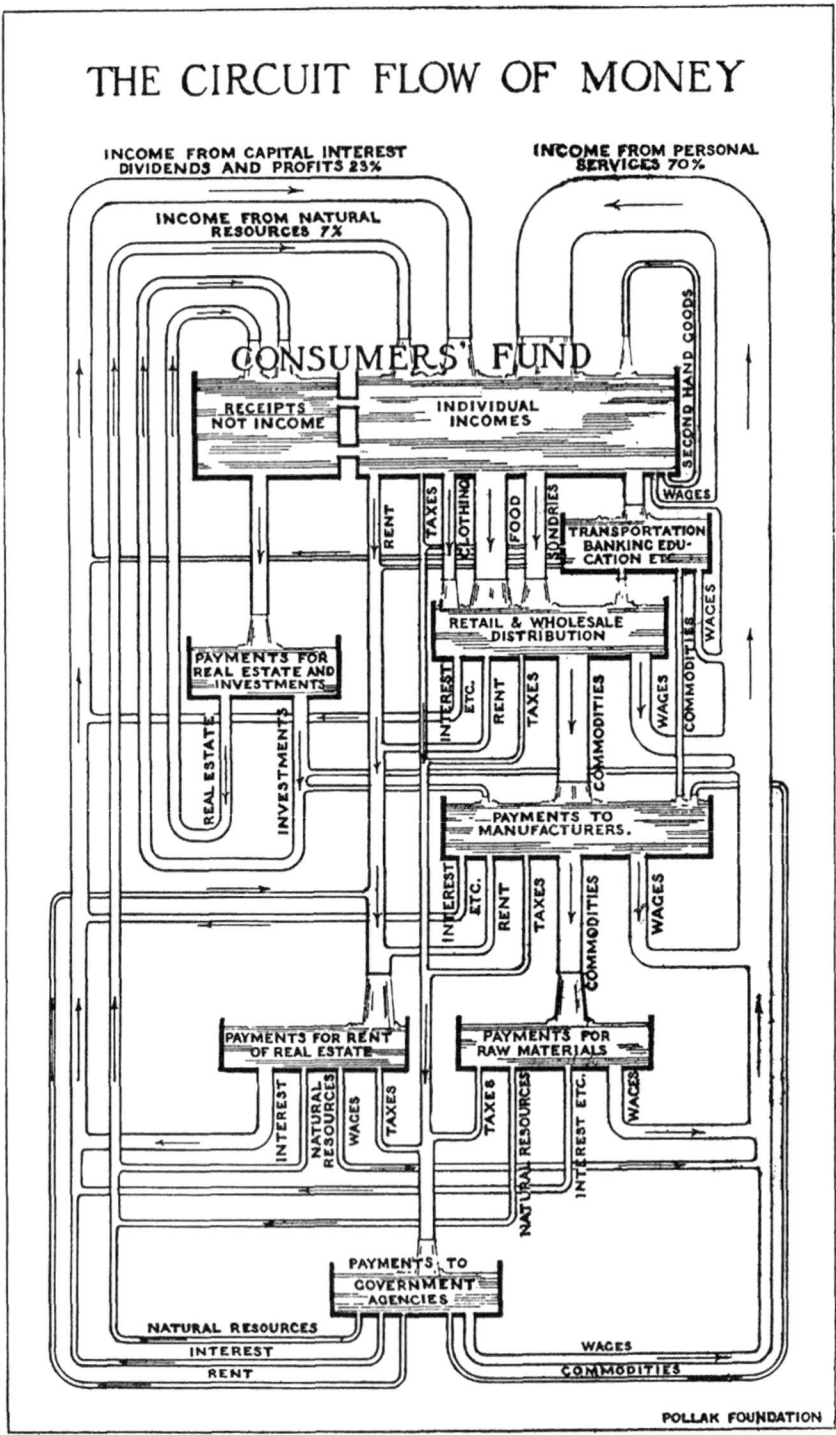

Figure 10.1 Granularity 100 years ago: The circuit flow of money.
Source: Foster (1922). Copyright American Economic Association;
reproduced with permission of the American Economic Review.

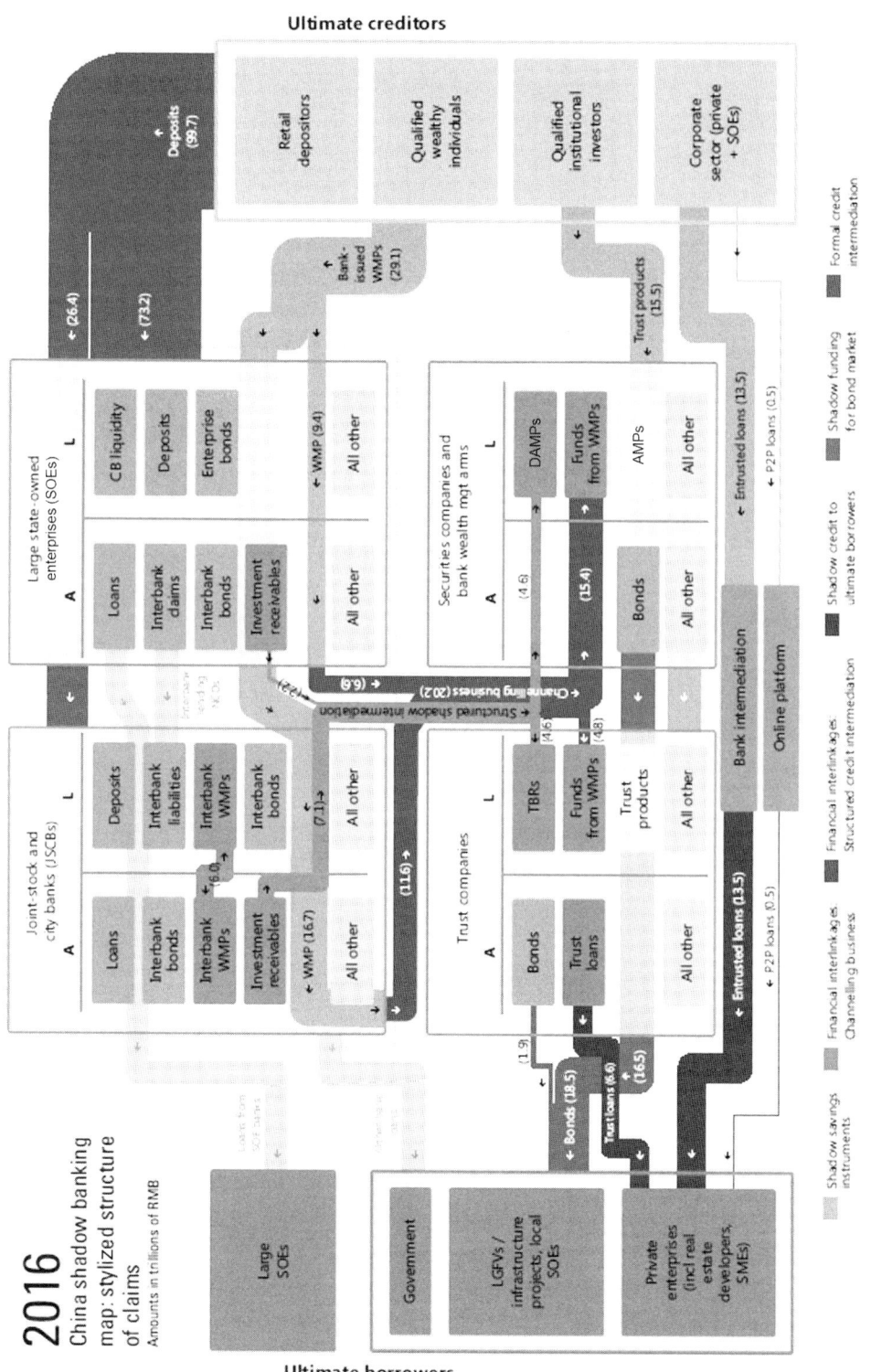

Figure 10.2. Granularity today: The Chinese financial system. AMP, asset-management product; CB, central bank; DAMP, directed asset-management product; LGFV, local government financing vehicle; P2P, person to person; SME, small and mid-size enterprise; TBR, Treasury bill rate; WMP, wealth management product. Source: Ehlers et al. (2018).

Data collection and analysis have been a core competency for most central banks and supervisors for many years. Central banks thus have a deep understanding of how to collect data from—primarily financial—economic agents and are thus well placed to capitalize on the current data trend. At the same time, they are faced with legacy systems that might lock them in inefficient solutions. Some authorities are looking for alternative ways to acquire data. In Austria, for example, direct reporting to the central bank has been replaced by reporting to a joint venture—owned by the bigger banks—tasked with data handling. This joint venture, in turn, provides the necessary reporting to the authorities (Kienecker et al., 2018). An alternative is plugging into the systems of private institutions directly. In theory, this would allow authorities to extract the precise data points at precisely the right time without imposing a material reporting burden. The practice is more elusive because firms have multiple information technology (IT) systems with unaligned definitions. For regulatory reporting, significant manual effort is required. Direct access to firms' data will not solve this but will push the necessary manual steps to the authorities.

The breadth of reporting has increased, especially in the area of OTC markets. In OTC markets, such as the interbank market or the derivatives markets, market conditions are more difficult to gauge because contracts are bilateral and generally private information. Recently, regulators have introduced new regulations to—at a minimum—introduce exposure reporting (without imposing limits or other constraints). The European Market Infrastructure Regulation (EMIR) is an example of such a regulation covering derivative markets trading. Under this regulation, all parties with significant dealings need to report their transactions centrally.[4] These central repositories are for-profit entities that collect the data and disseminate them to the relevant authorities. Another important initiative is AnaCredit, which became operational in the fall of 2018 and will provide a basic version of a European credit register. This will allow authorities to more comprehensively assess credit risk. Furthermore, AnaCredit data can be used to trace out the effects of monetary policy. A final example is Money Market Statistical Reporting (MMSR), which covers large banks and their counterparties in the European money market. It contains deal-level information on various short-dated money market contracts (less than a year) and is set up to construct a daily interest rate benchmark. An important side effect is that the MMSR data can be used to shed light on many other aspects of money market functioning. A relevant question could, for example, be how systemic or idiosyncratic stress affects the system.

Many data collections have become more granular, allowing for made-to-measure aggregations. Whereas traditionally the task of classifying a particular data point was put upon the reporter, a current trend is to request information on a much more granular level and leave the classification to the authorities. For example, reporting firms no longer need to determine the sector of the counterparty but only need to supply the Legal Entity Identifier (LEI). The LEI, which will be discussed in more detail later in the chapter, is a unique identification number for legal entities. Using the LEI, data can, for instance, be aggregated to a financial group. The LEI can also be used to generate sector aggregates. By leaving the aggregation process up to the data collector (i.e., the statistical agency, the supervisor or the central bank), consistent handling of the sector allocation is ensured. Furthermore, a change in the sector assignment of a particular firm can be applied consistently across all reported data.

---

[4] The US Dodd–Frank Act has similar stipulations.

A major advantage of data on a granular level is that they allow us to pinpoint the source of stress and the location of the most vulnerable parts of the financial system (Kwast et al., 2010). More granularity, however, also implies massive data sets and thus creates computational challenges. Moreover, if we are interested in feedback effects, we need to be able to trace such effects with sufficiently high frequency. If the timescale is too coarse, we will only be able to see end results once they have already materialized. For example, if overlapping portfolios are an important driver of contagion because they open up a channel to transmit stress, then we need timely information on securities holdings. However, holdings are often, if collected at all on a granular level, only collected on a monthly or quarterly basis (see, e.g., mutual fund data in the United States or the Securities Holdings Statistics (SHS) in Europe). In the period in between, positions can move in many ways, and in order to stress test, we need to make assumptions about how the portfolios change (Wang et al., 2018). In sum, examining the possibilities and challenges that new data and tooling bring in the stress test arena is useful, arguably not just for stress testing but for risk identification and management more broadly.

In stress testing, many of these new developments come together. Stress tests often require the ad hoc combination of data in order to be able to flexibly model unlikely scenarios. In the past, data availability dictated which scenarios were feasible to be stressed. For example, interbank exposures were only available as an aggregate without a split by counterpart. It was therefore very difficult to construct linkages from bank to bank. Without an accurate representation of the interbank network, tracing out the transmission of stress through the system is not possible.[5] Stress tests were thus restricted to investigating how aggregate risk measures reacted to stress. Furthermore, modeling individual actors was often not possible because available information was not detailed enough to map out utility functions or supply and demand curves. Increased data availability might thus warrant a reconceptualization of more traditional stress testing. A very promising avenue in this respect is heterogeneous agent models with network effects. These models are now computationally feasible and ready to deliver useful insights (Aymanns et al., 2018).

To be able to optimally conduct stress testing in this new environment, a few challenges need to be resolved. First, our data should be of sufficient quality. In particular, the associated metadata should allow us to connect and combine data sets as needed. Second, a conceptual framework on how to handle the data needs to be adopted. Third, we need to develop the capability to tackle complex stress tests. This involves both an adequate IT infrastructure and a data-friendly company culture. We will discuss each of these points in turn. Then we will discuss a case study with the SHS to show how granular data can provide the raw material for micro *and* macro stress tests.

2 Collecting Data: National Challenges and Opportunities

Typically, reporting requirements are developed in a "silo." These silos are defined by the country-specific legal frameworks that govern them. Mandates can be statistical, conduct of business (including market conduct), and micro- and macroprudential. Statistical mandates

---

[5] See van Lelyveld and Liedorp (2006) and van Lelyveld et al. (2011) for early efforts to construct network stress tests for the interbank and reinsurance markets, respectively.

have the longest history. The English King William the Conqueror ordered a stock-take of productive assets to determine the tax base as long ago as the year 1085 AD. Market-conduct mandates require data with the aim of ensuring fair and efficient trading. Microprudential reporting is primarily aimed at collecting data to assess an individual firm, whereas macro-prudential supervision takes wider effects into account. For the latter type of supervision, we are thus naturally more interested in data that can shed light on the interaction of different financial actors and the effects of financial markets on the wider economy. Each mandate has its own focus, and incentives to coordinate across regulatory domains are limited.

Depending on a country's institutional framework, the legal mandates are assigned to different agencies, potentially making sharing data difficult. Given the often-complex gover-nance, it is no wonder that reporting requirements are a patchwork of over- and underlapping components. Moreover, sharing data among agencies is often difficult because existing laws prohibit sharing confidential data. Even in the United States, a country with a long tradition of publishing a wide range of financial information, this seems to be the case. This point was reiterated by Ruth Judson (Federal Reserve Board) at a recent G20 workshop on data sharing (Judson, 2017). She noted that "The U.S. has problems similar to those of other countries with regard to sharing data across agencies and countries."

3 Collecting Data: Cross-Border Challenges

Reporting and sharing data across national jurisdictions is even more complicated than in a specific jurisdiction. For example, the Basel accords are a comprehensive set of rules for banks aiming to create a level playing field globally. The accords themselves are not legally binding, however. Basel Committee member countries have to put the accord in national legislation first. For the current accord, Basel III, this is in Europe, first and foremost, the so-called Capital Requirements Directive (CRD-IV) package. CRD-IV contains reporting requirements for licensed entities, that is, banks. There are generally very few reporting requirements for firms outside the prudential regulatory perimeter. Furthermore, there is little coordination of reporting requirements for, for example, insurance undertakings and banks.

In the field of statistics, various bodies aim to coordinate regulation and the collection of (regulatory) statistics across jurisdictions. Standards are set by international institutions such as the United Nations (UN), Organisation for Economic Co-operation and Development (OECD), IMF, and World Bank. Agreements reached are then either implemented on a best-efforts basis or cemented in national law.

4 Collecting Data: New Opportunities through Metadata and Code Distribution

As a consequence of how reporting requirements are typically designed, reporting of meta data is not well coordinated. Arguably, the precise reporting needed to achieve the mandate is best determined by the relevant authorities. For example, microprudential banking super-visors are best placed to determine what information is needed to safeguard a bank. They should thus decide on what data points to request and at what frequency. Macroprudential supervisors might need other information because the scope of their mandate is wider. The same holds for other supervisors (e.g., insurance-sector or financial-market supervisors).

Standardization across various reporting frameworks can be beneficial. Statistical agencies and standard setters have agreed on a long list of standards to ensure consistent measurement of social phenomena. For instance, there are International Organisation for Standardization (ISO) standards for country naming conventions. In financial markets, other conventions have been agreed on, such as the International Securities Identification Number (ISIN; ISO 6166) and the Committee on Uniform Security Identification Procedures identification number (CUSIP) used in the United States and Canada. These conventions do not yet have complete coverage in terms of products or global acceptance, but solid progress is being made. For example, if regional breakdowns are reported using standard ISO country codes, then we can easily combine information from, say, pension fund exposures to mutual fund investments in particular countries or regions. More complex issues arise if we want to unequivocally define economic activity across sectors: What constitutes a sector? What are the defining characteristics of a loan? If different sectors agree to use the same metadata frameworks, combining and validating data across different sectors can be simplified significantly. The choices in one domain can thus have an external effect and affect the usefulness and efficiency of reporting elsewhere. If banking supervisors, for instance, agree to universally adopt the LEI, other financial supervisors can follow; in that scenario, matching exposures of banks with those of, for example, insurers will become much easier. Berner and Judge (2019) support this line of thought and argue forcefully that developing standards jointly with industry increases social welfare.

An example of a key identifier in metadata is the LEI.[6] The advantages of the LEI are discussed in more detail by Bottega and Powell (2011). The LEI is a 20-digit alphanumeric code based on a standard (i.e., ISO 17442). It connects to reference information that enables clear and unique identification of legal entities participating in financial transactions. The LEI initiative is a joint effort initiated by the Group of 20, the FSB, and many regulators around the world. They have emphasized the need to make the LEI a broad public good. The LEI Regulatory Oversight Committee is at the highest level of LEI governance. Below this is the Global LEI Foundation, which brings together the information curated by local LEI granting organizations. These are typically credit registries and chambers of commerce. The number of LEIs is growing rapidly, fostered by new regulatory requirements, and now exceeds 1 million active identifiers (Legal Entity Identifier Regulatory Oversight Committee, 2018). The European Central Bank (ECB) estimates that the LEI covered at least securities with a total value of €95 trillion worldwide as of November 2017. This is a 25 percent increase since the end of January 2017. Although much has improved recently, some major challenges remain. For one, as shown in Figure 10.3, the coverage is neither even nor complete.

More technical challenges remain unresolved as well: for example, if an LEI lapses because the fees have not been paid, then a new LEI needs to be applied for. The entity will now have both an old inactive LEI as well as a new LEI. The old LEI might linger in reporting data for a long time if the entity entered into a long-running commitment, such as a long-dated swap.

---

[6] Other initiatives in the European System of Central Banks (ESCB) to improve data collection are the Banks' Integrated Reporting Dictionary (BIRD), the Integrated Reporting Framework (IReF), the Register of Institutes and Affiliates Data (RIAD), and the Centralised Securities Data Base (CSDB).

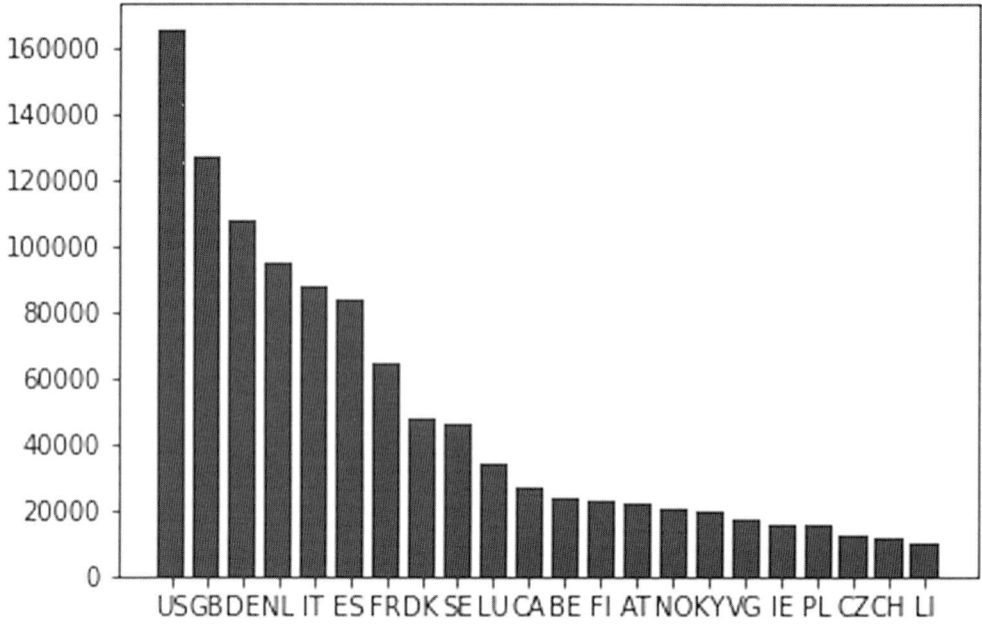

Figure 10.3 Number of LEIs per country.
Source: GLEIF (www.gleif.org). Parsing by Martien Lubberink.

The LEI makes it possible to start assigning activity and risks exactly to those subsidiaries that hold them. For example, for trading and holding of securities, it is clear to whom to assign the flow and stock of securities. With multiple jurisdictions, this allocation is of critical importance (Cerutti and Schmieder, 2012; Fang and van Lelyveld, 2014). A poignant example is the default of Lehman, where US authorities moved assets from Lehman's British subsidiaries before British authorities could react. Without information on the location of an asset (and how fungible it is), it is difficult to assess the robustness of a local institution.

To clarify how such allocation could work, we plot the legal structure of Rabobank in Figure 10.4. The information is taken from the Global Legal Entity Identifier Foundation (GLEIF) website and shows the structure and location of all the legal entities with LEIs. The different continents are shaded as explained in the legend. It is clear that even a relatively small player like Rabobank has a very complicated structure spanning multiple jurisdictions. Note that this is only a partial picture because not all the units have an LEI attached to them. Given this structure, authorities in different jurisdictions can allocate activities to these units and assess whether these activities pose a local or global threat to financial stability.

Being able to break down a reporting entity into its constituent parts is very useful in stress testing because it allows for pinpointing where and how stress would affect a firm (Cerutti and Schmieder, 2012). If a bank, for instance, extends mortgages in two subsidiaries in two distinct geographic areas with quite different business cycles, then being able to stress the mortgage book using two different cycles can be quite valuable. Alternatively, if different resolution schemes are in force for different parts of a bank, claims and liabilities used in the stress test need to be broken down by jurisdiction (cf. The Joint Forum, 2012), in particular on the role of internal guarantees. Although a unique identifier such as the LEI

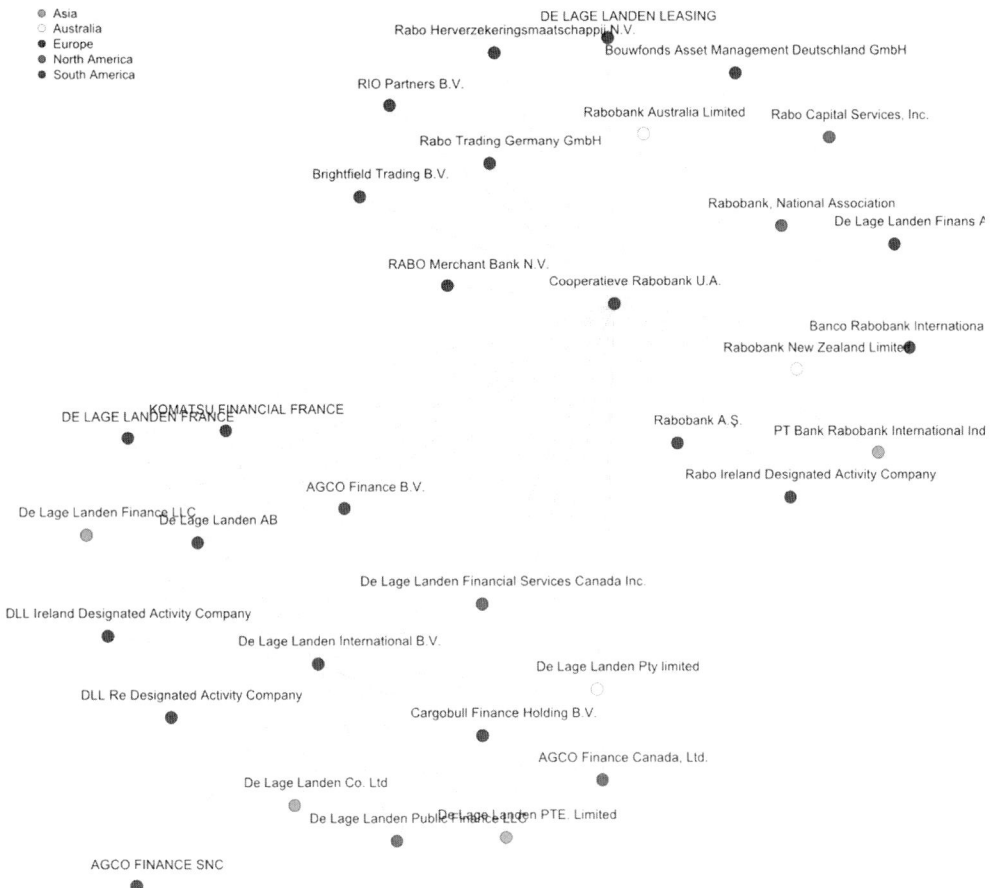

Figure 10.4 Legal structure of Rabobank.

is a necessary first step, we currently do not yet receive all the necessary information on a sufficiently disaggregated basis to allow for ad hoc choice of aggregation. That is, a stress test of a bank with the option that some of the subsidiaries are allowed to default still requires extensive data work. In principle, however, we can construct a reasonably precise picture of a large and complex internationally active bank, including a breakdown for the significant subsidiaries.

Ensuring a consistent reporting framework is quite a challenge, but some progress is being made. For example, the European Commission (EC) has undertaken an extensive study to assess whether the current reporting requirements are sufficiently harmonized (EC, 2019). Overall, the requirements are effective in meeting the objective of enabling supervisory authorities to fulfill their statutory tasks and mandates. Although the EC concludes that EU-level requirements are coherent in a broad sense, they are found to have a range of inconsistencies both across and within reporting frameworks. Many of these inconsistencies appear minor and purely technical in nature. They nonetheless put a burden on reporting entities and supervisors. Key inconsistencies identified include nonaligned definitions, different formats of data fields or templates, and inconsistent timing of mandated reporting. Note that

Table 10.1. Comparison of selected fields in EMIR, MiFIR, REMIT, and SFTR.

EMIR	MiFIR	REMIT	SFTR
Field identifier	Field identifier	Field identifier	Field identifier
Currency of price	Price currency	Price currency	Price currency
Report-submitting entity ID	Submitting-entity identification code	Reporting-entity ID	Report-submitting entity
Venue of execution	Venue	Organized marketplace	Trading venue

Source: EC (2019).

the EC's assessment did not back up the repeated claim by stakeholders that there are a significant number of duplicate requirements between different reporting frameworks. The EC did identify numerous cases of broadly or very similar data being requested. Authorities could thus consider if their information requirement merits multiple reports.

An example of the mundane but, in practice, quite real complications that arise in combining information is revealed by the differences for selected fields of the EMIR, Markets in Financial Instruments Directive (MiFIR), Regulation on Wholesale Energy Market Integrity and Transparency (REMIT), and Securities Financing Transactions Regulation (SFTR) requirements, as shown in Table 10.1.[7] The data fields shown are present in all four frameworks but carry different labels. This could, in principle, be resolved by making a mapping based on the definitions. However, in some cases, the definitions (or, equivalently, the admissible values) differ. For example, the data fields in the second row exhibit a validation inconsistency. Under EMIR, the status of the reported LEI shall be "Issued," "Pending transfer," or "Pending archival," whereas MiFIR reports are also accepted if the status of the LEI is "'Lapsed." For long-running swaps, this might increasingly become an issue because lapsed LEIs would then generate a missing value in the EMIR data—but not in MiFIR. Likewise, there is a partial inconsistency in terms of the content of the data fields in rows 2 and 3. Under REMIT, the reporting ID is not required to be an LEI308. Similarly, REMIT accepts a market identifier code (MIC), LEI, Agency for the Cooperation of Energy Regulators (ACER) code, or "XBIL" (for bilateral trades) "organized marketplace," whereas EMIR, MiFIR, and SFTR always require this field to be populated with an MIC code. Linking REMIT to the other data sets is thus not straightforward.

To overcome restrictions on sharing data, which could be used for more comprehensive stress tests, different approaches have been suggested—and in some cases executed as well. One way is to distribute code and only share the outcomes. Although the granular underlying data might be sensitive, the model outcomes—such as, for instance, regression coefficients—generally do not reveal the information encoded in the individual observations. In a project undertaken by the Basel Committee on Banking Supervision Working

---

[7] EMIR is a regulation on OTC derivatives, central counterparties, and trade repositories; MiFIR covers reporting of transactions in financial instruments that are traded on trading venues; REMIT deals with wholesale energy market integrity and transparency; and the SFTR provides securities financing transactions reporting. See EC (2019) and the references and links therein for further details.

Group on Liquidity Stress Testing, participating researchers ran the code on their respective financial network data and then fed back the results to a shared repository. This resulted in an overview of network characteristics and network vulnerabilities for a wide range of networks (Anand et al., 2018).[8] A similar approach to code and result sharing is followed by the International Banking Research Network.[9] An alternative is to assign a trusted party who handles data collection and analysis.[10] For example, supervisors of the largest banks in the world have drawn up a legal framework that governs the exchange of very granular, bank-specific information. These data are now being collected and analyzed in the International Data Hub (IDH) hosted by the Bank for International Settlements.[11] The detailed reports are then sent back to the data contributors. Over the years, the IDH has become more engaged in other regulatory workgroups, contributing to, for instance, analysis on the importance of central counterparties (CCPs), but external publication is still out of the question.

Data collection can have external effects. Although reporting requirements are almost always drafted with a single purpose in mind, the data collected can have multiple uses. For example, loan-level information on bank loans used in banks' internal credit-risk models is generally primarily collected for microprudential supervision. With such data, supervisors can assess the validity and robustness of a bank's credit-risk model. The data can also be used for other microprudential purposes, such as stress testing an individual bank. However, if such data are collected consistently *across* banks, then such data become potentially useful for macroprudential supervisors as well. The very same data can then also be used as input for an aggregate stress test. The aggregate stress test could in this case be micro-founded, with attention to the distribution in credit risk, and move beyond just examining credit aggregates. Alternatively, credit registry data—primarily collected for market transparency, statistical, and monetary policy purposes—could be used for micro- and macroprudential stress tests. At De Nederlandsche Bank (DNB), we are now investigating to what extent the information in the granular loan tapes, which we request ad hoc in on-site microprudential examinations, overlaps with information in AnaCredit, the ESCB's credit registry. The advantage of AnaCredit is timeliness and cost. For AnaCredit, banks have put in place a regular reporting schedule, which reduces the need for manual steps and hence costs. The disadvantage is that the definitions in place are not aligned with the supervisory definitions needed to assess compliance with, for example, Basel rules. At the very least, the AnaCredit information can provide a timely proxy for banks' credit risk.

To "recycle" data in other domains might dictate additional requirements that do not seem immediately useful to the reporting parties. For example, the definition of what precisely constitutes a "bad" loan might currently be left up to a reporting bank (because the definition ties into internal control procedures), and the definition of when a loan is past due is then firm specific. This firm-specific choice might be determined by accounting rules, regulations, or legacy IT systems. For understanding a bank's credit risk, the precise definition of past due

---

[8] The code for this project is available at https://github.com/imanvl/RTF_NTW_Horse. This approach is taken much further in the biomedical sciences, where similar security and storage concerns are at play. See, for instance, the Global Alliance for Genomics and Health (www.ga4gh.org).

[9] See www.newyorkfed.org/ibrn for more information.

[10] Yet another alternative is to encrypt the data in such a way that relevant information can be shared without revealing individual reporters' data. See Flood et al. (2013).

[11] The IDH is part of the FSB's Data Gaps Initiative. See FSB (2009) for more details.

is less important than that this definition is consistent over time and that the data are recorded accurately. To make the data useful for macroprudential analysis, however, it is important to use a single definition applied consistently across firms. Furthermore, if data are to be used for cross-country analyses, then these definitions need to be agreed upon internationally as well.[12]

The ECB has undertaken a pilot to see how the integration of highly granular data can be accomplished (Lauro and Traverso, 2018). The use case of this pilot was the analysis of the impact of the quantitative easing (QE) program on banks' balance sheets, on both the asset and the liability side. A key finding is that integrating two separate sets requires even higher data-quality standards of the contributing sets than in the case of separate reporting; the data not only need to be internally consistent but also must match across reporting frameworks. Authorities thus need to be "prepared for integration" by putting in place integrated data-management policies and adhering to best practices. Barbic et al. (2017) provide detail on how, based on microprudential information, the ESCB has developed a data set with consolidated banking data that can be used for macroprudential analysis.

5 Data Governance

To reap the benefits associated with combining different data sets on a regular basis, we first need a conceptual framework FOR how we handle data. One possible starting point is the four-quadrant model we employ at DNB (Figure 10.5). In this model, the quadrants are delineated on two dimensions. First, on the y-axis, we plot whether data are produced on demand (pull) or as part of a regular production cycle (push). An example of the former is the construction of a data set for a dashboard built by a business analyst. The analyst could, for instance, be interested in a holistic view of the risk of a bank and would like to see risk metrics coming from different reports all presented in a single view. In this case, the analyst is actively asking for the data. An example of the pushed data is the regular prudential reporting process. The collection of such data is mandated by law, and such data will be collected without analysts actively asking for it. This is not to say that analysts are not interested, merely that they do not have to instigate the process. The second dimension, delineated on the x-axis, is whether the data collection is a structured, systematic process or an ad hoc one. For instance, for producing macroeconomic statistics, well-developed processes are in place that have been honed over the years. For ad hoc research projects, in contrast, entirely new data sets or even methods to collect data need to be developed.

The combination of these two delineating dimensions results in four quadrants for which we require different data governance. To ensure that data in Quadrant I remains highly structured, we need to apply clear definitions and maintain high data quality. Generally, we define a logical data model (LDM). An LDM is an exact definition of a data set. It defines aspects such as the number of variables and the storage type(s) and provides a description of what the variable supposedly measures. It also describes the relationships between variables.

---

[12] Initiatives such as the development of the Financial Industry Business Ontology (FIBO) are very helpful. FIBO is an open-source business conceptual ontology of how financial instruments, business entities, and financial processes work across the global financial industry (see https://edmcouncil.org/page/fiboproductsaccessre).

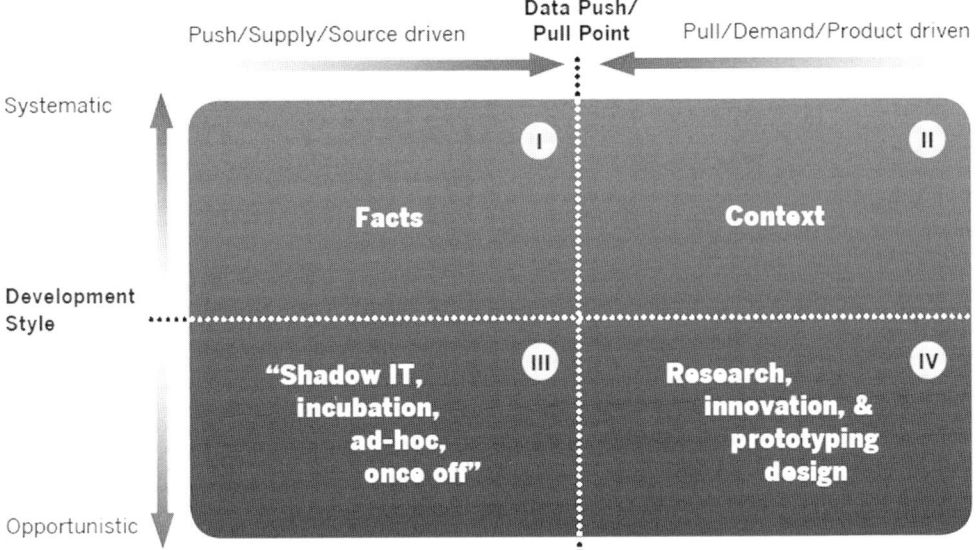

Figure 10.5 The four-quadrant model for data governance.

If reported data do not pass the data-quality tests dictated by the LDM, then one or more resubmissions are required until the data are of sufficient quality to be admitted to the production database. Quadrant I also requires an audit trail of data-point changes; otherwise, the integrity of the data cannot be guaranteed. For example, analysts could—unintentionally or otherwise—change important benchmark interest rates or inflation rates. Quadrant II holds those (intermediate) data sets that users can, for instance, visualize in a dashboard. As discussed previously, the analysts actively ask for data and are therefore also more involved in ensuring the quality of the data. Jumping ahead, Quadrant III covers those sets that require no centralized governance because they are not meant to be put onto the DNB data platform. Data sets in this quadrant are public or are local ad hoc data collections. The ownership—and hence the responsibility of properly maintaining the data—lies elsewhere (for public data) or with the local user.

For stress testing, Quadrant IV is the most important quadrant. For the most part, stress tests are bespoke exercises focusing on particular risks or scenarios. Although there might be a legal basis requiring periodic stress tests, the details are typically not spelled out. This implies the lower half of the *y*-axis of Figure 10.5. In some countries, stress tests are becoming an annual fixture, and hence the data requests might fit better in a Quadrant I process. However, to our knowledge, this has not happened anywhere. In terms of data governance, the best place for stress test analyses is Quadrant IV. Here the analyst can bring together the various data sources usually required in stress testing. Because stress test scenarios change regularly, the input required will also differ each time.

The choice of where to place a particular data set in the quadrant model depends on various factors. For example, the *importance* of the analysis using the data determines how much effort should be put into getting the collection process to the very highest level. With new data sources, this presents a problem because the universe of possible useful analyses is not immediately apparent to management, whereas the costs are real and immediate. *Repeated usage* might merit further investment in turning the reporting into a process. Another key factors is the *available budget*. Finally, there is some value in the *flexibility required*. Tying yourself to the mast with a very prescriptive data-intake process also implies that deviations from the process are costly. A deviation could, for instance, be requesting an accelerated delivery with lower data-quality standards because a crisis is unfolding and authorities value noisy timely information over perfect but information that arrives too late. In such data files, some checks or required resubmissions might have been foregone. The information will be available faster but will also be noisier.

A major advantage of promoting a data set to Quadrant I is that data quality is unambiguous. This makes, for instance, combining data straightforward; because identifying keys are clear, no time is lost in matching data points from different sources. Furthermore, issues with missing or tainted data should have been solved. It is up to the data owner to weigh the cost of promoting the data set to Quadrant I/II against the benefits of ease of use.

The challenge is to let the governance model create value in stress testing. As noted, stress testing often combines highly structured data (e.g., prudential call reports in the United States or Financial Reporting [FINREP] and Common Reporting [COREP] in the EU) with ad hoc data sources (e.g., banks' proprietary model parameters). Moreover, if the stress test covers multiple sectors, then information from different types of institutions (e.g., banks, insurers, and pension funds) needs to be shared by supervisors with different mandates. From Quadrant IV, the access to the high-quality structured data, which resides in Quadrant I, should be without impediments for those with the proper access rights. Given that the structure is clear, direct querying of databases using Structured Query Language (SQL) can be implemented without impinging on confidentiality. Such direct access will reduce data-wrangling costs, reduce errors, and potentially make it possible to deliver stress tests results in a more timely fashion.

Note that the quadrant model does not dictate the physical setup but primarily describes the governance of data. Traditionally, the physical setup (i.e., the hardware and the servers deployed on it) was tightly linked to the governance. Some servers were maintained in production with very tight governance, whereas others were research and development servers with a much looser regime. With the appropriate access-rights procedures and rights-maintenance software, this tight coupling can become looser—applications and data can be anywhere because their use can be controlled.

The questions to ask about data governance can thus be decoupled from the physical setup: What can we expect of the quality of the data? How important is traceability of data mutations? Who is allowed access? In practice, the quadrant model does still map to different types of workspaces. At DNB, we use primarily the tooling for our large databases (SAS), a structured business intelligence (BI) process, and strict governance for Quadrants I and II. For Quadrant IV, we have a research area network and a high-performance cloud environment where analysts have the freedom to deploy a much wider range of (open-source) tooling. With this comes the responsibility to maintain this tooling and vouch for the solidity of the analysis.

In an ideal world, all data would come with a completely fleshed-out LDM. An LDM defines the structure expected in the data. It will stipulate data types (e.g., string or float) and relations between data points (e.g., total assets should be equal to total liabilities). If the data complies with the LDM, a complete description of the data is ensured. Such a description can, for instance, be used to ensure consistency across jurisdictions and communicate with reporting entities. It also points out exceptions but does not in itself provide the "business rules" that need to be applied to find practical solutions to reporting errors. For example, if the LDM prescribes using two-digit ISO country codes, then a sensible business rule might be to also allow valid three-digit ISO codes even though, strictly speaking, these are a violation of the LDM. In the cleaning process, a mapping from the two-digit to the three-digit ISO codes could be applied.

In practice, we define LDMs for a limited set of reports. Defining an LDM is costly because it entails fleshing out all the possible admissible permutations in a reporting framework. This is not trivial because many relevant regulations are ambiguous in their definitions (because, more often than not, they are the result of negotiations, and the compromise has to reflect multiple views). Especially with ad hoc requests, there is insufficient time to work out a fully fledged LDM. Moreover, it might not be cost-efficient to define a complete LDM for a single-shot data request.

Sometimes the LDM approach might seem overbearing and overengineering for a single report. For instance, if we are designing an application to capture the contact information at supervised institutions (e.g., a mail address or phone number), it might seem overkill to work with an LDM. However, this contact information might feed into a larger system that, in the end, is meant to be shared through an internal or a public register. If the data source is not captured unequivocally, then the register data will be tainted. In practice, users across the institution will start to keep track of contact information, leading to inconsistent and incomplete information. The costs materialize in the relatively simple address application, whereas the benefits only materialize later for the organization as a whole. The same holds for more important information, such as the legal structure of supervised firms or the identifiers of financial instruments. Fairly attributing the costs of a local solution that is "too much" at the local level but has institution-wide benefits is a significant management challenge.

The challenge is to find a balance between LDM fetishism and "anything goes" reporting. Although adhering to an LDM ensures that the data we admit to the final database meet a specified standard, there can be good reasons to diverge. For instance, if a new data source needs to be available on short notice in a crisis situation, then defining an LDM will take too long. Alternatively, for small, one-off projects, the costs of definition might create too much overhead. Nevertheless, the concept of an LDM is worthy to be applied more widely. The adoption of LDMs is, however, hampered because among policymakers— mostly economists and lawyers by training—the concept is relatively unknown. Unfortunately, LDMs are thus generally an afterthought to reporting requirements that have been the subject of long-running negotiations. The structure and the content of the templates are then already set. Ambiguities that a rigorous approach such as an LDM definition process bring to the surface are then difficult to address. It would therefore be helpful if LDM-like thinking would be embraced from very early on in the reporting design. Furthermore, the external benefits that materialize only later and elsewhere in the organization should be incorporated in weighing the pros and cons. One way to accomplish this is to subsidize the drawing up of the LDM.

Acceptance of LDM-like thinking is hampered by, on the one hand, something we could term *LDM fetishism* and, on the other hand, the typical manner in which LDMs are presented to the noninitiated. LDM fetishism is sticking to the model no matter what. This is not helpful because policymakers and researchers generally prefer noisy information over no information at all. In some, if not most cases, a greatly simplified LDM is more than adequate. The second reason LDMs are not widely embraced is because of the way they are presented. The typical end user is used to thinking in terms of a template or an Excel sheet. The LDM, however, reduces these Excel sheets to their essence and strips common elements to be defined in a separate place. The logical structure is then presented graphically. These wall-size graphics, printed in tiny fonts and using, to the uninitiated, unintelligible symbols, are difficult to digest for most supervisors or policy officers. Modelers should put more effort into presenting the LDM in ways that appeal to the end user and that make clear how the LDM adds value.

6 Developing a Stress Testing Capability

Once all these governance challenges, technical difficulties, and conceptual issues have been resolved, we will be able to paint a complete picture of the financial ecosystem from all angles.[13] This requires a more or less comprehensive coverage of all relevant financial sectors. For some sectors, further detail is required to be able to identify systemically important entities. For example, data on lightly or nonregulated sectors are generally more sparse, and if financial activity in these sectors picks up and starts to become a factor influencing systemic risk, then new ways of unearthing this information need to be found. For instance, although currently FinTech firms' activity is only a fraction of the established financial sector (Frost, 2020), this might change in the future. For other sectors, a coarser aggregate can be sufficient. For instance, if contracts are linear, then stress test effects are less susceptible to tail scenarios, and we thus have less need for the full distribution of exposures. It also requires that we collect stocks of all relevant balance-sheet positions (including off-balance-sheet items and derivatives exposures). Furthermore, to detect fast-moving risk shifting, we will need to collect financial-market transaction data (i.e., flow data).

Ideally, the validity of the data is beyond reproach. In practical stress testing, significant effort is spent on discussions about whether the often-patchy data quality warrants the conclusion. However, discussions about the interpretation of stress test results should ideally not be about the validity of the bare facts; these discussions should focus on the conclusions and their policy implications. In practice, some of the data might indeed be noisy and—as some argue—this is inevitable. In stress testing, we often have no choice, however. We should be pragmatic and get the data up to acceptable levels with sensible business rules.

With data on the entire ecosystem, we will be able to deliver the raw material for both micro- and macroprudential stress tests. In a sense, we can seamlessly zoom from a highly aggregate macro level all the way down to the individual transactions of individual market participants. Moreover, we can also connect markets and see what market participants or sectors are doing in different instruments. Depending on the mandate, regulators and supervisors are interested in different scales of aggregation. For instance, a conduct-of-business supervisor will be interested in individuals trading with insider knowledge. A microprudential supervisor will be looking at the solvency of a particular firm, whereas a macroprudential

---

[13] See Bijlsma et al. (2018) for an example.

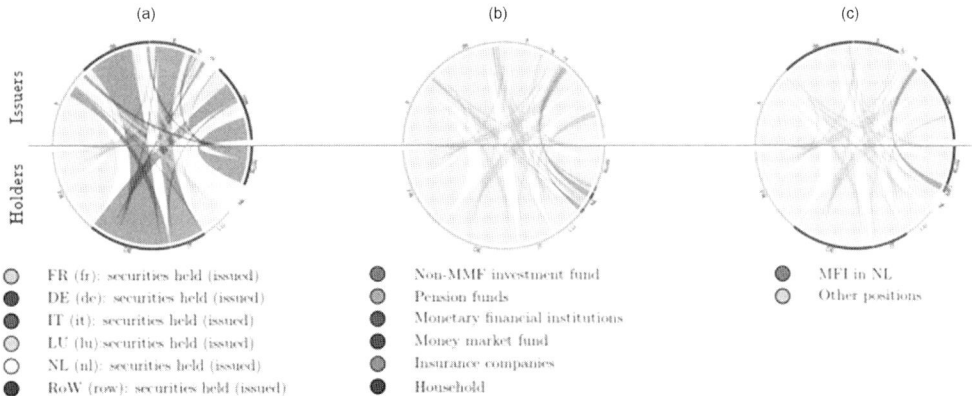

Figure 10.6 Securities data at different levels of granularity. MFI, monetary financial institution; MMF, money market fund.

authority will look for vulnerabilities at the macro level. With granular data, views for each of these can be generated from a single data source. To be able to zoom in on different timescales, we need the most granular level to start with.

To illustrate the notion that collecting granular data allows for flexible "zooming," we show data on securities holdings in the euro area. For compiling national accounts statistics, ESCB member banks collect securities holdings of euro-area residents combined in the SHS. Each central bank collects individual securities holdings from relevant entities in their jurisdiction. In the Netherlands, the reporting sample includes around 800 reporters. In principle, we can thus move from country aggregates all the way down to individual firms holding a particular instrument. We visualize this process in Figure 10.6 in three panels.[14] In each panel, we use the same data but focus on different levels of aggregation. For clarity, we collapse smaller countries to the category "rest of the world" (RoW) and show holder countries in the bottom half of the circle, with issuers are shown in the top half.

In panel (a), we show the total exposures of country holders to various issuing countries. This could be called the "macro" view. Here we see that France is the largest holder (8.3 trillion euro) and that its holdings have a significant home bias: 60 percent of its holdings are issued domestically, with 500 billion euro (6 percent) issued in Germany, a distant second. It is clear that home bias is an entrenched stylized fact. This level of aggregation is useful for macroeconomic analyses of international capital stock and flows and is most useful for macro stress tests.

Next, if we are interested in a particular sector, we can zoom in, as panel (b). We term this the "meso" view. Here, we show the exposures of the different sectors in the Netherlands to other issuing countries. The largest holdings are concentrated in the pension fund sector, with the monetary financial institutions (MFIs) or banking sector a close second. We see that for most sectors, the largest exposure is to Dutch firms (i.e., a self-loop).

Panel (c) highlights the exposures of a single sector, in this case the banking sector. Showing all sectors, as in panel (b), can sometimes conceal important exposures if we are

---

[14] See Sigl-Gloeckner (2018) for a survey of network visualization methods, in particular for the SHS data.

interested in just a single sector. Alternatively, we could use color to denote the issuing country. Such a view would allow us to focus on the source of foreign shocks.

Zooming in even further (not shown due to data confidentiality), we get to the "micro" view, where we could show to what country-sector one particular pension fund is exposed. This would be of interest to the pension fund's supervisor when assessing a fund's country risk. Finally, we can discern the "nano" view. Here, we could highlight the exposure of a single fund to a single instrument (identified by ISIN). As an additional flourish, we can show all related instruments issued by the same corporate group or, in the same vein, to a particular group of instruments (cf. Boermans and van Wijnbergen, 2018).

Having the ability to zoom is potentially extremely valuable for the efficiency of stress testing. The same source data can be used for stress testing from different micro and macro angles. For example, if we are worried about the country-risk exposures of a particular sector, we take a "meso" view and stress the valuation of the instruments held. This is generally a financial-stability concern. Conversely, if we are interested in a microprudential approach, we can just select the individual institution and apply the stress parameters to its portfolio items. The same data can thus serve different constituencies. Incidentally, having granular data at our disposal also means that macroprudential stress tests can have micro foundations, meaning that they can model/simulate shock amplification with much higher degrees of verisimilitude, potentially capturing otherwise-hidden risks.

With the underlying data of the SHS, national authorities can also move to the individual firm, be it a bank, a pension fund, or an insurance firm. The portfolios of the larger banking firms are available in full detail. On this level, we can currently add a lot of other firm-level data. For example, we can add public information from commercial data providers such as SNL and ORBIS. For research purposes, we can add prudential information from, among others, FINREP and COREP. With such information, we can, for instance, study how a bank suffering liquidity stress—as measured in the regulatory liquidity coverage ratio (LCR) reports—reacts by engaging in precautionary hoarding of liquid securities (Acharya and Merrouche, 2013) or is forced to fire sale (Diamond and Rajan, 2011).

Going forward, it should also become much easier to zoom using other cross-country data collections. For example, the EMIR, SFTR, and MMSR are all granular data collections that can be analyzed on different aggregation levels.

Consistent data across different markets will also allow for stress testing of connected markets. For example, a recent trend is that exposures in derivative markets are ever more mitigated by the exchange of collateral. This is driven by regulations mandating central clearing but also by firms' reduced appetite for counterparty credit risk. By exchanging collateral, counterparty credit risk is reduced significantly. However, because collateral requirements are very sensitive to market volatility, the reduction in credit risk is accompanied by an increase in liquidity risk. Clearly, sudden spikes in market volatility translate directly into collateral calls. And funding such calls can be particularly expensive or difficult at precisely these same times. To properly understand either the derivative or funding markets, such as the repo market, we should thus be able to analyze the two markets jointly (Adrian et al., 2013).

Collecting consistent information for an entire market might allow for better risk assessment than what financial markets can currently achieve. Financial markets are generally very efficient in digesting information. However, a financial market cannot price a risk it cannot observe. In many OTC markets, only a few core players can form a reasonably coherent view of all participants' positions. Such market-making parties can capitalize on the information embedded in order flow or knowledge—accumulated over time—of counterparties' positions (Duffie and Manso, 2007). In a dispersed market, however, such centralized knowledge is not available. The market can thus sometimes miss pockets of risks due to, for example, concentration risk. A well-known example is AIG, the American insurer that turned out to be a key player connecting the financial markets in the 2008 crisis. Information on the network as a whole can potentially improve on fragmented information. For example, Squartini et al. (2013) show that network information can show fragmentation in interbank network connections significantly earlier than market signals. Such market fragmentation can potentially hamper a market's ability to redistribute risks.

Note that stress tests are often kept simple on purpose in order to be able to tell causal stories (Drehmann, 2009). Some data-analytics methods do not fit well with such a strategy. The results from ML algorithms, for instance, are difficult to trace back to the actions of individual actors. This makes explaining the outcomes to mostly uninitiated users of the results (i.e., management or the general public) rather difficult. Without a comprehensible narrative, defending stress test outcomes becomes very difficult. One way forward would be to try to model the complexities in the tradition of Herbert Simon (1962). The understanding of the underlying processes would allow for the translation into causal stories. Another way forward would be to tell a broader story about how shocks to the system, depending on their nature and magnitude, can either be absorbed or are destabilizing (Wiersema et al., 2020). Finally, in some cases, contributions to particular outcomes can be attributed to particular agents or risk drivers using the concept of Shapley values (Joseph, 2019).

7 Challenges in a New Environment

The new possibilities for stress testing discussed so far also come with several challenges as well in terms of (a) governance, (b) the IT capacity to handle the data volume, and (c) the skill sets needed. We will discuss each of these in turn.

7.1 Governance

Collecting data on a more granular level implies more, often sensitive, detail. Moreover, the granular level sometimes contains information on natural persons. For example, loan-level information might contain natural persons' names, which are, in principle, not needed for stress testing (beyond being the primary key to link other information (e.g., tax records). Handling such data demands additional care. In this case, it might be useful to anonymize the loan-level data while preserving the ability to combine the data with other sources. In Europe, the General Data Protection Regulation (GDPR), which is being enforced as of May 2018, now comes with much tighter regulation of what firms and authorities can and cannot do with natural persons' data. In collecting data that contain or might contain natural persons' data, adequate thought needs to be given to ensure fair use while still adhering to the GDPR.

7.2 Volume

To collect, digest, and store larger volumes of data, authorities might need to explore unfamiliar terrain, for example, storing data outside their own servers. Although generally the costs of commercially available cloud services (i.e., computing and storage) have come down drastically, many government agencies are not yet ready to take advantage of this because this would mean outsourcing IT capacity in the cloud. The decision to outsource is an important strategic choice that also comes with operational, security, and legal risks. Arguably, security risks have economies of scale and could thus possibly be better handled by larger and more specialized outside parties. Note that in some cases, the data are not allowed to leave the jurisdiction, and in smaller jurisdictions, this limits the number of providers that are able to provide the services demanded.

To get a better understanding of the challenges, possibilities, and limitations of using cloud services, DNB started an initiative in early 2018 to bring confidential data to the cloud. Here, we can more easily apply advanced and computationally expensive techniques to data sets of several billion observations. Such volumes are already too big for most central banks and supervisors, and thus potentially useful insights are foregone. Moreover, one might even argue that the mandate to collect transaction-level data is rendered pointless if analysts do not have the IT infrastructure needed to study the data. One of the pilot projects undertaken to build on our capacity to analyze at scale is a computationally expensive stress test of margining requirements in the OTC interest rate derivatives market. The stress test looks into the liquidity risk of margin procyclicality, which has already attracted some attention in the academic literature (cf. Glasserman and Wu, 2018).

7.3 Skill Set

If an organization is to become truly data driven, then data should be embedded centrally in the business process. This requires that those involved should not be "data challenged." Some would argue that regulators are more likely to be "data challenged" than the regulated and that this creates its own pitfalls. Traditionally, market-conduct supervisors are trained lawyers, whereas central banks rely on economists. Both lawyers and economists generally have attractive outside options in the private sector. Authorities cannot remunerate on the same level as the private sector but try to compete by providing a better work–life balance and an appeal to civic duty. In addition, working in government for some time allows professionals to build a network and get a much better understanding of the regulatory process. This is very valuable for a career in the private sector, and the wages foregone can thus be seen as an investment in higher future income in the private sector. However, Luca Enriques has argued that such a trade-off might not be relevant for IT engineers and data scientists in RegTech or SupTech. Developing IT applications for regulatory compliance or supervisors is much less dependent on soft skills and localized knowledge. It will therefore be more difficult for government agencies to attract top-level IT engineers. Given the ample outside opportunities for people with these skills, it becomes even more important to offer interesting and rewarding jobs. Hopefully, the very rich data sets central banks and supervisors can use and the appeal of a meaningful job that offers rewards to society at large is sufficiently convincing to data scientists and IT engineers to accept a pay cut.

It is crucially important that data scientists interact with supervisors and policy officers. In doing so, data scientists will be inspired by real-world problems, whereas their colleagues in policy and supervision will get a better understanding of where central banking and supervision can benefit in practice from the data revolution. New techniques such as AI and ML will not drive out professional judgment, which, in terms of analyzing shocks and their impact on the financial sector, will remain essential for figuring out which shocks are feasible and for thinking outside the box more broadly. In terms of policymaking on the basis of stress testing, judgment calls will remain essential, for instance, in case of dilemmas where a choice is needed between two options that each affect a different group in society.

8 Discussion and Conclusion

To really make new opportunities in data pay off for stress testing requires hard work. The new and enticing possibilities are not a free lunch. To achieve them means to adhere to data-governance principles, for instance, by using metadata consistently, and on a large scale, collecting and cleaning new data sets, applying LEI and Unified Payment Interface (UPI) codes, and so forth. For a single stress testing project, the costs may seem disproportional. But usually there are substantial external benefits. As we go further down this road, new opportunities are emerging for an all-encompassing risk management approach on the basis of modeling, in a comprehensive way, the financial ecosystem.

In this chapter we have explored the opportunities that more granular data and reporting that is closer to real time can bring. Granular—even transaction-level—data allow us to aggregate the data in many ways. This is ideal for stress testing because here, flexibility is key; the strength of stress testing is that it is possible to entertain unlikely scenarios that have never materialized. Often these involve new markets or players for which we don't regularly collect data. These opportunities also come with challenges. Data collection is often organized in silos. For instance, insurance supervisors focus on insurance firms (e.g., Solvency II), whereas reporting for monetary policy operations (e.g., MMSR) primarily covers banks. In looking at the impact of interest rate setting on the wider economy, it might, however, be interesting to analyze how money market funding conditions are affected by insurance firms' repo transactions, and vice versa. Such analysis is hampered by inadequate coordination on reporting standards (frequency, metadata, etc.).

Having more and more granular data in-house also puts more weight on the governance of that data. For instance, who is the owner of the data? Dispersed ownership is an invitation to free riding and cost shifting. At DNB, the owner is therefore always a single person, although he or she is free to involve others for input or funding. Given that more granular data are often also useful more broadly, finding an organizational form that takes into account external effects is becoming more important. Furthermore, the owner should aim for maximally distributing the data to be able to realize its full potential. This implies a "share, unless . . . " policy, which should be accompanied by well-implemented access rights on the ground. Such a data policy should hold within the organization as well as for the interactions between the organizations involved—that is, the macroprudential supervisor (European Systemic Risk Board [ESRB] and Office of Financial Research [OFR]/Financial Stability Oversight Council [FSOC] in the EU and the United States, respectively) and supervisory authorities and central banks.

The flood of new data potentially also has reputational risk for authorities. For some parts of the precrisis system, authorities knew very little and were thus unaware of the build-up of risks. In the aftermath of the crisis, a massive increase in reported data ensued. The challenge authorities now face is to effectively use all the data and turn it into information and actionable analysis. Failing to incorporate the new data into analyses runs the risk of instigating a large public backlash the next time a crisis materializes. Some might conclude that authorities did have crucial information reported but failed to act on it.

In sum, new data and tooling have the potential to significantly improve stress testing, but challenges remain. The sheer volume of data requires an investment in data-processing capacity. More importantly, this requires a change in the way we use the data. Traditionally, those involved in stress testing have a supervisory or a financial stability background. Knowledge of database management and optimizing database queries is thus often insufficient to handle large volumes efficiently. Another challenge is how to communicate results to senior management and laypeople. Scrolling through the data in an Excel sheet to get a feel for the data is no longer feasible. In presenting results, more care should thus be given to supply appropriate dashboarding as well. The characterization of high-dimensional data with new concepts—such as the eigenvector centrality of a bank in a market—needs to be accompanied by sufficient explanation to instill a sense of comfort in decision-makers.

References

Abad, J., I. Aldasoro, C. Aymanns, M. D'Errico, L. Fache Rousová, P. Hoffmann, S. Langfield, M. Neychev, and T. Roukny (2016), "Shedding light on dark markets: First insights from the new EU-wide OTC derivatives dataset," ESRB Occasional Paper 11.

Acharya, V. V., and O. Merrouch (2013), "Precautionary hoarding of liquidity and inter-bank markets: Evidence from the sub-prime crisis," *Review of Finance*, 17, 107–160.

Adrian, T., B. Begalle, A, Copeland, and A. Martin (2013), "Repo and securities lending," Federal Reserve Bank of New York Staff Report 529.

Anand, K., I. van Lelyveld, A. Banai, T. Christiano Silva, S. Friedrich, R. Garratt, G. Halaj, I. Hansen, B. Howell, H. Lee, S. Martínez Jaramillo, J. L. Molina-Borboa, S. Nobili, S. Rajan, S. Rubens Stancato de Souza, D. Salakhova, and L. Silvestri (2018), "The missing links: A global study on uncovering financial network structure from partial data," *Journal of Financial Stability*, 35, 107–119.

Aymanns, C., C.-P. Georg, and B. Golub (2018), "Illiquidity spirals in coupled over-the-counter markets," Working Paper, available at http://bengolub.net/wp-content/uploads/2020/05/multilayer.pdf

Barbic, G., S. Borgioli, and J. Klacso (2017), "The journey from micro supervisory data to aggregate macroprudential statistics," ECB Statistics Paper Series 20.

Berner, R., and K. Judge (2019), "The data standardization challenge," ECGI Working Paper Series in Law 598.

Bijlsma, M., M. Castro Campos, R. Chaudron, and D.-J. Jansen (2018), "Building a multilayer macro-network for the Netherlands: A new way of looking at financial accounts and international investment position data," IFC Working Paper.

Boermans, M. A., and S. van Wijnbergen (2018), "Contingent convertible bonds: Who invests in European CoCos?" *Applied Economics Letters*, 25, 234–238.

Bottega, J. A., and L. F.Powell (2011), "Creating a linchpin for financial data: Toward a universal legal entity identifier," *Journal of Economics and Business*, 64(1), 105–115.

Cerutti, E., and C. Schmieder (2012), "The need for "un-consolidating" consolidated banks' stress tests," IMF Working Paper WP/12/288.

Diamond, D. W., and R. G. Rajan (2011), "Fear of fire sales, illiquidity seeking, and credit freezes," *Quarterly Journal of Economics*, 126, 557–591.

Drehmann, M. (2009), "Macroeconomic stress-testing banks: A survey of methodologies," in M. Quagliariello (ed.), *Stress-testing the banking system: Methodologies and applications*, Cambridge University Press.

Duffie, D., and G. Manso (2007), "Information percolation in large markets," *American Economic Review*, 97, 203–209.

Ehlers, T., S. Kong, and F. Zhu (2018), "Mapping shadow banking in China: Structure and dynamics," BIS Working Paper 701.

European Commission (2019), "Fitness check of EU supervisory reporting requirements," Commission Staff Working Document SWD.

Fang, Y., and I. van Lelyveld (2014), "Geographic diversification in banking," *Journal of Financial Stability*, 15, 172–181.

Financial Stability Board and International Monetary Fund (2009), "The financial crisis and information gaps," Report to the G-20 Finance Ministers and Central Bank Governors.

Financial Stability Board and International Monetary Fund (2017), "Second phase of the G-20 Data Gaps Initiative second progress report," Report to the G-20 Finance Ministers and Central Bank Governors.

Flood, M., J. Katz, S. Ong, and A. Smith (2013), "Cryptography and the economics of supervisory information: Balancing transparency and confidentiality," OFR Working Paper 11.

Foster, W. T. (1922), "The circuit flow of money," *American Economic Review*, 12, 460–473.

Frost, J. (2020), "The economic forces driving FinTech adoption across countries," DNB Working Paper 663.

Glasserman, P., and Q. Wu (2018), "Procyclicality in sensitivity-based margin requirements," *Management Science*, 64(12), 5461–5959.

Graff, Joshua (2018), "The real shape of Britain's post-Brexit talent brand," *Linkedin Pulse*.

Joseph, A. (2019), "Shapley regressions: A framework for statistical inference on machine learning models," Working Paper No. 2019/7, King's college London.

Judson, R. (2017), "Enhancing data availability: Recent U.S. experience with banking data," Workshop on Data Sharing, Deutsche Bundesbank.

Kienecker, K., G. Sedlacek, and J. Turner (2018), "Managing the processing chain from banks source data to statistical and regulatory reports in Austria," OeNB Statistiken 3.

Kwast, M. L., S. Holden, D. Jurcevic, and I. Van Lelyveld (2010), "Norges Bank stress testing of credit risks," Report of an external review panel.

Lauro, B., and R. Traverso (2018), "Data fitness for integration," Mimeo.

Legal Entity Identifier Regulatory Oversight Committee (2018), "The global LEI system and regulatory uses of the LEI," Progress Report.

Levels, A., R. De Sousa Van Stralen, S. K. Petrescu, and I. Van Lelyveld (2018), "CDS market structure and risk flows: The Dutch case," DNB Working Paper 592.

Sigl-Gloeckner, P. (2018), "Visualising financial systems," Thesis, Imperial College London.

Simon, H. A. (1962), "The Architecture of complexity," *Proceedings of the American Philosophical Society*, 106(6), 467–482.

Squartini, T., I. van Lelyveld, and D. Garlaschelli (2013), "Early-warning signals of topological collapse in interbank networks," *Nature Scientific Reports*, 3, 3357.

The Joint Forum (2012), "Report on intra-group support measures."

van Lelyveld, I., and F. Liedorp (2006), "Interbank contagion in the Dutch banking sector: A sensitivity analysis," *International Journal of Central Banking*, 31, 99–133.

van Lelyveld, I., F. Liedorp, and M. Kampman (2011), "An empirical assessment of reinsurance risk," *Journal of Financial Stability*, 7, 191–203.

Wang, D., I. van Lelyveld, and J. Schaumburg (2018), "Do information contagion and business model similarities explain bank credit risk commonalities?" DNB Working Paper 619.

Wiersema, G., A. M. Kleinnijenhuis, T. Wetzer, and J. D. Farmer (2020), "Scenario-free analysis of financial systems with interacting contagion channels," Institute for New Economic Thinking Working Paper No. 2019-10.

Stress Tests Disclosure: Theory, Practice, and New Perspectives[*]

Itay Goldstein and Yaron Leitner

1 Introduction

From the early days of stress testing banks in the wake of the financial crisis of 2008, disclosure has been a key issue of discussion. Although banking regulation and supervision were not very transparent over the years leading up to the financial crisis, the crisis created pressure not only to examine banks more closely and in a forward-looking way (which is done with stress tests) but also to reveal the results of the examination publicly. Initially, disclosure was constrained; for example, the first stress tests in Europe did not reveal much information. However, over time, an increasingly more frequent view among regulators and central bankers took hold, according to which they should follow the demand for more transparency and disclose much of the information revealed in the course of stress tests.

There are many controversies surrounding this view among practitioners, academics, and regulators. To give a flavor of the debate, in a *Wall Street Journal* article from March 5, 2012, former governor of the Federal Reserve Board Daniel Tarullo is quoted as saying that "the disclosure of stress test results allows investors and other counterparties to better understand the profiles of each institution." On the other side, the Clearing House Association is quoted as saying that this disclosure "could have unanticipated and potentially unwarranted and negative consequences to covered companies and U.S. financial markets." Such debates about the desirability of the disclosure of stress test results continued over the years.

Another issue, which has gained momentum recently and on which there is disagreement, is the transparency of the models that regulators use to conduct stress tests. Under the current policy, the Federal Reserve does not fully disclose its stress test models. Until 2018, the Federal Reserve only provided a broad description of its models (together with the publication of the results). Starting in 2019, the Federal Reserve moved to provide more details about its models, such as certain equations and key variables that influence the result. The Federal Reserve will also now illustrate how its models work on actual loans held by Comprehensive Capital Analysis and Review (CCAR) firms and on some hypothetical loan portfolios. In Europe, model disclosure has been more prevalent in recent years.

Economic theory suggests that public disclosure is not a panacea and that there are trade-offs that need to be considered when deciding how much information should be disclosed

[*] We thank Kathryn Judge, Til Schuermann, and other participants at the International Monetary Fund (IMF) "Rethinking Financial Stability: The FSAP at 20" conference for very helpful comments. We also thank Sophia Hua for research assistance.

publicly. Empirical evidence from various broad studies in economics and finance provide support for the possible downsides of disclosure. In an early policy paper commissioned by the Committee on Capital Markets Regulation after the introduction of stress tests, Goldstein and Sapra (2014) review various theories highlighting the costs and benefits of greater disclosure and link them to the context of stress tests. They conclude that although the benefits of disclosure are obvious, one has to be mindful of the costs and take steps to design disclosure policy to mitigate their impact.

As this chapter will suggest, we think that it is important for regulators to acknowledge that the disclosure of stress tests results and disclosure of the models behind the tests involve a trade-off and by no means can be viewed as a panacea. We believe it is important to think of a disclosure scheme that accounts for the costs and benefits and how they may evolve over time as the conditions in the economy and the financial system change. We caution against the current trend whereby increased stated policies for greater disclosure of results might just lead to weaker tests and where the resulting complacency in the financial system is not fully understood and accounted for.

The remainder of this chapter is organized as follows: In Section 2, we review the main costs and benefits of the disclosure of stress test results and their implications, as discussed by Goldstein and Sapra (2014). In Section 3, we review the implications of the unified self-contained frameworks, which were developed later on, dedicated to the analysis of disclosure of stress tests results. We focus on the framework of Goldstein and Leitner (2018) and other related works. In Section 4, we discuss the relation between implications for information disclosure and implications for stress test design and how the two are intertwined. In Section 5, we review the evolution of practice on the disclosure of stress test results over the years and caution against the link between greater disclosure and weaker tests. In Section 6, we discuss the disclosure of the models behind the tests. We conclude in Section 7.

2 Costs and Benefits of Disclosure

We review here the costs and benefits of disclosure discussed by Goldstein and Sapra (2014), their implications for stress tests, and some empirical evidence (see also Leitner, 2014). We also discuss some recent research that expands on the early work. We postpone discussion on recent work designed to build unified self-contained frameworks to analyze the disclosure of stress test results and the disclosure of the test models to Section 3 and Section 6.

A widely used argument in favor of disclosure of the test results is that it helps discipline banks. The idea is simple. If investors do not have information about the bank, the price of the bank's securities will not reflect the bank's true risk, and so banks may be induced to take more risk than is socially desired. However, if investors have information about the bank, prices will reflect the true risk, and risky banks will thus find it harder to raise money. This would induce banks to take less risk to begin with. More generally, the benefit of disclosure is that it allows market participants to make more informed decisions. This allows for better allocation of funds and also has a disciplinary effect on banks because they can be rewarded according to their actions. In a general corporate-finance context, various studies show that greater disclosure benefits firms by reducing information asymmetries and constraining managerial misbehavior (e.g., Leuz and Verrecchia, 2000; Greenstone et al., 2006).

Another benefit of disclosure is in allowing greater discipline and accountability of the regulators themselves. For example, disclosing information could make it easier for the regulator to commit to enforcement actions that are optimal ex ante but are suboptimal ex post. A recent empirical article by Kleymenova and Tomy (2019) provides evidence along these lines: when regulators are required to disclose enforcement actions, they are more likely to issue enforcement actions, as well as to rely on publicly observable signals to issue enforcement orders, suggesting a response to the increased public scrutiny of their actions.

However, the academic literature has also highlighted some problems with disclosure. First, public information might crowd out other sources of information. One form of crowding out is that the regulator may obtain less information from banks. A bank that has a temporary liquidity problem may benefit from government help, but it may be reluctant to reveal information to the regulator for fear of being stigmatized by the market. A bank may also conceal information for fear that it is revealed to competitors. Hence, greater disclosure of bank-specific information via stress tests might imply that banks will be less forthcoming in the information they provide to regulators, and this will limit the effectiveness of the stress tests to begin with. These issues are discussed by Prescott (2008) in the context of the disclosure of traditional bank supervisory information and by Leitner (2012) in the context of the disclosure of complex transactions that banks enter into.

Another form of crowding out of other information sources, which has been discussed much more widely in the general disclosure literature, is that the disclosure of public information can lead to a reduction in the private information that is produced and brought into prices by market investors. To the extent that market prices contain important information about banks' securities, which can help guide decisions made by creditors and others, this crowding out of market information is damaging and might even outweigh the direct positive effect of having more information disclosed via stress tests. A recent review of this literature is provided by Goldstein and Yang (2017). A general lesson is that the effect of information disclosure can vary significantly by the type of information that is being disclosed (Bond and Goldstein, 2015; Goldstein and Yang, 2019). It is always a good idea to disclose information about issues that decision-makers already know about and do not need to learn about from the price because this will encourage market participants to focus their attention on other things that decision-makers may still want to learn about, making prices more informative in a way that helps efficient decision-making. But revealing information, which might not be so precise, about things that the market may have a comparative advantage on might backfire. This will divert the attention of traders away from these dimensions and will make the price a less reliable source of information about them.

Second, as highlighted by Morris and Shin (2002), in the presence of strategic complementarities and coordination problems among bank counterparties, disclosing information about bank fundamentals might become a focal point, leading investors to put excessive weight on it and neglect their own private information. This is because public disclosure not only provides information on fundamentals but also helps investors guess what other investors will do. So even if the regulator is not much more well informed than private investors, these investors may end up overreacting to the regulator's announcement. This is a bad outcome from a social point of view, and it also undermines market discipline because it breaks the link between the bank's financial health and follow-up actions by investors (e.g., whether they extend credit to the bank). The implication of this is that disclosure is beneficial only if the regulator's information is sufficiently precise.

Third, greater disclosure might encourage bank managers to focus on short-term goals that are not necessarily aligned with long-term value creation. This insight arises from models such as that by Gigler et al. (2014), who show that frequent disclosure can backfire by diverting the incentives of managers. The point goes back to Stein (1989), who argued that managers in public corporations with frequent disclosure and public share prices often try to achieve short-term measurable gains at the expense of long-term value.

Goldstein and Sapra (2014) suggest that these downsides of disclosure arise mostly when bank-specific information is being disclosed and not so much for disclosure of aggregate outcomes. However, these downsides can be remedied in the case of bank-specific information disclosure to alleviate the potential damages. Here are some remedies. First, when disclosing negative results that might trigger a coordination failure among market participants, it would help if the negative results are followed up by a corrective action. Second, it may be better not to release public information on dimensions where the market may have a comparative advantage in providing information. Third, when the information being disclosed is very precise, the negative effects are diminished. Fourth, banks' investment choices should be carefully monitored to ensure that disclosure does not distort their incentives.

An important issue to keep in mind when evaluating the costs and benefits of disclosure is that their relative importance likely depends on the general environment we are in and in particular on whether we are dealing with stress tests in normal times or during a crisis. For example, the disciplinary role of stress tests is more relevant during normal times, when the main purpose is to ensure that banks do not take too much risk. On the other hand, the concern about public information getting too much weight due to a coordination problem, as in Morris and Shin (2002), is likely more acute during a crisis, when the objective is to avoid panic. Although the various theories provide insights along these lines, more work is needed to get a better understanding of how the purpose of a stress test changes along the cycle and how this affects the balance of the costs and benefits of disclosure.

In general contexts (i.e., outside the context of stress tests), several empirical articles have provided evidence for the downsides of disclosure mentioned here. Hertzberg et al. (2011) show how, in the presence of strategic complementarities among lenders, the revelation of a signal publicly causes them to put excessive weight on it. Jayaraman and Wu (2019) show that greater disclosure acts to crowd out information from financial markets and that this acts to reduce the ability of managers to make informed investment decisions. Agarwal et al. (2018) and Kraft et al. (2018) show that greater and more frequent disclosure lead to myopia by corporate managers, who choose to focus on short-term goals rather than long-term innovation and growth.

Following these discussions, recent work has attempted to find evidence of the negative consequences of the disclosure of stress test results using data from US stress tests between 2009 and 2015. Studies by Flannery et al. (2017) and Fernandes et al. (2017) do not find evidence consistent with negative effects. In particular, they do not find evidence for a reduction in other sources of information production or distortion in banks' investment decisions. However, as they note, their sample is limited to a few years that did not exhibit any major events, so it is not clear how these results will generalize to broader and more heterogeneous samples. Moreover, it is difficult to isolate the particular effect of disclosure from these samples, given that there is no clear counterfactual of what would have happened without disclosure.

Finally, a more recent argument that was brought forward against transparency in banks is developed by Dang et al. (2017). They argue that opacity on the asset side of banks is critical for the liquidity-provision role they perform on their liability side, allowing depositors to benefit from money-like assets. Clearly, the disclosure of stress test results would go against this objective of banks. Empirically, a recent article by Chen et al. (2019) provides evidence in support of this view. They show that banks that are more transparent and provide more precise information about upcoming losses suffer from a decreased ability to provide liquidity to depositors without relying on deposit insurance.

3 Unified Frameworks for Analyzing Stress Tests Information Disclosure

So far, we have discussed various costs and benefits of disclosure in isolation, as they were developed separately in the general disclosure literature and later tied to the context of stress tests. Following the introduction of stress tests, a few articles developed models to evaluate the question of stress test information disclosure using a self-contained framework in which costs and benefits arise in the model and unified policy prescriptions can be provided within the model. We describe a theory by Goldstein and Leitner (2018) that builds on the effects that disclosure will have on the risk-sharing arrangements attained by banks in interbank markets and financial markets more generally. We then describe a few other related works.

The idea that banks engage in risk-sharing arrangements among themselves, which is at the basis of the Goldstein and Leitner (2018) model, is strongly rooted in banking literature and practice (e.g., see the model of Allen and Gale, 2000). Going back to Hirshleifer (1971), it is known that more information might be harmful because it reduces risk-sharing opportunities for economic agents. Intuitively, if the realization of shocks is known, there is no room for mutually beneficial risk-sharing arrangements that ensure against these shocks. However, there are also forces that could make some disclosure desirable even when the objective is risk sharing. Indeed, during the 2008 financial crisis, interbank markets were not performing well, and there was a sense that some disclosure was necessary to prevent a breakdown in financial activity. Although market breakdowns are known to occur when market participants have asymmetric information (as in Akerlof, 1970), they can also occur when market participants share the same information, but the aggregate endowment is expected to be low (see Leitner, 2005), preventing risk-sharing arrangements from taking place. The model by Goldstein and Leitner (2018) captures these forces. We now provide a brief and simple description of it.

Suppose that the value of the bank's assets is $\theta + \varepsilon$, where θ and ε are independent random variables. The regulator learns θ during the stress test. We can think of θ as the regulator's forecast for the value of the bank's assets and of ε as additional noise that is unobservable at the time of the test and that has a mean of zero. Everyone knows the distribution of the random variables θ or ε. No one knows the realized value of ε, and only the regulator knows the realized value of θ. The regulator's objective is to minimize expected losses in the banking system. Suppose that the bank will suffer a loss if it ends up with value below 1 (e.g., because this will trigger a run on the bank). Banks can potentially guarantee that their values do not fall below 1 by selling assets for a price that equals the expected value. This is an example of a risk-sharing arrangement. However, the ability to get into such an arrangement depends on the overall conditions in the system (e.g., whether expected valuations are above 1) and on the disclosure policy employed by the regulator.

In the basic case in which the realization of the fundamental θ for individual banks is known only to the regulator, Goldstein and Leitner (2018) show that the optimal-disclosure policy depends on investors' prior beliefs regarding θ. If $E(\theta) \geq 1$, the regulator does not need to disclose any information because banks will be able to raise at least \$1 by selling their assets. In this case, disclosure will be harmful because of the Hirshleifer effect: banks whose fundamental is revealed to be below 1 will not be able to participate in risk-sharing arrangements and will not be able to guarantee themselves against losses. If instead $E(\theta) < 1$, then without disclosure, banks will not be able to raise \$1. In this case, full disclosure will lead to a better outcome than no disclosure because at least some banks (those with $\theta \geq 1$) will be able to protect themselves. However, partial disclosure will lead to even better outcomes. Specifically, it is optimal to separate banks into two groups. The first group includes all the banks with $\theta \geq 1$, as well as some banks with $\theta < 1$, such that the group's average θ equals the critical level 1. The second group includes all the other banks. The benefit of this partial disclosure is that it allows some of the banks with values below the critical level ($\theta < 1$) to share risk with banks that have values above the critical level (i.e., banks with $\theta > 1$). In some cases, the regulator can implement this partial disclosure via a simple cutoff rule: the regulator sets a cutoff below 1 and reveals whether the bank's θ is above or below the cutoff.

These results are in the spirit of the Bayesian persuasion literature (Kamenica and Gentzkow, 2011). Essentially, the regulator wants to implement an outcome in which the bank can protect itself against bad outcomes by selling its assets for a sufficiently high price. But because the regulator cannot force investors to purchase assets for more than what they think the assets are worth, the regulator needs to "give up" on some banks with low θ so that on average the remaining banks are above the threshold. Linking to the Bayesian persuasion literature, then, one can think of stress test disclosure rules as an exercise in information design: How should regulators design the disclosure of information in order to achieve their goal (which, in this case, is minimizing expected losses in the banking system)?

The model of Goldstein and Leitner (2018) then goes on to consider more complicated environments, in which not only the regulator knows the fundamental θ of an individual bank, but also the bank knows. Moreover, as is realistically the case, the regulator cannot force a bank to sell its assets. This case could lead to adverse selection because a bank that knows that the value of its asset is high will not agree to sell for only \$1. A strong bank (with high θ) would like to protect itself against its value falling below 1—but not at any price. In other words, each bank has a reservation price, a minimum price at which it is willing to sell in order to obtain insurance. Because the market anticipates this adverse-selection problem, they will not be willing to purchase at a price of \$1, and the partial-disclosure scheme will not work.

Goldstein and Leitner (2018) show that in this case, the regulator needs to disclose more information to maximize its objective function. This is expected: banks of higher type would not agree to be pooled with a group that has a much lower average, and disclosure will thus act to separate banks into more groups, effectively providing more information. The most realistic case, obtained in the model under some conditions, features optimal disclosure that separates banks into three groups based on their fundamental θ by defining two thresholds. Banks with low θ (below the lowest threshold) are excluded from risk sharing (i.e., they cannot protect themselves against value falling below the threshold), and it does not matter whether the regulator reveals or does not reveal their θ. Banks in the middle (between the two thresholds) are pooled together, and they all sell for the same price (same risk-sharing agreement). Because of the Hirshleifer effect, the regulator must not reveal the exact θ for banks in this group. The rest of the banks (those with θ above the second threshold) will sell for higher prices. For this group of banks, the purpose of disclosure is to ensure that they are willing to sell. If these banks are roughly the same, the regulator does not need to reveal their exact θ because even the highest bank in this group will be willing to sell at a price that reflects the group average. However, if banks in this group are very different from one another in terms of θ, the regulator will need to reveal more information because otherwise, a bank with a very high θ will prefer to take the risk that $\theta + \varepsilon < 1$, rather than get insurance by selling its asset as a very low price. There are even cases in which the only way to implement the optimal-disclosure policy is via full disclosure.

Summarizing the forces of the Goldstein and Leitner (2018) model, we see that the Hirshleifer effect is a force that pushes toward no disclosure, but if the regulator wants to guarantee that bank value is above some threshold, and the expected average value based on market priors is below the threshold, some disclosure is necessary. Adverse-selection problems will push for more disclosure. Adverse selection could be an issue if the bank knows the information that the regulator has and the regulator cannot force the bank to take corrective actions, such as raising capital, in order to protect itself. As these forces strengthen, full disclosure can emerge as the optimal outcome under this model.

The model has interesting implications for understanding how optimal disclosure may change with overall conditions in the financial system. Consider the case in which banks know their types, which are also revealed to the regulator in the course of the stress test. During booms, when there is no threat of bank runs and when every bank has sufficient capital (i.e., $\theta > 1$), the only relevant force is adverse selection, which could necessitate full disclosure. But if some banks are not in good condition (i.e., $\theta < 1$), but the overall state of the banking system is good (i.e., $E(\theta) > 1$), the Hirshleifer effect kicks in, which pushes toward less disclosure (i.e., some banks must be pooled together, but if there is sufficient adverse selection, the regulator may still need to disclose full information regarding the strongest banks). Finally, during a crisis, when the banking system as a whole is under-capitalized (which corresponds to $E(\theta) < 1$), some disclosure may also be necessary to separate some of the weak banks. When designing disclosure policies, regulators have to keep in mind how conditions might change and how this will affect how much they want to disclose.

Outside the model described so far, there can be other forces that push for more disclosure. For example, in a follow-up work, Orlov et al. (2018) develop a model in which the regulator learns from the stress test about systemic risks. In their model, the regulator wants to implement different outcomes across different banks, based on the expected performance of their risky assets in the bad state of the economy. In an optimal arrangement, banks with less fragile risky assets should hold more of them. The optimal-disclosure regime they derive out of this model also entails the separation of banks into different groups, guided by the intention to implement different outcomes for them.

Although this framework considers only the disclosure regime as a tool available to the regulator, governments can use other tools to alleviate fragility, such as directly intervening and injecting funds into banks. With this in mind, Faria-e-Castro et al. (2017) provide a model in which the optimal level of disclosure depends on the government's fiscal capacity, namely, its ability to inject money into banks and the cost of doing so. They show that governments that have more fiscal capacity can run more aggressive disclosure policies (i.e., they disclose more).

In their model, the benefit of disclosure is that it reduces adverse selection, allowing banks to raise capital to finance investment opportunities. The cost of disclosure is related to the Hirshleifer effect—it can induce runs on weak banks. Consistent with our previous discussion, the optimal amount of disclosure depends on the relative magnitude of each of these two forces. Now suppose the government can prevent runs by providing deposit insurance, but to finance this type of intervention, it will need to raise taxes later on, which is costly from a social point of view. Faria-e-Castro et al. (2017) show that if this cost is lower (i.e., if the government has more fiscal capacity), it is optimal to disclose more information. The intuition is simple. If the cost for the government to prevent runs is low, the government does not need to worry about runs, so the dominant force in its disclosure policy is reducing adverse selection.

This result can help explain the striking difference between the stress tests implemented in the United States and in Europe following the 2008–2009 financial crisis. Indeed, the first round of stress tests in the United States was considered to be very revealing, and for this it is always noted as a great success, whereas in Europe, the initial stress tests were weak and did not provide much information. The explanation here is that Europe could not afford to have highly informative tests because of its lack of fiscal capacity to intervene following bad results.

In Chapter 12 of this volume, Judge raises a similar point. Regulators may be hesitant in producing information unless they have the tools in hand to contain the fallout their findings might trigger. As a policy implication, she discusses the importance of designing safety nets, but she also discusses postcrisis reforms that scaled back the authority of financial regulators in the United States to provide crisis-time support. The Goldstein and Leitner (2018) model suggests that in some cases, a well-designed disclosure policy could supplement and possibly reduce the cost of having such safety nets because the central bank could rely not only on its own ability to inject funds into banks but also on cross-subsidies across banks that are pooled together. Reforms that weaken the regulator's authority to support banks during a crisis could potentially make disclosure policy even more relevant.

Finally, the frameworks just described start from the premise that regulators can disclose information according to some rules that are set in advance. This raises questions about commitment ability because ex post, the regulator may want to deviate from these rules. For example, in Goldstein and Leitner (2018), when disclosure is according to a cutoff rule, the regulator would like to say that θ is above the cutoff, even when it is below. This reflects the general idea that the regulator may want to hide bad news from the market. In the previous example, this is news about the health of an individual bank, but it could also be news about the health of the overall financial system: if investors did not know $E(\theta)$, the regulator might want to say that $E(\theta)$ is above the threshold, even when it is below. If investors expect this to happen, the stress test would lose its credibility. This issue is discussed by Bouvard et al. (2015).

4 Stress Test Design versus Information Disclosure

The Goldstein and Leitner (2018) article and the related literature discussed in the previous section are framed as analyses of optimal-disclosure schemes, by which regulators decide ex ante what rules to follow in presenting the information ex post. However, it is important to note that similar results can be obtained when regulators are committed to disclosing everything but have flexibility ex ante in how to design the test. In fact, the problem of stress test design is equivalent to the problem of information-disclosure design.

To see this point, suppose that according to the Goldstein and Leitner (2018) model, the optimal-disclosure rule entails no disclosure at all; that is, the circumstances are such that it is optimal not to reveal any information about the realization of the bank's type. For example, this would be the case in the model when the average fundamental of banks $E(\theta)$ is above 1 and the type is only revealed to the regulator and not to the bank itself. Analyzing a model of information design would suggest that after banks' types are revealed in the test, it would be optimal to assign all banks the same score, such that no information about their types is disclosed publicly. What we argue is that the same outcome could be achieved by designing a very weak test that does not identify meaningful differences between banks thus, even if the results of the test are fully revealed to the public, these results do not really differentiate among banks, and effectively, there is no disclosure.

Similarly, consider the case in which the optimal-disclosure rule in the model is to disclose everything, that is, to provide as much information as possible differentiating across the banks. This would be the case in the model if there are big differences across banks and they are aware of their types. Thinking about stress test design, such a solution can be obtained by designing a very strong test that will go to the root of the risks banks face, enabling sharp differentiation. In similar spirit, any disclosure rule prescribed by the model that features some form of partial disclosure (this is, of course, the most common outcome) can be achieved in a model of stress test design by planning a test that will highlight exactly the dimensions that are optimal to be revealed publicly.

Thinking about the problem through the lens of stress test design rather than information-disclosure design is also a way to address commitment problems of the type discussed previously. With information disclosure, one can ask whether the government has the ability to commit to follow ex post the ex ante optimal disclosure rule. A related concern is that political and legal constraints would make it difficult to lump together banks of different types ex post just to adhere to the ex ante optimal rule. Such concerns do not arise when we think about the problem as a problem of stress test design. The decision on stress test design is made ex ante, and there is no need to commit to it. Once the information is discovered during the stress test, it will be revealed publicly, but how informative it is will depend on the design decision that was made ex ante.

Overall, we can see here that the disclosure question cannot be fully separated from the question of the design of the stress tests. The two are intertwined. Ultimately, we care about what market participants will learn, and there are different ways to control this: either through disclosure or through the power of the test. This is an important observation, and we will come back to it when discussing the current practice in regulatory circles on stress test disclosure issues.

5 Current Thinking about Stress Test Information Disclosure in Regulatory Circles

Initially, following the introduction of stress tests, the views about the disclosure of their results were rather mixed among regulators; however, it seems that over the years, a growing consensus emerged that increased transparency of stress test results is desirable. Statements from policymakers reveal their belief in and commitment to greater disclosure, and the amount of information being revealed in both the United States and Europe has clearly increased.

So what kind of information is being provided following stress tests these days? In the United States, the Federal Reserve discloses various capital ratios and other accounting numbers, such as projected losses, revenue, and net income before taxes. The Federal Reserve also discloses loan-loss rates broken down by type of loan for each bank holding company (BHC) participating in the stress tests for both the adverse and severely adverse scenarios. Also disclosed are the Federal Reserve's decisions on capital plans for each BHC and the basis for the decisions. The European Central Bank's disclosure for European banks undergoing stress tests is much more detailed than that of the Federal Reserve. It is also becoming gradually more transparent over time, reaching 17,200 data points disclosed per bank in 2018. This is a striking development, given that in the first stress tests conducted in Europe, the results were kept rather confidential, and only aggregate numbers were provided. The information in Europe is also disclosed in a more user-friendly way than in the United States to help it be absorbed quickly into the financial market. In contrast, the Bank of England provides much less information for the stress tests it conducts.

The general view for greater transparency is clearly conveyed by the Federal Reserve and the European Central Bank in various policy papers and speeches. For example, for the Federal Reserve, the general philosophy is described in a supervisory staff report by Clark and Ryu (2013), who write: "Given a widely held view among supervisors and most third-party observers that the public disclosure of stress testing results enhances available information and supports market discipline, it will continue and it is perhaps even likely to be expanded over time." They then go on to dismiss several of the concerns raised in the academic literature and reviewed previously regarding the possible downsides of a greater-disclosure regime.

Although we agree that disclosure clearly has its benefits, as discussed earlier, we also think it is unrealistic to view disclosure as a panacea and dismiss the possible downsides. Ultimately, there is a trade-off, and theoretical frameworks help clarify the forces and provide policy prescriptions for given environments. For example, it is hard to ignore the fact that meaningful and precise disclosure will, in some cases, act to isolate banks with bad shocks from the market and disrupt the ability of the market to provide risk-sharing opportunities across banks. The question is whether regulators are prepared to accept such outcomes.

It should be noted that regulators have shown concern about such issues in various situations over the years. For example, one of the first uses of the Troubled Asset Relief Program (TARP) funds was providing capital to nine major financial institutions as part of the Capital Purchase Program, a program designed to infuse capital into healthy banks. During the audit, former Federal Reserve chair Ben Bernanke told the special inspector general for TARP that "there were differences in the nine banks in terms of strength and weakness, but that the selection was generalized in order to avoid stigmatizing any one bank as being a weak bank and creating panic" (Office of the Special Inspector General for the Troubled Asset Relief Program [SIGTARP], 2009). So, the question is whether the Federal Reserve or other policy institutions are prepared to ignore these kinds of considerations when they arise in the future and bear the costs that full disclosure entails.

Another important question to ask when evaluating the current regime of greater transparency is whether the stated goals and public statements about increased transparency are indicative of much valuable information being revealed. Consider the discussion in the previous section on stress test design versus information disclosure. Recall that one way to achieve less actual transparency is to design very weak and uninformative tests and then provide very detailed disclosure of their results, so effectively, no critical information is being revealed. Is it possible that the Federal Reserve, while committing itself to disclosing the results of the tests it conducts, might choose to weaken the tests, such that the negative consequences of greater disclosure are not incurred?

By now, there is some empirical research on the actual informativeness (as opposed to the stated transparency) of stress tests; albeit most of it still comes from earlier years. Although it is difficult to provide conclusive evidence on the informativeness of the tests, these studies do indicate that stress tests provided new valuable information to markets. Flannery et al. (2017) analyze stress tests in the United States between 2009 and 2015. They measure the market response to stress test disclosure using average absolute cumulative abnormal returns (CAR), to avoid positive and negative effects canceling out, and using abnormal trading volume. Absolute CAR is statistically significant around most disclosure dates. Moreover, they show a spillover effect of disclosure from tested banks to untested bank. Abnormal trading volume also spikes around disclosure. They conclude that valuable information is being provided. Similar results are obtained by Fernandes et al. (2017) on a similar data set. They provide evidence that information asymmetry (as proxied by bid–ask spread) increases with the announcement of a stress test and decreases following the release of the results. Importantly, they show that the market response to stress tests disclosures is attenuated after the initial rounds. This raises the possibility that stress tests are not revealing as much information more recently.

Related studies have been conducted on European data, especially following the 2011 European stress test, in which there was a jump in the scale of disclosure (including up to 3,400 data points for each of the 90 participating banks). Petrella and Resti (2013) find that the test result is not fully anticipated by the market, and hence stress tests play an active role in mitigating bank opacity. Ellahie (2015) shows that the detailed data on exposure to credit and sovereign risks disclosed in the 2010 and 2011 EU bank stress tests have significant predictive power for the cross-section of bank equity and credit returns (change in credit default swap [CDS] spread). High sovereign risk exposure implies a subsequent decline in equity return and a widening of the CDS spread.

The concern, however, is that more recently, stress tests have become much less informative, especially in the United States. Inspecting the latest rounds of stress tests in the United States, it is striking that all participating banks are predicted to do so well under unbelievably bad scenarios for the economy. Is this a reflection of the outstanding resilience of the banking sector or just an indication of weak tests being performed without revealing much critical information? In Chapter 9 of this volume, Sarin and Summers provide a detailed discussion, with references to other sources, of the flaws in the current system of stress tests in the United States. In particular, they highlight that market measures of banks' risks and fragility point to a much gloomier picture than the one provided by the regulatory stress tests. Hence, they call for an overhaul of the current methodologies and for greater reliance on market information when conducting stress tests. Glasserman and Tangirala (2016) provide related evidence in their analysis of the evolution of the information content from stress tests. They argue that the results have become predictable, with the information content declining over time.

Overall, the issue to pay attention to is that commitment to disclosure might come together with weaker tests. Although weak tests could be part of an optimal design, it would be troubling if the stress tests were claiming to identify all risks when, in fact, they were not designed to do so. First, current regulations are based to a large extent on stress test results, so capital disbursements and other measures loosening the positions of banks are triggered once stress test results suggest that risks are low. Second, although market information still seems to depict a different picture than the one portrayed by stress tests, there is a risk that the disclosure provided by the stress tests is affecting, or will affect in the future, the kind of information provided by the financial market and will make the whole system much more complacent. Third, weak tests could reduce the credibility of future tests in the next crisis because the market may lose confidence in the regulator's ability to assess risk. This issue is discussed by Judge in Chapter 12 of this volume. It is also related to the theoretical model by Morrison and White (2013), where crises spread across banks as a result of the common inference about the quality of the regulatory regime.

6 Disclosing the Model

As noted earlier, an important issue related to the disclosure of stress test results is the transparency about the models used by regulators to conduct the stress tests. Like in the case of disclosing the test results, we think there are costs and benefits in disclosing details about the underlying model, and one needs to have a framework in mind to evaluate the trade-off. It is also important to make a distinction between revealing the models to the public versus revealing them to the institution being tested (i.e., the bank).

In terms of disclosing models to the public, the benefits that come to mind are the following. First, disclosure enhances the credibility of the stress test by providing the public with information on the fundamental soundness of the models and their alignment with best modeling practices. Second, it helps the public understand and interpret the results of the stress test. Third, disclosure facilitates comments on the models from the public, including academic experts. These comments could lead to improvements, particularly in the data most useful to understanding the risks of particular loan types.

A possible downside of disclosing models to the public is that it could make it hard to maintain flexibility. The regulator will need to explain or consult with the public every time it wants to make changes in its models. The regulator may be afraid to experiment with new models that might contain mistakes, and this could lead to a situation in which the regulator sticks with old models that are "correct" but do not apply to the new environment. Frame et al. (2015) suggest that this force might have played an important role in the failure of the supervisory stress tests conducted by the Office of Federal Housing Enterprise Oversight (OFHEO) for Fannie Mae and Freddie Mac. This is an interesting experience to learn from because this is a previous attempt in the United States to tie stress test results to capital requirements, which clearly did not go well.

When it comes to disclosing models to the banks, the biggest disadvantage is that when the regulator reveals the model, banks can game the test. For example, they can use the models to make modifications to their businesses that change the results of the stress test without changing the risks they face. In the presence of such behavior, the stress test could give a misleading picture of the actual vulnerabilities faced by banks. This might become a game focused on passing the test rather than truly increasing resilience. In addition, if a specific asset is perceived as more advantageous to hold based on the regulator's model, disclosing the model could lead to correlated asset holdings across banks, further increasing systemic fragility. Finally, model disclosure could induce banks to use models similar to the regulator's model, rather than building their own capacity to identify, measure, and manage risk. As argued by Schuermann (2014), this "model monoculture," in which all banks have similar internal stress testing models, might miss key idiosyncratic risks faced by the banks.

On the positive side, revealing the model to the bank can help it understand the capital implications of changes to its business activities, such as acquiring or selling a portfolio of assets. Banks have constantly complained in the past about model secrecy, claiming that even their best efforts to prepare for a test could result in unexpected and costly failure, which leads to lack of trust and feeling of arbitrariness. This might have negative real implications, given evidence that regulatory uncertainty causes banks to reduce lending (see Gissler et al., 2016).

As noted earlier, it is useful to have a unified framework for assessing the trade-off involved in the disclosure of the underlying models. Leitner and Williams (2019) recently provide such a framework to evaluate the issue of sharing the models with the bank. In their setting, the bank has better capacity than the regulator to identify and measure risk, but there is a conflict of interest between the bank and the regulator: the bank wants to take more risk than is socially desirable. The outcome of this is that if the regulator discloses the stress test models, the bank games the test by overinvesting in assets for which it knows that the regulator's models underestimate the risk. Essentially, when the regulator reveals a model that shows that an asset is not very risky, it gives a "green light" for the bank to invest, so the bank may end up investing even when it knows that the asset is harmful to financial stability. Not revealing the regulator's model can mitigate this problem, but it opens the door to a new problem: the bank sometimes does not invest in risky assets even though it knows the assets are good from a social point of view. The reason is that although the bank knows that an investment is socially desirable, it may be worried that the regulator's model does not measure risks accurately, which could result in the bank failing the test. Revealing the test mitigates this problem. Hence, there is a trade-off. One implication of this is that revealing the test is preferred to not revealing it if the bank is too concerned about failing the test (e.g., if the bank's private cost of failing the test is sufficiently high). The bank's private cost of failing the test could represent the private cost for bank managers when the bank fails the test (e.g., the manager can get fired), the potential market reaction, the cost of altering the bank's capital plan, and so forth.

Leitner and Williams (2019) show that in some cases, the regulator can alleviate the bank's concern, thereby solving the underinvestment problem, by making the test easier. Specifically, the regulator can reduce the threshold for passing the test. If so, not revealing the test is preferred. However, this does not work if banks are very different from one another and the regulator must apply the same passing threshold to all banks. In this case, the regulator cannot calibrate the passing threshold to induce socially desirable investment by everyone because making the test easier to alleviate concerns by banks that are too cautious (i.e., those with a high private cost of failing the test) could induce the "reckless" banks (those with a low cost of failing the test) to take excessive risks. The implication of this is that if banks are very different from one another and the regulator must apply the same passing threshold to everyone, revealing the test is preferred to not revealing it.

Leitner and Williams (2019) also analyze the more complicated case in which the regulator can reveal only partial information (e.g., whether estimated losses from an asset, according to the regulator's model, are above or below some threshold). One of the results is that in some cases, some disclosure is optimal even if the regulator can set the passing threshold optimally and is not obliged to apply the same threshold to all banks.

This result reflects the general idea that in some cases, it is optimal to combine more than one regulatory tool to achieve a desired outcome. In Leitner and Williams (2019), the regulator wants to prevent gaming while still maintaining socially desirable investment, and under some circumstances, it is optimal to do so by combining partial disclosure with adjustment of the test difficulty (or capital requirements). Intuitively, each of these tools has costs and benefits, and a combination of the two tools helps minimize the total cost (i.e., maximize the total net benefit). Earlier, we saw that during crisis, when the regulator's objective is to prevent runs or ensure that each bank has a sufficient level of capital, the regulator can achieve this goal by combining safety nets and partial disclosure that pools together banks of different levels of stress.

Finally, a widely expressed concern is that disclosing the models could increase correlations in asset holdings among the banks subject to the stress tests. Leitner and Williams (2019) suggest that a similar effect could prevail if the regulator reveals the outcome of applying its models to hypothetical loan portfolios (e.g., under the new policy changes of the Federal Reserve from February 2019). The reason is that these portfolios could serve as a benchmark portfolio in which too many banks invest. In other words, a bank will overinvest in portfolios, for which there is less uncertainty as to how the regulator will measure risk, and it will underinvest in idiosyncratic investments, for which it is unclear how the regulator's models will work.

7 Conclusion

Stress tests have been proposed following the financial crisis as a tool to monitor bank risks in a forward-looking way. Unlike traditional regulatory monitoring and interventions in banks, which are typically kept confidential, the idea with stress tests is to provide more disclosure and transparency. This has become more prominent over the years because the amount of information being provided is now larger. In addition, regulators have taken steps to increase the transparency about the underlying models they are using.

We argue that disclosures of stress test results and of the underlying models involve trade-offs. We described various costs and benefits and also unified frameworks to evaluate such trade-offs. These forces represent issues that regulators have been concerned about in various situations and are likely to continue to play a role in the future. Hence, it is important to acknowledge that disclosure is not a panacea but, rather, that the disclosure regime has to be thought of carefully, taking into account the various forces and how they change over time.

We caution that if regulators do not acknowledge the trade-offs with disclosure, then commitment to full disclosure could just lead to weak tests which is something that might start happening already. If the system ends up producing such weak tests, while portraying them as strong and informative and basing regulatory actions on them, we might get to a point of extreme complacency, which is dangerous for future financial stability.

References

Agarwal, Vikas, Rahul Vashishtha, and Mohan Venkatachalam (2018), "Mutual fund transparency and corporate myopia," *Review of Financial Studies*, 31(5), 1966–2003.

Akerlof, George A. (1970), "The market for 'lemons': Quality uncertainty and the market mechanism," *Quarterly Journal of Economics*, 84(3), 488–500.

Allen, Franklin, and Douglas Gale (2000), "Financial contagion," *Journal of Political Economy*, 108(1), 1–33.

Bond, Philip, and Itay Goldstein (2015), "Government intervention and information aggregation by prices," *Journal of Finance*, 70(6), 2777–2811.

Bouvard, Matthieu, Pierre Chaigneau, and Adolfo De Motta (2015), "Transparency in the financial system: Rollover risk and crises," *Journal of Finance*, 70(4), 1805–1837.

Chen, Qi, Itay Goldstein, Zeqiong Huang, and Rahul Vashishtha (2019), "Bank transparency and deposit flows," Working Paper, available at www.bde.es/f/webpi/SES/seminars/2019/Fich/sie20190710.pdf.

Clark, Tim, and Lisa Ryu (2013), "CCAR and stress testing as complementary supervisory tools," Board of Governors of the Federal Reserve System Supervisory Staff Report.

Dang, Tri Vi, Gary Gorton, Bengt Holmstrom, and Guillermo Ordonez (2017), "Banks as secret keepers," *American Economic Review*, 107(4), 1005–1029.

Ellahie, Atif (2013), "Information content of mandated bank stress test disclosures," Working Paper, available at https://doi.org/10.2139/ssrn.2685919.

Faria-e-Castro, Miguel, Joseba Martinez, and Thomas Philippon (2017), "Runs versus lemons: Information disclosure and fiscal capacity," *Review of Economic Studies*, 84(4), 1683–1707.

Fernandes, Marcelo, Deniz Igan, and Marcelo Pinheiro (2017), "March madness in Wall Street: (What) does the market learn from stress tests?" *Journal of Banking and Finance*, forthcoming.

Flannery, Mark, Beverly Hirtle, and Anna Kovner (2017), "Evaluating the information in the Federal Reserve stress tests," *Journal of Financial Intermediation*, 29(C), 1–18.

Frame, Scott, Kristopher Gerardi, and Paul Willen (2015), "The failure of supervisory stress testing: Fannie Mae, Freddie Mac, and OFHEO," Federal Reserve Bank of Boston Working Paper No. 15–4.

Gigler, Frank, Chandra Kanodia, Haresh Sapra, and Raghu Venugopalan (2014), "How frequent financial reporting can cause managerial short-termism: An analysis of the costs and benefits of increasing reporting frequency," *Journal of Accounting Research*, 52(2), 357–387.

Gissler, Stefan, Jeremy Oldfather, and Doriana Ruffino (2016), "Lending on hold: Regulatory uncertainty and bank lending standards," *Journal of Monetary Economics*, 81, 89–101.

Glasserman, Paul, and Gowtham Tangirala (2016), "Are the Federal Reserve's stress test results predictable?" *Journal of Alternative Investments*, 18(4), 82–97.

Goldstein, Itay, and Yaron Leitner (2018), "Stress tests and information disclosure," *Journal of Economic Theory*, 177(C), 34–69.

Goldstein, Itay, and Haresh Sapra (2014), "Should banks' stress test results be disclosed? An analysis of the costs and benefits," *Foundations and Trends in Finance*, 8(1), 1–54.

Goldstein, Itay, and Liyan Yang (2017), "Information disclosure in financial markets," *Annual Review of Financial Economics*, 9, 101–125.

Goldstein, Itay, and Liyan Yang (2019), "Good disclosure, bad disclosure," *Journal of Financial Economics*, 131(1), 118–138.

Greenstone, Michael, Paul Oyer, and Annette Vissing-Jorgensen (2006), "Mandated disclosure, stock returns, and the 1964 Securities Acts Amendments," *Quarterly Journal of Economics*, 121(2), 399–460.

Hertzberg, Andrew, Jose Liberti, and Daniel Paravisini (2011), "Public information and coordination: Evidence from a credit registry expansion," *Journal of Finance*, 66(2), 379–412.

Hirshleifer, Jack (1971), "The private and social value of information and the reward to inventive activity," *American Economic Review*, 61(4), 561–574.

Jayaraman, Sudarshan, and Johana Wu (2019), "Is silence golden? Real effects of mandatory disclosure," *Review of Financial Studies*, forthcoming.

Kamenica, Emir, and Matthew Gentzkow (2011), "Bayesian persuasion," *American Economic Review*, 101(6), 2590–2615.

Kleymenova, Anya, and Rimmy Tomy (2019), "Observing enforcement: Evidence from banking," Chicago Booth Research Paper No. 19-05.

Kraft, Arthur, Rahul Vashishtha, and Mohan Venkatachalam (2018), "Frequent financial reporting and managerial myopia," *The Accounting Review*, 93(2), 249–275.

Leitner, Yaron (2005), "Financial networks: Contagion, commitment, and private sector bailouts," *Journal of Finance*, 60(6), 2925–2953.

Leitner, Yaron (2012), "Inducing agents to report hidden trades: A theory of an intermediary," *Review of Finance*, 16(4), 1013–1042.

Leitner, Yaron (2014), "Should regulators reveal information about banks?" *Federal Reserve Bank of Philadelphia Business Review*, Q3, 1–8.

Leitner, Yaron, and Basil Williams (2019), "Model secrecy and stress tests," Working Paper, available at www.basilwilliams.org/leitner.williams.secrecy.pdf.

Leuz, Christian, and Robert Verrecchia (2000), "The economic consequences of increased disclosure," *Journal of Accounting Research*, 38, 91–124.

Morris, Stephen, and Hyun Song Shin (2002), "Social value of public information," *American Economic Review*, 92(5), 1521–1534.

Morrison, Alan, and Lucy White (2013), "Reputational contagion and optimal regulatory forbearance," *Journal of Financial Economics*, 110(3), 642–658.

Office of the Special Inspector General for the Troubled Asset Relief Program (2009, October), "Emergency capital injections provided to support the viability of Bank of America, other major banks, and the U.S. financial system," Report 10-001.

Orlov, Dmitry, Andrzej Skrzypacz, and Pavel Zryumov (2018), "Design of macro-prudential stress tests," Working Paper, available at https://econpapers.repec.org/paper/redsed018/913.htm.

Petrella, Giovanni, and Andrea Resti (2013), "Supervisors as information producers: Do stress tests reduce bank opaqueness?" *Journal of Banking and Finance*, 37(12), 5406–5420.

Prescott, Edward Simpson (2008), "Should bank supervisors disclose information about their banks?" *Federal Reserve Bank of Richmond Economic Review*, 94(1), 1–16.

Schuermann, Til (2014), "Stress testing banks," *International Journal of Forecasting*, 30(3), 717–728.

Stein, Jeremy (1989), "Efficient capital markets, inefficient firms," *Journal of Finance*, 44(4), 1335–1350.

12

Stress Testing during Times of War

Kathryn Judge[*]

Introduction

In the spring of 2009, the United States was mired in the greatest recession it had faced since the Great Depression. In March, the Dow Jones Industrial Average had fallen to 6,594.44, a total decline of 53.4 percent from its peak in the fall of 2007. The official unemployment rate was over 9 percent and still trending upward, eventually exceeding 10 percent. With the support of Congress, the Federal Reserve (the Fed) and other financial regulators had launched an array of initiatives to contain the fallout of what had become a global financial crisis. These interventions, including a massive recapitalization of US banks and the effective elimination of large, independent investment banks, had succeeded in stabilizing much of the financial system, but full functionality remained elusive. The crisis had revealed significant deficiencies in the banks' risk management systems and the capacity of regulators to detect those weaknesses. Fear and distrust remained the order of the day.

Against this background, the Federal Reserve and other bank regulators took a gamble. On May 7, 2009, they publicly announced the results of the Supervisory Capital Assessment Program (SCAP). As then-chairman Ben Bernanke explained, "the SCAP marked the first time the US bank regulatory agencies had conducted a supervisory stress test simultaneously across the largest banking firms" (Bernake, 2013). The Fed further deviated from tradition in its decision to disclose the results of the SCAP. In providing an unprecedented level of detail regarding the methodology and inputs used in reaching those results, the Fed challenged the assumption that bank supervision should always be shrouded behind a thick veil of secrecy. Both gambles paid off. As Bernanke later observed: "The SCAP stands out . . . as one of the critical turning points in the financial crisis. It provided anxious investors with something they craved: credible information about prospective losses at banks" (Bernake, 2013).

Most policymakers, academics, and industry participants share Bernanke's positive assessment of the SCAP. Stress tests have now become a core part of the supervisory and regulatory toolkit and are one of the most important postcrisis regulatory innovations. The Dodd–Frank Act requires large banking organizations to undergo stress tests, and regulators on both sides of the Atlantic have come to see stress testing as a critical component of their ongoing efforts to prevent another financial crisis. These successes have been sufficiently

[*] The author would like to thank the editors of the *Handbook of Financial Stress Testing*, Til Schuermann, Doyne Farmer, Alissa M. Kleinnijenhuis and Thom Wetzer; Nat Benjamin; and other participants at the International Monetary Fund "Rethinking Financial Stability" for thoughtful feedback on earlier versions of this chapter.

great that other regulators, too, have embraced stress testing. There are even proposals for yet other ways that stress testing may be used to detect weaknesses in the financial system before they threaten the health or stability of that system. (See, for example, Chapters 28 and 30 in this handbook.) These are important developments, and broad-based, regular stress testing is appropriately here to stay. These postcrisis developments, however, have shifted attention away from the distinct value, and risks, of the SCAP as a crisis-time intervention.

This chapter shifts the focus back to crisis-era stress testing and other modes of just-in-time information production. Even with rigorous stress testing, the complexity and dynamism of the financial markets are such that regulators will always have incomplete information. These information gaps can prove particularly problematic during periods of distress. A lack of information can contribute to regulators' tendency to be too slow to recognize problems, exacerbating the size of a crisis and the long-term macroeconomic effects. Information gaps can also impede crisis-containment efforts once a crisis takes hold. Without accurate information about the size and location of capital and liquidity shortages, regulators often have little choice but to oversupply or mis-shoot in their efforts to combat dysfunction. This can exacerbate moral hazard and public outrage, as large sums of taxpayer money seem to flow to the very financial institutions that are perceived to be the cause of the crisis.

The COVID-19 crisis has made these issues timely in new ways. Unlike the 2008 crisis, the COVID-19 crisis grows out of changes to the real economy. The public health threat posed by the novel coronavirus, coupled with policy changes designed to slow its spread, caused massive, unanticipated changes to virtually every corner of the economy and society. Thanks in significant part to quick and aggressive actions by the Federal Reserve, boosted by aid from Congress, the early panic triggered by COVID-19 was quelled, and the reckoning going on in the real economy has not yet triggered a full-blown financial crisis. But it is far too early to declare victory. This chapter draws on lessons from past crises and expert analyses of when and how stress testing can aid supervision to explain the distinct value of crisis-time stress tests.

This chapter argues that stress tests can play a critical role in crisis containment, and more should be done in advance to enable crisis-time stress testing. Once things have started to go awry, stress testing provides an array of benefits. When fault lines emerge, regulators have a much clearer sense of the type of adverse developments that may threaten stability, where weaknesses may lie, and the specific fears that they must address to restore market functioning. Stress tests can provide much-needed information about the size and location of fundamental weaknesses in the system and the mechanisms through which those weaknesses may trigger broader dysfunction. Moreover, as the SCAP demonstrated, producing credible information and using that information to shape substantive policy interventions, like recapitalizations, can reduce the relative size of the amount of government support needed to restore faith and functionality.

The analysis here rests on the assumption that crises are in part information events. Coordination challenges may exacerbate dysfunction, but runs are rare when the entire financial system is safe and sound and everyone has credible information that it is so. Fragility—that is, dysfunction out of proportion to the triggering event—is in part the result of incomplete information about the ramifications or meaning of an adverse development. Stress tests can help address the unknowns that exacerbate fragility while providing guidance about how best to deploy other tools, such as guarantees and capital injections, that may be needed to calm fears and restore health. Even if these dynamics are not yet on full display with COVID-19, they may yet prove troubling.

Recognizing the importance of crisis-era information production reveals critical short-comings in the current toolkit policymakers have at their disposal. Most importantly, regulators will hesitate to tread into the unknown and generate new information when doing so could destabilize an already-fragile system. Without the passage of the Emergency Economic Stabilization Act, which empowered the Treasury to recapitalize weak banks if needed, regulators may well have lacked the will to take on the risks that the SCAP entailed. The Coronavirus Aid, Relief, and Economic Security (CARES) Act passed by Congress in March 2020 provides a much bigger fiscal hit for the economy but less discretion to regulators. Whether it will suffice as a backstop should panic set in remains to be seen.

Going forward, devising ways to vest regulators with the authority needed to contain, even if not resolve, a large-scale crisis is critical if we want regulators to produce the information required to reduce the size and scope of the next crisis. Time-limited guarantee authority, which I have advocated for elsewhere, is one method to encourage regulators to probe into the unknown when the system next starts to gyrate. There may well be others.

This chapter proceeds in three parts. The first part provides background. It briefly reviews, in informational terms, how the SCAP contributed to the crisis recovery, why informa-tion production is likely to remain important during periods of systemic distress, and the challenge of isolating information production from substantive interventions. The second part shifts the focus from the SCAP to consider the conditions under which just-in-time information production can aid crisis containment. It expands the analysis both temporally, to other stages in a crisis, and topically, to domains other than banks. The third and final part addresses the groundwork needed to aid crisis-era information generation while recognizing that some improvization is likely inevitable. It addresses the current situation but also the types of regulatory changes more broadly that may be warranted to promote just-in-time stress testing.

The most significant challenge moving forward lies in honing the relationship between the production of information and the capacity to address any weaknesses the analysis reveals. There is no easy way to resolve this tension. In a democratic system, fiscal decisions are usually reserved for the legislative branch or other elected officials, but these bodies are rarely equipped to act with the speed required to contain a growing financial crisis. In the United States, for example, it may well have been difficult to get Congress to act without the massive fallout triggered by the failure of Lehman Brothers, and even with that, it was not easy. Institutionalizing a guarantor of last resort is one way to bridge this gap because it would allow regulators to act swiftly to contain a crisis while still reserving to the legislature the question of which banks should be recapitalized and on what terms. In this way, it could provide regulators both the incentive and the capacity to generate information about the risks they are seeking to contain.

1 Information

1.1 The SCAP

The SCAP was a critical turning point in resolving the 2007–2009 financial crisis. The first round of stress tests provided high-quality information about the health of major financial institutions at a time when the lack of reliable information was continuing to impede market functioning. Subsequent analysis shows that the market already had a good sense of which banks were undercapitalized, but there was uncertainty about the degree of undercapitalization (Peristani et al., 2010). By using a standardized approach across the 19 financial institutions involved and providing market participants sufficiently detailed information that they could use to assess the credibility for themselves, the SCAP helped address lingering uncertainty about the health of major banks and their capacity to support the financial system. It also helped that the stress tests showed that the banking sector, as a whole, was healthier than expected.

Examining the success of the SCAP provides a foundation for considering the conditions required for information injections to aid crisis containment. Against this background, one factor distinguishing crisis-era stress tests from tests conducted under less adverse conditions is the nature of the assumptions market participants are using to address the inevitable gaps in the information they have about the health of one another and other macroeconomic factors that might affect their willingness to engage in risk sharing and other activities. At the time of the SCAP, banks and other market participants remained hesitant to work with one another in ways that were impeding the recovery.

Throughout the year and a half between the start of the crisis and SCAP, regulators and market participants had frequently been behind the ball. For the first year of the crisis, many regulators had underestimated and downplayed the scale of the problems plaguing the financial system. Banks, meanwhile, continued to pay dividends and were slow to recognize losses or to set aside adequate reserves for bad loans and liabilities related to mortgage-backed securities (MBSs), all while purporting to be well capitalized.

Information, or rather, a lack thereof, contributed to regulators' delayed recognition of the magnitude of the crisis they were facing (Judge, 2017). Among the challenges was that the crisis emanated initially not from banks, which are subject to prudential oversight, but rather from the "shadow banking system"—a market-based system of intermediation that was similar in size and even more complicated in scope than the formal banking sector. Although regulators were aware of many of the pieces of this system, and the system was closely interconnected at various points with banks subject to their oversight, its role in maturity and liquidity transformation and its corresponding exposure to the risks that come along with such activity were not fully appreciated by anyone before the crisis struck. As Richard Clarida, now vice chair of the Board of Governors of the Federal Reserve, has observed: "It would seem that the supervision and regulation of US investment and commercial banks during the great moderation was based on an assumption about how the financial system was supposed to work, not upon sufficient knowledge about how the financial system actually worked" (Clarida, 2010). Thus, a combination of insufficient information and a tendency to view the information available through an outdated lens seems to have contributed to the delayed recognition of just how bad things might get.

Eventually, as usually happens, problems that arose outside the banking sector made their way onto bank balance sheets. More than a year passed between when the crisis first hit and when Lehman Brothers failed, during which time there was a series of adverse developments, including the failure of Bear Stearns, that made it clear that the situation remained fragile. Nonetheless, it was not until the failure of Lehman Brothers that Congress passed the Emergency Economic Stabilization Act (EESA), giving Treasury broad, new authority to stabilize the growing crisis.

The Treasury soon put that new authority to good use, although not in the way originally envisioned in EESA. According to a Government Accountability Office (GAO) report
issued in January 2009, before the SCAP, the Treasury had already deployed $294 billion
of the funds it was authorized to use pursuant to EESA, with the bulk of that money going
into banks, pursuant to the Capital Purchase Program (GAO, 2009). The Federal Deposit
Insurance Corporation (FDIC) had also issued guarantees on a range of debt instruments
far beyond those that would normally be eligible for FDIC insurance. Thus, by the time of
the SCAP, the situation had already improved markedly—but only because of this series of
broad, risky, and politically unpopular government interventions.

Moreover, despite these myriad interventions, all was not necessarily well in the financial
system. Market participants continued to distrust banks' internal risk management capacities, and questions lingered regarding how the system as a whole might fare in the face
of a further adverse shock. It was at this time and in this environment that the SCAP,
which both recapitalized the banks that needed it and allowed market participants to evaluate
for themselves the accuracy of regulatory assessments of how much capital banks needed,
proved so useful.

1.2 Some Implications

Although it can be dangerous to extrapolate from a single example, the SCAP does suggest
a couple of lessons. One thing to note is that the environment into which the results of the
SCAP were released shaped their impact on market functioning. At the time of the SCAP,
fear and distrust remained the default positions of many market participants. The proxies on
which market participants had been relying before the crisis, from credit ratings to faith in
supervisors or their own capacity to assess the risk of other institutions, had proved wanting.
As a result, the default level of market discipline was excessively harsh relative to the actual
health of banking organizations. The credible information contained in the SCAP results
proved helpful, but this was largely because the gaps in what was known were being viewed
with fear, rather than neutrality or lack of concern.

A second important element of the SCAP was that the information produced and disclosed
addressed a real gap in what was known. The early stages of the financial crisis had revealed
banks' internal mechanisms for identifying and assessing risk exposures to be deficient, and
banks had lost faith in their own capacity to assess the health of counterparties. Subsequent
analysis shows that market participants generally knew which banks were undercapitalized,
but they did not know the size of the shortfalls or the types of circumstances that might
tip them over the edge. By providing that information about the size and reducing the
variance in market participants' assessments of bank health generally, the stress test results
addressed a gap in the information otherwise available to market participants. Had the
Federal Reserve merely provided information that was already incorporated into market
participants' assessments of bank health, the value of the information content of the stress
tests would have been negligible.

Third, and perhaps most obviously, the value of the information contained in the stress
test was contingent on market participants' assessment of the reliability of that information.
The importance of credibility was brought home by the European experience in Europe's

first round of stress tests in 2010. The 2010 European stress tests covered 91 banks spanning 20 countries, which, in the aggregate, held approximately 65 percent of the total assets in the European banking sector. In a spirit akin to the US stress tests, European authorities declared that "[t]he overall objective of the stress-testing exercise is to provide policy information for assessing the resilience of the EU banking system" (European Central Bank [ECB], 2010). To further this aim, they "decided to disclose a detailed report about the assessment of the resilience of the EU banking sector, the key results of the impact of the stress scenarios on each individual bank in the exercise, as well as their sovereign exposures, with a detailed breakdown between trading and banking book exposures" (ECB, 2010).

Unlike the experience in the United States, however, the disclosure of the results of the EU stress tests did not have the desired effect of restoring faith in the health of banks and market functioning more generally. Instead, the available evidence suggests that there was virtually no market response to the EU's disclosures (Alves, 2013; Ellahie, 2012). Moreover, subsequent adverse developments at a number of the banks that received a clean bill of health by the EU authorities suggest that market participants were right not to trust the results provided by EU authorities. At least one of the explanations for why the EU provided such rosy results despite the deep problems that clearly persisted was that—in contrast to the situation in the United States—regulators had no authority to undertake a massive recapitalization should they engage in a truly robust set of stress tests and should the results reveal the need for significant additional capital. Put differently, the EU was facing the prospect of setting off a crisis it did not have the authority to readily contain if it provided more accurate assessments of the situation, and everyone was aware of these limitations.

This suggests broader lessons. First, that information generation and disclosure are most likely to promote healthy market functioning when market participants are filling in information gaps with overly pessimistic assumptions or when, because of coordination challenges, liquidity hoarding, or psychological biases, they are excessively hesitant to transact. Significantly, this does not resolve—and may work against—disclosure when times are good. The reasons the optimal-disclosure policy may be state contingent are discussed in more detail by Itay Goldstein and Yaron Leitner in their contribution to this handbook (Chapter 11) and work cited therein. The considerations here are an extension of that already complicated matrix. (Goldstein and Leitner, 2018; Goldstein and Sapra, 2014; Flannery et al., 2017). Whether it is possible to make a change in disclosure policy at all is a difficult question, but it is clear that should there be any changes, they must go in the direction of more information—not less—as distrust increases. A related issue is that once a crisis takes hold, supervisory reputation is often already diminished, so disclosing not only results but also sufficient information to verify those results may be necessary for even the informational output to be seen as reliable. (Schuermann, 2016).

A second lesson suggested by the SCAP is that for information injections to be useful, the information generated must be otherwise unknown and must be credible. Given that there is likely permeability between what markets and governments know, the production of the information that is likely to be most useful may well entail the greatest risk, in that the process of generating that information might also yield bad news that could exacerbate market dysfunction if not accompanied by other stabilizing interventions.

1.3 Some Limitations

The focus thus far has been on SCAP as a mechanism for producing information. But SCAP was never just about producing information. Another important component of the SCAP's success was the requirement that banks found wanting would be required to recapitalize. Banks had the option of raising private capital if they could or accepting capital infusions from the Treasury if they could not. Either way, recapitalizing to a level that would allow each, and therefore all, of the largest banks to withstand further losses was mandatory. This was critical to overcoming a collective-action problem because the persistent weakness in the banking system had costs beyond those that any individual institution internalized as a result of its own undercapitalization (Dudley, 2011).

Given the multiple, interconnected mechanisms through which the SCAP promoted recovery, it is difficult to cleanly separate the information it generated from the changes it brought about. These dynamics make it difficult analytically to draw any clear conclusions regarding the value of the information SCAP injected into the market in improving conditions, apart from the other mechanisms through which it aided market functioning. Similar challenges arise with other examples of stress testing during periods of distress, like that undertaken by Japan in the 2000s, discussed further later in this chapter. In these and other instances, however, the information generation was critical to tailoring the intervention and enabling the intervention to have the desired effect of restoring faith in the banks and financial system. This reflects the distinct challenges and opportunities that arise when stress testing after conditions and confidence have already started to deteriorate, but it does not provide a reason to discount the value of information production during such periods.

2 The Life of a Crisis

Reviewing the SCAP provides a helpful introduction to why and when just-in-time information production can serve as a useful crisis-time tool. This section expands on that introduction by considering the information dynamics over the course of a financial crisis and how stress testing may change what is known at the various stages of a financial crisis. It first considers the background setting, suggesting information gaps are pervasive in modern finance. It then considers, in turn, the different ways that stress testing may facilitate strategies for crisis containment and resolution. Because the COVID-19 crisis is still evolving, it is addressed in the implications section that follows this one.

2.1 In the Beginning, Information Gaps

The importance of information generation during a crisis, as well as the particular type of information that a stress test can generate, depends a great deal on background information dynamics. Given space constraints, this chapter will do little more than summarize why unknowns loom large in modern finance and why the import of these gaps can change during periods of distress.

As a starting point, because of the costs of information generation, frictions in the transmission of information, and the inherent dynamism of financial markets and the economy more generally, market participants and regulators today often operate with radically incomplete information (Awrey and Judge, 2019; Gilson and Kraakman, 2014). Complexity is one of the greatest challenges. Even a cursory evaluation of the complexity of the largest banking organizations and the ever-evolving regime of market-based intermediation that continues to function alongside and interconnected with the banking system supports the view that massive information gaps are the new norm (Flood et al., 2017; Pozsar et al., 2010). The largest banking organizations often continue to have 1,000, and sometimes upward of 2,000, different legal subsidiaries. Moreover, these subsidiaries are often operating different lines of business, in different jurisdictions, and are subject to oversight by different regulators. (Carmassi and Herring, 2016). Instruments, too, remain remarkably complex, and a growing body of literature suggests reasons, from rent extraction to excess demand for information-insensitive assets, to expect this complexity to continue (Hanson and Sunderam, 2013; Holmstrom, 2015; Gorton, 2012). Simon Levin and Andrew Lo voice the view of many in their declaration that "the financial system has crossed a threshold of complexity where the system is evolving faster than regulators and regulations can keep pace" (Levin and Lo, 2015).

Accentuating the challenge and importance of these information gaps is that they do not arise arbitrarily but, rather, systematically show up in some of the most fragile spaces within the overall financial system. A growing body of literature documents the demand for assets that holders will hold and trade at face value with minimal due diligence into the value of the underlying assets. The overlapping concepts of "money," "safe assets," and "information-insensitive assets" are used to describe assets, of varying maturities, that have this quality (Krishnamurthy and Vissing-Jorgenson, 2012; Gorton, 2016; Holmstrom, 2015). The commonality across all of these is that the assets play a functional role, such as facilitatons transactions or serving as a tool for liquidity smoothing, which means holders are willing to pay a premium for these assets, above their risk-adjusted returns. Treasuries are a classic example, but when the supply of truly safe assets is insufficient relative to the demand, private actors step in to bridge the gap, often using structures that are intentionally opaque, such as debt on debt, to encourage holders to trade without having to worry that anyone else has private information (Dang et al., 2017). Information gaps are thus often greatest in precisely the domains undertaking the greatest liquidity transformation, making it all the more likely that information gaps will be large and will contribute to market dysfunction once holders start to ask questions about assets that had previously been treated as safe (Judge, 2019, 2017).

Both the size of information gaps and their distinct importance when things go wrong are exacerbated by the constant shape-shifting that is endemic to finance. Periods of stability induce behavior changes that change the structure of the financial system (Minsky, 1992; Shiller, 2001; Geanakoplos, 2010). Regulation also brings about change as market participants constantly seek out new ways to provide desired goods and services while minimizing the cost of regulatory compliance. Innovation, technological and otherwise, accentuates these dynamics and introduces an additional force toward change by sometimes enabling efficiency gains. As a result of this dynamism, even if regulators were to somehow develop a comprehensive picture of the entire financial system and all of the contingent obligations, explicit and otherwise, among the various bodies constituting that system, that picture would be outdated before it could be developed. Moreover, because regulatory arbitrage is one of the drivers of that shape-shifting, regulators in particular will almost always be working at an information disadvantage when signs of fragility first emerge.

The importance of all of these dynamics and the ways information gaps can exacerbate fragility were on display during the early stages of the last crisis. The aggregate value of subprime MBSs, for example, was relatively modest, but it nonetheless led to widespread runs on asset-backed commercial paper (ABCP; Covitz et al., 2013). ABCP was highly complex in ways that made it hard for holders to verify underlying asset quality, and given the low return, such effort was almost never cost justified. The patterns of runs on ABCP suggest that some proportion was likely due to a lack of high-quality information about how subprime assets were allocated across the system and what the downgrades of subprime MBSs might mean with respect to the value of other structured assets. Moreover, in contrast to the traditional bank setting, where bank managers and sometimes bank regulators understood the risks to which a bank was exposed, the complexity of the assets underlying many ABCPs created a real possibility that even plan sponsors had incomplete information about the value of the assets underlying a structured claim and the waterfall pursuant to which a particular instrument would be paid.

Although the specific series of events leading to the next crisis will inevitably look very different, there are broad patterns that do repeat. Fragilities often arise where liquidity transformation (and oftentimes, maturity transformation) is taking place, which is often the location in the financial system where assets and institutions have been structured to discourage information generation. Additionally, as economic historian Hugh Rockoff has shown, the majority of financial crises in the United States have first emerged not in the banking sector but in that day's version of a shadow banking system (Rockoff, 2018). Putting these pieces together suggests that there is a high probability that when the next crisis erupts, it will erupt in a space where both private actors and public regulators have only incomplete information about the value of assets and how risks are allocated. It further suggests that these information gaps will likely exacerbate the dysfunction and that filling in some of that missing information should be among the aims of regulators seeking to contain the fallout.

This assessment further suggests that even if there are questions about the capacity of stress tests to produce novel information when being performed during times of peace, they may well remain useful as a crisis-fighting tool. Because crises so often erupt in shadows and in domains where market participants had been relying on assumptions or proxies rather than high-quality information, the information gaps are often the largest in precisely those spaces where cracks first begin to appear.

The postcrisis reforms make modest progress in addressing the challenge of information gaps. There were numerous proposals on the table to substantially reduce the size and complexity of financial institutions or to try to scale back significantly on market-based finance. None of these proposals was adopted. There have been some attempts, through mechanisms like living wills, to scale back on the complexity of banks, and there have also been attempts, through efforts like centralized clearing, to improve transparency in other domains. Nonetheless, for the most part, Basel III, the Dodd–Frank Act, and other postcrisis reforms have not resulted in a massive simplification of the financial system, nor have they shut down on the types of dynamism that contribute to information gaps.

Shifting from the nature of the financial system to the rules governing it and the institutions through which those rules are implemented and enforced reveals further reasons that information gaps are the norm. Title 12 of the US Code, the provision governing banks and banking, is among—if not the single—most complex and convoluted areas of law in the United States today (Li et al., 2014). Moreover, as Andrew Haldane and others have explained, the postcrisis regulatory regime is even more complex than predecessors, leading to new types of uncertainty about how it will operate in different states of the world (Haldane, 2017). The regulatory architecture adds to the challenge. The United States has long had a particularly fragmented financial regulatory structure, one that not only separates the banking regulators from market regulators but also places multiple bodies in each of those categories.

The Financial Stability Oversight Council (FSOC) and the Office of Financial Research (OFR) were intended to overcome some of the information-generation challenges that arise from this siloed and fragmented regime. Similarly, the FSOC's authority to designate non-bank financial institutions as systemically significant and hence subject to oversight by the Federal Reserve was intended to enable the contours of federal prudential oversight to morph as institutions otherwise outside that perimeter evolved in ways that could threaten the stability of the financial system. These are helpful developments, but the intervening years have revealed them to be fundamentally incomplete. Other efforts that could help mitigate information gaps, such as data standardization, are similarly under way but lack in scale and scope relative to what would be needed to meaningfully reduce information gaps in the financial system (Berner and Judge, 2019).

The discussion here is descriptive, not normative. Given the incredible complexity of financial institutions and markets, the rate at which finance continues to evolve, and the costs and other frictions that impede information generation, there remain large swathes of information that are perfectly knowable and pertinent (at least in some states of the world) that are not currently known to any actor, private or public. Information gaps are part of finance as it exists today. Technology may change these dynamics, and FinTech and RegTech are already changing how information is produced and disseminated. Nonetheless, there has yet to be any indication that technology is on the verge of overcoming the numerous costs and other frictions that prevent anyone from having a complete picture of the financial system, how risks are allocated therein, and the myriad mechanisms through which problems in one domain can spread to others.

2.2 Stress Tests as Tool for Helping to Identify and Contain a Crisis

Although sometimes modeled with a single shock, financial crises often grow over time. The crisis had been under way for more than a year when Lehman Brothers failed in September 2008. And as Frederic Mishkin reminded his colleagues in August 2008, "in the Great Depression, when . . . something hit the fan, [laughter] it actually occurred close to a year after the initial negative shock. . . . We are now a year into this" (Federal Open Market Committee, 2008).

That crises are often under way for a meaningful period of time before things get really bad suggests this may be a window of opportunity for brave regulators. Among the key advantages of responding to the soft signals emitted during the early stages of a crisis is that those signals, those areas where the market response to news seems disproportionate to the news, can serve as a road map for the issues to investigate further. As Claudio Borio and coauthors emphasize in their own work that explains why crisis-time stress tests are more likely to succeed than peacetime efforts, when "the objective [of a stress test] is to support crisis management or resolution, the key risks are often apparent. For instance, if the crisis has originated in exposures to property markets, it is natural to stress them further" (Borio et al., 2014).

Borio et al.'s (2014) focus, which is helpful, is on the way the early signals of problems can provide useful guidance with respect to particular asset classes to watch. Currently, for example, there is growing concern about leveraged lending. A rise in corporate bankruptcies or a single corporate bankruptcy with large spillover effects on lending or the market for collateralized loan obligations (CLOs) could trigger stress tests focused on discerning the capacity of the financial system or firms within it to handle further deterioration in those markets.

Another way early indications of distress may helpfully inform stress testing is by identifying what it is that should be tested. As a number of the other chapters in this volume make clear, regulatory stress testing is not limited to banks. It can also be used to assess the viability of an array of institutional arrangements and, perhaps, the system itself. This type of effort will likely require coordination and communication among multiple regulators, but as more is learned about stress testing and techniques continue to improve, so, too, will the range of ways that stress testing might be used early in a crisis to provide new information about threats that may be imminent and foreshadowed before they become manifest.

A final point to highlight about stress testing in the face of trouble is that the information produced may be helpful not only in identifying specific problem areas but also in forcing market participants and regulators to acknowledge the nature and scope of the challenges that could well lie ahead. Although not talked about much in recent years, forbearance and its kin remain a real challenge. Regulators were not as quick to respond to the problems of 2007 as they could have been, and earlier intervention may have helped reduce the size of the crisis that followed (Judge, 2017).

More to the point, history abounds with examples of regulators being even slower to fully acknowledge and address capital deficiencies, with adverse effects on the macroeconomy. The US savings and loan crisis of the 1980s, Japan in the 1990s, and Europe emerging out of 2008 and the sovereign debt crisis that followed are just a few such examples. Although less dramatic than widespread panics, allowing a financial system to limp along can have disastrous macroeconomic consequences, leading to anemic or no growth for extended periods. Well-conducted stress tests can be a useful way to help avoid such situations by forcing an issue—making it plain that there are real problems or weaknesses that must be addressed sooner or later. Put differently, in addition to providing information that market participants and regulators appreciate the need for, stress tests can also force a change in mindset, compelling recognition that a situation may be more severe than anyone wants to admit. Hence, the final section examines the value of stress tests after things have stagnated for a while.

2.3 *Laying the Road to Recovery after the Situation Has Devolved*

The last section focused on the value of stress testing as an early-stage crisis-recognition and crisis-containment tool. SCAP represents a somewhat later-stage intervention, one component of a multicomponent, heavy-handed government effort to pave a road to recovery in the immediate wake of a panic. The other pattern that sometimes occurs is that regulators succeed in containing or averting a panic but lack the will or means to address the underlying problems, leading to a prolonged period of slow growth. Japan's lost decade (or two) starting in 1991 is a prime example and one that shows how here, too, stress testing can be useful.

The triggering event of Japan's prolonged malaise was the failure of Toho Sogo Bank, which revealed weaknesses that had been building in Japan's banking system for some time. Signs of trouble had emerged earlier, but a combination of wishful thinking and forbearance made it easier for banks to be too slow in recognizing losses. After first facilitating this wishful thinking by expanding deposit insurance and helping shield banks from market discipline, the government began to recognize a different tack was needed. The government started to intervene in more meaningful ways by injecting capital into *jusen*, specialized, nonbank housing loan companies, in 1995 and 1996. This was followed by a much more aggressive round of capital injections into banks in 1998. Even as regulators began to tackle the problem of capital deficiencies, unrecognized losses, and the need for more significant government intervention, however, they continued, for a while, to rely on banks' own, overly rosy assessments of their balance sheets.

It was not until the early 2000s that the government stopped accepting the banks' assessments of their health and future capital needs and instead started undertaking their own, more accurate and more dire, assessments of bank loans' likely performance. Like the SCAP, this horizontal supervisory exercise provided valuable new insights into where problems lay and what needed to be done to address them. Armed with these new insights, Japanese regulators finally forced banks to restructure or otherwise address troubled loans, thereby cleaning up the latent problems that had so long plagued the system and paving the way for future growth.

As with the SCAP, the production of more reliable information about the size and location of weaknesses was produced in conjunction with an effort to address those challenges, making it difficult to separate the value of the information production from the substantive actions. That the government had previously provided significant financial support without simultaneously compelling information production, and that such efforts failed to revive the system, nonetheless suggests that high-quality information is often a prerequisite to successfully bringing about a lasting inflection point once systemic distress takes hold.

Japan's lost decade also serves as a pointed reminder of the way information gaps can feed into tendencies to engage in forbearance and wishful thinking, patterns that can have lasting and detrimental effects on a country's economic health. In the face of stagnation or other indicators that a country's financial system is not performing as well as it could, targeted, forward-looking information generation could serve a helpful role in identifying the sites that require further attention and increasing confidence in those that do not.

Japan's experience is thus another illustration of the importance of accurate information about the location and size of losses in efforts to revitalize an ailing financial system. It is different in the sense that the production of information came far later into the crisis, when the situation was stable but still bleak. The role of the information was thus not to avoid runs by short-term creditors; rather, the information proved critical to breaking the feedback loop between the economy and financial system that was weighing heavily on both.

3 Setting the Stage for Success: COVID-19 and Beyond

The analysis thus far has explored the value of just-in-time information production when a crisis hits. During the early stages of a crisis, stress testing and related modes of information production can help regulators and others to recognize the magnitude of the change in state they are facing. As a crisis evolves, appropriately designed stress tests can provide valuable information about the actual capital deficiencies in the system. Stress tests can also be used to identify institutions that ought to be closed or even entire segments of the market that should be phased out. Furthermore, stress tests can be designed to identify interconnections or other mechanisms of contagion that must be addressed to restore calm. Depending on the stage when they are produced, these types of information can also facilitate engagement, allowing legislatures and the public to provide more informed feedback, processes that can be critical to the legitimacy and public acceptance of the crisis-management process.

This section first uses the lessons here to provide an early-stage assessment of the Fed's approach to stress testing in response to COVID-19. It then provides some general thoughts on the steps that ought to be taken to facilitate information generation during periods of systemic distress.

3.1 COVID-19

COVID-19 is a distinct type of crisis in many ways. The rapidity with which it hit the real economy and the speed with which both financial regulators, led by the Fed, and Congress responded with fiscal and other support are unprecedented. That so few were on the lookout for a global pandemic is itself a powerful illustration of why crisis-time stress testing will always be critical. There is simply no way that central banks can foresee in advance the array of exigencies that could threaten the health of banks and the broader financial system they help constitute.

Given the prompt response and the relative success of many of the efforts to stop widespread panic and contain the economic damage inflicted, it may be easy to think that the economic crisis will not morph into a financial one. But both the public health and economic crisis are still evolving, and the threat those developments pose to banks and other parts of the financial system remains far from clear.

Thus far, the US response has been decidedly mixed. In contrast to the United Kingdom, for example, the United States went forward with the stress tests when scheduled in the spring of 2020. But having recently made changes that incorporated the results of the stress tests into banks' capital requirements, the Fed opted to proceed with stress test scenarios that had been developed before the global pandemic took hold. The result was that even the severely adverse scenario contained assumptions that mapped poorly onto the dramatic pace of the contraction in global economic activity throughout the spring of 2020. It chose to complement those stress tests with "sensitivity analyses" that seek to capture how banks will perform in one of three possible scenarios—a quick return to economic health (V shape), a more prolonged downturn before returning to economic health (U shape), or a double-dip recession (the dreaded W). But it is not running a full stress test for any of the banks using those scenarios. Moreover, the Fed's vice chair for supervision, Randal Quarles, recently announced that the Fed is not planning to release the results of these sensitivity analyses for any specific bank. Rather, it will release only aggregate results.

The analysis here suggests that the Fed was wise to proceed with the stress tests and to complement the original scenarios with sensitivity analyses designed to provide insight into how banks may fare under some of the quite different possible economic scenarios. Running these additional analyses should provide the Fed more insight into which firms are most likely to run into trouble, and it may also help reveal sources of weakness or mechanisms of contagion that may not have been visible otherwise. Banks, too, should have better insight into the preparedness for what lies ahead, allowing them to plan accordingly.

At the same time, the analysis here also suggests that the Fed's response thus far has been far from sufficient. One of the reasons for not releasing individualized sensitivity analysis results, according to Quarles, is that these analyses are far less rigorous than the Fed's typical stress testing, raising questions about just how accurate the results are and what they might miss. Moreover, the refusal to provide information about how individual banks are expected to fare could yet stoke fear among bank creditors, including counterparties and depositors. Even if there is no immediate run, should any individual banks report results that raise red flags, market participants may become concerned about what else the Fed may be trying to hide by only providing aggregate results.

The Fed's decision to make only minor modification to its stress tests despite the onslaught of a global pandemic that changes so much is a sign that the Fed has not yet fully embraced the need to be responsive in the face of changed circumstances. On the whole, the Fed has been far more quick to roll out new programs and shift to a more accommodative approach to monetary policy than it was in 2007 and 2008. Nonetheless, staying ahead of a crisis requires good information. The Fed has not taken all the steps it could to understand bank vulnerabilities in the face of COVID-19, and there is limited evidence that it is working as closely as it could be with other financial regulators, other than the Treasury, to identify other potential sources of systemic weakness. More can and should be done.

3.2 Limited Safety Net

"If the primary objective [of a stress test] is to support crisis management and resolution, system-wide public-sector liquidity and capital backstops are essential. Without them, no exercise can be credible" (Borio et al., 2014). This declaration from Borio et al. reveals an important, even if unstated, assumption: regulators are going to be rationally hesitant to produce information that could exacerbate an already-fragile situation unless they have the tools in hand to contain the fallout their findings might trigger. Crises are, in part, information events. Bad news is often the trigger that leads to runs. It is unrealistic to ask financial regulators to produce timely information when the very process of doing so might trigger the crisis they are seeking to avoid. Hence, the most important policy takeaway is that there must be safety nets specifically designed to alleviate, or otherwise able to accommodate, the adverse systemic ramifications that could flow should the stress tests produce bad news about the state of the financial system or elements thereof.

A policy of nondisclosure might seem like a solution. After all, bad news must reach the market to trigger a market response. Given the long history of confidentiality in bank supervision, there may be some room for bank supervisors to expand their activities to produce some new types of information without that information getting out. Precisely because banks are so heavily regulated, however, the most pressing information gaps are unlikely to be information pertinent just to the health of individual banking organizations. And if the signals indicating a change in state reveal information gaps with respect to other sectors of the market, it is doubtful that bank supervisors will have the ability to devise a new form of stress test or related mode of information production without market participants getting a whiff that something is going on. Particularly given that for any exercise to be helpful, it will likely require significant information from and participation by the relevant financial institutions, the observation from Borio and colleagues (2014) seems apt: we cannot expect regulators to have the courage to produce new and potentially quite scary information with the hope that the information will not leak and that failure to disclose will not itself trigger concern.

This creates a very real chicken-or-egg problem. Information is key to developing a plan for crisis management, but regulators will have a hard time producing the information they need to develop a plan without first having meaningful support mechanisms already in place. This problem is accentuated by a structural problem, common to the United States and many democratic systems, that legislative approval—that is, approval by a diverse group of nonfinancial experts who are elected by the people—is required before the government can provide lasting fiscal support, such as capital injections. So this is the rub: How can regulators provide elected officials meaningful information about the location of weaknesses in the system and the threats that those weaknesses pose when the very process of producing that information could trigger the crisis they want to avert?

Regulators have found ways to try to navigate around these challenges, but all have limits and each can raise concerns regarding regulatory overreach. For example, most central banks have the ability to provide liquidity support to banks and sometimes nonbanks. But because the run point for firms (and the financial system) is breached well before firms become insolvent, liquidity support alone is insufficient to deter runs and restore stability in most instances. There is also the possibility that the bad news produced in a stress test might limit a central bank's legal authority to provide support where it is most needed. This practical challenge is not new. It has at times led to central banks stretching the bounds of their legal authority. It has also led to uninformed "emergency legislation" pursuant to which a legislature grants exceptionally broad authority to regulators to use their discretion to address the problem. Although either mechanism can overcome the need for legislative approval, neither suffices to achieve the type of broad-based buy-in that a legislative check is designed to provide and that can be critical to minimizing popular backlash.

There have been several postcrisis reform proposals that would help address this challenge. Eric Posner, for example, has argued that the authority of the Federal Reserve should be expanded to include recapitalizations under appropriate circumstances (Posner, 2017). The analysis suggests an additional advantage of such a reform: if regulators have the standing authority needed to address really bad news, they might be more willing to undertake innovative new forms of stress testing that are responsive to the crisis they are facing. The challenge—which is deeply rooted in the structure of the US government and other governments—is that such overt fiscal authority is hard to reconcile with the independence traditionally enjoyed by central banks, and central banks may well lose that independence if given such overt fiscal authority (Tucker, 2018).

In other work, I propose trying to address this challenge by attempting to separate crisis containment from crisis resolution. The aim here is to give regulators broad authority to take the actions needed to stabilize the financial system and absorb the shocks of further bad news while still protecting the legislature's prerogative to have a voice in how best to achieve lasting stability. One way to achieve this type of balance would be to vest the finance ministry, such as the US Treasury Secretary, with broad, time-limited guarantee authority. By allowing the Treasury Secretary to guarantee any debt claims, without pretending to know whether the underlying assets suffice to cover the claim, the guarantee regime would aim to keep private capital in the system while regulators produce the information needed to identify and address weaknesses. This approach would have the advantage of enabling regulators to undertake stress testing and other just-in-time information production anywhere in the financial system, even if those problems arise in domains other than banks or other entities subject to prudential oversight.

If appropriately structured, such a regime might go beyond enabling crisis-time stress testing to, in effect, mandating it. In placing an outside limit of 2 years on the duration of any guarantee scheme and requiring regulators to provide Congress a detailed report of where problems exist and how they should be best addressed, an emergency guarantee regime could also provide an incentive for crisis-time information production. Given how often regulators are slow to recognize the magnitude of the problems they are facing, this could be a distinct and related benefit of such a regime. This type of approach has numerous risks and challenges, including the inevitable moral-hazard concerns that arise from government intervention and legitimate concerns about vesting too much authority in a Treasury secretary beholden to the president. Nonetheless, starting with a realistic baseline of where things now stand, and recognizing the importance of producing new information once distress sets in and the practical challenges impeding regulators' willingness to produce such information, suggests the benefits may outweigh the costs (Judge, 2017).

Regardless of the approach taken, the inability to expect regulators to produce timely information about fragilities in the financial system without some capacity to address the fallout that information would trigger exemplifies what John Crawford has labeled the "moral hazard paradox of safety nets" (Crawford, 2015). If the "chicken" of broad recapitalization authority must precede information production about how that authority ought best be used, legislatures will likely have to provide much broader authority and impose fewer checks on how it is used than if regulators could first produce an "egg" in the form of information about the location and size of losses in the system.

The success of the SCAP and Japan's experience in the mid-2000s illustrate other ways that just-in-time information production brings this paradox to life. Whether the government provides broader support than is necessary or market participants run on more firms than the actual weaknesses would justify, indiscriminate behavior is at the center of crises and is a major factor contributing to the concerns about the moral hazard that crisis responses so often trigger. Crisis-time stress tests can produce critical, credible information about the location of weaknesses and the mechanisms via which they might propagate dysfunction. With such information in hand, market participants can tailor their responses, resulting in less widespread dysfunction, and regulators can better tailor their responses, resulting in fewer concerns about moral hazard. But none of these benefits can be realized if regulators are too afraid to produce that information, and without more authority to contain the fallout of bad news, there are reasons to expect that the frequency and rigor of crisis-time stress tests will be far below the level that is optimal.

3.3 *Preparing*

Financial crises, like war, require lots of advanced planning and a willingness to revise those plans and toss them out completely in the face of better information or evolving circumstances. Thus, the final broad point to highlight is that even though successful crisis-time stress testing will likely require some creativity and willingness to respond to the unique exigencies of the circumstances posed, there are steps that can be taken in advance to enhance the institutional capacity of the financial regulatory system to engage in successful crisis-time stress testing. Two areas where advanced effort could help merit particular attention.

3.3.1 *Regulatory Coordination*

That weaknesses may well first appear outside the formally regulated banking sector has important implications for the tools that will be needed for regulators to undertake appropriate just-in-time information production. Recall that the fragmented regulatory structure was among the features of the US financial system that were changed only modestly after the crisis. Moreover, the DNA of the various regulators has not changed in any fundamental way. The market regulators now have seats on the FSOC and a voice in crafting specific rules that are meant to address systemic stability, but their missions remain largely focused on protecting investors, combatting fraud, and facilitating capital formations—aims that are often orthogonal and sometimes contrary to effective crisis management. As reflected in the contentiousness and months of delay that characterized efforts for the Federal Reserve and US Securities Exchange Commission (SEC) to enter into a memorandum of understanding for sharing information once Bear failed and for the Fed to open two liquidity facilities to primary dealers subject to SEC oversight, crises cannot be trusted to produce "kumbaya" moments in which regulators magically overlook their differences and prerogatives for the sake of the collective good. In part because each often believes in the righteousness of their mission, and in part because no one has a view of the whole system and hence how weaknesses in one domain might spill over to others, effective coordination remains challenging.

In the United States, the creation of the FSOC makes some useful headway on these challenges. The regular meetings of the FSOC, for example, provide a setting for the heads of all of the federal financial agencies and some state representatives to get to know each other in an environment that does not have the stressful overlay that a crisis induces. These meetings, and the requirement that each member of FSOC attest to the completeness of the annual report that FSOC must provide Congress outlining potential threats to systemic stability (e.g., FSOC, 2017), also serve the useful function of ensuring that all of the regulators stay attuned to the possibility of a systemic event and consider whether other developments within the domain of their agency might have systemic repercussions that they might not otherwise have ignored. The OFR is similarly helpful in theory, given its broad authority to collect and standardize data and its orientation toward addressing systemic risk, but it is not currently on course to become a powerful mechanism for generating timely information not otherwise known about the health and structure of the financial system. Several other countries have gone further. The United Kingdom, for example, has formed a Financial Policy Committee and vested it with a range of complementary tools, including stress testing, to try to prevent, detect, and contain systemic disruptions.

Nonetheless, particularly in jurisdictions like the United States where the regulatory regime remains exceptionally fragmented, further progress in promoting coordination, communication, and trust among today's very different financial regulators could go a long way toward enhancing their capacity to work well together when indications of fragility again arise. The more this is embedded deeply into these organizations—such that it is not just the heads but long-time employees who are likely to remain with an agency through changing leadership—the more likely it is that they will have the goodwill and common language needed to work productively even when facing new challenges. One way to do this would be through regulatory overhaul, but more modest steps could also be useful. Of particular relevance to this volume, regular stress testing that requires the involvement of multiple regulators may be one way to help forge these relationships. Working together during peacetime to undertake macroprudential stress tests, for example, could help regulators develop shared understandings of how others approach data and testing issues in ways that may make it far easier to devise and implement crisis-specific stress tests when the time comes.

3.3.2 Skills and Credibility

Another factor that will affect the capacity of regulators to design and undertake stress tests during periods of systemic distress is whether they already possess the relevant skill set. The spread of stress testing as a tool for crisis preparedness and avoidance will likely help in this regard. Banks and their regulators are growing increasingly accustomed to, and seemingly sophisticated about, the process of conducting stress tests, improving the quality and reliability of the information they generate. As other chapters in this volume explain, other types of firms, such as clearinghouses, and other regulators, such as the Commodity Futures Trading Commission (CFTC), are also starting to use stress tests with greater frequency and could productively increase their use further for reasons apart from those addressed here (see Chapter 23 in this volume). Similarly, advances in macroeconomic stress testing and other innovative new forward-looking assessments of the capacity of firms, market structures, or the financial system to bear particular types of adverse developments could further expand the tools readily available to regulators, even if they then must be deployed in ways that address the particular types of risks that emerging threats make more likely.

That said, the spread of stress testing as a peacetime tool may be a mixed blessing. One of the reasons that the SCAP was so successful was that it employed a new technique for producing seemingly credible information about the health of large financial institutions and their capacity to withstand further adverse developments. Put differently, the information produced in the SCAP was credible in part because the accompanying disclosures allowed market participants to engage in some degree of verification but also because the process looked different from the supervisory oversight that had proved wanting.

Once stress testing, with disclosure, is the norm, then the process itself can be tainted and discredited. The same regulators, running effectively the same stress tests but with some tweaks to address new information, are not likely to produce information that the market will see as credible when earlier stress testing had provided overly rosy assessments. The efforts by the Office of Federal Housing Enterprise Oversight (OFHEO) to conduct risk-based capital stress tests for Fannie Mae and Freddie Mac prior to the crisis bring this challenge to life. The failure of those stress tests to reveal problems that later became evident had been there for some time—and had been disguised rather than uncovered by OFHEO's stress tests—discredited both the techniques the OFHEO used and the OFHEO itself. An autopsy of that failure by Scott Frame and coauthors demonstrates that errors with respect to model estimation frequency and specification and reliance on an insufficiently adverse house-price scenario contributed to a stress test that was considered "state of the art" when implemented becoming a "spectacular failure" (Frame et al., 2015).

To be sure, there were specific flaws in the OFHEO program that reduced its efficacy, and there have been attempts to learn from those mistakes. But as the stress tests of Iceland's banking system reflect (IMF, 2008), this was not an isolated failure (Borio et al., 2014). Stress testing will inevitably remain a work in progress and will inevitably sometimes produce assessments that underestimate growing risks. When this happens, tweaking assumptions will not suffice to restore faith in the process or outcomes. Thus, although regular stress testing on the whole is a positive development, one that reduces information gaps and expands the skill set of regulators to produce valuable information, further attention to these dynamics may be warranted to try to address the ways that a loss of regulatory credibility may exacerbate crises, creating yet more of a need for information that market participants can trust while reducing the capacity of regulators to provide it (Morrison and White, 2013).

4 Conclusion

When any crisis hits, regulators and market participants will inevitably lack some of the information that the events triggering the crisis reveal to be important. These information gaps can exacerbate market dysfunction and slow or otherwise impede appropriate regulatory response. Early in a crisis, stress testing can help compel regulators, as well as politicians, and market participants to acknowledge that they may be facing a bigger threat than they realize, leading to earlier intervention. After a crisis has taken hold, stress tests can provide helpful guidance about the location and size of capital deficiencies or other weaknesses impeding market functioning, allowing more tailored government interventions. Both dynamics may yet be important in connection with containing the fallout from COVID-19.

Stress testing can also provide market participants credible information that underlying problems have been addressed or that the government has a plan for rectifying them, further facilitating market functioning. However, none of the advantages that crisis-time stress testing can confer will be realized unless there is an adequate backstop in place to handle the fallout if the tests produce bad news. Time-limited guarantees, coupled with an appropriate affirmative investigation and reporting obligations, might be one way to create an incentive for and enable information production when it is most needed.

Regardless of whether such a step is taken, regulators should be on guard not to allow the results of stress tests or other supervisory efforts to make them too confident about the health of the financial system or institutions within it. This is one of the most critical lessons. The world is constantly evolving, and the financial system is shape-shifting faster than many domains. No test can fully reveal the vagaries of the interplays through which an unexpected adverse development can trigger effects elsewhere in the system. Looking out for and responding to soft signals that something might be amiss and being willing to ask hard questions can go a long way in helping regulators identify and address a crisis before it topples the entire economy.

References

Alves, Ricardo Jorge Santos (2013), "Information value of EU-wide stress tests: How did markets react to stress test results?" Master's Dissertation for the ISCTE Business School, Instituto Universitario de Lisboa, available at https://repositorio.iscte-iul.pt/bitstream/10071/9900/1/INFORMATION%20VALUE%20OF%20EU-WIDE%20STRESS%20TESTS.pdf.

Awrey, Daniel, and Kathryn Judge (2019), "Why financial regulation keeps falling short," Working Paper, available at https://scholarship.law.columbia.edu/cgi/viewcontent.cgi?article=3608&context=faculty_scholarship.

Bernanke, Ben S. (2013), "Stress testing banks: What have we learned?" Speech at the "Maintaining Financial Stability: Holding a Tiger by the Tail" financial markets conference sponsored by the Federal Reserve Bank of Atlanta, Stone Mountain, Georgia, available at www.federalreserve.gov/newsevents/speech/bernanke20130408a.htm.

Berner, Richard, and Kathryn Judge (2019), "The data standardization challenge." In Douglas W. Arner, Emilios Avgouleas, and Danny Busch (eds.)., Systemic risk in the financial sector: Ten years after the great crash, CIGI Press.

Borio, Claudio, Mathias Drehmann, and Kostas Tsatsaronis (2014), "Stress-testing macro stress testing: Does it live up to expectations?" *Journal of Financial Stability*, 12, 3–15.

Carmassi, Jacopo, and Richard Herring (2016), "The corporate complexity of global systemically important banks," *Journal of Financial Services Research*, 49 2–3, 175–201.

Clarida, Richard H. (2010), "What has – and has not – been learned about monetary policy in a low inflation environment? A review of the 2000s," Federal Reserve Bank of Boston Conference Paper No. 55, available at www.bostonfed.org/-/media/Documents/conference/55/papers/Clarida.pdf?la$=$en.

Covitz Daniel M., Nellie Liang, and Gustavo A. Suarez (2013), "The evolution of a financial crisis: Collapse of the asset-backed commercial paper market," *Journal of Finance*, 68(3), 815–848.

Crawford, John (2015), "The moral hazard paradox of financial safety nets," *Cornell Journal of Law and Public Policy*, 25(1), 95–139.

Dang, Tri Vi, Gary Gorton, Bengt Holmstrom, and Guillermo Oroñez (2017), "Banks as secret keepers," *American Economic Review*, 107(4), 1005–1029.

Dudley, William C. (2011), "US experience with bank stress tests," Remarks by Mr. William C. Dudley at the Group of 30 Plenary Meeting, Bern, Switzerland, available at www.bis.org/review/r110704c.pdf.

Ellahie, Atif (2012), "Capital market consequences of EU bank stress tests." www.newyorkfed.org/medialibrary/media/research/conference/2012/FinancialServices2012/Ellahie.pdf.

European Central Bank (2010, July 23), "Questions and answers: 2010 EU-wide stress testing exercise," Press Release, available at www.ecb.europa.eu/pub/pdf/other/euwidestresstestingexercise-qaen.pdf?860c12759915bf0e3a8cd5486cb595a4.

Financial Stability Oversight Council (2017), "2017 annual report," available at www.treasury.gov/initiatives/fsoc/studies-reports/Documents/FSOC_2017_Annual_Report.pdf.

Flannery, Mark, Beverly Hirtle, and Anna Kovner (2017), "Evaluating the information in the Federal Reserve stress tests," *Journal of Financial Intermediation*, 29, 1–18.

Flood, Mark D., Dror Y. Kenett, Robin L. Lumsdaine, and Jonathan K. Simon (2017), "The complexity of bank holding companies: A new measurement approach," Office of Financial Research Working Paper No. 17–03, available at www.financialresearch.gov/working-papers/files/OFRwp-2017-03_Complexity-of-Bank-Holding-Companies.pdf.

Frame, W. Scott, Kristopher Gerardi, and Paul S. Willen (2015), "The failure of supervisory stress testing: Fannie Mae, Freddie Mac, and OFHEO," Federal Reserve Bank of Boston Working Paper No. 15–4, available at https://pdfs.semanticscholar.org/048a/0dbc65109908151db462218be28e85aa89ea.pdf

Geanakoplos, John (2010), "The leverage cycle," Cowles Foundation Discussion Paper No. 1715R, available at https://cowles.yale.edu/sites/default/files/files/pub/d17/d1715-r.pdf.

Gilson, Ronald and Reiner Kraakman (2014), "Market efficiency after the financial crisis: It's still a matter of information costs," *Virginia Law Review*, 100(2), 313–376.

Goldstein, Itay and Yaron Leitner (2018), "Stress tests and information disclosure," *Journal of Economic Theory*, 177, 34–69.

Goldstein, Itay, and Haresh Sapra (2014), "Should banks' stress test results be disclosed? An analysis of the costs and benefits," *Foundations and Trends in Finance*, 8(1), 1–54.

Gorton, Gary (2012), *Misunderstanding financial crises: Why we don't see them coming*, Oxford University Press.

Gorton, Gary B. (2016), "The history and economics of safe assets," National Bureau of Economic Research Working Paper No. 22210, available at www.nber.org/papers/w22210.

Government Accountability Office (2009), "Troubled Asset Relief Program, status of efforts to address transparency and accountability issues," GAO Report GAO-09-296, available at www.gao.gov/new. items/d09296.pdf.

Haldane, Andrew (2017), "Rethinking financial stability," Speech by Andrew G. Haldane at the "Rethinking Macroeconomic Policy IV" Conference, Washington, DC, available at www.bis.org/review/r171013f.pdf.

Hanson, Samuel G., and Sunderam, Adi (2013), "Are there too many safe securities? Securitization and the incentives for information production," *Journal of Financial Economics*, 108(3), 565–584.

Holmstrom, Bengt (2015), "Understanding the role of debt in the financial system," BIS Working Papers No. 479, available at www.bis.org/publ/work479.htm.

International Monetary Fund (2008, August 19), "Iceland financial stability assessment update," available at www.sedlabanki.is/lisalib/getfile.aspx?itemid$=$6686.

Judge, Kathryn (2017), "Information gaps and shadow banking," *Virginia Law Review*, 103(3), 411–480.

Judge, Kathryn (2019), "The new mechanisms of market inefficiency," Columbia University and ECGI Law Working Paper No. 480, available at https://scholarship.law.columbia.edu/cgi/viewcontent.cgi?article$=$3577&context$=$faculty_scholarship.

Krishnamurthy, Arvind, and Annette Vissing-Jorgenson (2012), "The aggregate demand for Treasury debt," *Journal of Political Economy*, 120(2), 233–267. www.jstor.org/stable/10.1086/666526?seq$=$1#metadata_info_tab_contents

Levin, Simon A., and Andrew W. Lo (2015), "Opinion: A new approach to financial regulation," *Proceedings of the National Academy of Sciences*, 112(41), 12543–12544.

Li, William, Pablo Azar, David Larochelle, Phil Hill, and Andrew W. Lo (2014), "Law is code: A software engineering approach to analyzing the United States Code," *Journal of Business & Technology Law*, 10(2), 297–374.

Minsky, Hyman P. (1992), "The financial instability hypothesis," Jerome Levy Economics Institute Working Paper No. 74, available at https://papers.ssrn.com/sol3/papers.cfm?abstract_id$=$161024.

Mishkin, Frederic (2008), Statement at the Federal Open Market Commission Meeting, available at www.federalreserve.gov/monetarypolicy/files/FOMC20080805meeting.pdf.

Morrison, Alan D., and Lucy White (2013), "Reputational contagion and optimal regulatory forbearance," *Journal of Financial Economics*, 110, 642–58.

Peristiani, Stavros, Donald P. Morgan, and Vanessa Savino (2010), "The information value of the stress test and bank opacity," *Journal of Money, Credit and Banking*, 46(7), 1479.

Posner, Eric A. (2017), "What legal authority does the Fed need during a financial crisis?" *Minnesota Law Review*, 101, 1529–1578.

Pozsar, Zoltan, Tobias Adrian, Adam Ashcraft, and Hayley Boesky (2010, July), "Shadow banking," Federal Reserve Bank of New York Staff Report No. 458, available at www.newyorkfed.org/medialibrary/media/research/staff_reports/sr458.pdf.

Rockoff, Hugh (2018), "It is always the shadow banks: The regulatory status of the banks that failed and ignited America's greatest financial panics," in Hugh Rockoff and Isao Suto (eds.), *Coping with financial crises*, Springer.

Schuermann, Til (2016), "Stress testing in wartime and in peacetime," in Ronald Anderson (ed.), *Stress testing and macroprudential regulation: A transatlantic assessment*, Center for Economic and Policy Research.

Shiller, Robert J. (2001), "Bubbles, human judgment, and expert opinion," Yale Cowles Foundation Discussion Paper No. 1303, available at https://papers.ssrn.com/sol3/papers.cfm?abstract_id$=$275515

Tucker, Paul (2018), *Unelected power: The quest for legitimacy in central banking and the regulatory state*, Princeton University Press.

Part III

Microprudential Stress Testing

Stress Testing for Commercial, Investment, and Custody Banks[*]

Daniel Cope, Carey Hsu, Clinton Lively, James Morgan, Til Schuermann, and Evan Sekeris

1 Introduction

Stress testing is one of the oldest forms of risk management in banking. For example, banks have routinely run interest rate shocks through their books: a shock of 100 or 200 basis points (bps) up or down (assuming there is room for down), a steepening or flattening of the yield curve. Although simple and perhaps even simplistic, this approach manages to capture salient aspects of a fundamental risk type for banks: interest rate risk. Similarly, investment banks run series of single-factor shocks on their main risk exposures: equity market shocks, foreign exchange (FX), rates (e.g., interest rate), credit spreads, and so on; see Schuermann (2014) for a historical overview. So what is new?

The Global Financial Crisis (GFC) of 2008–2009 triggered a new and much more comprehensive stress testing approach. Instead of one or two risk factors experiencing a one-time shock, regulators designed full multifactor dynamic scenario paths including real economic (e.g., gross domestic product [GDP], unemployment, inflation) and financial (e.g., equity prices, rates, spreads, FX) variables evolving over a multiyear horizon. These scenarios are then translated to dynamically evolving balance sheets and the profit and loss (P&L) statements (or income statements) of banks. The counterfactual seems straightforward: How well would a bank survive the contemplated scenarios? And in particular, what would happen to its capital position at each time interval, be that at the end of each quarter or each year of the scenario? This in turn helps answer the central (microprudential) question of capital adequacy: Is a bank's current level of capital sufficient to withstand a range of severe but plausible outcomes and still perform its financial-intermediation function for the economy?[1] In this way, our treatment is focused on microprudential objectives.

Banks, by virtue of their intermediation role, are very closely tied to the business cycle and systematic risk factors. Schuermann and Stiroh (2006) show that banks are more correlated than any other sector to a standard set of risk factors. As an illustration, Figure 13.1 shows

[*] We wish to thank Tim Clark, Dick Herring, Akhtar Siddique, and Thom Wetzer, as well as participants at the the International Monetary Fund (IMF) "Rethinking Financial Stability: The FSAP at 20" conference for very helpful comments. All remaining errors are ours.

[1] Please see Chapter 2 in this handbook by Herring and Schuermann on objectives and challenges of stress testing, and for a historical perspective, see the chapters by Das, Dent, and Segoviano (Chapter 3) and Anderson et al. (Chapter 4).

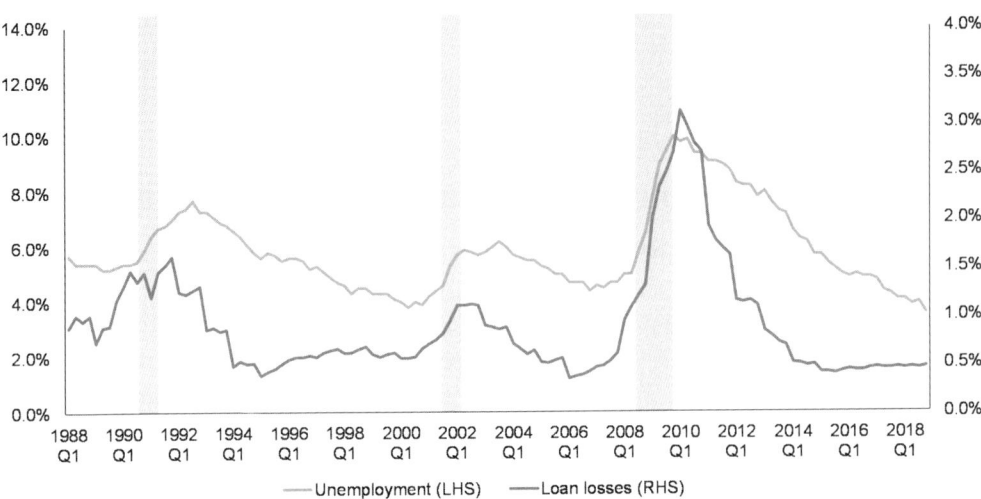

Figure 13.1 US unemployment rate and total US bank loan losses, Q1 1998–Q4 2018.
Total net loan charge-offs to total loans for banks—not seasonally adjusted.
Source: Federal Reserve Economic Data (FRED); authors' calculations.

the comovement of total bank loan losses in the United States with the unemployment rate, including National Bureau of Economic Research (NBER) recessions; the correlation is about 80 percent. Of course, not all risk factors affect all banking activities the same way. Consumer products such as mortgages and credit cards are heavily influenced by house prices and the unemployment rate. Commercial lending will depend on aggregate demand factors (e.g., GDP or industrial production), credit spreads, and sector-specific risk factors (e.g., oil prices and FX). The key, then, is to be able to identify the risk drivers facing the banks from the products and services they sell, the clients and customers they serve, and the markets and geographies they operate in. Good stress testing is not possible without first identifying and understanding the risks faced by the bank.

In this chapter we describe the stress testing process in banking across the major types of business models: commercial banking, including the wide range of retail and commercial products and services; investment banking and capital markets; and less common but equally important members of the banking ecosystem, trust and custody banks. Properly identifying and understanding the risks that arise from these business activities allows for appropriate scenario design; you have to know your risks and vulnerabilities to design a scenario to probe them. With the scenario(s) in hand, we go on to describe the translation or dynamic mapping of those scenarios to the banks' financials: balance-sheet evolution and P&L. Scenario translation involves a wide set of modeling approaches that depend heavily on data availability. The treatment and approach differ for the banking book, which is subject to accrual accounting, and the trading book, which is subject to mark-to-market/fair-value accounting. The presence of contingent, off-balance-sheet exposures such as commitments and derivative transactions brings further complications that must be accounted for.

Section 2 presents a discussion of risk identification and scenario design and gives an overview of the scenario-translation process. Section 3 is the heart of the chapter, where we discuss methodological approaches to stress testing the range of businesses and products in banking, considering losses, revenues, costs, and balance-sheet dynamics. We begin with commercial banking business lines and products, such as commercial and industrial lending and commercial real estate (Section 3.1); retail lending, such as mortgages, auto loans, credit cards, and personal loans (Section 3.2); investment banking and capital markets, including trading and counterparty credit risk (Section 3.3); the trust and custody business (Section 3.4); securities (Section 3.5); deposit behavior (Section 3.6); non-interest expenses ex operational losses (Section 3.7); nonfinancial risks resulting in (operational) losses (Section 3.8), and we close with a discussion of applications of stress testing to things other than capital adequacy, plus a forward-looking view (Section 4).

2 Risk Identification, Scenario Design, and Overview of Scenario Translation

The first step in any risk management program, stress testing included, is the careful and comprehensive identification of the risks faced by the businesses. Without a thorough understanding of those risks and vulnerabilities, it is not possible to design a set of scenarios that can test and probe the weaknesses to provide valuable risk management and mitigation insights. Flood et al. cover scenario design for regulators and other official-sector entities in Chapter 6 of this handbook, and Hopper provides details on risk identification and scenario design for financial institutions in Chapter 7.[2]

A key question banks seek to answer is whether they have adequate financial resources to withstand a severe shock with sufficient resilience to continue to provide financial intermediation services to the economy. We focus our attention here on capital adequacy, but liquidity adequacy would follow a similar logic and process. To answer that question, one first needs answers to two other questions: (1) How severe should the stress scenario be? (2) How much capital should the bank have post-stress? Both questions involve the risk appetite of the bank: The lower the risk appetite, the more severe the scenario should be to withstand and the higher the capital level post-stress. Regulators reveal their own risk appetite by providing stress scenarios and minimum post-stress capital levels; a bank's risk appetite may well be higher than the regulator's but will always be bound by the regulatory constraints.

Risks and risk exposures can be broadly divided into systematic and idiosyncratic sources. To fix ideas, consider the following general representation of a particular behavior i that would need to be projected for the bank at time t, denoted by y_{it}. This could be a revenue or cost item, a credit-risk component such as loss given default for a loan commitment, or a balance-sheet item such as deposit or loan volume:

$$y_{it} = f(x_t, z_{it}; \alpha, \beta, \gamma; \varepsilon_{it}),$$ (13.1)

where a common functional form is the linear regression,

$$y_{it} = \alpha_i + \beta_i x_t + \gamma_i z_{it} + \varepsilon_{it}.$$ (13.2)

We denote x_t to be a collection of systematic or common risk factors, such as GDP growth, unemployment, equity index returns, interest rates, and so on; z_{it} is a collection of risks that

---

[2] For a discussion of optimal scenario design, see Parlatore and Philippon (2019).

are not necessarily common to all banks; α is a fixed effect; and $\boldsymbol{\beta}$ and $\boldsymbol{\gamma}$ are sensitivities.[3] These might include risks to the systems and information technology (IT), legal risks, and key person risk—in short, a very wide collection of vulnerabilities that would affect the well-being of the enterprise. To be sure, not all elements in $\boldsymbol{x}_t$ need be economic or financial risk factors; one could include here an economy-wide cyberattack that, clearly, would affect all banks, albeit not all in the same way (hence activity- and bank-specific coefficients, $\boldsymbol{\beta}_i$). Both systematic and idiosyncratic risk factors might include lags. And finally, ε_{it} represents innovations that are uncorrelated with either set of risk factors.

The representation in equations (13.1) and (13.2) will serve as a guide throughout this chapter and help to illustrate the interwoven nature of risk identification, scenario design, and the specification of the translation models. In that model-building process, the business works together with risk management to determine candidate variables for $\boldsymbol{x}_t$ and z_{it}. These candidates, including lags and other transformations, then serve as the basis for the final model specification search. Because these are time-series regressions, and because there is a premium on sample length to include the financial crisis and prior recession experiences, stationarity testing is one of the core diagnostics performed by the modeling teams.

An important consideration is heterogeneity in client or customer behaviors for a given balance sheet or P&L item across different client or customer segments. For instance, mortgage delinquency behavior may vary depending on product type (e.g., fixed or variable rate), geography, house type (first or second), and so on. Some of these factors can be directly controlled for in the regression model (should regression be the chosen statistical methodology). Others may require separate models if there is meaningful heterogeneity in $\boldsymbol{\beta}$ and/or $\boldsymbol{\gamma}$. Finally, considerable attention is paid to the properties of the characteristics of the residuals, $\widehat{\varepsilon}_{it}$.

Correlations (and thus the potential for diversification) across business lines are captured in the joint distribution of the systematic risk factors $\boldsymbol{x}_t$ and the bank's sensitivity to those factors, business line by business line, product by product, as captured by $\boldsymbol{\beta}$. This relationship helps us to understand the extent of diversification in the business mix at the bank. A bank with a narrow business model, such as credit card monolines, has less potential for diversification across a range of shocks coming from $\boldsymbol{x}_t$ than a universal bank.[4]

The binding constraint at banks, unsurprisingly, is data availability. Because stress testing against economic and market-risk factors benefits from models calibrated against a long time series with several downturns, including, of course, the global financial crisis, banks are challenged in bringing to bear clean and long time series of internal data. Often, businesses have evolved significantly, especially since the financial crisis (the most prominent example being securitization); products have changed; client, customer, and market segments have been redefined—all of which leads to fragmented data. With these changes come behavior changes or, more accurately, changes in client and customer needs and preferences that drive change in the banking organization, resulting in changing parameters $\boldsymbol{\beta}$ and/or $\boldsymbol{\gamma}$. This source of model instability makes it very difficult to build lasting and robust models, that is, models robust to a wide range of possible shocks in the future.

[3] Bolded notation indicates more than one dimension or variable (vector or matrix); plain notation indicates a scalar.

[4] For a discussion of the limits of diversification for credit risk in particular, see Hanson et al. (2008).

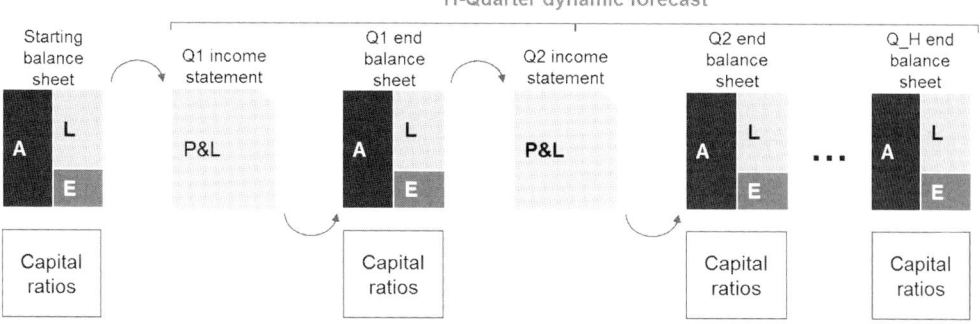

Figure 13.2 Dynamic projection of bank financials.

As a result, banks often complement the internal primary models with challenger or secondary models built on external data. This could be from stable sources, such as regulatory reports (in the United States, the call reports for banks and FR Y-9C reports for bank holding companies), or industry sources for particular products or businesses, such as residential and commercial mortgages, syndicated loans, debt and equity issuance, and so on. These data sources have the benefit of a long and relatively stable time series at the cost of missing bank-specific nuances. In this way, such secondary models play an important complementary role to the internal models, which typically have the opposite constraints: shorter time series but rich capture of bank-specific attributes.

It is not uncommon for large banks to have hundreds of models across the range of products and businesses. Clearly, the large universal banks have a wider model palette than smaller specialized firms, and if they are internationally active, country or regional heterogeneity must be accounted for, increasing the modeling challenge. All of these models must then come together to enable period-by-period (quarterly, annual) projections of the bank financials, conditional on the chosen scenario. This is illustrated in Figure 13.2. Typical horizons are 2 to 5 years; for the annual US Comprehensive Capital Analysis and Review (CCAR) program, it is nine quarters; for the European Banking Authority (EBA) biannual program, it is 3 years; and for the Bank of England's annual program, it is 5 years.

The balance sheet and income statement of a bank are highly interrelated and therefore need to be forecasted in an integrated manner. When forecasting based on a macroeconomic scenario, banks begin by developing a view as to which items are directly related to elements of the macroeconomic scenario, which elements are indirectly related to the scenario, and which elements are derived mechanically from other values. This process is sometimes referred to as developing "driver maps" because as the emphasis is on understanding which elements are driven by which effects. For example, loan balances could be forecasted directly, or they could be split into subcomponents: existing loan balance growth or reduction, new loan originations, and defaults. Balance growth or reduction could be further divided into draws on lines, repayment of principal, and full payoff or account closure; similarly, new loan originations could be directly forecasted or based on a forecast of market share and market size. Income-statement line items are similarly broken down into component drivers. Interest income would rely on the balance forecasting and interest rates, themselves split between a scenario-derived market rate (e.g., the prime rate) plus a margin.

Once each of the forecasted elements is derived, the balance sheet is projected for each forecast period, assembling the projected values per the logic in the driver maps. For example, new originations are the product of total market originations and a market-share projection, and this total is then to the outstanding loan balances. It may be useful to split market growth from one's market share to separately account for market versus bank-specific dynamics in a period of stress.

As the balance-sheet projection takes shape, some components of the balance sheet are derived from the remainder to ensure the balance sheet balances. Typically, these are funding elements such as wholesale funding and the securities portfolio, whose growth and contraction are determined based on funding needs. At every forecast period, the balance sheet has to balance.

The same logic is used to build up the income statement. Interest income and expense is typically derived from pricing forecasts, namely, rates and spreads. Non-interest income and expenses may be based on activity or balance drivers (e.g., the number of trades multiplied by a commission per trade) or deposit account fees collected per dollar of deposits. Non-interest expenses in the form of operational losses are typically derived in a separate operational loss modeling suite. The net income derived from these calculations is closed to the balance sheet prior to deriving the funding surplus or gap and the resulting assumed change in wholesale funding.[5]

Given the dependence of the stress testing process on models, it is perhaps no surprise that an important postcrisis control has been the focus on the management of model risk. Model risk as broadly defined is the risk of bad decisions resulting from incorrect or inaccurate estimates or misunderstanding of the modeled results.[6] A particularly complex and difficult aspect is the interaction of models and model errors across the often hundreds of models used by banks and regulators. For a more extensive discussion, see Canabarro's Chapter 16 in this handbook.

In the sections that follow, we elaborate on important considerations for the range of balance-sheet and P&L items across different products and business.

3 Model Consideration across Different Products and Business Lines

This section provides a basic description of each major business and product line in commercial banking to make clear how stress testing is implemented in those businesses. In each case, we discuss the loss, revenue, and cost, as well as balance-sheet components and common modeling approaches for each.

3.1 Commercial Banking Business Lines and Products

Commercial lending, which is a central component of many banks' businesses, is highly correlated with business cycles. The key dynamics to be captured in stress testing are primarily

---

[5] The Federal Reserve recently made public more detail about its supervisory model architecture, including functional forms and explanatory variables (i.e., the x_t and z_{it} in equations [13.1] and [13.2] above). See Board of Governors of the Federal Reserve (FRB, 2011).

[6] See Brown et al. (2015) for a historical discussion and FRB (2011) for a regulatory perspective.

around credit losses, which can be multiples higher from peak to trough of an economic cycle. Macro-sensitive elements of net revenue for commercial lending businesses consist primarily of net interest income from loan spreads, which expand and compress with credit conditions, and secondarily of fees associated with the provision of credit. As with other credit products, the evolution of balances combining credit, origination, and repayment effects is critical to estimating levels of both income and credit loss. In this section, we focus primarily on credit-loss stress testing approaches for commercial lending, with additional attention at the end on revenue stress testing.

In the years following the GFC, banks have refined a number of methods for commercial credit-loss stress testing. Key objectives of these approaches include capturing distinctions in subportfolio sensitivity to macro-factors—down to the loan level to the extent possible; modeling credit quality distribution changes through the stress testing horizon; and separating the three components of credit loss (probability of default [PD], loss given default[LGD], and exposure at default [EAD]), which can then be recombined in different ways for loss, balance-sheet, and capital evolution.

The first step for commercial credit stress testing analysis is portfolio segmentation to capture relevant heterogeneity and to group obligors and facilities anticipated to demonstrate common responses to macroeconomic drivers. Commercial and industrial (C&I) loans are typically broken down by the size of the obligor for PD purposes (e.g., large corporate vs. middle market), by collateral and seniority for LGD purposes, and with revolving facilities differentiated for EAD modeling. Corporates are further segmented by major industry group, although financial institutions, sovereigns, and subsovereigns are typically isolated for separate analyses. Commercial real estate (CRE) loans are typically differentiated by loan stage/purpose (e.g., income-producing vs. construction) and property type for both PD and LGD purposes.

Small business lending is between retail and corporate in both size and borrower behavior characteristics, and modeling approaches typically depend on how the bank handles the risk rating and credit process for the borrowers. For obligors with distinct corporate structures and reliable financial statements, versions of corporate models, as described in this section, tailored to these segments can be effective, whereas for obligors where repayment is dependent on the owner (and given the small size of loans, more data points are usually available), the retail techniques as described in the next section tend to work better.

Further differentiation of default sensitivity to macro factors is typically done at the level of portfolio credit quality, as represented by internal and external ratings, which synthesize a variety of obligor characteristics. Commercial borrowers are fewer in number and generally default less frequently than retail borrowers, meaning that data are generally insufficient for direct loan-level regression modeling of macro-sensitivity. Instead, the most common practice for C&I relies on transition models that are anchored to observed changes in historical ratings, including transition to default; see Belkin et al. (1998), Bangia et al. (2002), and de Bandt et al. (2013). Rather than regressing each ratings transition against macro factors, a transition matrix is typically reduced to a smaller number of factors using a weighting scheme, and these factors are then regressed against relevant macroeconomic variables. For C&I, LGD, and EAD, it is typical to differentiate by collateral and facility type and regress historical outcomes against relevant macro factors to build models. In CRE, ratings transitions may be used similarly, but structural models of default, where macroeconomic variables are regressed against income and valuation drivers, can be even more effective in deriving robust, integrated projections of the timing of default and magnitude of loss.

The previous description may raise the question of whether it is necessary to go through the step of modeling portfolio credit quality evolution as opposed to directly modeling losses. First, over a multiperiod forecasting horizon, the compounding effect of transitions to lower rating grades—often with varying magnitudes over the stress horizon—can result in substantially greater concentrations of low-rated loans over time, amplifying loss rates at times when the capital position is most vulnerable. Second, it is important to capture the impact of new originations on the evolution of the book; originations, although reduced during stress, typically have a meaningful moderating effect as they come in at higher levels of credit quality. And third, beyond directly modeling losses, it is important to be able to assess the impact of credit quality on capital requirements over time because PDs are direct drivers of capital levels under the Basel Advanced Approaches (or Advanced Internal Ratings Based) framework (Basel Committee on Banking Supervision [BCBS], 2006).

Revenue stress testing for commercial lending businesses involves the projection of interest income and fee revenues associated with lending balances. Interest income is a function of loan pricing and balances. As described at the beginning of this section, balance evolution of commercial lending portfolios requires combining credit, origination, and repayment effects to understand the size of the book and the distribution of pricing. Origination volumes through a stress period may be affected by both demand-side factors and banks' capacity for lending because losses may be eroding their capital base.[7] Origination demand projections may be informed by bank-specific or industry-level regressions of loan activity through different macroeconomic conditions.

Although many commercial loans have variable-rate pricing, which follows the fluctuations of interest rates through a stress scenario for the benchmark rate component, the spread is typically locked in at the time of origination. Regression modeling can be used to determine how spreads may vary under stress with different macroeconomic factors. The higher spreads for originations typically seen in stressed conditions, as reduced liquidity and increased risk aversion increase banks' pricing power, can be a meaningful offset to declines in volume and credit losses on the existing book.

3.2 Retail Lending: Mortgages, Auto Loans, Credit Cards, Personal Loans, and So Forth

Retail lending typically provides the most data-rich environment available for constructing loss forecasting and net income models. The products are standardized and usually well defined, with limited customization needed for individual customers; banks issue a great many of these loans, they do so frequently, and they regularly monitor performance through internal and external (e.g., credit bureau) channels. Furthermore, even when an individual institution may lack such granular data, they are often available from external data sources,

---

[7] Note: A macroprudential goal of supervisors through stress testing exercises is frequently to assure not only that banks are sufficiently well capitalized to sustain stress losses but also that they have sufficient capital to continue lending in this environment.

such as credit bureaus, securitization data vendors, or government or quasi-government sources. This has permitted the development of very granular regression models.

Because granular consumer credit data are typically proprietary, little academic research exists. Musto and Souleles (2006) use a consumer-level panel of credit bureau data to document systematic and idiosyncratic risk in consumer credit portfolios and the importance of heterogeneity. Butaru et al. (2016) use machine learning techniques on account-level credit-card data from six major commercial banks using consumer as well as macroeconomic variables. They find substantial heterogeneity in risk factors, sensitivities, and predictability of delinquency across banks. Wu et al. (2018) make use of credit bureau data for auto loans to show the importance of an aging effect in default risk as well as parameter instability, a cautionary finding given the desire to build models that are robust to a wide range of scenarios.

The modeling approach taken by banks is typically dictated by the size, complexity, and risk profile of the portfolio. At an aggregate level, one may directly project the net charge-off rate as a function of observables,[8] but in practice, banks have developed quite granular degrees of segmentation to capture relevant heterogeneity in products, markets, and customer characteristics. Moreover, charge-offs, or losses, are typically broken down into the typical PD, LGD, and EAD components.

Many banks have used age-period-cohort models for loss forecasting. These models seek to break down loss forecasting into a "maturation" effect capturing how loan default probabilities vary over "time on books," a "vintage" effect capturing how loan quality varies by vintage as a result of underwriting quality, and an "exogenous" effect capturing macroeconomic effects. Although popular in other loss-forecasting applications, their effectiveness has been mixed for stress testing because the vintage effect may mask or distort underlying macroeconomic effects, which are, of course, paramount for stress testing.[9]

The most complex and granular models in use are loan-level models. These models permit the integration of macroeconomic and loan-level characteristics within a single regression along the lines of equation (13.2) above. The models use a PD/LGD/EAD construct, and two methods of modeling the default component are state-transition models and hazard models.

State-transition models[10] attempt to predict transitions between "states" prior to default or write-off. These states may be defined by delinquency status or, in some cases, risk ratings, such as those derived from risk-rating tools based on a Basel internal ratings-based (IRB) approach. This approach requires a suite of individual models to capture the wide variety of possible transitions. By predicting the transition probabilities between individual states and then forecasting the flow of loans among those states, these model suites ultimately predict both the default or write-off events and the future distribution of loans across the modeled states. This additional detail may be used to feed elements of the forecast beyond the immediate loss forecast amount—for example, applicable risk weights for nondefaulted

---

[8] Hirtle et al. (2015) use the net charge-off approach in their CLASS model, which is based on Federal Reserve regulatory reports (FR Y-9C).

[9] The FRB (2013) notes that these models are in use while also noting the difficulty in separating the macroeconomic effects from the vintage effects.

[10] Loew and Crook (2014) describe a model predicting transitions between current, one payment behind, two payments behind, and three payments behind, which was defined as default.

loans may depend on risk rating or delinquency status. The drawback of this approach is that it is complex to estimate and execute forecasts.

A common alternative—the hazard model[11]—focuses on predicting the outcome of interest, such as the default or payoff event itself, directly rather than considering the complex network of state transitions. Hazard models establish a baseline hazard that captures effects such as loan age, forecast horizon, or time until maturity. This "baseline" effect represents the average impact of each factor to which loan-level and macroeconomic factors are added, thus allowing the baseline effects to adapt to individual loan characteristics and the scenario-specific macroeconomic conditions. Although this model is typically easier to estimate and avoids the challenges of estimating a system of regressions, it does surrender a level of granularity in the outputs by focusing on a single outcome of interest but ignoring the interim states on the path to default.

The LGD and EAD components are modeled separately and are typically less complex, although usually still modeled at the loan level. However, these may be a simple regression of LGD as a percentage on a set of loan-level and macroeconomic variables. And EAD models may be as simple as a mechanical calculation of expected amortization for term loans. More complex models are used where LGD or EAD may be subject to more heterogeneity— for example, more complex LGD models are common for collateralized obligations such as mortgages and auto loans, whereas more complex EAD models are common for credit lines such as credit cards and personal lines.[12]

3.3 Investment Banking and Capital Markets

A robust stress testing program for trading activity will use several—if not many—scenarios of market disruption in order to measure the vulnerability of the portfolio to stress emanating from different sectors of the markets and the global economy. For example, it would not be unusual to conduct market risk stress tests every day using a set of scenarios that would include financial market disruptions led by declines in housing values, corporate credit, leveraged credit, emerging markets, and commodity prices; the failure of a systemically important institution; and disruptions in the operating mechanics of markets (e.g., settlement mechanisms)—as well as others.

It is not uncommon that stress testing programs include both historical and hypothetical scenarios. There is a rich history of financial stress events over the past 30 years, starting with the stock market crash in October of 1987 through to the GFC and extending to the European sovereign crises of May 2010 and August/September of 2011, the "Taper

---

[11] Deng et al. (2000) describe early research on the use of competing hazard models in predicting mortgage default and payoff. Bellotti and Crook (2013) describe the use of discrete-time hazard models with macroeconomic variables in stress testing credit cards.

[12] Crook and Tony (2012) describe various methods for modeling credit card LGDs, finding the simples method using ordinary least squares (OLS) regressions the best. Leow and Crook (2016) describe a complex mixture-model approach to modeling credit-card balances throughout the life of the account and at default. Loew and Mues (2012) describe a two-stage LGD model, first predicting the probability of repossession and then predicting the haircut taken in a forced sale relative to the market value, finding that two-stage models outperform a single-stage model focused on predicting LGD directly.

Tantrum" in the US Treasury market in 2013, and the "Flash Crash" and the Turkish Lira in 2018, to name a few. Some of the historical events were global and systemic—such as the GFC—and will likely produce the largest projected potential loss of a stress testing program, given the severity of the scenario. Other historical events, however, were disruptions focused within specific sectors of the markets—such as an oil-induced price shock of the Kuwait invasion—and may be used to measure the vulnerability of the portfolio to disruptions in those sectors.[13]

Although the projected losses from these focused scenarios may not be the largest of the scenario set—and are therefore unlikely to be used for capital adequacy testing—they are likely to highlight specific vulnerabilities of the portfolio to potential events, similar to the historical event, which then provides the basis for a review of the positions and an assessment of appetite for loss given the likelihood of the scenario. For example, projecting the impact of an oil-price shock using the market events surrounding the invasion of Kuwait might highlight positions for discussion and review if a potential spike in oil prices is considered likely in the context of current events.

Complementing the historical scenarios are hypothetical scenarios for events that have not occurred but that are extreme, potentially plausible, and would break historical relationships. For example, a dominant characteristic of virtually all historical market disruptions is a flight to safety in US Treasuries and the US dollar. In fact, the canonical form of historical market stress can be modeled as follows: equities fall, credit spreads widen, volatility increases, and the US Treasury market rallies. A hypothetical scenario testing the failure of this canonical form would be essential for a stress testing program. For similar reasons, stress tests can be used to simulate the failure of pegged and managed currency programs, the diversion of funded and unfunded credit exposures, and other events rarely exhibited in history.

3.3.1 Business-Specific Modeling Approaches for Revenues

The investment-banking and capital-market businesses are a portfolio of businesses with different revenue streams that are typically sensitive to a range macroeconomic factors, but the modeling is complicated by market- and business-specific features. Examples include industry-wide market structure changes (e.g., increased electronification of trading businesses and the resulting impact on pricing, changes in securitization business models after the financial crisis), transfer of revenues between businesses (e.g., joint ventures where businesses share revenues, transfer pricing agreements to provide revenue to compensate one business for providing resources to another business), and bank-specific strategy decisions (e.g., strategic growth or cuts in balance-sheet allocation may affect the ability to generate revenue in any given macroeconomic environment).

Within each business, revenues can be modeled at the total revenue level or at the level of non-interest income versus net interest income. Some banks further segment non-interest income into client-revenue versus. market-revenue components or volume-based versus balance-based components. Client revenue and market revenue typically have different drivers; client revenue is linked to business activity, whereas market revenue from inventory is linked to a pricing index for the asset class.

---

[13] For an introduction to stress testing in a trading environment, see Jorion (2006).

Within the capital-market businesses, sales and trading (market-making) businesses are primarily fee based and are typically modeled as transaction volume × bid–offer spread per unit of volume. For the cash equities trading business, banks often use equity volatility (which is typically linked to trading volumes) and equity price indices as drivers. For the business involving trading of interest rate products, banks often use measures of the interest rate curve as drivers. For business involving credit products trading, banks often use measures of credit spread as drivers. Prime brokerage is a financing service that banks provide to hedge funds so that the hedge fund can borrow securities to execute its trading strategy. Banks often model revenues for this business as balances × spreads, with measures of interest rate spread often used as a driver for the financing-spread component.

The approach to modeling the fee-based advisory and underwriting businesses within investment banking is somewhat indirect but still follows the basic idea of volume × price. Because the market is relatively concentrated, banks compete for a share of investment banking transactions (mergers, acquisitions, equity and debt issuance, etc.). The first task is to project industry-wide volume using league tables of transaction activity conditional on observable risk factors (often broad measures of economic health, such as GDP growth and equity prices). Such industry-level data have the virtue of being available for a relatively long time series that included multiple stress periods (often starting from the mid-1990s and thus including data from both the dot-com boom-bust of the late 1990s and early 2000s and the GFC). Then, bank-specific assumptions for market share and fee margin can be applied to arrive at the bank's revenue projections (i.e., revenue for each investment banking business = industry-wide volume × bank's market share × bank's fee margin).

3.3.2 The Global Market Shock

A common approach to stress testing the trading book is to apply a so-called global market shock (GMS). The GMS measures the potential loss of commitments of the firm whose revaluation directly affects earnings and capital. Such commitments include securities, currencies, commodities, and derivative product positions held in (a) trading accounts whose values are functions of market-traded variables and whose change in value directly affects earnings and, possibly, (b) assets held for sale[14] or other investments whose change in value directly affects firm equity (e.g., in other comprehensive income). The shock of the GMS is a large and instantaneous change in global asset prices, interest rates, credit spreads, and exchange rates consistent with conditions of severe financial and credit-market distress.[15]

The purpose of using a large and instantaneous shock is to (a) trigger the impact of any adverse optionality or embedded leverage in portfolios—in which the rate of loss may increase as a function of the change in market variables (such out-of-the-money risk is, by definition, rarely within the focus of day-to-day trading risk management, given normal market price volatility and transactional liquidity) and (b) account for the potential inability

---

[14] Available-for-sale (AFS) positions are usually included in stress tests used for internal risk management purposes because their change in value, consistent with the GMS, directly affects capital. For US CCAR purposes, however, AFSs are excluded from the GMS and are modeled over the nine-quarter forecast horizon. The methodology for this is discussed more in Section 5.

[15] For a concrete example, consider the set of more than 20,000 parameters that constitute the Federal Reserve's GMS: www.federalreserve.gov/supervisionreg/ccar-2019.htm.

or unwillingness to take actions that would attenuate losses as a result of either (1) a severe loss in transactional market liquidity as market participants withdraw and become reluctant to commit capital (inability) or (2) an unwillingness by trading management to reduce positions (because such reduction will crystalize current losses); instead, management may retain positions in the hope that prices will revert (gambling for redemption). As a result, the risk positions held at the beginning of a severe market disruption are assumed to be held throughout the period of disruption for stress testing purposes—and are subject to instantaneous price shocks of size commensurate with long durations.

In concept, the mechanics of implementing trading stress tests are relatively straightforward. Once the scenario and associated price shocks have been determined, they are applied to the market-revaluation engines (models and systems) for trading positions to produce the stressed market values and change from initial value. For greatest accuracy, the stressed market values would be produced using all the machinery required for full revaluation of each position. This means that all pricing parameters must be shocked, and all pricing models must be rerun using the shocked values. This is a nontrivial operational task that can easily involve a large number of pricing variables and a large number of pricing models; the Federal Reserve's GMS has more than 20,000 parameters. Furthermore, the projections produced by such a complex multivariate system are not easily explained or understood because the key drivers of outcomes are difficult to identify in the wash of so many moving parts. This opacity created by complexity makes it difficult to use the process to control risk because the actions that must be taken to avert loss can be difficult to identify. In summary, the stress loss projected by full revaluation may be accurate, but the large number of positions and variables involved make it challenging to identify the small finite number of critical positions driving the outcome and translate those positions into actions that may be taken to control the outcome.

An alternative, or perhaps additional, process is to use the valuation engines of the firm to create the price sensitivities of the portfolio to the key market variable inputs. These price sensitivities collectively constitute, by definition, a projection function for the change in market value for the portfolio. The coefficients of the projection function are the price sensitivities. The function itself, as a parametric model of profit and loss, can then be used to produce stress outcomes. The benefit of this approach is that it is highly useful in clearly identifying the drivers of stress outcomes and the actions that would be taken to change those outcomes. The limitation of this approach is that as a parametric model, it provides an estimate of the actual outcome (otherwise provided by full revaluation). Much of the estimation error may be due to simplistic parametric estimation that fails to model curvature in the payoff function of a product (embedded optionality) or material cross-convexity in the payoff function, where the price sensitivity to one variable may change materially as another pricing variable changes (e.g., volatility and underlying price in options). These limitations can be reduced by further developing the sensitivities used, for example, measuring change in value for different gradients of change in price variables to address embedded leverage or specifically measuring the cross-convexity adjustment for the sensitivities of two price variables. The point is that there are at least two objectives of stress testing: the stress loss projection itself and the ability to control the stress loss outcome. These objectives may require different machinery: full revaluation for an accurate stress loss number and parametric revaluation for control.

3.3.3 *Counterparty Credit Risk*

Counterparty credit risk is the risk to each party of a contract that the counterparty will not live up to its contractual obligations. Counterparty credit exposure is the amount of profit on a contract that will be realized contingent upon the performance of the counterparty to the contract. The positions subject to counterparty credit stress testing are those for which the credit exposures are functions of traded market variables, including, for example, derivative products, repurchase agreements, reverse repurchase agreements, and securities lending transactions.

With regard to derivative contracts, the credit exposure to a counterparty is a function of the current market value of the set of derivative contracts collectively subject to a legally enforceable netting agreement with a counterparty. This legally enforceable "netting set" may result in a net receivable due from the counterparty or a net payable due to a counterparty, depending on the market values of the derivative contracts and collateral in the netting set. In the event there is a net derivative receivable due from the counterparty (a net profit across all trades and collateral provided and received within the netting set with a counterparty), the receivable is further adjusted for the creditworthiness of the counterparty. This adjustment is known as the credit-valuation adjustment (CVA) and discounts the derivative receivable for the credit quality of the counterparty (Canabarro and Duffie, 2003). CVA is analogous to the discount on a fixed income security for the credit of the issuer (the credit spread component of the security's yield). Stress testing derivative products involves shocking the underlying price variables referenced by the derivative contracts to produce new stressed market values and new net stressed receivables and net stressed payables. The CVA on each new net stressed receivable is then recalculated for the potential credit deterioration of the counterparty under stressed conditions. The change in CVA directly affects earnings. Under stress conditions, it is not uncommon both for the magnitude of the derivative receivables to increase and the credit quality of the counterparty to degrade. As a result, the change in CVA could be material and negative, with a nontrivial impact on earnings.

Significant exposure may stem from securities finance transactions, which include repurchase agreements, reverse repurchase agreements, and securities lending contracts. Credit exposures for these positions arise from the difference between the market value of a security and its repurchase price or the difference between the market value of securities lent and the value of collateral received. Under stress conditions, these values will likely change materially and may result in net unsecured credit exposures. For repurchase agreements, the security to be repurchased may increase in value well above the agreed repurchase price (a US Treasury, for example) resulting in unsecured credit exposure to the counterparty. For reverse repurchase agreements, the security purchased may fall in value well below its resale price. For securities lent and borrowed, the value of those lent may well exceed the value of those borrowed (the collateral).

The post-stress derivative receivables (with CVA) and the post-stress unsecured credit exposures of securities finance constitute a collective set of credit exposures that could be subjected to the credit-default models used in lending to arrive at stressed default-loss projections (see section on lending stress). The Federal Reserve's CCAR program avoids this complexity and instead simply defaults the largest net stressed exposure across all counterparty positions—derivatives, securities lending, repurchase and reverse repurchase agreements—with the loss added to the stress loss projection. This is known as the *largest counterparty default* and, along with the trading losses produced by the GMS, is assumed to be taken in the first quarter of the stress forecast horizon of the CCAR.

In addition to assessing capital adequacy, stress testing is used routinely to set controls upon and actively manage the risk of counterparty exposures. When conducting market-risk stress testing, it is not uncommon to project the largest stressed counterparty exposures resulting from the stress scenario. These stressed exposures are reviewed and assessed for potential wrong-way risk, in which the credit quality of the counterparty is likely to degrade as the exposure to the counterparty increases under the stress scenario. Actions may then be taken to change the exposure under stress. For example, if exposures to major European banks were seen to increase when simulating a stress event in the eurozone, specific trades may be initiated with those banks to attenuate the increase in exposure under the conditions of the stress scenario used. Stress testing outcomes are also used as control metrics for counterparty risk. In particular, it is not uncommon to place limits on unsecured net exposures resulting from stress scenarios for derivative counterparty risk and securities lending transactions. For securities finance transactions, stress testing may be used to assess the adequacy of the haircut on transactions to protect against periods of high price volatility and illiquidity.

There is a striking contrast in the valuation practices for derivative products as compared to securities finance transactions, and these different approaches specifically affect counterparty stress testing. For derivative products, the credit quality of a counterparty is accounted for by CVA, and the mechanics of CVA are replicated for stress testing. As a result, the stress loss projections for derivatives include the impact of both the change in receivable size (as underlying market variables change) and the potential credit degradation of the counterparty (as CVA changes).

In contrast to derivatives, many securities finance transactions are not marked-to-market for counterparty credit risk. The collateral—securities borrowed or received—is revalued, but any unsecured credit exposure neither appears as such on the balance sheet nor is valued for the credit quality of the counterparty (although such exposures are routinely tracked by internal systems for credit-risk management). As a result, the accounting mechanics for securities finance do not fairly report the size of credit exposures or the value at which such exposures would transfer in the marketplace (e.g., CVA-adjusted market value). For this reason, stress testing of securities finance using the standard accounting procedures will fail to account for potential loss in the fair market value of securities financing transactions or the change in the potential default of a counterparty. Perhaps this condition motivates the FRB to default the largest net stressed counterparty exposure as a proxy for the fair-value economic losses, which are otherwise unaccounted for.

3.4 Trust and Custody

Trust and custody banks generate significant income and expense from a range of services associated with securities processing, accounting, and administration. Although some of these services have linkages to the balance sheet and may involve directly taking credit and market risk, for the most part, they are highly operational in nature and do not expose the bank directly to these financial risks or their macroeconomic variability. However, they typically have significant net income sensitivity to macroeconomic factors because fees are linked to elements that change with market factors, such as the quantity of underlying assets under management or administration or trading volumes, whereas the cost base tends to have a larger fixed component.

Revenue and expense modeling for trust and custody fee-based businesses starts with the definition of driver maps that unpack the client pricing structure and how it relates to market-sensitive factors that can be modeled as part of the stress scenario. For example, if a fee arrangement has a component of pricing that varies with the assets under custody, the client assets can be grouped by asset class and representative index, and their volumes can be projected to shift with corresponding changes to the asset prices under stress. On the expense side, variable expenses such as sub-custodian fees need to be similarly modeled with macro-sensitivity in isolation from fixed cost-base elements (which may also be assumed to increase or reduce over a multiyear stress horizon through strategic actions but move more slowly).

From the perspective of balance-sheet and credit- and market-risk considerations linked to the core client business, trust and custody banks often have significant customer deposits; securities financing businesses; derivatives trading; and extension of short term credit, including through daylight and multiday overdrafts. Stress testing for securities financing and trading businesses follows similar approaches to those for other capital-market players, as described in the previous section.

In line with the greater concentration with a smaller number of large clients than retail or corporate business models, trust and custody banks' client deposits typically exhibit significant concentrations, which must be considered carefully in stress testing of liquidity because relatively few clients making large deposit moves can substantially affect total balances. The segmentation of deposits between operational deposits closely connected to the client's day-to-day cash management and trading activities, as opposed to more discretionary placements, is important to understand the vulnerability of deposits to reduction in a stress scenario (particularly one with idiosyncratic stress on a given institution).

The nature of the trust and custody business, which is centered around securities services and activities involving institutional clients, means large proportions of credit risk arise for various types of financial institutions, which typically have low default rates, as opposed to traditional corporate or retail borrowers. This places an emphasis on specialized ratings and credit stress testing models that capture the specific characteristics of these obligors. Additionally, counterparty stress testing (as described in Section 3.3.3) becomes important because there can be concentrated exposures in trading, securities finance, overdraft, and sometimes other credit businesses. Trust and custody banks frequently have larger securities portfolios than other banks and may take more credit risk in these portfolios because they have less credit risk from client-facing businesses; Section 3.5 on securities stress testing describes important considerations for stress testing of such exposures.

Trust and custody banks in many cases also have asset management arms. Net income stress testing for these follows mechanics similar to those described previously for other trust and custody fee businesses, where the evolution of assets under management driven by market factors needs to be considered in conjunction with pricing structures that may specify fees as a percentage of assets under management.[16] Although banks do not take the financial risk for the assets they manage through these businesses, market declines that adversely affect clients—particularly those that believed the funds they invested in carried a low risk of principal-value decline—may expose operational-risk issues or have reputational ramifications. In the GFC, some institutions provided noncontractual support to money market funds and other funds under their management. Thus, for funds that seek to sustain a stable net asset value, specific stress testing analysis is typically conducted to understand whether and how the stress scenario could create a loss in the funds, the magnitude of the shortfall, and mitigating factors.

3.5 Securities (Non-Trading)

Aside from trading inventory, securities are frequently held by banks as part of an investment portfolio (e.g., under AFS or held-to-maturity [HTM] classifications) and may be used for liquidity-management purposes or, in some cases, as long-term assets used to monetize low-cost liabilities.[17] For most banks, the scale of the investment portfolio is much smaller than their loan portfolio, but in some banks (particularly trust and custody banks, as discussed previously), the proportions are larger. Because trading-book revenue and risk-modeling considerations for securities were covered in Section 3.3.1, this section will focus primarily on the balance and income evolution and credit and valuation considerations for securities held in investment portfolios.

The total size of investment portfolios is typically a function of the evolution of other components of the balance sheet, rather than being driven directly by client demand considerations. As the size of the investment portfolio at different points in the stress horizon is determined, assumptions need to be introduced on how maturing securities will be reinvested and new purchases will be allocated—and to the extent sales are needed to affect a more rapid reduction or reshaping of the portfolio, how this will take place. The primary modeling complexities emerge around projecting principal and interest cash flows on individual securities, projecting credit losses, and projecting changes to valuation of securities.

Some securities, such as fixed-rate domestic sovereign bonds, may have no possibility of changes in their cash-flow profile that need to be modeled. However, the cash-flow profile of many securities can be subject to significant changes, including floating interest rates on many instruments; prepayments, particularly on mortgage bonds; default and recovery on credit-sensitive instruments; and availability of cash to different securitization tranches. Each class of security needs to be considered for these potential changes and addressed accordingly, usually with security-level differentiation of analysis. In some cases, the scenario-analysis capabilities of banks' core asset-liability management systems can directly handle interest rate—driven changes directly when fed with the appropriate interest rates—adjusting yields and calling on interest-sensitive prepayment models as needed to

---

[16] See also Chapter 14 in this handbook, where Longerstaey discusses stress testing of asset managers.

[17] Under US generally accepted accounting principles (GAAP), HTM securities are debt investments for which the management has both the intent and ability to hold them until maturity. The accounting treatment is amortized cost, with unrealized gains and losses disclosed in notes to the financial statements. AFS securities are those held neither for trading purposes nor HTM and are recognized and carried at fair value, with unrealized gains and losses included in other comprehensive income.

reestimate mortgage cash flows. For more complex impacts, such as the reestimation of securitization cash flows where underlying loan performance under stress and tranching structures must be considered, separate analytic modules are needed to project security-level cash flows. Once revised principal and interest cash flows reflecting the impact of relevant stress scenario factors are determined for all instruments currently in the portfolio, these can be combined with the assumptions for reinvestment and incremental new purchases and sales to determine the credit profile and income of the portfolio at each time step of stress testing analysis.

Some investment-portfolio holdings, including sovereign bonds and government-guaranteed mortgage-backed securities, do not have material credit-default risk to be modeled under stress. Evaluating credit-impairment risk for securities that represent a direct obligor exposure, such as corporate bonds and municipal bonds, follows similar analytic approaches to those used for loans in these classes of credit, as described previously in the section on corporate credit (Section 3.1). For securitizations, the credit stress testing analysis is more complex because it needs to look through to the underlying collateral (e.g., credit-card receivables, auto loans, residential mortgages, commercial mortgages, or corporate loans) and appropriately apply stress factors to the default, prepayment, and recovery characteristics of this collateral, then project these collateral cash flows through the deal's tranching structure to understand where changes in credit behavior may cause a credit loss.

For all securities, it is important in stress testing to be able to project changes in valuation, which can affect capital ratios or result in realized losses or gains if securities are sold. With cash flows forecasted under the stress testing scenario as described previously—utilizing stress scenario interest rate paths and other relevant factors—the remaining factor that needs to be projected is the appropriate spread to a benchmark rate to be used for discounting each security. During times of macroeconomic stress, risk aversion and liquidity premia typically increase in a "flight to quality" such that the prices of securities, even without anticipated credit losses, will decline substantially, implying that the spreads to the benchmark rate increase. Stress testing analytics for securities valuation typically involves using regressions to establish the relationships between macroeconomic factors and spreads for different types of securities, which can then be used to determine spreads to be used in a given stress scenario. Discounted cash-flow analysis then generates the appropriate fair-value prices for securities in the scenario. To the extent that a material amount of securities is anticipated to be sold during a stress period, for example, to meet projected liquidity shortfalls, further analysis is needed to determine the additional reductions to fair-value prices that are typically experienced when selling in volume in distressed markets.

3.6 Deposits

Modeling of deposits, a significant source of funding for banks, requires an understanding of both the evolution of deposit balances and the rate paid throughout various macroeconomic conditions. Typically, rate dynamics—such as deposit betas and spreads over reference rates—are captured in institutions' preexisting interest rate risk and balance-sheet forecasting tools, allowing the forecasting of rates paid on deposit balances based on the interest rates contained in the macroeconomic scenarios. However, these models and assumptions may be insufficient in some regimes, such as the liquidity crunch experienced during certain periods of the financial crisis.

The main incremental task therefore is projecting deposit balances. This is generally done using time-series models with the deposit balances or changes in the balances as the dependent variable. Explanatory variables include macroeconomic variables and indicator variables to capture structural effects such as seasonality, changes in deposit insurance rules, or other account characteristics. Explanatory variables may also seek to capture the attractiveness of deposits relative to other assets, such as the spread between deposit rates and actual or expected equity market returns, and an institution's own deposit rates relative to those of competitors. Further, given that stress testing occurs in a "nominal" world, impacts from changes in monetary policy, such as from the introduction and potential withdrawal of quantitative easing, should also be considered.

A key modeling decision is the choice of segmentation because different account and customer types exhibit different behavior during macroeconomic stresses. Common segmentation dimensions therefore separate retail and corporate clients; account types such as checking, money market, savings, and time deposit accounts; channel, such as online versus in-branch; and geographical markets. The insured status of the deposits is a key feature to incorporate that may be used separately or as an element of the account type.

3.7 Non-Interest Expenses (Ex Operational Losses)

Non-interest expenses are typically dominated by compensation (fixed and variable components) and headcount. Although it is tempting to drastically reduce headcount and variable compensation during a stress scenario, supervisors have been reluctant to fully accommodate such reductions because the exact length and depth of an actual crisis are unknown at the time, and premature staff reduction will make it difficult for firms to rebound once the crisis recedes, thus impairing franchise value. Other components include expenses for premises and fixed assets (rent, lighting, environment), IT expenditure, marketing expenses, expenses for outside professional and legal services, expenses related to other real estate owned, and so on. Although non-interest expenses, ex operational losses, are a significant contributor to net revenues, their modeling has historically received less attention.

3.8 Nonfinancial Risks

Nonfinancial risks (NFRs) in banking, as the name indicates, have historically been defined by what they are not: not credit, market, or other financial risks. This negative definition has created the perception that NFR is a catchall category of disparate risks with different root causes and behaviors, making it difficult to model them in any reasonable fashion.[18]

---

[18] The use of the term *NFR* rather than *operational risk* is a fairly recent evolution in the banking industry resulting from a desire to separate the field from its Basel II roots and its historical focus on capital modeling. Ultimately, the two terms refer to the same set of risks but tend to be associated with a different set of objectives. On the one hand, operational risk tends to be associated with Advanced Measurement Approach (AMA) modeling and often implies a focus on losses with a disregard for their drivers and risk management considerations. On the other hand, NFR is used when the emphasis is on risk management and the understanding of the root causes of the risk. In this chapter, we use the terms interchangeably.

The Basel Committee provides a more constructive definition (BCBS, 2011): "the risk of direct or indirect loss resulting from inadequate or failed internal processes, people and systems or from external events."

Stress testing of NFRs has proven to be significantly more challenging than stress testing of traditional banking risks, such as market and credit. This is primarily due to the more idiosyncratic nature of NFR losses and to the resulting weaker macroeconomic ties observed. Despite this weak observed relationship, there is a growing body of supporting evidence (Sekeris, 2012; Abdymomunov, 2014; Curti et al., 2019) that its impact is moderate or not well enough understood to be fully captured.

Early stress testing modeling efforts for NFR fell into two categories: regression models linking changes in operational risk losses to macroeconomic variables and Basel II AMA capital models repurposed for stress testing, usually by using a different (lower) percentile of the loss distribution than the 99.9th percentile used for capital purposes (BCBS, 2006). AMA capital models were often quite complex and fragile, resulting in unstable capital estimates (Opdyke and Cavallo, 2012; Ames et al., 2015).

There is ample anecdotal evidence of large NFR losses experienced by numerous banks resulting from the financial crisis. For example, the insider fraud committed by Jérôme Kerviel was exposed in 2008 through the market downturn. The financial crisis also triggered a plethora of mortgage-related settlements costing the industry tens of billions of dollars. Unfortunately, the significant time lags and the current state of NFR modeling have made it difficult to prove these relationships with any level of statistical significance. A number of financial institutions developed regression models attempting to link NFR losses with macroeconomic variables using either the frequency of NFR losses or the expected total loss. Attempts at linking the severity of individual events to the business cycle were abandoned early because of the poor performance of these models. Loss severity distributions are typically quite fat tailed, and samples are small at any single financial institution.

Those institutions that were able to successfully model their NFR losses quickly realized that these forecasts were ill-suited for stress testing purposes. In most instances, the projected losses were too small compared with those actually observed after crisis and in comparison to the much higher capital numbers that the industry had been accustomed to for operational risk under Basel II. This led institutions to resort to simpler approaches that, instead of attempting to formally link losses to macroeconomic conditions, asserted that link and tried to substantiate it with qualitative analysis.

These early approaches to stress testing of NFR are still in use today but usually as part of a broader framework that considers other elements to complement the projections. Beyond the weaknesses already reviewed, these approaches also have often been portrayed as ivory-tower exercises disconnected from the risk management activities at institutions and, most importantly, as failing to consider the constantly changing nature of NFR. The response to this criticism has been to increasingly rely on idiosyncratic scenarios as add-on losses meant to capture large losses that could materialize in the stress projection period, whether driven by the macroeconomic environment or not, and that have been determined to be of particular concern to the institution by a risk identification process.

The most commonly used framework for stress testing of NFR, illustrated in Figure 13.3, separates legal from nonlegal losses into three main modules: expected nonlegal losses, legal

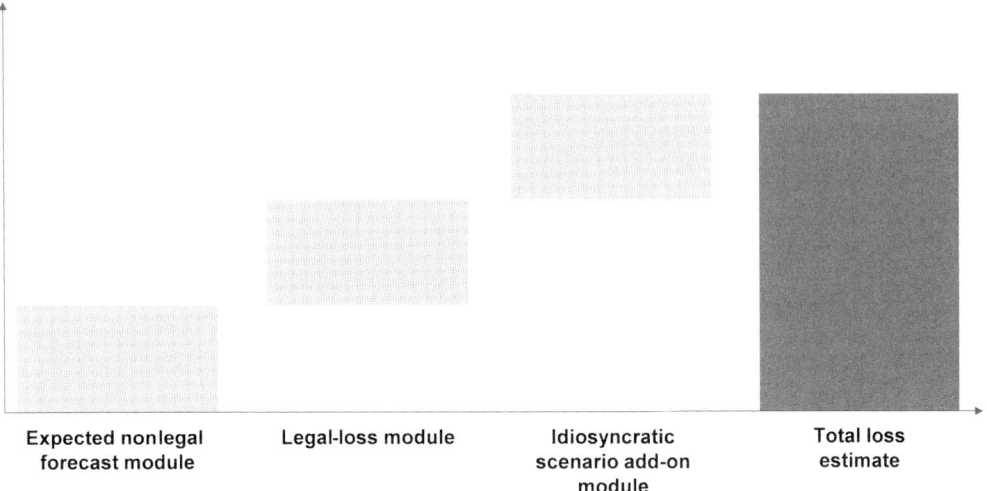

Expected nonlegal
forecast module
   Legal-loss module
   Idiosyncratic
scenario add-on
module
   Total loss
estimate

Figure 13.3 Typical NFR stress testing framework.

losses, and idiosyncratic scenario add-ons, bringing together qualitative and quantitative elements to produce a comprehensive NFR-loss forecast.

Each one of the modules plays a specific role so that the overall projection is comprehensive. The expected nonlegal loss forecast module is usually a candidate for more formal quantitative modeling. Material legal losses are excluded from the data set because of their unique characteristics, namely, the significant time lag (often several years) between the date on which an event occurs and the date on which a loss is registered.

The legal-loss module is typically a qualitative forecast of the losses an institution could experience given the current "portfolio" of legal material cases it is facing. This assessment is primarily driven by expert legal opinion and, at some institutions, by a history of similar cases. It is important to note that the definition of stress in this context is not necessarily just a macroeconomic stress but rather the expert assessment of what a stressful outcome would look like for the portfolio of legal cases. The outcome of cases that are currently being defended is usually not going to be influenced by the economic environment but rather by the idiosyncrasies of each case itself.

Finally, the idiosyncratic scenario add-on is meant to capture truly random tail events that could realistically affect an institution during the projection period. The scenarios are selected from a comprehensive risk identification process that identifies all extreme yet plausible events an institution is facing. The idea is to go beyond the already observed to the plausible yet severely adverse outcomes that the bank might face. Once the list of such risks has been compiled, a set of scenarios is then selected to be added to the NFR-loss forecast. The quantification of the impact of the chosen scenarios is usually done through workshops with the relevant stakeholders across the institution. The main drawback of this approach to quantifying scenarios is the highly subjective nature of such estimates and the difficulty in realistically minimizing it.

The combination of these three modules generates a total loss forecast in the spirit of a stress testing exercise by being reflective of both the expected behavior of NFRs in an adverse macroeconomic environment and the more idiosyncratic risks that are independent of the macroeconomic environment but are deemed to be of particular concern to an institution at that point in time. This approach addresses the problems raised by using quantitative techniques too narrowly focused on the macroeconomic elements of the forecast. Moreover, it casts the net widely in terms of information used and is more easily communicated to and understood by senior management and the board of directors.

4 Other Applications of Stress Testing

Aside from assessing the capital required to remain resilient in the face of severely adverse market and economic conditions, stress testing has found many other applications in banks. The US supervisory regime, in migrating from crisis (wartime) to ongoing peacetime stress testing, shifted quickly to incorporating this tool into capital planning. If the emphasis is on capital adequacy, then any capital distribution should be considered in the context of both business plans and possible adverse conditions. In this way, banks subject to the CCAR process increasingly make stress testing part of their strategic planning.

A natural application of stress testing is for limit setting narrowly and risk-appetite determination more broadly. At the top of the house, the risk appetite, set by the board of directors, provides the guardrails within which the bank is able to engage in risk-taking activities. Elements such as scenario severity and post-stress capital ratios, often called *capital goals*, are central to a risk-appetite statement. At a more granular level—business, portfolio, geography—stress test results can inform or even determine specific risk limits. To be sure, banks rarely take just one scenario as the determinant of a limit, lest they ignore other similarly plausible and severe scenarios. Rather, they consider a range of stress scenarios reflective of their risk profile to determine the limit-setting framework.

A second set of applications includes budgeting and performance measurement. The stress testing modeling infrastructure (following equation [13.1]) allows banks to separate contributions to losses or net revenues from systematic and idiosyncratic factors plus bank-specific contributions—effectively separating "beta" from "alpha." In this way, banks are able to budget and plan (on an ex ante basis) for contributions from markets and the economy (beta) to their future (planned or expected) performance versus what might be contributed by management effort (alpha). Ex post performance assessment of management can be similarly conducted.

Banks are subject to a range of capital and liquidity constraints. There is a growing set of regulatory metrics, including common equity Tier 1 (CET1), Tier 1, leverage, supplementary leverage, and so on. Depending on the business model and risk profile, different ratios may be binding under different scenarios, significantly complicating any capital planning and limit setting. As a result, banks are increasingly shifting their resources to industrializing their stress testing architecture to allow for rapid what-if analysis as they contemplate different risk-taking actions and their resilience to a wide range of stress scenarios.

As the crisis recedes into the past and the developed economies are enjoying a long, albeit tepid, stretch of economic growth, designing the right scenario becomes harder and harder. Borio et al. (2012) point out that it is much more straightforward to design an appropriate stress scenario in a crisis (wartime) than during apparently tranquil financial conditions (peacetime). Banks are increasingly developing capabilities to explore and run multiple (a dozen or more) scenarios to allow for a richer exploration of the relevant state space. These need not all be of similar severity because applications such as strategic planning and budgeting are concerned with performance in states of the world closer to the expected one.

References

Abdymomunov, Azamat (2014), "Banking sector operational losses and macroeconomic environment," Working Paper, available at http://ssrn.com/abstract$=$2504161.

Ames, Mark, Til Schuermann, and Hal Scott (2015), "Bank capital for operational risk: A tale of fragility and instability," *Journal of Risk Management in Financial Institutions*, 8(3), 227–243.

Bangia, A., F. X. Diedbold, A. Kronimus, C. Schagen, and T. Schuermann (2002), "Rating migration and the business cycle with applications to credit portfolio stress-testing," *Journal of Banking and Finance*, 26(2/3), 235–264.

Basel Committee on Banking Supervision (2006), "International convergence of capital measurement and capital standards: A revised framework, comprehensive version," available at www.bis.org/publ/bcbs128.pdf.

Basel Committee on Banking Supervision (2011, June), "Principles for the sound management of operational risk," available at www.bis.org/publ/bcbs195.pdf.

Belkin, B., S. Suchower, and L. Forest Jr. (1998), "A one-parameter representation of credit risk and transition matrices," *CreditMetrics Monitor* (3rd Quarter), 46–56.

Bellotti, Tony, and Jonathan Crook (2013), "Retail credit stress testing using a discrete hazard model with macroeconomic factors," *Journal of the Operational Research Society*, 65(3), 340–350.

Board of Governors of the Federal Reserve (2011, April), "SR 11-7: Guidance on model risk management," available at www.federalreserve.gov/supervisionreg/srletters/sr1107.htm.

Board of Governors of the Federal Reserve (2013, August), "Capital planning at large bank holding companies: Supervisory expectations and range of current practice," available at www.federalreserve.gov/bankinforeg/bcreg20130819a1.pdf.

Board of Governors of the Federal Reserve (2019), "Dodd–Frank Act Stress Test 2019: Supervisory stress test methodology," available at www.federalreserve.gov/publications/files/2019-march-supervisory-stress-test-methodology.pdf.

Borio, Claudio, Matthias Drehmann, and Konstantinos Tsatsoronis (2012). "Stress-testing macro stress testing: Does it live up to expectations?" BIS Working Paper No. 369.

Brown, Jeffrey, Brad McGourty, and Til Schuermann (2015), "Model risk and the great financial crisis: The rise of modern model risk management," in Douglas Evanoff, Andrew Haldane, and George Kaufman (eds.), *The new international financial system: Analyzing the cumulative impact of regulatory reform*, World Scientific Studies in International Economics.

Butaru, Florentin, Qingqing Chen, Brian Clark, Sanmay Das, Andrew W. Lo, and Akhtar Siddique (2016), "Risk and risk management in the credit card industry," *Journal of Banking & Finance*, 72, 218–239.

Canabarro, Eduardo, and Darrell Duffie (2003), "Measuring and marking counterparty risk," in L. Tilman (ed.), *Asset/Liability management of financial institutions*, Euromoney Books.

Crook, Jonathan, and Bellotti, Tony (2012), "Loss given default models incorporating macroeconomic variables for credit cards," *International Journal of Forecasting*, 28(1), 171–182.

Curti, Filippo, Marco Migueis, and Robert T. Stewart (2019), "Benchmarking operational risk stress testing models," FEDS Working Paper No. 2019–038, available at http://dx.doi.org/10.17016/FEDS.2019.038.

de Bandt, Olivier, Nicolas Dumontaux, Vinent Martin, and Denys Médée (2013), "Stress-testing banks' corporate credit portfolio," in *Débats économiques et financiers 2*, Banque de France.

Deng, Yongchen, John M. Quigley, and Robert Van Order (2000), "Mortgage terminations, heterogeneity and the exercise of mortgage options," *Econometrica* 68:2, 275–307.

Hanson, Samuel G., M. Hashem Pesaran, and Til Schuermann (2008), "Firm heterogeneity and credit risk diversification," *Journal of Empirical Finance*, 15(4), 583–612.

Hirtle, Beverly J., Anna Kovner, James Vickery, and Meru Bhanot (2015), "Assessing financial stability: The capital and loss assessment under stress scenarios (CLASS) model," Federal Reserve Bank of New York Staff Report No. 663, available at www.newyorkfed.org/medialibrary/media/research/staff_reports/sr663.pdf.

Jorion, Philippe (2006), *Value at risk: The benchmark for managing financial risk*, 3rd ed., McGraw Hill.

Leow, M., and Jonathan Crook (2014), "Intensity models and transition probabilities for credit card loan delinquencies," *European Journal of Operational Research*, 236, 685–694.

Leow, Mindy, and Jonathan Crook (2016), "A new mixture model for the estimation of credit card exposure at default," *European Journal of Operational Research*, 249(2), 487–497.

Leow, Mindy, and C. Mues (2012), "Predicting loss given default (LGD) for residential mortgage loans: A two-stage model and empirical evidence for UK bank data," *International Journal of Forecasting*, 28(1), 183–195.

Musto, David, and Nicholas Souleles (2006), "A portfolio view of consumer credit," *Journal of Monetary Economics*, 53(1), 59–84.

Opdyke, J. D., and A. Cavallo (2012), "Estimating operational risk capital: The challenges of truncation, the hazards of maximum likelihood estimation, and the promise of robust statistics," *Journal of Operational Risk*, 7(3), 3–90.

Parlatore, Cecilia, and Thomas Philippon (2019), "Designing stress scenarios," Working Paper, New York University.

Schuermann, Til (2014), "Stress testing banks," *International Journal of Forecasting*, 30:3, 717–728.

Schuermann, Til, and Kevin J. Stiroh (2006), "Visible and hidden risk factors for banks," Federal Reserve Bank of New York Staff Report 252, available at www.newyorkfed.org/medialibrary/media/research/staff_reports/sr252.pdf.

Sekeris, Evan (2012), "The new frontiers in advanced measurement approach modeling," in Ellen Davis (ed.), *Operational risk: New frontiers*, Risk Books.

Wu, Deming, Ming Fang, and Qing Wang (2018), "An empirical study of bank stress testing for auto loans," *Journal of Financial Stability*, forthcoming, available at https://ssrn.com/abstract=2591361.

Microprudential Stress Testing for Asset Managers

Jacques Longerstaey

1 Why Asset Managers Are Different

This chapter focuses on the applications of financial stress testing to firms that manage assets on behalf of third-party clients, either institutional or retail—the latter covering the wealth spectrum from mass affluent to ultra-high net worth. What makes asset-management stress testing different is that it applies to businesses that are not typically capital intensive. The focus therefore is not as much on testing the resilience of these firms' balance sheets, as is the case for traditional banking institutions, but more as tools to supplement traditional risk measures used in portfolio construction. As such, the stress testing involved is not only absolute—as in how much money can the asset manager lose in a particular market environment—but also relative, as in how an asset manager's portfolio will behave relative to the asset manager's investment objective in a stress scenario. Although stress tests are typically applied to individual products (e.g., pooled vehicles such as mutual funds or exchange-traded funds [ETFs]), they can also be extended to the aggregate portfolio managed on behalf of all clients to identify potential factor exposures that could damage the business franchise.[1]

If you want to consider how asset managers think about stress testing as opposed to their banking brethren—even though some are part of large banking institutions—you could say the following:

1. Banks perform these exercises as a way of shocking their *balance sheets*.
2. Asset managers stress test the following:
   a. Their client portfolios as an *off-balance sheet risk assessment*.
   b. Their aggregate assets under management (AUM) as a "revenue-at-risk" *income statement* exercise.

The need and methodological applications of stress tests for asset managers will also vary across the industry because many of the players differ along a number of dimensions:

1. *Ownership*: Asset managers are typically either independently owned (private or public) or owned by a bank or an insurance company. The ownership structure will have an impact on the scope of the stress testing activity required. For example, bank-owned asset managers will have to include their investment management activities within the

---

[1] Certain monoline or monoproduct asset managers are vulnerable to long periods of underperformance by the factors that their investment processes favor. In the United States, factor preferences such as value over growth, for example, can expose a manager to serious franchise risk.

framework of the Comprehensive Capital Analysis and Review (CCAR) exercise in the United States.

2. *Client Focus*: Asset managers can either be niche players, focusing on a limited client segment (institutional only, for example), or may focus on the full spectrum, which would include the broad range of retail and institutional clients.

3. *Product Type*: Certain managers will operate only separate accounts for their clients. Others may manage pooled vehicles, which may require stress testing for certain product types. For example, $1 net asset value (NAV) funds in the United States require stress tests under US Securities and Exchange Commission (SEC) rules, as do value-at-risk (VaR) types of undertakings for the collective investment in transferable securities (UCITS) funds under the applicable European directive.

4. *Assemblers and/or Manufacturers*: Most investment management firms create products from the ground up, but many are hybrids that both manage products and assemble strategies from a combination of internal and third-party products. This is particularly prevalent in the wealth-management business, where clients' requirements for exposure across a diversified set of asset classes and products may require using other firms' products to address market segments where one might not have the required expertise. An example of this might be a very traditional firm that relies on third-party products for alternatives. In the assembly business, it is important to understand the behavior of all the component products to a strategy as well as how they interact within the strategy. Think of an aircraft manufacturer, which will stress the tolerances of all the components of the airplane as well as the assembled airplane itself.

5. *Strategy Type*: The industry contains players who focus on a limited number of asset classes or investment philosophies, and these parameters will also define the use of stress testing. A manager of passive-type strategies whose aim is to mimic publicly available benchmark indices will have less use of stress testing than an active manager covering a broad spectrum of asset classes.

Some of the stress tests performed by asset managers are driven by regulatory requirements, such as the ones applied to US money market funds under SEC rules or the inclusion of broadly defined stable-value products in CCAR exercises for bank-owned businesses. But most of the applications focus on managing risk in client portfolios taken individually or books of business taken in aggregate. Because the business model of these firms is simple at it core, you can boil it down to this simple equation:

$$AUM \times Management\ fees = Revenue. \qquad (14.1)$$

Stress testing of the AUM provides management with valuable insights as to risks affecting business plans.

2 Managing Investment Risk

Stress testing of portfolios is one component of the process by which risk professionals monitor and manage investment risk. By virtue of their investment goals, asset-management clients are subject to investment risk.

> Investment risk is the potential for a client to fail to meet an investment objective, or lose principal, as a result of movements in equity prices, interest rates, credit spreads, commodity prices, foreign exchange rates, or even those variables not directly quoted or easily monitored and measured, such as liquidity.

Investment risk exists for financial products and portfolios and is inherent in the construction of investment portfolios. Investment risk is often confused with market risk, which is also concerned with the movement of financial market prices and rates but is direct on-balance-sheet risk with capital held as a buffer against adverse movements. A simple way to distinguish investment risk from market risk is that the former is agency, and the latter is deemed principal.

Although market risk and investment risk are closely related (think about investment risk as being market risk for your clients), the various sectors of the financial services industry have relied on different approaches to estimating risk, and these choices also affect how stress tests are performed, as we will see later in this chapter. Whereas banks will typically use security-specific VaR models, asset managers rely on factor models that map back to their investment philosophies. Asset-management clients typically have longer investment horizons (and hence risk horizons). Their managers may look at relative risk to a benchmark (e.g., tracking error) rather than absolute VaR-type measures. Their risk decomposition along factors will focus on components such as duration and credit for fixed income or momentum, value, growth, and capitalization size for equities.

The primary risk takers or managers are typically those individuals or teams that have the responsibility to manage assets for clients. These risk takers may be, among other descriptors, financial advisors, portfolio managers, or investment strategists. In addition, most firms have independent risk managers whose responsibility is to work with the primary risk takers to ensure that products and portfolios are managed on behalf of clients in a manner consistent with their objectives and expectations. This is broken into two components:

1. Risk estimation (Are the products or portfolios operating in the expected risk space?)
2. Risk decomposition (What are the risk components that make up the total product or portfolio risk?)

A good approximation for an investment-risk framework is typically guided by the following simple conceptual equation:

$$\text{Exposure} \times \text{Shock} = \text{Risk}.$$

A more general framework will address the limitation that although stress testing may inform on the response of portfolios to changes in inputs, it does not provide any information on the likelihood of any of this to happen. From this standpoint, in all generality, two equations would apply:

$$\text{Exposure} \times \text{Shock} = \text{Loss},$$
$$\text{Exposure} \times \text{Shock} \times \text{Likelihood} = \text{Risk (of loss)}.$$

That is, risk is a function of exposure to a particular driver or variable associated with financial markets and a shock, which may be broad-based (i.e., systematic) or security or product specific (i.e., idiosyncratic). Monitoring and measuring of risk are therefore inherently linked to understanding those factors to which a product or portfolio is exposed and gauging what shock could arise.

The inability to precisely assess future outcomes (i.e., to predict shocks) requires a risk framework to measure both exposures and risk metrics. Stress testing the latter ensures that you can answer the two following questions:

1. What risks might the model I am using have missed?[2]
2. What are the shape of the distribution of returns, and what do the tails of the distribution look like?

Exposure metrics can be monitored without the use of a risk model and may include concentration levels, top holdings, and other simple characterizations. Risk metrics incorporate shocks or variability. Risk metrics that estimate future shocks or variability require a model that aggregates exposures and shocks to derive a measure of shortfall for a stated investment objective. Absolute risk measures include the dispersion of returns, which is often called *volatility* and characterized by the standard deviation of returns. Relative risk measures include the dispersion of the squared differences in returns, which is often called *tracking error* and aimed at comparing portfolio returns to the returns of of a model or benchmark.

In addition to absolute or relative risk metrics, risk models also allow one to decompose a value to determine contributions to the aggregated measure. Although one may be interested in a simple point estimate of a risk measure, it is often the case that a distribution serves to shed light on the behavior of an array of products or portfolios. That is, as part of quality assurance, one may look at the standard deviation or tracking error for a particular product or portfolio, and one may also consider the distribution of the standard deviations and tracking errors for all products of a particular type or portfolios aligned with a certain model. The risk models used by asset managers also provide insights into how portfolios behave in historical stress scenarios.

Figure 14.1 illustrates how an asset manager might use a stress testing framework across various lines of business for both client portfolio-management purposes—and risk communication; we will discuss that aspect when we address risk-appetite considerations—and business-risk management.

---

[2] This assumes that the model used for general risk estimation and the scenarios for stress testing are independent. If the general model is missing certain explanatory factors, these may become evident in scenario testing.

Figure 14.1 Example asset manager stress testing framework.

3 Defining Investment Objectives and Risk Parameters

Managing assets on behalf of third parties requires firms to assess a client's tolerance for risk. For an institutional client hiring a manager on a mandate for a particular asset class, the total risk appetite would be set by the benchmark selected to measure the manager's performance (expected total volatility of equities or fixed income, for example). The relative risk appetite (often referred to as *active risk* or *ex ante tracking error*) would then be agreed upon with the client.

For a retail client, extracting the exact risk appetite is often more convoluted because it involves understanding the client's investment objective, often through the use of a questionnaire. The resulting investment objective is then mapped to a specific asset allocation from which the risk can be derived. The example in Table 14.1 shows a common set of nine investment objectives, and their associated risk and return expectations

In Table 14.1, we illustrate the metrics of how expectations of risk and return can be communicated to a retail client base. The table shows both long-term arithmetic and geometric expected annual returns (to cater to the requirements of various financial planning tools that use either metric as input), as well as annualized standard deviation (measure of volatility), yield, and downside risk. Downside risk is based on assumptions about average returns and the variability of returns. It represents the return that would be statistically likely in 95 percent of annual returns. In other words, in 19 out of 20 years, performance would likely be better than this figure, and in the 20th year, it would likely be worse. There is no guarantee that any particular 20-year period will follow this pattern.

Because we define investment risk as the probability of not meeting your investment objective, the risk measure used always relates to how that question must be answered:

- For an institutional investor who expects to achieve an objective consistent with the return of the chosen asset class (equities or fixed income, for example), measuring tracking error to the asset class benchmark could be the appropriate metric. Other risk metrics can also be used for different types of institutions. Pension funds, for example, could use the probability of falling below a target funding level as their benchmark metric.

Table 14.1. Hypothetical portfolio statistics

	INCOME			GROWTH & INCOME			GROWTH		
	CONSERVATIVE	MODERATE	AGGRESSIVE	CONSERVATIVE	MODERATE	AGGRESSIVE	CONSERVATIVE	MODERATE	AGGRESSIVE
Arithmetic expected return	4.3%	5.2%	6.0%	6.9%	7.6%	8.2%	8.7%	9.2%	9.6%
Geometric expected return	4.2%	5.1%	5.8%	6.5%	7.1%	7.5%	7.9%	8.2%	8.5%
Standard deviation	3.9%	5.4%	7.0%	8.7%	10.4%	11.9%	13.4%	14.5%	15.7%
Yield	3.0%	3.0%	3.1%	2.7%	2.7%	2.6%	2.6%	2.5%	2.4%
Downside risk	−2.0%	−3.4%	−5.0%	−6.8%	−8.6%	−10.2%	−11.8%	−13.0%	−14.1%

- For a retail client concerned about wealth preservation, a measure of absolute portfolio volatility would be appropriate.

Standard risk measures, such as volatility or tracking error, are useful when gauging a portfolio's behavior under normal conditions, essentially looking at the central region of the distribution of returns. What stress testing adds is (1) insights into the lower-left tails of portfolio return distributions and (2) potential scenarios that are not even contemplated in a historic distribution because these outcomes have never been recorded. With regard to the second of these additions, one must, however, constrain oneself to scenarios and outcomes that are legitimately within the scope of being reasonable.

The benefit of stress testing with regard to communicating with individual investors stems from the fact that it uses simple language. Whereas financial professionals understand the concepts of volatility, tracking error, and beta, most of their clients do not. And because gauging a client's risk appetite is a critical component of defining an appropriate set of investment objectives, we must refer to concepts that are simpler to communicate. Instead of telling a client that a particular portfolio would have an expected annualized return of 10 percent with a volatility of 15 percent, it might be better to describe the risk of the portfolio in terms of losses under scenarios that the client would be familiar with. That a particular portfolio might have sustained losses of 25 percent during the 2008 financial crisis is a much more vivid description than the implication that a 15 percent volatility with an expected return of 10 percent implies annual returns between certain thresholds that depend on your distributional assumptions. How much money a client may lose in a particular scenario is a much better way of getting at the risk-appetite question, notwithstanding the findings of behavioral finance that how people say they would react in a crisis and how they actually do often differs significantly.

Here again the notion of likelihood of such events is a separate factor. Note that these losses are unrealized unless the client makes an inopportune decision. In this case, the time it takes for the portfolio to recover with the market to half the level of loss may be helpful additional information to contrast the magnitude of the prospective loss and the degree to which its impact extends over time.

This is where stress tests are an important component to monitor adherence to risk appetite, although it is an indirect process. Because the monitoring of investment guidelines is typically done in the traditional expected volatility or tracking-error space, stress test results of various portfolios can be used to gauge a client's risk appetite or tolerance. From these, we can back out that a client wishing to minimize the probability of losing 25 percent of the value of his or her portfolio might be best suited with an asset mix displaying an annualized expected volatility of 8 percent for example. It is this volatility metric that will end up being monitored by the risk management function to ensure adherence to risk appetite.

4 Categories of Stress Tests

There are a few broad categories of stress tests that are typically used in the industry, and there are variations on their application, as discussed in the following subsections.

4.1 Historical Stress Tests

As their name implies, historical stress tests rely on past events to see how portfolios might have behaved had they been owned at that time. Many analytical software providers embed these historical scenarios in their applications, and these are therefore easy to access. Scenarios vary from equity market corrections, to interest rate surprises and emerging-market defaults.

Although these are useful applications, they also come with a few caveats, namely:

- History rarely repeats itself in the exact same manner. The changes in volatilities and correlations that occurred in the past are unlikely to be duplicated exactly because economic conditions are rarely identical, market structures change, and the players will not always act in a predictable fashion.
- Market structures evolve at many levels, both macro (e.g., the existence of a single currency bloc in the European Union will make pre-euro stress tests less meaningful) and micro (e.g., instruments used for leverage or hedging may have either existed or not, been regulated or not, required certain levels of capital or not) at different times in history.

When using historical scenarios, one must be acutely aware of the exact structural and market conditions that existed at the time and question whether a repeat is probable or even possible. For example, developed markets have not seen significant inflation in decades, so although we may have experienced credit crises in low-inflation periods, we have little to no data for such in inflationary periods.

4.2 User-Defined Stress Tests

Portfolio managers, risk managers, and other constituencies within asset-management firms can also devise their own scenarios. In the case of regulatory requirements, these scenarios may be imposed by a third party. User-defined stress tests allow practitioners significantly more leeway, from stress testing a single market factor to defining a comprehensive market scenario.

We typically separate user-defined scenarios between the following categories:

Simple Risk-Factor Stress: In this case, no correlations are used. Only the risk factors selected for inclusion will be varied by a given amount, and no other risk factor will be affected. This is the application of the economic precept "all other things being equal." In this variant, one might posit only a decline in a single equity market, for example.

The relevance and realism of this type of test must be kept in mind before forming conclusions.

Predictive Risk-Factor Stress: In this version, correlations across all risk factors in the stress test and portfolio are calculated based on settings of the user's choice (time-series period, return horizon, volatility, correlation forecasting methodology, etc.). For example, stressing a single point on the US Treasury curve will call on the correlation matrix to affect all other positions as well, based on history.

In the RiskManager® application, only significant nonhighly correlated risk factors are used. Risk factors that are too highly correlated impede the regression calculations required by the calculation engine.

4.3 Reverse Stress Tests

If historical and scenario stress tests can be viewed as a top-down exercise imposed on the portfolio, then reverse stress tests are the bottom-up equivalent. In this case, one is not asking, "What happens to my portfolio in this scenario?" but rather "What scenario can be most damaging to my portfolio?"

In addition to the scenario choice, which can be historical or predictive, reverse stress tests require that one identify which factors the portfolio might be most susceptible to. This is usually accomplished using regression analysis or factor-risk decomposition using a quantitative model. This component of the process is discussed in more detail in Section 6.1.

Reverse stress tests are an important component of a comprehensive approach because historical or scenario-based tests can often miss important exposures, particularly if they are imposed externally by a regulator, which may have designed them for other purposes. As an example, the CCAR stress tests designed for bank balance sheets are often not tailored to the exposures and behavior of stable-value or money market products sponsored by banks. In these scenarios, the funds actually perform quite well because their "safe-haven" and "flight-to-quality" characteristics make them a naturally preferred destination for money fleeing either declining equity markets or higher interest rates.

5 Stress Testing Model Portfolios

Model portfolios are often used by retail-oriented managers as a guide for field advisors or as a centrally managed option. Firms often use multiple sets of nine model portfolios that map to the nine investment objectives shown in Table 14.1 on hypothetical portfolio statistics. The various flavors can encompass those with three or 4 asset classes, using only passive or a combination of passive and active, as well as other factor focuses (use of environment, social, and governance (ESG), for example, has become quite common).

Table 14.1 shows how metrics of both expected return and expected risk increase as you move from less risky to more risky investment objectives (left to right). In Figure 5, we look at the risk distributions of these portfolios (prospective percentage loss of aggregate assets under management by investment objective) under a variety of scenarios to identify any potential deviations in the order of risk metrics. You would assume ex ante, for example, that a severe scenario for equities would have a greater impact on the model portfolio assigned to the most aggressive investment objectives. The opposite would be true for a negative event involving fixed-income securities. The advantage of running these historical scenarios rather than stressing a single factor at a time is that it reveals the complexities of behaviors of financial instruments across asset classes in times of stress. Volatilities go up, correlations change, and these events can inform portfolio managers about whether certain portfolios might look like outliers in certain circumstances.

Figure 5 shows the outcomes of a series of historical stress tests for nine model portfolios that correspond to the investment objectives shown in Table 14.1. It essentially confirms that the standard risk-metrics alignment in normal markets was confirmed during each of the chosen scenarios. On Black Monday in 1987, the conservative income (CI) portfolio lost around 6 percent, whereas the aggressive growth (AG) portfolio lost over 20 percent.

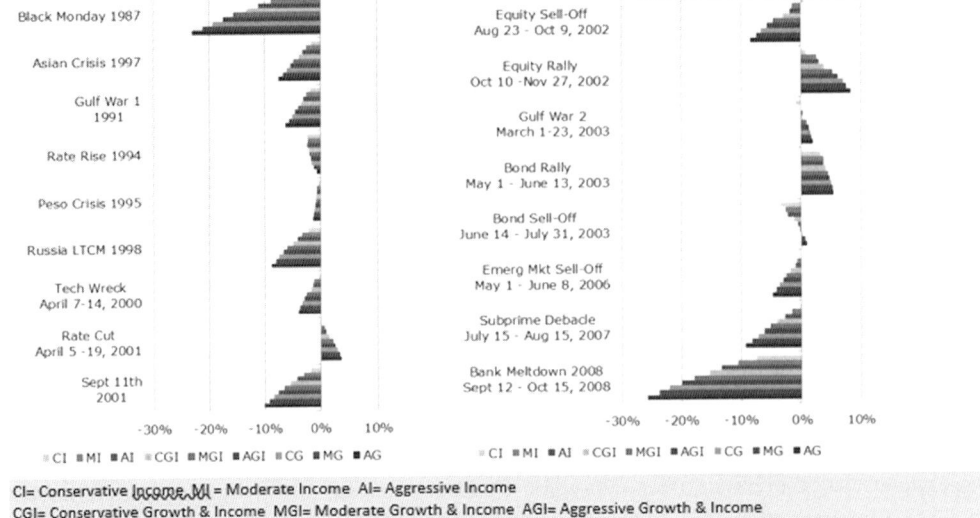

Figure 14.2 Data generated using RiskManager® application developed by MSCI.
Source: MSCI. The MSCI data contained herein is the property of MSCI Inc. and/or its affiliates (collectively, "MSCI"). MSCI and its information providers make no warranties with respect to any such data. The MSCI data contained herein is used under license and may not be further used, distributed, or disseminated without the express written consent of MSCI.

During the equity market sell-off of August to October 2002, only the CI portfolio eked out a small positive return, whereas any portfolio holding a more significant allocation to equities lost ground.

6 Stress Testing Client Portfolios

A portfolio report like that shown In Figure 14.3 is usually the starting point for an analysis of portfolio exposure and risk. It provides either the portfolio manager or the client with high-level statistics about overall and relative risk, asset-class decomposition, betas to individual risk factors, and total risk decomposition. It is, at any point in time, the model's best guess of how risky a portfolio is, given the current market environment.

This particular sample model portfolio follows the "moderate growth and income" investment objective—the middle of the road in terms of return and risk expectations (see Table 14.1)—and has an annualized volatility expectation of 5.5 percent, slightly below its benchmark target of 5.8 percent. It has an annualized expected tracking error versus its benchmark of 0.8 percent. It is invested 38 percent in US equities and 34 percent in US fixed income, with the remaining allocations diversified across non-US equities, fixed income, and cash. The model used for this example has been calibrated to a 3-month horizon.

Using the assumptions that underlie these multifactor models, we would expect that the portfolio's annual return would lie somewhere between 2.1 and 13.1 percent per annum, two-thirds of the time (using the annual expected return of the moderate growth and income portfolio of 7.6 percent shown in Table 14.1).

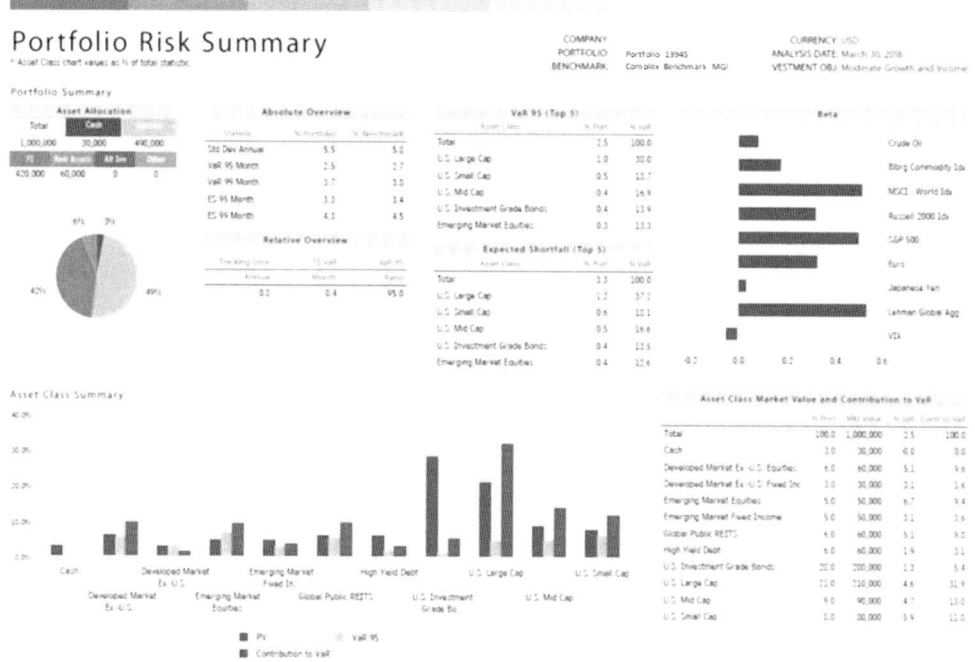

Figure 14.3 Data generated using RiskManager® application developed by MSCI.

Although these general statistics are useful in describing the risk profile of the portfolio, they are not sufficient to understand how the portfolio might behave in either stressful market conditions or under single-factor stresses that could correlate across assets. The portfolio stress test summary shown in Figure 5 provides insights into the potential losses that would have been incurred during historical periods going back over 30 years.

It would not be sufficient to describe the risk profile of this portfolio to a retail client by only focusing on the center of the distribution of returns, which, at the low end of 2.1 percent, might only elicit a yawning acknowledgment. It is imperative that managers speak to their clients about the potential for events such as 2008 bank meltdown, a 1-month period during which this portfolio would have lost 20 percent of its value, as shown in Figure 14.4. The same portfolio would have lost around 13 percent on Black Monday in 1997. It is important to note that the horizons for historical stress tests will vary because the models used pick events that vary in duration. For example, whereas Black Monday was a 1-day event, the bank-meltdown event lasted over a month.

Although financial professionals are usually comfortable with metrics such as volatility and tracking error, retail investors are more interested in the answer to the question, "How much money can I lose?" That's where historical stress tests provide a context to better understand a client's risk appetite.

6.1 Stress Testing "Assets under Management"

In addition to stress testing of individual client portfolios, many firms will also go through the exercise of creating an aggregate "AUM portfolio'," which entails nothing more than adding up all their clients' positions into a single set of holdings. Because the business model

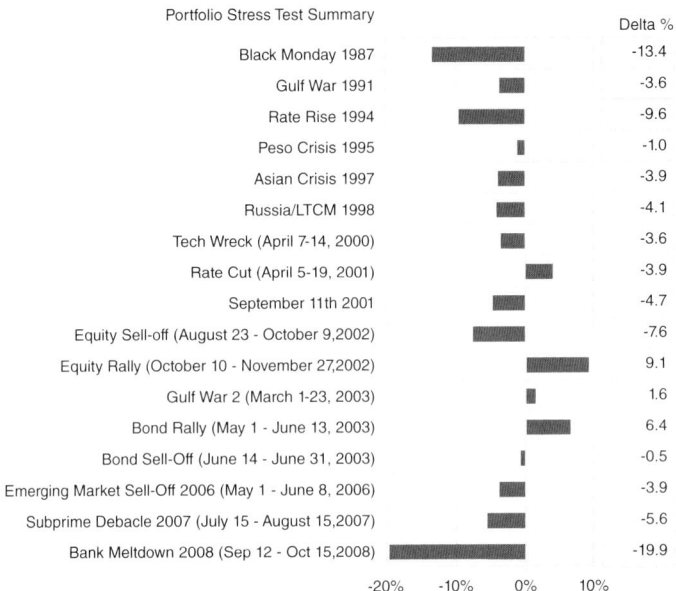

Figure 14.4 Data generated using RiskManager® application developed by MSCI.

of asset managers is reasonably simple—revenue is equal to the product of AUM times fees charged—this exercise allows the finance department to simulate changes to two important variables, the market-impact moves on the aggregate AUM and the potential changes in fees charged under various scenarios. Both of these will affect the firm's earnings.

Because asset managers do not face the same capital constraints as other financial institutions that act as principals, one can wonder why this exercise would be useful. It serves essentially as an expectation-management tool so that the leadership of these organizations can get a sense of the potential corrective actions that would need to be taken in response to major market corrections or competitive pressures on fee income. Many risk managers have also often raised the question of whether these exposures—at least the market ones—should be hedged because an asset manager has a structural imbalance between significant market exposure and a relatively fixed cost base. If your asset base goes from $100 to $50 because of a market event, your ability to cut expenses will be constrained by the fact that the work has not been cut by half.

The challenge here is that management must make a choice between being constantly hedged, which means giving up market appreciation in the long run and hedging only at times of higher perceived risk, which feels a lot like market timing, which no one is particularly good at.

One should not underestimate the data-management challenge of creating the "AUM portfolio," particularly in a large, multiasset, and potentially multi-boutique model. Getting all the client portfolios aggregated at the security level into a single portfolio is not a trivial exercise. One can find solace, however, in the fact that to be directionally correct, certain simplifying assumptions can be made. The look-through to the security level required for individual client risk analysis can be dispensed with and aggregate exposures mapped to representative asset-class benchmarks. For the purpose of a financial planning exercise, the adage "it's better to be approximately right than precisely wrong" definitely applies.

7 Factor Models and Stress Testing

One of the distinguishing characteristics of portfolio-risk measurement in the asset-management business is the widespread use of so-called "factor models." As described in equations 14.2 and 14.3, these models typically decompose expected volatility measures—whether absolute or relative—into a set of common explanatory factors F_k and provide the backbone of most quantitative measures of portfolio risk. They are essentially extensions of the simple equation that links a portfolio's volatility to the market's volatility (factor F_{Mkt}) with the slopes β_k ("betas") of the regression expressing the sensitivity to that systematic source of risk. In general, a term α_P ("alpha") is included to represent the contribution to the return (risk) that is not explained by the choice of factors, that is, idiosyncratic to the portfolio or product P—most typically active return (risk) relative to a benchmark. Factor models capture the exposures over a given period of time to a number of explanatory factors built as overall zero-investment portfolios. They are widely described in the financial literature (see, e.g., Ang [2014] for an engaging presentation of factor models). Specifically, equation (14.2) gives the four-factor Carhart (1997) model, an extension of the popular Fama and French model (1993) where value and growth (HML, high-minus-low book-to-market ratio), market capitalization size (SMB, small-minus-big), and momentum (Mom) are examples of factors that explain a significant percentage of the return and volatility of an equity portfolio. Factors such as duration, yield curve slope, and curvature are common factors used in fixed-income investing.

$$R_{P,t} = \alpha_P + \sum_k \beta_k \times F_{k,t} + \varepsilon_{P,t} \xrightarrow{\text{reg.}} \hat{R}_{P,t}$$

$$= \hat{\alpha}_P + \hat{\beta}_{\text{Mkt}} F_{\text{Mkt},t} + \hat{\beta}_{\text{HML}} F_{\text{HML},t} + \hat{\beta}_{\text{SMB}} F_{\text{SMB},t} + \hat{\beta}_{\text{Mom}} F_{\text{Mom},t}, \quad (14.2)$$

$$
\begin{aligned}
\text{var}\,(R_P) = \sigma_P^2 &= \text{var}\left(R_{P,t} - (\alpha_P + \varepsilon_t)\right) + \text{var}\,(\alpha_P + \varepsilon_t) \\
&\cong \underbrace{\sum_i \sum_j \beta_i \beta_j \times \text{Cov}(F_i, F_j)}_{\text{systematic}} + \underbrace{\text{var}(\alpha_P + \varepsilon_t)}_{\text{idiosyncratic}} \\
&+ \sim \sum_i \beta_i^2 \times \text{var}\,(F_i) + \text{var}\,(\alpha_P + \varepsilon_t).
\end{aligned}
\quad (14.3)
$$

Note how (1) in equation (14.2), estimated quantities resulting from the regression (reg.) of the data on the factors are designated with a caret, and (2) in equation (14.3) the factors in the last step may be approximated as independent of one another, although it is recommended to empirically verify this last point to detect multicollinearities.

It is worth pointing out that in this section, we are describing an ex post analysis. In the setting of active management and portfolio optimization, however, different models are used to forecast risk and the alpha term—the manager's return forecast (in excess of the benchmark)—because the manager may want to incorporate additional information not relevant in the risk model into the alpha forecast. The factors used in both models will mostly not be entirely different but show an overlap. For this reason, so-called "alpha factors" should be seen as consisting of one part that is a linear combination of the risk factors—with implied factor risk—and a part that is "orthogonal" to the risk factors, just as in equation (14.2).

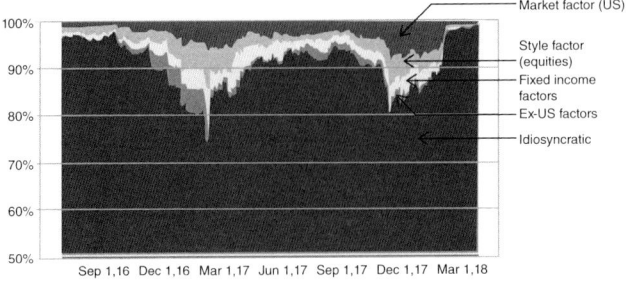

Figure 14.5 Explanatory power of the factors—R^2 decomposition.

Finding out which factors drive a portfolio's return and risk is an important part of the risk-measurement process (Figure 14.5). Portfolio managers may want to overweight factors that have positive expected long-term returns—the literature points to value, small size, and momentum as examples—while minimizing exposure to factors that may only add portfolio volatility. All of these factors display some amount of volatility; their returns show a distribution that can be modeled. Some of them can add excess return to a portfolio, whereas others add only risk.

Expanding equation (14.2), Figure 14.4 shows the contribution to the variance of the returns of a series of 10 equity, fixed-income, and international factors for a broadly diversified portfolio, rolling a 60-day regression window over a period between 2016–2018.

The explanatory power of these factors is extremely high, as the R^2 of the regression shows an average in excess of 96 percent and no dip below 90 percent.

Grouping the factors by categories—market and broad style (value/growth, size, and momentum) for equities, US interest rate curve and credit spread for fixed income, and their international counterparts—reveals comparable contributions, between 1 and 5 percent, on top of a portfolio overwhelmingly driven by the US equity market. The idiosyncratic contribution is similarly well under 5 percent most of the time.

Because we conclude that more than 90 percent of this portfolio's returns can be explained by the behavior of a relatively small number of common factors, we can turn our attention to these factors and analyze their behavior over time and in times of stress.

For illustrative purposes, Figure 14.6 shows a subset of the number of factors used in the analysis. We now focus the portfolio's factor (beta) exposures to four factors relative to its benchmark: market (US), US yield curve (term), value-versus-growth, and momentum. Over the period from June 2016 to March 2018, the portfolio shows positive if varying relative exposure to the US market, hovers between long and short in terms of value and momentum, and is short duration (negative term exposure). To test for the portfolio's performance in times of factor stresses, we also look at the individual behavior of the four factors in panel (b).

The factor returns display significant variability over the period, but the one we could most associate with a form of stress would be value-versus-growth, where the peak-to-trough rolling 60-day return between December 2016 and March 2017 is a more than

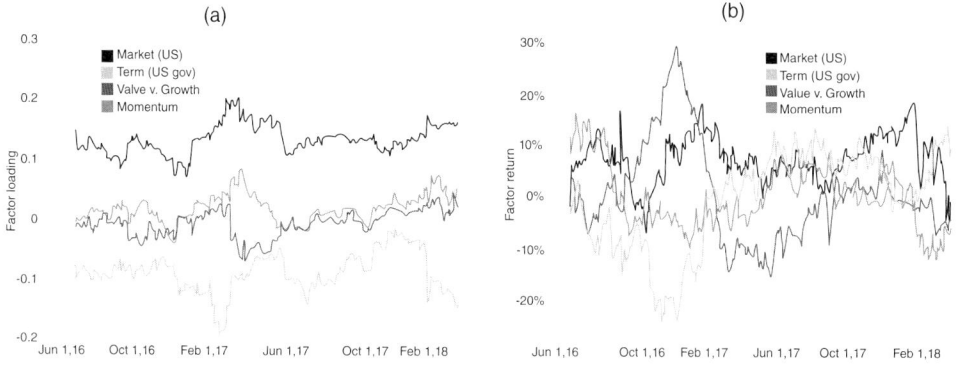

Figure 14.6 (a) Factor loadings (betas) and (b) returns relative to benchmark.

40 percent swing. If you recall, that period coincides with the US presidential election that brought the Trump administration to power. Value stocks definitely saw that as a stress.

The final part of the analysis is to combine the factor loadings with the factor returns and look at the portfolio's return profile over the period (Figure 14.6).

One of the most telling aspects of Figure 14.6 is that although one particular factor—value—may have encountered a period of relative stress in the latter half of 2016 and early part of 2017, the portfolio's return seemed relatively unaffected. In fact, the market impact dominated over the period and led the portfolio to show strong positive returns. Although the value-versus-growth dispersion can show significant return discrepancies over time, this particular portfolio did not have a meaningful exposure to it. On the other hand, a marked shock in the portfolio through its term contribution in Q1 2017 aligns with the sharp rise in US interest rates in late 2016 through the first half of 2017 but was mitigated by the short-duration position relative to the benchmark. This reinforces the need for any stress testing program to have robust and timely estimators of exposure so that one does not waste time and resources testing scenarios that would have little to no impact but rather inform contemplated tactical moves based on established drivers.

Figure 14.7 also tells us that although identified factors, whose exposures and returns can be modeled, drive the bulk of a portfolio's total return, there is always a residual or "remainder" that is unexplained by the model. This residual is either a factor you may have missed, a transient factor that shows up in markets infrequently and with insufficient intensity to be modeled effectively, or a pure idiosyncratic component.

Transient factors are an important thing to look out for. These tend to be factors that may affect portfolios over discrete periods of time and have little to no impact in normal markets. They are the quantitative representation of the fact that markets look at different things at different times and that the interpretation of a particular financial variable changes over time. Those of us who remember the 1980s will be reminded of the monthly impact on the value of the dollar of the publication of the US trade deficit at that time. The US external position has not meaningfully changed in 30 years, and yet foreign exchange traders no longer hold their breath on the day that the deficit comes out.

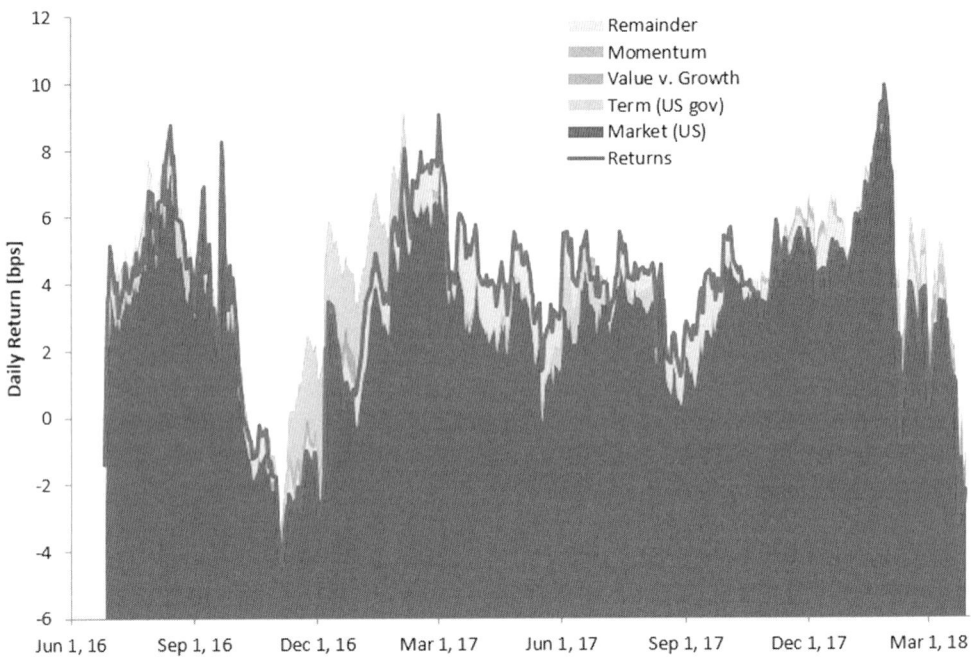

Figure 14.7 Contributions to portfolio returns. Rolling 60-day window.

8 Regulatory Stress Tests

Not being particularly capital intensive, asset-management businesses do not fall under the category of firms that need to be heavily stress tested for regulatory purposes. There are, however, a few exceptions worth mentioning, as detailed in the following subsections.

8.1 Stress Testing Stable NAV Funds for SEC Purposes

According to SEC Rule 2a-7(g)(8), a US-registered money market fund must engage in periodic stress testing that tests the fund's ability to have invested at least 10 percent of its total assets in SEC weekly liquid assets and the fund's ability to minimize principal volatility,[3] based on specified hypothetical events that include, but are not limited to, the following:

- Increases in the general level of short-term rates, in combination with various levels of an increase in investor redemptions
- A downgrade or default of particular portfolio security positions, each representing vari- ous portions of the fund's portfolio (with varying assumptions about the resulting loss in the value of the security), in combination with various levels of an increase in investor redemptions

---

[3] In the case of a money market fund using the amortized-cost method of valuation or the penny-rounding method, it is the fund's ability to maintain the stable price per share established by the board of directors for the purpose of distribution, redemption, and repurchase.

Money Market Fund Interest Rate Shock Analysis

March 31, 2016

Constant NAV Stress Test Summary

	A	B	C	D	E	F	G	H	I	J	K	L	M	N	O
				Curve Changes Resulting in NAV below 0.9975						Curve Changes Resulting in NAV below 0.995					
				No Investor Redemptions			20% Investor Redemptions			No Investor Redemptions			20% Investor Redemptions		
				Parallel Shift	Short Twist	Long Twist	Parallel Shift	Short Twist	Long Twist	Parallel Shift	Short Twist	Long Twist	Parallel Shift	Short Twist	Long Twist
Fund Description	NAV	WAM													
Fund 1	1.001140	21		500	570	3510	400	470	2840	960	1110	6820	780	900	5500
Fund 2	1.000106	21		490	570	3540	400	470	2860	960	1120	6930	780	910	5570
Fund 3	1.000164	17		650	740	4930	530	610	4000	1250	1440	9550	1010	1170	7700
Fund 4	1.000085	13		1190	1470	6260	850	1050	4430	2370	2920	NA	1790	2210	9370
Fund 5	1.000102	12		3620	4540	NA	2950	3700	NA	7090	880	NA	5750	7220	NA
Fund 6	0.999644	15		2270	2840	NA	1560	1960	7640	4910	6150	NA	3690	4630	NA

Figure 14.8 Sample report.

- A widening of spreads compared with the indexes to which portfolio securities are tied in various sectors in the fund's portfolio,[4] in combination with various levels of an increase in investor redemptions
- Any additional combinations of events that the advisor deems relevant

The approach to stress testing involves defining stress events consistent with the aforementioned guidelines detailed, evaluating the impact of incremental increases in each event's magnitude, and identifying the event magnitudes at which the NAV or SEC weekly liquidity level declines below threshold levels. For constant-NAV money market funds, the NAV threshold levels used are $0.9975 (i.e., warning track level) and $0.995 (i.e., break-the-buck level), and the SEC weekly liquidity threshold levels used are 30 percent (i.e., required minimum level) and 10 percent (i.e., mandatory liquidity fee level). For floating-NAV money market funds, the NAV threshold levels used are 99.75 and 99.5 percent of the daily average NAV over the last 12 months, and the SEC weekly liquidity threshold levels used are the same as those used for constant NAV funds.

Figure 14.8 shows an example of the monthly report produced to test for the implied market moves required for the NAV of each of the funds to breach either the $0.9975 or $0.9950 thresholds.

8.2 Stress Testing for CCAR Purposes

The CCAR is an assessment mandated by the Federal Reserve under the Dodd–Frank Act to determine whether large bank holding companies (BHCs) have sufficient capital to continue operations throughout times of economic and financial stress, as well as a robust, forward-looking capital-planning process that accounts for their unique risks.

Although the asset-management businesses of banks are typically the least capital-intensive businesses of these complex organizations and therefore not the main target of the regulation, stress testing requirements usually affect them in two ways:

1. Given their long history of supporting money market funds that had "broken the buck" in stressful environments, banks have had difficulties convincing regulators that their funds are not principal-risk exposures. Most banks therefore subject their $1 NAV funds and other stable-value products specifically to the required scenarios for separate analysis.

---

[4] A sector is a logically related subset of portfolio securities, such as securities of issuers in similar or related industries or geographic region or securities of a similar security type.

2. Client portfolios need to be stress tested in aggregate to gauge the impact on revenues of the regulatory scenarios. To do this effectively, banks need to estimate the factor exposures of the AUM portfolio (in essence, a roll-up of all client portfolios aggregated into a single portfolio) to the scenario parameters, as illustrated in Section 6.1. The factor sensitivities (or betas) to the various stressed parameters can then be applied to the scenarios to gauge the negative (or positive, in some cases) impact on asset-management revenues.

8.3 Stress Testing Requirements for UCITS Funds

The term *UCITS*, mentioned earlier in the chapter, refers to EU Directive 85/611/CE of December 20, 1985, the objective of which was to create a single European market for retail investment funds while at the same time ensuring a high level of investor protection. In Europe, stress tests are not required by regulation for all Common Investment Schemes (CISs), of which UCITS are one component, alternative investment funds (AIFs) being the other. For UCITS, the applicable European directive does not require the use of stress tests, except for funds that are VaR-type funds.[5] For AIFs, stress tests are mandated, regardless of their type. Readers can refer to the extensive regulatory documentation on the topic for more insights into the details mandated by the various regulators.

9 Conclusion

This chapter has focused on the uses of stress testing in the asset-management industry, where the focus is both on understanding the implications of downside market scenarios for firms' business models and providing add-on framework for understanding clients' risk preferences.

   The reader will have definitely noticed that there is no reference to stress testing in the actual portfolio-management framework because traditional risk measures, such as targeted volatility or tracking error, are the mainstay of that process.[6] Even nonquantitative managers embrace the framework of making portfolio construction decisions to "optimize" the return-over-risk ratio even if they do it less formally than "quants." In this process, the risk component is some measure of volatility, not a stress test. The main reasons for this are that, first, you cannot typically manage long-term assets in scenarios with no known probabilities. And additionally, markets display positive long-term returns. Even though building defensive portfolios might protect investors against such stress events—assuming you could forecast their timing—clients will never have sufficient patience to wait for a doomsday manager to maybe, potentially, be right.

References

Ang, Andrew (2014), "*Asset management: A systematic approach to factor investing*," Oxford University Press.

Carhart, M. M. (1997), "On persistence in mutual fund performance," *The Journal of Finance*, 52, 57–82.

Fama, E. F., and K. R. French (1993), "Common risk factors in the returns on stocks and bonds," *Journal of Financial Economics*, 33, 3–56.

---

[5] Funds that measure their global risk following the VaR approach.

[6] The one exception to this is low-volatility portfolios (money market funds and stable-value funds). Volatility and tracking-error-type metrics are less useful because the numbers are small. For these types of products, stress testing may be the primary risk management tool.

15

Stress Testing for Insurers

Brian Peters[*]

1 Background on the Insurance-Industry Business Model

The insurance industry comprises a diverse set of firms that provide economically useful risk-mitigating services to individuals and corporations. One definition of insurance is "an arrangement by which a company or the state undertakes to provide a guarantee of compensation for specified loss, damage, illness, or death in return for payment of a premium."[1]

Insurance companies assume contingent liabilities in exchange for a premium. These contingent liabilities may have a number of different characteristics but are generally a tranched risk; that is, there is an amount of loss that the insured typically assumes, and there is frequently also a cap on the amount of claims that can be paid. The insurer will estimate the funds needed to cover future claims and establish reserves to cover these estimated claims. These reserves (or funds) are then invested in financial-market investments designed to defease the expected payoff profile of the claims.

Insurance can be written to cover a wide variety of possible risks. At the highest level, insurance companies can be segmented into life, health and medical, property and casualty, title, reinsurance, and other. Frequently, a distinction is made between target customer channels: commercial insurance or consumer insurance. Commercial insurance is written for firms and can cover a wide variety of potential liabilities, such as negligence claims, errors, and omissions, or protect against claims should its operations injure a member of the public or damage someone's property in some way. Consumers purchase insurance to cover possible health-care and end-of-life needs, to insure goods such as automobiles and homes, or to provide a monetary benefit to another interested party who is hurt or injured for which an insured is held accountable.

As with other financial services, the bulk of the existing business is in developed, mature markets, but the area of more rapid premium growth is in emerging-market countries. The United States is the world's largest insurance market, accounting for more than $1 trillion in annual premiums, which is split almost equally between life/health and property/casualty.

The insurance business is highly regulated. The economic foundation for regulation is based on the concept of market failure. Market failures in insurance are due to imperfect information, information asymmetries, principal–agent problems, entry–exit barriers, and externalities. The basic premise underlying the need for regulation is that market failures

[*] The author would like to thank Jacques Longerstaey, Kathy Anderson, Brad Fischtrom, Chris Graff, Mike Gatenby, Alessa Quane, Carter Su, and Ming Zhang for helpful edits and suggestions.
[1] *Oxford English Dictionary*.

can diminish the efficiency and equity of market outcomes and harm the public interest. Many products and services are produced and sold only if adequate insurance is available. Insurance coverage may be a condition for engaging in a particular activity. Because of the high risk of new business failure, venture capitalists often make funds available only if tangible assets and the entrepreneurs' lives are adequately insured.

Regulation typically applies to individual subsidiaries on a national basis. Within the United States, regulation of subsidiaries occurs at the state level. Regulators may proscribe accounting treatments that differ from generally accepted accounting principles (GAAP) or International Financial Reporting Standards (IFRS), limit investments that back insurance-company reserves, approve or reject product design, and determine minimum capital requirements.

Insurance policies can be written to cover claims made within the coverage period or for events that occur during the coverage period even if the problems don't emerge for an extended period (e.g., asbestos claims).

Some insurance is obtained for risks that are quite predictable and emerge with high frequency, such as automobile accidents. Other times, insurance is written to cover highly remote events that could have a very severe impact on the insured; in these cases, the underlying risk may be quite difficult to predict, and the risk is best managed by pooling a large number of highly diverse claims.

Property and casualty insurance is short-term in nature, although certain liability and casualty lines can have "long tails" beyond the accident year as losses emerge; aggregate loss exposures are harder to model and more uncertain. Life insurance firms offer coverage for mortality and long-term disability risks. Contracts are long term in nature. Aggregate exposure can be measured with reasonable precision but is subject to long-term trends.

In addition, insurance in many countries is also provided as an investment vehicle, often to provide income protection in retirement. This may be done through the issuance of such products as fixed and variable annuities, through the guarantee of pension plans (pension risk transfer), or through guarantees that enable money market funds to assure their clients that they may withdraw their money at par. Many of these products are complex bundles of options sold to, and purchased from, the issuer of the product.

Reinsurance firms essentially provide risk-sharing services to primary insurance firms. They may take *pari-pasu* risk through quota-share treaties or provide out-of-the-money tail-risk protection, such as protection against excessive natural catastrophe losses. Reinsurers seek to avoid concentrations of risk by creating portfolios diversified by the type and layer of risk, geography, and counterparty.

Finally, some firms, known as *composite insurers*, participate in multiple insurance segments and lines.

The activity of insurance companies is different from that conducted by banks and investment banks. Whereas banks intermediate between depositors with short-term liabilities (e.g., demand accounts or overnight commercial paper) and intermediate term assets (bank loans and bond inventory), insurers' business activities tend to occur further out the maturity curve. At the extreme, such as in the life insurance business, the duration of the policyholder liability may be 50 years or more, significantly longer than the fixed-income assets available to defease the risk. Even for shorter-term liabilities, such as fixed and variable annuities in the retirement segment, prepayment penalties and minimum guarantees significantly reduce the incentive for insurance purchasers to surrender their policies, leading to liability durations of at least 2 to 3 years.

2 Literature Review

The actuarial profession has a long history and large literature associated with estimating specific insurance risks, including techniques for determining distributions and combining distributions. Stress testing was frequently proscribed for the evaluation of individual books of liabilities by both the actuarial societies and regulators.

There is less specific literature on firmwide, comprehensive stress testing in the insurance industry. Colquitt (1999) is among the first to discuss the role of risk management at insurers. Guo (2008) discusses both the theoretical and the application aspects of stress testing in managing enterprise risks. Heale (2013) and Moormann (2014) discuss advances by the insurance industry in its stress testing practices, spurred in part by insurance regulators through the Solvency II capital standards and Own Risk & Solvency Assessment (ORSA) requirements (see National Association of Insurance Commissioners, 2017). The International Actuarial Association (2013), Hunt (2013), and Creech (2016) discuss further advances in practice that have occurred in the insurance industry.

Campion and Schuermann (2017) assemble a unique data set of risk factors relevant for insurers and use this data set to estimate risk-factor correlations to better understand their dependence structure. They find that correlation between nonfinancial risk factors is very low (usually insignificant), that between financial risk factors is on the order of 30 to 50 percent, and there is a mix between the financial and nonfinancial risk factors.

3 Risk Identification

One defining feature that makes insurance companies different from other firms is that they bear more material and diverse risks on the liability sides of their balance sheets. Not only do insurers face traditional asset risks, common to most financial firms, but they also have material liability exposures to a wide variety of nonmarket factors. Where banks function to ensure that depositors can provide on-demand access to cash at par, insurers exist to pay for contingent claims that may be made in known and unknown amounts at some uncertain future date.

Asset risk is similar to that at most other financial institutions, at though the investment portfolio of insurers is frequently constrained by regulation. The investments that back

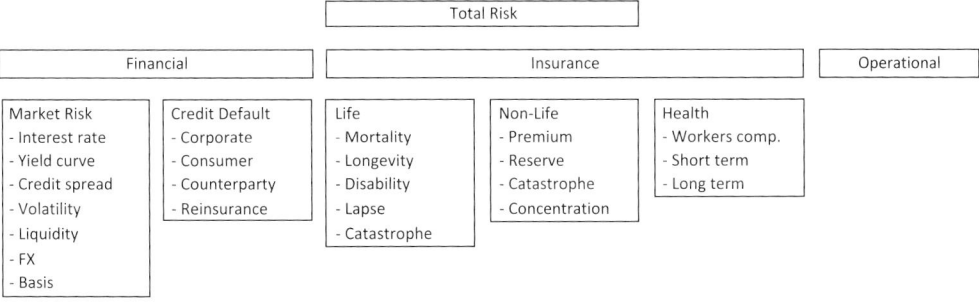

liabilities and surplus include investment-grade and high-yield corporate bonds; equity exposure through public equities, hedge funds, and private-equity investments; real estate in equity, debt and security form; foreign exchange (FX) exposure from both debt and equity positions as well as structural positions from owning the equity of overseas subsidiaries; and finally, to a lesser extent, commodity investments. Insurance companies with businesses that generate longer-term liabilities, such as life insurance and long-tail commercial lines, will have larger positions in less liquid investments, whereas businesses exposed to material catastrophe risk will have shorter-dated, more liquid investments.

For most asset risks, there is a reasonable amount of data to accurately estimate severe outcomes. In addition, under stress, most of these risks will move together in a correlated fashion; the "single systematic risk factor" will drive comovements of asset prices.[2]

The liability side of an insurer's balance sheet will reflect the actuarial reserves established; actuarial reserves represent the (frequently discounted) value of the future cash flows of an insurance policy, and the total liability of the insurer is the sum of the actuarial reserves for every individual policy. There are well-developed literature and practice in the methods used to establish these reserves.[3] The risk to the insurance company is that the level of reserves on the balance sheet may be judged in the future to be inadequate to cover known and anticipated claims. In these cases, the insurer will need to increase its reserves, through either income or a reduction in the surplus capital account.

The nature of the liability exposure at insurance companies will vary by business and product line. Overall, however, an insurer's liability stream has greater levels of idiosyncratic risk and lower levels of systematic risk than are observed on the asset side of the balance sheet. Because of the greater idiosyncratic risk, reserve estimation may be a noisy process; some insurers will carry some "redundancy" in their reserves to account for this noise. Insurers will typically use the amount of limit they are willing to extend, and facultative reinsurance, to limit the size of idiosyncratic losses that would affect reported earnings.

Additionally, where there are believed to be underlying systematic drivers of the risk in an insurer's liabilities, they may not be observable over short time horizons, have an observable data series, or be strongly correlated with other observed data. For example, changes in the litigious nature of the legal system, driven by changes in law and juror behavior, have led to changes in the frequency and severity of torts that affect certain insurance products. To control this risk, insurers devote time and effort to understanding how risks aggregate and how particular risks can affect a diverse group of insureds.

Some insurance liabilities are reasonably well behaved. They exhibit observable trending properties and have a reasonably low volatility around the estimate. These tend to be in areas where there is high frequency of claims (e.g., automobile insurance) or are thoroughly researched both by the industry and others in society (e.g., longevity or mortality risk). The downside of such observable, well-behaved business lines is that they are extremely competitive, with narrow margins, making deviations from expectations all the more

---

[2] This basic idea of a single latent systematic risk factor underlies, for instance, the Basel capital accord. See also Gordy (2003).

[3] For instance, the Actuarial Standards Board at www.actuarialstandardsboard.org, the American Academy of Actuaries at www.actuarialstandardsboard.org, or the Casualty Actuarial Society at www.casact.org.

important. Insurance is a business subject to strong adverse selection ("you win the business you underprice"), and it is important for insurers to understand the risk profile of their particular segment of the market.

Other types of insurance liability provide protection from infrequent, but very costly, events. Insurance that covers wind, storm, or flood damage from hurricanes, or structural damage from earthquakes, or protection from manmade catastrophes is an example of this type of insurance coverage. Here, the distribution of future outcomes is characterized by high variance, significant skew, and possibly fat tails. Category 5 hurricanes, or sizeable earthquakes that cause material damage, are not common occurrences in densely populated, high-value areas. A "bad CAT season" can bring claims significantly higher than budgeted or planned estimates. In these businesses, the risk is frequently estimated through the use of models or scenario analysis.

Finally, some insurance products, particularly those in the retirement segment that are designed to provide investment returns, are exposed to material customer behavior risk. Much like mortgages on the asset side of the balance sheet, where the borrower has the option to prepay their mortgage, investors in investment products such as fixed and variable annuities have the ability to surrender the policy. In both cases, customer behavior is studied in depth but changes over time. As interest rates and equity prices change, the incentive for the customer to surrender changes as well. This introduces optionality that the insurance firm can attempt to hedge or bear. A fundamental risk arises because insurance companies, based on experience, assume that policyholders will not uniformly exercise these options optimally. If policyholder behavior is more efficient than assumed, that is, if the insurer has misspecified an assumption in its model of customer behavior, the insurance company is at risk of loss.

Most firms will rely on internal data and claims-loss experience to appropriately estimate the risk arising from their liabilities. In some cases, it is possible to access historical data on some risk factors. For instance, the Society of Actuaries provides mortality experience studies on Americans residing in the United States, and the National Council on Compensation Insurance is a source of information on worker's compensation claims. However, for many types of insurance exposures, there is little publicly available information. This is particularly the case for new product coverage and emerging risks; examples include cyber-disruption, climate change, and changes to the social and political environment (trade wars, #metoo).

There is little literature on the causes of operational-risk losses in the insurance industry. Intuition would suggest that losses would be driven by a combination of asset-related and liability-related drivers. The most material operational-risk losses in the industry would relate to legal settlements and regulatory fines associated with mis-selling of consumer products or with insurance-treaty language, such as whether the attack on the World Trade Center represented one covered event or two.[4]

---

[4] In litigation, the leaseholder for the two World Trade Center towers claimed that he was entitled to recover $7.1 billion from the 22 insurers of the properties, twice the ostensible policy limit, on the ground that the attack of the center was two occurrences, not one. Two different trials ended with different interpretations of whether one or two events had occurred.

Finally, Campion and Schuermann (2017), using public data, find that "inter-risk correlation among financial factors is meaningfully positive at around 30 to 50 percent (this
is no surprise and already well known), among nonfinancial such as insurance factors is
relatively low, if at all significant, and between financial and nonfinancial it is mixed. There
are some instances of correlations of similar magnitude as among just the financial risk
factors, for instance P&C non-catastrophe with equity and credit losses, but most of the
pairwise correlations are low to insignificant." This aligns well with industry participant
intuition (hurricanes do not cause material life insurance losses or financial collapses, but
a financial collapse may, over time, cause an increase in worker's compensation claims
through heightened unemployment) and is an important factor in scenario design.

4 Scenario Design

Designing stress scenarios for insurance companies can be more difficult than for other
financial firms precisely because they are subject to wider sources of risk, because the data
on these risks may be less comprehensive or well understood, and because their intermediation role is performed further out on the maturity spectrum. It is important in this context to
understand what risks a stress scenario can illustrate and what it cannot illuminate.

As mentioned earlier, much of the intermediation that occurs at insurance companies is at
the medium to long end of the maturity ladder; stress scenarios are often best at illustrating
risks that may become apparent in the near term, say, the next few years. The design of
stress tests will vary both by the risks that the firm is trying to highlight and the nature of the
firm's activities; trading firms with capital-market-intensive business will emphasize short-
term scenarios that highlight leverage and market liquidity, whereas banking firms will have
longer horizons, typically measured over 3 to 5 years. Insurance companies' scenarios will
depend on the mix of business, but divergences and stresses over longer horizons will more
generally align with many insurers' exposures.

As with other financial firms, the design of the stress scenario will usually commence with
an assessment of the current economic environment, and a consideration of the potential
changes in the financial environment over the forecast horizon that would most challenge
the firm. For most insurance firms, the investment portfolio will be by far the largest source
of financial-risk sensitivity. In designing a scenario, firms will want to subject their most
concentrated asset position, say, commercial real estate debt and equity, to a recessionary
macroeconomic environment that specifically affects the concentrations.

For firms that participate in the provision of retirement services, the level and steepness of
the yield curve will be important drivers of their business strategy. Firms with large blocks
of business that provide high minimum guaranteed interest rates on retirement products may
want to illustrate this vulnerability by lowering rates below guarantee levels. Conversely, a
an environment with rapidly rising rates, perhaps associated with a flattening of the yield
curve, could be useful to evaluate policy-surrender behavior and new business volumes in
an environment of net-interest-margin compression.

For most firms, however, the primary financial scenario will involve some replication of
a severe financial downturn with a deterioration in both consumer and commercial credit
performance.

Single-theme stresses, such as a severe financial downturn, although having the potential to severely affect the financial performance and solvency of more highly leveraged monoline insurers, will rarely be material enough to threaten the solvency of major diversified commercial insurers precisely because insurers are managed, regulated, and capitalized to meet longer-term obligations. To meaningfully affect these firms will require stressing additional factors, such as including adverse reserve developments, catastrophe losses, or significant operational risk events, or meaningfully affecting new business or renewal volumes, such as through a material legal or reputational event.

Designing multifactor scenarios can be tricky because there may be multiple paths to a high-confidence-interval outcome. The more diversified the firm's businesses, the more sources of independent risk, and the more choices one needs to make in developing a scenario. The design of such a scenario can feel unrealistic; what is the chance that a severe earthquake destroys a large portion of San Francisco during a severe recession while the firm finds itself needing to increase its reserves for some long-tail line?

Finally, the choice of a time horizon is important. In general, the longer the scenario, the more operationally difficult it will be to run. Given that risk factors affect both the asset and liability sides of the balance sheet, projecting these balance sheets and (re)investing reserve balances in financial assets becomes a very dynamic exercise that is highly dependent on when certain events happen. As we shall see, this is particularly important when it comes to modeling subsidiary regulatory capital. Although the easiest assumption is that events happen simultaneously, there is a benefit to understanding the temporal diversification that occurs even over relatively short horizons (say, 2–3 years), such as when faced with uncorrelated events like hurricanes and earthquakes.

Many risks at insurance companies are not amenable to risk management strictly through short-term scenarios. A short-term scenario will not highlight whether pricing for long-tail risks is sufficient for emerging losses, nor whether the value accumulation required for longer-term payout requirements (life; income protection products) is sufficient; for these, other risk management techniques will be more important. It is always necessary to keep in mind the types of risk developments that scenarios can readily highlight and those that will require other techniques.

5 Translation of Scenario into Economic Outputs

One decision that insurance companies face when designing and running scenarios is the choice of whether to model results in purely economic terms, in aggregate under GAAP or IFRS accounting, or whether to model the results at the subsidiary level using the local regulatory accounting standards. This decision is particularly relevant in the United States for life and retirement products, where statutory accounting can differ materially from economic and GAAP accounting views.

Insurance is a heavily ring-fenced industry, generally at the national level, although even more granularly at the state level within the United States Ring-fencing aligns regulation, law, entity capital and reserves, and resolution funding with the risks and capital structure of individual legal entities. Regulators often seek to coordinate their activities and regulations across jurisdictions to minimize the costs of ring-fencing. The information created by running stress tests at the legal-entity level is often quite important in ensuring individual subsidiary solvency. This, however, can be extremely complex for firms operating in multiple jurisdictions under disparate regulatory accounting and solvency regimes. Legal-entity balance sheets will behave differently under a US statutory approach than a more market-sensitive IFRS approach.

If modeling subsidiary solvency, the firm is then faced with the challenge of projecting how required capital evolves over the time steps of the scenario as asset and liability exposures change. Deteriorating asset quality will raise statutory capital requirements in the United States, and the projected market-value changes of both assets and liabilities will affect the solvency ratios of firms operating under Solvency II or similar regimes. Further, if modeling solvency at the subsidiary level, the firm must establish decision rules around the minimum level of capitalization and around actions that will be taken should a subsidiary fall below a target capitalization level. In practice, although the measure of solvency may be a regulatory-defined minimum capital ratio, firms will usually target higher levels of solvency capital calibrated to their understanding of rating-agency and counterparty expectations.

Firms may choose to run the scenario under both GAAP and regulatory approaches.[5] This raises issues of reconciliation and the ability to explain the results of the stress test clearly for senior management.

Finally, as if this weren't complicated enough, the International Association of Insurance Supervisors (IAIS) is developing, under a mandate from the Financial Stability Board, an Insurance Capital Standard with its own set of rules that will apply at the consolidated level. In late 2019, the IAIS adopted the so-called "ICS 2.0" for a 5-year monitoring phase, during which international groups are expected to generate confidential estimates of the ICS ratio, in order to help the IAIS in assessing its design, methodology, calibration, and potential market impact.

Projecting asset values through the stress scenario is fairly straightforward. Insurance-company assets are financial instruments, although there may be a larger portion of less-liquid assets than one would find in other financial firms.

The liabilities of insurers are reserves set aside to cover possible claims expenses. The potential volatility of these reserves is what introduces risk. Volatility can be caused by the emergence of hidden drivers of loss or by misestimation of future claim estimates. Some liability risks are driven by economic factors, but many are not. Life and retirement firms will offer fixed- and variable-rate annuities, frequently with guarantees, whose value will be driven by the level of debt and equity markets, as well as options volatility. Some other liabilities will have no correlation with economic variables; earthquakes and hurricanes are not caused by changes in financial markets, and the evidence suggests that they have little influence on future financial returns. Finally, there are liabilities that have some degree of correlation to economic performance, although frequently weak and often with lags. Worker's compensation claims tend to rise with unemployment rates, and over the long term, they could be affected by medical inflation. Claims on directors' and officers' indemnity may be correlated with economic performance—but with a lower sensitivity and with a material lag. And, unfortunately, suicide-related deaths rise when economic conditions are poor.

---

[5] Firms may also need or want to model their performance under the guidelines published by the rating agencies because often these define the operating constraints for the firm.

In designing a scenario, the firm must decide how much idiosyncratic risk is assumed to occur; does the firm experience unexpected reserve development, or is it an abnormally bad year for natural catastrophes in the midst of a difficult financial environment? These decisions affect the outcome of the scenario, both by imposing losses and by requiring the firm to add additional financial assets to support increased reserve levels.

6 Complications in Scenario Execution

As previously mentioned, it is often critical for insurance companies to model the outcome of a stress scenario at the regulated-subsidiary level to ensure ongoing solvency or estimate the degree of parental support required.

Another complication prevalent within the insurance industry is the use of reinsurance. The insurance history has a long history of using reinsurance to manage its risk (facultative reinsurance) or obtain tail-risk protection (through reinsurance treaties with out-of-the-money attachment points). In this way, reinsurance has served as an important source of financing and capital for insurable risks.

Reinsurance may be acquired on a proportional basis, known as "quota share," or on a tranched basis, known as "excess of loss." A quota-share treaty is a pro rata reinsurance contract in which the insurer and reinsurer share premiums and losses according to a fixed percentage. Excess-of-loss reinsurance is a type of reinsurance in which the reinsurer indemnifies the ceding company for losses that exceed a specified limit. Excess-of-loss reinsurance is a form of nonproportional reinsurance.

This is also a potential source of counterparty risk because potential exposure to reinsurers may be among the larger credit exposures facing a firm.[6] In the main, purchased reinsurance will serve as a loss-absorbing resource for the firm; however, modeling of the timing of receipt of reinsurance recoverables and the effect on both capitalization and investments will complicate the execution of the stress scenario. Additionally, firms need to consider the effect of the stress on the reinsurance counterparty and its ability to honor its obligations.

Firms may also use internal reinsurance treaties to shift risk from one subsidiary to another through pooling. This allows individual subsidiaries to offer larger amounts of coverage to their customers than their standalone capitalization would permit. For instance, a firm may maintain a subsidiary in a foreign jurisdiction where a large international customer requires coverage for a particular project; the subsidiary could issue the policy and reinsure the risk back to the pool. The risk of the pool will be borne by its owners, one or more of the larger and better-capitalized subsidiaries within the group. On the whole, the proper use of internal reinsurance allows firms to more accurately align risk to capital. Here, too, although serving as a valid risk management tool, internal reinsurance can complicate the execution of the scenario, depending greatly on the specific design of the scenario. In a firm where the scenario results are executed in a decentralized manner, understanding how the specific scenario losses propagate through the internal reinsurance structure requires considerable attention to detail.

---

[6] As a result, one risk frequently considered in scenario design is the failure of a major reinsurer to pay on reinsurance receivables.

Understanding customer behavior on the liability side of the balance sheet is also a complicating risk. Just as banks must understand the behavior of their credit-card customers and mortgagors in response to changing economic conditions, so must insurers that provide life and retirement products. As stated earlier, many of the products offered to customers give those customers options. Frequently, there is an option to surrender the policy under changed economic conditions; in some cases, the insurance company will provide guarantees of either minimum income, a minimum interest crediting rate, or a future benefit. Oftentimes, these product benefits will be relatively recent innovations, and the firm will not have the benefit of having observed behavior through different economic conditions. In general, modeling these features for a portfolio of products could be very time consuming because the performance of the portfolio will depend greatly on its starting condition. For instance, a portfolio of policies guaranteeing a minimum interest rate will react differently depending on whether the guarantee is above or below current market rates, the size and duration of any prepayment penalty associated with the product, and nonfinancial factors such as tax status and stage of life. Additionally, customer behavior may exhibit behavioral patterns that are not economically optimal for them. As a rule, the nature of the guarantees offered under insurance policies is considerably more varied than one would find on assets such as mortgages.

In contrast to other financial firms, the linkage between the liability and asset sides of the balance sheet is more difficult for insurers to model. As a rule, and frequently as a regulation, insurers face strict requirements for cash-flow matching. Depending on how the economic scenario affects both policy surrenders and new originations, insurers could face demands to purchase or sell specific assets related to the targeted duration of the liability line in order to meet proscribed rules for cash-flow matching.

Many insurers seek to minimize the need to frequently rebalance their asset portfolios; this is done both to minimize transaction costs and to manage the recognition of capital gains and losses. This is particularly the case where the firm invests in less liquid assets. This introduces complications in modeling management action during the economic scenario. It requires some holistic thinking about the firm's balance sheet, and not just a line-by-line analysis and modeling of economic effects. For instance, a firm may decide that rather than selling assets in response to surrenders in a particular product line, it will offer a product with reduced profit margins to maintain the stability of its asset portfolio.

Both the tight linkage between liabilities and assets and the considerations around management action make the reinvestment decisions in a multiperiod scenario challenging for insurance stress scenario modeling.

As discussed earlier, idiosyncratic risks represent a large portion of the liability risk of insurance companies, and these risks generally do not have macroeconomic drivers. In modeling a scenario, firms must make a choice of whether to estimate these risks by embedding a statistically derived estimate of loss or whether to model specific events, such as a hurricane affecting an area of property insurance exposure.[7] Often, insurance firms will use smaller, specific scenarios to examine the "accumulation risk" in their portfolios that occurs from an event that affects multiple policies, often across multiple product lines.

---

[7] This would be done by selecting a particular path from the stochastically generated distribution of the desired risk.

A recent example of one such risk is coverage for cyber-risks, which will be covered in certain specifically written policies but where liability may also arise from other types of contracts. Another example would be exposures that arise from a costly potential new drug, which could extend clients' lives materially but requires considerable annual expenditures.

7 Stress Testing and Regulation

Regulators have found stress tests to be a useful tool for determining the risks to, and financial strength of, insurance firms, particularly following the financial crisis of 2008. Regulators have set out expectations for a firm's own stress testing processes, as well as requiring firms to conduct regulatorily proscribed stress testing scenarios.

Following the financial crisis, the IAIS in 2010 developed a set of Core Principles for the regulation of insurers. Insurance Core Principle 16 (ICP16), for example, established a requirement for insurers to conduct an ORSA. Although the specific requirements vary based on local regulatory implementation, the ORSA has established a global expectation that large insurers will adopt stress testing as an integral part of their risk management practices. Although neither ICP16 nor the ORSA specifies the form and nature of the stress test, local regulatory implementation documents may prescribe specific details, such as whether single-period shocks are acceptable or whether multiperiod scenarios are required.

In Canada, the Office of the Superintendent of Financial Institutions (OSFI) was among the first insurance regulatoras to promulgate guidance on stress testing governance and methodology in its 2009 publication on sound business and financial practices (Guideline E-18).

Overall, stress testing guidance has been issued by numerous national jurisdictions, including Austria, Bermuda, Canada, Czech Republic, Denmark, Germany, Japan, Singapore, Switzerland, the UK, and the United States.[8]

By virtue of its status as a BHC, the Federal Reserve subjected Met Life to its Supervisory Capital Assessment Program in 2009. This was a regulator-prescribed severe multiperiod financial stress scenario designed principally to ensure that banking firms hold sufficient capital to weather a repeat of the 2008 financial crisis.

The Committee of European Insurance and Occupational Pensions Supervisors (CEIOPS) conducted its first stress test exercise in 2010 of 28 major European insurers. Beginning in 2011, CEIOPS or its successor, the European Insurance and Occupational Pensions Authority (EIOPA), has conducted follow-up stress tests approximately every 2 years. Three stress scenarios were tested in the 2018 exercise, two scenarios combining market- and insurance-specific risks and one natural catastrophe scenario (EIOPA, 2018).[9]

Beginning in 2015, the Bank of England's Prudential Regulatory Authority (PRA) subjects regulated firms to a biennial stress testing requirement, the General Insurance Stress Test (GIST). The PRA's stress tests covered natural and man-made catastrophes as well as an economic shock. The PRA seeks to align the economic shock with the scenario also

---

[8] Jobst (2014) provides a detailed overview of regulatory stress testing regimes.
[9] See https://eiopa.europa.eu/financial-stability-crisis-prevention/financial-stability/insurance-stress-test?
_sm_au_=iHVW6krLjVHFrDV6k4ctvKsHftFMW.

proscribed for completion by banks. Most recently, the PRA's 2019 stress test sought to cover such topical risks as cyber-underwriting losses and climate-change risk.[10]

Stress testing has also been endorsed by the International Monetary Fund and adopted in its Financial Sector Assessment Program (FSAP). The purpose of FSAP stress tests differs from those conducted by firms or national regulators, which are generally designed to observe entity-level shortfalls because they are designed for macroprudential surveillance purposes.

8 Conclusion

Insurers face many challenges similar to those of other financial firms. However, the nature of insurance, with large idiosyncratic risks arising on the liability side of the balance sheet and a highly ring-fenced regulatory regime with multiple accounting lenses, adds considerable complexity to the design and execution of stress scenarios. No single stress scenario is likely to fully highlight the risks to a large multiline insurer, and the nature of the risks complicates the choices that need to be made in scenario design.

References

Campion, Scott, and Til Schuermann (2017), "Risk dependence, solvency and stress testing for insurers," Working Paper, available at https://ssrn.com/abstract=2906129 or http://dx.doi.org/10.2139/ssrn.2906129.

Colquitt, H. A. (1999), "Integrated risk management and the role of the risk manager," *Risk Management and Insurance Review*, 2(3), 43–61.

Creech, M. (2016, November). "Stress testing," *Actuary of the Future*, 39, 23–24.

European Insurance and Occupational Pensions Authority (2018). "EIOPA launches the fourth EU-wide insurance," available at https://eiopa.europa.eu/Pages/Financial-stability-and-crisis-prevention/Stress-test-2018.aspx.

Gordy, Michael (2003), "A risk-factor model foundation for ratings-based bank capital rules," *Journal of Financial Intermediation*, 12(3), 199–232.

Guo, L. (2008), "Effective stress testing in enterprise risk management," Working Paper, Society of Actuaries.

Heale, B. (2013, September), "Stress and scenario testing: How insurers compare with banks," *Moody's Analytics Risk Perspectives*; Stress Testing European Edition, Vol. 1.

Hunt, I. W. (2016), "Session 178 PD: The current and future state of stress testing," Presentation at the Society of Actuaries Annual Meeting, Las Vegas, NV.

International Actuarial Association (Association Actuarielle Internationale) (2013, July), "Stress testing and scenario analytics," available at www.actuaries.org/CTTEES_SOLV/Documents/StressTestingPaper.pdf.

Jobst, S. A. (2014), "Macroprudential solvency stress testing of the insurance sector," International Monetary Fund Working Paper WP/14/133.

Moormann, L. (2014), "Stresses and scenarios in the context," Munich Re Solvency Consulting Knowledge Series.

National Association of Insurance Commissioners. (2017), "NAIC Own Risk and Solvency Assessment (ORSA) guidance manual," available at https://content.naic.org/sites/default/files/publication-orsa-guidance-manual.pdf. www.naic.org/cipr_topics/topic_own_risk_solvency_assessment.htm.

---

[10] See www.bankofengland.co.uk/prudential-regulation/letter/2019/insurance-stress-test-2019.

Model Risk Management in Stress Testing and Capital Planning

Eduardo Canabarro

Introduction

Since 2009, *stress testing and capital planning* have been at the center front of the assessment of the capital adequacy of financial institutions in the United States. In the DoddFrank Act Stress Test (DFAST) and Comprehensive Capital Analysis and Review (CCAR),[1] the stress test is over a nine-quarter time horizon, and its context is a set of adverse and severely adverse economic conditions described by macroeconomic and market variables: gross domestic product (GDP) growth rates, unemployment rates, disposable income, inflation, interest rates, equities, commodities and foreign currency prices, market volatility, credit spreads, real estate prices, and so forth.[2]

The stress scenarios themselves, as well as their consequences for the banks' revenues, expenses, market/credit/operational losses, balance-sheet levels, capital requirements, and so forth, are estimated by many, sometimes hundreds of, econometric models. The heavy use of models renders the results of the stress tests substantially dependent on the quality of the models used to calculate the stress test by both the Federal Reserve and by the banks. Hence, it is not surprising that, concomitantly with the evolution of stress testing and capital planning, a new risk management discipline, called *model-risk management (MRM)*, has developed much and fast.

MRM establishes the risk management processes and controls to identify, contain, and report *model risk*, that is, the risk of incurring losses and/or making poor decisions because of the results of defective or misused models.

In finance and economics, the overall context of the application of models is distinct from the one in the sciences like physics, chemistry, and biology.[3] The laws of nature are *immutable*, and the scientific models (e.g., the ones built by Newton, Maxwell, Einstein, Dalton, Bohr, Darwin, etc.) aim at representing the natural laws that apply to the scope of their interest. The scientific models simplify reality by focusing on the most important features of the topic. The models are tested, retested, and then tested again against *immutable nature*. The models evolve and get better as the cumulative learning from their

---

[1] See the Federal Reserve Board DFAST's webpage at www.federalreserve.gov/supervisionreg/dfa-stress-tests .htm. See the Federal Reserve Board CCAR's webpage at www.federalreserve.gov/supervisionreg/ccar.htm.
[2] See the stress test scenarios developed by the Federal Reserve for the 2020 DFAST/CCAR cycle at www .federalreserve.gov/newsevents/pressreleases/files/bcreg20200206a1.pdf.
[3] See Derman (1996, 2012).

successes and failures in the tests leads to a better understanding of the laws of nature and to the improvement, and sometimes the full redesign, of the models.

The laws that govern economics and finance are not as precise and immutable as the laws of nature. They change over time and across different economic contexts. The data collected in the past may not be relevant to forecasting the economic features of the future. The *art of financial-economic modeling* is to identify the features that are resilient, sufficiently constant, and stable to have useful *predictive power* about the future. My own experience has been that financial economists and business practitioners tend to presume that the factors that drive economics and business are more stable and predictable than they are in reality. That is, they assume that the regularities they observe in the historical data are more repeatable than they are. As a result, their models are often caught in error, and they explain that their models failed because of the occurrence of "very extraordinary," "six-sigma" events or "black swans." The reality is that their models simply have not embodied the wide range of the potential outcomes, and they have underestimated the probability and severity of extreme events. The black swans appear too often, more often than they expect.[4]

> The aforementioned considerations are fundamental for the understanding and implementation of MRM in the context of stress testing.

By definition, stress tests examine exceptional economic and market circumstances. Models, which are calibrated to data obtained under *normal economic conditions*, are often used to estimate economic and market developments in *stressed conditions*. The working assumption is that the statistical relationships (e.g., the linear regression models) and regularities that exist in normal conditions will persist in stressed conditions. This is, more often than not, simply not the case.

> MRM in stress testing requires a fair amount of context and caution. The model builders and the model users need to understand the *big picture* of the limitations and uncertainty of their models and treat the models' results as subject to *substantial potential error*. As a result, the assessment of banks' capital adequacy based on stress tests needs to account for the inherent and unavoidable model uncertainty and apply the necessary allowances for conservatism.

This is the main theme of this chapter: **the model risk inherent to stress testing**. Herein, I will describe the risks and the risk management processes to mitigate it.

This chapter also covers the following:

- The evolution of the banks' use of models and of MRM as a discipline from the mid-1980s until now, helping to set the context.
- The *taxonomy of the models* used in stress tests and the different risks that each type of model faces.
- Guidance for model development, testing, and documentation.
- Guidance for model validation and governance.

---

[4] See Taleb (2007).

- The propagation of model risks through systems of interdependent and interconnected models.
- The presentation of the models' results in the context of their model risks and uncertainties.

Throughout, I will assume that the reader of this chapter is not a *quant* or a *modeler*, and I will use language that is accessible to a wide range of finance and economics professionals whose activities may involve the use of models. Also, and as usual, I will try to be concise and direct, avoiding lengthy text and financial jargon as much as possible.

1 A Historical Perspective

The 1980s were a challenging time for US banks. The decade started and ended with economic recessions, and it displayed an extraordinary upsurge in banks' credit losses. Several large banks failed during this decade, and the entire savings and loans industry eventually had to be rescued by the US government. Because of (a) the large losses associated with their lending, (b) the increased competition by nonbank institutions, and (c) the relaxation of protective regulations dated from the time of the Great Depression, US banks had to adjust their business models and strategies in order to generate income from sources other than their core business of supplying credit.[5, 6]

When I started my career in Wall Street in 1993, the use of models by banks was already extensive, and it was expanding very rapidly. The models were primarily used for trading and investments, not so much for risk measurement and management yet. It was a period of enormous innovation in finance and financial markets, with many new structures (financial derivatives) and a robust sense of confidence that the information technology available (including the models) enabled the financial institutions to manage the complexity and risks of their expanded set of activities.

The option-pricing model and the new finance theories published by Professors Fischer Black and Myron Scholes[7] put financial innovation into high gear. Their work was part of a prolific body of research[8] produced by a number of finance and economics professors since the late 1950s: Paul Samuelson, Harry Markowitz, John Lintner, William Sharpe, Merton Miller, Eugene Fama, Franco Modigliani, Robert Merton, John Cox, Stephen Ross, Mark Rubinstein, and many others. Their theories and techniques emerged just at the right time, when technology and institutions were ready for them, and they reverberated powerfully throughout Wall Street.

There were stumbles along the way, such as those in 1987, 1994, and 1998, but quick market recoveries ensued, and these episodes reaffirmed and strengthened the sense of confidence that finance technology (models and information processing) was evolving fast—getting better, more robust, and more prepared to handle the challenges of the future. With the exception of a very small number of financial institutions, MRM was still rudimentary or nonexistent in the 1990s: models were not well documented, and many models were designed, programmed, and managed by their users themselves. Also, there was very little

---

[5] See Federal Deposit Insurance Corporation (1997).
[6] See U.S. Congress (1980).
[7] See Black and Scholes (1973).
[8] See Merton (1997).

transparency on the methodologies. The models were mostly black boxes that only the very few people close to them could understand. Sometimes even these smart people did not fully understand the models, even though they thought they did, unfortunately. There were elements of competitiveness and secrecy too: if you are smart to construct these models, you should keep them secret so that your competitors cannot copy or replicate them.

Long-Term Capital Management (LTCM) is the poster child of this era.[9] A dozen or so super-smart Salomon Brothers' traders led by John Meriwether left the bank in 1993 to start their own hedge fund. Several of them were modelers with PhD degrees from MIT and other prestigious economics and business schools across the United States. They strongly believed in their statistical models, which were designed to detect abnormalities across the market prices, which they expected would (relatively quickly) revert to more regular historical patterns. Their trading strategies were wagers on *convergence* (i.e., return) *to regular market patterns*.[10] LTCM had an extraordinary run between 1994 and 1998, with the capital of its partners growing by a factor of four and then completely collapsing in the course of 3 months in the late summer of 1998 (see Figure 16.1). LTCM failed because its trades, and similar trades of other market participants that were emulating LTCM's successful strategies, had become so large in the marketplace that their size *warped the dynamics of the markets*, analogously to the way a large amount of mass or energy warps the texture of space-time. The convergence back to the patterns of the past did not materialize in the way and as fast as such convergences had in the past because the market participants were waiting for LTCM (and other large financial institutions) to liquidate their leveraged positions at a substantial loss. The LTCMs case was an unequivocal reminder of the model risk inherent to *financial-economic forecasting models*: the possibility that new (often endogenous) factors that had not yet been seen in the historical data suddenly appear and render the forecasts of the statistical models wrong, sometimes very wrong. In statistical terms, this means that the dynamics of the markets and the economy may not be stationary: market structures and players' preferences and actions change and adjust to new and developing scenarios in ways that are not represented in the historical data that were used to construct the models.[11]

Notice that I am not speaking of stationarity here in the sense that econometricians usually do. They examine properties of their data sets and draw conclusions about the regularities that they identify. I am saying that the data set itself may be highly incomplete and not representative of the complete dynamics and distribution of the potential outcomes of the economy. Thus, we get "surprises" all the time.

Similar problems with other econometric models had already been observed in 1987 and 1994. The stock market crash in October 1987 showed the vulnerability of the dynamic replication strategies implemented by *portfolio insurers*. The liquidity of the stock market vanished, and the hedges could not be rebalanced in time as the models required them to be. Mark Rubinstein and Hayne Leland, my finance professors at U.C. Berkeley, were among the main protagonists in that episode. In 1994, it was the defective forecasts of interest

---

[9] For a good description of the LTCM episode in 1994–1998, see Lowenstein (2000). This is the source of the graph in Figure 16.2.

[10] See Jorion (2000).

[11] See the analysis of the LTCM's debacle as written 10 years after the events: www.nytimes.com/2008/09/07/business/worldbusiness/07iht-07ltcm.15941880.html.

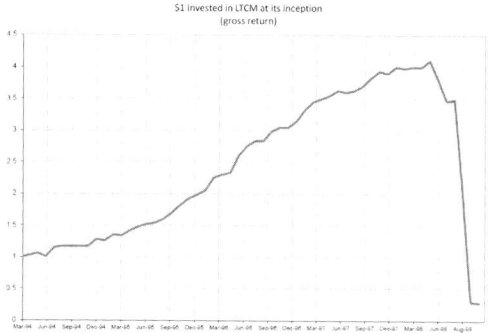

Figure 16.1 Value of a hypothetical $1 share of LTCM in the period 1994–1998: the appreciation from $1 to $4 and then the crash to $0.4 in the course of 3 months. Source: Lowenstein (2000), as cited in note 9.

rates and mortgage prepayments in an environment of fast-rising interest rates and their consequences to the valuation of leveraged bonds like structured notes and collateralized mortgage obligations (CMOs), such as principal-only (PO) and interest-only (IO) securities. The risk profiles of these securities changed massively and rapidly, and the losses were large. The debacle of Orange County (California) and its painful experience with its structured notes is a good example of the challenges of that time.[12] Then, similar problems would occur yet again in the Global Financial Crisis of 2007–2009, but much more severely this time around, now related to the *credit losses of mortgages* and their consequences to the leveraged structures: the collateralized debt obligations (CDOs) like asset-backed securities (ABSs) and commercial mortgage backed securities (CMBSs).[13]

What do all these situations have in common? Several features:

1. Extensive use of *forecasting models* that were based on the regularities contained in the past data.
2. The emergence of *new factors* that changed the market's dynamics observed in the past.
3. The catastrophic failure of *leveraged structures* (portfolio insurance, LTCM, IO/PO securities, CDOs, structured notes) that had used the results of the forecasting models to bet on the regularities of the past market patterns.

> A key lesson for model risk managers: the statistical models used in finance and economics are subject to large forecast errors, and leveraged bets on their predictions can lead to very large losses.

In other words, finance and economics are not like physics, where nature is precise and immutable. I will repeat this contrast many times in this chapter because many of the modelers who currently work in Wall Street have been trained in mathematics, physics, and other sciences. They tend to have more faith in econometric models than they should.

[12] See the report on Orange County's bankruptcy at www.nytimes.com/1994/12/08/business/orange-county-s-bankruptcy-the-overview-orange-county-crisis-jolts-bond-market.html.
[13] See Ashcraft and Schuermann (2008).

From 1980 to 2008, banking regulation of capital adequacy also became largely based on banks' internal models. The Market Risk Amendment (MRA) of 1996[14] to the first Basel accord (Basel I) allowed the banks to use their value-at-risk (VaR) models to quantify the risk of the changes in the prices of their trading positions in the calculation of the required regulatory capital. In the late 1990s, the Basel Committee on Banking Supervision (BCBS) initiated its Basel II project to expand the use of the banks' internal models to credit risk, derivatives counterparty risk, and operational risk. It was an exciting time for modelers in the risk management departments of the banks, and there was a strong and widely shared belief by bankers and regulators that models, despite all their already-observed shortcomings, were the most effective way to quantify and manage the multiple and complex risks faced by the largest and most sophisticated financial institutions across the globe.

In my view, the Basel MRA of 1996 was a main contributor to the Global Financial Crisis: it created strong economic incentives for banks to take and accumulate risk, especially credit risk, in their trading books by assessing insufficient regulatory capital on trading risks, much lower than the same risks would command if they were in the banking book. The trading risks were measured by the VaR models of the banks, the risks diversified away within the models, and the capital charges were low. The banks' trading books got loaded with credit risks that were marked to market, which created spirals of fire sales during the most acute run-to-the-exits phase of the crisis in September–October 2008.

In 2000, the Office of the Comptroller of Currency (OCC) issued a bulletin with "guidance to help financial institutions mitigate potential risks arising from reliance on computer-based financial models that are improperly validated or tested." The guidance outlined key model-validation principles and the OCC's expectations for "a sound model validation process." The guidance, OCC 2000-16, emphasized the importance of the independent and rigorous validation of models to ascertain that they were properly inventoried, documented, tested, and applied.[15]

In April 2011, after the Global Financial Crisis of 2008 and reflecting the model failures that occurred then, the Federal Reserve Board and the OCC jointly issued more comprehensive guidance on MRM: SR 11-7 or OCC 2011-12.[16] This guidance has been the cornerstone of the developments in MRM since then. DFAST and CCAR provided the setting for the implementation of SR 11-7, and they created a strong sense of urgency and the requirement for mandatory compliance with the guidance: banks that did not have good MRM failed the stress tests "qualitatively," and they were precluded from paying stock dividends and pursuing share buybacks.[17] Citibank was a prominent example of this when it failed CCAR for the second time in March 2014.[18]

---

[14] See BCBS (1996).

[15] See OCC (2000).

[16] The Federal Reserve's SR 11-7 (2011) is available at www.federalreserve.gov/supervisionreg/srletters/ sr1107.htm; OCC 2011-12 (2011) is available at www.occ.gov/news-issuances/bulletins/2011/bulletin-2011-12a.pdf.

[17] See Brown et al. (2015).

[18] See Federal Reserve Board (2014). The Federal Reserve wrote: "The Federal Reserve's objection to Citigroup's CCAR 2014 capital plan in part reflects significantly heightened supervisory expectations for the largest and most complex BHCs in all aspects of capital planning. While Citigroup has made considerable progress in improving its general risk management and control practices over the past several years, its 2014

Europe has lagged in MRM implementation. Mainly because of the required coordination across the many countries in the European Union, the Basel Committee's rules have been the preferred path to deal with banking regulation. Only in 2017 did the Bank of England issued a consultative paper with a set of principles for the MRM of the models used in its stress tests. The principles are aligned with the US guidance, and they were published in 2018 as the Prudential Regulation Authority's Supervisory Statement 3/18.[19]

In parallel with the MRM developments, the BCBS's regulatory capital frameworks took a new course after 2008. Differences in the capital calculated by the various and different banks' internal models led to a change in the direction toward more standardization of the regulatory capital framework. Regulators have become more prescriptive about the features of the internal methodologies (e.g., as in Basel 2.5), imposed floors on the capital levels calculated by the models, and completely eliminated the use of internal models in certain areas (e.g., in securitizations). Standardization has also been advanced for operational and credit risks.

Finally, summarizing the MRM context as I write this text in 2020:

- Models continue to be extensively used by banks, for valuation, risk management, and other applications.
- Stress testing and capital planning use models extensively, for both scenario design as well as calculation of the impact of the scenarios on the banks.
- Regulatory capital frameworks (Pillar 1) are becoming more standardized, based on specific prescriptions and less reliant on banks' internal models.
- MRM has emerged as new discipline in risk management at the same level of importance as the management of market, credit, operational, and liquidity risk.

2 Types of Models Used in Stress Tests and Their Risks

In this section, I present a taxonomy of the models used in stress tests:

- *Pricing models* used for the marked-to-market positions
- *Forecasting (statistical) models* used to obtain the following:
  - Multi-variate macroeconomic scenarios
  - Credit losses (impairment, charge-offs, and provisions)
  - Operational losses (litigation, execution, fraud, systems, employment, etc.)
  - Income (revenues and costs)
  - Balance sheet (loans, deposits, and other balances)
  - Regulatory capital requirements (risk-weighted assets [RWAs] for market, credit, and operational risks)

---

capital plan reflected a number of deficiencies in its capital planning practices, including in some areas that had been previously identified by supervisors as requiring attention, but for which there was not sufficient improvement. Practices with specific deficiencies included Citigroup's ability to project revenue and losses under a stressful scenario for material parts of the firm's global operations, and its ability to develop scenarios for its internal stress testing that adequately reflect and stress its full range of business activities and exposures."

[19] See Bank of England (2018).

Let me start by describing the fundamental distinction between *pricing* and *statistical* models.

Pricing models are *interpolators and extrapolators of prices* that exist in the markets.[20] The models are calibrated (i.e., mathematically fitted) to the market prices of the traded securities, and they are used as interpolators and extrapolators to obtain the prices of other securities whose prices cannot be directly observed in the markets. If the markets are *liquid, dense, and complete*, interpolation is the main determinant in the pricing process. If the markets are *illiquid, sparse, and incomplete*, the exercise is then extrapolation, and it relies heavily on the choice of the *functional form of the extrapolator*. In any case, pricing models are based on *current market data*, that is, a cross-section of the market data as opposed to a time series, and pricing is by reference to prices of *similar* securities that trade in these markets. The model risk of the pricing models comes from incomplete, sparse, and illiquid markets, where extrapolation based on the arbitrary choice of the functional form of the extrapolator is the dominant determinant of price.

Statistical models use historical data (i.e., time series) to forecast economic and market variables in the future. They implicitly assume that *the future economic/market dynamics will be similar to those of the past*. This is a brave assumption and, as mentioned before, has caused very substantial losses to financial institutions in the past, as in 1987, 1994, 1998, and 2008. The projection of economic and financial variables over multiple-year horizons is a very ambitious task. As a comparison, imagine forecasting the weather or the path of a hurricane even just a week ahead. In fact, financial forecasting and weather forecasting have much more in common than people realize, at least as far as forecast error and model uncertainty are concerned: we can predict relatively well over short time horizons, but we face large forecast risk over long horizons.

Economists, especially econometricians, are trained to build forecasting models. Thus, it is not a surprise that such models have become important components of the stress tests. The models are used to forecast the evolution of the economic variables over the stress test period, which usually stretches 2 to 5 years into the future, and then to infer from these economic variables how the losses, revenues, costs, balance sheet, and capital requirements of the banks will evolve. All the projections are based on the regularities and relationships of the economic variables observed in the past. Linear and nonlinear regression models, as well as other more sophisticated statistical models, are used extensively, with their R-squared values in the typical range of 30 to 70 percent.

In the more typical applications, econometric models are used to estimate a *central tendency*: the expectation with some potential statistical error around it (i.e., a confidence interval). The models are used to project the path of the future macroeconomic scenarios under circumstances that evolve naturally, gradually, and continuously from the current state of the economy. In stress tests, the models are used quite differently: they aim at projecting the behavior of the macroeconomic and market variables in extreme conditions, which are, in general, very different from the current state of the economy. Are the models good and robust enough to perform this? Probably not; they work much better under more moderate and gradual evolutionary conditions.

---

[20] For a good description of pricing models as interpolators and extrapolators of observed prices, see Breeden et al. (1978).

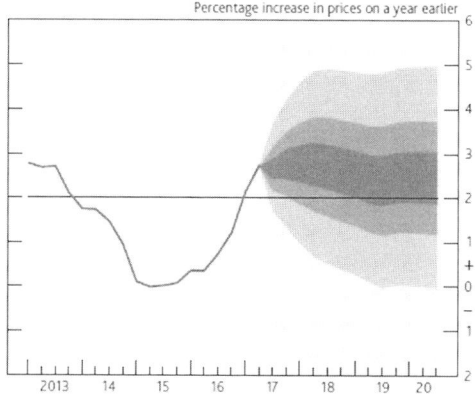

CPI inflation projection based on constant nominal interest rates at 0.25% (wide bands)(a)

Figure 16.2 Forecasts of UK CPI scenarios. Source: Bank of England (2017).

The main model risks in stress testing reside in the forecasting models (as opposed to in the pricing models). The errors around the projections of scenarios, income, and balance sheets can be large. Sensitivity analysis with respect to scenarios, assumptions, and model design is critical. Also, benchmarking of the model results against the results of other (so-called "challenger") models and other banks' and regulators' models is critical. There is no absolute *right or wrong* result in stress tests. There are ranges of plausible outcomes, which are often very wide. The modeler has to verify that the models are calibrated properly to obtain results within the plausible range and with the degree of conservatism that is required by the stress test.

Figure 16.2 shows the forecast of the UK Consumer Price Index (CPI) by the Bank of England in 2017. The dark-gray line is the historical, realized path of the CPI. The gray shaded area is the forecast period: 2017–2020. The colored shades correspond to 30 percent probability for each color and represent the forecast uncertainty. There is a residual 10 percent probability that the future CPI will be totally outside of the fan.[21]

This graphical representation is very useful to highlight the uncertainty of the forecast. The forecast is not a *line*; it is a *probability distribution*. Stress test results should be presented as a fan chart like this. If only a line is provided, regulators, senior management, and the board of directors should ask: What is the uncertainty around the numbers? More on this topic follows in Section 6.

3 Model Development, Testing, and Documentation

Model conceptualization and development combine multiple disciplines: business, economics, finance, mathematics, statistics, econometrics, probability theory, financial

---

[21] Other examples of similar forecast uncertainty representations can be found at www.federalreserve.gov/ newsevents/pressreleases/files/fomcfanchart20170303.pdf and bankofengland.co.uk/-/media/boe/files/ inflation-report/2017/fan-charts-aug-2017.

engineering, and computer science. Very importantly, they also require a full understanding of the context of the application of the models and a fair amount of pragmatism. Everything should be modeled as simply as possible, but not simpler than necessary. Unnecessary complexity and opacity should be avoided at all cost: Occam's razor should be applied at every step of the model-development process. Financial-economic intuition should always and constantly be assessed. It is possible, and these are good findings when they occur, that models may unveil unintuitive results, but these results should be scrutinized carefully and deeply to make sure that they are really correct. Models can help us correct flawed intuition. Intuition can help us detect models' flaws. The synergy between economic intuition and the rigor of modeling should be exploited at all times.

Pricing models (i.e., interpolators/extrapolators used to mark to market the trading positions) are a fertile ground of work for *mathematicians, physicists, and engineers*. These are called the *Q-quants*, where "Q" is the reference to the risk-neutral (i.e., cross-sectional pricing) probability measure. Financial-economic theory imposes structure and shape to the functional form of the interpolators, which often are solutions of the nonarbitrage equations that are derived from frictionless, continuous-time finance. Some of these functional forms are the solutions to partial differential equations that can only be obtained numerically by finite differences, finite elements, numerical integration, or Monte Carlo simulations. The testing of the pricing models should examine how tightly they fit the target set of market prices (the calibrating set). Stability and efficiency of computation are critical to enable the management of large portfolios of trades in real time. Historically, regulators have been concerned about the performance of pricing models in the stress test scenarios. These scenarios tend to be very different from the current state. Banks have to demonstrate that their pricing models perform properly in the severe stress scenarios. That is, they calibrate properly to the stressed market prices, and they interpolate and extrapolate correctly.

Forecasting models are the domain of *econometricians, statisticians, and actuaries*. These are called the *P-quants*, where "P" is the reference to the real-world (i.e., time-series) probability measure. The models are usually based on long time series of data, and the modeler attempts to capture the statistical regularities that exist in such data. The selection of the driving economic variables (e.g., GDP, unemployment rate, inflation, etc.), the functional form of the models (e.g., linear or nonlinear regressions, VAR models, etc.), the relevant data periods, and fitting methods are the key elements that have to be specified. The statistical tests usually examine the goodness-of-fit of the models to the input data, in-sample and out-of-sample accuracy of forecasts, the significance and stability of parameters, and so forth.[22] The modeler should keep in mind that the models will reflect statistical regularities that existed in *normal (regular) times*. Then they will be used to extrapolate to *stressed, unusual conditions*. This extrapolation introduces a substantial amount of model risk because we have no assurances that the statistical relationships that existed in normal times will persist in stressed situations. In fact, my own experience suggests that stressed economic and financial situations always present novel elements that challenge previous existing relationships. It is exactly these *surprises* that have caused some of the largest financial losses observed in the past.

---

[22] See Hong et al. (2018) and the references listed therein.

Clear and complete documentation of models is critical. It is a key requirement to enable the understanding of the models and their validation by independent parties. The model documents have to describe (a) the context and scope of the model application and use, (b) the goals of the model, (c) the rationale and processes used to select the model, (d) the choices made in the design and development process, and (e) the tests performed to confirm the validity of the model and the results obtained. The document has to tell the full story of the model and present the reasons and support for why the modeler believes that the model is appropriate to tackle the problem at hand. My experience is that it is very hard to obtain good model documents from model developers; they are much more concerned with model development, testing, and implementation than with writing model documentation.

> Thus, it is important to have a set of *MRM policies, procedures, and templates* establishing what is expected to be part of the model development and contained in the model documentation. These will guide the modelers through the model development and documentation.

4 Independent Model Validation

Banks have created independent model-validation groups to verify that their models are developed soundly and perform correctly to their intents. The validators examine the model documents, assess the strengths of the tests performed, perform additional and different tests, challenge the key assumptions, and determine if the support presented for the model is adequate and sufficient. But, and importantly, usually the validators know much less about the model, its use, and the context than the modelers do. This knowledge gap gives rise to a large number of iterations between modelers and validators until all the relevant aspects of the model are fully documented and understood by the validators. Validation is an intensive and potentially time-consuming process, and appropriate time should be allocated to it.

> The fundamental role of independent validation is to foster discipline in the model-development process as well as clarity and transparency in the description and explanation of the models. This, in turn, feeds back into the development processes, making them better, more effective, and more rigorous.

Contrary to the views of many, including of most regulators, I do not believe that model validators can match the skills and knowledge of the modelers. What the validators can do is to create a strong control and surveillance framework that fosters best practices by the modelers themselves. And by doing so, the validators make the entire model development process better, more robust, and more transparent.

5 Propagation of Risks in Systems of Models

The stress test exercises involve the use of many *interconnected models*. The outputs of some models feed into other models as inputs. It is important to understand the interactions of models and how model risks, deficiencies, and errors can propagate through the system. Total model risk compounds through the system of models, and it can produce a large amount of uncertainty around the final results of the stress tests: minimum Common Equity

Tier 1 (CET1), Tier 1 capital (T1), total capital (TC), and leverage ratios. It is important to assess and explain the model risk that exists around these key metrics of capital adequacy in order to provide the proper context for the users of the metrics (i.e., regulators, senior management, and board of directors).

Sensitivity analysis of the results to changes in the key assumptions and to different model choices can be informative. Also, it is important to benchmark the models' results against the stress test results of similar risks and portfolios of other banks and regulators that use similar models.

> It may be hard to come up with rigorous quantitative measures of the aggregate model risk. Qualitative assessments informed by sensitivity analysis and benchmarking are useful and better than not attempting to aggregate and understand model risk at all.

The published results of DFAST/CCAR over their eight runs from 2011 to 2019 provide useful data to assess the model risk of the stress test results: in each DFAST/CCAR run, each bank and the Federal Reserve calculate the minimum (stressed) CET1 of the bank, applying their own and different models. The analysis of the pairwise differences between the banks' and the Federal Reserve's CET1s over the eight CCAR/DFAST runs and with 20+ banks per run shows that the standard deviation of these differences is on the order of 1.3 percent of CET1.

This standard deviation suggests that if a bank calculates its minimum CET1 in stress as 7 percent, the 90 percent confidence interval (i.e., $+/-1.64$ standard deviations) around this estimate would be its CET1 being in the range of 5 to 9 percent. In other words, the model risk/uncertainty seems to be large.

6 Presentation of Results Including Model Uncertainty

Model results have to be presented in the proper context of the inherent model uncertainty. Human beings are not good at dealing with uncertainty—they prefer point estimates. It is important that model results be presented together with information about the model risks, deficiencies, and uncertainties and how these could affect the key metrics used to make capital adequacy decisions. This will make users of the stress test results more cautious and more realistic. It will also emphasize the need for conservatism in the actions and decisions that follow from the stress test results.

> "With the inclusion of our planned dividends and stock buybacks, our stressed CET1 is 5 percent. The threshold to pass the stress test is 4.5 percent. Should we increase our dividends and buybacks until the point that our stressed CET1 gets down to, say, 4.6 percent, just a bit above the passing threshold?" the chief finance officer (CFO) asks.

Some banks seem to be doing exactly this, as evidenced by the CCAR 2019 results.[23] But this completely disregards model risks and uncertainties. Do the Federal Reserve's models

---

[23] See Federal Reserve Board (2019).

really have the precision to allow the statement that 4.6 percent is greater than 4.5 percent CET1 and that it is a "real pass?" I do not think so. Banks and regulators should understand that the precision of the estimates is not that high. Again, based on the data available from 8 years of DFAST/CCAR, it appears that the standard error of the minimum CET1 estimate is around 1.3 percent of CET1. This number is a good starting point to think about model uncertainty in the context of stress tests of DFAST/CCAR.

How to handle the model risk in the stress test's results? One possibility is to have a variable multiplier of the stress test's CET1 drop that would increase as the minimum (stressed) CET1 approaches the threshold. For example, as the minimum CET1 falls to within the 90, 60 percent and 30 percent confidence zones, the multiplier would gradually increase.

---

7 Conclusions

In this chapter I provided the reader with the big picture to understand MRM in the context of stress tests and capital planning.

Most of the existing literature on model risk management tends to go straight into the technical modeling issues and quickly lose sight of the broader context of financial-economic modeling. The econometricians and statisticians talk about goodness-of-fit tests, in-sample and out-of-sample performances, significance tests, stationarity tests, and so forth. All these are relevant within their specific domains. But they will not eliminate the bulk of model uncertainty and model risks in stress testing. In my view, the most MRM can achieve is to establish a strong and robust framework that elicits sound model development, clear and transparent model descriptions, extensive model testing and benchmarking, and ultimately, the recognition that the results are point estimates with wide bands of uncertainty around them, especially over long stress testing horizons.

We are making progress; we are much better in MRM today than we were just a few years ago. But we have to advance even more, and we have to embed the MRM concepts and disciplines more deeply into the organizations that use models.

The Federal Reserve's stress test and capital planning programs have substantially strengthened the US banking system after the Global Financial Crisis of 2008.

The US banks have more than doubled their core equity capital. In situations of crisis and panic, it is only *capital and liquidity* (not *models or stress tests*) that will support the banking system. Thus, the fundamental preemptive function of the stress tests is to lead banks to retain more capital and liquidity reserves.

The fact that the Federal Reserve develops its own models and then assesses the capital adequacy of the banks using these models is a fundamental factor for the success of DFAST and CCAR.

If the banks' models were the only ones used, there would be great economic incentives to bias these models toward more benign numbers. This phenomenon has happened in the application of the Basel Committee's capital frameworks: banks' models have gradually become more benign, and model-based capital declined.

Banks have strong economic incentives to increase their leverage and invest in illiquid assets to maximize their current returns on equity (ROE). Banks' management is rewarded for achieving ROE targets and for beating the ROEs of their competitors. Rigorous stress tests and capital-planning programs, benchmarked by regulators' own models, lean against these strong economic incentives.

> The stress tests should be countercyclical; that is, the severity of the stress tests should increase when the economic and market conditions are good, especially if they have been good for a long period of time.

It is important to keep reality in sight: we will never get the stress scenarios and the stress losses right. It is futile to speak of the accuracy of the stress tests with respect to what will happen next. Regulators should use the stress tests as tools—very powerful tools—to set capital and liquidity at the proper levels across the banking system.

A strong, well-capitalized banking system is necessary for financial stability and sustained economic growth. This is the ultimate goal.

References

Ashcraft, A., and T. Schuermann (2008), "Understanding the securitization of subprime mortgage credit," *Foundations and Trends in Finance*, 2(3), 191–309.

Bank of England (2017), "Inflation report fan charts, August 2017," available at www.bankofengland.co.uk/~/media/boe/files/inflation-report/2017/fan-charts-aug-2017.

Bank of England (2018), "Model risk management principles for stress testing," Supervisory Statement SS3/18, available at www.bankofengland.co.uk/-/media/boe/files/prudential-regulation/supervisory-statement/2018/ss318.

Basel Committee on Banking Supervision (1996), "Market risk amendment," available at www.bis.org/publ/bcbs119.htm.

Black, Fischer, and Myron Scholes (1973), "The pricing of options and corporate liabilities," *Journal of Political Economy*, 81(3), 637–654.

Breeden, Douglas T., and Robert H. Litzenberger (1978), "Prices of state-contingent claims implicit in option prices," *Journal of Business*, 51(4), 621–651.

Derman, Emanuel (1996, April), "Model risk," Goldman Sachs Quantitative Strategies Research Notes, available at http://pricing.online.fr/docs/modelrisk.pdf.

Derman, Emanuel (2012), *Models behaving badly: Why confusing illusion with reality can lead to disaster on wall street and in life*, Free Press.

Federal Deposit Insurance Corporation (1997), "History of the eighties: Lessons for the future, Vol. 1," available at www.fdic.gov/bank/historical/history/vol1.html.

Federal Reserve Board (2014, March), "Comprehensive Capital Analysis and Review 2014: Assessment framework and results," available at www.federalreserve.gov/newsevents/press/bcreg/ccar20140326.pdf.

Federal Reserve Board (2019, June), "Comprehensive Capital Analysis and Review 2019: Assessment framework and results," available at www.federalreserve.gov/newsevents/pressreleases/bcreg20190627a.htm.

Hong, Han Deming We, and Qing Wang (2018, July), "A common pitfall in bank stress testing: Macroeconomic factors and model instability," available at https://ssrn.com/abstract=2896974.

Lowenstein, Roger (2000), *When genius failed: The rise and fall of Long-Term Capital Management*, Random House.

Merton, Robert (1997), "Applications of option-pricing theory: Twenty-five years later," Nobel Lecture, available at www.nobelprize.org/uploads/2018/06/merton-lecture.pdf.

Office of the Comptroller of the Currency (2000, May 30), "Model validation," Bulletin OCC 2000-16, available at https://ithandbook.ffiec.gov/media/resources/3676/occ-bl2000-16riskmodelvalidation.pdf

Taleb, Nassim (2007), *The black swan: The impact of the highly improbable*, Random House.

US Congress (1980), "Depository Institutions Deregulation and Monetary Control Act," available at www.federalreservehistory.org/essays/monetarycontrolactof1980.

17

A Supervisory Perspective on Stress Testing—the US Experience

Tim P. Clark

Since the global financial crisis of 2008–2009, there has been a widespread increase in the use of stress testing to inform forward-looking assessments of capital and liquidity needs at systemically important financial institutions (SIFIs). In addition to stress tests run by banks and supervisors to measure capital needs, supervisory assessments of banks' internal use of stress testing for capital- and liquidity-planning practices increasingly serve as a valuable lens through which supervisors view and assess many of the banks' key practices. This combination is arguably the most important and positive supervisory and risk management development in decades. It has promoted greater financial resiliency at the largest banks, which have greatly increased capital from precrisis levels, and promoted an improvement in the quality of capital through a shift toward common equity.[1] At the same time, it has led to much-needed improvements in risk management, internal controls, and corporate governance practices.

Importantly, these developments have also provided supervisors with a framework through which they can gain valuable perspectives on these practices at large banks while supporting both a deeper and broader comparative understanding of banks' financial condition as well as the strengths and weaknesses of risk management and control practices across groups of banks. The types of fundamental risk management, internal controls, and governance practices that were often assessed individually during traditional bank examinations are better assessed in the context of banks' internal stress testing and associated capital- and liquidity-planning processes. This provides a more meaningful contextual perspective for assessing banks' practices in a continuum from identifying and measuring risks to the execution and results of a stress test to understand how they evolve over stressful periods, to how critical decisions about capital and liquidity needs are informed by each of these factors. Through this more integrated approach, supervisors can better assess if banks' practices are effective and whether they provide the information decision-makers on boards of directors and in bank senior management need to make related decisions in a well-informed manner.

---

[1] The Supervisory Capital Assessment Program in the United States introduced *poststress* "Tier 1 common equity" as a capital definition and a supervisory requirement, a significant strengthening relative to total Tier 1 capital. This has evolved, and today the common equity Tier 1 definition of capital has been codified through Basel III.

Postcrisis supervisory approaches focusing on stress testing and capital and liquidity planning represent a significant and necessary divergence from past supervisory practices, which were exposed as inadequate for the largest banks by the crisis. Precrisis supervision of SIFIs—that is, those banks and others that can pose a threat to financial market functioning and the economic well-being of millions of people—was built on a strong belief that banks' informed self-interest, supported by market discipline, would constrain excessive risk taking. This proved to be badly misguided. For such banks, the potential to cause or exacerbate disruptions in the financial system and economy needs to be minimized, and the margin for error by supervisors is small. Given the incentives they face, neither the banks nor their investors and counterparties can be counted on to take actions that could protect the system if that necessarily reduces their return on equity, which higher capital requirements do, all else being equal. Indeed, as independent actors, they would not likely view protecting the system as their role. It is the role of supervisors of SIFIs to promote a safer system populated by financially resilient banks. Indeed, for prudential supervisors of SIFIs, this should be the primary objective. Focusing on their preparedness to withstand severely stressful events while being able to continue to operate provides a distinct path to that end.

Stress testing is not a "silver bullet" and will not provide the solutions to all of a bank supervisor's challenges, although this does not attenuate its value. There are no such silver bullets. The use of stress testing in supervision is a valuable tool in a business where good tools are scarce, and its prominence has created large benefits for supervisors, banks, and the stability of the financial system. Assessing the ability of SIFIs to withstand a period of severe stress, using both supervisory and internal bank practices and measures, should remain the critical area of focus for supervisors of SIFIs for the foreseeable future.

Underscoring the importance of promoting resiliency to stress, even with the progress that has been made to address large-bank resolution challenges, uncertainties surrounding the efficacy of proposed large-bank resolution regimes remain and are unlikely to be fully addressed in the medium term. Moreover, at least in the United States, some of the tools the authorities used to address past crises have been weakened or are no longer available. Despite some progress in resolution preparedness noted by supervisors over the past 5 years, the largest banks effectively remain "too big to fail" and will continue barring the requirement of more fundamental changes in bankruptcy regimes and the structures of and interrelationships within the largest banks. Trying to resolve one or more SIFIs during a time of stress is likely to prove too frightening for the authorities, and for good reason; there is simply too much uncertainty, and the potential for negative consequences is too great. Seeking to reduce the probability of facing such an event by requiring SIFIs to hold capital and liquidity to withstand severe stress will remain critically important. Supervisors should focus on ensuring that banks themselves are doing their best to do that and on informing the measurement of those needs.

1 The Evolution of Stress Testing in US Banking Supervision

Prior to the global financial crisis, bank supervisors and supervisory agency economists used scenario analysis and other stress testing techniques to analyze specific risks to banks and financial conditions in the banking sector. These were generally rough attempts to assess the possible impact of an event on the value of specific asset portfolios (usually loans, for which at least some limited data were available) and did not involve running a scenario through all aspects of banks' operations, as is commonly done today. Supervisors did not view stress testing as a tool they should use for assessing and making supervisory decisions about capital adequacy and liquidity sufficiency at banks. Indeed, the idea that a hypothetical, although possible, negative outcome could or should drive supervisory decision-making regarding capital and liquidity requirements was not widely supported by bankers or their supervisors.

The Basel Market Risk Amendment (1996) and Pillar 2 of the Basel II Capital Accord proposal (proposed in the United States in 2007) promoted the use of stress testing in risk management and internal capital adequacy assessment processes (ICAAPs), respectively. However, at least in the United States, the latter had not taken hold in a meaningful way precrisis and was not a significant focus of supervisory activities. Moreover, like the static point-in-time regulatory capital requirements that were used, even where Pillar 2 was in use, the economic capital measures (based on value at risk [VaR]) often used for ICAAP exercises did not identify concerns about capital in advance of the crisis, just as many other VaR-based measures of risk proved to be inadequate. In addition to challenges measuring risks in the far tail of the distribution, a backward-looking approach reliant on recent historical market performance, which had been quite positive for some time, made them inappropriate measures for assessing potential risks in a rapid and severe downturn.

Banks had long used stress testing for risk-measurement purposes but were not interested in "enterprise-wide" tests against severe outcomes to determine their capital needs. At some banks, management thought it was not a good use of time and that their boards of directors would not support using stress tests to inform their actions and decisions. They simply could not (or did not want to) believe things could get bad enough to threaten their survival.

The Supervisory Capital Assessment Program (SCAP), carried out in the United States in 2009, was the first time US banking supervisors used scenario-based stress testing to measure "required" poststress capital thresholds for large banks.[2] It was an attempt to reduce uncertainty by estimating and publishing potential losses and capital needs at the 19 largest US bank holding companies in the event of economic and asset-price deterioration that was worse than even the extremely negative outlook at the time. It was also aimed at reestablishing confidence in the banks by assessing whether capital at each of the banks was sufficient and, if not, requiring them to raise the capital they would need to absorb potential losses, meet the poststress threshold, and keep functioning and lending. Banks were required to go first to private investors for the capital, but the government provided a critical backstop in the event investors declined to step in. SCAP set a critical precedent for supervisory assessments of capital adequacy in the United States because it was deemed to be unsafe and unsound for banks to fall below a poststress capital threshold set by supervisors, paving the way for poststress capital measures to become a regular supervisory tool and a binding constraint.

The success of SCAP owed much to the breadth of its coverage of the banking system (roughly two-thirds of total US banking assets) and the transparency of the exercise. This level of transparency was a revolutionary change for bank supervisors and for banks. It proved an important factor supporting SCAP's credibility, and it informed views at the Federal Reserve about the value of transparency that were still in place at the time of this writing.

---

[2] For a detailed discussion of SCAP, see Hirtle et al. (2009).

As Ong and Pazarbasioglu (2013) observed with respect to the use of stress testing in addressing financial crises, "credibility is the bedrock of any crisis stress test."[3] If the work is not viewed as credible, it has little chance of being helpful and may prove counterproductive if it inspires views that the authorities are trying to hide something. The level of transparency supported SCAP's credibility by providing comparative data across firms that allowed the public to distinguish between the merely weak and the truly undercapitalized, allowing the public to assess them and make apples-to-apples comparisons across banks (Federal Reserve Board of Governors, 2009).

An important element of SCAP was the focus on stressed revenues. Most estimates of banks' capital shortfalls at the time were based on a revaluation of assets using current market-price-based "guesstimates" or other point-in-time measures, without consideration of the timing of potential losses and the revenues banks would be earning in the meantime to offset them as they occurred. Revenue generation was substantial across these banks, and ignoring that led to significant overestimations of capital shortfalls. Although SCAP estimated potential losses of roughly $600 billion, loss-absorbing resources in addition to capital (pre-provision net revenues + changes in loan-loss reserves) were estimated to be approximately $370 billion, which were available to offset losses as long as the banks continued to operate (Federal Reserve Board of Governors, 2009).

As a result of lessons learned during SCAP in the United States, there was a strong movement to resist a return to supervisory "business as usual" after the crisis. The failure of precrisis supervision and regulation was simply too evident. The crisis had made clear that substantially more focus needed to be put on events that posed existential threats to the banking system, particularly for systemically important banks. A critical postcrisis objective was to promote resiliency of SIFIs to enable them to better withstand severe outcomes and thereby reduce the chance they could cause or exacerbate a spiraling downturn. For SIFIs, forward-looking, tail-risk-sensitive, and dynamic ways to identify risk and assess capital and liquidity sufficiency were needed, even in so-called "normal" times.

Before the 2010 Dodd–Frank Wall Street Reform and Consumer Protection Act put in place a statutory requirement for stress testing in the United States, the Federal Reserve was already moving to make stress testing for capital adequacy a key piece of its bank supervision program for SIFIs. The first step was focusing on banks' internal capital adequacy assessments and requiring the use of scenario-based stress testing by the largest banks for this purpose. Supervisors continued to believe that the banks had to remain responsible for such decisions, but given recent history, they had little confidence they would make them in an informed way. The crisis had exposed woeful inadequacies in banks' data, risk management and controls, and governance processes. Those at the top of the banks who made critical decisions about risk appetite, business strategies, and associated risks, as well as about capital and liquidity needs stemming from those, had been doing so based on overly optimistic assumptions and incomplete and/or inaccurate information about their risks. Many banks' boards of directors and senior managers had been flying nearly blind with respect to the risks they were taking, and particularly how those risks would play out under stress.[4]

---

[3] See Ong and Pazarbasioglu (2013).

[4] See Citigroup Saw No Red Flags Even as It Made Bolder Bets, *New York Times*, November 22, 2008.

Supervisors needed a better way to assess how confident boards of directors should be that the decisions they make that affect banks' financial resiliency are well informed. To achieve this, supervisors need to know the answers to the following questions:

- Are the processes run by bank management allowing for a comprehensive and timely identification of the bank's material risks?
- Are the risk measures in place effective, and are they suitable for estimating potential losses under stressed conditions?
- Do risk managers have the information, practices, and stature they need to control risks and keep them in line with the board's stated risk appetite?
- Does the bank have robust internal controls that provide for a high level of confidence in the integrity of data and other information so that the numbers being used to inform decisions are a comprehensive and accurate representation of their positions?
- Is internal audit capable of assessing all the practices that support capital planning and of providing the board with an assessment of weaknesses in those practices and the implications of those?
- Is management providing the board with the information it needs, in a form that is digestible and understandable, so that it can actually use it to inform decision making?
- Is the board taking all of this into account when making decisions?

These are key questions that supervisors of SIFIs should be asking over and over again. The use of integrated assessments of capital and liquidity planning processes puts them in context and provides the most valuable framework for this.

2 The Supervisory Benefits of Stress Testing Programs

In the United States and elsewhere, the increased use of stress testing in supervision has been informed by lessons painfully learned from the crisis. Among the most important were the following:

- Regulatory capital regimes were neither well conceived to be dynamic and adaptable enough to rapidly changing business practices, operating environments, and risks, nor were they appropriately stringent for systemically important banks.
- Risk identification, measurement, and management capabilities at large, complex financial institutions had failed to keep pace with the growing scope, scale, and complexity of their activities, and banks generally did not have or were not using measures that were suitable for ferreting out tail risks or risks that only emerge under stressful conditions.
- Banks' internal controls practices were insufficient, given the complexity and massive size of the banks, creating great uncertainty as to whether those responsible for running the firm could have confidence that they understood what was going on in all parts of the firm.
- Corporate governance, including the role of independent risk management and internal audit functions, had either suffered a complete breakdown or had perhaps never functioned as assumed.
- Really bad things can and do happen to even the largest banks—the benefits of diversification had been overstated in deregulatory decisions, allowing banks to increase activities in a broad array of formerly restricted businesses.

- The negative consequences of badly run banks will go well beyond those SIFIs that make the gravest mistakes and can destabilize the entire system.

The Federal Reserve's capital stress testing program, the Comprehensive Capital Analysis and Review (CCAR), was launched with these problems in mind and has led to progress in addressing each of them. CCAR is supported by a regulation put in place in 2011 that heightened expectations for banks' capital-planning practices and defined the supervisory measurement of capital sufficiency based on a poststress analysis of capital (Federal Reserve Board, 2011). At a later date, requirements for liquidity stress testing by the banks and the implementation of the stressed-assumptions-based liquidity coverage ratio (LCR) were also put in place. To support these changes, the Federal Reserve changed its supervision program for the largest banks to increase the supervisory focus on banks' capital and liquidity planning and sufficiency under stress (Federal Reserve Board of Governors, 2012).

The Fed's CCAR program applies two complementary assessments. One is the supervisory stress test, independently run by the Fed, through which all of the firms are tested concurrently against the same adverse scenarios designed by the Fed and using the same models across the banks to maximize consistency and comparability. The second, the so-called "qualitative assessment," looks at firms' annual capital plans and the supporting practices that feed into them. These capital plans are required by regulation to incorporate internal stress testing using scenarios and models designed by the banks to best reflect and capture each bank's idiosyncratic business activities and associated risk profile. This combination seeks to minimize the potential for a "monoculture" to develop around risk modeling and scenario design, each of which benefit from a wide variety of differing perspectives and methods. As importantly, it maintains the correct emphasis on banks having the capacity to measure their own risks and associated capital needs, backstopped by a supervisory floor.

The use of an independent supervisory stress test to quantify supervisory expectations of capital needs was a momentous change from precrisis practices. US and other supervisors had long expected banks to hold more capital than minimum regulatory requirements, but there was no agreed-upon formal mechanism in place for how to measure this. In the United States and elsewhere, supervisors expected these capital levels to be determined by the banks and to be "commensurate with their risk profiles."[5] This turned out to be too vague to be of practical value and opened the door to inconsistent interpretations among and across banks and supervisors. Moreover, it was undermined by incentives for bank management and boards to keep regulatory capital as low as possible relative to the risks being taken to generate earnings, similar to the incentives that can undermine the effectiveness of Basel Advanced Internal Ratings Based (A-IRB) approaches to calculating risk-weighted assets (RWAs) for credit exposures.[6]

Poststress capital analysis using independent supervisory stress testing has a number of benefits as a measure of capital needs. It is explicitly forward-looking; has greater

---

[5] As noted elsewhere in this chapter, the Pillar 2 ICAAP had been proposed in the United States in 2007, and was being used in Europe, but was still in its early stages.

[6] A-IRB allows banks to use their own models to quantify required regulatory capital for credit risk. Banks can use the A-IRB approach only if their regulators have approved them for its use after determining controls are effective.

risk sensitivity than standardized approaches; and allows for greater confidence, cross-bank consistency, and comparability than the A-IRB approach of Basel II, through which banks measure their own risks and, as noted earlier, will try to minimize associated capital requirements. Also, and of critical importance, stress testing is a more dynamic measure of capital needs. It can be readily adjusted as banks' business models, products, risks, or operating environment change. Redesigning regulatory capital rules requires a lengthy process and is generally too slow to be effective in that regard, potentially leaving emerging risks unrecognized by regulatory capital requirements for an extended period of time.[7]

The qualitative assessment of banks' stress test-based capital planning allows supervisors to develop a comprehensive view of each of the key steps along the path from a bank's capacity to identify and measure its risks (the starting point), to designing the scenario and executing the stress test, to making decisions about potential capital needs to withstand stress (the ending point). Each of these steps aligns with supervisors' traditional set of safety and soundness concerns. Taken as a whole and on a bank-wide basis, this provides a valuable lens through which supervisors can assess a bank's risks and the practices it uses for identifying, measuring, and aggregating them; for attempting to ensure data integrity and other controls; for reporting risks to the board and management; and for making decisions about capital needs.

More traditional approaches to supervision generally focus on specific risks and associated controls in a silo-by-silo manner, with no direct link to assessments of capital or liquidity sufficiency. Such an approach is not wholly without value, but it can lead supervisors to forget or ignore the main reason strong risk management, internal controls and governance are needed: banks need to hold sufficient capital and liquidity to support the risks they take. When making decisions about future strategies, risk appetite, and capital and liquidity needs, boards of directors should have confidence that they are doing so based on accurate data and information. They need a full understanding of how their decisions could affect the banks' capacity to withstand stress and keep operating and what specific actions they can take to affect that. Supervisors, by assessing the practices that inform those decisions, can get an integrated view that is far more valuable than a risk-by-risk approach. And by doing so across a group of large banks, they can take advantage of perhaps the one key informational advantage they have relative to individual banks: the ability to assess and compare practices across many firms.

Scenario design and loss and revenue modeling generally get the most attention when discussing stress testing-related practices (and will be only briefly discussed in this chapter because they are the subjects of other chapters in this book). But often, the greatest challenges faced by the largest banks are much more fundamental. Banks struggle to effectively identify, capture, measure, and aggregate risks, including those that are difficult to observe because they may only manifest in a stressful environment. For very large banks, effective practices for risk management and internal control can be a particularly vexing challenge. The sprawl of their operations and the use of "legacy" information technology (IT) systems knit together after acquisitions and scattered across the globe complicate matters. It can be quite expensive to address these issues, which leads banks to move slowly to fix important

---

[7] For example, the full complement of proposed Basel III rules to address precrisis regulatory capital and liquidity deficiencies has still not been fully implemented more than 10 years after the crisis.

weaknesses, so the costs are more easily absorbed over time. A key benefit of required stress testing and capital planning in the United States has been that it pushed banks to quickly work to improve their processes for risk management and control processes and to collect/develop the data needed to assess risks, including potential risks under stress.

In addition to the benefits from a microprudential perspective, supervisory stress testing programs can serve macroprudential goals when applied to the most systemically important financial institutions and when capturing a significant share of the financial system. Most importantly, greater resilience across SIFIs reduces the likelihood that a downturn will spiral into a system-wide crisis. Stress testing also allows supervisors and others to look across the system and identify potential areas of emerging risks that may be of concern and might be overlooked if not considered in the context of a potentially stressed operating environment.

3 Operationalizing Stress Testing in Supervision

Two methods generally used to introduce stress testing into the supervision process are to require banks to use stress testing in risk management and decision-making related to capital and liquidity planning and for supervisors to use stress testing techniques themselves to measure banks' capital and liquidity sufficiency. Capital stress tests run by supervisors get the most public attention in the United States and elsewhere, but the requirement for SIFIs to use stress testing in capital and liquidity planning is equally important and more prevalent across different jurisdictions than the use of independent supervisory stress tests. Where supervisory stress tests are used, it should not be for supervisors to determine precisely how much capital and liquidity a bank needs but rather to set required regulatory floors, just as with any regulatory capital or liquidity requirement. It should always remain the responsibility of banks' boards and management to determine their own banks' capital and liquidity needs; supervisory emphasis should focus on qualitative assessments of banks' capacity for making such decisions, informed by effective risk management and control practices and governance and supported by clean data.

The use of stress testing in supervision continues to evolve, and design decisions can lead to significant differences that create different challenges. Two fundamental decisions must be made: (1) whether to require explicit minimum poststress capital thresholds or to use results to inform Pillar 2 capital assessments[8] and (2) whether to conduct an independent supervisory stress test where the banks provide data and the supervisors run the test, or to give the banks scenarios designed by supervisors, have them run the tests, and have supervisors evaluate the results and make adjustments where a bank has not addressed the risks appropriately.[9]

---

[8] For example, in the United States, the Federal Reserve's capital stress tests serve as a binding constraint on the largest banks because required capital ratios and capital distributions (dividends and share buybacks) can be temporarily restricted while a bank falls below the poststress threshold. In the eurozone, the primary emphasis is on the so-called Supervisory Review and Evaluation Process (SREP), through which supervisors may include a Pillar 2 add-on charge based on stress testing analysis as a component of a broader assessment.

[9] The United States is the only major jurisdiction that currently runs its own fully independent "bottom-up" stress tests concurrently across all major banks using detailed data collected from banks and stress testing models designed by the Federal Reserve. Other major jurisdictions often use a combination of banks running scenarios designed by the supervisors and "top down" or other supervisory modeling based on models created by supervisors.

In most jurisdictions, supervisors, at a minimum, require stress testing to be used by banks for the purposes of risk management and internal capital assessment, and these may be assessed periodically during examinations. Although an approach such as the US CCAR program that combines a supervisory stress test and an assessment of banks' stress testing–based capital planning is a particularly powerful tool, other approaches can yield benefits, as discussed in the prior section of this chapter, particularly when banks' practices are assessed holistically in an integrated review of banks' internal capital- or liquidity-adequacy assessments and planning.

4 The Use of Supervisory Stress Testing to Set Capital Requirements

The use of supervisory stress testing to set minimum capital requirements heightens the importance of a variety of challenges, and jurisdictions will need to decide whether the trade-offs involved are acceptable to them. These include but are not limited to (1) the need for consistency across firms when a one-size-fits-all approach will miss things at every bank, (2) the tension between a flexible and dynamic measure of capital and the industry's desire that regulatory requirements not present a "moving target," (3) the need for a scenario that represents an appropriately severe and plausible event and captures a broad swath of risks all banks face versus the use of a number of scenarios specifically designed for each firm, and (4) that the transparency necessary to support a credible stress testing program may have the potential to undermine the benefits if negative results lead to a bank facing destabilizing speculation about its financial condition. These challenges notwithstanding, the risk sensitivity and greater flexibility and dynamism relative to more traditional capital measures and more explicit forward-looking nature provide benefits that make it worth addressing these challenges, especially for SIFIs.

4.1 Consistency

As a practical matter, supervisors should strive to treat banks consistently. If two banks hold exactly the same asset and book it the same way—for example, in the banking book—required capital against that asset should be the same for both. This is especially difficult to achieve when banks themselves are calculating the stress test results using a scenario provided by the supervisors. It can also be an issue when supervisors run the stress test. In the latter case, consistency requires imposing supervisory standards for reporting positions across the banks and checking that banks are meeting those standards. Even working to enforce such standards cannot fully resolve this issue. Large banks often face challenges in capturing, measuring, and aggregating exposures in a timely and accurate manner, which can undermine comprehensiveness and consistency and pose a significant challenge for supervisors, who must rely on data reported by the banks.

These consistency challenges are not unique to stress test-based capital measures. The use of banks' internal models to calculate capital requirements from credit exposures can be materially inconsistent across firms and jurisdictions (*Basel Committee on Banking Supervision*, 2013), and although the imposition of Basel III "floors" for credit risk tries to address this, it cannot fully resolve the problem. More fundamentally, the incorrect classification of positions when using a standardized approach can also lead to inconsistent treatment that is difficult to observe without a strong process for reconciliation of banks' regulatory reporting back to the underlying positions, which is seldom done comprehensively by supervisors for the largest banks because these banks can have literally millions of positions.

4.2 A Dynamic Measure versus Year-to-Year Stability

Most would agree that in an uncertain world, a dynamic measure of capital will be a better method for determining capital needs than a static one. Indeed, the static nature of precrisis capital requirements was a key weakness in a rapidly unfolding downturn. However, the tension between the value of a dynamic measure that can be adjusted as things change and greater predictability desired by banks is an issue that continues to receive much attention. The industry has long argued that it is not reasonable, nor does it allow for effective capital planning, to have capital requirements that can vary every year. However, at the same time, other observers have found that the Federal Reserve's stress tests, for example, may have become too predictable and, for that reason, are becoming less effective.[10]

Attempting year-to-year consistency will encourage banks to adjust their portfolios to reduce potential losses under the hypothetical scenario. Moreover, they may do so in a herd-like manner, increasing risks to the system. The incentives for banks to do this are strong. A willingness to change scenarios each year can help to mitigate this problem but not eliminate it.

The concern that annually fluctuating capital requirements driven by changes in stress test results can complicate banks' planning efforts, although no doubt real, is often overstated. Running a SIFI responsibly is a complicated undertaking and should involve forward-looking attempts to adjust as the potential outlook changes. The world changes often and rapidly. A periodic change in forward-looking assessments of capital needs should result as much (or more) from banks' internal analyses as from a supervisory stress test. Banks' capital planning in 2005, 2006, and 2007 left the industry woefully unprepared for 2008, and banks failed to realize or acknowledge this until it was too late. Greater flexibility and dynamism may generate some limited uncertainty, but this is not a high price to pay for greater resiliency at SIFIs. Indeed, dynamism is a key positive feature (not a bug) of using stress-based measures of capital needs.

4.3 Scenario Design and Selection

There are a number of important considerations in scenario design and selection, including: (1) which and how many scenarios to use, (2) whether they should include "add-ons" to capture specific issues that may warrant further attention, and (3) the appropriate level of severity.

Scenario selection has been the subject of much discussion and critique. Some trade-offs are largely a function of the practical realities and operational challenges of using stress testing to measure capital sufficiency. In an optimal situation, if reducing the probability of default of a bank is the primary objective, one might run every conceivable and plausible scenario for each bank and, for example, always use the one that leads to the largest capital needs at each individual bank. This would allow for each bank to be tested against a variety

---

[10] See Glasserman and Tangirala (2015).

of scenarios that attempt to capture the idiosyncratic vulnerabilities associated with a bank's activities. But it may not be feasible today for supervisors to run such a multitude of different scenarios for each SIFI in a bank's jurisdiction. Nor is it likely to be seen as reasonable (or consistent) for supervisors to hold each bank to results from a different scenario that generates the largest capital needs.

This challenge gives rise to the real danger of too much focus being placed on the same scenario and has led to criticism of supervisors for "putting all their eggs in one basket" and not focusing on the "right scenarios." Because no one scenario can capture all risks any one SIFI faces, much less all material risks across a number of banks, supervisors should focus on "systemic" macro scenarios for supervisory stress testing. These consider the broad macroeconomic and financial market variables to which all large banks are sensitive—levels and changes in economic output, unemployment, interest rates, credit spreads, equity prices, and so forth. For banks concentrated in traditional lending and investment activities, this may be the most appropriate scenario type, and for most banks, it will be useful for a significant share of their activities. However, for banks active in complex trading activities, or in asset management, payments processing, or custodial activities, for example, the scenario needs to be augmented by shocks that capture specific risks stemming from those activities.[11]

A requirement that each bank run its own customized scenarios designed for its own activities and risk profile can help mitigate this concern. In the US CCAR program, each bank is tested against consistent supervisory scenarios and against ones of their own design, the latter of which are reviewed and assessed by supervisors to ensure they capture a bank's key risks.

Should scenarios include a specific focus on areas deemed of particular concern at the time, and how much and how often should those be refreshed? In order for stress testing to provide the full benefits of flexibility and dynamism, the answer to these questions should be yes and fairly often, respectively. The United States and the UK have each taken a different approach to keeping stress tests dynamic and allowing for the probing of specific issues that may arise.

The Federal Reserve uses the so-called "salient risks" approach, through which it may build into its broader macroeconomic and financial-market scenarios a specific issue that it finds worthy of probing in a given year. For example, if there were specific concerns about vulnerabilities in the eurozone, it could tweak the scenario to heighten the deterioration of exposures to that area. This augments its systemic macro scenario, and because it is built into the full scenario, it requires all tested banks to hold capital against systemic and salient risks.

The Bank of England also runs a systematic stress test every year using a fairly stable severely adverse macro scenario, and every other year, it includes a Biennial Exploratory Scenario (BES) to flesh out potential vulnerabilities in certain areas. The results of the

---

[11] For example, in the United States, all banks subject to Federal Reserve stress testing are tested against a broad macro scenario designed by the Fed, the globally systemic important banks (GSIBs) are also subject to an additional global market shock (the six largest trading GSIBs) that tests against a severe market disruption and/or a counterparty default scenario (all eight US GSIBs). Through these add-on shocks, the Fed is able to capture key potential sources of loss at these firms under stress while not subjecting those banks with relatively little relevant exposure to these add-ons.

BES are generally not made public but are used internally to identify and assess areas of potential vulnerabilities. This allows the Bank of England to probe for new areas of vulnerability while maintaining the annual test against the most likely sources of losses in a broad economic downturn. Banks are not required to hold capital against the biennial test. The value that could be gained from publicly disclosing the results of these tests is also foregone. This is a deliberate trade-off that allows supervisors to think freely and creatively about sources and channels of potential future crises without facing pushback from the industry that would result if they did require banks to hold capital against the more variable stress test.

Because one cannot assert that the results of a stress test will ever be "accurate" and given the importance of minimizing the probability of failure of a SIFI, supervisory stress tests used for measuring capital requirements should err on the side of conservatism when determining severity. This is not to say they should necessarily be extreme or unbound from the possible (e.g., the same interest rate cannot go up and down at the same time) but that when confronting a choice between two plausible scenarios, using the more conservative is reasonable given the context.

An important question that arises when considering scenario severity is about expectations of government action in a crisis. Historically, in most severe downturns, the government has provided support in a variety of ways. It is impossible to know with any precision how effective that support was and how much worse things would have gotten without it. Put another way, we cannot know how bad the global financial crisis might have been if there had not been massive government actions to support banks and market liquidity. Given the multitude of government programs to support the banks and financial markets in 2007–2009, we can assume with a high degree of confidence that, without such support, it would have been substantially worse than it was. Even if one agrees that the government should be there to correct for such disruptive "market failures," that does not mean banks should assume government support when identifying their capital needs. It is an open question how best to build this directly into the severity of a stress test scenario, although it does raise doubts about the legitimacy of the criticism from the industry that supervisors use scenarios that are "worse than what actually happened" in the global financial crisis.

A common critique of supervisory stress testing has been that it runs the risk of "putting all the eggs in one basket" and possibly missing important risks if the scenario is not perfectly designed. This critique appears to be based on a view that supervisors should be trying to predict the source and shape of the next downturn. This cannot be predicted, and that should not be the point of a stress test used to set capital requirements. A supervisory stress test is best seen as a way to calibrate capital needs against a range of possible severe outcomes. By setting the severity such that it is broadly representative of what the results of a severe downturn might be, it helps prepare banks for a range of possible negative events rather than attempting to specifically predict what the next downturn might look like.

It is a reasonable and commonly raised concern that regulatory capital regimes can shape the decisions made by banks and that this may have unintended consequences. But it is not a concern unique to stress testing. Any binding capital requirement creates an incentive for banks to adjust/structure positions in an attempt to reduce capital charges. It is not just stress testing that can lead to this behavior, although it may inform the specifics of the behavior in different ways than other capital requirements. The leverage ratio or an RWA approach will also drive decisions by banks if they are binding and given their static (and in the case of the leverage ratio, risk-insensitive) nature, perhaps in more dangerous ways. Stress testing should be seen as an important complement to other ways of measuring capital needs, not the only way. And it should be creative and dynamic, even (and perhaps especially) if that poses challenges for banks and supervisors. They operate in a volatile world.

4.4 Loss and Revenue Estimation

There is no one correct way to estimate potential losses and revenues under a hypothetical scenario—it is more art than science—but there are some bad ones. For example, a model (or other measure) that is not sensitive to the risk characteristics of a portfolio, including how those may change under stress, is of little practical value for stress testing purposes and assessing capital needs. Although this is an obvious point, it warrants noting because the use of such models has an unfortunate history. As one example, the experience of the Office of Federal Housing Enterprise Oversight (OFHEO) and government-sponsored entities (GSEs) provides strong evidence in support of the view that models should be revised and updated regularly to capture the changing risk characteristics of assets as the products that banks offer change. This may create some volatility in poststress capital requirements, but as the Fannie and Freddie examples illustrate, such volatility may well be appropriate and even necessary.[12]

One issue on which there is unusually broad agreement is that it would be a bad outcome if all banks and their supervisors estimated potential losses the same way. Although this comes up most often when discussing stress testing, it should be noted that this is precisely the effect of using a standardized approach to risk weighting of assets and/or a leverage ratio or setting binding floors for the Basel advanced approach to credit risk. With respect to the leverage ratio, not only is everyone using the same measure, but its indifference to variations in risk characteristics across asset classes makes it a particularly tricky one. Given the risk sensitivity and dynamism of stress testing, and the ability to update or revise models, stress testing could actually represent a significant mitigant to "model monoculture" relative to these measures (which are, by construction, a single fixed model).

A convergence in the methods used for estimating risks under stress may increase systemic risk by leading everyone to miss the same things, and in the same way.[13] Additionally, it will promote the design of products that carry risks in ways the models being used are not well equipped to capture. As noted earlier, this is the case with any regulatory capital measure that effectively (at least for a time) constrains banks' behavior. It is expected that banks will seek over time to turn every requirement into a compliance exercise, and in stress testing, the easiest way to do this in the United States is to simply copy the Fed's practices rather than put in the hard work of figuring it out for oneself—with the added parochial benefit of not risking estimating greater capital needs than the Fed requires. If the details of supervisors' models are not disclosed, substantial effort will be expended by banks in trying to reengineer them. The only credible way to push back against this predictable evolution is to continue to update and revise the scenarios and models on a regular and frequent basis.

---

[12] See Frame et al. (2015).

[13] For a good discussion of this issue, see Gutierrez Gallardo et al. (2015).

And to require the banks to develop and use their own models for stress testing and capital planning based on the specifics of their activities and risks and expect them to use them in practice rather than focusing solely on the supervisory stress test results.

5 Where Do We Go from Here?

Great progress incorporating stress testing into supervision has been made over the past 10 years, but there are still a number of issues to address. Most importantly, banks and supervisors need to keep pushing to expand the frontier of best practices, both with respect to stress testing and hopefully through the development of other compelling approaches that are as yet untried. Given the speed of changes in the financial sector, continuing evolution of practices is always needed, and resting on the successes of programs put in place to date would be a mistake. A short list of considerations for enhancing supervisory stress tests is as follows:

Expand the use of scenario analysis as a risk identification tool. The data now required to be submitted by participating banks to facilitate stress testing in many jurisdictions have provided a wealth of information that could only be dreamed of 10 years ago. Not only did supervisors not receive it in the past, but often banks themselves did not collect or save it. This information should be used to continuously probe for sources of possible vulnerability, both within individual banks and across the system. Although it may be difficult to then build this analysis directly into the capital regime, as discussed previously, the value of looking at many possible scenarios would be substantial, and it warrants further attention and effort.

Scenario design should remain creative. Related to the first point, scenario design should be directly linked to the continuous probing for emerging sources of vulnerability. Supervisors and banks need to resist the easy argument that greater predictability is a good thing. It distracts from the reality of ever-present uncertainty.

Better incorporation of liquidity and feedback effects. As has been noted by many, supervisory stress testing should explore the linkages between capital and liquidity needs under stress. It should also include a more direct incorporation of so-called feedback effects, which could include an iterative approach to better capture second-order implications. For practical reasons, these will both be challenging, but the potential value is great and worth the effort.

Better communication to minimize a false sense of security. Banks and supervisors should tone down the rhetoric that the results of supervisory stress tests (when positive) mean banks can necessarily weather a severe storm. Much of the potential outcome of a downturn for a bank will turn on the behavioral characteristics of its depositors, creditors, and counterparties. More transparency about the challenges of making such predictions would understandably make supervisors nervous—why should they say anything that might undermine one of their best tools? But the danger of creating a false sense of security is real. Moreover, overstating the value of stress testing poses risks to the reputations of both supervisors and stress testing as a practice. There is no magic bullet, just relatively better and worse ways to measure capital and liquidity needs.

Build stress testing directly into regulatory capital requirements for GSIBs. Poststress capital assessments should be built directly into regulatory capital requirements for all GSIBs, including the use of a poststress leverage ratio requirement that combines needed risk sensitivity and the beneficial simplicity of a traditional leverage ratio. If GSIBs are better prepared to withstand stress and can see that supervisors are holding their counterparts to the same high standards, the risk of a lack of confidence that can contribute to or exacerbate a downturn within and across banks should be at least somewhat reduced. This depends substantially on banks believing that the supervisors (and their also-supervised bank counterparts) are not missing critical risks or underestimating the impact of a downturn. This belief is strengthened by dynamic, robust, and conservative stress testing programs used by supervisors and banks and by strong capital and liquidity requirements informed by measuring the amount of each needed to increase the odds that a SIFI can weather a severe downturn.

Resist allowing stress testing to become just another compliance exercise. All supervisory requirements are, by definition, compliance exercises. This is a major challenge because if stress testing loses its dynamism, it will lose much of its value. Although banks' desire to improve efficiency and increase consistency is understandable, the process of routinizing assessments of capital and liquidity needs under stress can risk undermining the practice by decreasing necessary dynamism and reducing the direct engagement of key actors from across the bank. As we move forward, the emphasis should be on making stress testing more effective, with efficiency an important but second-order consideration that should not interfere with effectiveness.

After only ten years of using stress testing to measure capital adequacy and liquidity sufficiency, the field is still in its early years. Practices should and will continue to evolve. It will never be appropriate to proclaim current practices are sufficient. Stress testing, like all measures of capital and liquidity needs, is far from perfect, and the need to ensure resiliency at SIFIs in a world of uncertainty is a task with no end. Nonetheless, the strides made to date have strengthened the financial system by increasing resiliency at SIFIs and have provided supervisors with a framework through which to assess banks' capacity to run themselves in a safe and sound manner.

References

Basel Committee on Banking Supervision (2013), "Regulatory Consistency Assessment Programme (RCAP); analysis of risk-weighted assets for credit risk in the banking book," available at www.bis.org/bcbs/publ/d363.htm.

Federal Reserve Board of Governors (2009, May 7), "The Supervisory Capital Assessment Program: Overview of results," available at www.federalreserve.gov/newsevents/files/bcreg20090507a1.pdf.

Federal Reserve Board of Governors (2011), The Capital Plan Rule, 12 CFR Part 225.

Federal Reserve Board of Governors (2012), "Consolidated Supervision Framework for Large Financial Institutions," Supervision and Regulation Letter 12–17, available at www.federalreserve.gov/supervisionreg/srletters/sr1217.pdf.

Frame, W. Scott, Kristopher Gerardi, and Paul S. Willen (2015), "The failure of supervisory stress testing: Fannie Mae, Freddie Mac and OFHEO," Federal Reserve Bank of Atlanta Working Paper 2015–3.

Glasserman, Paul, and Gowtham Tangirala (2015), "Are the Federal Reserve's stress test results predictable?" Office of Financial Research Working Paper 15–02.

Gutierrez Gallardo, German, Til Schuermann, and Michael Duane (2015), "Stress testing convergence," Working Paper, available at https://ssrn.com/abstract$=$2636984.

Hirtle, Beverly, Til Schuermann, and Kevin Stiroh (2009), "Macroprudential supervision of financial institutions: Lessons from the SCAP," Federal Reserve Bank of New York Staff Report 409.

Ong, Li Li, and Ceyla Pazarbasioglu (2013), "Credibility and Crisis Stress Testing," IMF Working Paper WP/13/178.

18

Strengths and Weaknesses of Microprudential Stress Testing for Financial Institutions

Christine M. Cumming

1 Introduction

Modern capital and liquidity stress tests are transformative tools for the management and microprudential supervision of financial institutions. Modern stress testing has provided a more forward-looking and flexible framework than previously available to evaluate the resilience of financial institutions by quantifying the impact of a variety of scenarios, with a special focus on severely adverse scenarios.[1] Stress testing is transformative in its clear emphasis on scenario analysis; its capacity to focus on the far-left adverse tail of the distribution of a firm's financial outcomes; and its comprehensive, flexible, and modular structure.

Although stress testing has been practiced for decades, what we call *modern capital and liquidity stress testing* was developed by supervisors during the global financial crisis in 2008–2009 as an effort to determine the capital adequacy of large banking companies. Bank management and banking and securities supervisors faced compelling questions during the crisis. Which banks and securities firms were in a position to withstand the severe recession and distressed financial conditions being experienced in the fourth quarter of 2008 and first quarter of 2009, with signs that the distressed conditions could persist longer? If banks and securities firms could withstand the severe conditions, would they have sufficient capital (and liquidity) to conduct meaningful banking and securities operations for their customers and thereby support an economic recovery?

Modern capital and liquidity stress testing seeks to answer these questions with a projection of the amount and structure of capital or liquidity and funding available after a specified period of time. The extent to which a financial institution has adequate capital after a severe stress event defines what we will call its *insolvency risk*, the risk that capital falls below a threshold level that market participants or regulators view as necessary for the institution to meet effectively its customers' demand for banking or securities services. Similarly, the extent to which an institution has adequate stress- and post-stress-event liquidity defines its illiquidity risk, the risk that the institution—in reality or in market perception—can no longer generate sufficient cash flows to meet its financial obligations. Together, these two risks are key to determining a financial institution's *viability*, its ability not only to avoid

---

[1] The Board of Governors, Office of the Comptroller of the Currency (OCC), and Federal Deposit Insurance Corporation (FDIC) (2012) view capital and liquidity stress testing as equally important.

insolvency but also to provide services to the economy as laid out in its charter, license, or corporate mission statement.[2]

To understand the extent of insolvency and illiquidity risks, modern stress testing has characteristics that set it apart from previous stress testing, based on public-sector surveys of international bank practices.[3] Modern stress testing is severe, comprehensive, coherent, forward-looking, and flexible. It should also be frequent, transparent, and comparable over time.

Modern stress testing has the following characteristics:

- It utilizes severely adverse economic and financial scenarios to capture meaningful estimates of insolvency and illiquidity risks in financial institutions. Modern stress testing represents a decisive move to specific scenarios as a description of severely adverse conditions. The use of a scenario makes the overall severe shock concrete and plausible, although uncommon, and ensures that the overall behavior of the financial markets and economy is economically coherent. The need to capture the multiple dimensions of a severely adverse scenario, involving both financial and real economic disturbances, such as the 2008 financial crisis and Great Recession, requires specifying a significant number of key economic and financial drivers in an economically sensible manner. Prior to the global financial crisis, stresses tended to be moderate recessions or financial disturbances by post-World War II, pre-global financial crisis standards.[4] The stress was characterized by its percentile on the empirical distribution (e.g., 95 or 99 percent), often focused on a single driver, thus making the stress test both limited in scope and tethered to an estimated distribution from the precrisis period.[5]

- It strives to be comprehensive, in that it simulates the entire augmented financial balance sheet and income statement (and for liquidity, the entire cash-flow statement) over time. *Comprehensive* means both modeling the complete balance sheet and major off-balance-sheet items (together, the augmented balance sheet) *and* considering the full impact of the adverse scenario on the augmented balance sheet's evolution. For capital stress testing, the full impact includes not only the immediate capital losses and gains but also quarterly or annual income because the adverse scenario affects revenues and expenses, incorporating changes in the volume of activity, such as lending and fee-for-service businesses. Activity

---

[2] The use of the terms *insolvency risk* and *illiquidity risk* is meant to emphasize that the aim of modern stress testing is to probe deeply enough in the adverse tail to test viability. Viability is discussed in the Financial Stability Board's "Key Attributes of Effective Resolution Regimes for Financial Institutions" (2011).

[3] We use as a baseline for "modern" stress testing a recent study of the state of stress testing practices at banks by the Basel Committee on Banking Supervision (BCBS, 2017) and its updated principles of stress testing (BCBS, 2018) and the description and instructions for the Comprehensive Capital Analysis and Review, summarized by the Board of Governors (2018b). The baseline for the decade before the 2008–2009 global financial crisis is drawn from two studies of internal bank stress testing practices by the Committee on the Global Financial System (CGFS, 2001, 2005) and an extensive critique of internal bank stress testing methods (BCBS, 2009).

[4] The BCBS (2009) notes that "[p]rior to the crisis, 'severe' stress scenarios typically resulted in estimates of losses that were no more than a quarter's worth of earnings."

[5] Reports from the earlier period also note the use of sensitivity analyses, which evaluate the impact of changes, such as a parallel shift in the yield curve of +/−50, 100, or 200 basis points, that were often considered stress tests but usually fell well below the actual movement of rates in a rising cycle.

volumes change as customers and the financial institution respond to general financial and economic stress and the financial institution's apparent financial condition. Projecting preprovision net revenue (PPNR) is an especially important innovation because depressed revenues are often a significant limiting factor in rebuilding capital or liquidity once a stress event has occurred. The stress test should be coherent: the assumed impacts on individual income-statement or balance-sheet components should reflect an overall firm strategy over the scenario. Prior to the global financial crisis, stress testing was almost always confined to a segment of the balance sheet, not the whole, and largely focused on immediate capital gains and losses, without impacts on revenues or forecastable changes in customer or bank behavior, although the Basel Committee on Banking Supervision (BCBS, 2009) reports some exceptions.

- It is forward-looking and dynamic: forward-looking in that it builds on a narrative of what could happen and not just a replay of what has happened in the past, and dynamic, in that it allows management actions in each period to reflect the changes in the capital, liquidity, income, and balance sheet as the test progresses from one period to the next. To be dynamic, a stress test needs to incorporate the passage of time, facilitated by a structure that considers sequential discrete time periods. Although some stress testing regimes cover one or more annual periods, both greater periodicity and a longer horizon should be introduced in order to capture management actions, the impact of the ripple effects of an initial shock, and the initial stages of recovery. Ripple effects include depressed levels of economic activity, deepening borrower distress, and the spread of the initial shock to other sectors of the economy; they can persist well into recovery. Multiple periods also provide flexibility to consider a variety of scenarios: a single large blow to the financial and economic system (most common) or a sustained series of smaller shocks. Prior to the crisis, most stress testing was point-in-time testing, involving a single, instantaneous shock, with little attention to recovery from the shock.

- It should be frequent, transparent, and comparable over time. The results from stress tests, usually expressed as post-stress period capital levels, can provide valuable additional information about the financial condition of condition of financial firms. This was evident from the publication of the results from the first supervisory capital stress tests conducted during or just after the financial crisis until the present. To make these disclosures meaningful to investors and the process fair to financial institutions, supervisory stress test processes should be transparent. Investors and financial institution management should have a flow of periodic results sufficient to meaningfully assess the strengths and weaknesses of a financial institution's capital or liquidity position.

- It leverages the extensive improvements to financial data collection, review, and quality control; data integrity and safeguarding; and more stringent controls on data access that emerged in the 2000s. Modern stress testing relies on models that incorporate the necessary detailed data elements to capture a fuller set of characteristics of loans, securities, and trading positions than were previously available. The purpose of detailed data elements is not to analyze individual items but to group them into portfolios responding to the same essential drivers. The detail is often instrumental to achieving higher levels of precision and statistical power in stress test projections. These capabilities allow for greater comparability across bank loan portfolios because supervisors can potentially categorize loans by their price and nonprice terms and any underlying collateral using consistent definitions in the stress testing process. Stress testing in the 2000s generally relied on aggregates.

- It makes use of specific, theoretically grounded and disciplined economic and financial modeling to estimate the adverse scenario's impacts on all of the institution's significant portfolios. The models may be simple and parsimonious for some portfolios and more complex for others (e.g., mortgages, options). Disaggregated portfolios allow for differentiated models for the major elements of the augmented balance sheet, customer volumes, and revenues. In addition, an entire discipline has developed around model development, the use of assumptions, processes for vetting and testing models, documentation, and internal controls to safeguard model integrity. Because models may have known shortcomings, their use has been accompanied by similarly rigorous and well-documented methodologies for analytically based judgment, such as overlays. Prior to the financial crisis, the basics of model management were well recognized, but adoption was best characterized as gradual and evolutionary.
- It requires comprehensive, coherent management information systems. One effect of the push toward comprehensive capital and liquidity stress testing has been a dramatic surge of investment in management information systems by financial institutions, driven substantially by qualitative supervisory requirements. Meeting the goal of comprehensive and coherent management information systems is almost certainly incomplete, and the improved state of information systems is at risk if future investment is insufficient, especially as bank business models evolve. Prior to the crisis, information systems at many financial institutions grew by accretion through acquisitions. Few financial firms apparently saw the private benefits of integrating systems in mergers and acquisitions as exceeding the private costs of investing in common management information systems across the enterprise.[6]

The first modern capital stress test, the Supervisory Capital Assessment Program (SCAP) in 2009, embodied all these new features. As an innovation, it reflected both the urgent need to determine the insolvency risk for major banks in the United States and the new capabilities in data and model management. The urgency to answer the key question of solvency for major banks and securities firms helped break through two important barriers for financial institutions and financial supervisors. One barrier was psychological: the crisis swept away bank management reluctance to ask existential questions about what types and combinations of shocks could cause a large financial institution to fail and why. The second barrier involved resources. Urgency made it possible to leverage the increased data, modeling, and information systems capabilities into a resource-intensive stress testing regime. Breaking through these barriers was instrumental to achieving sufficient conceptual and empirical precision to form meaningful estimates of insolvency and illiquidity risk. Publication of the SCAP results validated the immediate solvency of the banks tested, even as 10 of the 19 banks failed the forward-looking quantitative stress test and were required to raise additional capital.

---

[6] The Senior Supervisors Group (2009) reported that banks assessed their information technology (IT) infrastructure as "ineffective" for risk data aggregation and exposure monitoring. In January 2013, a BCBS report noted that "making improvements in data risk aggregation capabilities...remains a challenge for banks" (BCBS, 2013, point 5).

The focus of these comparisons has been capital stress testing, but liquidity stress testing has also benefited from the same advances. The framework of an enterprise-wide liquidity stress test dates back to at least the early 1990s.[7] Liquidity stress testing simulates each important cash flow from one period's (augmented) balance sheet to the next period's, repeated over several periods. To be useful, liquidity stress testing generally needs to focus on shorter, more frequent time periods than the simulation structure of most capital stress tests. At this point, liquidity stress testing is generally not a formal supervisory test with substantial disclosures after each round. Liquidity stress tests are still largely run using supervisory scenarios on bank-designed models.

The questions of solvency and liquidity could not have been answered as fully as necessary in 2009 had it not been for at least three decades of applying statistical methods to financial and economic data, with a focus on the adverse tail of the distribution's behavior. The public-sector surveys of stress testing highlight important developments in the evolution of stress testing and make clear that from the beginning, regulators, central banks, and some banking companies set a goal of developing informative and actionable stress testing as comprehensive as today's, in order to supplement the rapidly evolving quantitative methods for risk measurement and management then being developed.

Credit, market, liquidity, and other risk disciplines at financial institutions have reached sufficient maturity to allow firms to combine them to answer difficult but critical questions about the viability of the firm and the sufficiency of its capital and liquidity during an event of severe financial distress and the ensuing recovery period. Management and regulators alike have found that effective capital and liquidity stress testing sets new prerequisites for the firm. Those prerequisites are stronger, more consistent, enterprise-wide application of risk disciplines and a strong risk and control mandate from the board of directors and senior management, supported in implementation by the "three lines of defense" in risk management and internal control.[8]

2 The Transformative Power of Modern Stress Testing

How is modern capital and liquidity stress testing transformative? The answers point to the strengths of modern stress testing. Principal areas of transformation are as follows:

- Modern stress testing explores a region of the distribution of capital and liquidity outcomes much closer to the possibility of financial institution failure than in the past. Two important shifts make that possible. The first moves from measuring losses to an explicit estimate of capital and liquidity levels, putting insolvency risk, illiquidity risk, and the viability of the financial firm at the center of analysis. The second moves from moderately adverse, largely single-threaded scenarios to severely adverse, multidimensional, and realistically complex scenarios based on an anchoring narrative emphasizing the dynamic behavior of economic agents and the economy under highly distressed circumstances.

---

[7] See BCBS (1992).

[8] See OCC (2014). The three lines of defense are set out in Section II, entitled "Standards of Risk Governance." It includes a discussion of the three lines of defense in risk and controls (business line, central unit, internal and external audit) as well as the role of the board of directors.

- Modern stress testing is more effectively forward-looking, in that the multiperiod time horizon of the projection is generally long enough to understand whether financial firms not only survive a severely adverse scenario but also have sufficient capital and liquidity to operate effectively during and after the scenario's trough.
- Modern stress testing can draw on an academic literature developed since the global financial crisis that seeks to explain why the combination of severe financial crisis and deep recession occurs, what makes the combination so damaging, and why recovery and rebound are so difficult. This literature has developed in parallel with the development of stress testing; both advances provide new and complementary insights. The literature demonstrates the importance of liquidity and capital, as well as the key role of leverage and liquidity mismatch in amplifying and extending severe downturns. Supervisory stress testing has provided a wealth of data and insights into bank business models and how capital influences bank behavior in normal times and in periods of stress.
- Modern stress testing is already providing transformative insights for investors in and counterparties of financial companies and institutions for which stress tests are disclosed. Similarly, microprudential stress testing has been transformative for financial supervisors of large financial institutions, increasing the emphasis and focus on insolvency and illiquidity risks in large financial institutions. The Federal Reserve, for example, currently states its key areas of supervisory focus for large financial institutions to be capital, liquidity, governance and controls, and recovery and resolution planning.
- Modern stress testing facilitates adding insolvency and illiquidity risks (tail risk) as a component of strategic decision-making at financial institutions, in addition to expected risk and return. The use of bank-internal stress testing models in strategic decision-making expands the range of financial metrics considered in strategic decisions and should improve long-term decision-making.

3 Further Discussion of the Strengths of Modern Stress Testing

3.1 Attention to Plausible, but Extreme, Adverse Outcomes and the Recovery Process

The attention to the extreme adverse tail of the distribution of financial outcomes and a bank's potential for resiliency and ability to recover is a transformative development, the most important analytical advance in the assessment of financial institutions to emerge from the global financial crisis. Modern stress testing stands alongside several post-global financial crisis assessment and supervisory tools that focus on plausible, but extreme, adverse outcomes, designed to address the resilience of financial institutions to historically high levels of financial and economic distress. Moreover, supervisors are looking for coherence between the bank's capital stress testing and its risk appetite and internal risk limit setting. Other such postcrisis developments are the implicit stress tests that support the setting of minimum bank capital requirements, especially the innovations of the Basel III capital conservation buffer, the cyclical risk buffer, and the systemic risk surcharge, the latter two applicable only to global systemically important banks (GSIBs). A liquidity stress test framework underlies the innovations of the Basel III liquidity coverage ratio (LCR) and net stable funding ratio (NSFR). Yet another set of postcrisis reforms focuses on near-failure and actual failure of large financial institutions: management-developed recovery and resolution planning (living wills), supervisory resolution planning, and the total loss-absorbing capacity requirement, a requirement that a financial institution must have sufficient senior and subordinated debt to recapitalize the bank in resolution, as well as sufficient liquidity, in order to operate successfully through a relaunch and recovery period. These developments, too, begin with a severely adverse scenario, sufficiently severe to test the financial institution's likelihood of near-term failure or inability to recover.

In considering the far-left tail of financial outcomes, modern stress testing is important in the same way that the introduction of mean and variance analysis and the identification of potential skewness and/or leptokurtosis (fat tails) have been in financial analysis and quantitative risk management practices. Attention to the far-left tail acknowledges that the left region of the distribution of capital outcomes is effectively truncated. As a large financial institution is deemed nonviable, the managers and shareholders of the institution forfeit control of the institution to the resolution authorities.

The modeling of the distribution in the region approaching nonviability can change sharply because the increasing likelihood of a financial institution's nonviability can and likely will substantially change its own behavior and that of its customers, creditors, and counterparties.[9] Behavior throughout the adverse tail is unlikely to be linear and may depend substantially on both macroeconomic conditions and the microeconomics of the firm's prospective financial condition. For large financial institutions, the point of nonviability is not fixed by regulation or statute and can only be estimated. For senior management, shareholders, and senior and subordinated debtholders, the move to resolution creates a largely irrecoverable loss.

3.2 Modern Capital and Liquidity Stress Testing Can Draw on New Postcrisis Theoretical and Empirical Research on Factors That Amplify Shocks and Cause Recessions to Persist

Since the 2008–2009 financial crisis, economists have made substantial advances in understanding how severe episodes of financial instability and recession develop. By integrating a more articulated model of the financial sector with a macroeconomic model (creating a macrofinance model), economists seek to illuminate how leverage and liquidity mismatches at financial institutions contribute to the severity of financial and economic distress. This literature provides an explanation for the difficulty financial institutions face in recovering after a decline in their net worth and posits the potential for multiple economic equilibria, including the possibility of a "bad" equilibrium in the form of a persistently depressed state.

Recent theoretical work (e.g., Brunnermeier and Sannikov, 2014; Adrian and Boyarchenko, 2018; Gertler et al., 2020; Kashyap et al., 2017) focuses on the financial mechanisms that cause amplification of shocks and persistence of their effects. Reinhart and Rogoff (2009) document amplification and persistence as characteristics of episodes of severe

---

[9] See Board of Governors of the Federal Reserve System (2018d, p. 17).

financial distress and economic underperformance. The recent theoretical literature focuses on asymmetry and nonlinearity in the behavior of financial institutions in crisis periods.[10]

Brunnermeier and Sannikov (2014) identify a crucial endogenous reaction to large negative shocks.[11] Occasional small negative shocks are usually absorbed if there is accumulated saving (net worth) or if assets can be sold with relatively little loss of value. Large shocks, however, can overwhelm buffers of accumulated savings or the sale value of assets. The large shocks set off a round of effects that drive down net worth beyond the initial shock by depressing capital-building and the value (price) of capital (equity) and thus net worth. The further decline in net worth sets off additional rounds of reactions, creating a (hopefully tightening) downward spiral until an effective bottom is reached. Brunnermeier and Sannikov identify this cycle of net worth declines, falling investment, and lower capital prices as an endogenous spiral that amplifies the shock and extends the persistence of its impacts.

The addition of financial markets and financial institutions to macroeconomic models highlights the importance of leverage and liquidity mismatch as further amplifying mechanisms in financial crises. Financial institutions such as banks and securities firms are inherently leveraged and often exploit the usually upward-sloping interest rate yield curve to fund longer-term assets with short-term liabilities. Losses in a leveraged balance sheet lead to more than proportional losses of net worth (equity). Large declines in financial institution net worth lead to constriction in the supply of credit in response to lower regulatory capital ratios. A lower credit supply creates the prospect of reduced economic growth, hurting revenue prospects and the value of capital and raising the potential for more borrower defaults. Net worth is revalued lower, providing fresh impetus to a continuing downward cycle. Similarly, liquidity mismatch can lead to an endogenous spiral as firms attempt to narrow the liquidity mismatch by liquidating assets (often the most liquid assets) in fire sales, driving down prices, and creating new liquidity and collateral demands on financial firms.[12]

Once such a financial crisis and severe recession begins, Brunnermeier and Sannikov (2014) posit that the economy may have one of two paths. In one, the endogenous spiral extends the economy's decline until the economy reaches a level where the powerful incentives of low prices and pent-up demand can generate a more-than-offsetting momentum toward recovery. In the other, the economy cannot generate sufficient momentum for recovery to offset the continuing endogenous spiral and remains in a depressed state. The notion of financial and economic instability leading to a possible bifurcation of economic equilibria is often associated with the writings of Minsky (1977).

---

[10] In a dynamic setting, downward-transition probabilities in outcomes from one period to another may rise sharply near the point of nonviability, higher than transition probabilities in the body of the distribution of outcomes.

[11] In an early version of their paper, Brunnermeier and Sannikov begin with a simple two-period model with at least two types of agents and no financial sector and gradually generalize the model to include illiquid capital, a financial sector, and financial markets.

[12] The authors note that financial institution size (i.e., the extent of concentration) can also amplify shocks. This nod to concentration suggests that concentrations form another class of potential amplifiers of external shocks.

Empirical investigation of the distribution of expected economic growth demonstrates observable nonlinearity across the distribution, with a striking long, fat left tail in the United States (Adrian et al., 2019). The resulting distribution of economic outcomes is asymmetric and variable in shape—positively skewed in years of higher expected growth, negatively skewed in years of low or negative growth. Although the region of expected positive growth above the mode is less leptokurtic than the region below the mode, the adverse tail becomes strongly leptokurtic in years of low to negative growth. Adrian et al. attribute the increasing leptokurtosis in the left tail to various financial frictions: elevated credit spreads, borrowing constraints, liquidity mismatches, and fire-sale dynamics, with the impact felt primarily in the financial sector. The importance of liquidity alongside capital, and not capital alone, is a key insight about the nature of combined financial and economic crises. This result mirrors the presence of asymmetry and nonlinearity in the adverse part of the distribution of financial outcomes as a result of financial frictions found in a model focused on financial-institution liquidity by He and Krishnamurthy (2013), among others.

This literature's insights reinforce the motivation for both capital and liquidity stress testing and for the disclosure of their results by pointing out the potential size and scope of economic spillover effects of illiquidity and capital distress in financial institutions. Such insights provide more support for regular microprudential capital and liquidity stress testing of financial institutions and for the importance of informed and significant market discipline through regular disclosure of stress testing results. The consequence of rising insolvency and illiquidity risks in financial institutions is the potential for a deeper economic downturn with greater difficulty in achieving recovery. Those consequences are relevant to all investors, and presumably rational and well-informed investors will seek to avoid such economic outcomes by disciplining financial institutions whose leverage or liquidity mismatch look to be headed beyond established norms. Thus, the outcome of capital and liquidity stress tests provides valuable information to the marketplace supplementary to standard accounting and risk disclosures.

4 The Value of Disclosure

Disclosure of stress testing results allows financial analysts and investors to understand the insolvency and illiquidity risks of a financial institution, its tail risks, and how the institution is earning the returns that allow it to meet its capital requirements. Thus, investors can better relate the firm's financial performance measures and its capital and liquidity strategies to its risks. Because supervisory stress tests are conducted around a common scenario using a common methodology, disclosure of stress test results allows a reasonably high degree of comparability of results across financial institutions. The modular structure of modern capital stress tests allows comparisons of the stress performance of key classes of assets across banks, as do the US DoddFrank Act Stress Test (DFAST) disclosures published by the Federal Reserve after the annual stress tests.[13]

---

[13] The DFAST differs from the Comprehensive Capital Assessment Review (CCAR) in that DFAST assumes that capital actions in the form of dividends, stock buybacks, and net changes in the stock of preferred stock and subordinated debt are the same as the previous year, whereas CCAR incorporates the proposed planned capital actions by bank management (Board of Governors of the Federal Reserve System, 2018a,c).

These important disclosures are largely limited to the results of microprudential capital stress tests of banking organizations. The results of microprudential liquidity stress testing of banks are not disclosed, even as aggregated results.

Financial institutions themselves disclose relatively little about their internal stress testing activities. What they report is mostly in a high-level discussion format, such as highlighting the importance of the firm's stress testing in its risk management, and the approaches to these disclosures are heterogeneous. An important exception is liquidity stress testing, where key factors influencing liquidity are usually summarized. Occasionally, a bank might describe the nature of the risk and the mitigation activities of the bank for a specific, high-profile risk, such as Brexit. In general, such specifics are infrequent, and no orders of magnitude are given.

Two other types of financial institutions, insurance companies and central counterparties, are subject to capital stress testing activities overseen by microprudential authorities, but the results of these tests are not disclosed. These capital stress tests illustrate the flexibility of the modern stress testing framework when applied to almost any financial company.

Insurance companies in Europe, the UK, and the United States, are subject to supervisory stress testing, often centered on testing of catastrophe risk, analogous to the severely adverse scenario for banks. One major difference is that catastrophic events are often (but not always) seen as having exogenous causes, in contrast to severely adverse events in banking, for which stresses are often substantially endogenous. Property and casualty insurance companies have maintained, for decades, good models of catastrophic events, such as earthquakes, hurricanes, and windstorms. Such tests often stop with the losses from the catastrophic event, however, and may not look at economic recession or other financial or economic disturbance. Catastrophe models are increasingly challenged by rapidly evolving demographics (life) and the combination of real estate and economic development and climate change (property and casualty). Insurance companies may stress test economic scenarios that affect both the substantial investments of surplus or reserves on the asset side and the impacts on the volume and pricing of premium-generating insurance contracts on the liability side. The US stress test allows insurance companies some flexibility in stress test design, and a company can include the combination of catastrophe losses and economic downturn in a format similar to that of bank stress testing. An assessment of capital also needs to take account of the well-organized markets in reinsurance for risk mitigation.

Central counterparties (CCPs), used for the clearing and settlement of derivatives, foreign exchange, and other traded instruments, are financial firms for which supervisors set high expectations for internal stress testing. The Committee on Payments and Market Infrastructures (2017) sets out that every CCP should conduct stress tests to determine the adequacy of its prefunded financial resources and its liquidity in a wide range of scenarios. The severity of the stress scenario involves multiple defaults of counterparties at a time of unusual market volatility, adverse price movements, and strained liquidity in asset markets.

As a matter of good risk management practice, the development of good stress testing models with scenarios of varying severity seems useful for any financial company that reaches critical mass. Where firms are seeking funding in the public or private markets, such stress tests should be an essential disclosure.

One of the major gaps in the financial reform efforts after the global financial crisis was the failure to take significant measures to improve the regulation and disclosure practices of shadow banks, less-regulated nonbanks.[14] A subset of nonbanks, shadow banks are involved in credit intermediation and have significant balance sheets that take on credit and market risks, including managing significant pipelines of assets destined for sale or securitization (Adrian and Jones, 2018). They include finance companies, mortgage banking firms, hedge funds, private equity firms, and some FinTechs. Shadow banks are at risk of credit loss and lack of funding.[15] Found in many countries, some nonbanks may have a relatively modest capital standard. Others, however, are regulated lightly or not at all. These types of entities are not subject to microprudential stress testing or required to provide stress test disclosures.

5 Stress Testing for Strategic Decision-Making and Assessment of Strategic Risk

The stress testing framework enables more sophisticated analysis of strategic risk and thereby better strategic decision-making.[16] Strategic risk is the potential that an organization could underperform (or even fail) as a result of misjudgments of strategy, assumptions, or capacity to execute. Virtually all strategic decisions involve consideration of measures of financial performance, such as net income, along with other dimensions, such as the risk profile of the firm. The stress testing apparatus can deliver measures of financial performance in baseline, moderately adverse, and severely adverse scenarios, along with accompanying measures of risk, notably including the tail risks of insolvency and illiquidity.

Early supervisory guidance noted the importance of linking private-sector stress testing to firm decision-making and to strategy decisions in particular.[17] As a tool for assessing shifts in strategic direction (mergers, acquisitions, entry into new businesses, changes of business mix, divestiture), the forward-looking, flexible, and dynamic nature of modern capital and liquidity stress tests and the alignment of stress test simulation results with the balance-sheet, income, and cash-flow disclosures by financial institutions facilitate its use in strategic decision-making.

---

[14] One reason for the shortfall in financial reform efforts vis-à-vis shadow banking is that the Financial Stability Board took a sectoral approach (e.g., money market mutual funds, securitization), focused on problematic sectors in the global financial crisis. In some sectors, nonbanks suffered high rates of failure during the crisis (US mortgage banks, mortgage conduits). In the current resurgence, nontraditional entrants, such as hedge funds, have become important in commercial lending markets, missed altogether in the sectoral approach. The US designation of systemically important nonbanks created controversy among its designees, and currently there are none. Systemically important financial market utilities are covered under another program.

[15] The definition of shadow banks follows Adrian and Jones (2018) and the article's focus on nonbanks carrying out credit intermediation.

[16] The BCBS (2017) notes that in its survey of bank stress testing practices, strategic decisions were mentioned by just 30 percent of its bank respondents, well behind assessing capital adequacy and liquidity (96 percent), recovery planning (55 percent) and related compliance and risk management issues. In contrast, Jamie Dimon, CEO of JPMorgan Chase, noted in a July–August 2018 *Harvard Business Review* interview, "We do more than 100 stress tests a week—related to geopolitics, capital downturns, recession, war."

[17] Principle 4 of the Board of Governors, OCC, and FDIC (2012) states: "Stress tests should be clear, actionable, well supported, and inform decision-making" (p. 7). It further states: "Stress testing should inform analysis and decision-making related to business strategies, limits, risk profile, and other aspects of risk management, consistent with the banking organization's established risk appetite." In the preamble to the principles, the regulators state that the stress testing framework enumerates benefits to management, including "contributing to strategic planning" and "enabling senior management to better integrate strategy, risk management, and capital and liquidity management decisions" (p. 4).

Such strategic analysis requires data as critical inputs, and the incompleteness or absence of available data points to the importance of sensitivity analysis to supplement scenario analysis in the stress testing framework. Data of sufficient quality, detail, and disaggregation comparable to what a financial firm possesses for its own operations may not always be available. The use of aggregates, imperfectly matching data series, and projections may be necessary. In such cases, the stress test structure provides a disciplined framework for sensitivity analysis, testing assumptions about revenues, costs, and risks and determining the sensitivity of net income and risk to rates of macroeconomic growth or changing production and input costs. In the stress testing framework, strategic decisions can be evaluated not only for their contribution to profitability and diversification of risk in normal times and in moderate recessions but also for their likely impact on overall results, capital, and liquidity in times of severe financial distress.

Senior management and, in turn, the board of directors can use stress test results to oversee business performance and to seek to detect the potential for large tail risks in the conduct of the business. Disaggregation to the business-line level helps management and the board understand the composition of the firm's net income and its risks. Management and the board can supplement their review of internal measures of performance and current measures of risk with a review of the potential for catastrophic loss or drains on liquidity at the business-line level. Financial institutions can make more nuanced judgments in establishing performance incentives, awarding compensation, and allocating capital and liquidity within the firm, particularly in those cases where strong financial performance is accompanied by substantial insolvency and illiquidity risks.

6 Development Opportunities for Microprudential Stress Testing

"Modern," post-2008, stress testing of capital and liquidity has not yet achieved its full potential and will require leadership from the public sector to do so. This section discusses three opportunities for advancing stress testing: developing an integrated capital and liquidity stress test; applying the stress testing framework to emerging and nonfinancial risks; and sustaining the momentum, vibrancy, and intellectual advances in stress testing efforts and disclosures by supervisors and financial firms. Microprudential supervisors made the initial investments in modern stress testing and are still best positioned to drive further progress. In some cases, they may need the support and assistance of other financial authorities.

6.1 Integrating Capital and Liquidity Stress Testing

The single most important near-term step forward would be integrating liquidity stress testing and capital stress testing to create a potentially powerful tool for executive management at financial companies and microprudential supervisors. Given the interaction of solvency and liquidity decisions in a financial firm, a single-track capital or liquidity stress test may be biased in estimating insolvency and illiquidity risk. In recently developed macrofinancial models, leverage and liquidity are determined simultaneously by the financial firm, influenced by the pricing and availability of capital and liquidity in the marketplace. Thus, a joint capital and liquidity stress test could be viewed as a "full" test of insolvency and illiquidity risks.

Adding urgency to integrating the tests is the experience of the global financial crisis. Many firms became illiquid, that is, unable to roll over or obtain new financing, and de facto failed well before they became technically capital insolvent, a condition called *liquidity insolvency*. Examples in the United States from the global financial crisis include investment banks, large commercial banks, thrifts (savings institutions), and a large insurance company. In Europe, the UK, and the United States, many institutions were spared liquidity insolvency by the combination of government capital injections and central bank extraordinary liquidity provision.

The risk of liquidity insolvency has risen somewhat. A principal reason is the tightening of regulatory restrictions on the use of (insured) deposit funding. These restrictions, such as Section 23a and Section 23b of the Federal Reserve Act and similar provisions in Japan, are intended to limit the use of insured deposits to fund "nonbank" activities, such as many capital-market activities. Since the global financial crisis, similar restrictions have taken effect in the UK and were proposed, but subsequently withdrawn, in Europe.[18] Such constraints have risen in impact because of the growth of capital-market activities, and the reliance on market funding creates vulnerability to a disruption of both funding and trading liquidity.

In the United States, existing restrictions have been tightened. In the 2008–2009 financial crisis, the Federal Reserve relaxed the Section 23a restrictions substantially; the 2010 Dodd–Frank Act narrowed the scope of the Federal Reserve's discretion. The Dodd–Frank Act also reduced the flexibility of the Federal Reserve's ability to lend to nonbanks in unusual and exigent circumstances.

The development of an integrated capital and liquidity stress test faces some significant hurdles. The first hurdle is developing an approach that retains the flexibility of a capital or liquidity stress test while dealing with the simultaneity of capital and liquidity determination. The simultaneity issue exists for perhaps three reasons. The first is that the firm has a single (augmented) balance sheet, and a decision to increase capital or liquidity will have an impact on both the solvency and the liquidity profile of that balance sheet.[19] The evolution of the balance sheet captured in stress tests is driven by the imperative that the income and cash-flow statements arrive at the same balance sheet at each period end.

The second reason is that some apparent substitutability exists between capital and liquidity in the market's overall gauge of an institution's financial condition. Empirical work such as Pierret (2015) and Schmitz et al. (2017) has found a statistical association between capital and liquidity levels in the cost of funding, suggesting a limited substitutability of capital and liquidity. Some theoretical models have explored potential simultaneity in the determination of capital and liquidity in a firm. Adrian and Boyarchenko (2017) demonstrate a

---

[18] In the UK, the authorities have "ring-fenced" traditional banking activities and their deposit funding from other activities in the banking company, which cannot be deposit funded, effective January 1, 2019. In Europe, the 2012 Liikanen Report recommended a similar approach, but that proposal was withdrawn in 2018.

[19] For example, an increase in capital funded by a sale of risky assets ceteris paribus will not only increase the firm's capital buffer relative to its credit risk, but it will also increase its liquidity by reducing its stock of assets. Similarly, an increase in liquidity (e.g., central bank balances) funded by a reduction of risky assets ceteris paribus not only increases the liquidity of the firm but also increases its capital buffer relative to its risk.

trade-off between capital and liquidity in reducing systemic risk.[20] Kashyap et al. (2017) have developed a theoretical model to understand how management of solvency and liquidity risk influences the macroeconomy.

A third reason is the only approximately understood role of financial frictions. Financial frictions (elevated credit spreads, borrowing constraints, collateral shortages, or declines in collateral value) are central concerns in liquidity stress tests.[21] Financial frictions are generally thought of at the macroprudential level as market-wide adverse developments, yet they can be firm specific, such as an increase in the credit-risk spread for a particular firm. At both the firm-specific and market-wide levels, financial frictions often seem to be driven by concerns about financial-institution credit quality and generally affect financial-institution liabilities (cost and availability of funding, access to equity and bond issuance, terms of capital market borrowing, availability of trading lines). Such costs and constraints associated with liquidity can force changes in the size and composition of the balance sheet and trigger fire sales of assets.

In addition to simultaneity, other headwinds to establishing integrated capital/liquidity stress testing include the following:

- Understanding the capital–liquidity nexus in activities that cannot be funded by deposits, such as capital-market activities. More work is needed to understand fully what kind of capital and liquidity requirements could enhance the stability of financial institutions with large capital-market activities.
- Understanding the dynamics of funding markets, that is, why funding and trading liquidity may appear to be available to a firm even when signs of deterioration in its creditworthiness are visible, and conversely (and less commonly), how credit may be available when liquidity is strained, until the apparent availability of one or the other suddenly gives way to run-like conditions.[22]
- Developing a stress testing strategy to deal with differences in the periodicity of changes in liquidity and capital: days, weeks and months for liquidity, quarters and years for capital. Estimating a system of capital and liquidity models, including imposing cross-system constraints arising from the single balance sheet, would be significantly more computationally intensive than separate capital and liquidity stress testing.

In the absence of an integrated capital and liquidity stress test, supervisors can pair a more formal microprudential liquidity stress test with the capital stress test and disclose the results,

---

[20] Their model allows for direct computation of the covariances in joint capital and liquidity optimization so that the policy analyst can determine the sign and size of those correlations. Interestingly, the trade-off between capital and liquidity is not necessarily present when considering consumption or economic growth, broad welfare measures.

[21] Adrian et al. (2019) link financial conditions to the nonlinearity in the left (adverse) tail of the distribution of expected gross domestic product (GDP) 1 year out, in contrast to their lack of impact in the body of the distribution. Related work is covered by the International Monetary Fund (IMF, 2017).

[22] An example of the latter occurred prior to the 1984 run on the Continental Bank. Its correspondent banks had largely abandoned Continental, whereas depositors in the euromarket vastly expanded their funding, relying on the market convention of the time that treated the top 10 US banks as having common credit characteristics.

enabling banks and investors to consider the capital and liquidity ramifications of a severely adverse stress scenario. Simultaneous testing of capital and liquidity could be an important first step toward an integrated capital and liquidity stress test.

7 Challenges in Quantifying Risk in Some Emerging Threats

A second area of opportunity for a financial institution's internal stress testing is the exploration of risks associated with nontraditional or emerging threats, many often initially described in nonfinancial or even nonquantitative terms. Many emerging and some traditional threats present the challenge of quantification of both the size of the shock produced by the risk and its financial impacts on the firm.

To make the discussion more concrete, consider the risk of cyberattack, often cited as a leading risk concern of bank management. The impacts of cyberattacks are thought of in terms of the hours or days needed to resume business (hours lost); remediation time and cost necessary to restore damaged or destroyed information, software, or equipment; and loss of customer trust. These impacts can be translated into the loss of short-run business revenue, the cost of remediation and the lost value of tangible and intangible assets, and reputational risk that drives customers and their long-term revenue away.[23] Each dimension can be translated, however imperfectly, into financial estimates.

The imprecision of these estimates is similar to those encountered when considering strategic decisions in the stress testing framework. As in that case, the lack of hard data suggests the value of conducting sensitivity analysis in addition to scenario analysis. Often, the realization of such emerging or nontraditional risks will create a dynamic situation to navigate. In a cyberattack, adjusting an existing stress test's assumptions and parameters for the specific and changing facts of an attack can be invaluable in navigating its operational and financial aspects. The stress testing framework and sensitivity analysis could also be compelling in working through and justifying preventative measures and business recovery strategies.[24]

More broadly, the capital stress testing framework offers potential to assist management in addressing emerging or complex risk challenges to a financial firm, such as geopolitical risk, trade tensions, business model disruption, and operational and reputational shocks. A starting point for geopolitical risk, for example, is to consider what revenues, costs, and asset valuations are at risk as a result of a particular potential flashpoint. Some financial services firms and vendors have developed data on operational and reputational problems, from their own experience and that of others, in order to make plausible estimates of the distribution of losses and remediation costs arising from such nonfinancial risks. Even a limited number of cases can be helpful in developing initial estimates.

---

[23] See, for example, Healey et al. (2018) and Warren et al. (2018) for discussion(s) of cyber-risk and financial stability.

[24] As Healey et al. (2018) point out, vulnerabilities such as leverage existing in the economy and financial system at the time of the attack could amplify the impact of a cyberattack.

7.1 Public-Sector Leadership in Sustaining and Expanding Stress Testing and Disclosure of Results

The modern capital and liquidity stress testing initiated by the microprudential supervisory community and the disclosure of microprudential capital stress testing results have catalyzed transformative efforts in the public and private sectors to measure and manage tail risk. That leadership should continue, with the support and assistance of other financial authorities.

7.2 Scenarios for Supervisory and Bank-Internal Stress Tests Need to Change over Time

Today's supervisory stress tests generally focus on a deep recession and financial distress commencing at the start of the stress test period, modeled on the global financial crisis and recession of 2008–2009. The simultaneous severity and undeniable plausibility of the scenario make it an attractive continuing benchmark for severely adverse stress. Accompanied by public disclosure, that benchmark affords comparability across the banking industry over time.

Yet two risks to banks and the overall financial system arise when applying the same stress scenario over time. As supervisory capital stress testing content and process have solidified, banks subject to the stress test have come to understand how the system of supervisory models, "the supervisory model," works.

The first risk is that banks might not be sufficiently attentive to how the changing economic and financial environment is altering their risks, or they may observe such changes but find it difficult to incorporate them into a sufficiently adverse scenario. The second risk is that banks seek to reverse-engineer stress tests in order to pack more risk into their business without increasing their capital needs under the stress test. Banks that engineer or "game" the stress test results treat the capital stress test as a short-term constraint on their efforts to maximize profits. As a result, the stress test becomes far less effective.[25]

Both risks can lead to an underestimation of risk at individual banks and an overestimation of their solvency and liquidity. If that underestimation of risk occurs broadly across the industry, it can become a source of systemic risk.

The implication is clear: supervisory stress testing should go beyond one anchor scenario to a variety of scenarios. The Bank of England, for example, conducts its capital-requirement-setting stress test of major banks annually for the seven largest banks, with a scenario more severe than the 2008–2009 financial crisis, but in roughly alternative years, it

---

[25] In promulgating Basel I in 1988, the BCBS was concerned that banks would seek to optimize against minimum requirements and attempted (unsuccessfully) to make clear that banks were expected by supervisors to operate well above the minimum capital requirements. In time, as financial institutions concentrated each loan category with the riskiest types of transactions allowed in that category, Basel I became viewed as ineffective and nonbinding, leading to the development of Basel II. This second risk reflects the Lucas critique that economic agents with an understanding of the economy learn how the public sector reacts to new data or a changed environment and anticipate its policy actions, reducing the effectiveness of the policy.

runs a stress test based on "exploratory" scenarios as an exercise to understand the vulnerabilities of the UK banking system without a formal evaluation of capital levels. In late 2019, the Bank of England sought input on three climate-change scenarios planned for 2020.[26] Other regulators could adopt similar practices to ensure that stress tests remain challenging and relevant. Some of those additional scenarios could thoughtfully incorporate new or emerging risks.

Other types of changes to scenarios could be considered. Instead of one large shock, a series of repeated smaller shocks or a double-dip recession could be considered.[27] Because even a small shock could trigger a severe economic recession and market seize-up through an endogenous spiral, research could explore how varying levels of leverage and liquidity mismatch, including in the external sector, could create other severely adverse macroeconomic scenarios.

It often requires imagination to see the potential for severely adverse scenarios in the contemporaneous economic situation. Financial and economic crises such as the Great Depression and the 2008–2009 global financial crisis result from multiple mistakes, misjudgments, and sometimes malfeasance in the private and public sectors, an uncomfortable aspect of scenario development. Some newly observed or newly understood risks pose challenges of analysis and quantification. But supervisors should see it as essential that they and financial institution management understand and address all substantial risks promptly and thoroughly.

7.3 Models Underlying Supervisory and Bank-Internal Stress Tests Need to Change over Time

Financial authorities collectively have an impressive history of scholarly leadership in the microeconomics of financial institutions, markets, financial stability, and stress testing. In addition, microprudential capital stress testing requires banks to collect, manage, and submit large amounts of their data to supervisory authorities.

These data represent a much-improved window into contemporary financial institutions and the credit and liquidity features of portfolios of different types of loans, securities, trading assets, and liabilities. The potential for substantial research on financial institutions, markets, and most especially, differentiated models of bank portfolios, still underexploited, could advance knowledge and help update stress testing models. Moreover, if liquidity stress tests involved data structure and collection similar to capital stress tests, the resulting data sets on capital and liquidity could dramatically further our understanding of these crucial buffers.

The challenge for the custodians of such highly sensitive stress testing data is to identify avenues to expand access to these data in a controlled manner that preserves the confidentiality of individual banking institutions. Access to these data could be made available to scholars in much the same highly controlled manner as US census data are made available for research.

---

[26] See Bank of England (2019).

[27] Brunnermeier and Sannikov (2014) demonstrate in their model that a series of small shocks aggregating to the same size as a single large recession/financial shock should have approximately the same effect once the series of shocks is completed.

Efficient quantitative methods to facilitate iterative models of interaction within stress testing models is another area for research. Making it possible to create scenarios in which financial-sector actions lead to changes in overall credit and liquidity supply to the economy would create a stronger link between bank actions in a stress event and the resulting macroeconomic environment in which they operate. Although largely a macroprudential stress testing exercise, it could have interesting insights into the microprudential health of individual institutions.

Such iterative quantitative methods could allow financial authorities to simulate intersectoral impacts of a shock. Like banks, nonbanks with substantial balance-sheet and credit activities can contribute to endogenous spirals that amplify and increase the persistence of a shock. The Bank of England (2015) discusses the possibility of a system-wide stress test of the financial system that includes sectoral interaction with every major financial-industry sector, largely for macroprudential purposes.[28]

The interaction between shadow banks and the banking and financial markets is especially important. Capital markets, short-term wholesale funding, and the banking system remain important sources of balance-sheet funding for shadow banks. Such firms are often at the leading edge of new risks. The shadow-bank sector has enjoyed a resurgence in recent years in many countries around the world. The shadow-bank sector should be drawn into macroprudential stress testing, paving the way for internal stress testing by those companies.

8 Expanding Stress Testing Disclosures

Capital and liquidity stress testing becomes transformative as it becomes a key element of financial analysis and thereby of market discipline. For that to occur, disclosure is imperative. Stress testing and public disclosure should be expected of any larger financial company seeking credit and funding in the capital and money markets, with private disclosure expected for investors in private markets and the banking system.[29]

The supervisory community can lead by expanding and innovating around disclosure of stress testing results and encouraging financial institutions to expand disclosures of their self-developed stress tests. Microprudential bank supervisors can begin by disclosing the results of liquidity stress tests. The time to do so is in a period when firms are strong and markets are liquid and open to a wide range of borrowers. If the stress test reveals problems, it forces institutions to address them with a meaningful plan. Once an economic downturn looms and markets are edgy, it inevitably appears too risky to initiate liquidity disclosures. A great value of regular disclosures of liquidity information is that they create a baseline against which investors and depositors can gauge information in more stressful times, rather than imagining the worst.

As supervisors conduct a wider range of capital stress tests, including those that envision unusual patterns of economic downturn and highlight emerging and nontraditional risks, supervisors should publish the results of the stress testing in at least a summary form with

---

[28] See Bank of England (2015, Box 5, pp. 30–31).

[29] In Chapter 11 of this volume, Goldstein and Leitner set out the benefits and costs of public disclosure and argue that disclosure policies should reflect a weighing of the benefits against the costs.

appropriate but informative quantitative content. Finding effective templates for such disclosures is not only helpful to the investors but also offers banks ideas for greater disclosures.

Supervisors should encourage banks to make fuller disclosures about their internal stress testing, including scenarios that are most relevant for them. Banks could provide more specific stress testing disclosures around the larger risks they consider in stress tests and seek to find meaningful ways to describe the results and actions they have taken to mitigate tail risks.

Providing this important information to investors in banking organizations may increase the demand by investors and creditors that larger nonbank financial companies, such as insurance companies and shadow banks, conduct stress testing and disclose the results. Securities regulators that usually oversee public disclosures could play an important role in setting expectations for stress testing disclosures. And financial stability authorities can seek to draw larger nonbanks and, where present, their regulators into macroprudential exercises to understand better the resiliency of the financial sector.

References

Adrian, Tobias, and Nina Boyarchenko (2018, July), "Liquidity policies and systemic risk," *Journal of Financial Intermediation*, 35(B), 45–60.

Adrian, Tobias, Nina Boyarchenko, and Domenico Giannone (2019, April), "Vulnerable growth," *American Economic Review*, 109P(4), 1263–1289.

Adrian, Tobias, and Bradley Jones (2018), "Shadow banking and market based finance," *Financial Stability Review*, 22, 13–24.

Bank of England (2013, October), "The Bank of England's approach to stress testing the UK banking system," available at www.bankofengland.co.uk/-/media/boe/files/stress-testing/2015/the-boes-approach-to-stress-testing-the-uk-banking-system.

Basel Committee on Banking Supervision (1992), "A framework for measuring and managing liquidity," available at www.bis.org/publ/bcbs10b.htm.

Basel Committee on Banking Supervision (2009, May), "Principles for sound stress testing practices and supervision," available at www.bis.org/publ/bcbs155.htm.

Basel Committee on Banking Supervision (2013, January), "Principles for effective risk data aggregation and risk reporting," available at www.bis.org/publ/bcbs239.pdf.

Basel Committee on Banking Supervision (2017, December), "Supervisory and bank stress testing: Range of practices," available at www.bis.org/bcbs/publ/d427.htm.

Basel Committee on Banking Supervision (2018, October), "Stress testing principles," available at www.bis.org/bcbs/publ/d450.htm.

Board of Governors of the Federal Reserve System (2018a), "Comprehensive Capital Analysis and Review Assessment framework and results 2018," available at www.federalreserve.gov/publications/files/2018-ccar-assessment-framework-results-20180628.pdf.

Board of Governors of the Federal Reserve System (2018b, February), "Comprehensive Capital Analysis and Review 2018 summary instructions," available at www.federalreserve.gov/newsevents/pressreleases/files/bcreg20180201a2.pdf.

Board of Governors of the Federal Reserve System (2018c, June), "Dodd–Frank Act Stress Test Results 2018: Supervisory stress test methodology and results," available at www.federalreserve.gov/publications/2018-june-dodd-frank-act-stress-test-supervisory-stress-test-results.htm.

Board of Governors of the Federal Reserve System (2018d, November), "Federal Reserve supervision and regulation report," available at www.federalreserve.gov/publications/2018-november-supervision-and-regulation-report-preface.htm.

Board of Governors of the Federal Reserve System, Federal Deposit Insurance Corporation, and Office of the Comptroller of the Currency (2012, May 14), "Guidance for stress testing for banking organizations with consolidated assets of more than $10 billion," available at www.federalreserve.gov/supervisionreg/srletters/sr1207a1.pdf.

Brunnermeier, Markus, and Yuliy Sannikov (2014), "A macroeconomic model with a financial sector," *American Economic Review*, 104(2), 379–421.

Committee on the Global Financial System (2011, April), "A survey of stress tests and current practice at major financial institutions," available at www.bis.org/publ/cgfs18.htm.

Committee on the Global Financial System (2005, January), "Stress testing at major financial institutions: Survey results and practice," available at www.bis.org/publ/cgfs24.htm.

Committee on Payments and Market Infrastructures (2017, July), "Resilience of central counterparties (CCPs): Further guidance on the PFMI—final report," available at www.bis.org/cpmi/publ/d163.htm.

Financial Stability Board (2011, October), "Key attributes of effective resolution regimes for financial institutions," available at www.fsb.org/wp-content/uploads/r_111104cc.pdf.

Gertler, Mark, Nobuhiro Kiyotaki, and Andrea Prestipino (2020), "A macroeconomic model with financial panics." *Review of Economic Studies*, 87(1), 240–288.

He, Zhiguo, and Arvind Krishnamurthy (2013), "Intermediary asset pricing," *American Economic Review*, 103(2), 732–770.

Healey, Jason, Patricia Mosser, Kathryn Rosen, and Adriana Tache (2018, October), "The future of financial stability and cyber risk," Monograph, Brookings Institution and Columbia University SIPA.

International Monetary Fund (2017, October), "Financial conditions and growth at risk," in *Global financial stability report*.

Kashyap, Anil K., Dimitrios P. Tsomocos, and Alexandros P. Vardoulakis (2007, August), "Optimal bank regulation in the presence of credit and run risk," Finance and Economics Discussion Series Divisions of Research & Statistics and Monetary Affairs, Federal Reserve Board, 2017–097.

Minsky, H. P. (1977), "A theory of systemic fragility," in E. I. Altman and A. W. Sametz (eds.), *Financial crises: Institutions and markets in a fragile environment*, John Wiley & Sons.

Office of the Comptroller of the Currency (2014, September 2), "OCC guidelines establishing heightened standards for certain large insured national banks, insured federal savings associations, and insured federal branches; integration of regulations," Docket ID OCC-2014-001.

Pierret, Diane, (2015, June), "Systemic risk and the solvency-liquidity nexus of banks," *International Journal of Central Banking*, 193–227.

Reinhart, Carmen, and Kenneth Rogoff (2009), *This time is different: Eight centuries of financial folly* Princeton University Press.

Schmitz, Stefan W., Michael Sigmund, and Laura Valderrama (2017, May), "Bank solvency and funding cost: New data and new results," IMF Working Paper 17/116.

Senior Supervisors Group (2009, September), "Risk management lessons from the global banking crisis of 2008," Monograph, available at www.sec.gov/news/press/2009/report102109.pdf.

Warren, Phil, Kim Kalvanto, and Dan Prince (2018), "Could a cyber attack cause a systemic impact in the financial sector?" *Bank of England Quarterly Bulletin*, Q4.

Part IV

A Macroprudential Perspective on the Financial System

19

The Structure of the Financial System: Implications for Macroprudential Stress Testing

Gonzalo Fernandez Dionis and Patricia C. Mosser

1 Introduction

What do financial systems do, exactly? A macroeconomist would say that the financial system is the mechanism through which those with savings (lenders) meet those with investment opportunities (borrowers) so that savings (S) = investment (I). A financial economist would say that a financial system is a place where financial claims and contracts are created, priced, and exchanged. And a risk management professional would say it is the place where risks (market risk, credit risk, liquidity risk, counterparty risk) are priced. Of course, financial systems do all those things (some of them better than others). In capitalist economies, however, all financial systems carry out these tasks in a very particular way, specifically via a set of highly interconnected institutions and markets, which *collectively* facilitate maturity and risk transformation with very significant amounts of leverage.

The widespread use of maturity and risk transformation combined with leverage and a high degree of interconnectedness makes financial systems inherently fragile. Large declines in asset prices (with attendant large losses on leveraged asset holdings) can cause loss of confidence by those who provide short-term financing, which—in the extreme—leads to system-wide fire sales of assets and runs during which the entire system can contract sharply. The macroeconomic, social, and political consequences of such collapses are enormous and have been well documented across decades, countries, and policy regimes. See Laeven and Valencia (2010, 2013), Schularick and Taylor (2012), and Reinhardt and Rogoff (2009). It is no accident that modern frameworks to monitor systemic financial risks focus on measuring these same vulnerabilities—leverage, maturity transformation, and interconnectedness. See Adrian et al. (2019), Aikman et al. (2018), McLaughlin et al. (2018), Adrian et al. (2015), and Blickle et al. (2019). For examples of financial stability reports, see Bank of England (2019), International Monetary Fund (IMF, 2019), Board of Governors of the Federal Reserve (2019), and Office of Financial Research (2019).

The rest of Part IV and Part V will present a set of examples and methodologies for macroprudential stress testing designed to capture such extreme left-tail financial events. At this stage, macroprudential stress testing is a work in progress, both in terms of modeling systemic risk and in terms of applications by central banks and regulators. Among the important open issues are how to directly incorporate spillovers across the system, particularly between funding and capital decisions of intermediaries; improved integration of stress testing for different parts of the financial system, particularly across key intermediaries and financial market infrastructures; modeling and measurement of dynamic interactions between the financial sector and the real economy; and research to develop new modeling techniques for stress testing that directly incorporates the dynamic behavioral interaction of actors across the financial system.[1]

2 Why Macroprudential?

The need for a macroprudential approach to stress testing reflects the fact that finance is a classic example of a complex, adaptive (dynamic) system. In general, such systems, whether physical, biological, or economic, are quite robust and respond in adaptive ways to adverse shocks and changes in the environment, allowing a system to recover and continue to function. See Haldane (2009). However, complex, adaptive systems also experience catastrophic crises in which the system rapidly collapses. These crises are rare but nearly impossible to predict in terms of their triggers or their timing. And when such crises happen, they devastate the entire system, not just those parts that are weakest or most directly affected by shocks.

In practice, financial crises can arise when leverage and maturity transformation are high and when financial risks are measured or perceived to be low (typically when some set of asset prices is particularly high). Indeed, it is the dynamic interaction of these factors and the adaptive behavior by a wide array of different institutions and markets that make financial-system risk endogenous. Low measured/perceived risks encourage more investment in risky assets, with greater leverage and more short-term funding, which leads to higher asset prices and even lower measured risk.[2] This endogeneity of risk is a major reason that financial systems are fragile. See Adrian and Shin (2011) and Danielsson and Shin (2003).

Standard statistical measures of financial risks are likely to be unhelpful in predicting catastrophic events and the fragilities that can cause crises. They are too procyclical, measuring low risk when asset prices are high, and vice versa. In addition, they typically cannot evaluate rare, large events far enough in the left-hand tail, in part because they are often estimated over time periods that do not include such an event. For example, widely used value-at-risk (VaR) measures are typically calculated using a few years of the most recent data on returns, and thus they only estimate the left tail of risk for that particular narrow time period. If that recent history includes an asset-price boom, then the VaR measure of "tail risk" may fall as the boom continues, encouraging greater risk taking and leverage, which can continue to feed the boom and reduce VaR again. Not only can VaR underestimate the potential losses of the bubble burst, but it can fall as systemic risks increase. The experience of the hedge fund Long-Term Capital Management (LTCM) is illustrative. A few months prior to its failure in 1998, LTCM significantly increased leverage and risk taking, in part because its measured VaR had fallen. See Jorion (2000).

Similarly, statistical measures are likely to be particularly unhelpful in measuring the risks of new products and markets. New products are more likely to be launched when asset prices are rising and profits are high, and thus statistical measures may include no

---

[1] See Chapters 4, 23–26, 28, and 30 in this handbook.

[2] This cycle in reverse—declining asset prices that cause losses, deleveraging, higher measured/perceived risks, loss of funding, dumping of assets, and further asset-price declines—typically happens more quickly and in the extreme will cause a crisis.

data reflecting their downside risks. In addition, standard statistical risk measures are often narrow, typically focused on a specific market or a specific category of risk within a firm. Because of these limitations, risk management professionals and financial regulators have increasingly used stress tests to assess how large far-left-tail events will affect the safety and soundness of individual firms as well as the broader financial system. See Chapter 2 in this volume by Herring and Schuermann for a detailed discussion of the comprehensive regulatory stress testing experience since the 2007–2009 financial crisis. In Chapter 3 of this volume, Das, Dent, and Segoviano provide a critical history of stress testing in finance by the private sector, international institutions and regulators.

Although stress tests have become a common tool for evaluating severe left-tail events, in practice, the most widely used stress testing techniques by firms and by regulatory agencies are not particularly macroprudential. Regulatory stress tests, although comprehensive in some dimensions,[3] remain largely microprudential in the sense that they evaluate the impact of a large negative tail event on a firm-by-firm basis. Modern regulatory stress testing regimes such as the Federal Reserve's Comprehensive Capital Analysis and Review (CCAR) and the European Banking Authority's EU-wide exercise, conduct stress tests by imposing a very broad set of very large negative shocks across all the financial risks—market, credit, counterparty, funding—faced by each firm. These shocks are calibrated out of a common, very dire macroeconomic and market scenario used by all firms.

In this sense, regulatory stress tests mimic the outcome of a systemic crisis but without actually capturing the dynamic interactions of firms and markets. They do not model firm behavior that can cause a negative-feedback loop of fire sales (falling asset prices, attempts to delever leading to a loss of funding, causing further asset price declines) or the dynamic game of a run. In short, they do not model interconnectedness across markets or the endogeneity of risk (and leverage) across the entire system, nor do they allow for feedback between real economic outcomes (employment and growth, for example) and financial outcomes. Macroprudential stress testing, which is the subject of this section of the handbook, aims to do just that.

In addition, most regulatory stress testing regimes are applied only to certain segments of the regulated financial system, most commonly to banks. The behavior of nonbank counterparties in the financial system is calibrated as a largely static set of shocks (e.g., to funding and counterparty risks of the bank), but the interconnections and dynamic feedback relationships between banks and nonbanks are modeled in a very limited way. Thus over time, bank-only stress test results will reflect changes in the behavior, business model, and risk profile of the banks being stressed but will only indirectly reflect similar changes in the structure of the rest of the financial system.

If the structure of the nonbank part of the financial system is relatively stable over time, such a segmented approach may be warranted. But the history of nonbank finance in many countries shows that the structure and the systemic risks posed by the various sectors of the financial system can and do shift considerably over time. In the United States, the rapid growth of money market mutual funds (MMFs), particularly prime funds, in the 1990s and 2000s facilitated rapid growth in the short-term financing of shadow-banking activities,

---

[3] For example, scenario design in regulatory stress tests typically takes into account the buildup of particular risks across the financial system.

particularly securitization. For example, asset-backed securities (ABSs) grew from USD 200 billion to USD 3 trillion from 1990 to the mid-2000s (19 percent compound annual growth rate [CAGR]). The short-term financing portion—asset-backed commercial paper (ABCP)—grew from USD 600 billion to USD 1.4 trillion over the same period (6.5 percent CAGR), much of it made possible by the investments from money market mutual funds, which grew from USD 450 billion in early 1990 to USD 2 trillion by the first quarter of 2005 (9.6 percent CAGR). The financial system overall experienced a very substantial increase in leverage and maturity transformation, facilitated by both nonbanks and banks, creating greater interconnections and dependencies among them. The increase in system-wide fragility was not well measured and so not apparent until the financial crisis of 2007–2009. The result was the sequence of runs in ABCP, repo markets, and ultimately a run of MMF shares themselves in the fall of 2008.[4] By not looking at fragilities across the entire financial system, the interconnected systemic risk was underestimated. Such an oversight may be more likely in portions of the financial system that are unregulated or lightly regulated or where the regulatory structure is very different. In such cases, information and data on the nature of financial risks or their linkages to the rest of the financial system may be incomplete or missing entirely.

The gap in both regulatory coverage and information leads to several important issues that should be addressed before truly macroprudential stress testing can be implemented. First, within the regulated segments of the financial system, more consistency of data on instruments, positions, business models, and their associated risks is needed. For example, banking and insurance companies clearly have very different risk profiles; see Cope et al. (Chapter 13 in this volume) and Peters (Chapter 15 in this volume). But harmonization of risk reporting where the two segments share common risks and strong interconnections is important in order for system-wide risks to be measured and stressed. For example, differences in both risk measurement and regulation contributed to a concentration of liquidity and credit risk held by some monoline mortgage insurers in the United States in the mid-2000s. The distress and failure of several mortgage insurers was an important contributor to contagion of mortgage distress to the rest of the financial system during the 2007–2009 financial crisis.

Second, because leverage and maturity transformation are key vulnerabilities that make the entire financial system fragile, more effort should be devoted to monitoring (and obtaining consistent data on) the products and activities that facilitate these vulnerabilities across the entire financial system. A good example is the secured financing market, including repo and securities lending products. Repo and many securities lending markets facilitate greater leverage and maturity transformation across a very wide variety of financial and nonfinancial firms, operating under different regulatory oversight regimes. From a macroprudential stress testing perspective, measuring secured financing markets is critical to understanding the risk of a contagious crisis that could impair a broad array of firms across the system.

Third, finance is constantly innovating, adding new instruments and markets, new institutions, and new technologies to finance economic activity. Although some of that new

---

[4] For additional analysis of the ABCP run, see Covitz et al. (2013); for analysis of repo runs, see Copeland et al. (2014); and for analysis of the MMF run, see McCabe (2010).

activity happens inside banks and other traditional regulated financial institutions, much of it does not. As a result, most new innovative financing techniques are not measured, and key information on potential systemic risks, such as leverage, is typically unavailable. In addition, it is particularly important from the macroprudential standpoint to understand whether financial activities are migrating outside the regulatory perimeter purely for regulatory arbitrage reasons or because of financial innovation.

The history of the Financial Products subsidiary of the American International Group (AIG-FP) is an interesting example of an unregulated entity that was an innovator and exploited its regulatory advantage. Although AIG-FP was an early innovator in credit default swaps, its extraordinary financial success in the mid-2000s was largely the result of it operating mostly outside the perimeter of the US regulations for financial institutions but adjacent to it. This allowed AIG-FP to use exceptionally high amounts of leverage combined with very large one-way (mortgage) credit risk and liquidity risk, which ultimately caused its failure and required a government bailout; see McDonald and Paulson (2015).

Even if regulatory stress tests can be made more standardized and dynamic, without information about the changing structure of the rest of the system, such efforts will necessarily become less accurate as the less regulated segments adapt over time. Furthermore, the transparency of stress testing regimes and exercises could become an incentive for regulatory arbitrage, shifting the largest risks (and the highest returns in good times) away from the regulated sector to avoid the measurement of potential negative stresses. Such an outcome could increase the likelihood that macroprudential stress testing regimes will cover a declining portion of the financial system over time and, in fact, could miss a significant proportion of systemic risk.

Fourth, to conduct true system-wide macroprudential stress testing, consistent scenarios across different segments of the financial system are needed. Stress testing regimes applied to insurance companies and asset managers typically differ from those applied to banks. In this volume, Cope, Hsu, Lively, Morgan, Schuermann, and Sekeris (Chapter 13) provide a related discussion for banks, including commercial and retail lending, capital markets (investment banking, sales, and trading), and trust and custody activities; Longerstaey (Chapter 14) focuses on asset managers and potential tools to supplement traditional risk measures used in portfolio construction; and Peters (Chapter 15) discusses key considerations for stress tests in the insurance industry.

In a market-based financial system, such as the United States, greater consistency in stress testing regimes would involve a very large coordination effort across the regulators of banking, insurance, securities firms, financial market infrastructures, mortgage and securitization companies, real estate investment trusts (REITs), and asset management and finance companies, to name a few. Such coordination efforts may be smaller for economies with more bank-based financial systems, for example in the euro area, but nonetheless it would need to go well beyond any previous efforts to summarize and aggregate systemic risk. Moreover, the 2018 "Global Monitoring Report on Non-Bank Financial Intermediation" from the Financial Stability Board (FSB) makes clear that the share of market-based finance, particularly via investment companies, continues to rise while shares of banking, insurance, and pensions decline; see FSB (2019). The report estimates nonbank financial intermediation to have grown for 6 consecutive years to USD 184.3 trillion in assets, representing 48.2 percent of total global financial assets at the end of 2017.

Finally, macroprudential stress testing would ideally also include scenarios with explicit feedback between financial risks and macroeconomic outcomes. This is a particularly tall order because macroeconomic models, such as the dynamic stochastic general equilibrium (DSGE) models used to evaluate the impact of macro and monetary policies, have great difficulty incorporating realistic financial interactions in a tractable way. The current stress testing frameworks are challenged to add behavioral spillovers effects within just the financial sector, for example, the dynamic response of funding to capital losses and asset-price declines. The addition of macroeconomic dynamics and multipliers is even more aspirational.

The remainder of this chapter will discuss in more detail the structure of financial systems, focusing particularly on the roles of the banks and nonbanks in financial intermediation, their interconnections, diversity of institutions and markets, and the regulatory perimeter. It will also highlight areas where additional information is needed to (a) follow innovation and its associated risks, (b) fill in gaps in information needed to assess systemic risks (with or without stress tests), and (c) discuss the role of stress testing and other regulatory efforts in making progress on both.

3 Financial System Structure: Banking versus Market-Based Finance

The financial system plays a critical role in society by pricing assets and bringing together borrowers and savers. In practice, this financial intermediation takes place through two main channels: banks (bank-based finance) or through a series of financial instruments created and transacted by financial and nonfinancial companies (market-based finance). In both cases, the intent of intermediation is to allow excess capital to flow to where it is needed. And in each of them, it involves a certain degree of risk as the credit, liquidity, and maturity of the financial assets provided are transformed. Credit transformation relates to how default risk is shifted and allocated from the original borrower to the bank, investor, or other intermediary. Liquidity transformation allows hard-to-sell, less liquid assets to be transformed into instruments that can be easily traded and priced. And through maturity transformation, long-term assets are (ultimately) financed through short-term money-like liabilities.

3.1 Bank-Based Finance

In its simplest form, banks conduct credit, liquidity, and maturity transformation by borrowing via short-term deposits to issue long-term loans. Given its inherent fragility, trust in a banking system is reinforced by a long-standing regulatory environment that typically includes minimum capital standards, activity restrictions, reserve or liquidity requirements, and deposit insurance. The last reflects the fact that required capital is a fraction of the total volume of liquid assets, so in the event of a loss of confidence, bank runs can occur.

Recent regulation of bank-based finance (particularly Basel III; see Basel Committee on Banking Supervision [2019]) has focused on improving the ability of banks to withstand significant stress by harmonizing and tightening the definition of capital, making capital requirements risk sensitive, introducing more robust liquidity requirements, and improving

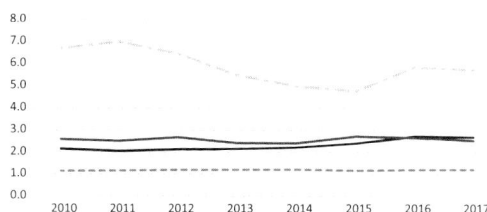

Figure 19.1 Traditional banking system assets* as a ratio to gross domestic product (GDP).
*Follows the FSB definition. Banks include all deposit-taking institutions and other equivalent
instruments. Assets refer to all financial assets on an unconsolidated basis.
**Europe only includes data for Belgium, France, Germany, Ireland, Italy, the Netherlands, and
Spain. Source: FSB (2019). World Bank GDP data.

banking supervision, including the use of regulatory stress testing.[5] Although recent regulatory changes made the banking sector more resilient, they have also come with a cost to the
flow of bank credit and its impact on output; see Fender and Lewrick (2016).

This raises the issue of incentives. Is modern bank regulation and supervision creating
incentives to move intermediation activity out of banks and into a less regulated market-
based finance, as suggested by Pozsar et al. (2013)? For example, postcrisis regulatory
reform has increased both capital and liquidity requirements for banking firms, particularly
those classified as systemic, by significantly more than the requirements for nonbank
financial companies, including insurance companies, REITs, and securities firms. As shown
in Figure 19.1, financial activity has since rapidly migrated to nonbanks in many countries.
Even in the United States, where the nonbank financial sector contracted sharply during the
financial crisis, it is now increasing while traditional banking activity stagnated at least until
the onset of the 2020 pandemic. The use of regulatory stress testing is important here: as
a practical matter, the binding capital requirement for many large banks is that needed to
"pass" regulatory stress tests; see Cetina et al. (2017). But regulatory stress tests for other
types of financial companies—if done at all—are less likely to be the binding constraint on
regulatory capital or are structured in ways that make comparisons difficult. In addition, the
final Basel III reforms introduce modifications to stress testing practices and risk weighting
that could ultimately have significant consequences for traditional bank-based credit
intermediation. A recent EBA report estimated that the impact of the final Basel III reforms
on EU banks could be an average increase in Tier 1 capital of 19.1 percent;[6] see EBA (2019).

3.2 Market-Based Finance

Market-based finance uses financial instruments traded in capital markets to price risk and
allocate capital between borrowers and savers. Public equities, corporate bonds, syndicated

---

[5] Basel IV proposals are somewhat less risk sensitive than Basel II or Basel III as a result of the inclusion of
more standardized approaches to risk weighting.

[6] See https://eba.europa.eu/-/eba-publishes-updated-impact-of-the-final-basel-iii-reforms-on-eu-banks-
capital-and-updates-on-the-compliance-with-liquidity-measures-in-the-eu.

loans, commercial paper, secured funding instruments such as repo, and the many types of securitizations are just a few of such instruments used to finance business and household borrowing through markets. Purchasers (lenders) with different risk profiles buy these assets, ranging from mutual funds and insurance companies to hedge funds and other leveraged investment vehicles, to pension funds, to banks. These institutions in turn often use additional financing instruments (typically short-term borrowing, such as repo or commercial paper) to fund at least a portion of their purchases. In addition, they regularly use derivative instruments such as swaps, options, and futures to hedge interest, credit, and foreign exchange risks. Much of this activity depends directly on borrowing, backstop liquidity arrangements, and counterparty relationships with banks.

The role of securitization in providing market-based intermediation services is an important example. By pooling a group of assets (typically illiquid individual loans) into securities, financial markets can carry out credit, liquidity, and maturity transformation. For example, many securitizations create tranched securities, with each tranche ranked by seniority and therefore risk. The higher tranches are typically designed to minimize some aspect of risk—credit risk or interest rate risk—which improves their liquidity. Lower-rated tranches are typically structured with more leverage: for a given size loss to the underlying illiquid loans, the decline in the price of mezzanine tranches can be significantly larger than that for senior tranches. Other securitization structures, for example, pass-through mortgage securitization in the United States, include credit protection and are highly standardized, so the securities are traded frequently in liquid markets. Finally, many investors (both banks and nonbanks) purchase securitizations on a levered basis, financing their investments through market instruments; some of this in the form of short-term funding.

This deconstructed intermediation process is the basis of market-based finance or so-called "shadow banking." The FSB publishes yearly estimates to quantify the total volume of assets that are involved in this type of credit intermediation. The FSB approach defines a monitoring universe of nonbank financial intermediation (MUNFI) that broadly incorporates insurance corporations, pension funds, financial auxiliaries, and other financial institutions. At the end of 2017, this monitoring universe was estimated to comprise \$184 trillion in assets across 21 countries and the euro area,[7] or on aggregate, roughly half of total financial assets in the countries in the report. In addition, the FSB has a narrow definition of nonbank financial intermediation that includes the assets of nonbank financial entities whose bank-like activities could pose a risk to financial stability. The definition includes entities based on five economic functions: (1) management of collective investment vehicles with features that make them susceptible to runs (MMFs and other funds), (2) loan provision that is dependent on short-term funding (finance and credit companies), (3) intermediation of market activities that is dependent on short-term funding or on secured funding of client assets (broker-dealers), (4) facilitation of credit creation (credit insurance, financial guarantors), and (5) securitization-based credit intermediation and funding of financial entities (structured finance vehicles, ABSs).

---

[7] Argentina, Hong Kong, Saudi Arabia, Australia, Indonesia, Singapore, Brazil, India, South Africa, Canada, Japan, Switzerland, Cayman Islands, Korea, Turkey, Chile, Mexico, the United Kingdom, China, Russia, the United States, and the euro area.

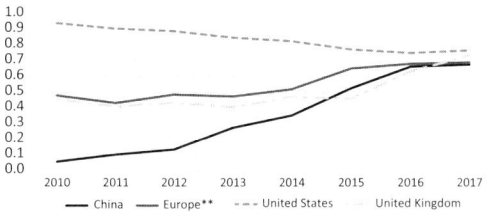

Figure 19.2 Narrow measure of nonbank financial intermediation as a share of GDP. Narrow definition includes entities under the following categories: (1) management of collective investment vehicles with features that make them susceptible to runs, (2) loan provision that is dependent on short-term funding, (3) intermediation of market activities that is dependent on short-term funding or on secured funding of client assets, (4) facilitation of credit creation, and (5) securitization-based credit intermediation and funding of financial entities. *Europe only includes data for Belgium, France, Germany, Ireland, Italy, the Netherlands and Spain. Source: FSB (2019). World Bank GDP data.

The FSB's narrow measure of shadow banking accounts for close to $52 trillion in assets,[8] or just under one-sixth of total financial assets. As shown in Figure 19.2, since 2010, the FSB's narrow measure of shadow banking—nonbanking financial intermediation as a share of economic activity—has grown significantly in many countries, with the notable exception of the United States.

The FSB's definitions highlight the role of key markets and nonbank institutions in financial intermediation. They include the following:

- Open-market paper and repurchase agreements. Open-market paper includes commercial paper and short-term unsecured borrowing issued by (highly rated) financial companies. Repurchase agreements (repos) are secured short-term funding, typically secured by longer-term bonds or equity. Both traditional and shadow banks are large users of both types of financing, often borrowing and lending with each other, leading to a high degree of interconnectedness.
- Saving vehicles like mutual funds or hedge funds, which channel savings to buy financial assets. Leverage across such funds varies widely, with MMFs and some hedge funds being highly levered but significantly lower leverage levels in other types of funds. Savings vehicles provide liquidity to the financial-intermediation activity by buying and selling securities, including bonds and securitizations. Mutual and exchange traded funds (ETFs) also provide liquidity transformation. They allow their investors the ability to sell shares daily or even intraday, even if underlying fund assets are illiquid risky assets, such as high-yield bonds or ABSs. ETFs have not only grown incredibly rapidly in recent years, but they also have a complicated liquidity profile that could cause significant disruptions and large losses in a stress event. ETFs secondary-market trading is done in highly liquid markets, typically on exchanges. However, primary-market creation (and destruction) of ETFs

---

[8] Argentina, Hong Kong, Saudi Arabia, Australia, Indonesia, Singapore, Brazil, India, South Africa, Canada, Japan, Switzerland, Cayman Islands, Korea, Turkey, Chile, Mexico, United Kingdom, China, Russia, the United States, and the euro area (only includes Belgium, France, Germany, Ireland, Italy, Luxembourg, Netherlands, and Spain).

is done by a small number of very large financial companies, which have the capacity to regularly buy and sell the underlying (often less liquid) assets in securities markets (and arbitrage in ETF market). In a large negative stress event, an inability (or high cost) to sell could cause a disconnect between the ETF prices and the values of underlying assets and could potentially result in a withdrawal of market-making intermediaries; see Blackrock (2015). Because ETFs are now used as collateral for funding instruments such as repo and securities lending, such liquidity tail risk could also be transmitted to (typically more fragile) funding markets as well; see Burne (2019).

- Special government-related financial institutions, such as the government-sponsored enterprises (GSEs) in the United States, which often play a key role in the wholesale financing markets. Companies like Freddie Mac and Fannie Mae provided credit enhancement to the securitization process by effectively securing the mortgages and asset pools used to issue the new securities. Their securitizations were highly liquid in part because of standardization of the securitization process but also because of an implicit (since 2008, explicit) backing by the US government. Their combination of implicit government backing, favorable tax status, and very low capital requirements led to significant migration of mortgage credit away from the US banking system. For the same reasons, they entered the portfolio-management business and became the largest investors in their own securities. The portfolio was funded by short-term money market borrowing and engaged in exceptional maturity transformation at very high leverage.
- Insurance companies and pension funds are large-scale investors in market-based finance, with very large portfolios of equities, corporate bonds, and securitizations, financed through customer premiums and annuity payments. Insurance companies use significant leverage and a smaller amount of maturity transformation, for example, through lending their securities for cash in the securities lending market.

US Shadow-Banking Sector

Market-based finance has been an important component of the US financial system for the last century; see Gordon and Judge (2018). Starting in the 1980s, US nonbank financial intermediation grew very rapidly, and by the mid-2000s, it was the dominant channel for financial intermediation, larger than the US banking sector. Despite retrenchment after the 2007–2009 financial crisis, market-based intermediation in the United States remains nearly as large as bank-based intermediation (Figure 19.3).

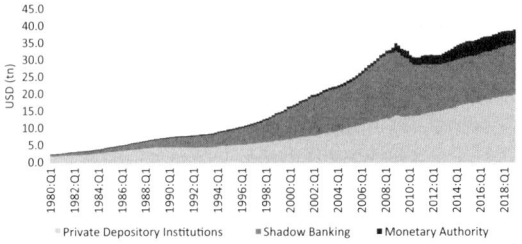

Figure 19.3 US financial system assets (USD trillions).
Source: US Flow of Funds. Analysis by authors.

In addition, a very sizable proportion of US market-based finance is shadow banking in the sense that it is funded on a levered basis, often with short-term liabilities, most of which have no ex ante access to government backstops or central bank liquidity. Figure 19.4 gives a breakdown of US shadow banking by type of financing rather than by asset type. The focus on liabilities reflects the key fragility that can lead to extreme financial stress, namely, the runability of many types of wholesale funding.

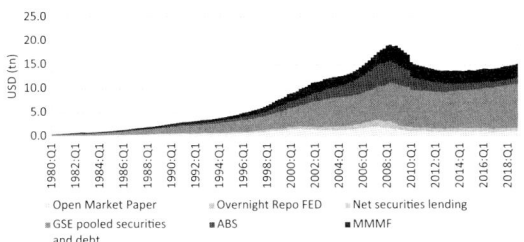

Figure 19.4 US shadow-banking short-term liabilities (USD trillions).
As of the first quarter of 2010, the introduction of accounting rules Financial Accounting Standard (FAS) 166 and 167 result in differences in the consolidation of special purpose entities (SPEs), producing a discontinuity in the GSE and ABS series.[9]
Source: US Flow of Funds.

Market-based finance has several benefits relative to traditional bank finance. The process of decomposing banking activity can provide credit-intermediation services at a cheaper price because risk can be diversified throughout a series of agents and investors according to their preferences and risk appetite. Overall, there is increased information in the financial system because disclosure requirements around securities such as corporate bonds and securitizations mandate a higher degree of transparency than simple banking loans. Allen and Gale (2000) provide a more detailed comparison of market-based versus institution-based financial systems, noting that neither system has a clear overall dominance. They note that the strengths and weaknesses of the two systems are different with respect to risk sharing, pricing risk, efficiency, and competition.

In practice, the largest globally active universal banks (bank holding companies in the United States) are very large participants in market-based finance activities in addition to their traditional banking business of funding loans with deposits. They underwrite and trade securities and create securitizations, finance books of the same securities using money

---

[9] As a result of two new accounting rules, FAS 166 and 167, the assets and liabilities of some SPEs have been moved to the balance sheets of the US chartered commercial bank sector (Table L.110), the GSE sector (Table L.124), and the finance-company sector (Table L.127). The consolidated assets and liabilities were removed from the agency- and GSE-backed mortgage pool sector (Table L.125) and the issuers of ABSs sector (Table L.126). Almost all of the consolidations resulting from the new accounting rules occurred in the first quarter of 2010. In the Flow of Funds Accounts, these changes are treated as discontinuities that affect the levels of outstanding assets and liabilities in the relevant sectors but not the flows (from Flow of Funds).

market instruments, run very large derivatives businesses, and manage mutual funds. The interconnections between banks and shadow banks are widespread and complex.

As a result, the end-to-end intermediation process in market-finance systems (and between banking and market finance) is hard to monitor. Although there may be greater transparency about individual securities and markets, the risk to the overall system is quite opaque and complicated to measure. Risk estimates and particularly correlations and exposures across different markets and products measured in normal conditions will almost certainly under-estimate systemic risk. In a crisis or stress event, contagion across institutions and markets will increase significantly.

4 Financial System Structure: Defining the Perimeter

As discussed earlier, the complexity and interconnections between financial intermediaries across the financial system are a key challenge for creating a truly macroprudential stress testing regime. Differences in regulatory requirements and standards, data inadequacy and inconsistencies, and innovations outside the traditional financial industry are persistent chal-lenges to understanding where risk is housed (and intermediated) and in developing a macro-prudential approach to stress testing. We discuss these issues in turn.

Because of its diffuse nature, regulatory oversight of shadow-banking activities is also typically diffuse, different regulatory agencies and Basel Committee on Banking Supervi-sion and subject to oversight by rules for different types of financial institutions and markets and very limited oversight. As a result there is wide variation in regulatory of the entire intermediation process and the risks the collective activity may pose to the financial system. Despite efforts to improve surveillance of the entire system by central banks, financial stability authorities, and international institutions, consistent regulatory treatment of similar risks in different types of institutions remains elusive; see BCBS, 2018).

Since 2008, there has been a series of regulatory changes to improve the supervision of the shadow-banking system by regulators. In particular, there have been important reforms to short-term funding market instruments, including commercial paper markets (Canada,[10] France[11]), repo markets (United States[12]), and MMFs in both the United States[13] and the European Union.[14] These reforms have primarily focused on reducing run risk and improv-ing the structural resiliency of these markets during times of financial stress. In addition, broad international agreement on the regulation, reporting, and central clearing of deriva-tives transactions was a signature global financial market reform with the same purpose.

However, efforts to level the regulatory playing field in some dimensions can shift finan-cial risk to less transparent or less regulated venues. For example, changes in the capital treatment of mortgage servicing rights in Basel III have led US banks to sell servicing rights, largely to hedge funds and private equity funds, shifting large quantities of highly

---

[10] See https://albertasecurities.com/-/media/ASC-Documents-part-1/Regulatory-Instruments/2018/10/5057743-v1-CSA-Notice-and-Annexes-attached-SAFI.ashx.

[11] See www.banque-france.fr/sites/default/files/dossier-technique-en-reform-of-the-negotiable-debt-securities-market.pdf.

[12] See www.newyorkfed.org/banking/tpr_infr_reform.html.

[13] See www.sec.gov/rules/final/2014/33-9616.pdf.

[14] See www.ecb.europa.eu/stats/financial_markets_and_interest_rates/money_market/html/index.en.html.

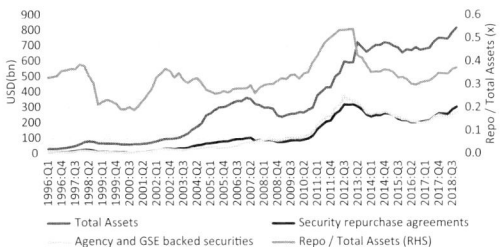

Figure 19.5 REITs. Source: US Flow of Funds.

leveraged interest rate risk out of the regulated banking system; see Covas and Fernandez (2019). Similarly, in the syndicated loan market, tighter capital standards for banks have shifted corporate credit risk, particularly for the lower-rated leveraged lending, to nonbank syndicate members, which typically face lower capital charges; see Board of Governors of the Federal Reserve, Federal Deposit Insurance Corporation, and Office of the Comptroller of the Currency (2019).[15] And tighter Federal Housing Finance Agency (FHFA) regulation of Fannie Mae and Freddie Mac after their 2008 conservatorship included requirements to unwind their very large mortgage-backed security (MBS) investment portfolios and to sell off some of the credit risk from their securitizations. A major purchaser of both has been REITs, particularly MBS REITs (see Figure 19.5), which have grown rapidly in the last 10 years and finance their purchases with significant amount of short-term financing such as repo. In effect, REITs, which do not have minimum capital or liquidity standards, are financing significant amounts of long-term US mortgage risk with short-term debt.[16]

Like regulation, consistent measurement of the degree of leverage, maturity, credit, and liquidity transformation in market-financed intermediation is a challenge. For example, the definitions of traditional and shadow banking in the previous section appear clear, but in practice, the line between them is easily blurred. Is shadow banking just a matter of the institutions that carry out credit, maturity, and risk transformation, or is it dependent on the activity itself? Because regulation particularly for safety and soundness is largely done at the institutional level, it is not surprising, that most measurement is done institution by institution. Ideally, of course, a focus on activities, specifically credit, liquidity, and maturity transformation, might be preferable, but at present, most regulation and data collections are not consistent with that approach.

Using US and European definitions, shadow banking seems to be defined as the process of credit, maturity, and liquidity transformation performed outside the traditional banking system, specifically without explicit access to central bank liquidity and without ex ante public-sector backstops.[17] From a macroprudential stress testing perspective, this definition

---

[15] See www.bloomberg.com/news/articles/2019-05-20/leveraged-lending-isn-t-posing-crash-threat-fed-chairman-says.

[16] As of the end of 2018, these had reached over $800 billion in assets, of which almost $275 was in agency and GSE-backed securities (US Flow of Funds L.129 Estate Investment Trusts [REITs]).

[17] In practice, of course, these two characteristics were not true during the financial crisis of 2007–2009. This has likely increased systemic moral hazard in markets-based finance, inducing investors to consistently underprice risk. Pozsar et al. (2013) provide an analysis of the emergency lending facilities from the Federal

is useful because it focuses directly on measuring intermediation activities and contagion channels, which may be fragile.

Unfortunately, measurement of nonbank financial intermediation is incomplete and done differently across jurisdictions. For example, the US Flow of Funds offers an immense array of data going back to 1945 but does not include information on the maturity of liabilities financing longer-dated assets in the different sectors of the financial system. Therefore, understanding when and where maturity, credit, and liquidity transformation are taking place becomes an arduous task. Similarly, measuring the degree of public-sector support is nearly impossible if guarantees are implicit rather than explicit. The role of GSEs and the moral hazard attached to too-big-too-fail nonbanks are obvious examples.

In Europe, the European Systemic Risk Board (ESRB) includes in its broad measure of shadow banking a very different breakdown: MMFs, non-MMF investment funds, financial vehicle corporations (FVCs), non-FVC special-purpose entities, and other financial institutions (OFIs); see ESRB (2018). Here, too, however, the data available are insufficient. First, some categories (in particular, OFIs) are very broad and do not allow a granular understanding of the underlying risks. In some cases, the data are available only on a country-by-country basis, and the collection does not capture short-term liabilities (under 1 year) but instead uses a maturity standard of under 2 years.

International efforts to improve the quality and consistency of data on financial risks, across both banks and nonbanks, have made some progress in recent years. As noted earlier, the FSB publishes the "Global Monitoring Report on Non-Bank Financial Intermediation." Through this report, the FSB aims to standardize an approach to measure shadow banking across jurisdictions while acting as a supervisory tool to track and detect potential economic imbalances created by nonbank financial intermediation. In addition, international reporting standards for secured financing transactions (securities lending and repo) and derivatives transactions and more complete information on international banking flows and exposures have been added. Nonetheless, major gaps remain, including a lack of consistent information on counterparty exposures (interconnectedness) across the financial system and still-limited information on nonbank flows and exposures.

A final macroprudential challenge is how to classify and include risk assessments of new products and institutions appearing through financial innovation. Traditional financial institutions are regularly challenged as new technology is deployed in new or more efficient ways of carrying out financial transactions, creating and pricing assets, and providing funding. The new instruments and firms—as well as the ways that existing banks and nonbanks adjust their business models in response to innovation—will change both the structure and perimeter of the financial system, creating or at least shifting risks and intermediation activities. A few recent examples are as follows.

- The volume of technology-based (FinTech) mortgage originators such as Rocket Mortgage from Quicken Loans increased by around 30 percent annually from 2001 to 2016 reach 8 percent of US mortgage lending, despite higher fees than traditional lenders.

Reserve, almost 10 of which were either created or were expanded to assist nonbanking credit intermediation. The authors conclude: "Upon the complete rollout of the liquidity facilities and guarantee schemes, the shadow banking system was fully embraced by official credit and liquidity puts and became fully backstopped, just like the traditional banking system."

Such FinTech mortgage companies offer an automated application process completely online and strictly for securitizations, typically through the Government National Mortgage Association (GNMA) or the mortgage GSEs. Fuster et al. (2018) show the companies have improved the efficiency of mortgage intermediation in the United States, with a 20 percent faster processing time for mortgage applications (for the same default risk) and faster adjustments to changes in mortgage demand. However, they are also largely unregulated, with very modest capital.

- Venmo, a cell phone app that moves money directly between users, transferred more than $60 billion in 2018.[18] The app is both a transfer mechanism and a store of value, allowing users to keep money in an e-wallet, which they can then use to make payments or send to their bank accounts at a later date. The total amount of money stored in Venmo accounts at any given point in time is unknown, as is its risk profile.[19] These e-wallet transactions can seem to function as a substitute for bank deposits, but they occur outside the traditional safeguards of the banking system.

- The well-known rise of cryptocurrencies has been explosive. The regulatory approach has been mixed, with some countries, such as China, forbidding financial institutions from dealing with Bitcoin while others are more permissive.[20] In the United States, market regulators are debating whether these financial products should be treated as currencies or as securities with a higher degree of regulation.[21] Regardless of the outcome, financial and tech companies are looking at ways to structure these products to avoid market regulations. A careful assessment of the liquidity and market risks of these instruments is warranted.

- Technological advances in processing speed and data analytics have transformed trading in many markets. Although the share of total assets held by firms engaged in high-frequency-trading strategies are very small, they account for the bulk of trading volumes in equities, futures, Treasury, and foreign exchange markets.[22] Given these changes, market risk and operational risk are almost one and same, at least in terms of how shocks may be transmitted within and across markets.

The point of these examples is not to suggest that any one of them poses a systemic risk to be incorporated in current stress testing regimes. But monitoring the risks they pose—and how they and other innovations may interact with the rest of the financial system during periods of stress—is an important ongoing task for a macroprudential stress testing framework.

5 Conclusion

This chapter has highlighted key dynamics and cross-system risks that should be included in macroprudential stress testing regimes. The challenges are many, including fragmentation

---

[18] PayPal 2018 Annual Report.

[19] In February 2018, Venmo settled with the Federal Trade Commission over a misrepresentation of its ability to transfer funds and its privacy settings.

[20] See www.loc.gov/law/help/cryptocurrency/china.php.

[21] The definition of a security stems from a 1946 court case on contracts for lands with citrus groves, called the "Howey Test" in honor of W. J. Howey, owner of the lands. The test aims to determine if the financial product represents an investment in a common enterprise with the expectation of a profit.

[22] See www.sec.gov/divisions/investment/private-funds-statistics/private-funds-statistics-2018-q3.pdf.

and incompleteness of both regulation and data to measure underlying exposures and risks. One cannot stress what is not measured. Greater data and regulatory coordination prioritized toward addressing key systemic fragilities throughout the entire financial system would seem the logical place to start. This could include consistency in regulatory approach and data collection in areas such as short-term funding risks, leveraged institutions and instruments, and key interconnections such as financial market infrastructures. The constant migration of financial activity in response to regulation, innovation, and competitive forces calls for regular monitoring and reevaluation of how (and through what markets and institutions) financial contagion may spread. Doing so will require regular adjustments to both scenario design and the technical implementation of regulatory stress tests. Finally, technical and modeling improvements to directly incorporate feedback effects and behavioral spillovers—across the financial system and between the financial system and the real economy—remain the great research and technical challenge for future stress test frameworks. All of these suggest that the future of macroprudential stress testing should be closely tied to financial stability monitoring and macroeconomic stability, as well as to the kinds of tools and techniques that are discussed in the coming chapters.

References

Adrian, T., D. Covitz, and N. Liang (2015), "Financial stability monitoring," *Annual Review of Financial Economics*, 7(1), 357–395.

Adrian, Tobias, Dong He, J. Nellie Liang, and Fabio Natalucci (2019, August), "A monitoring framework for global financial stability," IMF Staff Discussion Note SDN/19/06, available at https://www.imf.org/en/Publications/Staff-Discussion-Notes/Issues/2019/08/23/A-Monitoring-Framework-for-Global-Financial-Stability-46645.

Adrian, Tobias, and Hyun Song Shin (2011), "Financial intermediary balance sheet management," Federal Reserve Bank of New York Staff Report No. 532.

Aikman, David, Jonathan Bridges, Stephen Burgess, Richard Galletly, Iren Levina, Cian O'Neill, and Alexandra Varadi (2018, July), "Measuring risks to UK financial stability," Staff Working Paper No. 738, Bank of England, available at www.bankofengland.co.uk/working-paper/2018/measuring-risks-to-uk-financial-stability.

Allen, Franklin and Douglas Gale (2000), "*Comparing financial systems*," MIT Press.

Bank of England (2019, December), "Financial stability report," available at www.bankofengland.co.uk/-/media/boe/files/financial-stability-report/2019/december-2019.pdf.

Basel Committee on Banking Supervision (2018), "Implementation of Basel standards: A report to G20 leaders on implementation of Basel III regulatory reforms," available at www.bis.org/bcbs/publ/d453.htm.

Blackrock (2015, July), "Bond ETFs: Benefits, challenges, opportunities," available at www.blackrock.com/corporate/literature/whitepaper/viewpoint-bond-etfs-benefits-challenges-opportunities-july-2015.pdf.

Blickle, Kristian, Fernando Duarte, Thomas Eisenbach, and Anna Kovner (2019, December), "Banking system vulnerability: Annual update," available at https://libertystreeteconomics.newyorkfed.org/2019/12/banking-system-vulnerability-annual-update.html.

Board of Governors of the Federal Reserve (2019, May), "Financial stability report," available at www.federalreserve.gov/publications/2019-may-financial-stability-report-purpose.htm.

Board of Governors of the Federal Reserve, Federal Deposit Insurance Corporation, and Office of the Comptroller of the Currency (2019), "Shared National Credit Program: 1st and 3rd quarter 2018 examinations," available at www.federalreserve.gov/newsevents/pressreleases/files/bcreg20200131a1.pdf.

Burne, Katy (2019, June), "ETFs to join the collateral party?" available at www.bnymellon.com/us/en/locale-assets/pdf/markets-group/aerial-view-magazine/etfs-to-join-the-collateral-party.pdf.

Cetina, Jill, Bert Loudis, and Charles Taylor (2017), "Capital buffers and future of stress testing," Office of Financial Research Brief 17–2.

Copeland, Adam, Antoine Martin, and Michael Walker (2014), "Repo runs: Evidence from the tri–repo market," *Journal of Finance*, 69, 2343–2380.

Covas, Francisco, and Gonzalo Fernandez Dionis (2019), "Ways to curb nonbank activity in the mortgage market and reduce systemic risk," Bank Policy Institute, available at https://bpi.com/ways-to-curb-nonbank-activity-in-the-mortgage-market-and-reduce-systemic-risk/.

Covitz, Daniel, Nellie Liang, and Gustavo A. Suarez (2013), "The evolution of a financial crisis: Collapse of the asset-backed commercial paper market," *Journal of Finance*, 68(3), 815–848.

Danielsson, Jon, and Hyun Song Shin (2003), "Endogenous risk," in P. DeGrew (ed.), *Modern Risk Management: A History*, Risk Books.

European Banking Authority (2019), "Basel III monitoring exercise—results based on data as of 20 June 2018," available at www.eba.europa.eu/sites/default/documents/files/documents/10180/2551996/a8f383db-79f5-4d41-a8f8-3fd76f0dba30/Basel%20III%20Monitoring%20Exercise%20Report%20-%20data%20as%20of%2030%20June%202018.pdf?retry=1.

European Systemic Risk Board (2018), "EU shadow banking monitor No. 3," European Systemic Risk Board Report.

Fender, Ingo, and Ulf Lewrick (2016), "Adding it all up: The macroeconomic impact of Basel III and outstanding reform issues," BIS Working Papers No. 591.

Financial Stability Board (2019), "Global monitoring report on non-bank financial intermediation 2018," available at www.fsb.org/2019/02/global-monitoring-report-on-non-bank-financial-intermediation-2018/.

Fuster, Andreas, Matthew Plosser, Philipp Schnabl, and James Vickery (2018, February), "The role of technology in mortgage lending," Federal Reserve Bank of New York Staff Report No. 836.

Gordon, Jeffrey N., and Kathryn Judge (2018), "The origins of a capital market union in the United States," ECGI Working Paper Series in Law No. 395.

Haldane, Andrew G. (2009, April 28), "Rethinking the financial network," Speech at the Financial Student Association.

International Monetary Fund (2019, October), "Global financial stability report: Lower for longer," available at www.imf.org/en/Publications/GFSR/Issues/2019/10/01/global-financial-stability-report-october-2019.

Jorion, Philippe (2000), "Risk management lessons from long-term capital management," *European Financial Management*, 6, 277–300.

Laeven, Luc, and Fabian Valencia (2010), "Resolution of banking crises: The good, the bad, and the ugly," IMF Working Paper 10/146.

Laeven, Luc, and Fabian Valencia (2013), "Systemic banking crises database: An update," IMF Working Paper 12/163.

McCabe, Patrick (2010), "The cross section of money market fund risks and financial crises," FEDS Working Paper 2010–51.

McDonald, Robert L., and Anna Paulson (2015), "AIG in hindsight," *Journal of Economic Perspectives*, 29(2), 81–106.

McLaughlin, Joe, Adam Minson, Nathan Palmer, and Eric Parolin (2018, March), "The OFR financial system vulnerabilities monitor," OFR Working Paper 18–01, available at www.financialresearch.gov/working-papers/2018/03/28/ofr-financial-system-vulnerabilities-monitor/.

Office of Financial Research (2019, December), "Annual report to Congress," available at www.financialresearch.gov/annual-reports/2019-annual-report/.

Pozsar, Zoltan, Tobias Adrian, Adam Ashcraft, and Hayley Boesky (2013), "Shadow banking," *Economic Policy Review*, 19(2).

Reinhart, Carmen M., and Kenneth S. Rogoff (2009), "The aftermath of financial crises," *American Economic Review*, 99(2), 466–472.

Schularick, Moritz, and Alan M. Taylor (2012), "Credit booms gone bust: Monetary policy, leverage cycles, and financial crises, 1870–2008," *American Economic Review*, 102(2), 1029–1061.

20

Holistic Bank Regulation

By Charles A. E. Goodhart

1 Introduction

At an International Monetary Conference held in Atlanta in June 2011, Jamie Dimon, chief executive officer (CEO) of JP Morgan Chase, directly asked Ben Bernanke, then president of the Federal Reserve Board, "Has anyone bothered to study the cumulative effect of all these things [new regulations]?" To this, Bernanke replied, "Has anybody done a comprehensive analysis of the impact on—on credit? I can't pretend that anybody really has. You know, it's—it's just too complicated. We don't really have the quantitative tools to do that" (CNN Money, 2011).

It is difficult and complex enough to try to estimate the quantitative effect of any one regulatory initiative taken in isolation, let alone the cumulative effect of several separate regulatory requirements. In contrast to the vast outpouring of studies on the real effects of monetary policy instruments on the real economy, studies on the effects of bank regulation on wider macro developments have been relatively scarce (although there is more literature on those instruments, such as loan-to-value [LTV] limits, aimed directly at the housing market).

There are a few aspects of the effects of bank regulation on the wider economy that have been reasonably well documented. One example is the impact of tougher risk weightings on banks' holdings of marketable securities on liquidity in bond markets, where the Federal Reserve Bank of New York has done some excellent work; see Adrian et al. (2017a,b).

But the main instrument of modern regulation has been to enhance capital adequacy requirements (CARs), and here the literature has been rather more sparse. Thus, Thakor (2014), one of the main economic writers on this topic, states, "Let us now turn to the relationship between bank capital and bank value. There has been surprisingly little work done on this issue" (p. 36). And Kanngiesser et al. (2019) open their abstract with this statement:

> How do changes in bank capital requirements affect bank lending, lending spreads and the broader macroeconomy? The answer to this question is important for calibrating and assessing macroprudential policies. There is, however, relatively little empirical evidence to answer this question in the case of the euro area countries.

That said, there is now an increasing number of such studies; see, for example, Budnik et al. (2019), "The Benefits and Costs of Adjusting Bank Capitalisation: Evidence from Euro Area Countries."

Moreover, such earlier work was often plagued by identification issues. Thus, more successful banks (and banking systems), with a more profitable conjuncture and business plan, will find it simultaneously easier to meet their CARs, plus a sizeable buffer, and at the same time expand loans and deposits. So in a cross-section, and even perhaps a time-series, exercise regressing measures of bank credit expansion on equity capital, there is almost bound to be a positive relationship. But that would tell us almost nothing about the aggregate macro effects of raising bank capital requirements, either transitional or longer run. Nor should studies showing that banks reduce their lending when they experience negative exogenous shocks to their capital (e.g., Peek and Rosengren, 1997; for a review of similar studies, see Hanson et al. [2011]), lead one to the conclusion that an increase in CARs would also cause more lending.

However, the evidence, as reviewed by Hanson et al. (2011), that even a very large increase in capital requirements is likely to have a rather modest effect on loan interest rates, via its impact on the bank's weighted average cost of capital, is more compelling. Their estimate is that if capital requirements go up by 10 percentage points, bank loan rates are likely to increase by 25 to 45 basis points. The effect of this on lending will depend on the price elasticity of loan demand. As Thakor (2014) states, these kinds of studies are useful because they are beginning to address calibration issues, on which the theories on bank capital structure are mostly silent but in which regulators are particularly interested. But even here, a modest effect could exist side by side with a much stronger transitional impact.

Nor can we turn easily to event studies. The process of designing, deciding on, and implementing bank regulation involves such a long, drawn-out, semi-public process that it is hard to point to clear events (e.g., the Dodd–Frank Act), although Eickmeier et al. (2018) is a good effort. Among the only examples of studies that can circumvent this problem are those by Aiyar et al. (2012) and Meeks (2017), who examine banks in the UK, where regulators have deployed time-varying bank-specific minimum capital requirements, which are not made public; Alyar et al. document that regulated (UK-owned) banks reduced lending in response to higher capital requirements, whereas unregulated banks (resident foreign branches) increased lending, and this effect was significant. Also see Jiménez et al. (2017) for a similarly constructed exercise on Spanish banks.

In any case, this chapter is *not* primarily focused on the macro effect of any single instrument, whether it be CARs, stress tests, liquidity ratios, or whatever, but on the fact that banks are simultaneously affected by *all* of them, at least to some degree, although some may, on occasion, not be binding. When it comes to empirical studies of the joint, interactive effect of the whole set of such regulatory instruments on banks and hence on the real economy, I feel on stronger grounds to claim that there are virtually no such empirical exercises.

Indeed, the idea that it might be possible to quantify the joint effects of the whole set of such measures is, to say the least, far-fetched. But the thesis of this note is that the regulatory authorities have not given much, if any, consideration to the other regulatory initiatives being undertaken when they initially set up their new, or revised, measures. Moreover, this has often been the case for initiatives within categories of regulation (e.g., capital or liquidity regulation), as well as being generally so between such categories. In other words, regulators have tended to work in silos, disregarding what is being done in the regulatory field by other bodies.

Besides the attempt to justify this claim about regulatory tunnel vision, I shall also suggest that there is a further category of regulation, on margins, that has been completely ignored, perhaps out of excessive enthusiasm for the benefits of competition.

2 Capital

The most beneficial regulatory initiative, devised in the aftermath of the Global Financial Crisis (GFC), has been the much greater reliance on stress tests, following on from the successful application of the Troubled Asset Relief Program (TARP) exercise on US banks in May 2009. But how, if at all, do such stress tests fit together with the regular CARs propounded by the Basel Committee on Banking Supervision (BCBS), that is, Basel I, II, and now III? Sometimes the effective constraint on commercial bank expansion is a current, or recent, stress test; sometimes it is a Basel CAR. Does this dual jeopardy matter?

It is commonly stated that foreseen bank losses, for example, from credit or interest risk, should be covered by having sufficiently high interest margins, or spreads (discussed later in the chapter), whereas capital should be held to cover unforeseen or unforeseeable risks. But if they are unforeseeable, how does one know how much capital is required? Moreover, some risks, such as nuclear war, or a Tokyo earthquake for Japanese banks, or the collapse of the euro, although clearly feasible, are so encompassing that they could not be satisfactorily met by self-insurance via higher capital; the state would have to step in, as, for example, it did in London in August 1914 (Roberts, 2013).

Thus, although in principle, capital requirements should be based primarily on stress tests, the difficulty of forecasting the unforeseeable and the complications of trying to assess endogenous second, and subsequent, contagious rounds and effects, despite considerable technical strides in improving such analysis (see, e.g., Alla et al., 2018), mean that a belt-and-braces approach is almost inevitable. Stress tests have to be supported and accompanied by CARs. Nevertheless, the two approaches should, in general, be complementary. If a particular bank appears to be strong on one of these two approaches but weak on the other, then that should be a signal for some detailed reconsideration of the reasons for such a discrepancy. Is that bank's business model unusual, or do the parameters of either the risk weightings of the stress test or the CAR need reconsideration? I am not aware of any such cross-casting between stress testing and CARs being done. I hope that it has been, but if so, it has been behind the scenes.

Another problem of stress testing is that it is meant to explore bank resilience to feasible, yet quite extreme, tail events. The implication must therefore be that in such tests, several banks may fail them. If such failure is then made public, this could generate a run on such failing banks, unless there is a ready source of additional capital that can be rapidly put in place. But if the failing banks are not sufficiently profitable, as is likely, this will not easily come from the private sector. There are two obvious potential answers to this, but both have severe disadvantages.

The first is for the regulatory authorities not to publicize the results of the stress tests, which also has the advantage of making the banks less pressured to try to manipulate the results. This is now how the Australian Prudential Regulation Authority (APRA) operates in

Australia. The disadvantage in this case is that this goes exactly counter to the proposed need for transparency, accountability, and openness that is likely to unite the press, the media, and academics in opposition to such a closed system. The second alternative is to have public-sector funding available and in place, as was done with the TARP exercise in the United States in May 2009, to meet such cases as could not be met from private-sector sources. The disadvantage here is that such public-sector funding would be excoriated as a bailout, and it is now proclaimed that (virtually) all such bailouts will cease and be replaced by bail-ins.

So, if the results are to be made public, and public-sector funding cannot be applied, the only way to be reasonably confident of avoiding having the exercise instigate a run and a subsequent crisis is for the authorities to manipulate the exercise so that no bank seems to be required to raise an unmeetable capital funding requirement. But if so, as it was on occasion in Europe, the exercise will be met with much skepticism and will lack credibility. This conundrum has not yet been fully solved.

Finally, the contracyclical add-on to Basel III and stress testing would, or should, be overlapping. If the economy and/or asset markets are booming, then one would expect the plausible tail downside in the current stress test to be greater—another form of belt and braces. Once more, one can ask how far the authorities review the coherence between changes in the contracyclical add-on and the severity of the scenario(s) being applied in the current stress test.

One field where stress testing and another area of regulation should be capable of complementarity would be in the initiation of recovery exercises by weak banks; stress tests should provide much greater content to the recovery arm of the Bank Recovery and Resolution Directive (BRRD) of the EU.

3 Liquidity

As Cecchetti and Schoenholtz (2017) have already noted:

> Over the course of the past decade, the Basel Committee has responded with the development of two liquidity requirements. The first, the liquidity coverage ratio (LCR), is aimed at ensuring a sufficient quantity of liquid assets; and the second, the net stable funding ratio (NSFR), aims at controlling the level of maturity transformation. . . .
>
> In looking at these, it is natural to ask why we need *two* liquidity requirements, rather than one. . . .
>
> To see why the two requirements are likely to be redundant, start with a simple example. Assume that we are looking at a bank with assets that are either liquid or illiquid and liabilities that are either runnable or stable. Assume also that there are no off-balance sheet exposures.
>
> In this simple case, the bank's balance sheet looks like this:

Simple Bank Balance sheet	
Assets (A)	**Liabilities (L)**
Liquid	*Runnable*
Illiquid	*Stable*

Each of these assets and liabilities corresponds to a category in the one of the two regulations. Liquid assets are the high-quality liquid assets (HQLA)—primarily central bank deposits and short-term domestic sovereigns—in the LCR. Illiquid assets are the long-term loans and securities that require stable funding in the NSFR. Runnable liabilities are the funding sources that generate outflows for the LCR. And, finally, stable liabilities are the available stable funding in the NSFRs.

In this simple case, the two liquidity requirements look like this:

$$LCR: \qquad Liquid \geq Runnable \qquad \Rightarrow \qquad Liquid - Runnable \geq 0,$$
$$NSFR: \qquad Stable \geq Illiquid \qquad \Rightarrow \qquad Stable - Illiquid \geq 0.$$

Now, note that the bank also has a balance sheet identity. This means:

$$Balance\ Sheet\ Identity: Liquid + Illiqid = Runnable + Stable$$
$$\Rightarrow Liquid - Runnable = Stable - Illiquid.$$

From this, we can see that, in this very simple case, the two requirements are the same! Obviously, the assumptions needed to reach this conclusion are not very realistic. . . .

To be bound by the NSFR, instead of the LCR, requires that a bank have a large quantity of long-term assets with relatively high-risk weights that are funded by unsecured wholesale borrowing. This seems unlikely.

This exercise leads to the critical conclusion that if a bank meets the LCR, then it is very likely the bank will meet the NSFR. This assessment is consistent with the evidence in the BCBS quantitative impact study for end-2016: of the 105 largest global banks (those with tier 1 capital in excess of € 3 billion), 91% met the fully phased-in LCR and 94% meet the fully phased-in NSFR. . . .

The practical implication of this complementarity is profound. It means that we really only need one liquidity requirement, and it should be the LCR. If there is concern that the shadow NSFR is overly lax, allowing too much short-term wholesale funding of long-term illiquid assets, then there is a simple fix: increase the horizon of the LCR from 30 days, with run-off rates that slowly decrease.

Thus, the regulators failed to take notice of the interrelationships between the two requirements within the field of liquidity. The problem of failing to note such interrelationships is even more glaring when we turn to taking account of them across different regulatory fields, such as between capital and liquidity.

4 Cross-Field Relationships

It is depressing, but remains the case, that I am not aware of a single directive or regulatory document on capital adequacy that reviews whether capital holdings might be adjusted to take into account an intermediary's relative liquidity, or vice versa, a directive or document on requiring liquidity that takes into account relative capital holdings. Again, a partial exception is stress testing. Insofar as second- and subsequent-round effects are taken into account in such stress tests, more liquidity will lead to fewer fire sales and hence protect capital valuation.

As a generality, more liquidity protects capital. Almost by definition, it reduces interest rate risk and protects against runs. By much the same token, greater equity capitalization protects the liquidity position of a financial intermediary. The more strongly capitalized is such an intermediary, the more easily it can meet a net cash outflow by borrowing in wholesale markets, and the less likely would be a panic run on the bank in the first place.

Indeed, one of the factors that led to the diminution of concern about sufficient liquidity, prior to the GFC, was the belief that commercial bank adherence to the Basel CARs would so buttress their apparent solvency that they could always borrow cash in the now wider and more efficient wholesale money markets (see Goodhart, 2011, Chapter 9, especially p. 330).

Thus, a bank that is more liquid has no countervailing benefit from a lower required CAR, nor, of course, vice versa. The two are assessed in separate silos. A recent but, alas, typical example of this is to be found in the recent December 2017 BCBS discussion paper "The Regulatory Treatment of Sovereign Exposures," which discusses potential capital risk weights for sovereign exposures without relating such weightings to the role of such exposures in providing HQLAs. This is mentioned in the paper's Chapter 3, "The Holistic Role of Sovereign Exposures," but then it seems to play no active part in the key Chapter 5, "Potential Ideas Related to the Regulatory Treatment of Sovereign Exposures." I wrote a comment on this paper, which is reproduced, subject to a few minor changes, in the Appendix.

5 Bank Profitability and Margins

In our chapter "Towards a Measure of Financial Fragility," Chapter 11 in Goodhart and Tsomocos (2012), we (Aspachs et al.), argued that there were two key factors characterizing financial fragility. These were banks' probabilities of default, on the one hand, and their profitability, on the other hand. Even if a bank was following a risky business model, it could survive if it was (expected to be) consistently profitable because current running profits could repair balance-sheet losses, and expected future profits could enable a stricken bank to raise additional equity capital.

A major factor in determining bank profitability is the margin, or spread, between (bank) borrowing costs and lending rates. When such margins, and spreads, become narrowed (e.g., by competition), not only is bank profitability directly adversely affected, but bankers will have an incentive to take on a riskier portfolio, to reach for yield, in order to maintain a desired return on shareholder equity (ROE). Thus, contemporaneous analysis of the interwar financial crisis and resulting depression ascribed much of this to excessive competition, with the resulting remedy being the encouragement of cartelization and an oligopolistic setting of rates and margins. Montagu Norman was particularly keen on this.

The liberalization of such oligopolistic systems, as in Competition and Credit Control in the UK in 1971, led in many cases to a drastic narrowing of margins, a rapid expansion of credit and money, and a financial boom/bust. Of course, narrowing margins, a concomitant of greater competition, provide a boon to bank clients. But the question that regulators need to ask themselves is whether such narrowing is broadly sustainable, for example, a result of a more efficient approach, perhaps a better technology, or is the result of a rush for growth at all costs. As Chuck Prince neatly put it, when the competitive music is playing, everyone is induced to dance. Sometimes the dance is led by public-sector banks with less immediate concern for profitability.

Yet the concerns of yesteryear about undue compression of margins and potentially excessive competition currently seem to be cast aside. Competition is judged to be always good. Policies that serve to restrict bank margins, such as purposeful flattening of the yield curve, as in QE2 and 3, are introduced without a backward glance at their effects on bank profitability. When has macro-monetary policymaking incorporated a prior assessment of its implications for bank profitability? When have regulators queried a bank's business model because of the effect more widely on banking margins and profitability?

It is, at least, arguable that the most stable banking structure is an oligopoly. Countries that restricted competition from foreign banks and were totally dominated by a few local bank (e.g., Australia, Canada, India, Sweden) did best in the GFC. There is, surely, a trade-off to be found between stability and competition, but there should be some recognition that more (of some kinds of) competition will lead to less stability. That recognition currently seems largely lacking.

6 Conclusions

The theme of this chapter is that regulators tend to have tunnel vision when introducing or amending financial regulation. They consider each proposal in isolation, without attempting to see how it might fit, holistically, into the broader canvas of the whole set of financial regulations affecting banks. A secondary theme has been that the authorities have been reluctant to appreciate the trade-off between stability and competition and are rarely prepared to consider whether some features of unrestrained competition may be inconsistent with financial stability.

In principle, the most helpful way of moving toward more holistic regulation would be further development of, and emphasis on, stress testing. Such tests, both at the micro and systemic levels, should, in theory, be able to take account of relative capital, liquidity, and profitability concerns, all together in a single package. Moreover, the events of a failure of a stress test and the trigger for starting a recovery program need to be integrated, which has not yet been done but should be. But there is a long way to go before we can rely largely on stress testing procedures to provide adequate resilience to banks to ensure sufficient financial stability, and how should one calibrate "adequate" and "sufficient"?

Such stress tests now concentrate on the banking sector. If these, and regulation more broadly, become more onerous, there is a greater incentive for financial intermediation to move toward nonbanks, "shadow banks," tech companies, peer-to-peer facilitators, and so forth. This is generically known as the *border problem*, on which Rosa Lastra and I have already written (Goodhart and Lastra, 2010). As a generality, the more that such nonbank financial intermediation partakes of, and takes over, credit expansion and the provision of means of payment, the more that such new channels need to be brought within the ambit of stress testing procedures. But this is easier said than done.

Appendix: The Regulatory Treatment of Sovereign Exposures

A Comment on the BCBS Discussion Paper, December 2017

A.1 Introduction

There are many praiseworthy features of this paper, but it is an academic's role to criticize; I shall do so under four headings. First, there is a degree of dissimulation within the paper. The suggestion is made that the problems are general and economic, whereas actually, they are primarily political and specific to European concerns. Second, the main purpose for holding sovereign debt is to provide safe, liquid assets. But there is no attempt within the paper to integrate the need and requirements for HQLAs with the proposed possible adjustments to capital requirements. Third, all claims on sovereigns are treated in the paper as if they are identical in all characteristics. In practice, they are not, but no account is taken of this. Fourth, no mention has been made of the history of the treatment of sovereign exposures by the BCBS, despite the fact that the present problems are markedly similar to those met at the outset, in the 1980s, of trying to deal with this issue.

A.2 Some History

Let me start with the final criticism. I have, of course, some personal interest in this because I wrote that history in my 2011 book, on *The Basel Committee on Banking Supervision*. The history of how this problem arose is set out at some length in Chapter 6, "Capital Adequacy and the Basel Accord of 1988." To recapitulate, the US representatives put forward a logical idea, which was that in each country, commercial banks could treat the sovereign debt of their own separate central government as riskless but that it should use credit ratings to assess the relative risk of all other sovereign debt. This was shot down by the Europeans and the European Commission (EC; ibid., p. 154, and Section D, pp. 181–194), on the grounds that the Treaty of Rome allowed no discrimination between member states, so all the sovereign debt of all EC members must be equally given zero risk weight, irrespective of which banks held the debt. This led to the notorious "club" approach, which has caused problems ever since.

The current problems of sovereign exposures simply involve a continuation, and further complication, of the original decisions made in the mid-1980s.

A.3 Modeling Sovereign Risk

The suggestion is made, on page 25 of the text of the BCBS report, "Removal of the IRB Approach for Sovereign Exposures," that it is particularly difficult "to robustly model risk parameters for sovereign exposures." With respect, this is largely nonsense. Although the

377

number of actual defaults has been quite low, there has been a great deal of practice and experience in credit risk modeling for sovereigns, and the credit-rating agencies have a lot of experience and practice in this respect. Such risk modeling is a great deal easier than in the case of dealing with financial instruments when they are new and untried, such as collateralized debt obligations (CDOs), or industries introducing novel innovatory technologies, such as the tech giants.

The problem is not economic analysis but political concerns. Credit-rating agencies are under frequent considerable pressure not to downgrade their ratings for important countries. There are also threats from some countries that if such downgrades did occur, the country would subsidize its own national credit-rating agencies, which its own institutions would then have to use. That political problem would be far worse for internationally active global systemically important financial institutions (G-SIFIs). For these, the sovereigns of most countries are among their most important clients. The political pressures on such G-SIFIs or national systemically important financial institutions (N-SIFIs) would be intense to compete for business by giving overstated ratings. Thus, the conclusion that one cannot rely on banks to model such ratings accurately is correct, but the reason given is misleading.

A.4 Currency Unions

A major reason why sovereign debt is regarded as riskless in nominal terms is that, in extremis, the government can order its own central bank to create additional money to pay off any required coupon or principal repayment. A problem with a currency union is that it is no longer within the power of members of that union to force the central bank to do so. Thus, Greece could not require the European Central Bank (ECB) to create euros on its behalf in order to pay off the holders of Greek debt. This puts members of the euro area in a quite different position from those of sovereigns controlling their own individual central bank. In a sense, the same is true of countries with a currency board, rather than a fully operative central bank, such as Hong Kong or Panama. But at least in these latter cases, it is technically feasible for such countries to shift their currency board into central bank form, whereas within the EU, a member country, such as Greece, cannot reestablish a central bank of its own while still being a member of the EU currency union. On page 23, subheading ii, where governments with a central bank are treated as exactly equivalent to those "of a currency union to which that state belongs" is consciously misstating a crucial difference.

A.5 Some Principles

To my mind, the basic problem is as follows: We want commercial banks to hold a sufficiency of liquid assets (HQLAs) to support and protect their businesses, and notably so in each country and currency in which they operate. On the other hand, an excess (beyond those holdings that can be described as an appropriate liquid base for operations) at some point represents a potentially speculative bet on the relative successful economic outcome of that country. This raises another point. All G-SIFIs will be operating in many countries and with many currencies. In each country/currency business where they are operating, they should be allowed and encouraged to hold sufficient liquid assets to support that activity. Thus, there is a suggestion that other sovereign entities should have a relatively high risk weight. But if the commercial bank under consideration has a significant business in that country/currency, then it would be entirely appropriate for it to hold, as supporting liquidity, adequate exposure to the sovereign debt of that country/currency at zero weight.

Thus, my own suggested approach would be to require all such internationally active banks to record the proportionate weights of their country/currency business. Then, to provide sufficient liquidity for all such currency business, they would be allowed and encouraged to hold the relevant sovereign debt up to some proportion of that business. The question of appropriate proportion has already been considered in devising the proposals for HQLAs. But at some point, the need for HQLAs in the relative country/currency would pass over into the field of speculation rather than appropriate liquidity holding. After that point, one might envisage increasing capital ratio requirements.

There are obvious severe practical problems, but the point that I am trying to make here is that commercial bank holdings of sovereign debt ought to be proportionate to the business transacted within that country/currency area. This argument about the need for proportionality is based on the fact that the need for assets that appear to be liquid and riskless in any country/currency pair is related to the need to maintain a sufficient liquidity base for that business. Decisions taken on appropriate capital ratios should be closely and carefully integrated with the need for liquidity, and equivalently, regulation on capital and liquidity needs to be treated holistically, rather than independently and separately, as is the case in this latest paper.

A.6 Claims on Sovereigns Are Not Homogeneous

Throughout the discussion paper, bank exposures to sovereigns are treated, implicitly, as if they were all identical in all respects. This is clearly not so. During the GFC, the actual experience of credit loss from default on the senior tranches of collateralized mortgage obligations (CMOs) was actually quite low. But during the panic and crisis, the market value of such assets collapsed. In many cases, they were marked to market, and in other cases, analysts could work out what the loss of value was. It was the loss of market value, rather than actual default, that in most cases drove many banks toward insolvency. The same experience holds for sovereign debt. It would be the loss of market value, rather than the actual default, that in many cases would lead to increasing risk of bank insolvency.

Clearly, the risk of loss of market value is far greater with long-duration bonds. Both liquidity and solvency considerations should require that the allowance for banks to hold short-dated sovereigns is far greater than the allowance for them to hold long-dated sovereigns. In general, it is believed that banks tend to be well behaved in this respect, mostly holding only short-dated sovereign exposures. Nevertheless, any measures to impose liquidity or capital requirements should take specific account of the expected volatility in market prices of such exposures. So, there should be some increasing discount in the allowance of such holdings, depending on their duration. The higher the average duration of such exposures, the less it should be allowed under liquidity considerations, and the more it should be potentially penalized by greater capital ratios as a speculative holding.

Again, there is no discussion of such issues in the paper.

References

Adrian, T., N. Boyarchenko, and O. Shachar (2017a), "Dealer balance sheets and bond liquidity provision," *Journal of Monetary Economics*, 89, 181–194.

Adrian, T., M. Fleming, O. Shachar, and E. Vogt (2017b), "Market liquidity after the financial crisis," *Annual Review of Financial Economics*, 9, 43–83.

Aiyar, S., C. Calomiris, and T. Wieladek (2012), "Does macro-pru leak? Evidence from a UK policy experiment," NBER Working Paper No. 17822.

Alla, Z., R. A. Espinoza, Q. H. Li, and M. Segoviano (2018, March), "Macroprudential stress tests: A reduced-form approach to quantifying systemic risk losses," IMF Working Paper WP/18/49.

Aspachs, O., C. A. E. Goodhart, D. P. Tsomocos, and Zicchino (2012), "Towards a measure of financial fragility," in C. A. E. Goodhart and D. P. Tsomocos (eds.), *Financial stability in practice: Towards an uncertain future*, Edward Elgar.

Basel Committee on Banking Supervision (2017, December), "The regulatory treatment of sovereign exposures," BCBS Discussion Paper, Bank for International Settlements.

Budnik, K., et al. (2019, April), "The benefits and costs of adjusting bank capitalisation: evidence from euro area countries," European Central Bank Working Paper No. 2261.

Cecchetti, S. G., and K. Schoenholtz (2017 November), "Regulatory reform: A scorecard," Centre for Economic Policy Research, Discussion Paper DP12465.

CNN Money (2011), "Jamie Dimon gripes to Bernanke," June 7, 6:20 p.m. EST.

Eickmeier, S., B. Kolb, and E. Prieto (2018), "Macroeconomic effects of bank capital regulation," Deutsche Bundesbank Discussion Paper No. 44/2018.

Goodhart, C. A. E. (2011), *The Basel Committee on Banking Supervision: A history of the early years, 1974–1997*, Cambridge University Press.

Goodhart, C. A. E., and R. Lastra (2010), "Border problems," *Journal of International Economic Law*, 13(3) 705–718.

Goodhart, C. A. E., and D. P. Tsomocos (eds.) (2012), *Financial stability in practice: Towards an uncertain future,* Edward Elgar Publishing Limited,   Cheltenham, UK.

Hanson, S., A. Kashyap, and J. Stein, (2011), "A macroprudential approach to financial regulation," *Journal of Economic Perspectives*, 25(1), 3–28.

Jiménez, G., S. Ongena, J.-L. Peydró, and J. Saurina (2017), "Macroprudential policy, countercyclical bank capital buffers, and credit supply: Evidence from the Spanish dynamic provisioning experiments," *Journal of Political Economy*, 125(6), 2126–2177.

Kanngiesser, D., R. Martin, D. Moccero, and L. Maurin (2019), "The macroeconomic impact of changes in economic bank capital buffers," European Central Bank Working Paper No. 8.

Meeks, R. (2017), "Capital regulation and the macroeconomy: Empirical evidence and macroprudential policy," *European Economic Review*, 95, 125–141.

Peek, J., and E. Rosengren (1997), "The international transmission of financial shocks: The case of Japan," *American Economic Review*, 87(4), 496–505.

Roberts, R. (2013), *Saving the city: The great financial crisis of 1914*, Oxford University Press.

Thakor, A. V. (2014), "Bank capital and financial stability: An economic trade-off or a Faustian bargain?" *Annual Review of Financial Economics*, 6, 185–223.

21

Leverage and Macroprudential Policy

John Geanakoplos

1 Introduction

At the World Econometric Society Congress of 2000, I proposed a theory of booms and crashes: the leverage cycle, caused by a buildup of too much leverage and then a faster deleveraging.[1] I argued that leverage is an endogenous variable that can vary widely. Rising leverage leads to rising asset prices and increasing economic activity. But this rise paradoxically makes the economy more vulnerable, so that eventually, a little bit of "scary bad news" can cause leverage to plummet and trigger a great crash.

Between 2000 and the 2007–2009 crisis, leverage did indeed rise in the balance sheets of banks and households, and so did prices for housing and mortgage-backed securities. Then leverage and asset prices collapsed. After 2010, leverage and asset prices eventually recovered.

Changes in leverage are often caused by changes in the perceived downside risk of assets serving as collateral for loans. There is a feedback between lessened down risk, caused, for example, by more optimism or lower asset volatility, increased leverage, and a rise in asset prices. This feedback loop becomes dangerous when more perceived down risk leads to lower leverage and falling asset prices. The culmination can be an economy-wide margin call leading to a crash.

Traditionally, macroeconomic cycles have been attributed to insufficient or excessive aggregate demand for investments and assets, that is, to fluctuations in demand caused by the "animal spirits," irrational exuberance, risk appetite, or precautionary savings of investors. The leverage cycle begins with the observation that much of demand is facilitated by borrowing. When uncertainty rises, investors see more downside risk but also more upside potential. Lenders, on the other hand, don't share in the upside; they see bigger downside losses. Lenders therefore have a bigger incentive than borrowers to change the terms of the contract. They ask for more collateral.

In Section 2, I lay out the logic of my original leverage-cycle story. One crucial element is the heterogeneity of agents. In my original work, this stemmed from different priors. It could be due to different risk tolerance or to many other differences between agents. But it implies that different agents have different marginal propensities to buy the asset when their wealth changes. Leveraging the purchase of an asset is a bet on the price of the asset. Good news makes everyone value the asset more, but the resulting rise in its price also shifts wealth toward the agents who had been leveraging the asset. These beneficiaries of

---

[1] See Geanakoplos (2003). See also Geanakoplos (1997, 2010a).

the price rise likely are the agents with the higher marginal propensity to buy it, increasing demand still more. The wealth redistribution from news thus reinforces the news, creating more asset-price volatility than would occur with homogeneous agents.[2]

On top of that, if the news itself also changes the forward view on the asset volatility, this will change leverage. For example, calming good news will lead lenders to worry less about collateral values and thus to lend more easily. The same buyers with a high propensity to spend will then be able to leverage more (gaining access to more of their future wealth), and this will drive the asset prices still higher, greatly amplifying the fundamental news.

The crash in the leverage cycle occurs when scary (volatility-increasing) and bad (expectation-reducing) news come together *while* the economy is already leveraged. The bad news lowers the price because of the direct effect of bad news on asset value. The prevailing high leverage then lowers the price more by magnifying the wealth redistribution stemming from the falling price. The scary news then lowers the price still more because old holders are forced to deleverage by selling the asset, and new buyers cannot leverage as much as the old buyers had previously because of tightened lending standards.

In Section 3, I point out that leverage is a special kind of asset tranching that has been extended by all sorts of financial innovations, which themselves in effect create even more leverage. One can thus speak of a financial innovation cycle. In Section 4, I point out that there can be multiple leverage cycles, where each one triggers the next one until they all come crashing down together. In Section 5, I argue that leverage or collateral is just one of many credit terms, albeit the most important, that the lenders require of borrowers, all of which vary over the cycle. When one of them, say, FICO gets looser, even if leverage stays the same, effectively they have all gotten looser, ceteris peribus. The multiplicity of credit terms is captured by the credit surface.

The leverage cycle and leverage-cycle crash are effectively a risk scenario. The leverage-cycle story thus lends itself perfectly to stress testing, either by a regulator or by a risk manager. The leverage-cycle apparatus of the credit surface also leads to a more sophisticated kind of monetary policy, in which the interest rate is no longer regarded as the only lever to manage credit. In Section 6, I briefly describe the leverage-cycle implications for monetary policy and macroprudential policy, including stress tests.[3]

2 The Leverage Cycle

The conventional view in macroeconomics had long been that cycles are caused by fluctuations in aggregate demand. These can be smoothed over by raising the interest rate when demand is too high and lowering the interest rate when demand is too low. The trouble with this demand-centric and interest rate-centric view of macroeconomics is that it ignores the lenders. It leaves unanswered what we mean by "tight credit," if not just a high interest rate. When businesspeople talk about tight credit, they don't mean that the riskless interest rate set by the Federal Reserve is too high. They mean that at the going riskless

---

[2] This wealth redistribution is proportional to the change in the total value of the purchased assets. If each agent is spending a small fraction of his or her wealth on the asset, the amplification will be small. But if the asset is, say, housing, then the wealth redistribution is much more important.

[3] This essay is adapted from part of Geanakoplos (2019).

interest rate, or anything close to it, they cannot get a loan, because lenders are afraid they might default. Default is what is missing in the traditional macroeconomics theory.

Once default is recognized as a possibility, we should expect lenders to require additional terms for a loan, such as a maximum ratio of debt to income (DTI), or a minimum credit score (FICO). The most important requirement is usually collateral, and I concentrate on collateral here.

If an $80 loan requires collateral of $100 (or enables the purchase of a $100 asset that serves as collateral), then we say that the collateral rate is 125 percent, the loan to value (LTV) is 80 percent, the margin or downpayment is 20 percent, and the leverage is 5 because $20 cash can allow for the purchase of an asset worth $100. All of these amount to the same thing. It has been known for centuries that more leverage leads to more risk. If the collateral falls in value to $99, and the $80 loan is paid off, the borrower is left with $19 out of the original $20. A 1 percent fall in the collateral price leads to a 5 percent fall in investor capital, which are in the same ratio as the leverage.

The new idea in the leverage cycle is that more leverage causes higher collateral prices. The only precedents for this seem to be in the work of Minsky (1977) and the economic historian Kindleberger (1978). Neither of these authors used a mathematical model to express his ideas, and neither had collateral explicitly in mind (Minsky was talking about a firm borrowing money, and by "leverage" he meant a ratio of debt payments to income). Both of them made the extrapolative (irrational) expectations of borrowers the linchpin of their theories.[4]

There are three mechanisms that drive the leverage cycle and three more mechanisms that create the leverage-cycle crash. The first driver is that leverage can be made endogenous via the credit surface. The second is that leverage increases when perceived down risk decreases, either because of greater optimimism or reduced volatility. The third is that higher leverage makes for higher asset prices, all else being equal. All three mechanisms can be described by precise mathematical theorems in the binomial uncertainty world. The leverage cycle typically moves from good news that reduces volatility to higher leverage, then to higher asset prices. Eventually, the news worsens and uncertainty rises, which leads to lower leverage and falling asset prices.

The downward trajectory of the leverage cycle is sometimes much more violent than the slower upward trajectory. A highly leveraged economy is vulnerable to crashes stemming from small shocks that create more uncertainty, which I call "scary bad news." A little bit of scary bad news can topple a highly indebted economy through three leverage-cycle-crash mechanisms, which all sometimes come together in a margin call.

The scary news leverage-crash mechanism occurs when scary news abruptly increases uncertainty, which in turn abruptly reduces leverage, leading to margin calls. This mechanism is the same as the volatility-leverage mechanism. It just moves much faster in the downward direction.

The bad news liquid wealth crash mechanism occurs when bad news reduces a collateral price, leading to a margin call even when leverage is unchanged. The margin call reduces

---

[4] Collateral appears in formal macroeconomic models first in Bernanke and Gertler (1986) and then simultaneously in Kiyotaki and Moore (1997), Holmstrom and Tirole (1997), and Geanakoplos (1997). One difference between my approach to leverage and the rest is that I emphasized the endogeneity of leverage and changes in leverage, whereas they did not.

the liquid wealth of the asset's leveraged owner, potentially forcing him or her to sell even though there is no conventional wealth effect, just when the lower price might have incentivized him or her to buy. Finally, when debts are high, and large sales are necessary to repay the debts, as in a margin call, the economy can be very fragile because of a wealth-redistribution effect if the marginal propensity to spend on the collateral is higher for its leveraged owners than for the new buyers. I call this last mechanism the *debt-fragility crash mechanism*.

Each mechanism again corresponds to a precise mathematical theorem. Panic plays no necessary role in the mechanisms. Of course, panic might exacerbate the mechanisms. But they are powerful even without the added froth of panic. If the leverage-cycle theory of crashes had to be stated in one or two words, it would not be *panic*; it would be *margin call*.

2.1 The Three Mechanisms of the Slow Leverage Cycle

2.1.1 Endogenous Leverage: The Credit Surface

As a graduate student, I had never heard the word *collateral* mentioned in any course I took, even in macroeconomics and finance. But when I worked in the fixed-income department at Kidder Peabody in the late 1980s and early 1990s, collateral came up in almost every conversation. I began to think about collateral as a theorist, and I was immediately struck by a puzzle. How can one supply-equals-demand equation for a loan determine the price (or interest rate) on the loan and also the collateral rate or leverage or LTV on the loan? It seemed impossible that one equation could determine two variables. This same problem becomes even worse when one considers all the other terms of a loan, such as FICO and DTI.

I resolved this puzzle for collateral when I realized that I should be thinking about a different price for each different level of leverage. A loan should be defined by a pair (promise, collateral), not just by the promise, and each pair must have its own separate price. Fixing the collateral, bigger and bigger promises give rise to higher and higher leverage. At first, the loans are so small that the collateral fully protects the lender. But after a certain point, the loans are not fully protected and might default. They get riskier and riskier, and the interest rises. The surface generated by the interest rate corresponding to each level of leverage is what I called the *credit surface*.[5] See Figure 21.1. More generally, one could imagine a credit surface with independent axes including LTV, DTI, and FICO and a vertical axis giving the corresponding interest rate charged to a loan with any combination of those three characteristics.[6] I shall be content to mostly stick with the collateral credit surface in this chapter, although I return to the more general case in Section 5.

Borrowers and lenders each choose where they want to be on the credit surface. In equilibrium, for each level of leverage, there is a separate supply-equals-demand equation and a separate price. At many leverage levels, there may be zero supply and zero demand. The most interesting borrowers are not the ones on the flat part of the credit surface, who are able to borrow unconstrained quantities at the riskless interest rate, as in the old style of macroeconomics. The agents who are at point A and beyond are often the pivotal drivers

---

[5] See Geanakoplos (1997).

[6] See Geanakoplos (2016). One could also imagine different axes corresponding to different kinds of collateral, depending on the precise legal rights for the confiscation of the collateral.

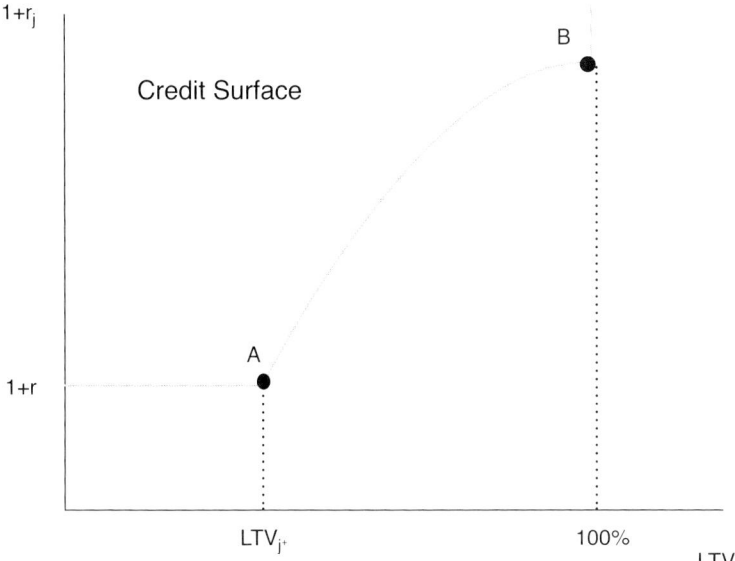

Figure 21.1 Credit surface.

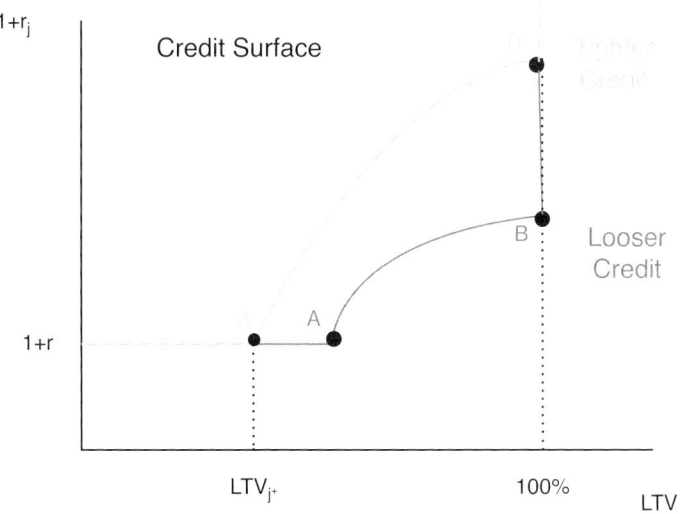

Figure 21.2 Credit surfaces, with tight and loose credit.

of fluctuations in economic activity, and they are constrained because each time they try to borrow more (on the same collateral), they face a higher interest rate.

The credit surface also clarifies the meaning of tight credit. It is not the height of the riskless rate per se but also the steepness of the credit surface that renders credit tight. Thus in Figure 21.2, the bottom credit-surface line is looser than the top credit-surface line even though the riskless interest rate is the same.

For binomial economies with financial assets, Ana Fostel and I proved that the only leverage level that would be positively traded in equilibrium is the maxmin loan, which promises the maximum without any risk of default. This is point A in Figure 21.1.[7] The theorem guarantees that in equilibrium, the credit surface rises sufficiently fast beyond point A that nobody will choose to trade there. Leverage is completely endogenous, chosen freely by borrowers and lenders at any point, but the theory predicts exactly where it will end up. Of course, the binomial assumption, that only two things can happen, is very unrealistic.[8] (It approximates reality best with very short-term loans, such as repurchase agreements [repos]). But the conclusion does not depend on the preferences of the agents or their endowments or their probability assessments of the future states, or whether lenders' probabilities differ from borrowers' probabilities.[9, 10]

2.1.2 Leverage and Down Risk or Volatility

The binomial no-default theorem has an immediate consequence for leverage, which Fostel and I called the *binomial leverage theorem*. Geometrically, it is clear that point A is defined by the worst-case scenario. With a little bit of algebra, we showed that in binomial models with financial assets, equilibrium LTV is equal to the worst-case gross return, divided by the gross riskless rate of interest:

$$\text{LTV} = \frac{1}{1+r} \frac{\text{worst collateral payoff}}{\text{price of collateral}}.$$

I emphasize that leverage rises when the down risk abates, that is, when the world gets safer. One way this can happen is if expectations become so optimistic that returns on the collateral are anticipated to be higher in every future state.

Feeling optimistic and feeling safe often go hand in hand. But sometimes they can be quite different. If agents think there is more upside in just the best state or just higher probabilities of the best state, leverage will not rise.[11]

When risks are symmetric, the worst case is worse if volatility is higher. This shifts the credit surface to the left and up, as indicated in Figure 21.2. The theory then predicts that leverage will go down for assets whose expected volatility goes up. The great advantage of expected volatility as a marker of safety is that it is observable, either through the implied volatility of option prices or through recent volatility, which is highly predictive of expected volatility. And indeed, margins (in, say, the commodities markets) almost always go up when

---

[7] See Fostel and Geanakoplos (2015). Financial assets give no direct utility for holding them (like a painting would), and their future dividends do not depend on who holds them. Think of a share of GE stock or of a mortgage-backed security.

    Thus, in binomial economies with financial assets, leverage to the right of point A will never be observed. This is not true for houses or paintings, or for financial assets with trinomial states, where the most interesting borrowers might indeed be to the right of point A. Loans to the left of point A are overcollateralized. If we ignore the irrelevant extra collateral, we could say those loans are maxmin loans on a smaller collateral base.

[8] Everybody has to know that only two things can happen and agree on which two things can happen, although not necessarily on their probabilities.

[9] In Geanakoplos (2003), I had proved the same theorem but only under the additional hypothesis that agents are risk neutral.

[10] The question of what expectations lead to more leverage in cases where there are more than two states is quite complicated. The biggest advance was made by Simsek (2013). See also Phelan (2017).

[11] If the probability of the good state rises, the price of collateral will likely rise. By the binomial no-default theorem, the equilibrium promise will remain the same. So leverage will fall.

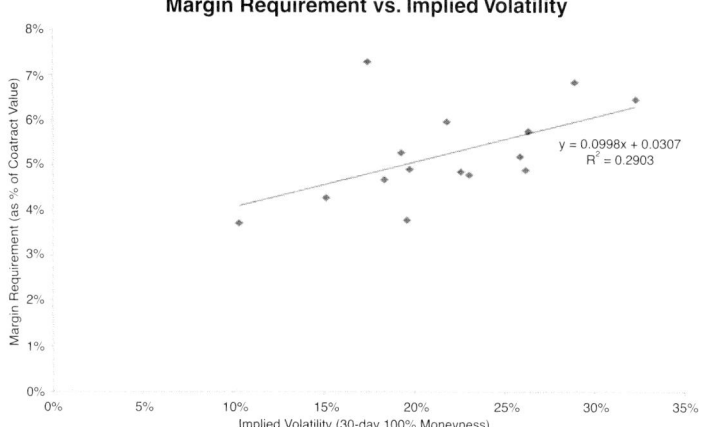

Figure 21.3 Volatility and leverage.

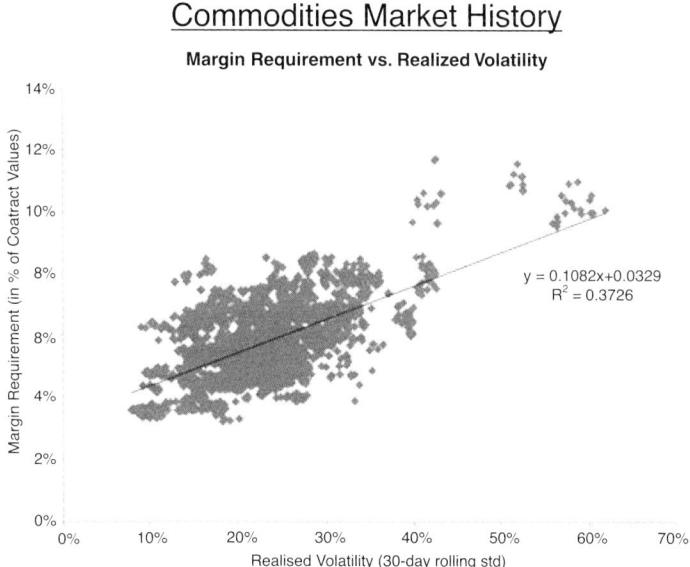

Figure 21.4 Historical volatility and leverage.

either kind of volatility goes up. By contrast, it is notoriously difficult to quantify optimism about the mean of tomorrow's price; in stock prices, yesterday's direction does not predict today's direction.

Figure 21.3 shows the connection between margins in the commodities futures markets and the implied volatility of the underlying prices.

Similarly, Figure 21.4 shows the connection between margins in the commodities futures markets and the recent volatility of the underlying prices.

2.1.3 Leverage and Asset Prices with Agent Heterogeneity

The third key mathematical idea is that all else equal, more leverage increases asset prices. The reason is almost self-evident, yet it had not really been examined in the literature. With a smaller required downpayment, more buyers can express their demand for the collateral (houses or mortgage-backed securities, etc.), and the same buyers can buy more units, leading to greater demand and a higher price, provided there is heterogeneity in the valuations agents place on the asset.[12] Fostel and Geanakoplos (2014) proved that in any binomial model with financial assets, constraining leverage below the equilibrium maxmin value, for example, by prohibiting leverage altogether, always lowers the value of an asset, assuming that the risk-free interest rate does not change.[13] The magnitude of the price effect brought on by the change in leverage depends on the heterogeneity of agent valuations. The more the heterogeneity, the bigger the leverage–price effect.

The link between leverage and asset prices contradicts the famous Modigliani–Miller (M-M) theorem, which asserts that prices should be unaffected by leverage. One difference is that Modigliani and Miller did not explicitly discuss collateral. They did have in mind a firm, which, to be sure, might be thought of as collateral for its bond issuances. But they overlooked that their argument depends on the reliability of nonfirm debt as well. Their argument, as clarified by Stiglitz (1969), is essentially the following. Suppose a firm issues a debt promise of D and raises the rest of its money by issuing equity of value E. Suppose it does not default on D in any state of nature. If the firm were restricted to sell a promise $D' < D$, then it would have to issue more equity E'. The bondholders who had previously purchased the promises $D - D'$ would be disappointed at losing access to riskless debt, and the equity holders would be forced to absorb more equity, and tamer (less leveraged) equity, possibly reducing their expected returns. The M-M theorem is proved by noting that the equity holders could themselves issue the missing debt $D'' = D - D'$, thereby giving the market the same debt it had before and at the same time releveraging the equity E' so that it becomes just like E. In essence, the reduced leverage at the firm level is compensated by increased leverage at the investor level.

One flaw in this M-M proof is that collateral is not generally transferable; just because the firm can be used as collateral does not necessarily mean the equity can be used as collateral. The equity holder might have a different propensity to repay, perhaps not as reliable as the original firm, so D'' would not be treated by the market as a perfect substitute for D. When leverage goes down for the economy as a whole, there are real consequences.

---

[12] Imagine all the buyers arrayed on a vertical corresponding to their valuation of the asset. The marginal buyer is the agent whose valuation is equal to the price. The higher-valuation agents will be buyers, and the lower-valuation agents will sell the asset. As the buyers get access to more borrowing, a fewer number of them can buy all the assets, creating a higher marginal buyer and thus a higher price. This is essentially the demonstration given in Geanakoplos (2003) that higher leverage leads to higher collateral prices.

[13] If the interest rate rose as agents leveraged more, agents would discount the cash flows from the asset more harshly, and so their lower valuations would party offset their gain in purchasing power, leaving the final collateral price ambiguous.

Leverage and Collateral Prices

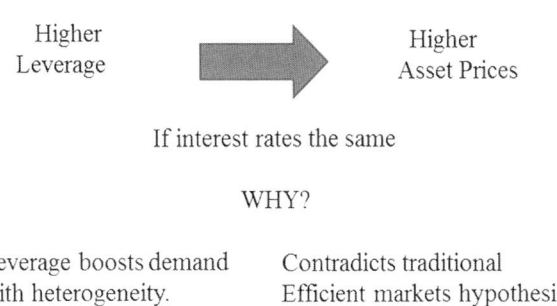

Higher Leverage → Higher Asset Prices

If interest rates the same

WHY?

Leverage boosts demand with heterogeneity.

Contradicts traditional Efficient markets hypothesis

Figure 21.5 Leverage and asset prices.

For example, consider a new homeowner who is limited (say, by regulation or by a worse down risk in housing prices) to taking out a mortgage at a smaller LTV. The homeowner would simply have to come up with a bigger down payment because taking out a second loan would not be permitted by the regulation or by the worse down risk. There is no outside agent who could use the homeowner's increased equity to increase his or her leverage. The drop in debt will necessarily have real consequences, for the economy and for the price of the houses. This same argument applies word for word to the purchaser of any asset, such as a mortgage-backed security. The only situation in which the M-M logic partially applies is the one they had in mind. The buyer of firm equity could indeed use the equity as collateral for a further loan, thus compensating for the lower debt-to-equity ratio at the firm level. But the flaw emerges here as well if we go one step deeper. If increased firm down risk reduces firm debt, the equity may increase in size, but it will not support a compensating increase in debt, so total leverage will go down, and the firm price will fall. See Figure 21.5.

The lead-up to the 2007–2009 financial crisis, the crisis itself, and also its aftermath give some evidence for the leverage-cycle connection between collateral prices and leverage. In Figure 21.6, we see the connection between the price of a portfolio of AAA subprime bonds and the LTV lenders offered to the hedge fund Ellington Capital on similar bonds. As the figure indicates, leverage and prices fell together during the crisis and eventually rose together in the recovery.[14]

In Figure 21.7, we see the connection between the Case–Shiller housing index and the leverage on nongovernment mortgages from 2000 to 2009. Again, leverage and the asset

---

[14] This graph first appeared in Geanakoplos (2010) and is updated here. For a graph of margins on repo just after 2007, see Gorton and Metrick (2012).

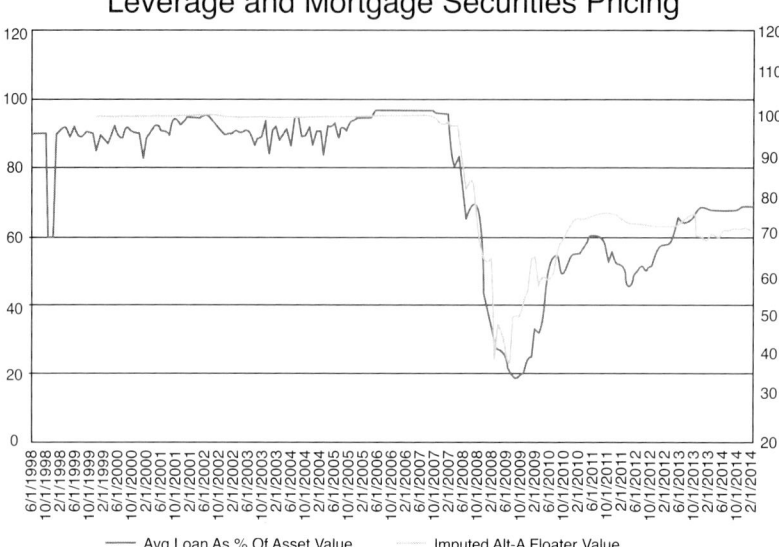

Figure 21.6 Leverage and mortgage securities pricing.
Source: Geanakoplos (2009), updated to 2014.

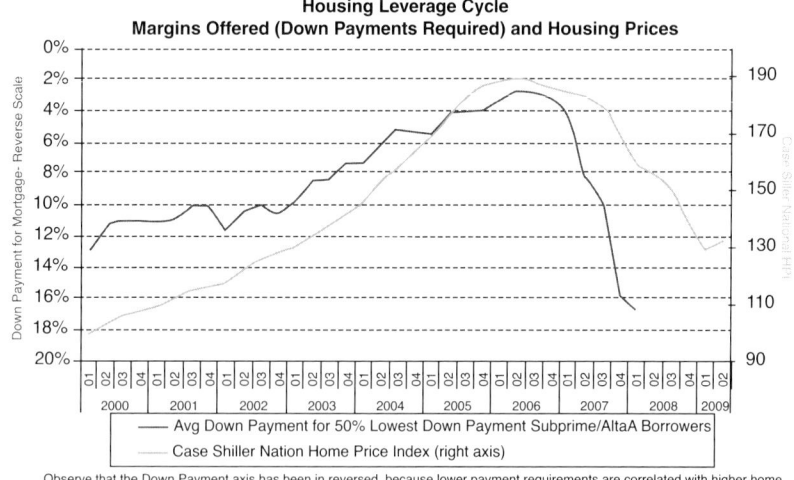

Observe that the Down Payment axis has been in reversed, because lower payment requirements are correlated with higher home prices.

Note: For every AltA or Subprime firstloan originated from Q12000 to Q1 2008, down payment precentage was calculated as appraised value (or sclae price if available) minus total mortgage debt, divided by appraised value. For each quarter, the down payment percentages were ranked from highest to lowest, and the average of the bottom half of the list is shown in the diagram. This number is an indicator of down payment required: clearly many home owner put down more than they had to, and that is why the top half is dropped from the average. A 13% down payment in Q1 2000 cores ponds to leverage of a about 7.7, and 2.7% down payment in Q2 2008 corresponds to leverage of about 37.

Figure 21.7 Leverage and housing prices.

price move together.[15] The great crisis of 2007–2009 was the culmination of a double leverage cycle, in mortgages and in mortgage securities.

2.1.4 The Slow Leverage Cycle

The leverage cycle typically moves from good news that reduces volatility and flattens the credit surface, which leads to higher leverage, and then to higher asset prices. Eventually, the news worsens and uncertainty rises, which leads to a steeper credit surface, lower leverage, and falling asset prices. This is shown in Figure 21.8, which puts together the three leverage cycle mechanisms. I call this the slow leverage cycle because it does not yet explain why the downward trajectory might be faster than the upward trajectory.

Needless to say, a steady alternation between good news and bad news will create a cycle in asset prices. The novel contribution of the leverage cycle is that an alternation between low uncertainty and high uncertainty will also create a cycle in prices because leverage will rise and fall. This latter price oscillation can occur even if every individual's expectation of the mean of future payoffs remains constant. The amplitude of the price cycle is magnified when both kinds of news occur together because good news and calming

---

[15] Adelino et al. (2018) argue that housing mortgage leverage did not rise leading up to the crisis or fall afterward. They announce the surprising nature of their findings as "contrary to popular beliefs," including of most of Wall Street. Let me mention several ways in which their interpretation of the data strikes me as wrong. First, Adelino et al., acknowledge that DTI rose dramatically leading up to the crisis, as emphasized by Greenwald (2018). So credit terms manifestly became looser. The credit surface is multidimensional, as I mentioned at the outset and as I shall emphasize more in Section 5. LTV is not a standalone variable. By inverting the credit surface and writing LTV as a function of interest rate and other credit terms, such as DTI and FICO, the LTV surface got looser according to their own analysis. Second, LTV is a subtler measure when talking about a long-term loan (e.g., a mortgage) as opposed to a short-term loan (e.g., a repo). With a more appropriate measure of LTV, they would have found a big rise during the mid-2000s. Goodman (2019) constructs a risk index of mortgages and finds that it rose substantially in the mid-2000s. The mid-2000s were famous for introducing riskier mortgages, such as interest-only mortgages, negative amortizing mortgages, and floating-rate mortgages (whose initial interest is lower, especially with teaser rates for the first 2 or 3 years). These became 30 to 40 percent of all the originations in that time period. They defaulted much more frequently in the crisis than conventional mortgages with comparable borrowers. The reason these mortgages are regarded as riskier is that the borrowers pay less over the early years of the mortgage. Mortgage default is much more likely on a day in the third year than on the very first day. The mortgage payments through the third year are a smaller fraction of the original debt with the riskier mortgage. A measure of LTV that corresponds to the scheduled LTV in the third year (assuming stable housing prices) would have risen in the mid-2000s. Third, Adelino et al., focus their attention on new mortgages. But of course, mortgage refinancing, especially cash-out refinancing, was notorious for higher LTVs during the mid-2000s, in part because the appraisal value used in the LTV calculation was widely viewed, even at the time, as inflated. It is simply not true that LTV on refinanced loans does not affect housing prices. Homeowners in need of cash can sell their homes, but if they can borrow more without selling, then the homes do not go on the market. An important reason housing prices fell rapidly during the crisis is that homeowners could not refinance their loans and were forced to sell. Fourth, and most importantly, by their own measures, Adelino et al., show that private lending standards, including LTV, did indeed get looser in the run-up to the crisis of 2007–2009, consistent with the diagram in Figure 21.6. Adelino et al., say that the looser private standards were compensated by stricter government lending during the same time. (By "government," they do not mean Fannie and Freddie, which themselves were delving for the first time into subprime-like loans, but rather Federal Housing Administration loans.) They lose track of this distinction when, later, they emphatically declare that lenders did not change their LTV standards. The economy is much more vulnerable when the private sector is holding high-LTV loans and the government is holding low-LTV loans than it is in the reverse situation.

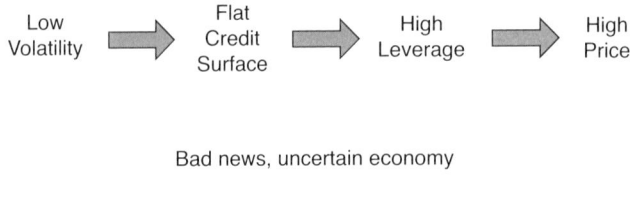

Figure 21.8 The leverage cycle.

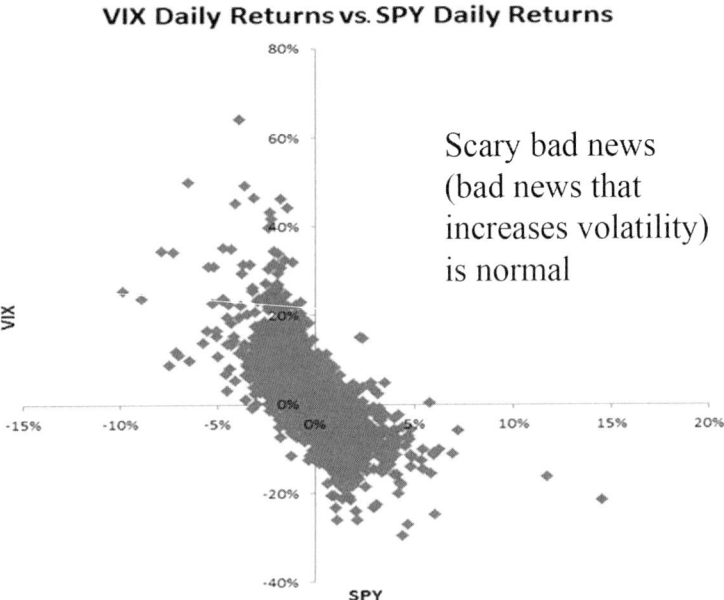

Figure 21.9 Bad news and high volatility.

news raise asset prices for different reasons, the first directly and the second through higher leverage. In particular, an increase in expected outcomes that also reduces the maximum percentage shortfall below the expectation will increase prices by much more than the increased expectation.

The coincidence between good news and reduced uncertainty, or bad news and increased uncertainty, is not rare. As Figure 21.9 shows, volatility usually does go up (as measured by the Volatility Index [VIX]) when news is bad (as measured by a fall in the Standard & Poor's [S&P] 500).

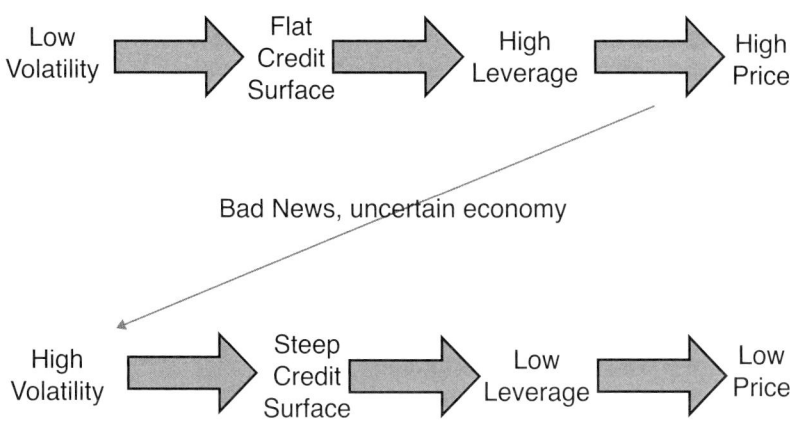

Figure 21.10 Dynamic leverage cycle.

2.2 The Three Mechanisms of the Leverage Cycle Crash

Sometimes in the leverage cycle, prices can go down much faster than they go up. The slow leverage cycle is just a comparative statics exercise. It does not take into account that the economy is more vulnerable *while* it is highly leveraged. Bad news then transfers wealth away from the leveraged owner, who has effectively made a bet on the asset price rising by leveraging. The key idea of the leverage cycle crash is that with agent heterogeneity, this transfer of wealth causes a further price decline. The transfer of wealth can be even more consequential because it is liquid wealth; some agents with high future wealth may nonetheless be forced to respond to the disappearance of their liquid wealth. The down cycle is thus more severe when following the up cycle than it would be on its own. This dynamic aspect of the leverage cycle is emphasized by the amendment to Figure 21.8 shown in Figure 21.10.

Paying closer attention to the leveraged starting point of the downward stage, we uncover three more mechanisms that can combine to cause a crash. All three come into play in a margin call: a situation in which a leveraged holder of an asset has to repay his or her debt and would like to reborrow it (i.e., to "roll it over") but finds that he or she can reborrow less than he or she must repay. The first two precipitate the margin call, and the third accelerates the downward price spiral once it starts. The first and third rely on an added hypothesis, that the leveraged buyer has a higher marginal propensity to spend on the collateral asset than the lenders do.

2.2.1 Bad News–Liquid Wealth Mechanism

The first margin-call mechanism arises from high leverage and the debt coming due, even if there is no change in leverage and even if the leveraged buyer did not intend to sell the collateral. This kicks in when bad news leads to a fall in the asset price and a loss of equity for the leveraged holder. Normally, if a commodity declines in price by $1, an owner

Bad News—Liquid Wealth Mechanism

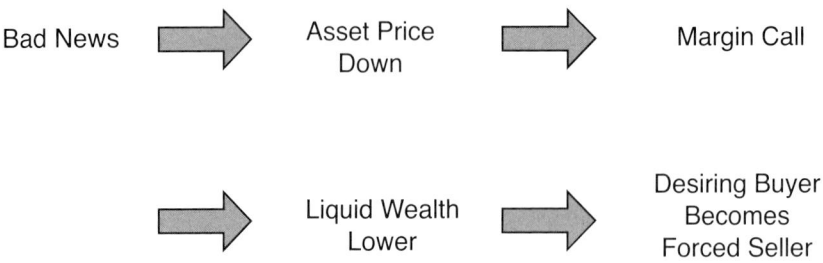

Figure 21.11 Bad news—liquid wealth crash mechanism.

who is not planning to sell or buy it faces no loss of purchasing power. In conventional microeconomic theory, where there is no issue of default, there would be no wealth effect.[16] The situation is different for a leveraged owner of an asset. A leveraged owner whose debt is coming due and who plans to roll over the debt at the same LTV without trading the asset faces a margin call of LTV $\times$ 1 if the asset price drops \$1. This loss in liquid wealth gives the owner an incentive to sell the asset, despite its drop in price, in order to restore his or her liquid wealth. Of course, the lender who receives the margin payment now has more wealth to buy the asset herself. But typically, the lender has a much smaller marginal propensity to spend on the asset out of wealth than the owner.

In fact, the owner has a liquid-wealth-effect incentive to sell a great deal of the asset. If the owner's marginal propensity to spend on the (downpayment for the) asset out of each liquid dollar is m, and if he or she is leveraging the asset λ times, then the owner will want to sell LTV $\times m \times \lambda = m \times (\lambda - 1)$ dollars worth of the asset on account of the liquid wealth loss of LTV $\times$ 1 dollars.[17] If the owner has no buffer of liquid wealth and no other assets to sell or borrow against, then $m = 1$. If $m = 1$ and $\lambda = 4$, the owner will sell \$3 worth of each asset he or she owns for every \$1 fall in the asset price. By contrast, the unleveraged owner of the asset, who has $\lambda = 1$, has no liquid wealth incentive to sell.

The substitution effect normally stabilizes the price by propping it up when it falls by inducing bigger demand. Had the asset been owned free and clear ($\lambda = 1$), the owner might have wanted to increase his or her holdings at the lower price. By contrast, the liquid wealth effect for the leveraged owner of the asset is to sell after the price falls, causing the price to fall further. This destabilizing effect makes for a more fragile economy. See Figure 21.11.

2.2.2 Scary News—Leverage Mechanism
The second mechanism involves an abrupt change in the anticipation of down risk, which I call *scary news*. It will prevent holders of collateral whose debts are coming due from rolling

---

[16] The conventional wealth effect is the product of the intended net trade and the change in price. When the intended net trade is zero, there is no conventional wealth effect.

[17] Recall that LTV $= (\lambda - 1)/\lambda$.

Scary News—Leverage Mechanism

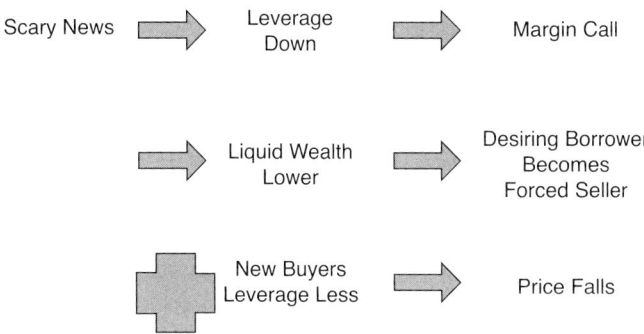

Figure 21.12 Scary news—leverage crash mechanism.

over all the debt, creating a similar liquid wealth reason to sell as in the first mechanism, causing a fall in price even if there had been none before. In addition, the new buyers will not be able to borrow as much, which will cause the asset price to fall more. The scary part of the news is potentially more powerful than the bad part of the news. See Figure 21.12.

2.2.3 Debt-Fragility Mechanism

Debt crises have always been linked to fragile economies. Historically, in times of debt troubles, politicians often make speeches about restoring confidence. President Roosevelt said you have nothing to fear but fear itself. Bernanke and Geithner said similar things about restoring confidence, as did Prime Minister Tsipras of Greece. All of them seemed to believe that by changing expectations, they could move the outcome a long way. In other words, they thought the economy was fragile: a small push could cause a big shift. So why does high debt make for fragile economies?

The third mechanism arises from high levels of debt in periods when the debt needs to be paid off. Paying off the debt requires the sale of commodities or assets. Large sales themselves are not necessarily a sign of fragility because they are matched by large purchases. But the fragility arises if the marginal propensity to spend on the traded asset, out of the last dollar of wealth, is higher for the sellers than the buyers. I call this a *propensity-to-spend reversal*, because the agent with the higher propensity to spend is the seller. This propensity-to-spend reversal is not the norm, except when selling leveraged assets to pay off debt. The agents with a high marginal propensity to spend leveraged to buy the assets. A margin call forces them to sell the asset, thus causing the propensity-to-spend reversal.

The propensity-to-spend reversal causes fragility through the old microeconomic dichotomy called the *income and substitution effect*. When the price of a good Y goes down, the substitution effect is that agents will try to buy more of it because, all else equal, it is more attractive by virtue of being cheaper. This tends to stabilize prices. But if an agent is already selling Y, then all else is not equal. There is an additional income effect. The lower price makes the seller poorer, which means he or she might want less of everything, including Y. In more dramatic words, the further the price goes down, the more he or she might have to sell. The usual stabilizing effect of lower prices raising demand can be reversed for sellers. In the language of demand theory, the income effect counteracts

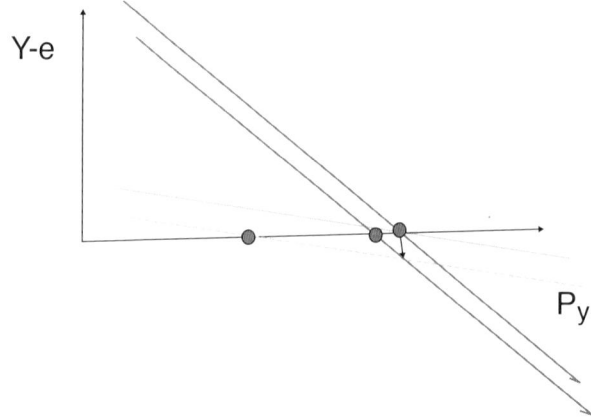

Figure 21.13 Equilibrium with flatter excess demand gives rise to fragile equilibrium.

the substitution effect for the sellers. On the other hand, the income effect reinforces the substitution effect for the buyers. As the price goes down, they effectively get richer, and for that reason, they want to buy more, beyond their pure substitution effect. The crucial observation is that if the marginal propensity to buy Y (out of an additional dollar of wealth) is higher for the sellers than for the buyers, then the sellers' income effect will be stronger.[18] A drop in the price of Y effectively redistributes income from the sellers to the buyers, in proportion to how much is sold. If the marginal propensity to spend on Y out of income is higher for the sellers, then their income-induced drop in the consumption of Y will be greater than the buyers' income-induced increase in consumption. In aggregate, the income effect will tend to reverse the substitution effect. Unlike the income effect, the substitution effect is invariant to the quantity sold. Hence for bigger sales, the aggregate income effect diminishes the stabilizing aggregate substitution effects more. With big enough sales, the aggregate demand curve for Y will be close to flat.[19] But a flat demand curve means that equilibrium prices will have to move dramatically to restore equilibrium after a small shock. The economy is fragile. Thus, a little bit of bad news can have a big effect on prices in an economy with large sales of some good.[20]

See Figure 21.13, which illustrates how the same vertical downward shock will produce a small change in the equilibrium price of an economy with steep demand (i.e., with a dominant substitution effect) but produce a large change in the equilibrium price of an economy with flat demand (i.e., with a dominant income effect)

---

[18] The famous Slutsky equation says that the income effect is the product of the marginal propensity to consume and excess demand. Because in equilibrium, the excess demand of the sellers of Y must be the negative of the excess demand of the buyers, the aggregate income effect on Y is the product of the difference between the sellers' and the buyers' marginal propensities to consume Y and the excess demand of the sellers for Y.

[19] With still bigger sales, the income effect will reverse the substitution effect, and demand will be increasing. But that means there are multiple equilibria.

[20] This is worked out in Ben Ami and Geanakoplos (2017).

Leverage – Debt – Propensity-Reversal – Fragility Mechanism

Fragility emerges through the income-substitution effect
because of the propensity-to-spend reversal

Figure 21.14 Debt—fragility crash mechanism.

When leverage rises, asset prices rise, so borrowers are borrowing a higher percentage of a higher number. Thus, with higher leverage, borrowing goes up for a squared reason, so debt can skyrocket. When there is a large debt that is coming due, then there must be a large sale, either of some good or of more promises, to pay the debt. If collateral is scarce, and if leverage becomes low, then there is a hard cap on the sales of new promises, and so there must be sales of some good or asset. The borrowers who accumulated the large stock of that asset presumably did so because they had a high marginal propensity to spend on it. When a margin call forces them to sell, we get a propensity-to-spend reversal. If the marginal propensities to spend are markedly higher for sellers than buyers, then the economy is fragile. Economies that have large short-term debts therefore are perpetually in a vulnerable situation because they perpetually have enormous sales.

Putting all this together, the debt-fragility mechanism is really a leverage–debt–propensity-reversal–wealth-redistribution–fragility mechanism. See Figure 21.14.

2.2.4 The Leverage-Cycle Crash

Putting the volatility–leverage mechanism together with the debt–fragility–income-redistribution mechanism, we see that a little bit of scary bad news can have a huge effect on asset prices. The leverage cycle I described in 2003 combines all six mechanisms and goes like this: A long period of low volatility leads to a flatter credit surface and thus increased leverage, as well as laxer credit standards generally (for the same reasons). That raises asset prices and increases activity. But it also makes the economy more vulnerable because of the double boost to new debt of higher asset prices and higher leverage. A little bit of bad news decreases everybody's valuations and lowers prices a little. But as we saw at the outset, the most leveraged buyers will lose the highest fraction of their wealth from the price drop. They are likely to be the highest-valuation, highest-marginal-propensity-to-spend buyers, and their disappearance (or reduced purchasing power) further reduces asset prices, from the income effect discussed earlier. If the news is scary, as well as bad, the increased uncertainty steepens the credit surface and lowers leverage. Thus, asset prices drop for three reasons: the bad news; the wealth transfer away from high-leverage, high-valuation, high-marginal-propensity-to-spend agents; and the final drop in leverage reducing old and new buyers'

Marginal Buyer Theory of Price

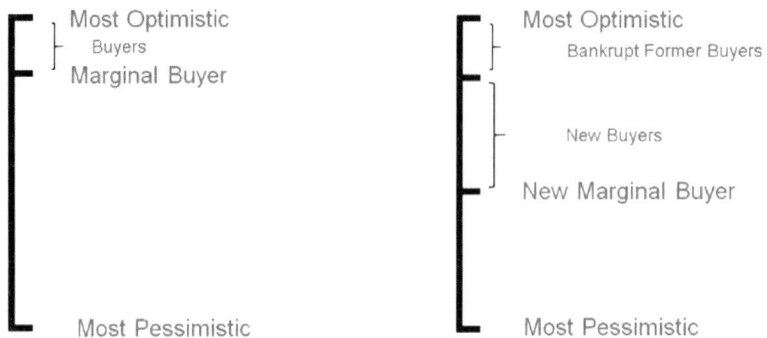

Figure 21.15 Leverage cycle and crash.

demand for assets.[21] Asset prices and activity will stay low as long as uncertainty remains high and the credit surface remains steep. And as I added in 2008, if the debt is too high relative to the lower asset prices, full repayment may become impossible. With a big enough disparity, partial forgiveness may be the only way out of the recession.

The leverage cycle and crash can be described in a simple special case that illustrates all six mechanisms and gives rise to another diagram that uses the idea of a marginal buyer. Suppose that we array all the agents in the economy in a vertical line according to their valuations of an asset, with the highest valuation at the top and the lowest valuation at the bottom. (For simplicity, think of a continuum of agents, each valuing the asset at a level independent of how much he or she buys). The valuation heterogeneity could have many causes. Some agents might be more risk tolerant. Some might get higher utility out of holding the asset or could use it more productively. Some might be more optimistic about the future value of the asset. The heterogeneity is important, not the source. In my 2003 paper, from which this example is taken, I assumed differences in optimism. See Figure 21.15.

Whatever the asset price, some agent, whom I call the *marginal buyer*, will think it is fair. More optimistic agents will buy the asset, and less optimistic agents will sell it. In the expansionary part of the leverage cycle, when volatility is low, leverage will be high and set by the down risk (mechanisms 1 and 2). Few agents will be needed to buy the asset because each one can buy a lot using borrowed funds. With few buyers, the marginal buyer will be high up the line and because his or her valuation is equal to the price, the price will be high, as indicated on the left side of Figure 21.15 (mechanism 3).

When bad news comes, every agent, including the marginal buyer, will value the asset less, so the asset must fall at least a little in price. But the old buyers will be forced to sell the asset in order to pay back their loans (mechanism 4). The initial fall in the asset price

---

[21] Another driver of the crash is the sudden emergence of the credit default swap (CDS). A CDS is a way for pessimists to leverage their short selling of the asset. For the same reason that leverage increases asset prices when buyers can leverage more, so, too, does increased access to leverage by the short sellers of the asset lower its price. I had not anticipated CDSs in 2003 but added them to the story in 2010. See Section 3 on the financial innovation cycle.

causes them to lose more wealth, especially because they are so leveraged. On the right-hand side of Figure 21.15, we consider an extreme case where the old buyers lose so much wealth that they can no longer buy any assets. The new marginal buyer is necessarily further down the line and more pessimistic. So the asset price falls for a second reason, caused by the loss in wealth of the original buyers (mechanism 6).

If the news is scary, anticipated volatility will be high, and leverage will drop. The new buyers will not have access to as much borrowed funds (mechanism 5). Thus, even at a lower price, there will need to be many more buyers than previously, and the gap down from the original marginal buyer to the new marginal buyer will be very large. The price will then reflect the valuation of a much more pessimistic buyer. The fall in price is more attributable to the change in marginal buyer, occasioned by the wealth losses of the optimists and the curtailment of leverage, than it is to the bad news itself.

The leverage-cycle crash is related to so-called "fire sales." For a good account of the important literature on fire sales, see Shleifer and Vishny (2011). There are, however, several differences. The most important is that the leverage cycle injects the critical element of varying and endogenous leverage. The fire sale literature misses the overvaluation and buildup of debt due to the soaring leverage and the sudden transition from high leverage to low leverage, which plays a vital role in all crashes. It also misses the aftermath during which the credit surface is still steep and new borrowing remains low. The fire-sales literature addresses part of the middle game, without discussing the opening or the endgame. The more recent fire-sales literature uses language like *deleveraging* without actually endogenizing asset leverage. It does, however, include the idea of heterogeneous buyers and the loss in price when high-valuation buyers are forced to sell to low-valuation buyers.[22]

2.3 Expectations and the Leverage Cycle

The leverage cycle, as I told it, does not rely on extrapolative or irrational expectations. It can perfectly well occur with rational expectations. The story does depend on heterogeneity in asset valuations and marginal propensities to buy out of income. I motivated this heterogeneity by different priors, meaning my agents were all completely rational Bayesians, aware of all the possible states of nature, but always put different prior probabilities on the down state. There could have been other ways to explain the heterogeneity, including differences in risk aversion or differences in utility for the assets, as I have demonstrated in later work. My early leverage-cycle work had rational updating; the cycle would have gotten even more dramatic with extrapolative expectations.[23]

I do not discount the importance of irrational expectations but merely note that they are not needed for the story. Different beliefs are not indispensable for the story either, but they can play an important role, as they did in my model of 2000–2003, dating from before the 2007–2009 crisis. What characterizes the run-up in the model is that the risk of falling prices is low and is recognized as low by everyone. Indeed, Bernanke dubbed the era as the "great moderation." With little perceived down risk, even the pessimistic agents can

---

[22] More subtly, the fire-sales literature conflates valuation with the marginal propensity to spend out of income, although to be sure, the two often go hand in hand, such as when there are linear utilities.

[23] This is explored in Thurner et al. (2012).

participate in the bubble stage by lending so much, while the more optimistic agents exercise disproportionate influence over prices by borrowing so much. When bad news hits and down risk increases in everyone's reckoning at the same time (scary bad news), leverage falls, and prices fall to reflect the views of a different and more pessimistic class of agents.

After the crisis, a number of authors sought to explain the 2000–2006 price surge as a bubble stemming from bubbly expectations. According to this view, in 2000, everybody began to think that the future demand for housing was going to be high, and this persisted until 2007, when everybody mysteriously began to think the future demand would be low.[24] Anabtwawi and Schwarcz (2011) pointed out that optimistic expectations would lead lenders to give out loans with more leverage, which would push prices up. In that sense, the (irrational) exuberance story and the leverage cycle are similar. Indeed, Robert Shiller, who famously recognized the housing bubble as it was beginning, advised me to interpret the leverage cycle as irrational exuberance by the lenders. I am grateful to him for saying so simply what he took to be the innovation of the leverage cycle, that asset prices could be pushed up by the changing beliefs of lenders, even with no changes in the beliefs by buyers. But I find the great-moderation (i.e., reduced-volatility) explanation for increased leveraged lending more aesthetically pleasing than the story of irrational exuberance (about price trends), although, as I said, both could be right.[25]

---

[24] See, for example, Kaplan et al. (2017).

[25] Kaplan et al. (2017) take a more extreme view, that changing expectations from 2000–2007 about future housing demand caused the boom and the crash, and yet that the changing lending standards (that they acknowledge) played absolutely no role in moving housing prices. The reason they give for the latter is that homeowners could always rent instead. In their view, borrowing constraints do not affect the total demand for housing but merely redirect it from owning, which requires borrowing, to renting, which does not.

Although very interesting, I find the Kaplan et al. (2017) story far-fetched. First, it takes an unprecedented shift in expectations alone to justify a 90 percent increase in housing prices in the 6.5 years from 2000 to 2006, as measured in the Corelogic Case–Shiller housing index. From where did this change in outlook come? Interest rates are fixed in the Kaplan et al. model, so those don't explain the change. At the same time, Kaplan et al. must assume a simultaneous and completely exogenous shift in credit standards. How convenient these two happened at exactly the right time, and together.

By contrast, there had been years of stability leading up to the 2000s, which not even a foreign attack on American soil could shake, that led people to call it the great moderation. There is no mystery as to why people might have rationally believed in the 2000s that down risk was lower (without thinking that things were going to rapidly improve). In the leverage cycle, the volatility assumption endogenously leads to laxer credit conditions, which in turn endogenously produce price appreciation.

Shiller might argue that the Greenspan–Fed drop in interest rates in early 2000s got housing prices going up, and that extrapolative expectations kept pushing them further. That could also happen in a more sophisticated model of dynamic expectation revision, as in Bordalo et al. (2018). But they recognize the importance of expectations on credit conditions as a channel for affecting asset prices.

Second, although Kaplan et al. (2017) incorporate changing credit conditions into their model, their "proof" that credit terms do not influence housing prices depends on the ability of entrepreneurs to convert owner-occupied housing to rental housing at low cost. It also requires the use value of home ownership to be not much greater than the use value of home rentals.

Finally, and most importantly, the paper ignores, by assumption, the obvious heterogeneity in the population. If some agents are more optimistic than others, they will prefer to buy rather than rent, and as credit conditions ease, they will want to spend still more on buying. Their total demand for housing very much will depend on credit conditions.

3 The Financial Innovation Cycle

Half a century ago, more and more goods became usable as collateral for leveraging. Thirty years ago, securitization and tranching, especially of mortgage-backed securities, emerged and grew dramatically. Finally, over the last 10 to 15 years, the CDS mortgage market suddenly blossomed at the end of the securitization boom. After the crisis of 2007–2009, the complexity of these instruments declined, but it is now on the rise again.

In Fostel and Geanakoplos (2014), we argued that there is a financial innovation cycle that follows and boosts the leverage cycle. The financial innovation cycle made the crash of 2007–2008 bigger than it would have been otherwise.

In periods of quiet, financiers innovate to stretch the available collateral. When a single asset can be used to collateralize multiple loans, it is stretched. When collateral backs promises that are in turn used as collateral to back further promises, which I call *pyramiding*, the original collateral is effectively reused and thus stretched. Leverage can be thought of as buying an asset while simultaneously borrowing. But it can equally well be thought of as a way of cutting the collateral into two pieces, a bond and a risky junior piece. Cutting the bond into still more pieces, which involves pyramiding and tranching, is a more advanced financial innovation, requiring more complex record keeping, a more sophisticated court system, and accomodating tax laws. By skillfully cutting the collateral into appropriate pieces, entrepreneurs can sell the pieces for more than the original collateral. Competition then bids the whole collateral price up to the sum of its parts. The search for profits from scarce collateral through financial innovation makes collateral more valuable, over and above its payoff value. Leverage raises the prices of assets, and tranching raises their prices still more. And they rise higher because the financial innovation comes in stages, not all at once. Once the prices get high enough, which, unfortunately, is the moment when the indebted economy is becoming especially vulnerable, another financial innovation, the CDS, is introduced, which enables the pessimists to bet against the asset. This tends to lower asset prices. A little bit of bad news can then lead to a great crash.

The run-up to the crisis of 2007–2009 fits the pattern of the financial innovation cycle perfectly. Throughout the later 1990s and 2000s, higher-LTV loans, called *subprime loans*, began to be initiated by the private sector. These in turn were collected into pools, which were then tranched. The subprime market grew from almost nothing in 1990 to over \$1 trillion in 2006. Housing prices skyrocketed from 2000 to 2006. At the end of 2005, a small group of investors who thought housing prices and mortgages were overvalued pushed to get the indexed subprime mortgage CDS market established so that they could bet against the subprime mortgages. The indices stayed high for about 11 months but then cracked at the end of 2006 on the release of delinquency reports for subprime mortgages. The housing market tumbled soon afterward. Had the CDS been trading robustly from the beginning, prices might not have gotten so high.[26]

---

[26] A similar story unfolded with Greek sovereign debt. After Greece gained entry into the European Monetary Union in 2000, it was able to borrow more money. Eventually, Greek banks were buying Greek sovereign bonds, at very high LTV, because the capital requirements for sovereign debt were so low. As the ratio of debt to Greek gross domestic product (GDP) rose, investors became more jittery. When the Greek crisis started just after the revision of Greek deficit numbers, Prime Minister Papandreou blamed it all on the CDS market.

Krishnamurthy et al. (2018) similarly describe the exponential rise of Bitcoin price and the subsequent crash as the result of financial innovation.

4 Multiple Leverage Cycles

Many kinds of collateral exist at the same time; hence, there can be many simultaneous leverage cycles. Each one has its own credit surface. Collateral equilibrium theory not only explains how one leverage cycle might evolve over time, but it also explains some commonly observed cross-sectional differences and linkages between cycles in different asset classes, such as flight to collateral and contagion.[27]

It is commonly observed that in times of crisis, some assets retain their value (or even rise in value) while the others lose value. This situation is often called a *flight to safety*. Another way to describe the situation is a *flight to collateral*. The safe assets, with low volatility, turn out to be the assets that can be leveraged more.[28]

A second commonly observed phenomenon is that when bad news hits one asset class, the resulting fall in its price seems to migrate to other assets, even if their payoffs are statistically independent from the original crashing asset. There are two reasons for this contagion connected to crossover investors. As we saw in Section 1, the leverage cycle in one asset amplifies the bad news and creates wealth redistribution away from the most optimistic buyers of the asset. If these buyers are also crossover holders of a second asset, their losses in the first asset might force them to raise money by selling the second asset. Moreover, the leverage-cycle price decline in the first asset will make these buyers feel there is a special opportunity there, leading them to withdraw even more money from the second asset to take advantage. These two reasons to withdraw demand for the second asset lead to price declines there.

Two examples of this kind of spillover into seemingly unrelated markets involve the mortgage market and emerging markets bonds in 1997–1998 and in 2007–1908. In 1997, a crisis started in Asian and Russian emerging markets and was followed within 6 months or a year by a sudden downturn in mortgages. (See the leverage graph in Figure 21.6.) In 2007, a crisis in mortgages then seemed to migrate to emerging markets.

As I mentioned earlier, the great crisis of 2007–2009 was the culmination of a double leverage cycle, in two separate but interrelated markets, mortgages and mortgage securities. Declining cash flows in one asset induce lenders to tilt the credit surface in the other. George Soros's (2009) principle of reflexivity includes the proposition that historical crashes invariably involve disasters in two separate but interrelated markets. Although he didn't apply this insight to housing and mortgage securities, the mortgage crisis fits. Leverage rose in housing and in mortgage securities together. Trouble with mortgage delinquencies depressed mortgage securities prices, which led to cutbacks in housing leverage, which depressed housing prices, which indicated future default losses, which reduced mortgage security prices.

5 The Credit-Terms Cycle and Central Bank Policy

As I mentioned at the outset, leverage is just one of many terms that come with loans, besides the interest rate. In boom times, many credit terms get relaxed, not just leverage. It is important to keep track of all of them. The general credit surface is the loan interest rate as

---

[27] In this section, I follow Fostel and Geanakoplos (2008).

[28] In the language of Fostel and Geanakoplos (2008), they have more collateral value.

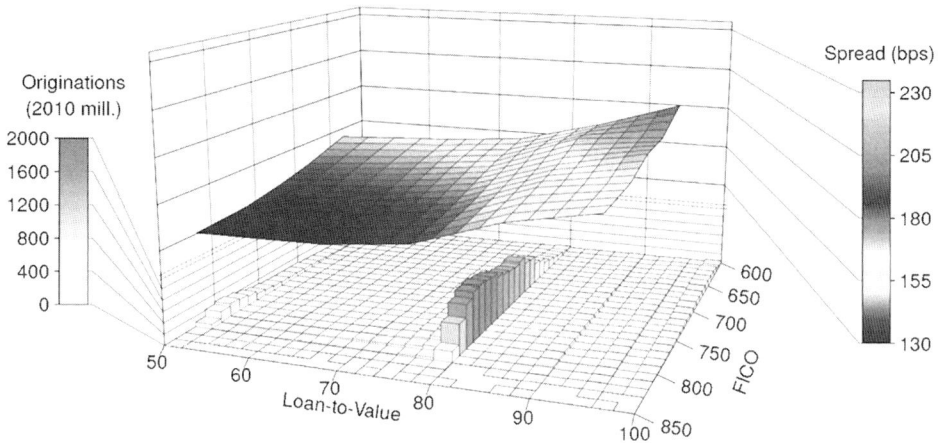

Figure 21.16 Q2 2006 mortgage credit surface.
Source: Geanakoplos and Rappoport (2019) using Black Knight Financial Services and BLS.

a function of its various terms, including LTV, DTI, FICO (or credit score), and of course, maturity. Not all of these terms can be displayed easily in the same picture. By picking any two credit terms, such as LTV and FICO, the Washington Federal Reserve has worked with me to produce credit surfaces like that shown in Figure 21.16.[29]

Figure 21.16 shows the average interest rate charged on all fixed-rate Federal National Mortgage Association (FNMA) and Freddie Mac mortgage loans in the second quarter of 2006 as a function of LTV and FICO. Loans with the highest FICO and lowest LTV, in the southwest corner, are the safest loans. Loans with the highest LTV and lowest FICO, in the northeast corner, are the riskiest loans. Even for the conforming group of households that passed many hurdles to get into the government programs, there is a difference in interest rate depending on credit characteristics. But the curve is generally quite flat, indicating a loose credit surface. The rectangular blocks below the surface give the volume of loans at the point on the surface just above. One can see that the bulk of the loans had less than 80 percent LTV. But there is a significant number with LTVs close to 100 percent and FICOs around 650.

Consider next the mortgage credit surface in the last quarter of 2008, after the crisis had started. It is much steeper, and the number of low-FICO, high-LTV loans is much less. See Figure 21.17.

In Figure 21.18, we see the corporate bond credit surface for 2007.[30] As it was for mortgages in 2006, the corporate bond credit surface is very flat.

---

[29] See Geanakoplos and Rappoport (2019).

[30] There are complications in presenting simple interest rates for different bonds at different times, if, for example, some of the bonds are callable and others are not. For corporate bonds, we replace the interest rate with something called the *option-adjusted spread*, which adjusts for the option value of the bonds. I do not have space to go into these details here, but I refer the reader to Geanakoplos and Rappoport (2019).

2009Q1, 30-year conventional purchase mortgages

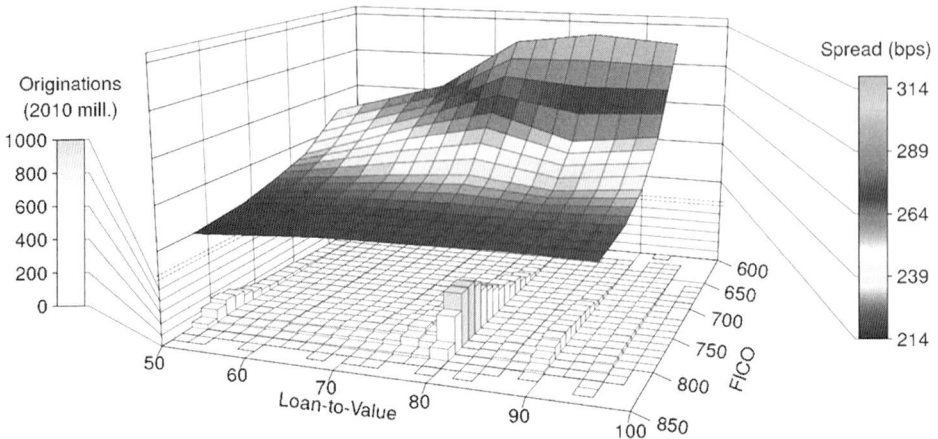

Figure 21.17 Q4 2008 mortgage credit surface.
Source: Author's elaboration using Black Night Financial Services and BLS.

2007Q2, 7-10-year corporate bonds

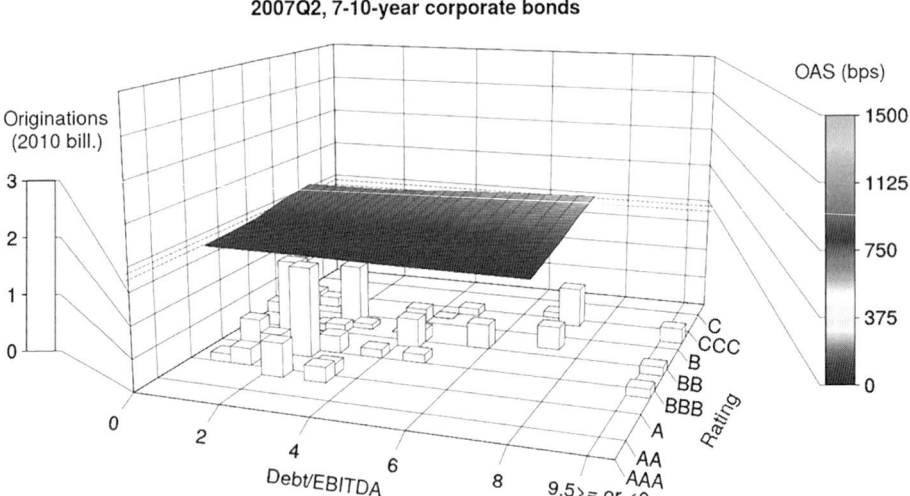

Figure 21.18 Q2 2007 corporate bond credit surface.
Source: Geanakoplos and Rappoport (2019) using ICE, Bond Indices, Mergent FISD,
CRSP/Compustat, Compustat, and BLS.

In the fourth quarter of 2008, the credit surface became remarkably steeper, making it much more difficult to borrow. The reader should be aware that the credit surfaces do not always move in tandem. Today (in late 2019), for example, the corporate credit surface has again become very flat, but the mortgage credit surface has not. These differences should be taken into account by the Fed in its deliberations. See Figure 21.19.

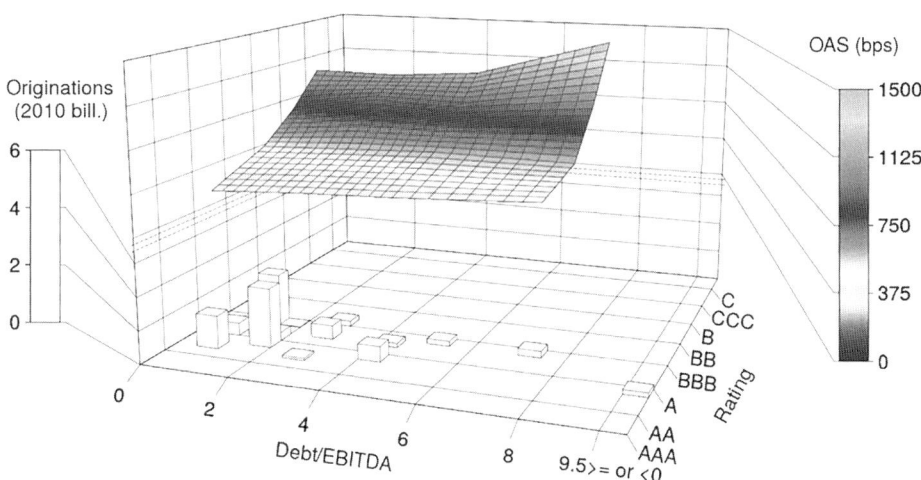

2008Q4, 7-10-year corporate bonds

Figure 21.19 Q4 2008 corporate bond credit surface.
Source: Author's elaboration using Bank of America, bond indices, Mergent FISD,
CRSP/Compustat, and BLS.

6 Monetary Policy and Macroprudential Policy

The policy implications of the leverage cycle are that central banks should smooth the cycle, restraining leverage in booms and propping up leverage in the acute stage of the crisis.[31] How to do so involves monetary policy and macroprudential policy. Indeed, an implication of the leverage cycle is that it is hard to separate the two. See Figure 21.20.

One crucial question is always, Where in the leverage cycle are we? This question can be addressed by monitoring and by stress tests. Stress tests are most important when we get closer to the top of the leverage cycle. Of course, by then, it is difficult to stave off a crash, even if we know it is coming. The interesting thing is that knowing that we are approaching the precipice, we should reverse the monetary policy that we had been using to keep us from getting too far up the mountain.

A second crucial question is whether the upswing in the leverage cycle is due to an unusual investment opportunity or to a market in which borrowers and especially lenders are taking bigger risks. Here again, monitoring and especially stress tests can give valuable information.

6.1 Monitoring the Leverage Cycle

In my opinion, the Federal Reserve should produce credit-surface pictures for the general public each quarter. Not only that, but they should be produced for the unsecured consumer

---

[31] I have also argued that if in the aftermath of a leverage-cycle crash, depressed asset prices are too low relative to debts, debt must be partially forgiven. See Geanakoplos and Koniak (2008, 2009) and Geanakoplos (2010b, 2019).

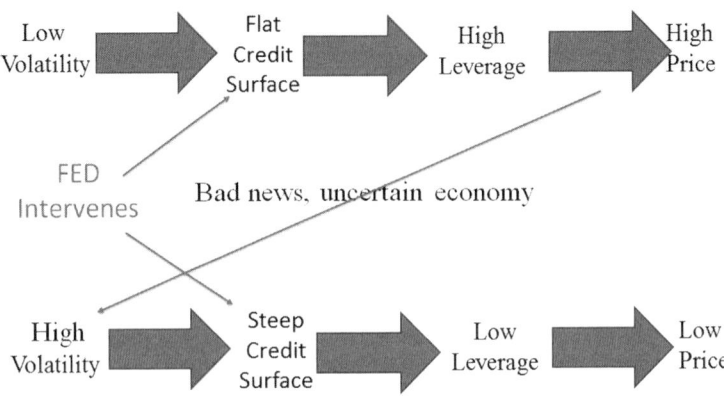

Figure 21.20 Monetary policy and macroprudential policy.

loan credit surface and other surfaces, as well as for the mortgage credit surface and the corporate bond credit surface. This will give economists, businesspeople, macroprudential regulators, and the Fed itself a much better picture of credit conditions in the economy.

The credit surface also includes the number of loans made at each point of the surface. The Fed can keep track, for example, of the percentage of new loans made at over 90 percent LTV, or the median LTV of the 50 percent of loans with the highest LTV, as in Figure 21.7. Similar numbers can be recorded for FICOs, DTIs, and so forth.

The Fed should also be keeping track of the evolution of asset prices. Each credit surface that corresponds to a collateral type should be associated with the price of the collateral (or the index of the collateral type). At the same time, the trailing volatility of the collateral price should be recorded. (Better still would be a trailing down volatility.) The Fed might also keep track of activity such as production in the sectors associated with each loan type.

The Fed should carefully note when there is a financial innovation in the loan market corresponding to a collateral type.

The hallmark of the upswing of a leverage cycle is a flattening credit surface, increasing leverage (as given by the volume data), increasing numbers of low-FICO loans, increasing asset prices, and low down volatility. Once the cycle has progressed unusually far, the Fed should intervene with monetary policy to slow things down. How to tell when things have gone too far will require some more dedicated historical research. The situation is analogous to the traditional Fed role in monitoring the Phillips curve (connecting output or employment to inflation), except that asset prices replace commodity prices. Indeed, in my opinion, traditional monetary policy could be substantially improved by incorporating leverage-cycle concepts.

In conducting monetary policy, the Fed should be aware of how the changes it makes in the riskless rate (in the southwest corner) affect the whole credit surface of each type. Perhaps the changes move every credit surface rigidly upward or downward, or perhaps the risky end of the credit surface moves less than the safer end, blunting much of the power of conventional monetary policy.[32] Do Federal Reserve risky asset purchases (called *quantitative easing*) have similar effects, or are they better at tilting the credit surfaces? In my opinion, the Fed should use the language of the various credit surfaces to explain its policy aspirations. Is the hope to shift or steepen the mortgage credit surface or the corporate credit surface?

Finally, if there are parts of some credit surface the Fed or macroprudential regulators wish to target, then they should use unconventional tools to specifically affect those areas alone. For example, at the current time, the mortgage credit surface is still very unkind to low to medium-FICO borrowers. If the Fed wanted, it could purchase loans of that type, which would bring down their rates. If the Fed thought that housing prices were rising too rapidly, it might declare, as Stanley Fischer did in 2010 as head of the Bank of Israel, that no mortgage loans could be issued with more than 60 percent LTV.[33]

6.2 Stress Tests

Stress tests are designed to judge whether crucial intermediaries could survive a severe downturn and to warn of increasing risk of a severe downturn. Monitoring the leverage cycle displays the preconditions for a crash. Stress tests are indispensable to uncovering whether the preconditions are creating genuine risk.

The leverage cycle is an archetypal scenario, which recurs in crisis after crisis, that in broad strokes lays out the critical stress test scenarios. The defining signature of the leverage-cycle crash is bad news and simultaneously much tighter credit and higher volatility. Looking at the portfolio of loans a bank or other intermediary faces, what happens if the loans start to default (say, because unemployment rises or activity and profits decline), and at the same time the loans cease to prepay because borrowers' access to credit has been curtailed? And at the same time, the banks themselves lose access to their credit? And at the same time, what if the prices of the securitized loans they own also fall further than the reduction in their underlying loan cash flows warrant because leverage on the securities declines? Do the intermediaries have enough capital to maintain their positions, or will they be forced to sell? Will they default?

Of course, if we assume enough simultaneous bad shocks, every bank will collapse. The point of the leverage cycle is that it describes plausible simultaneous shocks. The scary scenario described in the last paragraph is plausible according to leverage-cycle principles, and it happened in the crisis of 2007–2009. The simultaneity of shocks emerged because leverage tightened for both securities and loans at the same time.

Turning to the economy as a whole, the leverage cycle becomes systemic when the assets that function as collateral are similar across the whole economy or because the same group of borrowers owns several asset classes that together are large. Prices declining in two unrelated industries, such as mortgages and emerging markets, can happen because crucial leveraged buyers are crossover investors in both industries. The macroprudential regulators therefore

---

[32] I analyze just this question in Geanakoplos and Rappoport (2019).

[33] There is some doubt about whether the US Federal Reserve has such powers because it has not exercised them. It certainly does have the power to regulate margins on stocks.

need to keep track of the portfolio composition of all highly leveraged crossover investors in order to see which assets are linked through common ownership.

Stress tests should not only examine the effects on bank capital from changes in cash flows but also from price declines. And the price declines should not simply be taken as exogenous but themselves be generated from stress tests with data-driven inputs. For example, a crucial determinant of the size of plausible price declines is the heterogeneity of the valuations of different potential buyers. When some assets are propped up in value by a small class of investors who are highly leveraged and who (at that leverage) value the assets much more than anybody else, there is risk of steep price declines.[34] In order to judge the resilience of the system, the macroprudential regulators must know which big players (who themselves have high valuations) have enough free capital to absorb assets that might be disgorged by leveraged holders facing margin calls.

References

Adelino, M., A. Schoar, and F. Severino, (2018), "The role of housing and mortgage markets in the financial crisis," *Annual Review of Financial Economics*, 10, 25–41.

Anabtwawi, I., and S. Schwarcz (2011), "Regulating systemic risk: Toward an analytical framework," *Notre Dame Law Review*, 86(4).

Ben-Ami, Y., and J. Geanakoplos, (2017), "Debt, fragility, and multiplicity: Thinking outside the Edgeworth box," Working Paper, Yale University.

Bernanke, B., and M. Gertler (1986), "Agency costs, collateral, and business fluctuations," Proceedings, Federal Reserve Bank of San Francisco.

Bordalo, P., N. Gennaioli, and A. Shleifer (2018), "Diagnostic expectations and credit cycles," *Journal of Finance*, 73(1), 199–227.

Fostel, A., and J. Geanakoplos, (2008), "Leverage cycles and the anxious economy," *American Economic Review*, 98(4), 1211–1244.

Fostel, A., and J. Geanakoplos (2014), "Endogenous collateral constraints and the leverage cycle," *Annual Review of Economics*, 6(1), 771–799.

Fostel, A., and J. Geanakoplos (2015), "Leverage and default in binomial economies: A complete characterization," *Econometrica*, 83(6), 2191–2229.

Geanakoplos, J. (1997), "Promises, Promises," in W. B. Arthur, S. Durlauf, and D. Lane (eds.), *The economy as an evolving complex system, II*, Addison-Wesley.

Geanakoplos, J. (2003), "Liquidity, default, and crashes: Endogenous contracts in general equilibrium," in Mathias Dewatripont, Peter Hansen, and J. Turnovsky (eds.), *Advances in Economics and Econometrics: Theory and Applications, Eighth World Conference*, Volume II, Econometric Society Monographs, Cambridge University Press.

Geanakoplos, J. (2010a), "The leverage cycle," in D. Acemoglu, K. Rogoff, and M. Woodford (eds.), *NBER Macroeconomics Annual 2009*, Vol. 24, University of Chicago Press.

Geanakoplos, J. (2010b), "Solving the present crisis and managing the leverage cycle," *Federal Reserve Bank of New York Economic Policy Review*, 101–131.

Geanakoplos, J. (2016), "The credit surface and monetary policy," in International Monetary Fund and Massachusetts Institute of Technology (eds.), *Progression and Confusion: The State of Macroeconomic Policy*, MIT Press.

---

[34] The leveraged buyers are often the banks, which, because of their access to sticky deposits and sophisticated valuation models, can find extra value in assets, so long as they don't cause large regulatory capital charges. In fact, these assets might be so valuable to the banks because they are not assessed for high capital charges while most other assets are.

Geanakoplos, J. (2019), "Leverage caused the 2007–2009 crisis," in D. W. Arner, E. Avgoules, D. Busch, and S. L. Schwarcz (eds.), *Systemic risk in the financial sector: Ten years after the great crash*, McGill-Queen's University Press.

Geanakoplos, J., and S. Koniak, (2008, October 20), "Mortgage justice is blind," *New York Times* op-ed.

Geanakoplos, J., and S. Koniak (2009, March 4), "Matters of principal," *New York Times* op-ed.

Geanakoplos, J., and D. Rappoport (2019), "Credit Surfaces, Economic Activity, and Monetary Policy," Working Paper, Society for Economic Dynamics.

Goodman, L. (2021), "Housing credit availability index," available at www.urban.org/policy-centers/housing-finance-policy-center/projects/housing-credit-availability-index.

Gorton, G., and A. Metrick (2012), "Securitized banking and the run on repo," *Journal of Financial Economics*, 104, 425–451.

Greenwald, D. (2018), "The mortgage credit channel of macroeconomic transmission," Working Paper, MIT Sloan.

Hale, G., A. Krishnamurthy, M. Kudlyak, and P. Schultz (2018), "How futures trading changed Bitcoin prices," FRBSF Economic Letter, Federal Reserve Bank of San Francisco.

Holstrom, B., and J. Tirole (1997), "Financial intermediation, loanable funds, and the real sector," *Quarterly Journal of Economics*, 112, 663.

Kaplan, G., K. Mitman, and G. Violante (2017), "The housing boom and bust: Model meets evidence," Working Paper, New York University.

Kindleberger, C. (1978), *Manias, panics, and crashes: A history of financial crises*, Basic Books.

Kiyotaki, N., and J. Moore, (1997), "Credit cycles," *Journal of Politcal Economy*, 105(2), 211–248.

Minsky, H. (1977), "A theory of systemic fragility," in E. D. Altman and A. W. Sametz (eds.), *Financial crises: Institutions and markets in a fragile environment*, John Wiley and Sons.

Phelan, G. (2017), "Collateralized borrowing and increasing risk," *Economic Theory*, 63(2), 471–502.

Schleifer, A., and R. Vishny (2011), "Fire sales in finance and microeconomics," *Journal of Economic Perspectives*, 25(1), 29–48.

Simsek, A. (2013), "Belief disagreements and collateral constraints," *Econometrica*, 81(1), 1–53.

Soros, G. (2009), "*The crash of 2008 and what it means: The new paradigm for financial markets*," Public Affairs.

Stiglitz, J. (1969), "A re-examination of the Modigliani-Miller theorem," *American Economic Review*, 59, 784–793.

Thurner, S., J. D. Farmer, and J. Geanakoplos (2012), "Leverage causes fat tails and clustered volatility," *Quantitative Finance*, 12(5), 695–707.

22

Monetary Policy and Financial Stability

William B. English*

1 Introduction

Since the 2007–2009 financial crisis, there has been great interest in possible policies to dampen growing financial imbalances that could lead to financial crises with large macroeconomic consequences. One set of responses is microprudential—for example, higher capital ratios and liquidity ratios to strengthen the safety and soundness of individual institutions.[1] A second set is macroprudential—that is, policies that are intended to strengthen the financial system as a whole. Some macroprudential policies are static—for example, even higher capital and liquidity requirements for systemically important firms to further increase the resilience of the system—and some are countercyclical, such as loan-to-value limits that could be introduced or adjusted to dampen growing imbalances. Stress tests of key institutions are another macroprudential tool. In a sense, they are countercyclical, in that they can be used to address perceived buildups of risk in particular areas. In addition, as implemented in the United States, the stress scenarios are more adverse in good times than in bad.[2]

In addition to these microprudential and macroprudential tools, some have noted the potential use of tighter monetary policy as a countercyclical tool to push back on growing financial imbalances. As a general rule, it would be preferable to use micro- and macroprudential tools to directly address growing risks, rather than monetary policy, which may have significant spillover effects on the broader economy. However, policymakers may not know how to appropriately calibrate the use of such tools or be confident that they will be effective because the experience with them is relatively limited.[3] Moreover, in some

---

* I thank Til Schuermann and Steve Cecchetti for helpful comments on an earlier draft. I thank Lars Svensson for useful discussions. All remaining errors are my own.

[1] See Basel Committee on Banking Supervision (2010) for evidence on the effectiveness of higher capital and liquidity ratios. If the effects of such policies are large enough, and the perimeter of the regulated sector is broad enough, these policies may reduce the probability of a crisis substantially, even in the absence of macroprudential policy. In that case, the role of monetary policy in combating financial instability may be greatly reduced.

[2] In particular, the adverse scenarios assume a larger rise in unemployment if the current situation is one of low unemployment. See Federal Reserve (2013).

[3] Tight enough micro- and macroprudential standards would presumably be effective in reducing the odds of a financial crisis, but they might come with adverse side effects. Hence, there may be room for monetary policy to be used as well. See Cecchetti and Kohler (2014) for a discussion of the potentially overlapping roles of monetary and macroprudential policy and the resulting possible need for policy coordination.

countries, the use of macroprudential tools may be limited owing to a lack of legal authority.[4] Thus, policymakers may choose to use monetary policy to "lean against the wind;" that is, they could respond to growing imbalances that cannot be addressed effectively with micro- and macroprudential policies by raising interest rates, with the aim of discouraging excessive risk taking (Smets, 2014). Moreover, some have argued that tighter monetary policy could affect the incentives to employ leverage in a range of markets and so, by "getting in all the cracks," reduce risk-taking by those outside the perimeter of micro- and macroprudential regulation (Stein, 2013).

In this chapter, I will outline a modeling framework for thinking about how and when monetary policy might be used to combat growing imbalances, look at some calibrations of such models, and then discuss some historical examples where monetary policymakers considered whether it was appropriate to use monetary policy to lean against the wind. I'll end with a discussion of the implications for the design of stress tests.

2 A Conceptual Framework

In order to consider these issues, it is helpful to have a simple model of the factors involved. A useful starting point is the model in Woodford (2012). Woodford begins with a standard New Keynesian model and adds the possibility of financial crises. A crisis causes a widening of risk spreads in credit markets, which reduces aggregate demand and supply, and so reduces output and has an ambiguous effect on inflation.[5]

Crises are assumed to follow a Markov process. In "normal times," there is no crisis, and the economy has standard dynamics; in "crisis times," risk spreads widen by a fixed amount, independent of both economic conditions at the time of the crisis and of monetary policy—assumptions to which I will return later in the chapter. There is assumed to be a fixed probability of the economy transitioning from crisis times back to normal times, regardless of the policy followed by the central bank. By contrast, the transition probability from normal times into crisis times is a function of "leverage," which can be thought of as an index of the fragility of the financial sector. The probability of transition into a crisis is assumed to rise with leverage and at an increasing rate. Leverage, in turn, is assumed to depend in part on the level of economic activity, with higher activity associated with higher leverage.

To assess the implications of this model for optimal monetary policy, Woodford assumes that the goal of the central bank is to minimize a quadratic loss function reflecting deviations of output from its natural level, deviations of inflation from its target level, and the level of risk spreads. The first two terms are standard in the monetary policy literature, and the final term captures the direct inefficiencies resulting from wide spreads in credit markets as a result of a crisis. Thus, in this model, the central bank cares about crises both directly and also (and probably more importantly) indirectly through the potentially large effects of crises on output and inflation. Woodford shows that optimal policy in this model is characterized by a price-level targeting rule under which policymakers balance deviations of output from its

---

[4] For example, in the United States, regulators do not have the authority to impose economy-wide loan-to-value or debt-to-income limits.

[5] See Woodford (2012) for a complete discussion.

natural level, deviations of the price level from its desired level, and the effects of monetary policy on the probability of a crisis.

This targeting-rule result indicates that the central bank should "lean against" high leverage. If there were no risk of a crisis, then the central bank would aim to balance the marginal costs of deviations from its desired levels of output and prices, as in a standard inflation-targeting model. For example, if the price level is below the target value, then output should be above its target value by a compensating amount. Otherwise, the central bank could ease policy to move both output and the price level higher and so be better off. However, if the marginal effect of tighter monetary policy on the probability of a crisis is negative, as assumed by Woodford, then there will always be an incentive, at the margin, to operate with the price level and output (or more precisely, the linear combination of the two in the targeting rule) at least slightly below the levels that would otherwise be desirable because doing so would reduce the risk of a crisis. However, as Woodford (2012) notes, this broad result is likely an overstatement because in some periods and for some levels of leverage, the marginal effect of tighter policy on the probability of a crisis may be essentially zero, implying no significant effect on monetary policy.

This theoretical result provides no quantitative information on the extent to which monetary policy should lean, and it leaves aside a number of complications that could matter in practice. First, the model of the economy has only forward-looking dynamics, and there are no limits on the ability of the central bank to provide accommodation. For example, if the effective lower bound on nominal interest rates binds, the central bank can use forward guidance to provide the desired level of accommodation. In practice, of course, the dynamics are more complicated, and the effective lower bound is a more significant problem.[6] As a consequence, the central bank may not be able to get the economy to a point that balances the marginal costs of deviating from its output and inflation objectives and so generates the optimality of leaning against the wind.

A second complication is that the model of crises and their economic effects is so simple, with the probability of a crisis depending only on leverage and the effects of a crisis fixed exogenously. Presumably, the effects of monetary policy on the probability of a crisis will depend on the specific factors that are undermining financial stability. Moreover, the effects of higher rates on the probability of a crisis may depend on the horizon because higher interest costs and a weaker economy could increase the pressure on indebted firms and households in the near term (Svensson, 2017a).[7] Furthermore, the adverse effects of a crisis may depend on the extent of the financial imbalances prior to the crisis. Thus, leaning against the wind may be helpful not just because it reduces the odds of a crisis but also because it makes the crisis less costly.[8]

3 Some Calibrations

In order to help inform policy decisions, this theoretical result needs to be calibrated to assess the extent to which policymakers should lean against the wind in different economic and

---

[6] See Del Negro et al. (2012), McKay et al. (2016), and Woodford (2018) for a discussion of the "forward guidance puzzle" and the limits to the effectiveness of forward guidance.

[7] See the discussion in Svensson (2017a). Indeed, a new paper by Schularick, ter Steege, and Ward (2021) suggests that tighter monetary policy increases the probability of a financial crisis.

[8] This possibility is discussed further later in the chapter. For evidence along these lines, see Jorda et al. (2017). For additional discussion on this point, see Adrian and Liang (2018) and Gerdrup et al. (2017).

financial situations. A number of authors have provided such calibrations, and I summarize a few key efforts here.

Ajello et al. (2019) provide a simplified two-period empirical implementation of the Woodford model that yields a numerical assessment of the extent to which policymakers should lean against the wind as a function of credit growth. Their model calibration uses fairly standard values for the macroeconomic parameters. To calibrate the equation governing leverage, they use US data on bank credit and estimate a dynamic equation for credit growth based on both the output and inflation gaps. They use a somewhat simplified model along the lines of Schularick and Taylor (2012) to calibrate the probability of a crisis as a function of average credit growth over the previous 5 years, and they calibrate the macroeconomic impact of a crisis based on the actual outcomes in the Great Recession. To assess optimal policy, they assume that the central bank cares equally about deviations of inflation from its target and output from its potential. Thus, they assume that the central bank does not care directly about the deterioration in credit market conditions occasioned by a crisis but only about the effects of the crisis on output and inflation.

In calculating optimal policy in the model, Ajello et al. (2019) note a significant question regarding the modeling of expectations. Under rational expectations, agents would perceive the risks posed by high levels of credit growth, and they would respond by reducing spending and inflation in anticipation of the possibility of a sharp recession. Those actions, in turn, would reduce the growth rate of credit and so reduce the odds of a crisis even without any leaning against the wind by policymakers. Indeed, with the assumption of rational expectations, the authors find that it may be optimal to lean *with* the wind—that is, cut rates when credit growth is high in order to offset the effects of the risk of a crisis on agents' behavior. Given this counterintuitive result, and the observation that such anticipation effects are not seen in practice, the authors assume that agents in the economy perceive the risk of a crisis as very low (essentially zero) and that those expectations are unaffected by credit growth or monetary policy.[9]

With their baseline calibration, Ajello et al. (2019) find that for credit growth at its average value in the United States over the past 50 years, it is optimal for the central bank to lean against the wind, but only very slightly—approximately 3 basis points (bps). Such a small change in policy has almost no effect on output and inflation, as well as on the probability of a crisis. However, given the considerable uncertainty around many of the key parameters related to the probability of a crisis, the authors go on to consider several alternative calibrations. First, if tightening is assumed to be considerably more effective in curbing credit growth and credit growth is assumed to have a considerably larger marginal impact on the probability of a crisis, then leaning against the wind by approximately 25 bps could be appropriate. And if credit growth is at the high end of its historical range and the costs of a crisis are assumed to be similar to those seen in the Great Depression rather than the Great Recession (as assumed in the baseline), then leaning against the wind by approximately 75 bps would be optimal in the model.

---

[9] Adam and Woodford (2018) find a similar result when considering the optimality of leaning against house prices. If agents are rational, such leaning is not optimal, but if agents' expectations deviate from rationality, then monetary policymakers should lean against high house prices.

The simple two-period model used by Ajello et al. (2019) includes two key assumptions that have significant implications for their results. First, like Woodford (2012), they assume that monetary policy decisions in a given period are made after learning if a crisis has occurred in that period. Thus, if a crisis occurs in period 2, monetary policy will not lean against the wind in that period and, indeed, will presumably be eased in response to the crisis. As a consequence, the unemployment rate in period 2 in the event of a crisis is not affected by the decision to lean against the wind in period 1. Svensson (2017a) emphasizes that this is an unrealistic assumption because leaning against the wind prior to a crisis would presumably imply weaker aggregate demand and higher unemployment when the crisis begins, and that weakness would affect the economy in subsequent periods, given the lags with which monetary policy affects the economy. As a consequence, leaning against the wind will make unemployment in the crisis higher and thus boost the central bank's loss by a potentially significant amount if the loss is assumed to be proportional to the square of the deviation of the unemployment rate from its natural level.[10]

The second key assumption is that there are only two periods. With this assumption, the effects of tighter monetary policy in the first period on credit, and thus on the probability of a crisis in the second period, are, in a sense, permanent. But in an infinite-horizon model with a stationary process for credit, the effects of monetary policy on credit will be transitory, with credit ultimately returning to its long-run level. If the probability of a crisis depends on credit *growth*, then any reduction in the probability of a crisis in the near term will be countered by a higher probability further into the future as the effects of tighter monetary policy on credit unwind. As a result, the benefits of leaning against the wind in the near term will be offset, at least in part, further in the future. The same can be said for benefits of leaning related to the costs of crises if they occur.

Svensson (2017a) provides a simple approach to incorporating model calibrations that address both of these issues. He uses an infinite-horizon model to evaluate a temporary period of leaning against the wind in an economy in which leaning against the wind increases the unemployment rate for a time, reflecting lags in the effects of monetary policy. As a result, leaning against the wind can boost unemployment in a subsequent crisis if the crisis occurs before the effects of tighter monetary policy on employment have passed. In addition, he considers both stationary and nonstationary credit dynamics. Svensson's calibration of the macroeconomic effects of tighter policy in normal times is based on the Riksbank's Ramses model. He calibrates the effects of a crisis on unemployment based on International Monetary Fund (IMF, 2015) and Riskbank (2013), and he sets the benchmark probability of a crisis starting to 0.8 percent per quarter, with crises assumed to last eight quarters. He calibrates the effects of debt growth on the probability of a crisis using the results in Shularick and Taylor (2012) and the effects of debt growth on the size of a crisis using the work of Floden (2014). Finally, he uses estimates from Riksbank (2014) to calibrate the effects of monetary policy on real household debt.

---

[10] As Svensson (2017a) points out, in the simple case where crises are purely exogenous events, this effect should lead policymakers to run an easier monetary policy in normal times to soften the blow of a crisis, should one occur—that is, to lean with the wind. By contrast, in Ajello et al. (2019) or Woodford (2012), if crises are exogenous, policymakers will behave in normal times as they would in an economy without crises.

With these calibrations, Svensson's (2017a) model suggests that the threshold for leaning against the wind is a high one. The benchmark results show a cost of leaning against the wind that is roughly five times the benefit. In part, this strong result reflects the assumption that leaning against the wind, by weakening the economy in advance of a crisis, makes economic performance in a crisis worse than it would be without leaning against the wind. In addition, Svensson's estimates of the marginal effects of leaning against the wind on the probability of a crisis and the cost of a crisis are both quite small. He estimates that the effect of a temporary 1-percentage-point increase in the policy interest rate reduces the probability of being in a crisis by a maximum of 0.4 percentage points and reduces the unemployment rate in a crisis by a maximum of 0.06 percent.

Svensson (2017a) shows that this negative result is robust to a range of alternative calibrations. In his model, leaning against the wind remains undesirable even if monetary policy has permanent effects on the level of credit, as one might expect if monetary policy helped to counter an excessive buildup in debt. Perhaps more surprisingly, he shows that the result remains even in a credit boom that significantly raises the probability of a crisis or boosts the size of a crisis if one occurs. These counterintuitive results reflect the increase in unemployment in a crisis caused by leaning against the wind. Because the loss associated with unemployment is assumed to be proportional to the square of the unemployment rate, increased leaning against the wind actually makes the crisis more costly. Indeed, Svensson finds that both the costs and benefits associated with leaning against the wind rise proportionately with the probability of a crisis, rendering leaning against the wind less attractive as the probability increases. However, if the economic effects of a crisis are large enough, then leaning against the wind can be desirable because it reduces the probability of such bad outcomes. However, the crisis effects must be very large; Svensson estimates that, given his other calibrations, an effect on the unemployment rate of more than 30 percentage points would be required—an even larger increase than was seen in the Great Depression.

Other researchers, however, have made different modeling assumptions, and their results suggest that a case can be made for leaning against the wind. For example, Gerdrup et al. (2017) use a calibrated open-economy dynamic stochastic general equilibrium (DSGE) model with endogenous switching between a "normal" state and a "crisis" state to consider the possible benefits of leaning against the wind. The model is broadly similar to the Woodford (2012) model discussed earlier, with the effects of debt growth on the probability and virulence of crises estimated based on Organisation for Economic Co-operation and Development (OECD) data. The other model parameters are chosen based on work by Justiniano and Preston (2010). Like Ajello et al. (2019), the authors assume that agents do not respond to the probability of a crisis, but the central bank can respond to elevated credit growth—captured in their model by a simple monetary-policy rule augmented with a credit-growth term and calibrated to minimize a standard loss function. The results suggest that leaning against the wind by as much as 50 bps can help reduce losses, primarily by reducing the adverse effects of crises when they occur. As noted by Svensson (2017b), this result reflects in part that the DSGE model used has a relatively short and front-loaded effect of monetary policy on the unemployment rate relative to its effects on the probability of a crisis and losses in a crisis. Consequently, the unemployment rate is less likely to be elevated if a crisis occurs after a temporary period of leaning against the wind, reducing the costs of the leaning.[11]

Finally, Gourio et al. (2017) follow a broadly similar approach, using a calibrated New Keynesian model in which the probability of a crisis is a function of "excess credit"—that is, the excess of credit over the level it would have if there were no price distortions and financial shocks in the model. They consider whether monetary policy rules based on either excess credit or on inflation, the output gap, and excess credit can outperform policy rules based only on inflation and the output gap. Critically, they assume that the effects of a crisis on output are large (10 percent of gross domestic product [GDP]) and permanent, implying a much higher cost of a crisis than assumed in the other studies. As a result, they find that leaning against the wind can have substantial welfare benefits even though its effects on the probability of a crisis are small. In their model, leaning against the wind leads to greater variation in output and inflation, but those negative effects are outweighed by the reduced odds of a crisis and its associated massive costs.

Of course, there is considerable debate over the persistence of the adverse effects of financial crises. There is little doubt that these effects have generally been long-lived in the past (see Reinhard and Rogoff, 2009). And evidence from the recent financial crisis suggests that those economies that were worst hit by the crisis saw larger subsequent reductions in estimates of potential output and even the growth rate of potential output, suggesting either permanent or very persistent effects (Ball, 2014). On the other hand, some researchers have claimed that these long-lived effects reflect insufficient policy responses rather than an intrinsic effect of the crises themselves (Romer and Romer, 2017).

To summarize the empirical results on leaning against the wind, conventional monetary models augmented with conventionally calibrated effects of monetary policy on crises, along the lines of Ajello et al. (2019) and Svensson (2017a), suggest that the bar for leaning against the wind is high. But the modeling of financial imbalances and financial crises, as well as the effects of monetary policy on such imbalances and thus on the probability and virulence of financial crises, is still a new area of research. As a result, different plausible modeling approaches can lead to different conclusions about the utility of leaning against the wind. As a practical matter, at least for now, the decision of whether to lean against the wind will have to depend on policymakers' judgments, based on the available information, regarding the likely costs and benefits of leaning in a particular case.

4 Case Studies

To explore how central banks make such judgments, this section examines three cases where central banks have considered whether to adjust monetary policy to combat elevated

---

[11] Filardo and Rungcharoenkitkul (2016) also find that leaning against the wind can be worthwhile. In part, their results appear to reflect a relatively small impact of tighter monetary policy on unemployment. They find that an 80-bps lean implies an output loss of only 0.31 percent—or, given Okun's law, a roughly 0.15-percentage-point higher unemployment rate. By contrast, in Svensson's (2017a) calibration, leaning against the wind is assumed to involve a percentage-point increase in the policy rate, and this leads to a rise in the unemployment rate of approximately 0.5 percentage points—a multiplier approximately three times as large.

financial stability risks.[12] These cases can shed light on what considerations and data have informed such judgments, and they may also allow some conclusions to be drawn regarding how central banks can improve their decision-making in such cases in the future.

The decision-making process for the most-discussed case of leaning against the wind—that of the Riksbank in 2010–2012—is hard to evaluate for two reasons. First, agreement is lacking on the extent to which policy rates were increased for traditional macroeconomic reasons rather than as a result of a desire to push back on rapid growth in household debt.[13] Second, the evidence considered by the Riksbank prior to its decision to lean against the wind has not been published. Although a number of papers and analyses of the decisions have been published subsequently, it is not clear whether the information they contain captures the discussion at the policy meeting when the decision to lean against the wind was taken.[14]

Given these complications, the Riksbank example is not considered here. Instead, I consider three examples where there is more information on the arguments and analysis that lay behind the decision: Norges Bank's move to lean against the wind starting in 2012, and two decisions by the Federal Reserve to not lean against the wind, one in 2005 and the other in 2012. Norges Bank provided considerable information in 2012 regarding its decision to lean against perceived imbalances in the housing sector, as well as the empirical approach used in judging the appropriate path for policy.[15] The Federal Reserve decided not to take monetary policy action to address possible excesses in the housing market in 2005, and in the fall of 2012 it decided to go forward with new asset purchases and forward guidance despite concerns that additional policy accommodation might undermine financial stability over time. In both cases, transcripts of Federal Open Market Committee (FOMC) meetings and associated briefing materials provide details on the arguments and conclusions of policymakers regarding the potential costs and benefits of using monetary policy to address financial stability concerns.

4.1 Norges Bank in 2012[16]

Perhaps reflecting, in part, the debate over leaning against the wind in Sweden, Norges Bank provided considerable information on its decision to lean against the wind starting in 2012. Although the information provided to the monetary policy committee and the details

---

[12] Note that the focus here is on leaning against the wind—that is, raising rates to address a growing imbalance prior to a possible financial crisis. There is little doubt that monetary policymakers respond to changes in financial conditions that could affect the economy. For evidence of the latter, see Gerlach-Kristen (2004) and Peek et al. (2017). There is even evidence that policymakers may adjust monetary policy to offset the macroeconomic impact of bank capital regulations (see Cecchetti and Li, 2008).

[13] See the discussion in Jansson (2014) on this point. He suggests that higher rates were chosen by the Riksbank primarily for standard macroeconomic reasons and that leaning against the wind mattered only at the margin. Critics of the Riksbank have seen a desire to lean against household debt growth as a more important consideration.

[14] Work by Svensson suggests that concerns about debt growth were not sufficient to justify leaning against the wind in terms of his simple cost–benefit test (Svensson 2014, 2017a).

[15] See Norges Bank (2012), Evjen and Kloster (2012), and Olsen (2012).

[16] I thank Karsten Gerdrup for helping me navigate the Norges Bank's monetary policy communications. The interpretations, of course, are my own.

of its deliberations are not made public in Norway, several inflation reports and associated staff documents laid out the conceptual framework used. Those documents explained how leaning against the wind could improve economic outcomes and reported on how financial stability is incorporated into the Norges Bank's simulations of the policy interest rate.

Norges Bank changed its approach for characterizing the appropriate policy path in 2012. Prior to that time, the bank had indicated that the policy path should achieve the inflation target over time; be flexible, in the sense that it should take account of capacity utilization; make gradual changes in rates consistent with past policy responses; and, "as a cross-check," limit any significant deviations from simple policy rules. In practice, the path shown in the Monetary Policy Report was based on a loss function that minimized squared deviations of inflation from target, squared deviations of output from potential, squared changes in the policy interest rate, and squared deviations of the policy rate from the prescriptions of a simple policy rule (Norges Bank, 2011). However, in 2012 the last two criteria were changed to indicate that the appropriate path should be "robust," in the sense that policy should mitigate the risks of a buildup of financial imbalances and also should provide acceptable outcomes under a range of assumptions about the economy (Norges Bank, 2012). To implement this new approach, Norges Bank modified the loss function used in its projections in two ways. First, it marked up the weight put on output deviations, on the argument that output persistently above potential could lead to financial excesses, and, second, it changed the final term of the loss function to account for deviations of the policy rate from its longer-run normal level, reflecting concern that persistently low levels of the policy rate could also encourage financial excesses.

As shown in Figure 22.1, the result was a policy path that involved significant leaning against the wind. The policy rate was more than half a percentage point higher than would have been suggested by the output and inflation objectives alone, and nearly 2 percentage points higher than would have been appropriate under strict inflation targeting. As a consequence, output was projected to overshoot potential by less, but inflation was projected to remain below target throughout the projection period rather than closing the gap over a few years.

A 2012 Norges Bank Staff Memo laid out the theoretical justification for this policy path (Evjen and Kloster, 2012). The authors note that macroprudential tools are not likely to be fully effective, leaving the possibility that monetary policy can be helpful in addressing remaining financial stability risks. They point to the Woodford (2012) paper discussed earlier, emphasizing that policymakers may be able to identify when financial stability risks are elevated even if they cannot identify with great certainty when a crisis is likely to occur. And they suggest that the marginal crisis risk might be captured by high output gaps and low interest rates relative to historical norms, both of which might be expected to contribute to a buildup of credit.

The staff memo also provides some empirical evidence for the Norges Bank's approach. The authors first note that capacity utilization and credit appear to be correlated in Norway, particularly in terms of their medium-term cycles. They also provide evidence that Norwegian banks' asset growth and leverage are positively correlated. If their assets are assumed to be related to real activity, this relation might also point to a link between activity and bank leverage. Finally, they point to research showing that high levels of house prices, equity prices, capital formation, and credit relative to trend can help forecast financial crises

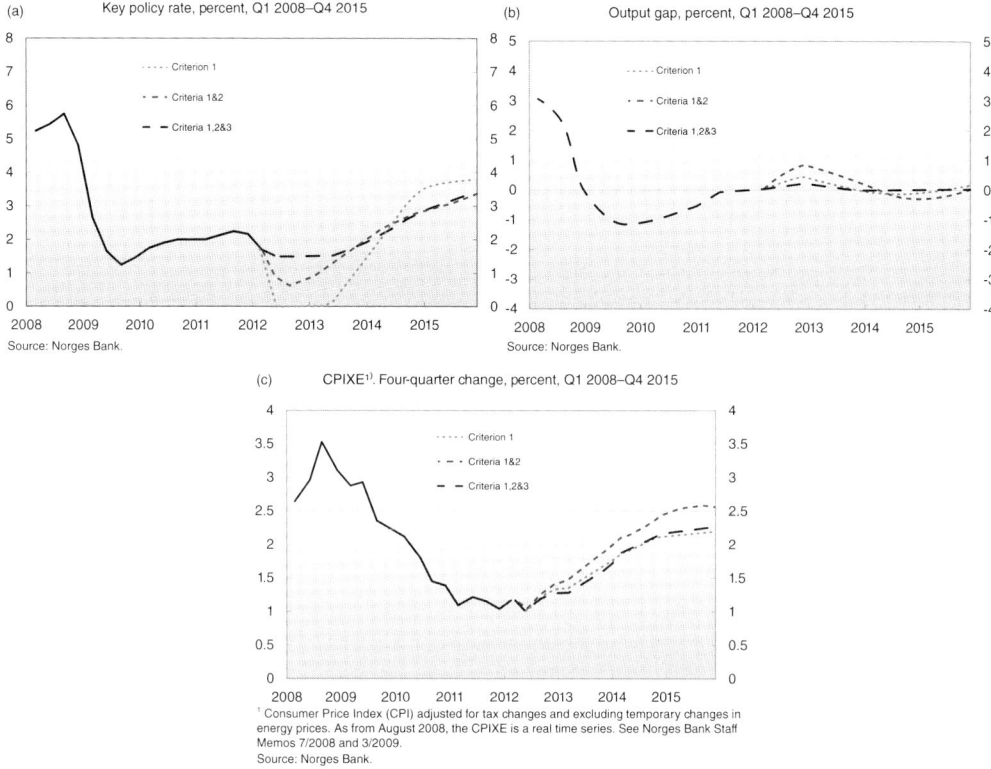

Figure 22.1 Leaning against the wind in Norway in 2012.
Note: Criterion 1 is based only on the inflation target. Criteria 1 and 2 are based on the inflation target and a goal of keeping output and employment stable. Criteria 1, 2, and 3 also involve leaning against the wind, as described in the text. The baseline policy forecast in the Monetary Policy Report was the one based on Criteria 1, 2, and 3.
Source: Norges Bank (2012).

in Norway. They argue informally that these correlations can justify the increased weight on the output gap term in the loss function used in policy simulations. The authors also suggest that easier policy, as measured by low levels of the policy rate relative to its longer-run normal level, could increase risk taking via a risk-taking channel of monetary policy and also by encouraging investors to reach for yield. The authors note that the empirical evidence for these effects is mixed, but they argue that there is at least some support for such effects, which could justify the addition of the new interest rate gap term to the loss function used in Norges Bank simulations.

Although this public explanation of Norges Bank's approach is commendable in its transparency, it is only qualitative. Norges Bank did not explain how the weights on the output gap, policy rate change, and policy rate deviation terms in the loss function were determined. In particular, it would be useful to show that the loss function can be calibrated based on the empirical evidence regarding the effects of monetary policy on debt growth and other forms of financial excess. In addition, as Norges Bank noted, leaning against the wind has costs—in the absence of a financial crisis, inflation and output will be further from their desired levels, and if a crisis does occur, those variables could be further from desired levels as well. Without an assessment of the links between monetary policy and financial stability and the possible costs of using monetary policy to address financial stability risks, it is hard to be sure that the policy paths resulting from Norges Bank's revised loss function raise expected welfare.

Of course, the Norwegian authorities likely based their policy decisions on other judgments as well. For example, they may have had insight based on their market monitoring regarding the specific risks that they faced and the possible consequences of the debt growth they observed.[17] And the decision to lean against the wind may have reflected broader concerns about downside risks to the outlook even in "normal" recessions, not solely in crises. For example, Governor Olsen has emphasized that recessions following rapid debt growth tend to be deeper and more protracted than other recessions (Olsen, 2016). This difference presumably includes results from crisis periods but may also suggest a more general dynamic along the lines of calculations of "GDP at risk" (IMF, 2017). If so, Norges Bank may have judged that leaning against the wind, by reducing GDP at risk, might have been desirable.

Norges Bank continued to lean against the wind until 2017, noting in the final Monetary Policy Report of that year that "the key policy rate has been set somewhat higher in recent years than the projections for inflation and the output gap in the coming years would in isolation imply" (Norges Bank, 2017). But there was a correction in house prices in 2017, and Norges Bank stopped, or at least reduced, its leaning against the wind as a result. Thus, in the same Monetary Policy Report, Norges Bank states, "The risk of a further build-up of financial imbalances therefore appears to have diminished somewhat. Norges Bank's overall judgment suggests that the interest rate path is adjusted up somewhat less than the changes in the outlook for inflation and the output gap alone would indicate" (Norges Bank, 2017).

4.2 The Federal Reserve in 2005

In contrast to Norges Bank's decision to lean against the wind in 2012, the Federal Reserve decided against taking such action in 2005, despite evidence of growing excesses in the housing market. At its June meeting that year, the FOMC considered possible monetary policy responses to the ongoing housing boom in the United States.[18] The meeting included extensive staff briefings and a substantial committee discussion of these issues.

Two staff briefings addressed the issue of house prices. The first presented information showing that house prices appeared to be very high relative to rents. As shown in Figure 22.2, empirical analysis suggested that house prices could be overvalued by as much as 20 percent, although the timing and pace of a return to more normal valuations were quite uncertain. A second briefing noted complications in the measurement of house prices, suggesting that the effects of various upward biases might have led to an overstatement of the rise in house prices over the previous several years. Having two briefings taking the two sides on this issue suggests that there was some uncertainty about the assessment, but

---

[17] See the discussion of this point in Jorda et al. (2013).

[18] The transcript of the meeting is provided in Federal Reserve (2005a), with briefing materials in Federal Reserve (2005b).

(a) Price-rent ratio and real carrying costs.

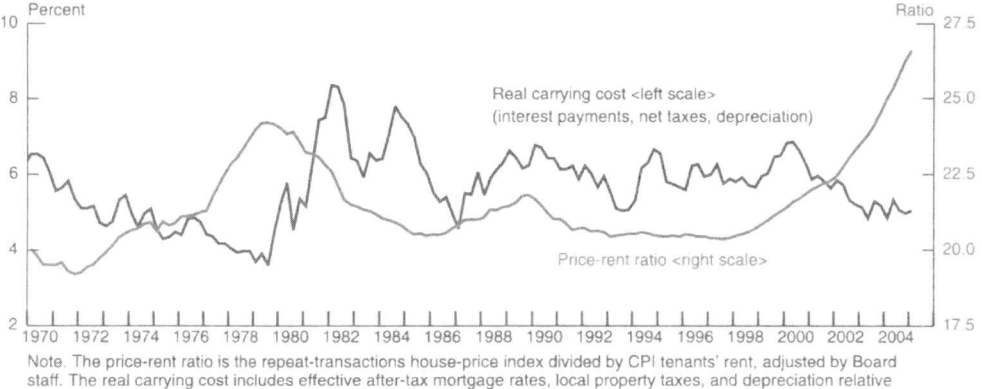

Note. The price-rent ratio is the repeat-transactions house-price index divided by CPI tenants' rent, adjusted by Board staff. The real carrying cost includes effective after-tax mortgage rates, local property taxes, and depreciation relative to 10-year inflation expectations from the Philadelphia Fed survey.

(b) Price-rent ratios and subsequent changes in real prices.

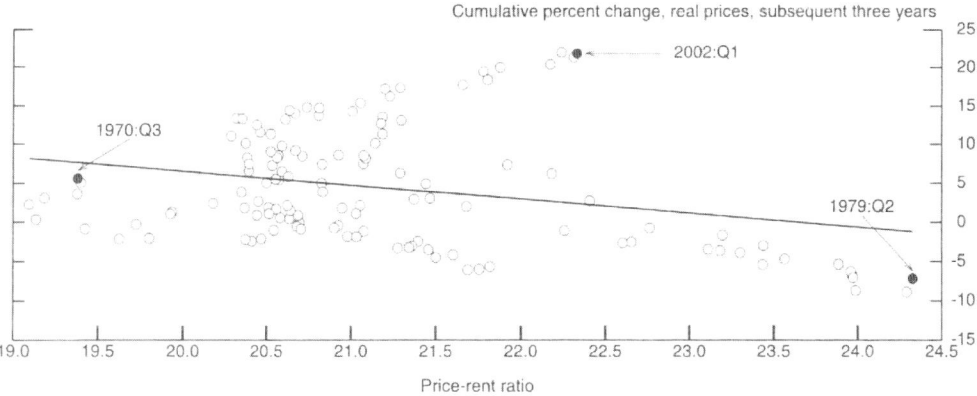

(c) Estimated loan-to-value distribution of outstanding mortgages.

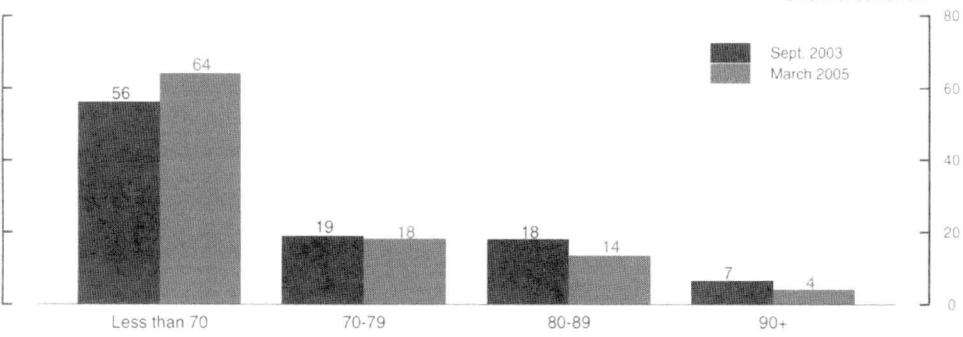

Figure 22.2 Not leaning against the wind in the United States in 2005.
Source: Federal Reserve (2005b).

the subsequent discussion showed that committee participants generally thought prices and activity in the housing sector were excessive and that a reversal was highly likely.

Another staff briefing considered the possible financial risks posed by the increased use of unconventional mortgage products, such as interest-only loans. The briefing suggested that such mortgages posed only limited risks for several reasons. First, because house prices had been growing strongly for some time, the distribution of loan-to-value ratios for outstanding mortgages had moved down somewhat, and delinquency rates were not elevated (Figure 22.2). Second, the borrowers with interest-only mortgages were not seen as especially risky. Third, securitization was thought to have transferred risk to investors who were willing and able to take it on, and financial institutions appeared to be strong enough to manage any losses they might face. Although the staff briefing was fairly sanguine about the risks associated with unconventional mortgages, the briefer ended by noting that "a historically unprecedented drop in nationwide nominal house prices ... will ... no doubt expose weaknesses that will only be obvious in hindsight."

A final staff briefing considered the extent to which monetary policy could be used to address the effects of a disorderly unwinding of the housing boom—that is, how effective the FOMC might be in cleaning up after a possible bust in the housing sector. The briefer presented three scenarios of increasing severity, with monetary policy assumed to follow a Taylor rule or to be calculated optimally. The scenarios were optimistic compared with the actual outcome: even under the worst scenario, optimal policy kept unemployment from rising significantly, and a Taylor rule kept the unemployment rate below 6.25 percent.[19] Given these projections, as well as uncertainty around the likely effectiveness of preemptive rate hikes in damping the excesses, the staff indicated that an ex post response was preferable.

The committee discussion after the briefings followed a broadly similar contour. Although participants generally appeared to see house prices as too high, and so expected a decline would come at some stage, they were fairly confident that the financial sector was sufficiently robust to manage the resulting strains, and also that the Federal Reserve could address the resulting economic weakness successfully, avoiding a deep and persistent downturn.

As one participant summarized:

> On the whole, the financial institutions seem to be in pretty good shape. The role of securitizing mortgages is to lay off risks to parties who are willing and able to bear the risks. Capital levels of the financial institutions are relatively high, so it appears that these markets are performing their roles well. And in the event of a sharp drop in housing prices, the odds of a spillover to financial institutions seem limited. And as I mentioned, it was helpful to hear the suggestion that monetary policy can be effective in responding to a sharp drop—if there is one—in housing prices. So I come away somewhat less concerned about the size and consequences of a housing bubble than I was before. (Federal Reserve, 2005a, pp. 47–48)

Although some committee participants were less confident in their assessments, this view seemed to be fairly broadly held, and the committee did not choose to try to use monetary policy to lean against the housing boom.

Given the care and depth of the committee discussion, it is interesting to consider how the conclusions regarding the likely outcome of the housing boom could have proved so wrong. Although the committee appeared to anticipate a significant decline in house prices and a resulting downdraft for the economy, participants did not understand the fragility of

---

[19] That said, the staff also noted that under the most challenging scenarios, the zero bound could become a significant constraint on the effectiveness of monetary policy.

the financing of riskier mortgages, in particular the run dynamic that could get underway in the shadow-banking sector once investors anticipated significant losses on mortgage-related assets. The result was a far sharper tightening of financial conditions than had been anticipated and greater deterioration in the real economy. It appears that the possibility of runs on major financial firms and markets simply was not seen as a possibility by the staff or the committee. This apparent lack of imagination proved very costly and points to the need for policymakers to think broadly about possible risks and take steps to learn about new institutions, instruments, and business models that could undermine financial stability.

4.3 The Federal Reserve in 2012

Another case in which the Federal Reserve considered taking financial stability risks into account in monetary policymaking arose in 2012, when the FOMC was discussing the possibility of initiating a new asset-purchase program.[20] In effect, in this case the question of leaning against the wind was reversed, as the FOMC considered whether providing additional accommodation to a very weak economy posed unacceptable financial stability risks.

In the summer of 2012, the economic recovery appeared to be stalling, with the unemployment rate leveling out above 8 percent and inflation falling further below the FOMC's 2 percent target. Against this backdrop, there were discussions at several FOMC meetings over the second half of the year regarding the possible pros and cons of providing additional monetary accommodation. A risk that some committee participants emphasized was the possibility that pushing interest rates lower via additional asset purchases could encourage investors to "reach for yield"—that is, invest at longer maturities or with greater credit risk in order to achieve higher expected returns. If that occurred, then investors subsequently could be wrong-footed by an increase in rates, causing outsized losses and potentially spillovers that could lead to renewed financial strains and an economic downturn.

To address these concerns, the staff provided a briefing at the September 2012 FOMC meeting on risks to financial stability.[21] The briefing mentioned risks related to the crisis in the euro area and possible resolutions to the "fiscal cliff" then confronting US policymakers, as well as possible funding vulnerabilities in the banking, broker-dealer, and money market mutual fund sectors. However, the main risk discussed was the extent to which low interest rates appeared to be spurring excessive risk taking. The staff evaluated evidence on the impact of persistent low interest rates on valuations in a range of financial markets, on the leverage of financial institutions, and on measures of risk taking by such firms. The overall conclusion was that there was "little evidence that the low rate environment has fostered additional vulnerabilities to date," with the staff noting that asset valuations did not appear stretched and that increased risk taking did not appear to be widespread. In their discussion following the briefing, committee members generally appeared to agree that existing policies

---

[20] The committee also changed its forward guidance at the December 2012 meeting, but the postmeeting statement indicated that the new guidance was intended to be consistent with the earlier language (Federal Reserve, 2012d).

[21] The transcript of the September meeting is provided in Federal Reserve (2012a), with briefing materials in Federal Reserve (2012b).

had not generated significant financial stability risks, but some continued to express the concern that additional asset purchases could lead to financial stability risks going forward that would be hard to counter with other tools.

Staff also provided simulations at the September and December 2012 meetings showing the estimated positive economic effects of additional asset purchases and changes in forward guidance (Federal Reserve, 2012b,d). These projections showed that the additional monetary accommodation could lead to a return of unemployment to its longer-run normal level and inflation to its 2 percent goal by 2016—2 or 3 years sooner than without the additional policy action (Figure 22.3).

Given this evidence on the possible costs and benefits of further action, the committee adopted a new mortgage-backed security (MBS) purchase program in September and a new Treasury purchase program in December. Taken together, purchases under these programs totaled $85 billion a month, and the committee indicated that the purchases would continue until the outlook for the labor market had improved substantially "in a context of price stability" (Federal Reserve, 2012e).

An important factor in reaching this decision was that without these additional actions, monetary policy would not provide sufficient accommodation to move the economy back close to the committee's objectives over a reasonable horizon. Indeed, even with the additional purchases, the committee did not expect to achieve its objectives for some time: in the December 2012 Summary of Economic Projections, the median forecast showed core PCE inflation only returning to its 2 percent target after 3 years and the unemployment rate still well above its longer-run normal level at that horizon (Federal Reserve, 2012f). The inability to achieve better macroeconomic outcomes reflected the depth of the recession and the limited strength of the FOMC's policy tools. Against this backdrop, the marginal cost of leaning against the wind—that is, not providing the additional monetary accommodation in order to avoid boosting financial stability risks—was very high because doing so would leave both employment and inflation even further below their desired levels for some time.

In addition, most committee participants appeared to find the arguments regarding the risks associated with greater accommodation unconvincing. As Chairman Bernanke noted in the policy discussion at the December 2012 FOMC meeting, the effects of easier policy on financial stability cut two ways: asset purchases increase the liquidity of financial institutions and also strengthen the balance sheets of nonfinancial firms. More broadly, he emphasized that simply pointing to possible risks was not sufficient, given the economic situation. Instead, he suggested that it was necessary to assess those risks quantitatively. Thus, those making the argument to hold back needed to be explicit about which markets and firms might be affected by additional accommodation, and to what degree; which firms would bear losses, and to what extent; and what the spillovers might be to broader markets and the economy. He noted that the popping of the tech bubble of the late 1990s had only a limited macroeconomic impact, so not all financial imbalances would necessarily lead to problems that monetary policy could not address reasonably effectively (Federal Reserve, 2012c).

Although it went forward with the new purchases, the committee did note in its post-meeting statement, "In determining the size, pace, and composition of its asset purchases, the Committee will, as always, take appropriate account of the likely efficacy and costs of such purchases" (Federal Reserve, 2012e). That language was intended to communicate that the committee could adjust the size of the program as evidence was obtained on its efficacy

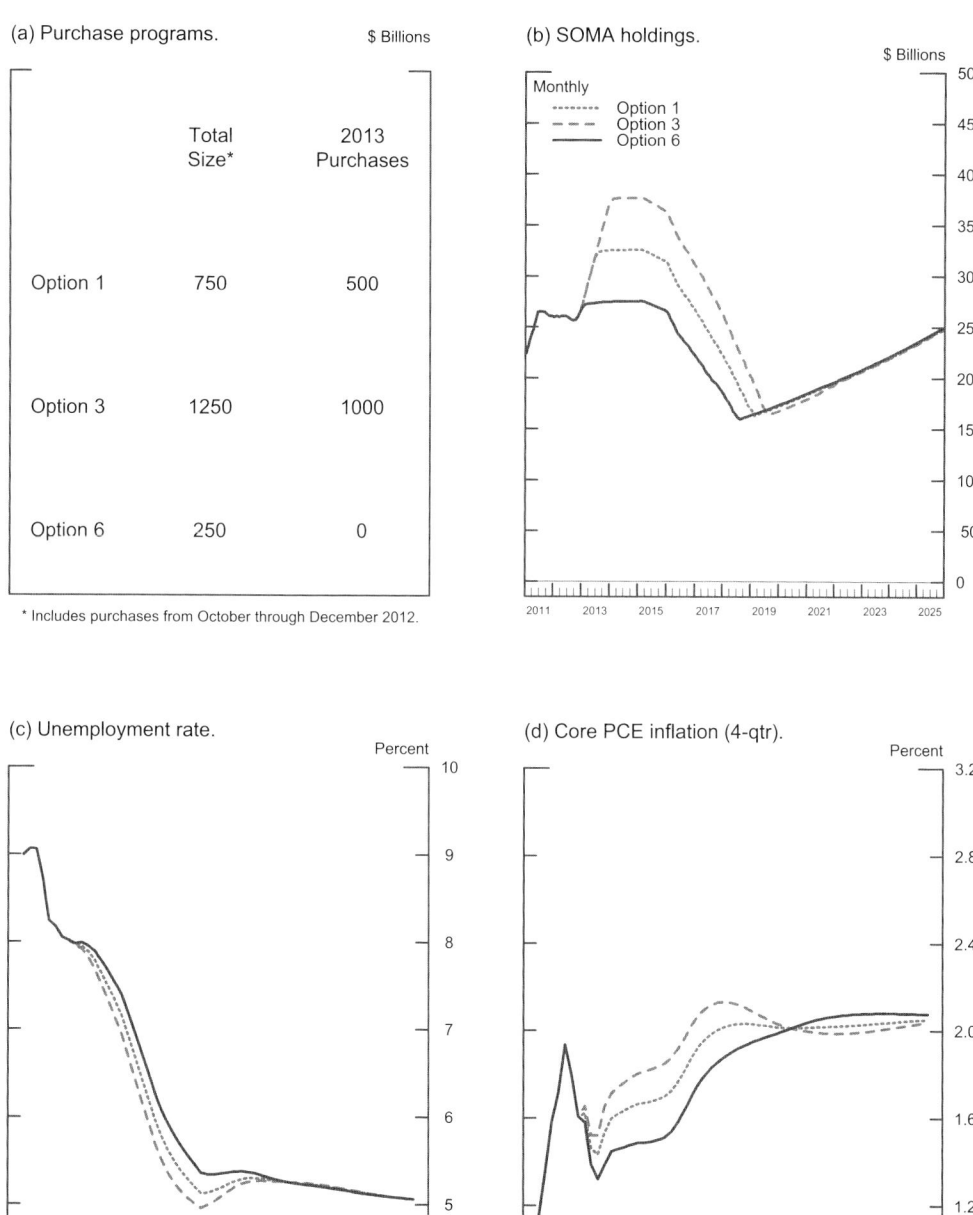

Figure 22.3 Not leaning against the wind in the United States in 2012.
Note: Options for new purchase programs provided by staff at the December
2012 FOMC meeting. PCE, personal consumption expenditure;
SOMA, System Open Market Account.
Source: Federal Reserve (2012d).

and costs (Federal Reserve, 2012c). In particular, the program could be wound down sooner if risks to financial stability clearly emerged. The chairman also called for regular updates from the staff on both the efficacy of the purchases and on financial-market developments, including any evidence that risks were building in the financial sector. Such reviews would give the committee the information it needed to assess the ongoing benefits and costs of purchases and thus judge the appropriate size and duration of the program.[22]

Ultimately, the committee judged that the purchases did not have adverse effects on financial stability. As then-governor Jerome Powell stated in a speech in July 2013, "The concern would be that... our policies, could be encouraging irrational expectations.... [But] there is little basis for arguing that markets show excessive optimism.... Nonetheless ... markets will need careful monitoring" (Powell, 2013). As a consequence, the Fed continued its purchases until the end of 2013, with the purchases under the program totaling approximately $1.7 trillion—more than twice the initial expectation of the staff. When the purchases ceased, the unemployment rate was 6.7 percent—a very clear improvement relative to the start of the program, although core PCE inflation was still only 1.6 percent.

4.4 Bottom Line

As these cases show, decisions regarding leaning against the wind are difficult. The difficulty reflects both a lack of information regarding the nature of imbalances that might be building and uncertainty about the likely effects of monetary policy on such imbalances. Even with considerable information, it is hard for policymakers to see risks that can turn out to be critical because the source of the risks may be so unexpected. And even if policymakers see the risks, they may not be confident that they could lead to a crisis bad enough to justify the costs of leaning against the wind. Moreover, even if policymakers are confident enough about the risk of a crisis, it may be very difficult to know if tighter monetary policy will help or hurt.

5 Implications for Stress Testing

The possibility of leaning against the wind to address growing financial-sector imbalances has potential implications for the design and use of stress tests. First, if the central bank believes that some leaning against the wind may be desirable, then building such a policy into the stress test would clearly be appropriate to ensure that banks are prepared for that possibility. Second, the central bank could use stress tests to obtain information on how leaning against the wind could affect the banking firms, and potentially the wider financial sector and the economy, along the lines discussed in Parlatore and Philippon (2018). This information would be much more granular than that provided by aggregate models linking credit aggregates or leverage to the probability and size of financial crises, potentially allowing the central bank to better calibrate the degree of leaning against the wind that would be most effective in fostering its economic objectives, or even whether leaning against the wind would be appropriate at all.

---

[22] Similar concerns arose in the UK, and the Monetary Policy Committee delegated the assessment of the financial stability risks posed by monetary accommodation to the Financial Policy Committee (Bank of England, 2013).

Including an assumption of leaning against the wind in a stress test scenario would be particularly appropriate if the central bank judged that such a policy could lead to a more challenging financial environment for banking institutions. For example, if there was an ongoing boom in a particular sector, driving rapid loan growth and temporarily reducing credit losses, then an assumption of tighter monetary policy would presumably reduce projected loan growth and boost projected loan losses in the stress scenario. Moreover, interest rates would be higher at the start of the stress scenario and so would likely fall by more as conditions deteriorated. The resulting larger swing in interest rates could impose losses on some banks' trading positions. It would be good to capture such effects in the stress test, both to see the extent of the resulting losses and also to encourage banks to place greater weight on those risks in their planning.

The results of such stress tests could also be helpful to the central bank in its decision about whether to lean against the wind, and if so, by how much. Stress tests involving a scenario with leaning against the wind could potentially lead banks to build capital and liquidity, which could allow the central bank to lean against the wind without putting the stability of the banking system at risk. More broadly, such stress tests would provide information on the likely impact on banking firms of leaning against the wind. For example, the central bank could learn about the effects of leaning against the wind on bank lending behavior in a stress environment. Such information could allow the central bank to better judge the extent of leaning that would best balance the costs and benefits. Finally, stress tests along these lines could also help the central bank to respond more effectively in the event of a crisis because it would have a better understanding of the likely behavior of banks and the operation of the monetary-policy transmission mechanism in such an outcome.

6 Concluding Remarks

Given the costs of the Global Financial Crisis and subsequent recession, there is great interest in policies that could be used to respond to growing financial imbalances. Macroprudential policies seem the most focused response, but such policies remain relatively new and untested, and authority for them is limited in some jurisdictions. Thus, in some cases it may be appropriate for central banks to use monetary policy to lean against such imbalances. But doing so is surely not a panacea. Monetary policy is a blunt tool in such situations (Bernanke, 2010), and the work of Svensson (2017a) suggests that central banks need to be careful in judging the costs and benefits of leaning against the wind.

To do a better job of making such judgments, policymakers need better information on the specific risks to financial stability at a given time, as well as the effectiveness of monetary policy in addressing those risks. The empirical models developed to assess the links between monetary policy and financial stability have generally been based on broad measures of debt. However, vulnerabilities in an economy's financial sector can build in a variety of ways and in a range of financial institutions and markets. As a result, policymakers need to consider a range of financial stability indicators and also look for the emergence of new institutions, instruments, and business models that could increase the vulnerability of the financial sector over time.[23] A great deal of additional research will be required to allow

---

[23] See Adrian et al. (2014) for a summary of the Federal Reserve's approach to financial stability monitoring, and see Aikman et al. (2015) for a structured approach to assessing financial stability risks in the United States.

quantitative assessments of the risks to financial stability posed by particular constellations of financial stability indicators and the likely effects on those risks of changes in monetary policy.

Policymakers also need to take account of the possibility that changes in monetary policy could trigger the financial crisis that they are trying to avoid. For example, at the February 1994 FOMC meeting, Chairman Greenspan expressed the concern that the first rate hike following a long period of monetary accommodation could "crack" markets (Federal Reserve, 1994). But running easier monetary policy to avoid triggering a crisis could lead to a bigger crisis later on. The result may be a complicated dynamic effect of tighter monetary policy on the probability and size of a crisis.[24]

With research on these issues still at an early stage, central bankers will have to use their best judgment, given the information they have, when deciding whether it is appropriate to use monetary policy to lean against the wind. In doing so, they should be explicit—both with themselves and with the public—about the judgments they are making. If central bankers choose to lean against the wind, then they should be clear about the imbalances that they are aiming to address and their judgments regarding the beneficial effects of tighter monetary policy that lead them to the decision. Such communication would not only improve transparency and accountability but also may help make the policy more effective because tighter monetary policy, and the surrounding communications, should provide a clear signal that the central bank views financial imbalances as a significant risk. That signal, along with any associated regulatory or supervisory actions, should cause financial market participants to think more carefully about the risks that they are taking and the likelihood of significant losses in the event of a crisis. Adding leaning against the wind to stress test scenarios could also send such a signal, and the results of such stress tests should provide information that helps central banks to decide on whether to lean against the wind, and if so, by how much.

References

Adam, Klaus, and Michael Woodford (2018, May), "Leaning against housing prices and robustly optimal monetary policy," CESifo Working Paper No. 7971.

Adrian, Tobias, Daniel Covitz, and Nellie Liang (2014, February), "Financial stability monitoring," Federal Reserve Bank of New York Staff Report No. 601.

Adrian, Tobias, and Nellie Liang (2018), "Monetary policy, financial conditions, and financial stability," *International Journal of Central Banking*, 14, 73–131.

Aikman, David, Michael T. Kiley, Seung Jung Lee, Michael G. Palumbo, and Missaka N. Warusawitharana (2015, June), "Mapping heat in the U.S. financial system," FEDS Working Paper No. 2015–059.

Ajello, Andrea, Thomas Laubach, David Lopez-Salido, and Taisuke Nakata (2019), "Financial stability and optimal interest-rate policy," *International Journal of Central Banking*, 15, 279–326.

Ball, Laurence (2014), "Long-term damage from the Great Recession in OECD countries," *European Journal of Economics and Economic Policies: Intervention*, 11, 149–160.

Bank of England (2013), "Minutes of the Monetary Policy Committee Meeting held on 31 July and 1 August 2013."

Basel Committee on Banking Supervision (2010, August), "An assessment of the long-term economic impact of stronger capital and liquidity requirements," available at www.bis.org/publ/bcbs173.htm.

---

[24] See IMF (2015) for estimates of the dynamic effect of credit growth on the probability of a financial crisis.

Bernanke, Ben S. (2010, January), "Monetary policy and the housing bubble," Speech at the Annual Meeting of the American Economic Association, Atlanta, GA.

Cecchetti, Stephen G., and Marion Kohler (2014), "When capital adequacy and interest rate policy are substitutes (and when they are not)," *International Journal of Central Banking*, 36, 205–231.

Cecchetti, Stephen G., and Lianfa Li (2008), "Do capital adequacy requirements matter for monetary policy?" *Economic Inquiry*, 46, 643–659.

Del Negro, Marco, Marc Giannoni, and Christina Patterson (2012), "The forward guidance puzzle," Federal Reserve Bank of New York Staff Reports No. 574 (Revised 2015).

Evjen, Snorre, and Thea B. Kloster (2012), "Norges Bank's new monetary policy loss function—further discussion," Norges Bank Staff Memo No. 11/2012.

Federal Reserve (1994), Transcript of the February 1994 meeting of the Federal Open Market Committee.

Federal Reserve (2005a), Transcript of the June 2005 meeting of the Federal Open Market Committee.

Federal Reserve (2005b), Briefing materials for the June 2005 meeting of the Federal Open Market Committee.

Federal Reserve (2012a), Transcript of the September 2012 meeting of the Federal Open Market Committee.

Federal Reserve (2012b), Briefing materials for the September 2012 meeting of the Federal Open Market Committee.

Federal Reserve (2012c), Transcript of the December 2012 meeting of the Federal Open Market Committee.

Federal Reserve (2012d), Briefing materials for the December 2012 meeting of the Federal Open Market Committee.

Federal Reserve (2012e), Postmeeting statement from the December 2012 meeting of the Federal Open Market Committee.

Federal Reserve (2012f), Summary of economic projections from the December 2012 meeting of the Federal Open Market Committee.

Federal Reserve (2013), "Policy statement on the scenario design framework for stress testing," 12 CFR 252, Appendix A, Section 4.2.2.

Filardo, Andrew, and Phurichai Rungcharoenkitkul (2016, December), "A quantitative case for leaning against the wind," BIS Working Paper No. 594.

Floden, Martin (2014), "Did household debt matter in the Great Recession?" [Blog post], available at on Ekonomistas.se.

Gerdrup, Karsten R., Frank Hansen, Tord Krogh, and Junior Maih (2017), "Leaning against the wind when credit bites back," *International Journal of Central Banking*, 13, 287–320.

Gerlach-Kristen, Petra (2004), "Interest-rate smoothing: Monetary policy inertia or unobserved variables?" *Contributions to Macroeconomics*, 4, article 3.

Gourio, Francois, Anil K. Kashyap, and Jae Sim (2017, December), "The tradeoffs in leaning against the wind," Mimeo, Booth School.

International Monetary Fund (2015, September), "Monetary Policy and Financial Stability," IMF Staff Report.

International Monetary Fund (2017, October), *Global Financial Stability Report October 2017: Is growth at risk?*

Jansson, Per (2014, December), "Swedish monetary policy after the financial crisis: Myths and facts," Speech at the SvD Bank Summit.

Jorda, Oscar, Bjorn Richter, Moritz Shularick, and Alan M. Taylor (2017), "Bank capital redux: Solvency, liquidity, and crisis," NBER Working Paper No. 23287.

Jorda, Oscar, Moritz Schularick, and Alan M. Taylor (2013), "When credit bites back," *Journal of Money, Credit and Banking*. 45(suppl. 2), 3–28.

Justiniano, Alejandro, and Bruce Preston (2010), "Can structural small open economy models account for the influence of foreign disturbances?" *Journal of International Economics*, 81, 61–74.

McKay, Alisdair, Emi Nakamura, and Jon Steinsson (2016), "The power of forward guidance revisited," *American Economic Review*, 106, 3133–3158.

Norges Bank (2011), "Criteria for an appropriate interest rate path," *Monetary Policy Report 3/11*.

Norges Bank (2012), "Response pattern of monetary policy and criteria for an appropriate interest rate path," *Monetary Policy Report 1/12*.

Norges Bank (2017), "4. Monetary policy analysis," *Monetary Policy Report 4/17.*

Olsen, Oystein (2012, September), "Monetary policy in turbulent times," Address at the Centre for Monetary Economics (CME)/BI Norwegian School of Management.

Olsen, Oystein (2016, October), "A flexible inflation targeting regime," Speech at the Centre for Monetary Economics (CME) / BI Norwegian Business School.

Parlatore, Cecilia, and Thomas Philippon (2018), "Designing stress scenarios," 2018 Meeting Papers 1090, Society for Economic Dynamics.

Peek, Joe, Eric S. Rosengren, and Geoffrey M. B. Tootel (2017), "Does Fed policy reveal a ternary mandate?" Federal Reserve Bank of Boston Working Paper No. 16–11.

Powell, Jerome H. (2013, June). "Thoughts on unconventional monetary policy," Speech at the Bipartisan Policy Center, Washington, DC.

Reinhart, Carmen M., and Kenneth S. Rogoff (2009), "The aftermath of financial crises," *American Economic Review*, 99(2), 466–72.

Riksbank (2013, July), "Financial imbalances in the monetary policy assessment," *Monetary Policy Report.*

Riksbank (2014, February), "The effects of monetary policy on household debt," *Monetary Policy Report.*

Romer, Christy, and David Romer (2017), "New evidence on the aftermath of financial crises in advanced countries," *American Economic Review*, 107, 3072–3118.

Schularick, Moritz, Lucas ter Steege, and Felix Ward (2021), "Leaning against the wind and crisis risk," *American Economic Review: Insights*, 3, 199–214.

Schularick, Moritz, and Alan M. Taylor (2012), "Credit booms gone bust: Monetary policy, leverage cycles, and financial crises, 1970–2008," *American Economic Review*, 102(2), 1029–1061.

Smets, Frank (2014), "Financial stability and monetary policy: How closely interlinked?" *International Journal of Central Banking*, 10(2), 263–300.

Stein, Jeremy C. (2013, February), "Overheating in credit markets: Origins, measurement, and policy responses," Presentation at the "Restoring Household Financial Stability after the Great Recession: Why Household Balance Sheets Matter" research symposium sponsored by the Federal Reserve Bank of St. Louis, St. Louis, MO.

Svensson, Lars E. O. (2014, June), "Inflation targeting and leaning against the wind: A case study," Memo, Stockholm School of Economics.

Svensson, Lars E. O. (2017a), "Cost-benefit analysis of leaning against the wind," *Journal of Monetary Economics*, 90, 193–213.

Svensson, Lars E. O. (2017b), "Leaning against the wind: Costs and benefits, effects on debt, leaning with DSGE models, and a framework for comparison of results," *International Journal of Central Banking*, 13, 385–408.

Woodford, Michael (2012), "Inflation targeting and financial stability," *Sverges Riksbank Economic Review*, 7–32.

Woodford, Michael (2018), "Monetary policy analysis when planning horizons are finite," in Martin Eichenbaum and Jonathan A. Parker (eds.), *NBER Macroeconomics Annual 2018*, Vol. 33, National Bureau of Economic Research.

Stress Testing Networks: The Case of Central Counterparties

Richard B. Berner, Stephen G. Cecchetti, and Kermit L. Schoenholtz*

1 Introduction

Complex and diverse connections among financial institutions and market participants and across financial activities and borders are a defining characteristic of the modern financial system. Indeed, they are what make it a system, rather than merely a collection of individual firms and markets.

Interconnections can add to financial-system resilience and efficiency. As Janet Yellen, then chair of the Federal Reserve, noted in 2013, the more complete is the network of interconnections, the more resilient it may be: "The principle behind this result is familiar and basic to economics: Diversification reduces risk and improves stability." Diverse interconnections also allow for efficient resource allocation that can reduce the costs of transactions.

However, interconnections can also act as channels for transmitting and amplifying shocks. Liquidity or credit shocks in one part of the financial system may spread to other parts, resulting in runs and fire sales, with adverse consequences for economic activity. Furthermore, there is a tendency for networks to be both complex and opaque. The combination can blind market participants to the nature and extent of exposures and can trigger individually rational but collectively destabilizing behavior, amplifying the effects of an initial shock. As Yellen also observed, more concentrated, highly interconnected systems with a few key players or critical nodes in the network can be particularly vulnerable to shocks unless these institutions are highly resilient. Even where direct linkages are sparse, common exposures can generate herding, amplifying shocks.

Two aspects of interconnectedness illustrate these benefits and costs. First, linkages in networks for payments, clearing, and settlement offer clear efficiency and risk management benefits. Second, those same linkages in periods of stress could amplify and transmit financial shocks across networks and from one part of the financial system to another, increasing costs.

In this chapter, we evaluate how to use stress testing to assess the resilience of central counterparties (CCPs)—key nodes in the payments, clearing, and settlement network and arguably the most interconnected of all financial institutions. CCPs have a host of

---

* A longer version of this chapter appears as NBER Working Paper No. 25686 (March 2019). We would like to thank Robert Engle, Richard Haynes, Til Schuermann, and Bruce Tuckman for helpful discussions; Martin Scheicher for helpful comments; and the New York University (NYU) Stern Volatility Lab for providing firm-level systemic-risk data.

counterparties, some of which are large and interconnected and some of which are highly leveraged; they conduct an enormous volume of transactions; and they have potentially sizable liquidity needs, especially under stress. Some CCPs also provide critical services for which there are no immediate substitutes. Hence, studying the vulnerabilities of these *financial-market utilities* is essential for assessing and informing policies aimed at securing financial stability: What makes a CCP a source of systemic vulnerability? How might we determine the potential for these network hubs to transmit and amplify shocks? How can we mitigate their vulnerabilities?

Following this introduction, we start with a description of stress testing, including a discussion of its value for determining financial system resilience, as well as its limitations. We then move to a discussion of the topology of networks and the challenge of testing their fragility. Finally, we turn to a discussion of the unique characteristics of CCPs, suggesting a high-frequency metric for assessing the extent to which they would likely transmit and amplify shocks. Unlike most studies of CCP resilience that focus on the adequacy of pre-funded resources, we consider the impact on the financial system in the event that these resources run out.

Interconnections among CCPs and their counterparties can be direct, indirect, or both. For example, a CCP's clearing members have direct connections through mutual counterparty exposures and indirect connections to other firms that are the clearing members' counterparties. Similarly, through holdings of correlated portfolios of assets and through common funding and resolution mechanisms, firms face similar risks.

In either case, analyzing and monitoring interconnectedness, and the benefits and vulnerabilities it creates, requires a system-wide approach. Researchers suggest two broad methodologies for addressing this challenge. One relies on the estimation of models from market-price data and perhaps some details on quantities. Such a nonnetwork approach relies on the validity of the model and its applicability to stress events. Examples include the development of so-called *cross-sectional measures* like covariance (CoVaR), the default insurance premium (DIP), marginal expected shortfall, and systemic risk (SRISK).[1] The other approach begins by mapping the network and then turns to graph-theoretic tools to assess the strength of transmission mechanisms through both direct and indirect exposures.[2]

In principle, mapping CCPs' connections is straightforward, so they would seem to be natural entities for the application of network analysis. In practice, however, creating such a map requires that researchers and policymakers have detailed, consistent, timely data on counterparty exposures (including derivatives transactions) to assess and monitor interconnectedness. Challenges remain in collecting the necessary high-quality granular data, in ensuring the confidentiality and security of those data, and in developing methodologies to reconstruct the full network from partial information when full data are unavailable.[3] For pragmatic reasons, analysts customize network analysis to the special characteristics of each

---

[1] There are other measures of institutional risk in addition to SRISK, including credit default swap (CDS) spreads, expected default frequency, and CoVaR. It would be useful to test the robustness of results when using these alternative measures of institutional risk. We discuss SRISK pros and cons later in the chapter.

[2] For further discussion, see Kara et al. (2015).

[3] In cases such as in the market for CDSs, complete data are available to official entities; see, for example, Paddrik et al. (2016).

financial network, including its topology, the heterogeneity of its participants, and the speed with which it can become vulnerable.

2 Stress Test Preliminaries

Stress tests applied to firms have three broad objectives. First, they help evaluate potential losses under adverse conditions and thus help determine the adequacy of resources for self-insurance, such as capital or liquidity buffers. Second, they help ensure the use of robust processes, data, and information systems.[4] Third, they help make informed decisions about a firm's resilience in the face of shocks to net worth and liquidity.

These *micro*prudential stress tests differ from the *macro*prudential kind. The former measures losses that a public or private guarantor would suffer following the collapse of an *individual institution* that results from a firm-specific shock. By contrast, macroprudential tests provide information about a *system's* resource adequacy in the face of an aggregate shock that hits all institutions simultaneously, accounting for the spillovers to the system resulting from shocks that may create material distress or failure of an individual firm.[5]

To be effective, all stress tests should be severe, flexible, and not overly transparent.[6] All three attributes matter: First, if the scenarios are insufficiently dire, there is no point to the test. Second, flexibility is essential, lest the tests become a mechanical exercise. Third, there are limits to transparency: premature disclosure of the scenarios invites gaming, allowing firms to manage to the test.[7]

In addition, stress tests must also be frequent, comparable over time, and coherent. Infrequent tests may miss material changes in system health. Without comparability, tests are difficult to interpret. Incoherent tests may be implausible.[8] Specifically, the combination of assumed changes in factors (e.g., employment, inflation, and asset prices) that affect the results needs to fit together.

Existing stress testing methodology has two limitations. First, there is no obvious way to account for interconnectedness and feedback, either adverse or beneficial, when testing a group of institutions (a point we will return to shortly). Second, there is no straightforward way to determine the behavior and impact of institutions (e.g., banks, in the case of CCPs) that are outside the scope of the test. Assessing system-wide or network resilience requires that we address these limitations.

---

[4] See Counterparty Risk Management Policy Group (2008).

[5] Such tests can help calibrate a system's resilience consistent with society's tolerance for using taxpayer resources to address systemic shocks. See the discussion in Cecchetti and Tucker (2015).

[6] See Clark and Ryu (2015).

[7] See Cecchetti and Schoenholtz (2018a). Note that the precrisis stress tests of Fannie Mae and Freddie Mac—the government-sponsored enterprises (GSEs) that serve as mortgage lenders and guarantors—failed on all three counts. Although the GSEs had always passed the annual government stress tests, they collapsed in September 2008. Why? First, the stress scenario was insufficiently severe: house prices *rose* for the first 10 quarters of the test, before falling only modestly over the full 8-year horizon. Second, they lacked flexibility: from year to year, the parameters and the macroeconomic conditions were unchanged. Third, the tests were too transparent: their regulator published the models and scenarios prior to initiating the tests.

[8] See the discussion in Liang (2018).

3 Conceptual Issues

To assess stress tests and CCPs in the context of financial networks, our discussion proceeds in three steps. We begin by outlining the basic tools of network analysis. Next, we examine how to test networks, such as one created by the combination of a CCP, its clearing members, and their customers. Finally, we discuss the need for *high-frequency* stress tests that, given the speed with which asset prices and exposures change, can complement traditional tests.

3.1 Basics of Network Analysis

A financial *network* is the combination of a group of financial entities (the *nodes*) that transact with one another, and the set of *connections* among these nodes. Risk flows in both directions between the nodes; for each transaction, there is a payer and a payee; for each obligation, there is a claimant and an obligor.[9]

In theory, network analysis can help us to assess what makes a financial network resilient or fragile and how interconnectedness can amplify or dampen shocks to the system. Is network health simply an aggregation of the resilience of its nodes? Or is a network as brittle as its weakest node? How do resilience and the resistance to contagion depend on factors such as node strength, distance between nodes, and network structure and growth?[10]

In principle, fragility in simple network structures is easy to spot. In a *ring* of payment obligations, for example, a shock to a single entity can trigger a total collapse.[11] In a hub-and-spoke topology, the central node must be an effective shock absorber. But even for relatively simple networks, heterogeneity among the nodes can create sudden, large shocks.[12]

In practice, financial networks are complex and heterogeneous, making it difficult to assess their resilience. For example, Figure 23.1 maps the connections in 2017 across 15 jurisdictions among 26 CCPs and their top 25 clearing members (CMs).[13] The figure reveals the concentration both of credit risk in a few CCPs (the large dark-gray circles) and of prepositioned resources in a somewhat larger number of CMs (the large light-gray circles).

Indeed, Glasserman and Young (2016) conclude that even in relatively straightforward networks—such as the market for interbank lending—empirical work "has not yet produced a compelling link between traditional network measures and financial stability."[14] So analysts are fashioning tools that exploit partial data in an effort to gauge a financial network's structure and vulnerabilities.[15]

---

[9] Private cryptocurrencies present an exception to this generalization because there is no clear obligor for Bitcoin, Ether, or the like.

[10] Classic references in the vast and growing literature on financial networks include Allen and Gale (2000), Elliott et al. (2014), and Acemoğlu et al. (2015).

[11] See Allen and Gale (2000).

[12] See Covi et al. (2019).

[13] See Basel Committee on Banking Supervision et al. (2018).

[14] Analysts employ a range of metrics in assessing financial networks, such as the *degree distribution*—the number of connections for each node—and the scale of exposures for each node; measures of *centrality*, hypothesizing that central nodes are critical for resilience; and *node depth*, which measures the amplification of losses arising through network connections. See the discussion in Glasserman and Young (2016).

[15] See Anand et al. (2018).

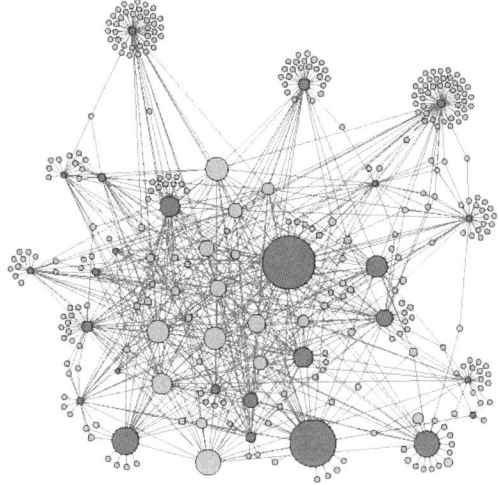

Figure 23.1 Global CCP network.
The size of each CCP node (dark-gray circle) is a proxy for the credit exposure to its clearing members (CMs), whereas the size of each CM node (light-gray circle) reflects the scale of its resources posted at the network of CCPs.
Source: Basel Committee on Banking Supervision et al. (2018).

For example, Glasserman and Young (2015) focus on the importance of key nodes, stressing not only size but also their leverage and interconnectedness. These authors place upper bounds on *node depth* as a proxy for the contribution of particular nodes to systemic risk (even in the absence of full information about the network topology). They conclude that contagion is most likely when the nodes vary in size and a shock emanates from a highly leveraged and interconnected node.

3.2 Network Analysis and Stress Tests

Despite the challenges, a network perspective can alert us to important sources of vulnerability in a financial system. Focusing on the nodes, for example, one obvious concern is the resilience of those serving critical functions that lack substitutability. When a key node is irreplaceable, its failure can lead to a collapse of the system. The most prominent instances of limited substitutability in the financial system lie in the payments, clearing, and settlement functions that CCPs and other financial-market utilities provide. Accordingly, Section 4 of this chapter focuses on stress tests for CCPs as nodes in financial networks.

To be sure, the central node of a financial network is neither the only nor always the main source of system vulnerability. To see this, consider Figure 23.2, which maps the financial network of the CDS market involving US entities in 2015.[16] At the center of this hub-and-spoke system is a CCP. Its CMs, denoted by gray circles, are in the innermost ring around it. The outer ring is composed of the counterparties of the CMs (information the CCP itself may not possess). The arrows denote the direction and the size of the additional variation margin a particular node is obligated to post following an asset-price shock. The mapping reveals

---

[16] See Paddrik et al. (2016). One CCP clears virtually all US entities' CDS trades.

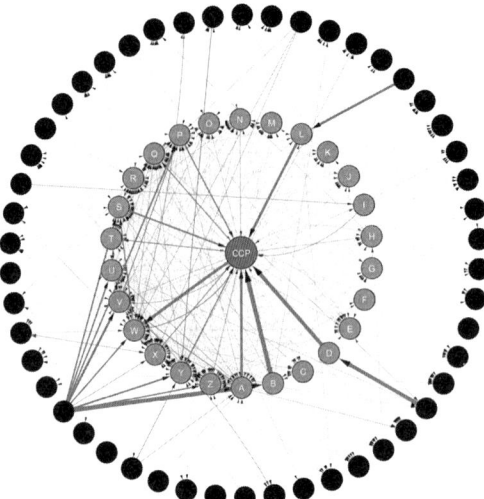

Figure 23.2 Variation margin payment network after an asset-price shock.
Source: Paddrik et al. (2016).

that the CMs' collective exposure to a single counterparty that is *not* a CM is larger than the CCP's exposure to any individual CM. Importantly, only by mapping the *system* beyond the CMs can we can see that the network faces a vulnerability to stress that is outside of the direct CCP–CM system.

Indeed, the transmission of vulnerabilities may work in both directions. In discussing the risks associated with structures like the one in Figure 23.2, Faruqui et al. (2018) note the potential for a destabilizing feedback loop from the interplay of CCPs' risk management and banks' risk taking. The authors consequently propose *jointly stress testing banks and CCPs* to capture these interactions.

In practice, most stress tests focus precisely on key nodes as triggers, channels of transmission, or amplifiers of shocks. The key nodes—such as CCPs—may administer their own stress tests on a daily basis.[17] Although supervisory stress tests are only annual at best,

---

[17] The Principles for Financial Market Infrastructures (see Committee on Payment and Settlement Systems and Technical Committee of the International Organization of Securities Commissions [2012]) specify: Stress tests should be performed daily using standard and predetermined parameters and assumptions. On at least a monthly basis, a CCP should perform a comprehensive and thorough analysis of stress testing scenarios, models, and underlying parameters and assumptions used to ensure they are appropriate for determining the CCP's required level of default protection in light of current and evolving market conditions. A CCP should perform this analysis of stress testing more frequently when the products cleared or markets served display high volatility, become less liquid, or when the size or concentration of positions held by a CCP's participants increases significantly. A full validation of a CCP's risk management model should be performed at least annually. In conducting stress testing, a CCP should consider the effect of a wide range of relevant stress scenarios in terms of both defaulters' positions and possible price changes in liquidation periods. Scenarios should include relevant peak historic price volatilities, shifts in other market factors such as price determinants and yield curves, multiple defaults over various time horizons, simultaneous pressures in funding and asset markets, and a spectrum of forward-looking stress scenarios in a variety of extreme but plausible market conditions.

they typically impose the same test across entities, so they are directly comparable across institutions and likely better capture network effects.

Reflecting the international accord on CCP resilience, supervisory stress tests require a CCP to withstand the default or nonpayment of its two largest CMs and affiliates. For two reasons, such tests may overstate network resilience, as well as that of key nodes. First, as the Systemic Risk Council (2019) points out, they do not assess the full implications of the default of a systemically important financial entity. Such a default today would only occur if that firm could not be resolved, and in those circumstances, the distress of the entity and the loss of credibility of the regulatory regime likely would trigger broad contagion across financial markets.

Second, network effects are in fact likely to be substantial. Research by Paddrik and Young (2017) examines the direct and indirect effects of nonpayment by members and/or their clients through the full network of exposures and takes account of network spillover effects in the CDS market. They conclude that the CCP that clears this market "is potentially more vulnerable to default than conventional stress tests would suggest."

Importantly, we can interpret the adverse scenarios of supervisory bank stress tests—such as the Fed's Comprehensive Capital Analysis and Review (CCAR) test—as a large financial network where, following an episode of severe stress and multiple rounds of network amplification, the connections all freeze. In this distressed state, the CCAR banks are no longer able to sell assets or raise new equity funding. Consequently, each institution is completely isolated, and information about its well-being is unavailable to others. Moreover, the write-downs arising from the asset-price changes implied by each scenario determine both individual and collective capital adequacy. In these circumstances, if *every* institution passes the test—implying a sufficient pre-shock level of self-insurance in the form of equity finance—the system is as safe as the sum of its parts.[18]

It would, of course, be desirable to model the entire process for all markets from start to finish, including the destabilizing feedback mechanisms that amplified the initial asset-price plunges. However, constructing such a fully specified model remains beyond the current capacity of technical experts inside and outside the official sector.[19]

3.3 High-Frequency Stress Testing of Financial Networks

The low frequency of the current supervisory stress tests is a concern. Exposures of key nodes can change extremely quickly, swiftly rendering point-in-time observations obsolete.

---

[18] We note that such ad hoc effects are part of most stress tests. A test might specify system-wide parameters that determine the severity of real estate losses or that limit a firm's share of funding coming from deposits during the stress period. Assessing scenario plausibility across CCPs also includes carefully considering the relationships between asset classes and within asset classes; see Tompaidis (2018).

[19] We should note that *reverse* stress tests are also useful to assess resilience; they are standard in some firms and are a feature of recent Commodity Futures Trading Commission (CFTC) and forthcoming European Securities and Markets Authority (ESMA) CCP stress tests. The purpose of such a test is to gauge the size of a shock necessary to precipitate the failure of a key node or set of nodes. Combined with an estimate of the probability of such a shock, policymakers could determine whether the resilience of the system is consistent with social objectives. In the case of a CCP, how large a decline in key asset prices would trigger the default of one or more CMs and be sufficient to deplete the institution's resources?

To monitor the system effectively, supervisors need high-frequency tools and indicators. To this end, some authorities and researchers use high-frequency financial market data to construct heat maps and financial stress indexes that can alert supervisors to emerging risks.[20] Others construct real-time measures of institution and systemic stress using publicly available information. One example is the NYU Stern Volatility Lab's weekly publication of SRISK, a forward-looking measure of each firm's estimated shortfall of capital (relative to a stated norm) in a bad state of the world (e.g., a 40 percent plunge over a 6-month horizon in a broad aggregate of stock prices).[21] The notion is that a firm's contribution to systemic risk reflects the externalities (e.g., fire sales and liquidity hoarding) that arise from its capital shortfall precisely when the system as a whole is short of capital.[22]

To be sure, SRISK has limitations. First, it is only available for listed firms, so it excludes those held privately or held by a government. Second, comparison of SRISK measures across types of institutions (e.g., as banks and insurers) and regions with different accounting standards (e.g., generally accepted accounting principles [GAAP] and [IFRS]) requires some caution. Third, market-based measures like SRISK tend to be procyclical. Fourth, although SRISK's reference leverage ratios aim to exceed the minimum capital thresholds necessary to avoid runs in a crisis, their calibration is imprecise.[23]

That said, we see market measures like SRISK as among the most promising high-frequency tools to help guide supervisors in quickly refining and modifying stress tests. With that in mind, we now turn to a discussion of CCP stress testing from a network perspective.

4 Central Counterparties Structure, Risks, and Stress Testing

Systemic stress associated with bilateral over-the-counter (OTC) derivatives markets prior to and during the crisis led authorities to push for an increase in central clearing.[24] Clearing is the posttrade process of transmitting, reconciling, and confirming transactions prior to settlement. In 2009, the G20 leaders mandated clearing of standardized OTC derivatives through CCPs.[25] As a consequence, central clearing grew dramatically: In 2018, CCPs cleared approximately three-fourths of OTC interest rate derivatives, triple the proportion just a decade earlier.[26] Central clearing involves significant economies of scale and scope,

---

[20] For example, the Office of Financial Research has produced stress and vulnerability indicators.

[21] For a description and analysis of SRISK as a measure of resilience of the financial system, see Brownlees and Engle (2017).

[22] The idea of using market measures as a guide to financial resilience is not limited to SRISK. Sarin and Summers (2016), for example, use an array of indicators—including volatility, implied volatility, out-of-the-money put options, correlations, CDS spreads, price–earnings ratios, and preferred stock prices and yields—to assess the riskiness of the largest US banks.

[23] We view the fact that these reference ratios currently exceed regulatory requirements—8 percent in Asia and the United States and 5.5 percent in Europe—as a feature of the SRISK calibration, rather than a flaw.

[24] See Gorton (1984) and Bernanke (2011) for discussions of the history of clearinghouses in the American financial system. The first US clearinghouse—the New York Clearing House Association established in 1853—followed the model of the London Clearing House of 1773.

[25] See paragraph 13 of G20 Leaders (2009).

[26] See Faruqui et al. (2018).

combined with network externalities.[27] As a result, CCPs tend to become extremely large; for example, as of December 2019, LCH Clearnet (the largest CCP in Europe) had over $340 *trillion* in gross notional contracts outstanding.

Central clearing has numerous benefits. First, by substituting the CCP as the buyer to every seller and the seller to every buyer, central clearing *mutualizes* and can—with appropriate margining, trade compression, position liquidation procedures, and reporting—*reduce* the counterparty risks that arise from the lag between the initial transaction and the ultimate payment.[28] To manage its risk, a CCP runs a matched book and can absorb losses both through overcollateralization of exposures and by setting appropriate initial and variation margins. Correspondingly, a CCP's resilience depends importantly on the liquidity, quality, and quantity of the collateral it holds. Second, central clearing facilitates enforcement of *uniform* margining standards.[29] Third, through multilateral netting and trade compression, CCPs economize on the use of collateral. Fourth, should a clearing member (CM) default, central clearing facilitates the orderly liquidation of that member's positions and provides for some sharing of the default costs. Finally, mandated trade reporting makes risk concentrations transparent.

While central clearing may reduce some aspects of systemic risk, it creates others. A change in asset prices that triggers a margin call transforms market risk into counterparty risk. Should a CM default, the web of complex counterparty risks can become credit and liquidity risk *at the CCP*. If meeting a CCP's obligations requires significant sales of securities, and if a defaulting clearing member's collateral is either illiquid or funded in the repo market, the CCP will face further risks as it works to raise the necessary cash. The CMs also play an important role in limiting the losses during the auction process of defaulted collateral.

Ultimately, experience shows that when a CCP fails, financial markets become severely impaired. In the past, the spillover of these collapses, as in the case of the 1974 failure of the Paris commodities futures market, was limited by the size of the market or by the lack of cross-border financial integration. However, given the development of the global financial system since 1980, a failure today could have a much broader impact.

The services that CCPs provide are inherently systemically important; central clearing creates *concentration* and *substitutability* risk. This makes it essential that a CCP be resilient and have sufficiently rigorous risk management systems. Put differently, the resolution of a failed CCP must aim to sustain critical services that lack substitutes. Each CCP must confront risks arising from the default of a CM and from operational failures. Although we agree that cyber-risk and other operational risks are among the biggest current threats to CCPs and to the financial system as a whole, our focus here is on CM default.[30]

4.1 The Default of a Clearing Member

What if one or more large counterparties to the CCP defaults? Bell and Holden (2018) describe the Lehman default in 2008 and the default in 2018 of a Norwegian CM trading

---

[27] See the discussion in Alexandrova-Kabadjova et al. (2018), especially Section 3.3.

[28] See Duffie et al. (2015) and Tuckman (2015).

[29] See Duffie and Zhu (2011).

[30] See Cecchetti and Schoenholtz (2018b).

Nordic power spreads on a Sweden-based Nasdaq exchange.[31] In both cases, the CCPs involved proved resilient and were able to continue operating. However, in the recent Nasdaq episode, the exchange called on its remaining CMs to replenish their prepositioned guarantee funds rapidly, an action that could have transmitted and amplified distress in a financial system in which CMs face a broad capital shortfall.

To be sure, postcrisis reforms to bank resolution regimes, requirements for increased capital and liquidity, and improvements in risk management should meaningfully reduce the likelihood of the default of a large CM—typically operating subsidiaries in banking groups.[32] However, in light of the 2018 Nasdaq episode, prudence requires preparing for one or more such defaults.

In the event of a CM default, the CCP has access to resources in a particular order known as a *waterfall*.[33] Figure 23.3 provides a stylized view of such a waterfall.

The waterfall works as follows: The CCP's first line of defense against default is *margin* (collateral). To enter into a derivative contract, both parties need to supply *initial* margin. Whenever the price of the contract changes, the losers must pay the winners in *variation margin calls*. When a CM's margin is exhausted, its contribution to a CCP guarantee fund is the second line of defense. If both are exhausted, the process next looks to the CCP's own capital, and then to the guarantee fund contributions of other CMs. And finally, the CCP can call on its CMs—in ex post *assessments*—once all the prefunded backstops are depleted.[34] Will the nondefaulting CMs be able to meet such a capital call or not?[35]

Table 23.1 shows the various components of the waterfall buffer that protects the integrity of operations for the CME and LCH.[36] In both cases, margin is by far the largest piece. At the end of September 2018, for the CME and LCH, the margin was 11.0 and 13.7 times the potential assessment, respectively. That said, the CCPs' buffers appear small relative to the size of their exposure. For LCH, the margin is less than 6 basis points (bps) of open

---

[31] The most important lesson from the 2018 Nasdaq exchange incident is this: *set margins commensurate with risk.* Failure to do so can lead the CCP to call on the resources of its clearing members in a period of stress when doing so can transmit the shock broadly across the financial system. Another concern is governance: Do the clearing members that contractually share in the losses faced by the CCP have the authority to impose discipline on the risk management of the CCP? In the absence of such discipline, they bear risk without meaningful control.

[32] Cecchetti and Schoenholtz (2018c) discuss the reforms to the resolution regimes for systemic intermediaries. For a detailed analysis of CCP resolution, see Duffie (2015).

[33] For a description of the usual waterfall, see Cecchetti and Schoenholtz (2017).

[34] Based on Table 23.1 for LCH, the margin, prepaid guarantee fund, and CCP's own contribution are 0.06 percent, 0.004 percent and 0.00003 percent of notional gross open interest, respectively. The ex post assessment can supplement these prepositioned resources. We cannot calculate these ratios for the Chicago Mercantile Exchange (CME), which does not report the value of open interest.

[35] Clearly, the assessment acts as a transmission mechanism and amplifier of financial stress elsewhere in the system. We return to this shortly. Faruqui et al. (2018) provide a detailed discussion of how the waterfall functions.

[36] Note that assessments depend on each CCP's individual rules. More generally, waterfall design can vary widely among CCPs, with important consequences for the allocation of losses between the CCP and its CMs, and for the resilience of the CCP. For example, a larger guarantee fund relative to margin would increase the CCP's resilience and lower systemic losses but increase the costs to the CMs. See Ghamami et al. (2019).

Table 23.1. Margin and default resources (end September 2018, billions of US dollars)

	CME	LCH
Gross margin posted	$122.89	$150.17
CCP's own capital	0.25	0.08
Prepaid guarantee fund	7.33	10.73
Commitment/further assessment	11.19	10.73
Open interest	NA	251,797.02
Gross notional outstanding	NA	332,500.00

Source: CPMI-IOSCO quarterly disclosures:
www.cmegroup.com/clearing/cpmi-iosco-reporting.html,
www.lch.com/resources/rules-and-regulations/ccp-disclosures, and
www.lch.com/services/swapclear/volumes; and authors' calculations.
Note that the CME does not disclose the value of either open interest or
gross notional amount outstanding, whereas LCH reports both.

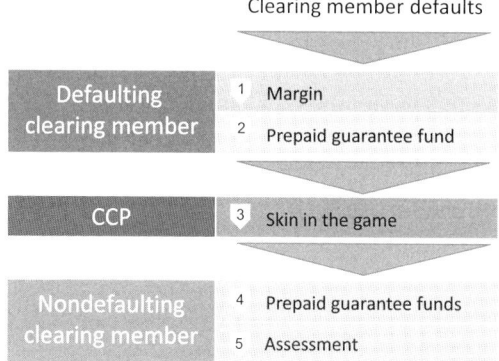

Figure 23.3 Typical CCP default waterfall.
Source: Office of Financial Research (2017).

interest, whereas own capital, the prepaid guarantee fund, and potential assessments sum to less than 1 bps.[37]

As mentioned previously, CCPs also face liquidity risk. Following a CM default, the CCP needs cash to make timely payments to settle contracts that it has guaranteed. Raising cash could necessitate the sale of non-cash collateral in potentially stressful circumstances. Indeed, CCPs may encounter difficulty in selling securities, and the attempt to do so could aggravate market stress.

One way of mitigating liquidity risks is to expand central clearing. Including asset managers and other buy-side firms in the CCP's committed liquidity arrangements (particularly

---

[37] Although open interest likely overstates CCP exposure (especially for short-maturity instruments), these ratios still strike us as small. The data are available at www.cmegroup.com/clearing/cpmi-iosco-reporting .html and www.lch.com/resources/rules-and-regulations/ccp-disclosures. See Table 6.1 for margin and Table 18.3 for the percentage posted by the five largest clearing members.

those firms that are less reliant on short-term liquidity for their own business) could increase and diversify the CCP's sources of liquidity. In 2015, for example, the Options Clearing Corporation and eSecLending established a prefunded, committed repurchase facility with the CalPers pension fund.

To provide liquidity to solvent CCPs under stress, some central banks offer backstops.[38] In the United States, the Federal Reserve can establish deposit accounts at the Fed for the eight designated financial-market utilities (DFMUs).[39] The Fed, with a majority vote of the Board and in consultation with the Treasury Secretary, also can lend to solvent DFMUs in times of market stress (what are termed "unusual and exigent circumstances"). Whether these arrangements are sufficient is far from clear.[40]

To assess the liquidity vulnerabilities of CCPs, stress tests can incorporate critical liquidity risks. For example, an individual CCP test would assess whether it is able to convert collateral into cash in the event of the default of the top one or two CMs under stressed market conditions. A collective test, potentially involving the risk of fire sales and liquidity runs, might identify the cumulative size and timing of the market transactions that a group of CCPs would undertake to meet its liquidity needs under a common stress scenario.[41]

4.2 CCP Stress Testing

Both CCPs and their supervisors conduct regular stress tests. CCP managers need to know whether their books will remain matched following a shock, as well as where stresses might emerge. Supervisors look to an official internationally agreed-upon framework for supervisory stress testing of CCPs. This framework provides for governance arrangements and disclosure standards, as well as protocols for the development of stress scenarios and metrics for evaluation.[42]

Authorities have translated these standards into action. For example, the ESMA and the CFTC have both conducted supervisory stress tests. The purpose of these is to assess the ability of the CCP to survive a shock with the resources it has on hand, without recourse to further assessments. Their first tests, described in Box 23.1, suggest that CCPs were resilient at the time of the tests.

It is surely encouraging that authorities are performing periodic stress tests on CCPs. But as is the case with bank supervisory stress tests, they are costly for both the financial

---

[38] The central banking community decided in 2012 to ensure that there are no technical obstacles in the way of providing emergency liquidity assistance to CCPs that are distressed but fundamentally sound. See Appendix II of Financial Stability Board (2012, p. 48).

[39] See www.federalreserve.gov/paymentsystems/designated_fmu_about.htm.

[40] To take one obvious case where matters remain unsettled, imagine a circumstance in which a US-based CCP would require liquidity in a currency other than dollars. Without access to the appropriate central bank, it is unclear how such a problem could be resolved. Conversely, a CCP based outside the United States that needs to liquidate dollar collateral in a crisis may need access to a central bank that is able and willing to provide dollar funding against high-quality collateral.

[41] See Powell (2017) for a discussion of liquidity risks and policy implications, and see Anderson et al. (2018) for a useful taxonomy of micro and macro liquidity risks in the context of stress tests.

[42] See Committee on Payment and Market Infrastructure (CPMI) and the Technical Committee of the International Organization of Securities Commissions (IOSCO, 2012).

institutions and the authorities. One reason is that they require vast amounts of granular, detailed, and timely information. Furthermore, the tests are quite time consuming, taking months to complete. Partly as a result, testing is infrequent compared with the rapid changes in market and institutional conditions that can affect the vulnerability of CCP networks, and the tests are typically published several months after their reference date. The CFTC conducted new tests in the spring of 2019, and ESMA has published the methodology for new tests to be completed in Q2 2020. Both include reverse stress tests that align with a goal of this chapter: to assess the impact on the financial system in the event that CCPs' prefunded resources run out. And the timing of both is welcome; as we noted earlier, infrequent tests may miss material changes in system health.

Box 23.1   ESMA and CFTC stress tests

ESMA's (2016) first EU-wide stress test exercise assessed the resilience of 17 CCPs (including all authorized EU CCPs) for dates in the final quarter of 2014. The test calculated the counterparty credit risk that EU CCPs would face if multiple CMs default in the face of simultaneous market-price shocks. It did not consider liquidity and other risks. The goal was to determine whether extreme but plausible circumstances would produce losses that exceed the prefunded or unfunded resources of the CCPs.[a]

The CFTC (2016) stress test in 2016 assessed the ability of each of five CCPs to satisfy the resilience standard under 11 extreme but plausible scenarios depicting stressed market conditions. The exercise focused solely on credit risk arising from scenarios in futures, options on futures, and cleared swaps. It included contracts based on financial products and contracts based on physical commodities. To incorporate interconnectedness, the tests focused on firms that serve as CMs at more than one CCP from a universe of the 15 largest CMs at each CCP. Reflecting the overlap in memberships among the CCPs, the exercise included 23 CMs. CFTC staff analyzed both the house accounts and the customer accounts of these CMs at each clearinghouse. Because two of the CCPs have separate guaranty funds for some asset classes, the exercise covered eight different guaranty funds.[b]

The tests revealed that under a number of severely adverse scenarios and across a wide range of products and instruments, all five CCPs had the financial resources to meet the standard in which the two top-ranked CMs default. Furthermore, under nearly two-thirds of the stress scenarios, clearinghouses could meet a standard in which all of the CMs in the exercise that incurred a loss defaulted. In addition, CM risk was diversified among the scenarios and across the CCPs. No single scenario accounted for more than 19 percent of the worst outcomes, and risk was not concentrated among a few firms. Furthermore, the scenarios did not cause CMs to incur losses at all CCPs.

Following this first round of CCP supervisory stress testing, both ESMA (2018a,b) and the CFTC (2017a,b,c) undertook tests that addressed liquidity (in addition to credit risk), publishing them in 2017 and 2018. In both tests, all CCPs could meet liquidity needs, but EU CCPs would have required external resources to do so.

Finally, the CFTC (2019) conducted a third set of tests based on data for 2018 and published in 2019. This two-part report describes the results of a reverse stress test of two large CCPs' resources and an analysis of stressed liquidation costs. The reverse stress test results indicate that it would take a 1-day scenario that exceeded the experience since 2008 to exhaust all prefunded resources. Of course, market shocks are unlikely to last only a day. The liquidation test results indicate that prefunded resources could cover market losses and twice the historically observed liquidation costs in extreme but plausible scenarios. And ESMA (2019) has published extensive details on the methodology for its third round of tests to be published in Q2 2020. They will cover all 16 EU-authorized CCPs (including the three UK CCPs, unless a no-deal Brexit takes place). The new tests will assess resilience under a combination of market price shocks and clearing-member defaults for credit and liquidity; assess liquidation costs derived from concentrated positions; like CFTC, include reverse stress tests; and conduct other analyses aimed at assessing concentration risks and spillovers.

[a] ESMA employed an array of scenarios: CM defaults, market-price scenarios developed from history and models, and reverse stress tests. The default scenarios assume the default of each CCP's top-two CMs ranked by exposure. Furthermore, the test assumed cross-default: if a CM defaults, it does so in all CCPs in which it is a member. That is, the top-two CMs in any CCP default across all CCPs. Combined with the worst-case market-price scenarios, this creates an extremely severe test (and in ESMA's view, results in an implausibly high number of entities simultaneously defaulting at EU-wide and CCP levels). The results in these scenarios trigger CCP assessments for resources that were not prefunded and still leave a small residual of uncovered losses (less than €100 million).

[b] The CFTC methodology involves assuming a large move in prices; calculating the resulting variation margin requirement; checking that against the sum of initial margin, financial contributions of CCPs, and guaranty funds; and determining whether available resources were adequate without resorting to further assessments. The scenarios were extreme, representing the largest price moves of the past three decades, a 50 percent increase in the implied volatility of all options on the futures contracts, and a significant increase in cross-asset correlations in the exercise.

5 Complementing Supervisory Stress Tests Using SRISK

In Section 3 of this chapter, we discussed the use of market information to construct a high-frequency measure of the contribution to systemic risk of any publicly traded financial institution. Here we apply a related method to gauge the vulnerability of the financial system to an assessment by a CCP on its CMs. That is, we focus on the impact on the financial system in the event that a CCP's prefunded loss-absorbing resources are depleted, either partially or completely.

This matters for the simple reason that CMs are subject to assessments to replenish a depleted prepaid guarantee fund. In the 2018 Nasdaq case, where the guarantee fund was partially depleted, assessments occurred relatively quickly—within days of the loss. Looking back at Figure 23.3, this means that CMs are liable to a capital call at the fourth step of the waterfall as well as at the final step.[43] In addition, if a large aggregate shock depletes capital levels throughout the system, the potential for capital calls to trigger further stress is high.

---

[43] Although there are obviously examples of the first of these, the second is extremely rare. In their discussion of the three known CCP failures, Faruqui et al. (2018) suggest that the closest historical example to the case in which a CCP drew on assessment powers of nondefaulting members was the failure of the Hong Kong Futures Guarantee Corporation in 1987.

We estimate systemic vulnerability arising from a CCP assessment as the weighted average of the SRISK of the CMs. Recall that SRISK is a market-based measure of the capital shortfall (relative to a benchmark) of a financial institution in the event of a large decline in global equity prices over a 6-month period. For our illustration, we use the NYU Stern Volatility Lab's default definition of a large decline: a 40 percent drop. A simple way to think about SRISK is that the higher an institution's leverage and the bigger its exposure to the aggregate market when market prices plunge (the higher its *conditional* beta), the more sensitive its capitalization is to a severe event. Furthermore, the bigger the weighted-average capital shortfall, the bigger the stress created for the financial system as a whole by a CCP call on its CMs' capital.

Ideally, we would measure the CCP's potential for transmission and amplification by weighting CM SRISK based on the size of its potential assessment. Unfortunately, that information is not publicly available.[44] In lieu of that, we use two types of information: measures of the gross notional derivatives exposure for a CM's parent and information on a CM's client margin posted at the CCP. The first alternative, available for 2017, comes primarily from the Basel Committee on Banking Supervision's disclosure of the high-level indicators used to compute capital surcharges for global systemically important banks (GSIBs).[45] The second comes directly from the CCP and is only available for the CME.[46] Gross notional derivatives exposure almost surely overstates the counterparty risk of a CCP.[47] The client margin weightings are likely closer to the ideal measures based on potential assessment, but they are imperfect primarily because they exclude the CM's own exposure.

We construct an SRISK index for both the CME and LCH using weights computed from the gross notional derivatives exposures of their CMs. For the CME, we have information for 32 of the 65 CMs with independent parents.[48] LCH has 68 firms with independent parents; of these, we have data on 51. Importantly, 28 of the firms (or their subsidiaries) are CMs of both CCPs. This large overlap is unsurprising because CMs frequently use these CCPs to clear different products: CME clears futures and swaps based on a variety of financial instruments and commodities, whereas LCH mainly clears interest rate swaps. Figure 23.4 plots SRISK indexes for the CME and LCH using these weights.[49] In each case, the index is set to 100 at its peak (February 2009). Two conclusions emerge from this exercise. First, for both CCPs, largely as a result of the improved market capitalization of the CMs, the weighted SRISK index is less than half of what it was a decade ago. Second, reflecting the large overlap in CMs, the SRISK indexes are nearly identical: the correlation is 0.998.

---

[44] Perhaps surprisingly, CCP clearing members generally do not know each other's assessment obligations.

[45] See www.bis.org/bcbs/gsib/gsib_assessment_samples.htm. For CMs too small to be in the Basel Committee's sample, we took information from individual banks' annual Pillar III disclosures.

[46] We draw these from the Financial Industry Association's reporting of data from the CFTC. See https://fia.org/fcm-comparison-table.

[47] See Haynes et al. (2018).

[48] Although we know neither the individual shares nor the collective share of total activity at the CME accounted for by the 32 firms for which we have data, we do know that these firms account for more than 90 percent of posted customer margin.

[49] We base all of these calculations on the presumption that a group will not choose to allow a CM subsidiary to default if there are adequate resources at the group level. Although such a default may be legally possible in some jurisdictions, we suspect that the reputational consequences would be sufficiently dire that a firm would only take this action if the alternative were holding company insolvency.

Figure 23.4 LCH and CME SRISK index (February 2009 = 100), January
2004—November 2018.
Note: The plot shows the weighted sum of SRISK of the publicly traded parent firms of CMs
at the CME and LCH using weights based on 2017 gross notional derivatives exposure.
Source: Authors' calculations based on data from the NYU Stern Volatility Lab, Basel
Committee on Banking Supervision, and various firm disclosures.

The immediate implication is that when the system is vulnerable to a CME call, it is also
vulnerable to an LCH call (and vice versa), and that a capital call from one transmits stress
to the other.

In Figure 23.5, we disaggregate the CME SRISK index by region. Here we use weights
based on posted client margin, which are probably a better proxy for the CCP's assessments.[50] This exercise helps us to understand how a CCP can transmit and amplify a shock
across borders. The striking result here is that the portion of the CME's vulnerability emanating from Europe (the light-gray-shaded area) ranges from 30 to 50 percent of the total,
with the most recent reading near the high end of the range. Although the CME surely

---

[50] The CME's CPMI–IOSCO disclosure reports that between 75 and 80 percent of margin is posted by
customers of the CMs, rather than the CMs themselves (see CME, 2018, Table 6.1). For the SRISK indexes
that we construct, the change in weighting from gross notional outstanding to client margin makes very little
difference because the correlation between the total in Figure 23.5 and the CME index (black line) in
Figure 23.4 is 0.993.

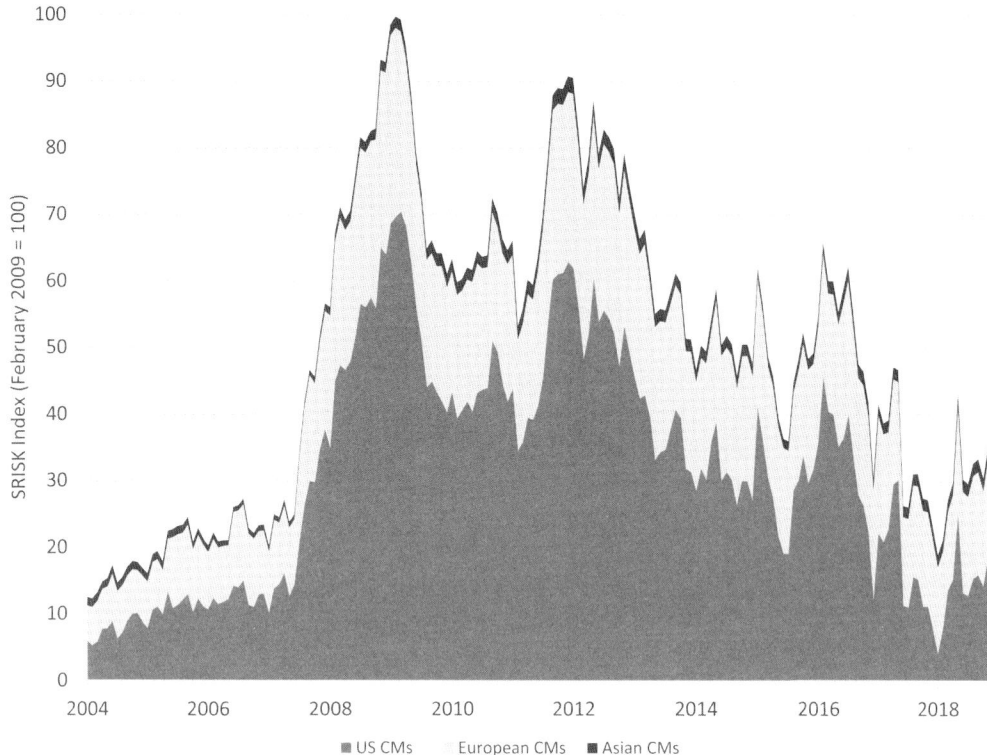

Figure 23.5 CME SRISK by region (February 2009 total = 100), January 2004–November 2018.
Note: The index of client-margin-weighted shares of CME SRISK is decomposed by world region. The fraction accounted for by Canadian clearing members is never more than 0.6 percent of the total, so it is included in the US total.
Source: Authors' calculations based on data from the NYU Stern Volatility Lab and the Futures Industry Association (FIA) (https://fia.org/fcm-comparison-table).

remains exposed to domestic shocks, through capital assessments, it can transmit shocks from defaulting European members to other CMs inside and outside the United States.

A word of caution is in order. We do not intend our CCP SRISK index as a tool for dynamic policy implementation. If policymakers were to use heightened SRISK as a trigger for altering margin or collateral requirements, their response and the effect would be pro-cyclical. Instead, we interpret the large historical variation in the CCP SRISK index as an indicator of the need for additional steady-state buffers. The fact that the default resources appear small relative to indicators of exposure (see Table 23.1) raises the question of whether the CCP's capacity for loss absorption—including its margin, capital, and prepaid guarantee components—is adequate.[51]

---

[51] The debate over the adequacy of CCP resources and who will contribute to them is far from resolved. See, for example, Allianz Global Investors et al. (2019).

6 Conclusion

Although stress testing of banks has a long history, testing of nonbanks is far less common. The empirical application of network analysis to financial systems also is relatively new and remains limited to well-defined networks where data availability significantly exceeds the usual norms.

In this chapter, we highlight the benefits of mapping financial networks to identify sources of vulnerability and of applying customized stress tests to these mapped networks to quantify those vulnerabilities and track them over time. We use CCPs as an example, both because regulators have made some progress in mapping these networks and because CCPs now concentrate risk in ways that can transmit and amplify stress both domestically and across borders. As a result, they are arguably the most important nodes in modern financial networks.

We emphasize the importance of using high-frequency market-based indicators as tools for informing and complementing less frequent, but comprehensive supervisory tests. As an example, we show how to use to SRISK, a measure of capital shortfall, to see how CCP assessments of their CMs can make the financial system more vulnerable to cross-border activity in derivatives markets.

Given the interest of financial regulators and the increased effort to map and monitor key global financial networks, we expect significant further progress in the stress testing of networks that, like CCPs, are amenable to mapping. In the case of CCPs, regulators can utilize these tests to assess margin requirements, the adequacy of prepositioned guarantee resources, and access to liquidity. They can further enhance the impact of their stress tests by making the process and results increasingly transparent to CCPs, CMs, and their stakeholders. So long as CCP governance allows, such transparency would allow healthy CMs to impose greater market discipline on the CCPs with which they transact business.

We view such network-based, customized stress tests as a promising means to assess the vulnerability of the global financial system to shocks originating in key network nodes that lack short-run substitutes.

References

Acemoğlu, Daron, Asuman Ozdaglar, and Alireza Tahbaz-Salehi (2015, February), "Systemic risk and stability in financial networks," *American Economic Review*, 105(2), 564–608.

Alexandrova-Kabadjova, Biliana, Evangelos Benos, Jo Braithwaite, Jorge Cruz Lopez, Ronald Heijmans, Mark Manning, David Murphy, and Francisco Rivadeneyra (2018), "FMIC 2 special issue introduction: A policy view on developments in the field of financial market infrastructures," *Journal of Financial Market Infrastructures*, 6(2/3), 1–20.

Allen, Franklin, and Douglas Gale (2000, February), "Financial contagion," *Journal of Political Economy*, 108(1), 1–33.

Allianz Global Investors et al. (2019, October 24), "A path forward for CCP resilience, recovery, and resolution," available at www.jpmorgan.com/solutions/cib/markets/a-path-forward-for-ccp-resilience-recovery-and-resolution.

Anand, Kartik, Iman van Lelyveld, Ádám Banai, Soeren Friedrich, Rodney Garratt, Grzegorz Hałaj, Jose Fique, Ib Hansen, Serafín Martínez Jaramillo, Hwayun Lee, José Luis Molina-Borboa, Stefano Nobili, Sriram Rajan, Dilyara Salakhova, Thiago Christiano Silva, Laura Silvestri, and Sergio Rubens Stancato de Silva (2018, April), "The missing links: A global study on uncovering financial network structures from partial data," *Journal of Financial Stability*, 107–119.

Anderson, Edward, Fernando Cerezetti, and Mark Manning (2018, December), "Supervisory stress testing for CCPs: A macro-prudential, two-tier approach," Finance and Economics Discussion Series 2018–082, Board of Governors of the Federal Reserve System.

Basel Committee on Banking Supervision, Committee on Payments and Market Infrastructures, Financial Stability Board, and International Organization of Securities Commissions (2018, August 9), "Analysis of central clearing interdependencies," available at www.fsb.org/2017/07/analysis-of-central-clearing-interdependencies/.

Bernanke, Ben (2011, April 4), "Clearinghouses, financial stability, and financial reform," Presentation at the 2011 Financial Markets Conference, Stone Mountain, GA.

Bell, Sarah, and Henry Holden (2018, December), "Two defaults at CCPs, 10 years apart," *BIS Quarterly Review*.

Brownlees, Christian T., and Robert F. Engle (2017, January), "SRISK: A conditional capital shortfall measure of systemic risk," *Review of Financial Studies*, 30(1), 48–79.

Cecchetti, Stephen G., and Kermit L. Schoenholtz (2017, October 17), "Resolution regimes for central clearing parties," available at www.moneyandbanking.com.

Brownlees, Christian T., and Robert F. Engle (2018a, January 22), "Ensuring stress tests remain effective," available at www.moneyandbanking.com.

Brownlees, Christian T., and Robert F. Engle (2018b, July 16), "Cyber instability," available at www.moneyandbanking.com.

Brownlees, Christian T., and Robert F. Engle (2018c, December 3), "E pluribus unum: Single vs. multiple point of entry resolution," available at www.moneyandbanking.com.

Cecchetti, Stephen G., and Paul M. W. Tucker (2015, November), "Is there macroprudential policy without international cooperation?" in R. Glick and M. Spiegel (eds.), *Policy challenges in a diverging global economy, proceedings of the Asia Pacific Policy Conference*, Federal Reserve Bank of San Francisco.

Chicago Mercantile Exchange (2018), "CPMI-IOSCO quantitative disclosures, CME_DataFile_2018Q3," available at www.cmegroup.com/clearing/cpmi-iosco-reporting.html.

Clark, Tim P., and Lisa H. Ryu (2015, June 24), "CCAR and stress testing as complementary supervisory tools," Working Paper Federal Reserve Board, June 24, 2015.

Committee on Payment and Settlement Systems and Technical Committee of the International Organization of Securities Commissions (2012, April), "Principles for financial market infrastructures (PFMI)," available at www.bis.org/cpmi/info_pfmi.htm.

Commodity Futures Trading Commission (2016, November), "Supervisory stress tests of clearing-houses," available at www.cftc.gov/sites/default/files/idc/groups/public/@newsroom/documents/file/cftcstresstest111516.pdf.

Commodity Futures Trading Commission (2017a), "Proposed amendments to the swap data access Provisions of Part 49 and certain other matters," RIN Number 3038-AE44, *Federal Register*, 82(15).

Commodity Futures Trading Commission (2017b, July 10), "Roadmap to Achieve High Quality Swaps Data," available at www.cftc.gov/sites/default/files/idc/groups/public/@newsroom/documents/file/dmo_swapdataplan071017.pdf.

Commodity Futures Trading Commission (2017b, July 10), "Evaluation of clearinghouse liquidity," available at www.cftc.gov/sites/default/files/idc/groups/public/@newsroom/documents/file/dcr_ecl1017.pdf.

Commodity Futures Trading Commission (2019, May 1), "CCP supervisory stress tests: Reverse stress test and liquidation stress test," available at www.cftc.gov/system/files?file=2019/05/02/cftcstresstest042019.pdf.

Counterparty Risk Management Policy Group (2008, August 6), "Containing systemic risk: The road to reform," available at www.crmpolicygroup.org/.

Covi, Giovanni, Mehmet Ziya Gorpe, and Christoffer Kok (2019), "CoMap: Mapping contagion in the euro area banking sector," available at www.ecb.europa.eu/pub/pdf/scpwps/ecb.wp2224.en.pdf.

Duffie, Darrell (2015), "Resolution of failing central counterparties," in Thomas Jackson, Kenneth E. Scott, and John B. Taylor (eds.), *Making failure feasible*, Hoover Institution.

Duffie, Darrell, Martin Scheicher, and Guillaume Vuillemey (2015, May), "Central clearing and collateral demand, *Journal of Financial Economics*, 116(2), 236–256.

Duffie, Darrell, and Haoxiang Zhu (2011, December), "Does a central clearing counterparty reduce counterparty risk?" *Review of Asset Pricing Studies*, 1(1), 74–95.

Elliott, Matthew, Benjamin Golub, and Matthew O. Jackson (2014, October), "Financial networks and contagion," *American Economic Review*, 104(10), 3115–3153.

European Securities and Markets Authority (2016, April 29), "EU-wide CCP stress test 2015," available at www.esma.europa.eu/sites/default/files/library/2016-658_ccp_stress_test_report_2015.pdf.

European Securities and Markets Authority (2018a, February 2), "EU-wide CCP stress test 2017," available at www.esma.europa.eu/sites/default/files/library/esma70-151-1154_eu-wide_ccp_stress_test_2017_report.pdf.

European Securities and Markets Authority (2018b, February 2), "Questions and answers (Q&A) on ESMA's EU-wide stress tests for CCPs," available at www.esma.europa.eu/sites/default/files/library/2016-665_qa_ccp_stress_test.pdf.

European Securities and Markets Authority (2019, April 3), "Methodological framework: 3rd EU-wide central counterparty (CCP) stress test exercise," available at www.esma.europa.eu/sites/default/files/library/esma70-151-2198_framework_for_the_2019_ccp_st_exercise.pdf.

Faruqui, Uman, Wenqian Huang, and Elöd Takáts (2018, December), "Clearing risks in OTC derivatives markets: the CCP-bank nexus," *BIS Quarterly Review*.

Financial Stability Board (2012), "OTC derivatives market reforms—third progress report on implementation," available at www.fsb.org/2012/06/third-implementation-progress-report-on-over-the-counter-otc-derivatives-market-reforms/.

G20 Leaders (2009), "G20 Leaders statement: The Pittsburgh Summit," available at www.treasury.gov/resource-center/international/g7-g20/Documents/pittsburgh_summit_leaders_statement_250909.pdf.

Ghamami, Samim, Mark Paddrik, and Simpson Zhang (2019, November 18), "Central counterparty default waterfalls and systemic loss," Working Paper, available at www.financialresearch.gov/working-papers/2020/06/18/central-counterparty-default-waterfalls-and-systemic-loss/.

Glasserman, Paul, and H. Peyton Young (2015, January), "How likely is contagion in financial networks?" *Journal of Banking and Finance*, 50(C), 383–399.

Glasserman, Paul, and H. Peyton Young (2016, September), "Contagion in financial networks," *Journal of Economic Literature*, 54(3), 779–831.

Gorton, Gary (1984), "Private Clearinghouses and the origins of central banking," available at https://fraser.stlouisfed.org/title/business-review-federal-reserve-bank-philadelphia-5580/january-february-1984-557614/private-clearinghouses-origins-central-banking-523625

Haynes, Richard, John Roberts, Rajiv Sharma, and Bruce Tuckman (2018, January), "Introducing ENNs: A measure of the size of interest rate swap markets," available at www.cftc.gov/sites/default/files/idc/groups/public/@economicanalysis/documents/file/oce_enns0118.pdf.

Kara, Gazi, Mary Tian, and Margaret Yellen (2015, July 31), "Taxonomy of studies on interconnectedness," FEDS Notes, available at www.federalreserve.gov/econresdata/notes/feds-notes/2015/taxonomy-of-studies-on-interconnectedness-20150731.html.

Liang, Nellie (2018, February 2), "Well-designed stress test scenarios are important for financial stability," Working Paper, Brookings Institution.

Office of Financial Research (2017, March), "New public disclosures shed light on central counterparties," Viewpoint Paper 17–02.

Paddrik, Mark, Sriram Rajan, and H. Peyton Young (2016, December), "Contagion in the CDS market," Office of Financial Research Working Paper 16–12.

Paddrik, Mark, and H. Peyton Young (2017, November), "How safe are central counterparties in derivatives markets?" Office of Financial Research Working Paper 17–06.

Powell, Jerome H. (2017, June), ''Central Clearing and Liquidity'' Presentation, at the Federal Reserve Bank of Chicago Symposium on Central Clearing, Chicago, IL.

Sarin, Natasha, and Lawrence H. Summers (2016), "Understanding bank risk through market measures," *Brookings Papers on Economic Activity*, 57–127.

Systemic Risk Council (2019, March 18), "Note to the Financial Stability Board on CCP resolution."

Tompaidis, Stathis (2018), "Measuring systemwide resilience of central counterparties," *Journal of Financial Market Infrastructures*, 6(4), 41–54.

Tuckman, Bruce (2015, September 29), "In defense of derivatives: From beer to the financial crisis," Cato Institute Policy Analysis No. 781.

Yellen, Janet L. (2013, January 4), "Interconnectedness and systemic risk: Lessons from the financial crisis and policy implications," Speech at the American Economic Association/American Finance Association Joint Luncheon, San Diego, CA.

Part V

Macroprudential Stress Testing

24

Enhancing Stress Tests by Adding Macroprudential Elements[*]

William F. Bassett and David E. Rappoport

1 Introduction

Since the Global Financial Crisis (GFC), many jurisdictions have begun to conduct stress tests on a regular basis (Basel Committee on Banking Supervision [BCBS, 2017]). In advanced economies, these tests are increasingly being used to inform macroprudential policy, such as increasing the resilience of the financial system during expansions and better understanding the financial spillovers and amplification dynamics within the financial sector or between the real and financial sectors. In this chapter, we discuss several macroprudential elements that can be included in solvency stress tests to better capture spillovers, as well as the pros and cons of various modeling approaches. To provide a concrete example, we discuss the use of an add-on approach that incorporates reduced-form relationships between the cost of funding and the average capital ratio of the banking sector to simulate a funding shock that dynamically adjusts to changing vulnerabilities in the banking sector and the macroeconomic scenario.

For this chapter, we define a macroprudential policy as one that aims to reduce systemic risk—the risk of widespread disruption to the provision of financial services that can negatively affect the real economy (International Monetary Fund [IMF] et al., 2016). This policy goal is then pursued through three distinct intermediate objectives: (1) increasing the resilience of the financial system to shocks, (2) leaning against the buildup of systemic vulnerabilities over the financial cycle, and (3) limiting structural vulnerabilities that arise from the interconnectedness of intermediaries or the critical role of individual intermediaries.

Not all regulatory stress tests explicitly consider the amplification of shocks from spillovers either within the financial sector or between the financial and real sectors, but most do contain elements that achieve important macroprudential policy goals.[1] Stress

[*] We thank Grace Brang and Candy Martinez for excellent research assistance and are grateful to David Arseneau, Ken Heinecke, Andreas Lehnert, Lisa Ryu, Jason Schmidt, Alex Vardoulakis, Cindy Vojtech, participants at the conference "Rethinking Financial Stability: The FSAP at 20," the editors, and an anonymous referee for useful comments. All errors herein are ours. The views expressed in this chapter are those of the authors and do not necessarily represent those of Federal Reserve Board of Governors or anyone in the Federal Reserve System.

[1] One exemption is the Bank of England modeling of funding and fire-sale spillovers. In addition, some academic papers have proposed frameworks that incorporate spillovers. For example, He and Krishnamurthy (2019) develop a model with financial spillovers that amplify the effect of initial shocks and can be used to compute the probability of reaching a systemic risk state. We review other academic proposals to model fire sales in Section 5.

tests most clearly build resilience in the banking sector against the materialization of severe macroeconomic or financial shocks by requiring banks to limit capital distributions or raise more capital if the results show weaknesses. However, that resilience may be insufficient if the scope of the tests is too microprudential. Stress tests that include macroprudential elements may also be more effective in leaning against the buildup of systemic vulnerabilities over the financial cycle (Adrian et al., 2014; see also Chapter 4 in this volume). The clearest example of such a leaning-against-the-wind policy would be taking steps to make the stress test more severe in the expansionary phase and less severe in the contractionary phase of the financial cycle.

Indeed, the considerable severity of the scenarios used in most jurisdictions' stress tests implicitly incorporates estimates of the damage to financial markets and the economy caused by a seizure of funding markets or the distress of large financial institutions.[2] Nonetheless, given the potential for financial spillovers to disrupt the financial system, the macroprudential goals of stress tests could be significantly enhanced by incorporating elements that can explicitly project spillovers conditional on current vulnerabilities and a given scenario (Haldane, 2009; BCBS, 2009; IMF, 2012; Baudino et al., 2018; see also Chapter 4 in this volume). Endogenizing the reaction to the stress through reduced-form or structural models can also enhance financial stability monitoring and risk identification by making stress tests a tool to analyze more primitive shocks—such as an energy price spike—and creating a better understanding of how those shocks amplify vulnerabilities and generate financial spillovers.

The important macroprudential objectives achieved by the aforementioned elements already in place raise the question of how to incorporate additional macroprudential goals in existing frameworks. Some macroprudential elements can be incorporated into existing stress test frameworks, either as add-ons or by integrating them as part of the suite of scenario variables, with the associated losses projected using either reduced-form or structural models. However, other combinations of macroprudential elements and existing stress test frameworks would require supervisors to revisit existing assumptions, develop new frameworks, or collect additional data to incorporate models of financial spillovers.

We provide two examples of how to incorporate funding of spillovers into stress tests through an add-on approach. The first example uses a simple exogenous shock to the cost and availability of short-term wholesale funding (STWF), which we call a "prudential shock." One feature of this add-on is that the funding stress affects different banks than the macroeconomic stress considered in the Dodd–Frank Act Stress Test (DFAST). The second example uses a reduced-form model of spillovers between the health of participating banks and the cost of STWF to create a dynamic shock that responds to the evolution of bank capital and growth in gross domestic product (GDP) over the projection horizon. Thus, funding stress is greater when the starting capital ratios are lower or losses from the macroeconomic shock are larger.

---

[2] These mechanisms seem to have fueled, in large part, the sharp reduction in credit availability and the depth of the GFC (see, e.g., Kashyap et al., 2008). In fact, theories of macroprudential regulation developed since the GFC have focused on the interactions within the financial system and the largest financial firms and the costs they impose on the broader economy and other financial institutions when those largest firms are distressed (Brunnermeier et al., 2009; Hanson et al., 2011).

We further discuss other models of spillovers, such as fire sales and the feedback between the macroeconomy and the financial system. Next, we discuss vulnerabilities where the use of a structural model or highly granular data seems to be necessary. For instance, reduced-form relationships are poor approximations of exposures arising from bilateral interactions, such as counterparty default.

Finally, we argue that the benefits of promoting new macroprudential goals need to outweigh the costs of increased complexity or the potential to inhibit the goals that are promoted by existing macroprudential elements. Moreover, it is important to remember that neither reduced-form nor structural models can accommodate all of the macroprudential elements discussed herein at the same time. For instance, a stress test that requires banks to meet a macroprudential goal of maintaining credit supply by growing the size of their balance sheet during the test would be inconsistent with one in which fire-sale dynamics also play an important role. Therefore, jurisdictions should remain flexible in their uses of macroprudential stress testing.

The rest of the chapter is organized as follows. In Section 2, we review macroprudential elements currently in place in selected jurisdictions. In Section 3, we discuss how funding stress can be incorporated into existing stress testing frameworks. In Section 4, we describe our approach to incorporating macroprudential funding spillovers into existing frameworks. In Section 5, we discuss how our approach relates to other proposals to incorporate other spillovers. Finally, Section 6 provides some conclusions.

2 Macroprudential Elements Currently in Place

Current stress test frameworks incorporate several macroprudential elements that help to limit systemic risk through some of the three intermediate objectives of macroprudential policy previously mentioned.

First, stress tests have become an important input into prudential capital requirements in many jurisdictions, helping to build economy-wide resilience against the stresses considered in the test. For example, results from the European Banking Authority (EBA) stress tests are used by bank authorities in their Supervisory Review and Evaluation Process (SREP), which assesses capital, among other risk dimensions (EBA, 2020). In 2014, the results of the EBA stress test required approximately 20 banks to issue new capital.[3] Likewise, the Federal Reserve has used the results from its Comprehensive Capital Analysis and Review (CCAR) to object to a firm's capital plan if a firm fails to demonstrate its ability to maintain the required minimum capital ratios over the planning horizon or if it finds the bank's capital-planning processes to be unreliable.[4] An objection by the Federal Reserve usually has resulted in the bank being required to limit its dividends and share repurchases to no more than its previous year's capital distributions.[5]

---

[3] See www.eba.europa.eu/risk-analysis-and-data/eu-wide-stress-testing/2014/results.

[4] See Board of Governors (2019). Additionally, even if the Federal Reserve does not object initially to firms' capital plans, firms generally have been required to request prior approval of a capital distribution if the dollar amount of the capital distribution will exceed the amount described in their capital plan.

[5] In its 2016 report, the US Government Accountability Office (GAO) stated that on a macroprudential level, some firms felt "the stress tests have led to higher capital levels and improved risk management that have contributed to the stability of the financial system." See the GAO Report "Additional Actions Could Help Ensure the Achievements of Stress Test Goals," page 28, available at www.gao.gov/assets/690/681020.pdf.

In addition, some jurisdictions use stress test projections to calibrate macroprudential capital buffers. For example, the Bank of England and Swiss National Bank (SNB) use the projections from their stress tests as an input in their countercyclical capital buffer (CCyB) framework.[6] A different approach is to link stress-loss projections with capital conservation buffers (CCoBs). For instance, the Federal Reserve has requested public comment on a rule that will make CCoBs time varying and risk sensitive by conditioning them on individual bank losses in the test.[7]

Second, the use of common scenarios and the simultaneity of the exercise enable regulators to examine the resiliency of the system to certain shocks and to consider the appropriate policy response. For instance, the IMF's Financial System Assessment Program (FSAP) includes stress tests of banks representing 60 percent or more of total bank assets, and in some jurisdictions FSAPs have also considered stress tests of the insurance and pension funds sector (Jobst et al., 2013; see also Chapter 26 in this volume). Moreover, scenarios can also be tailored to examine the resilience of the system to emerging vulnerabilities, such as the negative interest rates included in the severely adverse scenario of the 2016 DFAST exercise.

Third, increasing the severity of the hypothetical stress scenario when the economy is strong, together with the use of the results in setting prudential capital requirements, can lean against the inherent procyclicality of the financial system (see Chapter 4 in this volume).[8] In this way, when times are good, stress tests can induce banks to build up more capital than they otherwise would in order to increase resilience in a subsequent downturn. The US DFAST and CCAR do this in several ways, most clearly through the calibration of the unemployment rate shock, which is described as typically rising 3 to 5 percentage points over the scenario horizon, to a minimum of 10 percent (12 CFR 252). Hence, with US unemployment falling below 4 percent, CCAR scenarios since 2018 have seen unemployment rise 6 percentage points or more, a full percentage point larger than the upper end of the typical range.[9] Similarly, the Bank of England calibrates the severity of the stress scenario to policymakers' assessment of the state of the financial cycle, as well as the business cycle (Bank of England, 2015).

Fourth, stress tests have been tailored to address the critical role and interconnectedness of the largest financial institutions. From a macroprudential perspective, the failure of a

---

[6] In the UK, the CCyB is set by the Financial Policy Committee (Bank of England, 2015); in Switzerland, it is set by the Federal Council on a proposal by the Swiss National Bank (SNB) after consultation with the Financial Markets Supervisory Authority (IMF, 2019).

[7] See the proposal for the "stress capital buffer" at www.federalreserve.gov/newsevents/pressreleases/bcreg20180410a.htm.

[8] This inherent procyclicality manifests in the banking system in several ways. First, lending standards tend to loosen as the last crisis fades into history (Berger and Udell, 2004). Second, investors' overreliance on models can lead to excessive risk taking in the run-up to financial crises (Haldane, 2009; Daniellson, 2011; see also Chapter 4 in this volume). Third, loss rates in US and EU stress tests have tended to decline as time has passed since the 2007–2009 financial crisis.

[9] However, such features will probably prove insufficient to achieve macroprudential goals, in part because larger increases in unemployment and steeper declines in asset prices become harder to justify as they move further outside the range of historical precedent. In addition, marginal changes in scenario variables (e.g., a 5-percentage-point increase in unemployment versus a 6-percentage-point increase) may not lead to proportional changes in loss estimates.

major financial institution has greater implications for financial stability than the failure of a smaller institution (see Lorenc and Zhang, 2018). The size, interconnectedness, and complexity of large financial institutions make them sources of instability should an adverse situation in financial markets materialize. Thus, stress test scenarios can differentially address the risks posed by the largest institutions, which should help mitigate the additional and unique risks that these types of institutions pose to financial stability. For example, CCAR considers a market shock and the default of the largest counterparty to stress the trading activities and interconnectedness of the largest and most complex participating firms. We come back to the modeling of these financial spillovers in Section 5.

Of course, the efficacy of the previous four elements, which aim at building resilience or leaning against the wind, is limited by the scope of regulatory stress tests within the financial system.[10] This limitation is more important the larger the amount of credit intermediation and liquidity transformation that takes place in less regulated institutions and markets, the so-called "shadow-banking system." However, this limitation can be mitigated by regular surveillance of the entire financial system, which can be used to identify salient risks and spillovers from nonbank intermediaries to banks and the real economy and incorporated into the scenarios used in the bank stress tests.

Fifth, the extent of public disclosure of the results is an often-overlooked element of stress tests that provides macroprudential benefits. Indeed, the public disclosure of the results of the US stress test in 2009 contributed to restoring confidence in markets by providing much-needed transparency during a period of financial uncertainty, and similar effects have been found for the European Comprehensive Assessment and Asset Quality Review (Georgescu et al., 2017). Although there is not consensus on the optimal level of information disclosure, there is consensus that publicly announcing stress test results can promote financial stability in turbulent times.[11] Some contend that the credibility of supervisory stress tests is undermined by the lack of market information in their design (Archarya et al., 2014; see also Chapter 9 in this volume). Yet other observers have pointed to the credibility of US supervisory stress tests as one factor for the relatively strong market performance of US banks amid adverse market events early in 2016.[12] Aspects of tests that have been cited as contributing to market confidence are credible estimates of potential losses (e.g., through

---

[10] Supervisory stress tests tend to focus on the sector, but several jurisdictions have also stressed the activities of nonbank intermediaries, such as central clearing counterparties (CCPs), insurance companies, and asset managers. For example, the FSAP for Japan and Sweden in 2017 included the insurance sector (IMF, 2017a,b), and the FSAP for Belgium in 2018 included the insurance sector and financial conglomerates, comprising asset managers (IMF, 2018). In the Belgium FSAP, insurance companies and banks used the same scenarios, but in other cases, the scenarios applied to different institutions can be different, limiting some macroprudential benefits of the exercise. Additional examples are supervisory stress tests where multiple CCPs are evaluated under common scenarios. For instance, stress tests on CCPs are conducted by the Commodity Futures Trading Commission (CFTC) in the United States (see www.cftc.gov/system/files? file=2019/05/02/cftcstresstest042019.pdf) and by the European Securities and Markets Authority (ESMA) in Europe (see www.esma.europa.eu/press-news/esma-news/esma-publishes-results-second-eu-wide-ccp-stress-test).

[11] See Faria-e-Castro et al. (2017), Goldstein and Leitner (2018), and Chapter 11 in this volume.

[12] See www.reuters.com/article/us-britain-eu-usa-banks-idUSKCN0ZD0OP. Jobst et al. (2013) argue that the credibility of the Supervisory Capital Assessment Program (SCAP) exercise in 2009 was useful to restore market confidence in the US financial system.

the use of independent supervisory models rather than banks' own estimates), transparency regarding individual institutions' projections, and a credible plan for dealing with institutions found to be undercapitalized or insolvent.

The development of independent models by supervisors to calculate each bank's projected losses enhances the efficacy of several of the aforementioned macroprudential elements. But the development and credibility of these models to a large extent depend on the existence of robust data-collection programs designed to support stress tests, such as the national credit registers maintained by many Basel member countries.[13] Likewise, the Federal Reserve has supported its postcrisis capital stress testing program by introducing the FR Y-14 data collection, which provides obligor-level data for securities, business loans greater than $1 million, credit cards, and residential mortgages, as well as finely segmented data for other types of loans.

Nonetheless, blind spots, especially for publicly available data, continue to exist.[14] For instance, detailed and timely data on bank funding sources and relationships are available only for the small number of very large banks that are subject to the liquidity coverage ratio (LCR) and are confidential. In part, these blind spots remain in the United States because of statutory requirements to balance reporting burden with potential benefits, as well as firms' concerns about the potential inadvertent disclosure of proprietary information, such as trading strategies. One way to increase stability while reducing the burden of more adequate reporting could be for banks to be incentivized to make larger investments in data infrastructure. In the following discussion, we also highlight some specific improvements in data collection that would facilitate the implementation of macroprudential elements to capture funding stress.

3 Adding Funding Stress

Evidence from the GFC points to a credit squeeze in STWF markets and liquidity hoarding as key amplification mechanisms within the financial system. Strains in STWF markets intensified starting in the second half of 2007 (Gorton and Metrick, 2012; Acharya et al., 2013). As uncertainty rose from mid-2007 through the early part of 2009, the cost of these funding instruments spiked, and some markets previously thought to be very low risk, for example, auction-rate securities, closed entirely (McConnell and Saretto, 2010). Although banks identified as financially weak faced acute funding constraints, even banks with strong financial positions saw their funding costs increase. The resulting widespread stress created negative spillovers, a generalized pullback by credit providers, and liquidity hoarding.[15]

---

[13] It is worth pointing out that expanded data availability itself constitutes an element of the macroprudential policy toolkit. In fact, Bassett et al. (2011) have argued that a lack of data contributed to the inability of regulators, particularly in US mortgage markets, to understand the buildup of vulnerabilities in the financial system between 2003 and 2008.

[14] The absence of data on derivative counterparties led to the failure of regulators to appreciate the central role that AIG Financial Products Group was playing in the credit default swap (CDS) market for subprime mortgage-backed securities (MSBs) in the run-up to the GFC (Kroszner and Strahan, 2011).

[15] Ashcraft et al. (2008) present evidence of liquidity hoarding in the US interbank market. Similarly, Heider et al. (2015) present evidence of liquidity hoarding in the euro-area interbank market. Finally, Acharya and Merrouche (2013) present evidence of liquidity hoarding by large settlement banks in the UK.

These spillovers are interrelated and can cause negative-feedback loops. The pullback by credit providers incentivizes banks to hoard liquidity and exacerbates the credit squeeze (Diamond and Rajan, 2005; Gale and Yorulmazer, 2013). This is especially the case for STWF, which reprices quickly because sophisticated and risk-averse wholesale credit providers respond swiftly to the first signs of distress. In response to the lessons of the financial crisis, many jurisdictions have adopted the LCR and net stable funding ratio (NSFR), both of which are calibrated to a common stress scenario and designed to allow firms to self-insure against a loss of funding for up to 1 year. Some also have implemented liquidity stress tests, such as the Comprehensive Liquidity Assessment and Review (CLAR) in the United States. Therefore, adding funding stress to solvency stress tests may seem to be duplicative of enhanced liquidity regulation.

However, we argue that the addition of a funding shock to capital stress tests would be complementary because the experience of 2007 and 2008 shows that access to liquidity and capital adequacy are inherently linked. Although the LCR and NSFR account for the potential for short-term funding to run as banks experience losses that erode their capital position, they do not account for the impact that restricted access to such funding would have on the funding costs and hence the profitability of even healthy banks once markets are disrupted. Thus, unless the capital stress test captures these additional costs adequately, it will underestimate the capital shortfall associated with the supervisory scenario. In fact, several jurisdictions have included funding shocks as a driver of stress in solvency stress tests.[16] Moreover, explicitly modeling the negative spillovers and amplification dynamics under stress helps to build resilience against these higher-round effects, improves risk identification, and leans against the buildup of financial vulnerabilities.

The relationship between system-wide capital and the average costs of STWF is most clear during the previous financial crisis. To study the dynamics of these variables, we use information from FR Y-9C regulatory reports for US bank holding companies (BHCs) that participated in DFAST 2018.[17] We summarize the system-wide capital ratios with the unweighted geometric mean of common equity Tier 1 (CET1) capital ratios for those banks. The use of the geometric mean is motivated by the fact that extreme stress at a small number of institutions may precipitate widespread concern about the banking industry that results in contagion to otherwise-healthy institutions.

For the cost of STWF, we consider three different proxies but focus primarily on the TED spread. The TED spread is the difference between the interest rate on 3-month US Treasuries and the interest rate on interbank loans, measured by the 3-month London Interbank Offer

---

[16] The IMF country FSAPs have included funding costs as a driver of losses under stress for France, Germany, Japan, New Zealand, Poland, the UK, and Sweden (Jobst et al., 2013; see also Chapter 26 in this volume). In addition, the Bank of England in its 2019 stress scenario considers an increase in the cost of wholesale funding implemented as an increase of 2 percentage points in the 5-year senior unsecured bond spread.

[17] To be precise, we consider the 34 BHCs for which the Federal Reserve Board published the results of DFAST 2018 and that report nonnegative CET1 ratios in FR Y-9C. For the list of DFAST 2018 banks, see www.federalreserve.gov/newsevents/pressreleases/bcreg20180621a.htm, as opposed to the 38 BHCs that were initially subject to DFAST (see www.federalreserve.gov/newsevents/pressreleases/bcreg20180201a.htm) before the Crapo Banking Bill became law in May 2018 (see www.banking.senate.gov/newsroom/majority/president-signs-crapo-banking-bill-into-law).

Rate (LIBOR) based in US dollars.[18] LIBOR has been a widely used benchmark for funding costs over our sample period, so it is expected to be a reasonable gauge of banks' STWF costs. But an important limitation of LIBOR is that it is based on bank surveys.[19] For robustness, we use the spread between 3-month US Treasuries and the interest rate on 3-month commercial paper (CP) issued by financial firms with high credit ratings.[20] Relative to the TED spread, the CP spread offers the advantage that it is based on market data, but it only represents a single STWF source, as opposed to a broader measure of bank funding.

Finally, we compute a comprehensive cost of STWF using the FR Y-9C regulatory reports, following Bassett et al. (2020). STWF is defined as in the US implementation of the capital surcharge for globally systemically important banks (GSIB rule), which includes (1) secured funding transactions, (2) unsecured wholesale funding, (3) covered asset exchanges, (4) short positions, and (5) brokered deposits (Board of Governors of the Federal Reserve System, 2015). These categories are mapped as closely as possible to data on liability balances and interest expenses reported in FR Y-9C forms,[21] supplemented with CP rate information.[22]

The calculation proceeds as follows. For each bank participating in the 2018 DFAST, the weighted-average effective cost of each STWF source is calculated using the ratio of the relevant interest expense item to the average balance of the corresponding liability for each quarter in which those banks reported on the Y-9C. Second, for each STWF source in each quarter, the distribution of the estimated rates paid is trimmed at the 10th and 90th

---

[18] We use the quarterly average of the daily TEDRATE, retrieved from FRED, Federal Reserve Bank of St. Louis (https://fred.stlouisfed.org/series/TEDRATE).

[19] LIBOR's reputation was undermined by accusations over rigging, and it is currently being phased out; see www.fsb.org/wp-content/uploads/P181219.pdf.

[20] In all our calculations, we use the 3-month Treasury constant maturity rate (DGS3MO), retrieved from FRED, Federal Reserve Bank of St. Louis (https://fred.stlouisfed.org/series/DGS3MO). We use the quarterly average of the 3-month AA financial commercial paper rate (DCPF3M), retrieved from FRED, Federal Reserve Bank of St. Louis (https://fred.stlouisfed.org/series/DCPF3M). Both series are sourced from the Board of Governors of the Federal Reserve System (US).

[21] The exact liability items from FR Y-9C are as follows (codes reported in parentheses): federal funds and repos (HC-K.2), trading liabilities (HC-D.13.a-b), CP (HC-M.14.a), other borrowed money with a remaining maturity of 1 year or less (HC-M.14.b), brokered deposits with a remaining maturity of 1 year or less (HC-E-M.1), uninsured time deposits with a remaining maturity of 1 year or less (HC-E-M.3), and foreign deposits with a remaining maturity of 1 year or less (HC-E-M.4). The exact interest expense and liability items used to compute cost rates are as follows: foreign deposits (HI.2.a.[2] and HC-K.7), federal funds and repos (HI.2.b and HC-K.2), uninsured time deposits (HI.2.a.[1][b] and HI.2.a.[1][a] before Q4 2016, and HC-E.1-2.e), brokered deposits (HI.2.a.[1][a] and HI.2.a.[1][b] before Q4 2016, and HC-E.1-2.d), trading liabilities and other borrowed money (HI.2.c and HC.15-16). Note that interest expenses sometimes correspond to less granular categories than the categories identified as STWF components in the FR Y-9C.

[22] The CP rate corresponds to the average of short-term CP rates, weighted by quarterly issuance, where CP rates are calculated as the quarterly average of daily series. The series, from the Board of Governors of the Federal Reserve System (US) and sourced from FRED, Federal Reserve Bank of St. Louis, are as follows: 1-month AA financial commercial paper rate (DCPF1M), 2-month AA financial commercial paper rate (DCPF2M), and 3-month AA financial commercial paper rate (DCPF3M). The respective weights are as follows: total value of issues, with a maturity between 21 and 40 days used in calculating the AA financial commercial paper rates (FIN2140AAAMT), total value of issues, with a maturity between 41 and 80 days used in calculating the AA financial commercial paper rates (FIN4180AAAMT), and total value of issues, with a maturity greater than 80 days used in calculating the AA financial commercial paper rates (FINGT80AAAMT).

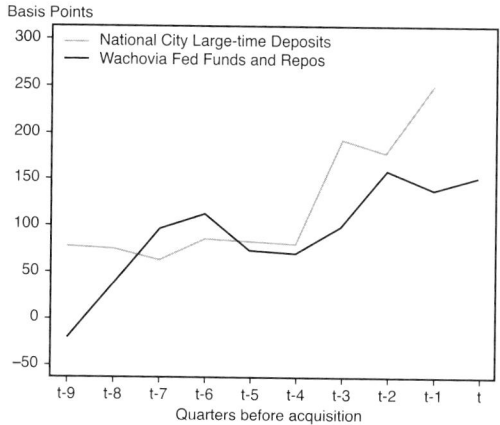

Figure 24.1 Spread of STWF for banks with idiosyncratic stress.
Note: Spread over a group of peer institutions that did not experience acute stress.
Source: Authors' calculations based on Call reports.

percentiles. Then, the average rate for each source-quarter is computed using the trimmed cost rates, weighted by average liability balances. Finally, the time series of average rates for each source are averaged, weighted by liability balances, and expressed at an annual rate. Thus, this measure of STWF costs represents the weighted-average effective rate paid by a representative bank for wholesale funding in that quarter, and the spread is calculated relative to the 3-month Treasury. Relative to the other spreads measuring STWF costs, the aforementioned STWF spread has the advantage that it represents the actual average funding cost paid during the quarter, encompasses a broader set of STWF sources, and corresponds to the set of DFAST 2018 banks that we analyze. However, some of the liabilities that comprise the STWF spread reprice at frequencies lower than a quarter, so it reflects both current and past funding conditions.[23] Therefore, it fluctuates around the more market-sensitive spreads and takes negative values when market interest rates are rising, as was the case in 2006 and 2018.

The individual components of this comprehensive measure can be used to show the relationship between banks' financial health and the cost of STWF at the individual bank level for banks that approached failure and then entered into merger agreements under extreme distress. This was the case for two large banks before the height of the GFC: National City and Wachovia. Figure 24.1 presents the spread of the effective interest rate paid on a representative liability at these two banks for the nine quarters before their failure relative to the cost of that funding source at a sample of banks of similar size.[24] The figure

[23] This issue is particularly problematic with FR Y-9C data because interest expenses are not reported for short-term maturities. Instead, cost rates for STWF sources are imputed from cost rates of similar liabilities for all maturities.

[24] For National City, which was acquired in October 2008 and had assets of $151 billion in Q3 2008, we considered banks with assets between $150 and $500 billion; for Wachovia, which stopped independent operations in September 2008 and had assets of $675 billion in Q3 2008, we considered banks with assets between $300 billion and $1 trillion.

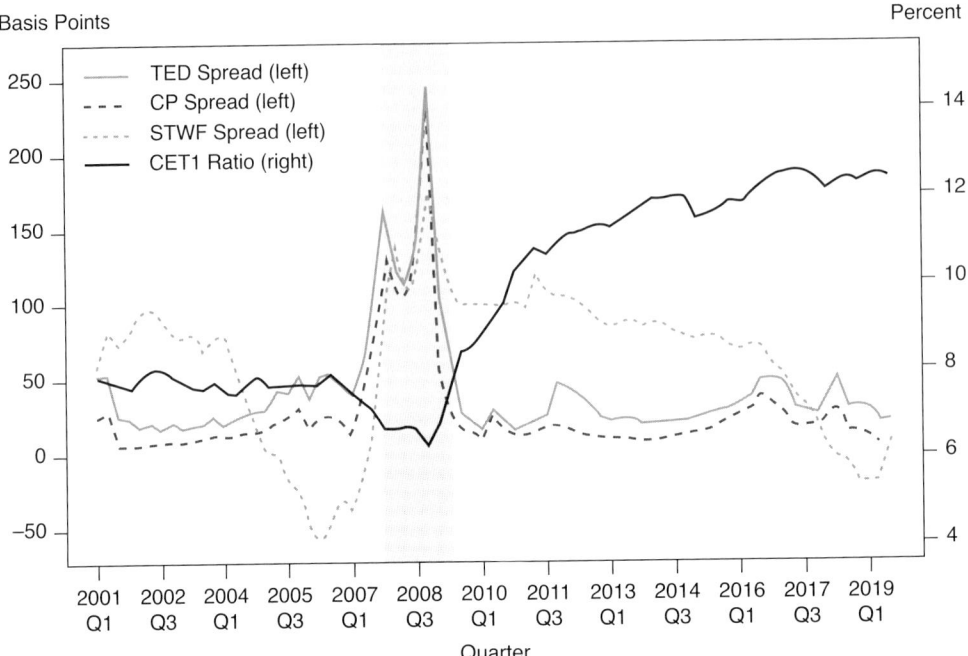

Figure 24.2 DFAST 2018 banks: Average capital ratios and STWF spreads.
Note: Average CET1 ratios correspond to the geometric mean of CET1 ratios for DFAST 2018 banks. TED, CP, and STWF spreads are relative to 3-month US Treasury bills. Shaded area denotes National Bureau of Economic Research (NBER) recession quarters.
Source: Authors' calculations based on forms FR Y-9C, FRED, US Bureau of Economic Analysis, and Board of Governors of the Federal Reserve System (US).

shows that the spreads of their funding costs over those of similarly sized banks not (yet) experiencing acute stress increased steadily in the quarters ahead of their failure or merger. National City's spread on large time deposits (in dark gray) peaked at approximately 250 basis points (bps), whereas Wachovia's spread on federal funds and repos (in light gray) peaked at approximately 150 bps.

Figure 24.2 depicts the time series of average capital ratios and the three proxies for the cost of STWF. The figure shows spikes in the cost of STWF to more than 200 bps across all three proxies in late 2008, when the financial crisis intensified and average capital ratios declined to their minimum levels. Thus, when markets were most disrupted, the TED spread and CP rates, which are more representative of banks' marginal funding costs than the STWF spread, were also consistent with the more comprehensive measure of banks' overall cost of funding. As governments introduced unprecedented support for the financial system and capital ratios started rising in 2009, overall STWF costs dropped back to approximately 100 bps, but the TED spread and CP spread dropped further, to below 50 bps. The close convergence of the three measures during the financial crisis and its immediate aftermath provides important validation for the choice of proxies. Therefore, the rest of this chapter uses the TED spread as the baseline measure of STWF costs, reflecting the balance of representativeness and responsiveness in its crisis and postcrisis behavior.

Nonetheless, the limitations of each of the three proxies highlight that a jurisdiction implementing a funding shock would likely want to revisit its information collections to better capture the quantity and pricing of STWF in a systematic way over time. For instance, the FR Y-15 data that support the calculation of the GSIB surcharge have a complete accounting of STWF, but these data are not fully granular and do not include the interest rates paid on those liabilities. The new FR 2052A data used for monitoring liquidity requirements has a granular breakdown of STWF instruments, but they are only collected from the largest banks. The new FR 2420 data collect interest rates paid by banks on certain wholesale funding sources and are a somewhat larger panel. All three of these sources may become valuable for modeling purposes but have only been available for a limited time.

3.1 Prudential Funding Shock

The most straightforward way to add a component to the stress test that simulates a funding shock is to impose a significant exogenous increase in banks' costs of funds over the projection horizon.[25] To illustrate, we follow Bassett et al. (2016) and consider a 100-bps shock over the first four quarters of the stress scenario. That shock represents a generalized increase in the costs of all STWF (as defined previously in calculating the comprehensive STWF spread) that hits all banks equally. Keeping with the current design of the DFAST and CCAR framework, we consider that banks do not change their mix of liabilities (between STWF and other sources) or shrink their balance sheets in response to the funding shock.[26]

This simple calibration helps to illustrate the effect of a funding shock and is within the range of the evidence described previosuly. That is, if s_{wt} denotes the TED spread in basis points at an annual rate, the increase in the cost of STWF Δs_{wt} equals 100 if $1 \leq t \leq 4$ and 0 if $t \geq 5$ Mechanically, an increase of 100 bps in the cost of funds for a fraction of a bank's liabilities will have a commensurate effect on interest expenses, dampening projected net interest income.[27] The impact of the STWF shock on poststress minimum capital ratios then corresponds simply to the original projection of capital ratios minus the cumulative decline in net income as a fraction of risk-weighted assets (RWAs).

Table 24.1 presents estimates of the effect of this 100-bps shock on bank minimum capital ratios for BHCs that participated in DFAST in 2018 based their STWF balances in Q4 2017. The estimated effect of this shock to STWF averages 35 bps. That is, the 100-bps increase in the cost of STWF over the first four quarters of the exercise reduces net income by 35 bps of initial RWAs. Compared with the average net interest income as a percentage of RWA of approximately 360 bps at large commercial banks in 2018, these

---

[25] This approach is consistent with rationing by price or by quantity. Regulators could achieve this spike in funding costs by specifying an exogenous increase in the interest rate for a specific liability or set of liabilities in the stress test scenarios or by specifying the closure of a particular funding market that forces banks to obtain alternative funding at a higher cost, resulting in a 100-bps increase in the overall cost of STWF.

[26] In stress test models with a dynamic balance sheet, we expect funding costs to have two opposing effects on profitability. On the one hand, as banks delever, they may avoid paying higher funding costs at the margin. On the other hand, credit balances may shrink, lowering interest income.

[27] This calculation abstracts from the impact of taxes, which would somewhat reduce the magnitude of the findings. In this implementation, we also do not credit banks with additional interest income for their investments in floating-rate assets tied to the same STWF rates.

Table 24.1. Effect of funding and macroeconomic shocks for DFAST 2018 banks

Shock	Mean	Standard Deviation	10th Percentile	Median	90th Percentile
Prudential funding	35	33	7	27	80
Macroprudential funding	6	6	1	4	13
Macroprudential funding (2-percentage-point lower initial CET1 ratios)	50	48	10	38	115
Severely adverse DFAST 2018	402	187	183	390	657

Note: Funding shocks are described in the text. The severely adverse scenario in DFAST 2018 includes macroecnomic and global market shock.
Source: Authors' calculations based on forms FR Y-9C, FRED, US Bureau of Economic Analysis, and Board of Governors of the Federal Reserve System (US).

additional losses are nontrivial. Further context for these projected losses can be obtained by comparing them with the 400-bps decline in capital ratios experienced in the severely adverse macroeconomic scenario in DFAST 2018 presented in Table 24.1 and Figure 24.3.[28]

Figure 24.3 presents the history of the TED spread (solid; light-gray line) and the historical geometric mean CET1 ratio for banks that participated in DFAST 2018 (solid black line). The dashed dark-gray lines represent the projected path of these variables in the DFAST Supervisory Severely Adverse scenario for reference. The DFAST severely adverse macroeconomic scenario does not include the TED spread in its variable inventory; for simplicity, we assume projections consider a constant spread at its jump-off value. As we describe in the Appendix, we use public disclosures of the starting and ending values of CET1 capital ratios and scenario variables to impute the path of CET1 capital over the stress horizon for each participating bank. The dashed black line in Figure 24.3 shows the small decrease in the geometric mean of the individual bank projections after accounting for the effect of the shock to the TED spread.

The estimated effect of the prudential STWF shock exhibits ample dispersion among banks, as evidenced by the large standard deviation relative to average losses shown in Table 24.1. In addition, for 10 percent of banks, this shock represents losses of no more than 7 bps of RWAs. By contrast, the 10 percent of firms with the largest losses from this shock experience declines of CET1 of at least 80 bps of RWAs. Thus, for some banks that had otherwise ended the test near the regulatory minimum, this additional decrease in capital buffers could have led to a reduction in their dividends or share repurchases.

---

[28] Throughout this chapter, we consider the projected losses under the DFAST. The latter corresponds to the supervisory stress test required under the Dodd–Frank Wall Street Reform and Consumer Protection Act (12 U.S.C. 536[i]; 12 CFR part 252) and is related to but distinct from CCAR. In particular, DFAST projections assume constant dividends equal to the average of the four quarters prior to the beginning of the exercise and no buybacks, instead of the planned capital distributions considered in CCAR. Given that our simulations envision additional levels of stress, the generally much smaller distributions assumed in DFAST seem the more appropriate comparision.

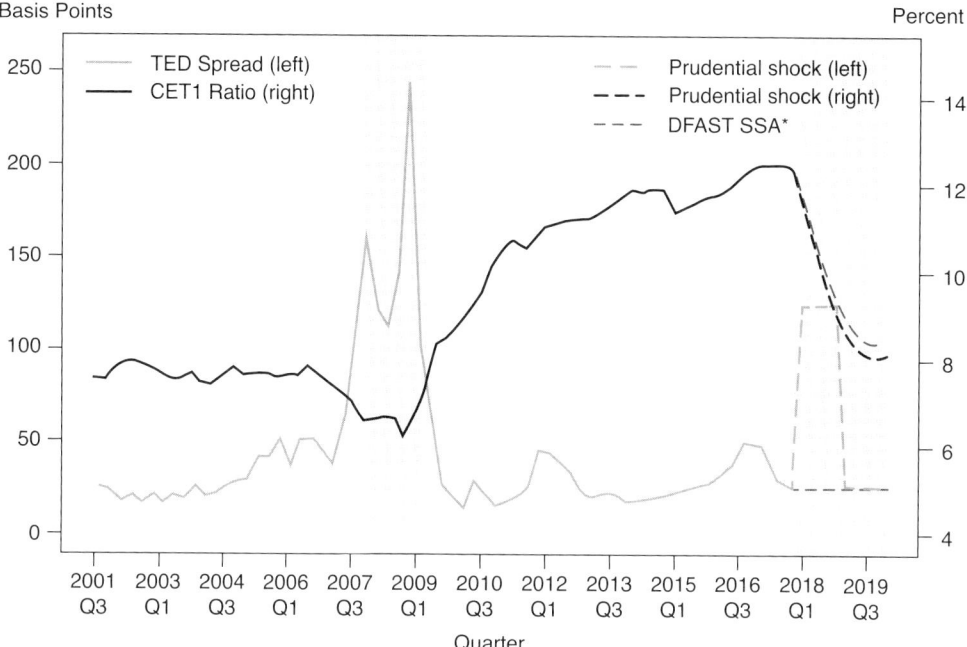

Figure 24.3 DFAST 2018 banks: Prudential funding shock.
Note: DFAST SSA corresponds to the severely adverse scenario in DFAST 2018.
Source: Authors' calculations based on forms FR Y-9C, FRED, US Bureau of Economic
Analysis, and Board of Governors of the Federal Reserve System (US).

Finally, the banks affected most by the funding-cost shock are different from those affected most by the macroeconomic shock in the DFAST 2018 exercise. In fact, the correlation between losses from the shock to STWF and the decline in capital ratios from the macroeconomic shock is only 0.54. Figure 24.4 depicts a scatterplot of the decline in capital ratios under prudential funding stress and macroeconomic stress. The solid line depicts the best-fitted linear relationship between these declines. The dispersion of points away from this line reveals the different distributional impact of the two shocks. Thus, a prudential funding shock seems to capture a missing risk dimension in stress tests rather than being duplicative of factors already included in the tests.

4 Adding Funding Spillovers

This section extends the analysis of a funding shock to model financial spillovers by creating a dynamic relationship between the assumed increase in funding costs and the deteriorating health of the banking sector and slowing economic activity.

As in the previous case, we consider a cost shock to all STWF sources and hold banks' balance sheets constant for the simulation. But in contrast with the previous case, the magnitude of the STWF shock in each projection quarter depends on the projected health of all

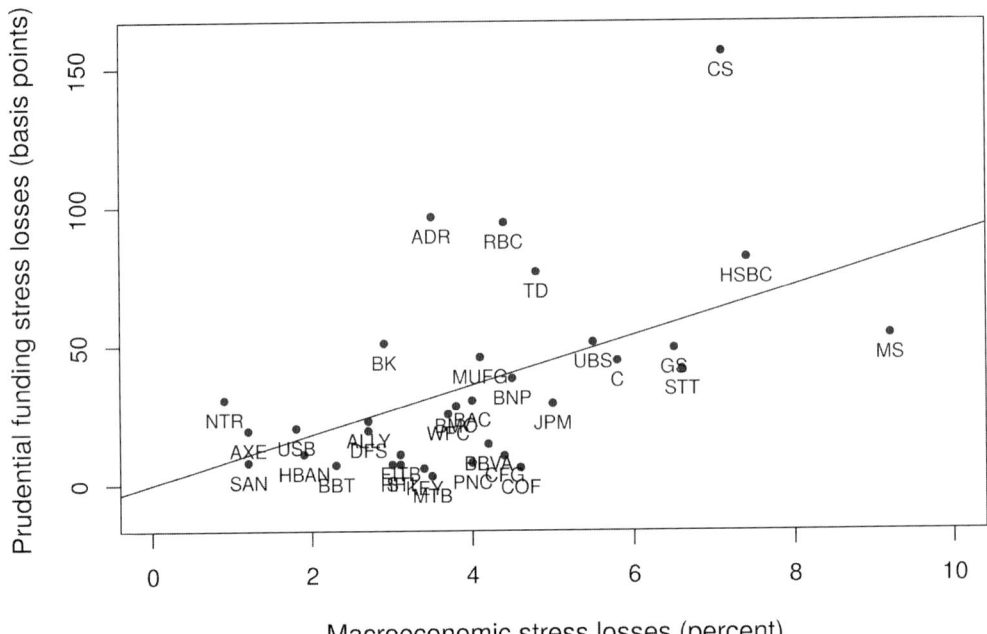

Figure 24.4 DFAST 2018 banks: Losses from prudential funding versus macroeconomic stress.
Note: Losses expressed as a fraction of RWAs. Losses from macroeconomic stress correspond to
the losses under the severely adverse scenario in DFAST 2018. The solid line corresponds to a
fitted linear model of prudential funding losses on macroeconomic losses ($\beta = 8.94$).
Source: Authors' calculations based on forms FR Y-9C, FRED, US Bureau of Economic
Analysis, and Board of Governors of the Federal Reserve System (US).

participating firms and the state of the economy.[29] From the perspective of each participating
institution, this form of funding shock provides an incentive to tune its capital plan to the
overall health of *all* participating firms.[30] For instance, when the average starting capital
levels are high, firms will not necessarily need to account for industry health in their capital
plan. In contrast, when overall starting capital levels are low, even firms that are themselves
strongly capitalized will have incentives to reduce planned capital distributions. Moreover,
this specification of a dynamic funding shock furthers the macroprudential benefit of the

---

[29] Of course, such estimates may underestimate the relationship because the unprecedented level of government
support deployed during the financial crisis likely attenuated the rise in funding spreads and reduced the
duration of the spikes. In addition, most developed-country recessions have not had an associated financial
crisis but rather were associated with significant declines in short-term interest rates stemming from
flight-to-quality flows and monetary-policy accommodation. Therefore, parameter estimates that incorporate
multiple benign business cycles and one crisis period may also generate smaller projected shocks.

[30] The 2017 Japan FSAP includes a stress test scenario where projected funding costs are a function of an
aggregate funding shock and idiosyncratic funding shocks that are a function of projected individual bank
capital (IMF, 2017a). When projected funding costs depend on the financial health of individual firms, the
funding shock will incentivize firms to steer away from capital plans that scrape the minimum capital
requirements during the scenario, because they would see their funding costs increase when that occurs.

stress tests because it explicitly models spillovers and amplification dynamics within the financial sector and between the real and financial sectors.

4.1 Relationship between STWF Spreads and Banks' Capital

We postulate the following relationship between STWF costs, the health of participating banks, and the real economy:

$$s_{w,t} = f\left(\bar{k}_t, \text{GDP}_t\right), \tag{24.1}$$

where $s_{w,t}$ denotes the STWF spread, $\bar{k}_t$ denotes the (geometric) average CET1 ratio for participating banks, and GDP_t denotes the annualized quarterly growth rate of nominal gross domestic product (GDP).[31] A more general specification of this relationship could account for bank heterogeneity, as indicated by the experiences of National City and Wachovia presented earlier. However, the lack of reliable data on individual banks' marginal cost of STWF prevents us from being able to estimate a specification that accounts for this heterogeneity. In addition, STWF costs depend mechanically on past financial and economic conditions because some liabilities have fixed interest rates and terms longer than a quarter. This approach could accommodate this dependence on past conditions if appropriate time series of STWF sources and prices were available, but it would complicate the simulation.

We estimate this empirical relationship between these variables by fitting GDP growth and a second-order polynomial of average CET1 ratios to the STWF spread, using the following regression:

$$s_{w,t} = \alpha + \beta_1 \bar{k}_t + \beta_2 \bar{k}_t^2 + \beta_3 \text{GDP}_t + \varepsilon_t, \tag{24.2}$$

where α, β_1, β_2, and β_3 are coefficients to be estimated, ε_t are mean-zero innovations, and $\bar{k}_t^2$ corresponds to (geometric) mean capital ratios squared.[32] From an econometric perspective, the regressors in equation (24.2) are endogenous, so the estimated coefficients should not be interpreted as the causal effect of changes in bank capital and economic activity on spreads. Rather, equation (24.2) should be thought of as summarizing the relationship of these variables, which we will use to calibrate an STWF shock that is sensitive to the scenario and vulnerabilities in the banking sector. Moreover, from the perspective of model risk management, the relationship in equation (24.2) offers the advantage of being both intuitive and "as simple as possible, but not simpler" (see Chapter 16).

We expect the signs on the coefficients of equation (24.2) to represent a decreasing function of STWF costs on economic activity and bank capital over most of its range, that is, $\beta_1 + 2\beta_2\bar{k}_t \leq 0$ and $\beta_3 \leq 0$. Moreover, structural models of credit default in the tradition of Merton (1974) imply a nonlinear relationship between capital and funding spreads, motivating the inclusion of the second-order polynomial. In particular, funding costs are expected to be more sensitive to bank capital as capital becomes insufficient to protect bank creditors from credit losses (Aymanns et al., 2016; Schmitz et al., 2017). This convexity

---

[31] GDP growth corresponds to US Bureau of Economic Analysis, GDP (A191RP1Q027SBEA), retrieved from FRED, Federal Reserve Bank of St. Louis (https://fred.stlouisfed.org/series/A191RP1Q027SBEA).

[32] In unreported analysis, we considered orthogonalized capital ratios; that is, the linear capital term is mapped to the interval $[-1, 1]$, whereas the quadratic term equals $\left(3\bar{k}_t - 1\right)/2$. Results were qualitatively and quantitatively robust.

Table 24.2. STWF spread, bank capital, and economic conditions

Dependent variable: TED spread				
	(1)	(2)	(3)	(4)
CET1	−1.651***	−1.624***	−1.185***	−5.871
	(0.211)	(0.187)	(0.395)	(8.674)
CET1^2	0.082***	0.077***	0.050***	0.293
	(0.011)	(0.010)	(0.018)	(0.485)
GDP growth	−0.045***	−0.018	−0.059***	0.032
	(0.010)	(0.012)	(0.012)	(0.014)
Constant	8.512***	8.773***	7.517***	29.406
	(0.965)	(0.863)	(2.128)	(38.766)
Sample period	Q1 2001–Q3 2019	Q4 2007–Q3 2019	Q1 1996–Q3 2019	Q1 1996–Q3 2007
Capital ratio	CET1	CET1	T1CE	T1CE
F-statistic	43.21***	61.98***	20.14***	12.35***
($\bar{k}_t = \bar{k}_t^2 = 0$)				
Observations	75	48	95	47
Adjusted R^2	0.651	0.818	0.338	0.371

Note: *** significant at 1 percent, ** significant at 5 percent , and * significant at 10 percent. T1CE, tier 1 common equity capital.
Source: Authors' calculations based on forms FR Y-9C, FRED, US Bureau of Economic Analysis, and Board of Governors of the Federal Reserve System (US).

in the relationship between funding costs and bank capital is then key to modeling the amplification from funding spillovers.

Table 24.2 presents the estimates of equation (24.2). The first column considers the full sample from Q1 2001 until Q3 2019. The estimated coefficients imply a decreasing relationship between the TED spread and GDP growth, which is significant at the 1 percent confidence level. For bank capital, we consider the joint magnitude and significance of the first- and second-order coefficients of CET1. These results show that funding costs are decreasing in the health of the banking sector over much of the observed range of CET1 capital ratios. However, the positive coefficient for the quadratic capital term indicates that the fitted relationship is convex. That is, as banks approach their minimum CET1 ratio, the cost of funds becomes more sensitive to bank capital, as predicted by credit-risk models. Conversely, the marginal benefit of capital on STWF costs declines as capital ratios reach very high levels. This equation achieves its inflection point when the CET1 ratio is above 10.1 percent, a value that is in the lower end of estimates of optimal bank capital ratios (Firestone et al., 2017).

Column 2 of Table 24.2 reports the results of the regression in equation (24.2) when the sample is split at the beginning of the GFC in Q4 2007. In the postcrisis period, the convex relation between bank health and STWF costs is similar to the full sample, although the statistical significance of GDP growth dissipates. The precrisis sample is too short to reliably estimate the relationship between STWF costs and capital and GDP growth.

The estimated relationship between these variables appears robust to the choice of capital ratio and a somewhat longer time horizon. In columns (3) and (4) of Table 24.2, we proxy bank health by the geometric mean of Tier 1 capital ratios that are available since Q1 1996.

Table 24.3. Heterogeneous STWF spreads, bank capital, and economic activity

Dependent variable:	(1) TED spread	(2) CP spread	(3) STWF spread
CET1	−1.651***	−1.500***	1.125***
	(0.211)	(0.210)	(0.418)
CET1^2	0.082***	0.075***	−0.058***
	(0.011)	(0.011)	(0.022)
GDP Growth	−0.045***	−0.040***	−0.103***
	(0.010)	(0.010)	(0.020)
Constant	8.512***	7.582***	−4.140**
	(0.965)	(0.932)	(1.914)
Sample period	Q1 2001–Q3 2019	Q1 2001–Q3 2019	Q1 2001–Q3 2019
F-statistic ($\bar{k}_t = \bar{k}_t^2 = 0$)	43.21***	33.46***	3.63**
Observations	75	75	75
Adjusted R^2	0.651	0.593	0.265

Note: *** significant at 1 percent, ** significant at 5 percent, and * significant at 10 percent.
Source: Authors' calculations based on forms FR Y-9C, FRED, US Bureau of Economic Analysis, and Board of Governors of the Federal Reserve System (US).

The results using this longer sample and the alternative measure of bank capital are quite similar to those in the primary regression specification (column [3]). Namely, STWF costs depend negatively on economic activity, and the relationship with Tier 1 capital ratios is convex, with these relationships being statistically significant at the 1 percent confidence level. The inflection point for the Tier 1 capital ratio in this regression is 11.7 percent, consistent with the result for CET1 capital. Using this longer sample, we are also able to obtain qualitatively similar results for the precrisis period. However, the only statistically significant estimates are, jointly, the two terms associated to capital (column [4]).

To further explore the robustness of the relationship between STWF costs, banks' health, and economic activity, Table 24.3 reports regression results for equation (24.2), for the two alternative proxies for STWF costs, as previously described. To facilitate the comparison, column (1) of Table 24.3 presents our baseline results from Table 24.2. Column (2) considers the CP spread as a proxy for STWF costs. The results are very similar to our baseline, with a convex relationship between the CP spread and capital ratios that shows an inflection point at 10 percent. The effect of GDP growth on the CP spread is negative and statistically significant. Finally, column (3) presents the results when funding costs are measured by the STWF spread, computed as previously described. But in this case, the function that relates capital ratios to STWF costs, although statistically significant, is concave. This counterintuitive result seems to be associated with the additional persistence of the STWF spread, relative to the forward-looking spread measures, because adding lags of the dependent variable in equation (24.2) restores a convex relationship. Nevertheless, the ability to use a more parsimonious functional form for this relationship reinforces our choice of the TED spread as the primary variable for this exercise.

Figure 24.5 plots the combination of TED spreads, CET1 ratios, and the fitted values from model (2), holding GDP constant (0 percent growth). As noted, the estimated relationship, $\hat{f}\left(\bar{k}_t, \text{GDP}_t\right)$ implies that funding costs are higher when banks' capital increases above

TED spread (basis points)

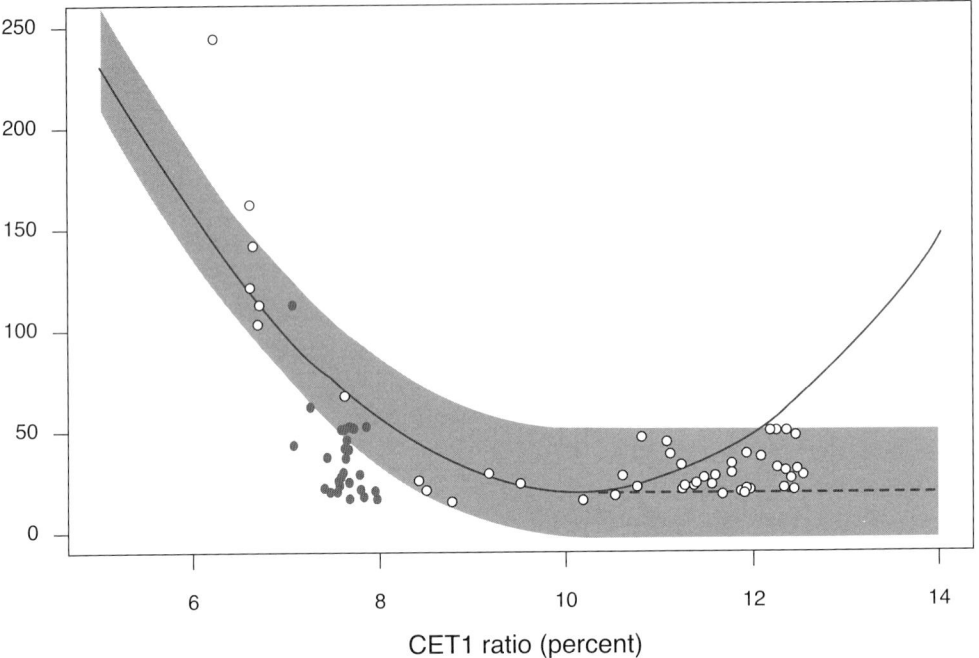

Figure 24.5 DFAST 2018 banks: Average CET1 ratio and TED spread.
Note: Postcrisis (dark-gray dots) corresponds to the period from Q4 2007 to Q4 2017
and precrisis (white dots) corresponds to Q1 2001 to Q3 2007. The solid gray line
is the projection of the TED spread from the regression in column (1) of Table 24.2.
The dashed line represents the transformed value when the CET1 ratio is greater than
$\hat{\kappa} = 10.1$ percent based on equation (24.3).
Source: Own calculations based on forms FR Y-9C, FRED, U.S. Bureau of Economic
Analysis, and Board of Governors of the Federal Reserve System (US).

10.1 percent. To avoid this counterintuitive prediction, we set funding costs equal to the
prediction of the second-order polynomial when bank capital is below 10.1 percent and at
the value corresponding to a capital ratio of 10.1 percent when bank capital is higher. That is,

$$s_{w,t} = \alpha + \beta_1 \min\{\bar{k}_t, \kappa\} + \beta_2 \left(\min\{\bar{k}_t, \kappa\}\right)^2 + \beta_3 \text{GDP}_t, \tag{24.3}$$

where $\kappa = -\beta_1/(2\beta_2)$ corresponds to the minimum of the second-order polynomial on bank
capital.

The dashed gray line in Figure 24.5 depicts this fitted relationship; the gray bands correspond to the range of values that the spread can take in the simulation, given the minimum
and maximum projected GDP growth over the DFAST 2018 severely adverse scenario. The
CET1 ratio was well below 10 percent for the entire precrisis period but moved above 10
percent at the end of 2010 and has hovered around 12 percent since 2013. Therefore, this
specification would appropriately conclude that funding-cost shocks are less likely and less
disruptive when banks enter a downturn with substantial capital buffers, as was the case
in 2019, than if they were to enter a downturn already impaired. However, the average

CET1 ratio falls well below 10 percent during the DFAST stress tests, providing variation in projected losses.

4.2 Macroprudential Funding Spillovers

The empirical specification for $f\left(\bar{k}_t, \text{GDP}_t\right)$ together with the estimates of the evolution of capital over the stress scenario, $\left\{\bar{k}_t\right\}_{t \in \{1,\ldots,9\}}$, and projected GDP growth under the DFAST 2018 severely adverse scenario simulate the effect of a macroprudential funding shock. The cost of STWF over the stress scenario is assumed to be $s_{w,t} = \hat{f}\left(\bar{k}_{t-1}, \text{GDP}_t\right)$, where we use the lagged value of average capital ratios to project STWF costs. This timing assumption allows us to simulate the macroprudential funding shock without violating consistency requirements. The challenge with using an add-on considering the contemporaneous relationship is best illustrated with an example. Suppose that in the first projection quarter of the test, the poststress average CET1 is 8 percent. Then the estimated relationship between STWF funding costs and average capital implies that STWF costs will be higher, generating additional losses. We could stop there, but without an additional assumption that would generate a fixed point on a subsequent iteration, the updated path of capital will imply additional projected costs of STWF, which in turn imply lower projected capital, higher STWF costs, and so on.

The final trajectory of average CET1 ratios together with the estimated relationship between CET1 and funding costs imply STWF spreads, $s_{w,t}$, increase by 25 bps to 51 bps in the second projection quarter and then decline to 23 bps by the ninth projection quarter (dashed light-gray line in Figure 24.6). Specifically, this spread in Q4 2017 stood at 26 basis points and we denote it with $s_{w,0}$. Thus, the effective shock to STWF costs in quarter t is given by $s_{w,t} - s_{w,0} = \hat{\alpha} + \hat{\beta}_1 \min\left\{\bar{k}_{t-1}, \hat{\kappa}\right\} + \hat{\beta}_2 \left(\min\left\{\bar{k}_{t-1}, \hat{\kappa}\right\}\right)^2 + \hat{\beta}_3 \text{GDP}_t - s_{w,0}$. It follows that the macroprudential STWF shock increases to 25 bps in the second quarter and then declines to -3 bps by the ninth quarter.

Table 24.1 presents the projected effect of this funding shock on bank minimum capital ratios for DFAST 2018 banks. The estimated average effect on net income of the macroprudential shock to STWF averages 6 bps of RWAs. Although not close to the level of the prudential shock, the estimated effects of the macroprudential funding shock still exhibit some dispersion among banks, reflecting the substantial heterogeneity in banks' reliance on STWF. For 10 percent of banks, the funding shock represents losses of no more than 1 bps of RWAs. By contrast, the 10 percent of firms with the largest losses from this shock experience declines of CET1 of at least 13 bps of RWAs.

Average projected losses and the dispersion of outcomes are much smaller than with the prudential funding shock described previously. This follows directly from the empirical relationships between the TED spread and bank capital. As depicted in Figure 24.5, the high starting values of bank capital in the 2018 exercise prevent material increases in funding costs unless the geometric average capital ratio for participating banks falls below 10.1 percent. Moreover, GDP growth recovers over the latter half of the projection horizon, which causes STWF spreads to decline and end marginally below their jump-off value (Figure 24.6). By contrast, the prudential shock of 100 bps results in an STWF cost that jumps to 126 bps over the first four projection quarters and then returns to the jump-off value.

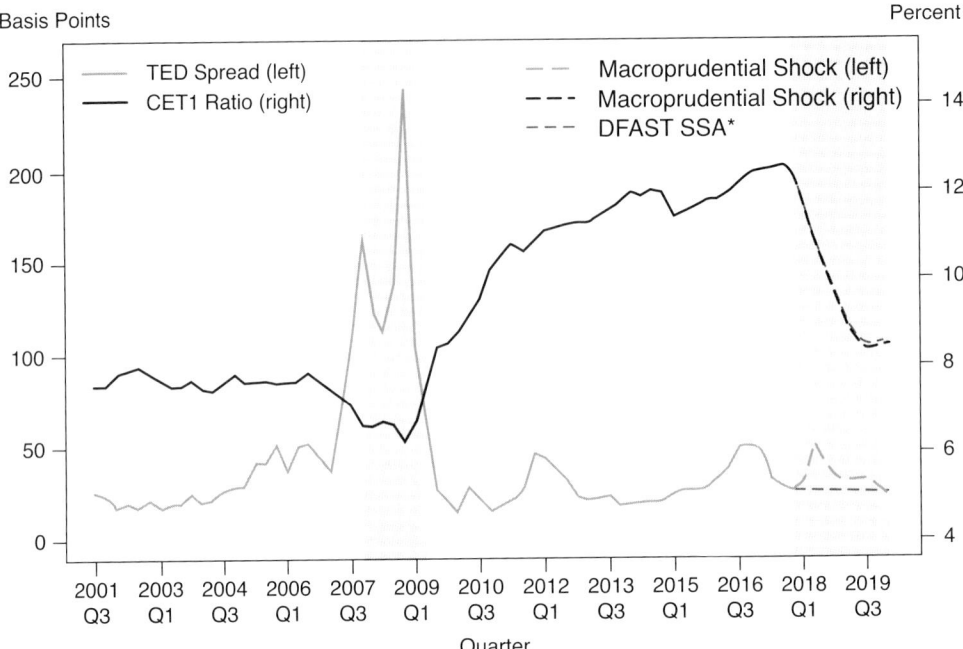

Figure 24.6 DFAST 2018 banks: Macroprudential funding shock.
Note: DFAST SSA corresponds to the supervisory severely adverse scenario in DFAST 2018.
source Own calculations based on forms FR Y-9C, FRED, U.S. Bureau of Economic
Analysis, and Board of Governors of the Federal Reserve System (US).

4.3 Macroprudential Funding Shock and Banks' Capital

To better highlight the nonlinear interaction between the severity of the projected funding
shock and the financial health of participating firms at the beginning of the exercise, we
consider a counterfactual exercise where DFAST 2018 firms are assumed to start the exercise
with CET1 ratios that are 2 percentage points lower than their values at the test's jump-off
point.[33] We assume that even with these lower starting capital levels, banks experience the
same losses as projected under DFAST, so the trajectory of CET1 ratios over the projected
horizon shifts down by the same amount as the jump-off values.

With banks' financial health consistently worse over the projected horizon, our estimated
relationship projects much larger STWF spreads. The 26-bps jump-off spread is projected
to increase between 4 and 113 bps to between 30 and 139 bps (Figure 24.7). Table 24.1
presents our estimates of the effect of this funding shock. The projected effects on net
income of the counterfactual macroprudential STWF shock average 50 bps of RWAs, which

---

[33] Indeed, several of the largest US banks have announced medium-term targets for the CET1 ratios that are 1
percentage point or more lower than their current ratios. See the recent transcript of earning calls from Bank of
America, www.fool.com/earnings/call-transcripts/2020/01/15/bank-of-america-corp-bac-q4-2019-earnings-
call-tra.aspx; Wells Fargo, www.fool.com/earnings/call-transcripts/2020/01/14/wells-fargo-co-wfc-q4-2019-
earnings-call-transcrip.aspx; and Capital One, www.fool.com/earnings/call-transcripts/2020/01/21/capital-
one-financial-corp-cof-q4-2019-earnings-ca.aspx.

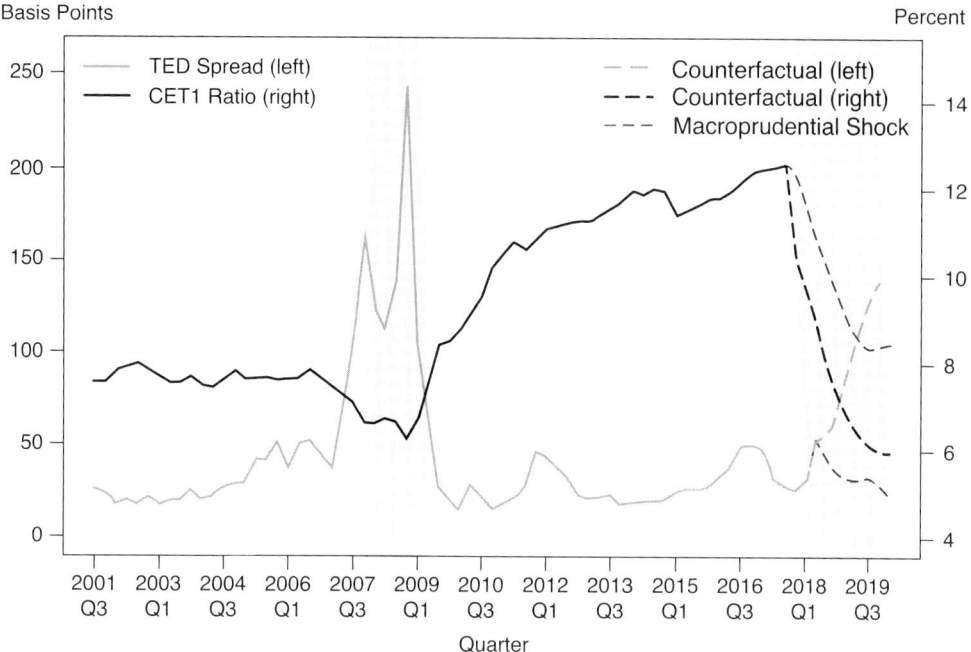

Figure 24.7 DFAST 2018 banks: Leverage and funding spillovers.
Source: Own calculations based on forms FR Y-9C, FRED, US Bureau of Economic Analysis,
and Board of Governors of the Federal Reserve System (US).

is considerably larger than the average 6 bps projected using the current CET1 ratios at the jump-off. It is instructive to note that the additional funding stress is concentrated among banks that rely more on STWF. In fact, for the 10 percent of banks with smaller losses, those losses are smaller than 1 bps, whereas for the 10 percent of firms with the largest losses, those losses reach at least 115 bps of RWAs (Table 24.1).

Thus, modeling the financial spillovers from funding shocks can significantly increase the severity of projected funding stress, but only when banks enter the test with somewhat weaker capital ratios than currently prevail. This feature improves the ability of the test to identify conditions that can put participating banks at risk and build resilience against those shocks before economic conditions actually deteriorate. However, in stress tests where amplification of financial spillovers is explicitly modeled, the initial supervisory scenarios should be constructed so as to avoid double counting of these effects.

In addition to illustrating the way that amplification dynamics are brought into the test through financial spillovers, our macroprudential funding simulations underscore the data and modeling requirements to incorporate financial spillovers as add-ons, using reduced-form relationships. In our simulation, ratios of capital and STWF to RWAs represent the *bank variables* involved in the spillovers, and the spread of STWF represents the *market outcome* involved. Data-collection programs are needed for accurate measurement of bank variables and market outcomes. In the case of our funding spillover, we use FR Y-9C forms and the TED spread from FRED to approximate these variables. As previously noted, additional projection accuracy would be obtained if more granular data on STWF balances and costs were part of data-collection programs.

There are two model requirements. One is a model that links bank variables with market outcomes, which we model using the reduced-form relationship between average CET1 ratios and the spread of STWF (equation [24.3]). The other is a model to update the projection of bank variables conditional on projected market outcomes. In our case, we use a bottom-up approach to approximate additional income losses based on the simulated spread of STWF and an individual bank's ratio of STWF to RWAs. Finally, we note that the development of supervisory models to update bank variables is required for practical considerations. The alternative would involve a sequence of bank submissions and updates to the test's scenario, which seems impractical.

5 Adding Other Financial Spillovers

Financial spillovers in periods of stress include other channels besides funding spillovers. In this section, we review three of these channels: fire sales, macrofinancial feedback loops, and network effects.

One important source of spillovers of financial distress is fire sales (see, among others, Shleifer and Vishny, 2011; Allen et al., 2012). In fact, theoretical analyses draw a link between fire sales, or market liquidity, and conditions in credit markets, or funding liquidity (e.g., Geanakoplos, 2003; Brunnermeier and Pedersen, 2009). For instance, a bank that suffers losses and experiences a decline in its CET1 capital ratio might adjust by selling assets to reduce its RWAs and funding needs, rather than facing higher funding costs as assumed in the simulations in the previous sections. These asset sales under already-stressed conditions create further stress in markets, especially for less liquid assets. As a result, the prices received for those assets are depressed, which reduces the mark-to-market value of similar assets at other institutions and reduces the prices for other sellers.[34] By including potential fire-sale losses in banks' capital buffers through stress testing, prudential authorities can hope to control vulnerabilities at the beginning of a downturn, rather than contributing to a downward spiral.

Several frameworks exist to simulate the spillover effects of fire sales. Greenwood et al. (2015) develop a linear model of fire-sale spillovers, under three key assumptions: (1) upon a negative shock, banks sell assets to return to their initial leverage ratio; (2) banks' asset sales are proportional to their initial holdings; and (3) the loss on fire-sold assets is proportional to aggregate quantities sold. The advantage of these stylized assumptions is that they yield simple linear expressions for the effect of an initial shock to banks' leverage, which are used to measure individual banks' contribution to systemic risk.

Subsequent authors adapted this approach to construct an index of aggregate vulnerabilities to fire-sale spillovers for euro-area banks (Cappiello and Supera, 2014) and for large US banks (Duarte and Eisenbach, 2018; hereafter DE). DE relax the assumption that banks aim to restore their original leverage and instead assume that banks have a latent target, where both the latent target and the speed of adjustment are time varying. The speed of adjustment

---

[34] Mark-to-market losses, under the Basel III implementations in most jurisdictions for advanced-approaches banks, are included in accumulated other comprehensive income that affects the accretion of CET1.

and target leverage ratios are jointly estimated, fitting a dynamic model of capital structure with partial adjustment toward a leverage target.

Cont and Schaanning (2017) use this framework to model fire sales in a macroprudential stress test of European banks. These authors assume that banks adjust their leverage ratios when they breach their regulatory constraints and target 105 percent of the minimum regulatory leverage ratio (other regulatory ratios for capital or liquidity may also be considered).[35] This adjustment rule implies that fire sales only occur when losses exceed bank-specific thresholds. Moreover, these authors discuss how price-impact functions that are linear in the value of assets sold can be calibrated using market depth, which varies by asset class.

The fire-sale approach has the merit of modeling bank actions, closer to a structural approach, but each of the previously described frameworks could be implemented as an add-on. Nevertheless, there is little historical evidence to inform the modeling of banks' adjustment.[36] Moreover, a structural approach to modeling fire sales would necessarily need to consider nonbank intermediaries and nonbank liquidity providers that participate in these asset markets, especially in markets where those firms are not subject to prudential capital and liquidity regulation.[37]

Finally, note that allowing banks to adjust their balance sheets over the projected scenario represents a key methodological departure from assuming static balance sheets, as in the DFAST exercise (Baudino et al., 2018). Such a change would require that supervisors explicitly model balance-sheet dynamics as an element of the framework. In addition, the assumption of a constant balance sheet has an important macroprudential impact: it prevents banks from meeting the requirements of the test by "shrinking to health," which somewhat reduces the likelihood of a credit-supply shock that could exacerbate the downturn if an adverse macroeconomic scenario were to materialize.

Another important source of amplification and spillovers of financial distress is the feedback loop between macroeconomic and financial conditions. Most of the current stress testing frameworks use a very severe macroeconomic and financial-market scenario in order to embed likely spillovers and financial-accelerator effects within the single round of stress testing.[38] However, at the end of such an exercise, banks are typically in significantly weaker condition than when they began the test, and policymakers would benefit from studying whether they would continue to be a drag on economic performance in that state.

Therefore, directly modeling the feedback loop between the financial system and the real economy could better advance the macroprudential goals of stress tests.[39] As banks' ex ante

---

[35] Cont and Schaanning (2017) define the leverage ratio as the ratio of assets to equity and consider that banks set a target of 95 percent of the maximum regulatory asset-to-equity ratio.

[36] Some authors even call into question whether banks deleveraged after the financial crisis for the purpose of adjusting to their desired leverage or capital ratios (Begenau et al., 2018).

[37] For evidence on fire sales by nonbanks, see Khandani and Lo (2011), Manconi et al. (2012), and Jotikasthira et al. (2012).

[38] For an example of a stress testing framework modeling macrofinancial linkages, see Andersen et al. (2019).

[39] In a macroprudential framework, one would prefer the modeling of the feedback between financial and macroeconomic variables to be informed by both the health of participating banks and the projections for nonbank financial intermediaries. Information on the latter is intrinsically sparse, but reduced-form models can make use of simple statistics, such as the share of nonbank intermediated credit.

financial health improves, we expect that their supply of credit will be more resilient to the same real economic shocks. By contrast, as the share of poorly capitalized banks rises, a negative economic shock will limit the supply of credit, and macrofinancial spillovers will be stronger. Thus, supervisors may be able to glean important insights from studying the reaction of the economy and banking system to a smaller initial shock and then accounting for the endogenous response of banks and the economy by iterating through higher-round effects. These spillovers require modeling macroeconomic and financial variables jointly, suggesting integrating these channels into the test design rather than using an add-on as we do for funding spillovers. Although a reduced-form approach may be suitable in some instances, advances in modeling of macrofinancial feedback loops will likely require structural models with robust financial sectors.

Finally, we discuss some approaches to address counterparty defaults in a macroprudential framework. Counterparty defaults are an important source of amplification and spillovers of financial distress, with the spillovers from the default of a large financial institution being widespread. Stress tests in some jurisdictions require banks to contemplate the default of one of their major counterparties.[40] By contrast, a more macroprudential approach to model default risk might consider the same scenario for all banks (i.e., the default of the same counterparties), and the second-round effects of spillovers from such defaults on the surviving counterparties. This is challenging to achieve without considering the structure of the network of individual counterparty relationships.

Some structural frameworks have been developed to incorporate financial spillovers from counterparty default into stress tests. For example, Eisenberg and Noe (2001) show how the payments after the default of counterparties are jointly determined by all the bilateral relationships and their relative priority in a network representing a single clearing mechanism. All of those factors also contribute to the repricing of nondefaulted derivative contracts, and larger margins from derivatives counterparties including CCPs.[41] This result underscores the necessity of the structural approach to modeling counterparty defaults.

In order for supervisors to conduct a macroprudential large-counterparty default test, the quantity and quality of data collected will need to be improved. Improved data would aid in the development of structural loss models that would be used to project losses. The reporting of such highly confidential data would require an understanding among firms and regulators about the ways in which the data would be protected and could be used. Another option would be to require banks to use their own models and data to project initial losses and then spillovers, although that may require multiple rounds of estimates as supervisors evaluate results across the system. A complication of the bank-run option is that the information that

---

[40] For example, in the DFAST severely adverse scenario, credit default losses (CDLs) are calculated using the following approach. The largest trading firms are required to rank their counterparty stressed losses and report the largest, excluding G-7 sovereigns and CCPs. These losses are calculated while including securities financing and derivative exposures. See www.federalreserve.gov/supervisionreg/files/bcreg20180201a1.pdf.

[41] Understanding the role of CCPs is important, given their increased prominence in handling counterparty risk within financial markets after the recent financial crisis and given recent large defaults that have tested these institutions. For instance, on September 17th, 2018, the default of two large traders depleted around two-thirds of Nasdaq's mutual default fund (see www.bloomberg.com/view/articles/2018-09-17/central-counterparties-mean-derivatives-risk-hasn-t-gone-away).

banks would need to run second-round effects could be used by banks to determine the confidential counterparty exposures of their competitors.

6 Conclusions

In this chapter, we discuss several macroprudential elements that can be included in solvency stress tests to better capture spillovers, as well as the pros and cons of various modeling approaches. We provide two examples that incorporate funding spillovers into solvency stress tests through an add-on approach. First, we consider a "prudential shock," where bank losses are a function of an exogenous shock to the cost and availability of STWF. Second, we use the relationship between the health of participating banks, economic growth, and the spread of STWF relative to short-term government securities to simulate spillovers from the financial health of banks to their costs of short-term wholesale funds. This reduced-form relationship can be combined with banks' current capital position and STWF intensity to project a dynamic shock that is responsive to current and projected levels of bank capital and GDP growth. This approach achieves an additional macroprudential goal because the effective shock to the spread of STWF is negatively correlated with starting capital levels and the severity of the macroeconomic scenario.

We show how our approach to modeling funding spillovers relates to existing models of fire-sales spillovers, many of which can also be implemented as reduced-form add-ons to existing frameworks. We also discuss models of spillovers between the macroeconomy and the financial system; however, we conclude that those elements are more effective when integrated into a comprehensive framework, as opposed to incorporated as an add-on. Finally, we discuss the challenges faced by policymakers who desire to move beyond the add-on approach for modeling bilateral interactions in financial markets, such as counter-party defaults. In order to incorporate second-round effects of counterparty defaults, stress test designers will need to choose between developing a structural model to project losses, which also requires enhanced data collection, or relying on bank-provided loss estimates. One limitation of the latter option is that it would require banks and regulators to iterate and might require regulators to disclose information that can then be used to back out confidential bilateral exposures.

Appendix: Projected Trajectories of CET1 Ratios

The public disclosures of DFAST results do not include the full projected path of CET1 ratios; rather, the disclosures provide the ending and minimum poststress CET1 ratios and the jump-off CET1 ratio.[42] Using the publicly available information, we estimate the trajectory of CET1 ratios over the severely adverse scenario to simulate the macroprudential funding shock as follows.

First, we modify the approach of Durdu et al. (2017) to calculate severity based on unemployment and house prices. In particular, we construct a severity index over the nine projection quarters, as the average between the normalized severity score for the projected unemployment and house-price paths. Formally, let u_t and HPI_t denote the projected paths for unemployment and house prices for quarter $t \in \{0, \ldots, 9\}$, where 0 represents the jump-off point and the other 9 quarters represent the 9 projection quarters in the test. We calculate the severity index in projected quarter t as

$$\text{severity}_t = \frac{1}{2} \left[\frac{(u_t - u_0)}{\max\limits_{j \in \{1, \ldots, 9\}} (u_j - u_0)} + \frac{(\log HPI_0 - \log HPI_t)}{\max\limits_{j \in \{1, \ldots, 9\}} (\log HPI_0 - \log HPI_j)} \right]. \quad (A.1)$$

Second, we assume that the capital loss in percentage points of RWAs during the projected scenario is proportional to the severity index. Using the minimum CET1 ratio for a given bank i, k_{i,t^*}, we can determine the constant of proportionality between projected CET1 ratios and the severity index, $\alpha_i = (k_{i,0} - k_{i,t^*}) / (\text{severity}_{t^*})$, where t^* denotes the quarter where the severity index peaks. Our assumptions imply that the projected path of capital reaches its minimum when severity peaks and provide an estimated path for each bank's CET1 ratio, $k_{i,t} = k_{i,0} - \alpha_i \text{severity}_t$. Using individual banks' projected paths, we compute the geometric mean of CET1 ratios for participating firms in each quarter, $\bar{k}_t$.

References

Acharya, V. V., R. Engle, and D. Pierret (2014), "Testing macroprudential stress tests: The risk of regulatory risk weights," *Journal of Monetary Economics*, 65, 36–53.

Acharya, V. V., and O. Merrouche (2013), "Precautionary hoarding of liquidity and interbank markets: Evidence from the subprime crisis," *Review of Finance*, 17(1), 107–160.

Acharya, V. V., P. Schnabl, and G. Suarez (2013), "Securitization without risk transfer," *Journal of Financial Economics*, 107(3), 515–536.

Adrian, T., D. M. Covitz, and N. Liang (2015), "Financial stability monitoring," *Annual Review of Financial Economics*, 7(1), 357–395.

Allen, F., A. Babus, and E. Carletti (2012), "Asset commonality, debt maturity and systemic risk," *Journal of Financial Economics*, 104(3), 519–534.

[42] See www.federalreserve.gov/publications/files/2018-dfast-methodology-results-20180621.pdf

Andersen, H., K. R. Gerdrup, R. M. Johansen, and T. Krogh (2019), "A macroprudential stress testing framework," Norges Bank Staff Memo 1/2019.

Ashcraft, A., J. McAndrews, and D. Skeie (2011), "Precautionary reserves and the interbank market," *Journal of Money, Credit and Banking*, 43(Suppl. 2), 311–348.

Aymanns, C., C. Caceres, C. Daniel, and L. Schumacher (2016), "Bank solvency and funding cost," International Monetary Fund Working Paper 16/64.

Bank of England (2015), "The Bank of England's approach to stress testing the UK banking system," available at www.bankofengland.co.uk/-/media/boe/files/stress-testing/2015/the-boes-approach-to-stress-testing-the-uk-banking-system.

Bank of England (2019, March), "Stress testing the UK banking system: Key elements of the 2019 annual cyclical scenario," available at www.bankofengland.co.uk/news/2019/march/key-elements-of-the-2019-stress-test.

Bank for International Settlements (2018, April), "Framework for supervisory stress testing of central counterparties (CCPs)," available at www.bis.org/cpmi/publ/d176.htm.

Basel Committee on Banking Supervision (2009, May), "Principles for sound stress testing practices and supervision," available at www.bis.org/publ/bcbs155.pdf.

Basel Committee on Banking Supervision (2017, December), "Supervisory and bank stress testing: Range of practices," available at www.bis.org/bcbs/publ/d427.pdf.

Bassett, W. F., G. B. Brang, D. E. Rappoport, and D. Schwindt (2020), "Banks' short-term wholesale funding: Evidence from regulatory reports," Mimeo.

Bassett, W. F., S. Gilchrist, G. C. Weinbach, and E. Zakrajšek (2011), "Improving our ability to monitor bank lending," in M. Brunnermeier and A. Krishnamurthy (eds.), Risk topography: Systemic risk and macro modeling, National Bureau of Economic Research.

Bassett, W. F., K. Heinecke, D. E. Rappoport, and J. Schmidt (2016), "How and why capital stress tests might incorporate a funding shock," Presentation at the Fifth Annual Stress Test Modeling Symposium, Federal Reserve Bank of Boston, available at www.bostonfed.org/-/media/Documents/2016-modeling-symposium/Panel_4_William_Bassett_Symposium_2016.pdf?la=en.

Baudino, P., R. Goetschmann, J. Henry, K. Taniguchi, and W. Zhu (2018), "Stress-testing banks—a comparative analysis Bank of International Settlements," available at www.bis.org/fsi/publ/insights12.pdf.

Begenau, J., S. Biggio, and J. Majerovitz (2018), "Data lessons about bank behavior," Mimeo.

Berger, A. N., and G. F. Udell (2004), "The institutional memory hypothesis and the procyclicality of bank lending behavior," *Journal of Financial Intermediation*, 13(4), 458–495.

Board of Governors of the Federal Reserve System (2015, July 20), "Calibrating the GSIB surcharge," White Paper, available at www.federalreserve.gov/aboutthefed/boardmeetings/gsib-methodology-paper-20150720.pdf.

Board of Governors of the Federal Reserve System (2019), "Comprehensive capital analysis and review 2019: Assessment framework and results," available at www.federalreserve.gov/publications/files/2019-ccar-assessment-framework-results-20190627.pdf.

Brunnermeier, M. K., A. Crocket, C. Goodhart, A. D. Persaud, and H. Shin (2009), *The fundamental principles of financial regulation*, International Center for Monetary and Banking Studies Centre for Economic Policy Research.

Brunnermeier, M. K., and L. H. Pedersen (2009), "Market liquidity and funding liquidity," *Review of Financial Studies*, 22, 2201–2238.

Cappiello, L., and D. Supera (2014), "Fire-sale externalities in the euro area banking sector," *ECB Financial Stability Review*, 99–108.

Cont, R., and E. Schaanning (2017), "Fire sales, indirect contagion and systemic stress testing," Norges Bank Working Paper 02/17.

Danielsson (2011), "Risk and crises: How the models failed and are failing," VoxEU.org, available at https://voxeu.org/article/risk-and-crises-how-models-failed-and-are-failing.

Diamond, D., and R. Rajan (2005), "Liquidity shortages and banking crises," *The Journal of Finance*, 60(2), 615–647.

Duarte, F., and T. Eisenbach (2018, October), "Fire-sale spillovers and systemic risk," Federal Reserve Bank of New York Staff Reports No. 645, revised June 2018.

Durdu, B., R. Edge, and D. Schwindt (2017, May 5), "Measuring the severity of stress-test scenarios," FEDS Notes, available at www.federalreserve.gov/econres/notes/feds-notes/measuring-the-severity-of-stress-test-scenarios-20170505.htm.

Eisenberg, L., and T. Noe (2001), "Systemic risk in financial systems," *Management Science*, 47(2), 236–249.

European Banking Authority (2020), "2020 EU-wide stress test: Methodological note," available at https://eba.europa.eu/file/228066/download?token=3mU425ga.

Faria-e-Castro, M., J. Martinez, and T. Philippon (2017), "Runs versus lemons: Information disclosure and fiscal capacity," *Review of Economic Studies*, 84(4), 1683–1707.

Firestone, S., A. Lorenc, and B. Ranish (2017), "An empirical economic assessment of the costs and benefits of bank capital in the US," Finance and Economics Discussion Series 2017–034.

Gale, D., and T. Yorulmazer (2013), "Liquidity hoarding," *Theoretical Economics*, 8(2), 291–324.

Geanakoplos, J. (2003), "Liquidity, default, and crashes: Endogenous contracts in general equilibrium," in *Advances in Economics and Econometrics: Theory and Applications, Eighth World Conference*, Vol. II, Econometric Society Monographs.

Georgescu, O. M., M. Gross, D. Kapp, and C. Kok (2017), "Do stress tests matter? Evidence from the 2014 and 2016 Stress Tests," ECB Working Paper No. 2054.

Goldstein, I., and Y. Leitner (2018), "Stress tests and information disclosure," *Journal of Economic Theory*, 177, 34–69.

Gorton, G., and A. Metrick (2012), "Securitized banking and the run on repo," *Journal of Financial Economics*, 104(3), 425–451.

Greenwood, R., A. Landier, and D. Thesmar (2015), "Vulnerable banks," *Journal of Financial Economics*, 115(3), 471–485.

Haldane, A. (2009), "Why banks failed the stress test," Speech at the Marcus Evans conference on Stress Testing.

Hanson, S. G., A. K. Kashyap, and J. C. Stein (2011), "A macroprudential approach to financial regulation," *Journal of Economic Perspectives*, 25(1), 3–28.

He, Z. and A. Krishnamurthy (2019), "A macroeconomic framework for quantifying systemic risk," *American Economic Journal: Macroeconomics*, 11(4), 1–37.

Heider, F., M. Hoerova, and C. Holthausen (2015), "Liquidity hoarding and interbank market rates: The role of counterparty risk," *Journal of Financial Economics*, 118(2), 336–354.

International Monetary Fund (2012, August), "Macrofinancial stress testing—principles and practices," available at www.imf.org/external/np/pp/eng/2012/082912a.pdf.

International Monetary Fund (2017a), "Japan: Financial sector assessment program," available at www.imf.org/~/media/Files/Publications/CR/2017/cr17284.ashx.

International Monetary Fund (2017b), "Sweden: Financial sector assessment program," available at www.imf.org/~/media/Files/Publications/CR/2017/cr17307.ashx.

International Monetary Fund (2018), "Belgium: Financial sector assessment program," available at www.imf.org/~/media/Files/Publications/CR/2018/cr1869.ashx.

International Monetary Fund (2019), "Switzerland: Financial sector assessment program," available at www.imf.org/~/media/Files/Publications/CR/2019/1CHEEA2019003.ashx.

International Monetary Fund, Financial Stability Board, and Bank for International Settlements (2016), "Elements of effective macroprudential policies: Lessons from international experience," available at www.imf.org/external/np/g20/pdf/083116.pdf.

Jobst, A. A., L. Lian Ong, and C. Schmieder (2013), "A framework for macroprudential bank solvency stress testing: Application to S-25 and other G-20 country FSAPs," IMF Working Paper WP/13/68.

Jotikasthira, C., C. Lundblad, and T. Ramadorai (2012), "Asset fire sales and purchases and the international transmission of funding shocks," *Journal of Finance*, 67, 2015–2050.

Kashyap, A., R. Rajan, and J. C. Stein (2008), "Rethinking capital regulation," in *Maintaining stability in a changing financial system*, Federal Reserve Bank of Kansas City.

Khandani, A. E., and A. W. Lo, (2011), "What happened to the quants in August 2007? Evidence from factors and transactions data," *Journal of Financial Markets*, 14(1), 1–46.

Kroszner, R. S., and P. E. Strahan (2011), "Financial regulatory reform: Challenges ahead," *American Economic Review*, 101(3), 242–246.

Lorenc, A. G., and J. Y. Zhang (2018), "The differential impact of bank size on systemic risk," Finance and Economics Discussion Series 2018–066.

Manconi, A., M. Massa, and A. Yasuda (2012), "The role of institutional investors in propagating the crisis of 2007–2008," *Journal of Financial Economics*, 104(3), 491–518.

McConnell, J. J., and A. Saretto (2010), "Auction failures and the market for auction rate securities," *Journal of Financial Economics*, 97(3) 451–469.

Merton, R.C. (1974), "On the pricing of corporate debt: The risk structure of interest rates," *Journal of Finance*, 29(2), 449–470.

Schmitz, S., M. Sigmund, and L. Valderrama (2017), "Bank solvency and funding cost: New data and new results," International Monetary Fund Working Paper 17/116.

Shleifer, A., and R. Vishny (2011), "Fire sales in finance and macroeconomics," *Journal of Economic Perspectives*, 25(1), 29–48.

Accounting for Amplification Mechanisms in Bank Stress Test Models at the Bank of Canada[*]

Grzegorz Hałaj and Virginie Traclet

1 Bank Stress Testing at the Bank of Canada Is Focused on Assessing Systemic Risk

The Bank of Canada does not have a supervisory mandate regarding financial institutions.[1] Rather, as part of its commitment to promote the economic and financial welfare of Canada, the bank conducts analysis and research to identify and mitigate systemic risk, that is, the risk that the financial system becomes impaired and the provision of key financial services breaks down, to the point where the real economy may be materially affected.[2] Stress testing models are an important component of the analytical toolkit used at the Bank of Canada to assess systemic risk and the resilience of the financial system.[3] Given the Bank of Canada's focus on systemic risk, it would be ideal to have a comprehensive systemic-risk stress testing framework that would consider the following:

- all financial system entities (banks and nonbanks) and their interlinkages (direct and indirect) to account for the propagation and amplification mechanisms that characterize systemic risk;
- first- and second-round effects affecting these financial system entities; and
- the regulatory standards that apply to these entities. These standards are designed to safeguard the soundness of entities, but if entities have difficulties satisfying these requirements under severe stress, they may be enticed to take management actions that seem desirable from their individual perspective but could have negative externalities.

---

[*] The views expressed in this chapter are those of the authors. No responsibility for them should be attributed to the Bank of Canada. We would like to extend special thanks to Sofia Priazhkina for her expertise on the Macro-Financial Risk Assessment Framework (MFRAF) and her ongoing support and advice in the writing of this chapter and Nicolas Whitman and Andisheh Danaee for their technical assistance. We also thank Marc-André Gosselin and Miguel Molico for their comments on this chapter.

[1] In Canada, the Office of the Superintendent of Financial Institutions (OSFI) regulates and supervises federally regulated financial institutions, including banks.

[2] The analysis of financial system vulnerabilities—that is, preexisting conditions that can amplify and propagate shocks throughout the financial system—is at the core of the Bank of Canada's framework to assess financial stability risks. For details on the framework, see Christensen et al. (2015).

[3] For instance, the Bank of Canada has developed the Framework for Risk Identification and Assessment (FRIDA), a suite of models used to quantify the impact of financial stability risk scenarios on aggregate macrofinancial variables, as well as the household, business, and banking sectors. For details, see MacDonald and Traclet (2018).

For instance, to satisfy leverage requirements, institutions may sell assets, potentially putting downward pressure on the price of these assets, thereby affecting other entities that hold similar assets.

Such a comprehensive stress testing framework would require overcoming significant methodological and data challenges (Anderson et al., 2018). This is why, about a decade ago, the Bank of Canada, like many other central banks, started by focusing its stress test modeling efforts on banks, given their central role in the financial system (lending, market-making, financial market infrastructure [FMI] participants, etc.).[4] The Global Financial Crisis (GFC) clearly demonstrated that the banking sector can be severely affected by second-round effects (e.g., fire sales, funding liquidity disruptions) resulting from the propagation and/or amplification of the initial shocks. These second-round losses can arise "mechanistically" (e.g., because of contractual exposures to entities that default) but also from the behavioral responses of entities. For instance, entities affected by the initial shock may sell securities to deleverage, in turn affecting other entities holding the same securities via mark-to-market (MTM) accounting.

Given the Bank of Canada's focus on systemic risk, we have put the emphasis on capturing these amplification (or second-round) effects in our bank stress test models.[5] This, however, raised significant challenges, notably because it requires modeling the behavior of financial institutions under stress, an area where the literature was (and remains, to a large extent) under development. The evolution of our ability to model behaviors under stress is reflected in the two banking-sector stress test models described in this chapter. In the Macro-Financial Risk Assessment Framework (MFRAF), our older model, bank behavior is modeled in a rather simple, mechanistic manner (e.g., deleveraging through securities fire sales). However, MFRAF innovates by modeling endogenously the interactions between funding liquidity and solvency to capture the type of dynamic observed during the GFC, something that was typically not considered in bank stress test models at the time (International Monetary Fund [IMF], 2014).[6] But banks do not have an optimization objective in MFRAF, which implies that the way they adjust their balance sheets in response to shocks and regulatory constraints likely misses on some channels that could contribute to systemic risk. The Bank Dynamic Balance Sheet model (BDBS), our most recent model, innovates significantly in this regard by explicitly modeling banks' decisions to adjust their balance sheets in response to shocks and regulatory constraints under a profit-oriented optimization objective. This is achieved by modeling banks' behaviors following the principles of optimal portfolio choice theory and asset and liability management, thus capturing the objectives of

---

[4] More recently, the Bank of Canada has developed a stress test model of open-ended mutual funds to quantify the impact of these funds on market liquidity (Arora et al., 2019). For an application, see Bank of Canada Financial System Review, May 2019.

[5] Tarullo (2016) notes that the Federal Reserve's review of its stress programs highlighted the importance of integrating indirect risks to bank capital, such as funding and fire sales, to expand the macroprudential dimension of stress testing. Other central banks (e.g., Bank of England, European Central Bank) have also expanded the coverage of their stress testing models to capture a wider range of systemic risk drivers. See Dees and Henry (2017) and Farmer et al. (2020).

[6] For instance, Dees and Henry (2017) note that because solvency shocks may affect the liquidity situation of banks via several channels, capturing the solvency–liquidity interactions in the European Central Bank (ECB) STAMP€ modeling framework was a necessary extension of that framework.

profitability and survival. The development of our models has been made possible thanks to an increasingly rich set of bank regulatory data over the last decade.[7]

Bank stress test modeling remains a dynamic field, with ongoing progress at central banks, at supervisory agencies, in academia, and in the private sector, as illustrated in this handbook. In this chapter, as we describe MFRAF (Section 2) and the BDBS (Section 3) and explain our modeling choices, we focus on the practical considerations from the perspective of the Bank of Canada for these choices (with a few comparisons with modeling approaches used by other central banks for illustrative purposes) as opposed to providing an exhaustive view of the various modeling options offered in the literature. Section 4 concludes.

2 The Macro-Financial Risk Assessment Framework

2.1 Objectives and Model Overview

The Bank of Canada started developing MFRAF in the late 2000s, with a view to capturing the impact of the various sources of risks, and their interactions, affecting banks in order to understand and quantify systemic risk. Doing so required modeling both first-round losses (i.e., credit losses and market losses) and second-round losses. We put the emphasis on three sources of second-round effects that played a prominent role in the GFC, contributing to significant losses for many banks worldwide and even the failure of some of them: (1) funding liquidity risk and its interactions with solvency risk, (2) indirect price-mediated contagion due to common asset holdings (i.e., market liquidity risk from fire sales), and (3) direct contagion due to interbank exposures.

Stress test modeling was relatively new when MFRAF development started, especially for risks other than credit risk. But as the field evolved rapidly over the last few years, MFRAF has capitalized on this progress, with major improvements to the model since its first version (described by Gauthier et al. [2010] and Anand et al. [2014]). For instance, the modeling of feedback-loop effects between solvency and funding liquidity risk has been significantly enhanced over time: initially, MFRAF considered only the implications of deteriorating solvency on funding liquidity risk (i.e., run risk), whereas it now accounts also for the effects that rising funding costs and costly liquidation of assets have on solvency risk (Fique, 2017). MFRAF has also been modified to account for changes in banking regulation. For instance, a leverage module was added to account for the Basel III leverage ratio and the associated potential for leverage-constrained banks to engage in asset fire sales.[8] Making these changes to MFRAF has been facilitated by its modular structure (Figure 25.1), which allows for the modification of targeted aspects of the model over time as we leverage research at the Bank of Canada and progress in the literature. It is important to note, however, that the modules are not independent from each other: they are linked by the dependencies between the different sources of risk.

---

[7] In Canada, OSFI, the Bank of Canada, and the Canadian Deposit Insurance Corporation (CDIC) have access to bank regulatory data. The three agencies work together to develop and enhance bank regulatory returns.

[8] It should be noted, however, that MFRAF does not model all drivers of capital ratios: income and risk-weighted assets (RWAs) are generated outside of MFRAF in satellite models. For details, see MacDonald and Traclet (2018).

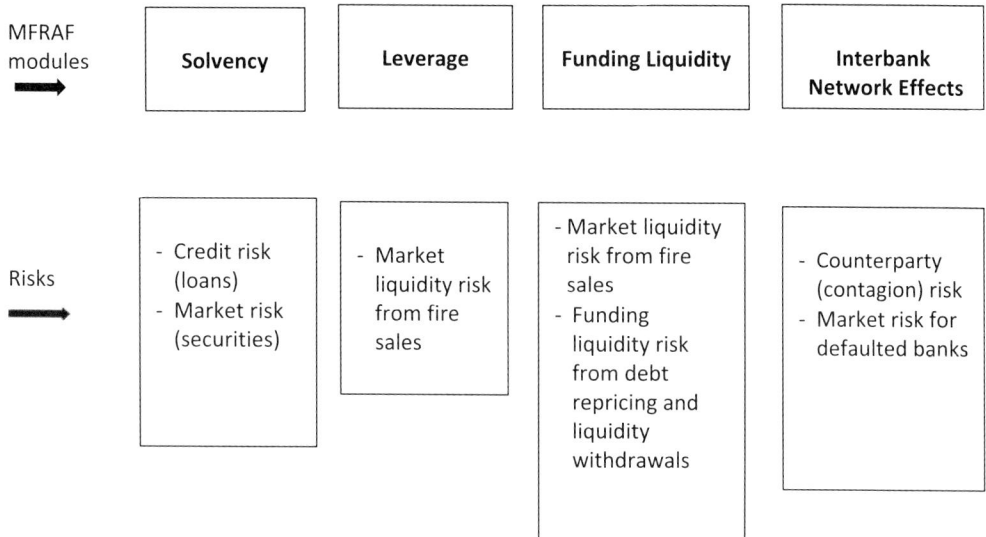

Figure 25.1 MFRAF modular structure and types of risks.

Because a core objective of MFRAF is to explain and decompose the drivers of systemic risk in the banking sector, the different sources of risk captured in MFRAF are all quantified in terms of their impact on one measure: banks' common equity Tier 1 (CET1) ratio. Decomposing and quantifying systemic risk drivers with a unified metric allows for the identification of vulnerabilities in the system and can inform policy discussions. Nevertheless, the model also generates other outputs (e.g., leverage ratio, liquidity coverage ratio [LCR]), thus allowing a better understanding of how stress episodes would affect banks relative to the multidimensional regulatory standards in Basel III.

MFRAF is a two-period banking model with distribution of credit and income losses in both periods. In addition to considering the behavior of banks, the model accounts for the behavior of rational creditors who provide funding to banks conditional on the uncertainty of banks' solvency and liquidity positions. Because creditors' behavior is based on both existing financial stress and expectations about future stress, MFRAF is modeled as a two-period model with an interim funding-cost decision and sequential income and credit losses. Figure 25.2 illustrates the timelines of events in MFRAF, and Figure 25.3 illustrates the balance sheet of a bank.

Stress testing typically relies on the assumption of a static balance sheet. Under this assumption, banks cannot actively modify their assets and liabilities via management actions to deal with the impact of stress scenarios, although the size and composition of their balance sheets evolve over the stress horizon because of the effects of the scenario (e.g., defaults on loan exposures, securities price shocks, etc.).[9] In MFRAF, we rely on the static-balance-sheet assumption in the solvency and network modules. However, we made the choice

---

[9] The static-balance-sheet assumption implies that assets and liabilities that mature over the stress horizon are replaced with similar instruments in terms of type, currency, credit quality, and original maturity as at the beginning of the exercise.

$t = 0$	$t = 1$		$t = 2$		
Adverse scenario	- Credit losses $_{t=1}$ - Initial market shocks - Income shocks $_{t=1}$ - RWA shocks $_{t=1}$	Leverage-driven fire-sales	- Liquidity effects: ▪ Rising funding costs ▪ Fire sales to repay funding withdrawals - RWA shocks $_{t=2}$	- Credit losses $_{t=2}$ - Income shocks $_{t=2}$	Interbank losses

Figure 25.2 Timelines in MFRAF.
Note: Italics refer to capital drivers that are generated in satellite models.

Assets	Liabilities
Cash (existing funds and retained earnings)	Maturing liabilities (composed from weighted short-term and long-term debt exposures)
Securities	Nonmaturing liabilities
Loans	
Other assets (partially liquid and illiquid)	Equity

Figure 25.3 Bank balance sheet in MFRAF.

to partially deviate from this assumption in the leverage and liquidity modules because we wanted to capture the potential for fire-sales contagion among banks stemming from banks actively engaging in asset liquidations to restore compliance with the leverage ratio and to cover maturing liabilities, respectively. The GFC clearly illustrated that the fast-paced impact of fire sales and associated liquidity spirals can rapidly lead to significant losses and contagion among banks, hence making this channel particularly important for a stress test model that aims to capture sources of systemic risk. Moreover, in the funding-liquidity module, banks that can access funding markets replace their maturing liabilities with unsecured short-term wholesale funding, implying a change in their liability structure. When modeling balance-sheet adjustments, MFRAF uses an approach based on behavioral heuristics: if banks breach regulatory requirements in the second time period, banks do not respond by taking actions to restore compliance.

2.2 Key Model Characteristics[10]

Because MFRAF captures different sources of risk, it combines several modeling techniques chosen to be best suited for each of these risks.

2.2.1 Solvency Module

The solvency module produces two types of first-round capital losses that are explicitly applied to the profits of banks: market losses and credit losses.[11]

---

[10] This section provides a high-level description of MFRAF and the rationale for the modeling approaches used in the model. For more details, including a technical description of the model, see the MFRAF Technical Report at www.bankofcanada.ca/wp-content/uploads/2017/09/tr111.pdf.

[11] As mentioned earlier, income and RWAs are generated using satellite models. For income, we use simple reduced-form econometric models that capture the relationships between a set of macrofinancial variables and

Market losses are calculated simply by applying the market shocks from the stress scenario to banks' initial securities holdings.[12] Because banks can have different securities portfolios, market losses can vary among banks under a common stress scenario.

Credit losses are equal to the probability of default (PD) of loans multiplied by the corresponding loss given default (LGD) and exposures at default (EAD) for each of the sectors to which banks lend. Although this approach is standard, we consider the entire distribution of credit losses as opposed to a single value. This approach was chosen to consider not only the most expected but also the less likely realizations of the credit shocks. This is important, first, to account for the fact that stress scenarios typically consider low-probability events associated with high uncertainty, making a case for considering the entire distribution of losses. Secondly, the uncertainty associated with credit losses in the solvency module plays an important role in the second-round effects in the model: the magnitude of the second-round effects varies with the higher moments of the distribution of PDs (e.g., variance, skewness, and kurtosis). Although tail events associated with very low probabilities are less likely to happen, they are associated with higher credit losses and hence a more pronounced decline in capital positions. As we explain later in the chapter, second-round effects materialize because of observed and expected deterioration in capital positions; consequently, higher credit losses lead to larger second-round losses.

The multivariate distribution of PDs is positively skewed and allows for stress-sensitive correlations between the different economic sectors. Therefore, banks' credit losses are correlated because banks are exposed to the same sectors of the economy and credit risk is correlated across economic sectors. The stochastic component of credit losses comes exclusively from the distribution of PDs, which are generated using satellite models accounting for cross-sector correlations. In contrast, LGDs and EADs are deterministic and calibrated consistently with the stress scenario considered.[13]

In addition to market and credit losses, shocks applied to off-balance-sheet exposures affect both the leverage ratio and the risk-weighted capital ratio.

2.2.2 Leverage Module

As banks' capital positions worsen because of initial solvency shocks, banks also face a deterioration in their leverage ratios, to the point where one or several banks could even

---

interest and noninterest income, respectively, thus ensuring that income is consistent with the stress scenario. Retained earnings (i.e., net income after taxes minus dividends) can be calculated using various dividend rules, including constraints on dividend distributions associated with the Basel III capital conservation buffer. The RWA model "mimics" the Basel III RWA framework such that the evolution of credit RWAs is consistent with the probabilities of default that enter MFRAF's solvency module. For details on these satellite models, see MacDonald and Traclet (2018).

[12] Because of data availability, we make the simplifying assumption that all securities are MTM, whereas in practice, MTM accounting applies only to trading book securities and available-for-sale (AFS) securities in the banking book, not to held-to-maturity (HTM) securities. This is because the data on securities holdings do not distinguish between these categories. Moreover, we do not account for hedges or other market-risk mitigating strategies. As a result, we may overestimate market losses.

[13] Although it would be appealing to also consider distributions for LGDs because one can expect that LGDs would also be affected by uncertainty, there is little in the literature to guide the development of LGD models, and LGD data availability is scarce. However, sensitivity analyses can be conducted to assess the sensitivity of the results to the LGDs considered.

breach the regulatory leverage requirement. In MFRAF, we consider that banks would engage in deleveraging by selling securities to return to a leverage ratio that satisfies requirements.[14] We measure the leverage ratio using Tier1 capital and exposure measures consistent with regulatory standards. Because the exposure measure includes on-balance-sheet bank exposures, the liquidation of securities may affect both the numerator and the denominator of the capital ratio, eventually leading to a lower leverage ratio. We assume that banks account for this effect and will sell assets only if the sale improves their regulatory ratio.

The deleveraging actions of leverage-constrained banks could lead to MTM losses, both for the banks that sell and for other banks that hold the same securities. Common securities holdings can thus create an indirect price-mediated contagion channel. This could in turn result in more banks breaching the regulatory requirement, ultimately leading to fire sales as more banks would engage further in liquidations to restore their leverage ratios.

MFRAF uses a mechanistic approach to model banks' deleveraging actions. First, banks sell securities if they breach the regulatory leverage requirement and the fire sales improve leverage. Banks do not form expectations regarding their future leverage ratios and do not consider preemptive actions to avoid regulatory breaches.[15] Secondly, deleveraging banks liquidate securities with the objective of minimizing the impact of MTM losses on solvency. In doing so, they focus exclusively on the impact of their own actions; that is they do not form expectations about the actions of other banks and their potential price impact.

The module is implemented in the following iterative way. First, we compute the leverage ratio following the realization of credit and market losses in $t = 1$ for each bank and each credit shock draw; if no bank breaches the leverage-ratio requirement, the module terminates (i.e., there are no fire sales). If at least one bank breaches the leverage-ratio requirement, the recursive deleveraging process described in Figure 25.4 takes place.[16] In this process, banks assess the solvency impact of selling each type of security and execute the sale(s) that minimizes losses. Because of MTM accounting, the new securities price (post-sale) is applied to the remaining holdings of these securities for the liquidating bank and for all banks that also hold these securities. Other banks can then become leverage constrained because of these MTM losses. The fire-sales process stops when no bank is leverage constrained. We allow banks to sell securities up to a certain amount at a time, such that a bank needs to conduct multiple rounds of sales to liquidate a large number of securities. This feature reduces the bias that can arise as a result of the first-mover advantage given to the first randomly chosen bank. It also makes the simulation more consistent with the length of stress horizons because banks are modeled to sell assets over multiple months.

---

[14] Although deleveraging actions could include selling a broader range of assets in practice, we restrict the model to the sale of securities for two reasons. First, selling of securities can be done relatively expediently, which makes it an attractive option to banks. Secondly, from a practical viewpoint, it is easier for us to determine an asset-selection process and to calibrate prices for securities than for other types of assets that are less liquid (e.g., loans).

[15] To "mimic" preemptive sales in the model in a simplistic manner, it is possible to consider that banks start selling securities when their leverage ratio breaches thresholds that are higher than the regulatory requirement. Sensitivity analyses can be conducted for different leverage-ratio thresholds.

[16] If several banks are leverage constrained, we randomly select one of these banks to start the deleveraging process. The bank order may affect the results.

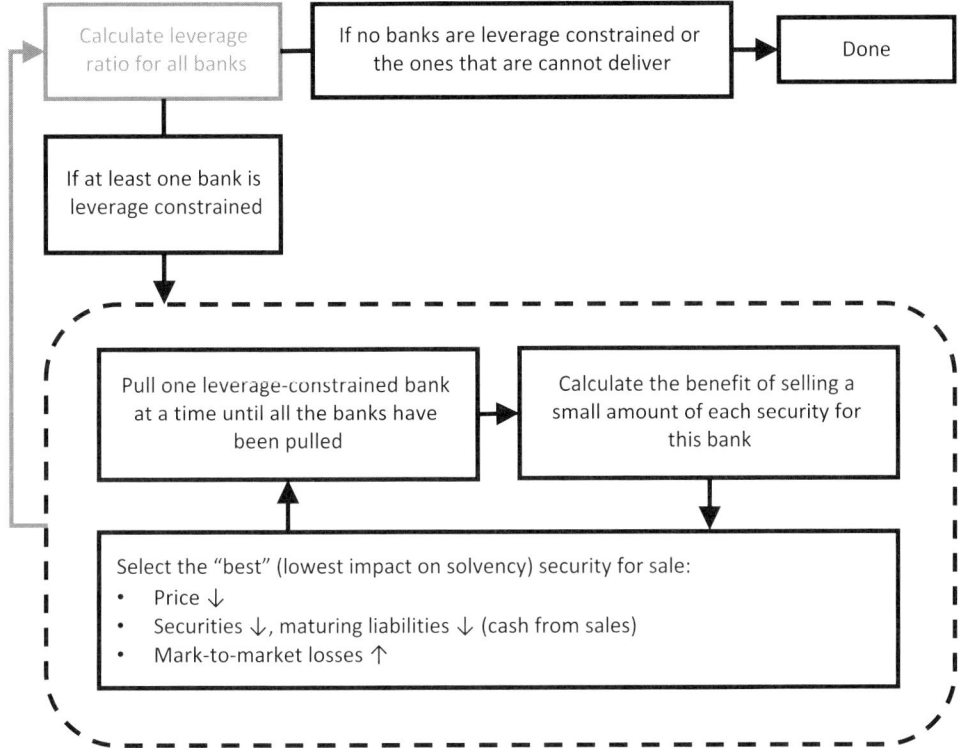

Figure 25.4 Schematic description of the leverage module and associated fire sales.

The market-price impact of securities sales depends on the stress scenario and the market characteristics of the securities (e.g., the more liquid the securities, the lesser the price impact). The calibration of price curves is scenario specific and is based on market expertise and supporting models. Although MFRAF models banks only, the calibration of price curves indirectly accounts for other types of players in these markets because it considers both the fundamental characteristics of the securities and the different participants in these markets, including how they would be affected by the stress scenario considered (e.g., they could be sellers or buyers, thus either amplifying or offsetting the sales conducted by banks). Although this approach relies heavily on expert judgment, it allows for the consideration of a range of impacts and the sensitivity of stress test results to a range of assumptions regarding market dynamics under stress. Figure 25.5 shows illustrative price curves for three types of securities.

2.2.3 Funding Liquidity Module

The approach for MFRAF's funding liquidity module was chosen to capture the key characteristics of bank liquidity risk highlighted during the GFC. More specifically, we wanted to capture the interplay between bank solvency, funding, and liquidity risks. When MFRAF was originally developed, these interactions were generally not considered in the stress testing models used by central banks. An exception was the Risk Assessment Model of Systemic Institutions (RAMSI), the Bank of England stress test model at the time

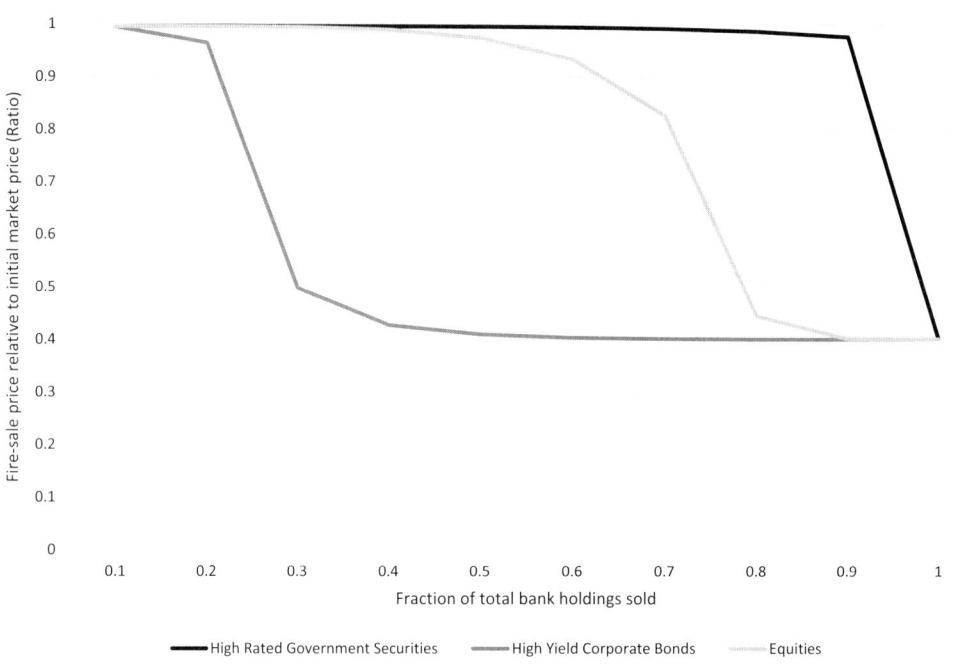

Figure 25.5 Illustrative price curves.

(Aikman et al., 2009; Burrows et al., 2012). RAMSI considered that banks would lose access to funding markets in response to perceived solvency risks using a simple scoring approach based on calibrated thresholds for a range of bank indicators (expected future capital ratio, past profitability, reliance on unstable sources of funding, etc.). In contrast, we wanted the interplay between solvency and liquidity to be informed by the microeconomic foundations of the behavior of banks' creditors, thus allowing for an endogenous mechanism in the model.

- The GFC illustrated the mutually reinforcing deterioration between banks' funding liquidity and the liquidity of their asset holdings, which ultimately led to liquidity spirals (Brunnermeier and Pedersen, 2009). To capture this interplay, MFRAF accounts for the interaction between banks' liability structures and the composition of their assets. This was made possible thanks to very detailed regulatory data on liabilities, their maturity profile, and securities holdings.[17]
- MFRAF follows Morris and Shin (2016): the decision of a bank's creditors to roll over or withdraw funding is driven by their concerns about the bank's future solvency

---

[17] The data come from the Net Cumulative Cash Flows (NCCF) return, which OSFI implemented as part of its Liquidity Adequacy Requirements (LAR) Guideline. The LAR Guideline establishes the framework within which OSFI assesses whether banks maintain adequate liquidity (www.osfi-bsif.gc.ca/Eng/fi-if/rg-ro/gdn-ort/gl-ld/Pages/LAR_index.aspx) and details the various liquidity metrics used by OSFI to assess the liquidity adequacy of banks. The NCCF is one of these metrics and is aimed at measuring an institution's cash flows beyond the 30-day horizon (up to 12 months) to capture the risk posed by funding mismatches between assets and liabilities.

(thus linking solvency and liquidity risk), the extent to which the bank relies on unstable funding sources, and the quality of its assets. It also accounts for the strategic behavior of creditors demonstrated by Rochet and Vives (2004): the rollover decision of banks' creditors is also influenced by their beliefs about the proportion of creditors who withdraw funding.

In MFRAF, a fraction of banks' liabilities matures in $t = 2$, and banks need to replace these liabilities.[18] We assume that they do so by accessing the unsecured short-term wholesale funding market at the end of $t = 1$. As shown during the GFC, some banks were unable to access funding markets altogether, which led them to conduct massive asset liquidations at distressed prices, whereas other banks were able to access funding markets but at higher prices than usual, which had a negative impact on their capital positions via the profit-and-loss account. Given our focus on the mechanisms through which systemic risk arises, our objective was to model access to funding markets and funding costs jointly in a manner that accounts for solvency risk.

The liquidity module follows the following timeline. At the end of $t = 1$, banks' capital positions deteriorate because of first-round effects that increase both solvency risk and funding costs. At that time, banks offer short-term debt contracts to replace the share of liabilities that will mature in $t = 2$. Creditors can redeem this debt claim at an interim date, thus exposing banks to rollover risk (i.e., the risk that a fraction of its creditors decide to withdraw).[19] If a large fraction of creditors withdraws, the bank may be forced to liquidate assets, which results in fire-sales losses. If cash reserves held on balance sheet and funds generated from the liquidation of securities are not sufficient to repay creditors, banks sell loans and other nonliquid assets at highly discounted prices.

In the model, the withdrawal decision of creditors depends nonlinearly on the bank's solvency. To understand this, consider the perspective of an individual creditor. If the bank was solvent in $t = 2$ regardless of the fraction of creditors that withdraw, each individual creditor would have no incentive to withdraw. Conversely, if the bank was insolvent in $t = 2$ for all realizations of income and credit shocks, and hence unable to meet its obligations, individual creditors would have no incentive to provide funding. However, if there are positive probabilities of a bank defaulting and staying solvent, the bank faces only partial withdrawals. It is assumed that a certain proportion of creditors withdraw under stress because they face external liquidity shocks or the return on loans promised by the bank is not enough to compensate for risk and opportunity cost. In this case, the bank either repays the withdrawals using available cash equivalents or liquidates assets to face its obligations. The asset sales lead to MTM losses or early liquidation costs and, hence, a decline in its

---

[18] The fraction of maturing liabilities is calibrated exogenously using liability data from the NCCF return by applying runoff rates that account for both the fundamental characteristics of the various funding sources (i.e., higher runoff rates for less stable liabilities) and the features of the stress scenario considered (e.g., higher runoff rates on foreign funding sources for Canadian banks in a Canada-centric stress scenario than in a global stress scenario).

[19] The rollover decision is considered at the individual-bank level, which implies that banks are assumed to have different creditors. Moreover, these creditors are "third-parties"; that is, banks are not the creditors of each other in MFRAF.

Table 25.1. Creditor payoff

	Fraction of withdrawing creditors $\leq T$	Fraction of withdrawing creditors $> T$
Roll over	r^B if bank is solvent	Depends on bank's liquidation process
Withdraw	r^F	r^F

capital position in $t = 2$.[20] If the share of creditors that withdraw is above a threshold T, which we call the *illiquidity threshold*, the bank becomes unable to honor its obligations.

The payoff structure for an individual creditor is as follows:

- If the creditor withdraws at the interim date (irrespective of whether the bank survives or not), he gets the risk-free rate r^F.
- If the creditor rolls over, he gets r^B ($> r^F$) if the bank remains solvent in $t = 2$ and only a partial repayment otherwise.[21] The funding cost r^B is determined endogenously by the bank. The uncertainty about the credit and income shocks in $t = 2$ makes creditors uncertain about the future solvency of the bank. Therefore, investors may roll over funding to a bank that will default in the following period.

Table 25.1 summarizes the payoff for an individual creditor depending on the fraction of creditors that withdraw.

In MFRAF, stressed funding costs are determined using a two-stage game: the bank proposes the funding rate in the first stage, and investors play a rollover game in the second stage. The model is solved by backward induction. In the rollover subgame, the behavior of creditors is conditional on the banks' future solvency, which is modeled using a coordination game with strategic complementarities. Because a bank's solvency in $t = 2$ is unknown in $t = 1$, creditors make their funding decision based on expectations. They receive a noisy signal regarding the additional credit shocks that the bank will face in $t = 2$. The strategic complementarities in the game arise from the fact that if one investor decides to withdraw, other investors also have more incentives to withdraw. As shown by Ahnert et al. (2018), parametrization exists that guarantees the existence and uniqueness of the equilibrium withdrawals in the rollover subgame.

The equilibrium illiquidity threshold T is thus determined endogenously, depending on the following:

- The fire-sales losses faced by the bank when it liquidates assets to offset funding withdrawals by creditors. These losses depend on the composition of a bank's asset holdings and market-liquidity conditions. If a bank holds a limited share of liquid assets, it is more likely to have to sell assets at fire-sale prices, thus generating higher

---

[20] In addition, MTM losses on securities also affect other banks that have similar securities holdings, like in the leverage module, thus contributing to contagion among banks.

[21] The exact amount that will be repaid to the funding providers depends on the amount of liquidity that a bank can extract by selling securities and illiquid assets in the state of default.

losses.[22] The higher these losses are, the less likely it is that a bank will be able to access unsecured funding.

- The bank's solvency prospects. These depend on the bank's initial solvency position, the magnitude of credit and market shocks, and their profitability. The better these prospects, the more likely a bank will be able to access unsecured funding.
- The degree of conservatism of banks' creditors.[23]
- The cost of funding offered by banks to creditors and alternative investment offers available.

MFRAF follows a simplifying approach by assuming that a bank will offer the minimum possible funding rate to keep the withdrawing investors. Given higher credit risk in a stress environment, the stressed funding cost is above the prestress levels. The illiquidity threshold and the cost of funding are jointly determined in equilibrium. But there can be cases where there is no solution, which implies that banks cannot access unsecured funding markets. Therefore, the model provides insights regarding the conditions under which banks can or cannot access markets and the price at which they can do so.

With this approach, we can disentangle the different components of funding liquidity risk (i.e., fire-sales losses versus rising cost of funding), which contributes to quantifying and explaining systemic risk.

2.2.4 Interbank Network Module

The final module in MFRAF captures the potential for banks to face counterparty credit losses as a result of their interbank exposures.[24] Following the deterioration in their capital position resulting from credit, market, fire-sales, and liquidity losses, some banks may be unable to fulfill their interbank obligations (in part or in their entirety). If this is the case, banks face credit counterparty losses on these exposures, leading to a further decline in their capital positions. This may, in turn, make them unable to fulfill their own interbank obligations, thus leading to a cascade of losses.

To calculate these network effects, we rely on the widely used clearing-payment vector developed by Eisenberg and Noe (2001), whereby interbank exposures are subordinate to all other debts and banks repay their interbank counterparties a sum that is proportional to the amounts due. The Eisenberg and Noe (2001) model is adjusted to account for additional MTM shocks for defaulting banks to reflect that the balance sheet of a defaulted bank may

---

[22] MFRAF does not account for the regulatory constraint imposed by the LCR, which requires banks to hold a certain proportion of liquid assets to cover funding withdrawals under stress over a 30-day period. However, the fact that banks liquidate assets to face funding withdrawals in stressed conditions in the model is consistent with the notion of liquidity-buffer usability embedded in the LCR. Although banks can liquidate all types of assets (liquid and illiquid), they have an incentive to sell liquid assets (consistent with the LCR) to minimize losses and their negative impact on capital ratios.

[23] In the model, the decision of creditors is delegated to risk managers whose remuneration depends on making the correct choice between lending to a bank and the bank remaining solvent, which results in inflows for risk managers, versus withdrawing while the bank remains solvent, which generates outflows. The more conservative risk managers are about this trade-off, the less likely the bank can obtain unsecured funding.

[24] The interbank exposures considered in MFRAF include deposits, secured and unsecured lending, bankers' acceptances, reverse repos, debt holdings, and over-the-counter (OTC) derivatives. The data come from the Interbank and Major Exposures Return (www.osfi-bsif.gc.ca/Eng/fi-if/rtn-rlv/fr-rf/dti-id/Pages/imer.aspx)

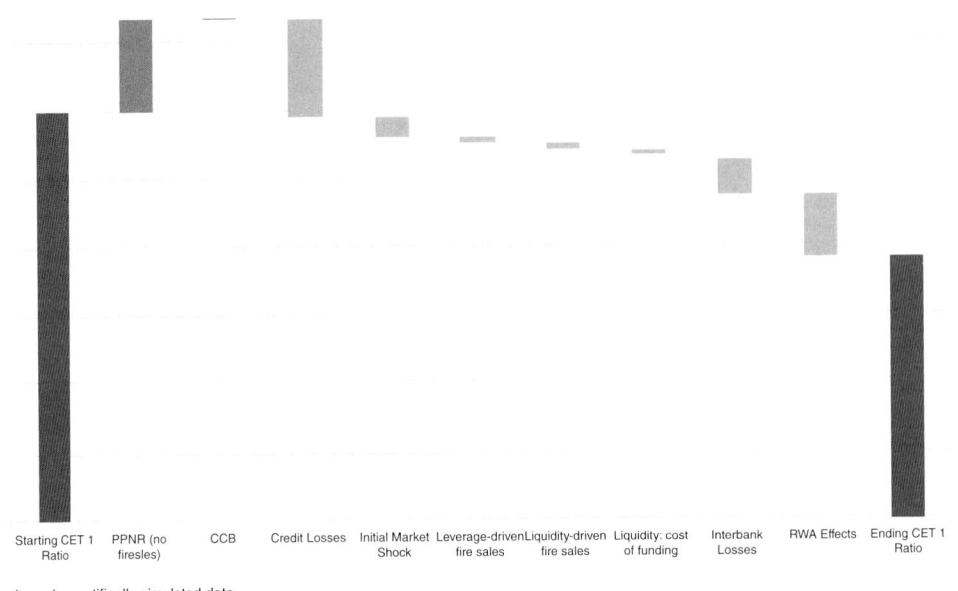

based on artifically simulated data

Figure 25.6 Decomposition of CET1 ratio drivers.
Note: The light-gray bars represent the positive contributors to capital positions, and the
dark-gray bars represent the negative contributors.

be further discounted as a result of liquidation costs, thus affecting its ability to repay other
banks. We chose to follow Eisenberg and Noe for two reasons. First, we assume that the
interbank contagion module would resolve instantaneously, given the concentrated nature
of the Canadian banking sector. Therefore, a clearing payment in equilibrium is a more
appropriate representation of the interbank contagion mechanism than an iterative default
cascade. Secondly, MFRAF leverage and liquidity modules already capture second-round
effects arising from common exposures and MTM effects, potentially mitigating the risk of
underestimating second-round effects (Battiston et al., 2016; Roncorini et al., 2019; Barucca
et al., 2020).

2.3 Model Outputs—Illustration

All figures presented in this section are illustrative and based on artificially simulated data.[25]
Because all the losses captured in MFRAF translate back into impact on the CET1 ratio, we
can decompose the drivers of the evolution of banks' capital position under stress in terms
of the various risks that affect banks (Figure 25.6).

Because MFRAF generates the entire distribution of the CET1 ratio, we can consider the
range of impact that a stress scenario could have on banks (Figure 25.7). This is an attractive
feature, given the uncertainty associated with tail events considered in stress tests.

---

[25] For details on data sources and model calibration, see Fique (2017).

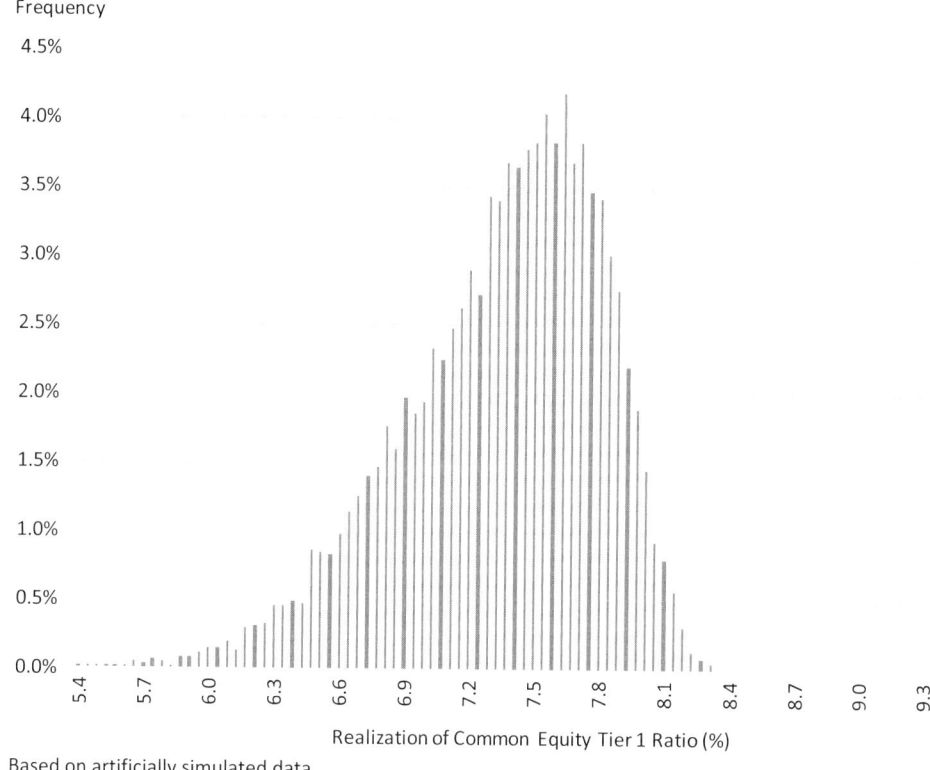

Based on artificially simulated data

Figure 25.7 Distribution of final CET1 ratio.

The main rationale for incorporating second-round effects in our model is that neglecting them can lead to a biased assessment of banks' resilience. This is illustrated in Figure 25.8: when second-round effects are not accounted for, the worst realizations of the impact of stress scenarios on capital positions are ignored. The distribution of the CET1 ratio with second-round effects displays a fatter tail than the one without second-round effects, which is consistent with systemic risk.

Although MFRAF captures a range of second-round effects, thus enhancing our quantification of systemic risk in the banking sector, banks do not have an optimization objective in the model, and management actions are limited to fire sales, which are modeled mechanistically. Consequently, MFRAF does not account for the behavioral component of systemic risk described in Section 1. This led us to invest in developing another model to assess these effects.

3 Accounting for Banks' Behavioral Responses to Stress under Multiple Regulatory Constraints and a Profitability Objective in a Stress Test Model

3.1 Objectives and Modeling-Approach Overview

The behavior of financial entities can amplify initial shocks, especially in turbulent market conditions, thus contributing to systemic risk. Entities tend to take action either

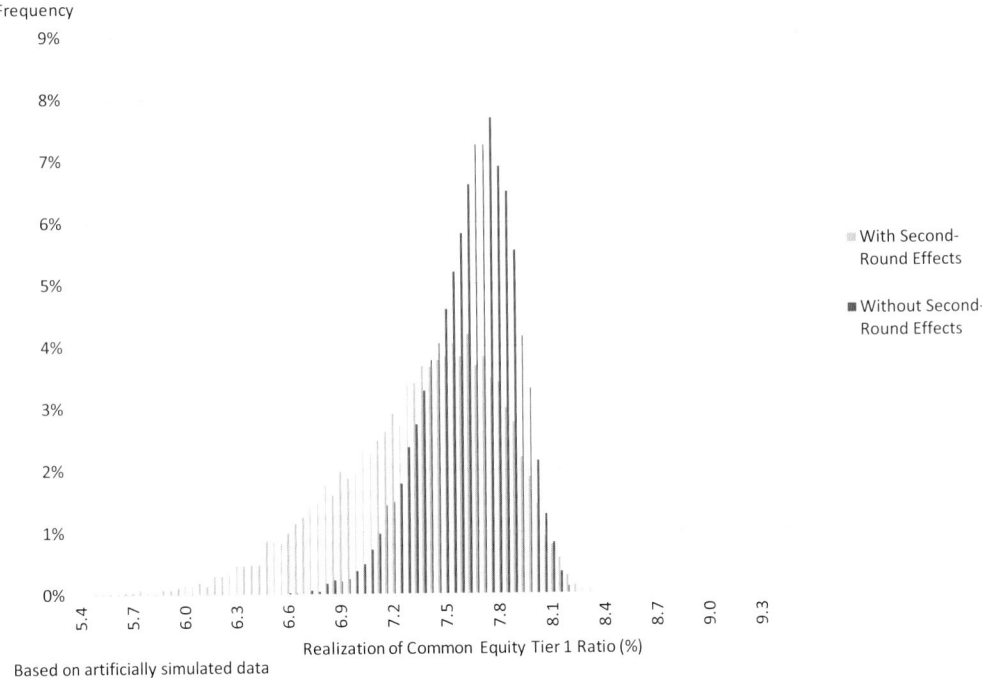

Figure 25.8 Distribution of CET1 ratio with and without second-round effects.

pre-emptively to protect themselves from shocks or ex post to counteract their impact. Although these actions may seem desirable from their individual viewpoints, they may not have the intended effect if they ignore the actions taken by other entities and could even lead to negative externalities for the system. Consequently, accounting for behavioral responses in stress test models is important to capture the potential for banks' strategic responses to contribute to systemic risk. As explained in Section 2, banks do not have an optimization objective in MFRAF, and the limited actions considered in the model are modeled mechanistically.[26] Given the Bank of Canada's focus on systemic risk, accounting for management actions constitutes the next step in our efforts to develop bank stress test models that can be used to assess systemic risk and financial-system resilience. Accounting for behaviors in stress testing models is an area of research in its infancy and faces significant challenges, thus calling for innovative approaches.

   In practice, banks could consider a broad range of behaviors under stress, ranging from changes to staff compensation to liquidation of entire business lines. To simplify the modeling task, we decided to consider a specific subset of actions related to asset, funding, and equity management informed by our stress testing experience and high-level discussions on the content from banks' recovery plans. Moreover, we decided to model banks' behaviors by following the principles of optimal portfolio choice theory and asset and liability management. This allows us to capture the objectives of profitability and survival, according

---

[26] Other examples of mechanistic dynamic balance sheets in stress testing are Burrows et al. (2012) and ECB (2017).

to which banks would consider management actions that increase their chances of withstanding shocks while limiting the associated costs (i.e., banks take profitability considerations into account) and respecting regulatory requirements.

Our objective is to resolve the cost–benefit trade-offs of different management actions to select those that balance risk and return subject to multiple regulatory requirements. This requires having a model to study the interplay of the multiple criteria considered by banks when deciding on management actions that dynamically affect their balance sheets. There is no such model in the literature on bank stress testing. However, we can adopt modeling techniques from finance and operations research to create a stress testing framework that considers strategic responses to shocks. The main modeling trade-off we had to address is about the granularity of banks' balance sheets and the completeness of the set of management actions and the tractability of the model to process comprehensive stress scenarios. Therefore, we had to compromise on the criteria that banks may consider in their choice of specific management actions. Moreover, we adopted a rather conservative assumption that banks do not raise new capital from the private sector, even though it is an admissible alternative management action to generated capital; see OSFI (2018).

Techniques related to optimal portfolio choice developed in the field of mathematical finance can inform our modeling strategy.[27] Kusy and Ziemba (1986) started applying these techniques to bank asset and liability management using stochastic programming techniques from operations research. Nevertheless, it remains difficult to directly transpose the optimal portfolio choice approach to the question of banks' balance-sheet management for at least three reasons that we need to address in our model:

- First, banks' balance sheets are made of a significant proportion of illiquid assets (e.g., loans, high-yield corporate bonds), whereas the assumption of high liquidity in finance models of optimal portfolio choice is essential to obtain closed-form solutions. To deal with this issue, we follow Kuhn and Luenberger (2010) and Ang et al. (2014) and use the infrequent-trading assumption, according to which loans are perfectly illiquid. This assumption is consistent with the observation that in times of stress (when there can be a high level of uncertainty on asset quality), it is difficult to deleverage loan portfolios.[28]
- Second, in making decisions about balance-sheet management, banks must take into consideration the market impact of these decisions. More specifically, their decisions must account for the following:
  - The price impact of traded volumes (related to liquidity conditions). Returns on assets are sensitive to trading activity in terms of frequency, timing, and volumes.[29] In the

---

[27] There is a rich strand of research on risk–return trade-off in finance, which followed Markowitz's (1952) breakthrough in financial risk management and Merton's (1969, 1971) dynamic model. This provides a quantitative approach to mathematically formulate and solve the investment decision-making problem faced by agents.

[28] Another approach used in the literature is to consider transaction costs to introduce stickiness in the rebalancing of the portfolio structure (Davis and Norman, 1990; Hilberink and Rogers, 2002). However, this approach typically generates computationally intractable problems even for a very small set of asset classes, thus making it nonviable for bank stress test models where capturing a relatively large set of asset classes is important to account for their different risk characteristics.

[29] See Dufour and Engle (2000), Easley and O'Hara (1987), and Baranova et al. (2017).

model, we factor into the trading strategy the relationship between securities prices and changes in balance sheet-exposures.

   o The price effects of demand and supply in the loan market (Obizhaeva and Wang, 2013; Bassett et al., 2014). In the model, we assume that the parameters of a bank's loan portfolio (e.g., return on loans and default rates) depend on the volumes offered by all banks competing in a given market.
   o The limited capacity of funding markets to provide financing for balance-sheet growth. We embed into the model the relationship between funding costs and increasing leverage of banks.
- The third difficulty stems from the need to account for the multiple regulatory requirements faced by banks, as well as self-imposed risk limits.

3.2 The Bank Dynamic Balance Sheet Model: Description

In the BDBS model, banks seek to maximize profitability, which, in combination with regulatory requirements, leads them to optimize their balance sheets as financial conditions evolve. More specifically, the following hold in the model:

1. Banks maximize risk-adjusted return to capital.
2. Banks are subject to three regulatory constraints: risk-weighted capital ratio, leverage ratio, and LCR.
3. Banks can consider management actions related to (1) asset management (securities and loans), (2) funding management, and (3) equity management.[30] Consequently, balance sheets are dynamic.
4. The coexistence of multiple regulatory requirements and profitability objective leads to a complex interplay that affects the selection of management actions.

3.2.1 Maximization Objective

In the model, return on capital is calculated as net income divided by capital.[31] Net income combines interest income (from loans and securities) and expenses (from deposits and wholesale funding), corrected by recognized gains and losses from traded securities and credit losses incurred in loan portfolios and some sources of noninterest income.[32] We consider a 1-year period as the basis to calculate accrued profits and losses. Because the components of net income are uncertain, banks' objective is to maximize the expected return on capital, adjusted by the variance of the return, which measures its riskiness over a 1-year horizon. It can be conveniently written down as a constrained mean-variance optimization, like a simpler model of Hałaj (2016) and advocated by Adam (2007) as one of the advanced approaches for asset and liability management. A caveat of using this

---

[30] At this stage, equity management is not modeled yet.

[31] Return on capital is a standard and commonly used concept in financial analysis embedded into the risk management frameworks of banks via risk-adjusted return on capital (or risk-adjusted capital). The approach is an extension of Hałaj (2016).

[32] At this stage, noninterest income captures fees and commissions related to lending. There are other sources of noninterest income (e.g., wealth-management income) that we do not model explicitly. However, they can enter the optimization function as other income depending on the changes in the total balance-sheet sum. These other sources of income will be calibrated in future versions of the model.

maximization objective is that banks may not be able to attain their planned optimum and, consequently, may operate suboptimally from a theoretical perspective. They may also have a different objective that does not conform with the mean-variance modeling paradigm of portfolio choice theory.

3.2.2 Regulatory Constraints

We consider the major regulatory constraints faced by banks. Because banks are required to be adequately capitalized on both a risk-sensitive and a risk-insensitive basis under Basel III, in times of stress, they would not only reduce capital-intensive exposures (i.e., high RWAs) but also the size of their balance sheets. Regulatory capital risk weights (hence capital requirements) are sensitive to some parameters of the exposures.[33] The second aspect of asset and liability management—liquidity risk—is captured in the constraints of the optimization via the LCR requirement.

Should banks expect deteriorating economic or market conditions, they would adjust their exposures well before breaching regulatory requirements. Capturing banks' risk sensitivity relative to regulatory constraints in our optimization-based approach is therefore an important element to account for in our model. In the current literature, it is often assumed that banks react early to deteriorating regulatory ratios by trying to hit their internal long-term leverage and capital targets (Greenwood et al., 2015). However, this strand of literature sets these self-imposed targets exogenously and independent from conditions. In the BDBS model, we go further by allowing banks to determine their capital and liquidity buffers endogenously, depending on their risk aversion.[34] This allows us to avoid situations where, for example, a bank would disinvest from a profitable portfolio to hit a long-term target significantly above the regulatory ratio.

3.2.3 Management Actions

Although our goal is to model management actions related to assets, funding, and equity, at this early stage of development, given the complexities associated with modeling management actions, we focus on modeling asset-management actions.

The asset-management actions considered include changes in the holdings of securities, maturing loan exposures, and cash transfers. Notably, we allow for both buying and selling of securities and deleveraging and expansion of loan portfolios. In the current version of the model, we assume that banks update their funding structure in a mechanistic manner to replace repaid liabilities and finance the acquisition of new assets. Banks finance the expansion of their balance sheets with a fixed, prescribed proportion of different funding categories. The choice of proportions would be contingent on the scenario. For instance, if funding markets are under stress, in the simulations, banks would be allowed to use only some segments of secured funding that imply high expenses. Therefore, the proportions are model parameters that can be used to control the level of stress on the funding side of

---

[33] For instance, under the internal-risk-based (IRB) approach where banks use their internal rating systems for credit risk, risk weights associated with loan portfolios vary with the expected PD and LGD: the higher the PD, the larger the risk weight.

[34] Because risk aversion is not observed, it must be calibrated. For instance, it could be inferred from the distance between the regulatory requirements for capital and liquidity and the actual levels maintained by the banks.

the balance sheet. Moreover, the model accounts for the correlation between asset returns and funding costs in addition to behavior-based spread adjustments (i.e., pass-through). Consequently, even though the funding side of the balance sheet is not directly optimized, the funding mix changes endogenously following the evolution of the balance sheet.

3.2.4 Feedback Effects and Market Equilibrium

The complexity of the behaviors modeled increases when the feedback effects of banks' decisions are accounted for. The sensitivity of asset prices to changes in the supply of loans and securities renders the individually optimal response to a shock suboptimal when all banks' responses are aggregated. Similarly, the funding cost of each bank is affected by the liquidity demands of other banks, allowing for endogenously formed liquidity shortage and corresponding increases in funding costs. Therefore, in the model, banks react strategically not only to shocks but also in anticipation of other banks' reactions. We model this pre-emptive behavior by embedding a Nash equilibrium concept into the portfolio-optimization problem, and we compute the equilibrium recursively.[35] Specifically, banks take the supply of assets provided by other banks as given and optimize their balance sheets as explained before. Based on the outcome, they repeat the process of updating their expectations about the supply-driven prices and reoptimizing the balance-sheet structure until convergence (i.e., when expectations stabilize across banks).

   To guarantee the existence and uniqueness of the equilibrium balance sheets, we make a few simplifications in our optimal portfolio-based model of banks' behaviors under stress. This also allows us to better understand the main drivers of banks' decisions compared with a fully fledged model. First, we adopt mechanistic rules governing the evolution of the funding side to capture the main principle of risk-adjusted optimizations and the market-wide feedback effects of the decisions.[36] Second, we choose the risk–return characteristics such that the resulting optimization problem becomes a linear-quadratic program, thus making it a simple extension of the Markowitz optimization framework. To this end, we assume that the price-impact functions of changing exposures are linear. Third, given that we calculate the regulatory ratios with respect to the initial capital, not considering potential future losses, the constraints of the optimization are linear functions of the decision variables (i.e., management actions). Otherwise, the nonlinearities would be related to the losses affecting not only the RWAs in the numerator of the capital ratio but also the capital figure in the denominator.

---

[35] This approach provides a well-established framework for assessing the system-wide consequences of banks' actions under profit maximization. However, it may be argued that banks could act differently in crisis times, for instance, by trying to exploit the first-mover advantage or simply by adopting herding behavior. This may lead to substantially different outcomes in terms of capital and liquidity conditions. These types of behaviors could be considered in agent-based models.

[36] The rules dictate by how much each wholesale-funding category changes if the balance sheet changes by a given value; the changes are expressed as fixed proportions of the value. For instance, if the rule says 60 percent unsecured interbank funding and 40 percent short-term nonfinancial corporate deposits, then a C\$ 1 billion increase in the size of the balance sheet would be financed by C\$ 600 million of interbank deposits and C\$ 400 million of short-term nonfinancial corporate deposits.

ASSETS		LIABILITIES	
L1	Mortgage loans	Retail demand deposits	D1
L2	Consumer loans	Retail term deposits	D2
L3	Corporate loans	...	
L4	Loans to financial institutions	Unsecured wholesale funding	W1
...	...	Secured wholesale funding	W2
S1	Government securities	Repo	W3
S2	HR Corporate bonds	Subordinate debt	W4
S3	LR Corporate bonds	...	...
S4	ABS		
S5	Equity	Capital	E
...	...		

Figure 25.9 Illustrative balance sheet in the BDBS model.

3.3 Model Illustration for a Stylized Stress Scenario

Because we know the risk drivers of the balance-sheet categories underlying the calculation of net income, we can apply any stress scenario to study the behavior of banks on an individual-bank basis first, then for all banks taken together. As shown in the illustrative balance sheet in Figure 25.9, assets are broken down into loans (illiquid but partially maturing) and securities (liquid), with the two categories broken down further into more granular subcategories. Liabilities are made of three types of funding sources: retail deposits (most stable), wholesale funding (higher rollover risk), and capital. Exposures are characterized by the expected interest rate charged and its volatility, their maturity profile, and regulatory parameters (i.e., risk weights for capital adequacy and runoff rates for liquidity). Loan categories exposed to credit risk are additionally described by default probability and LGD default parameters. Finally, the balance-sheet categories are correlated in the sense that the parameters describing their returns are correlated.

A stylized illustration of the balance-sheet reoptimization process for an individual bank is shown in Figure 25.10. In our example, we consider a stress scenario where the expected PD for loans to a specific sector of the economy (L3) increases. The bank responds by reoptimizing its balance sheet, specifically by reducing credit provision to this economic sector. However, the impact of the stress scenario goes beyond the bank changing its exposures to L3. Because of the correlation structure of dependencies between balance sheet categories, it is also optimal for the bank, from a riskreturn trade-off perspective, to increase its exposure to securities S2 from 10 to 15 units. The correlation of returns and costs, leverage, and wholesale funding costs implies a rebalancing of the funding mix by decreasing the bank's reliance on wholesale funding of type W1 to 10 units, increasing its reliance on funding type W3 to 15 units, and getting new funding of type W4 equal to 10 units.

However, banks do not reoptimize their balance sheets in isolation but, rather, internalize the market-wide consequences of their decisions. This is schematically illustrated in Figure 25.11 for a stylized system of three banks. For instance, an increase in the expected default probability for one of the loan categories induces bank 1 to change its asset structure, notably by modifying its loan portfolio. A change in loan supply implies new price conditions in the loan market. Consequently, all banks active in the segments affected by the reoptimization by bank 1 would reconsider the volume of loans to originate in the market. The resulting sequential feedback mechanism of aggregate loan provision and loan prices would stabilize in a Nash equilibrium, provided that all banks know the response function of

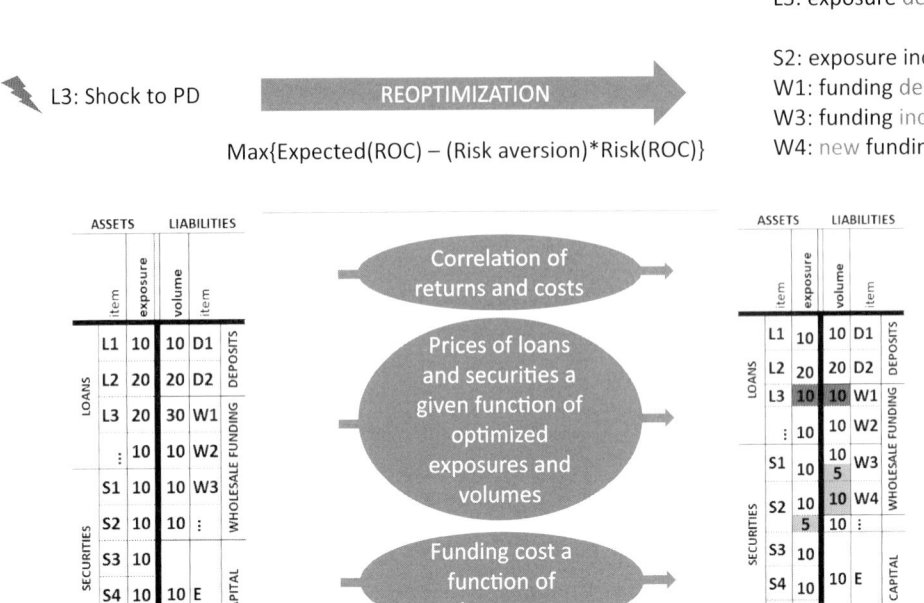

Figure 25.10 Schematic view of management actions for a stylized balance sheet following a shock to the PD of loan category L3.

loan prices to changes in loan volumes. The Nash equilibrium represents the supply of loans such that each bank would be worse off deviating from the committed loan origination.

We run a shock simulation in a stylized setup of the banking system. We assume that there are two identical banks in the system. They invest in four categories of securities (S0–S4) and three categories of loans (L0–L3). Expected returns on all asset categories are heterogeneous, sampled from a uniform distribution of [0.01,0.04] (i.e., expected returns can be as high as 4 percent and as low as 1 percent).[37] The risk of return is the same for all assets and equal to 0.01. The correlation between returns is positive and equal to 20 percent. Loans are sticky, which means that only 20 percent of their volumes mature within a horizon of reoptimization and can be considered for reinvestment (i.e., 80 percent of their volumes cannot be changed). Those stylized parameters can be obtained from supervisory disclosures.

The interactions of the balance sheets are captured by sensitivities of expected returns of assets to changes in the aggregate supply of the reoptimized assets. For simplicity, we assume that an increase of the exposure to any asset category by 10 percent of the outstanding volume in the market leads to a decline of the expected return on this asset category by 10 basis points (bps). In the application of the model in Hałaj and Priazkhina (2021), the sensitivities are estimated based on a time series of balance sheets and their parameters. It is achieved by addressing the following question: For what values of the sensitivities does the

[37] The exact numbers do not matter for the illustration of the mechanism of the game. Hałaj and Priazkhina (2021) present a fully fledged analysis based on a consistent macroeconomic scenario.

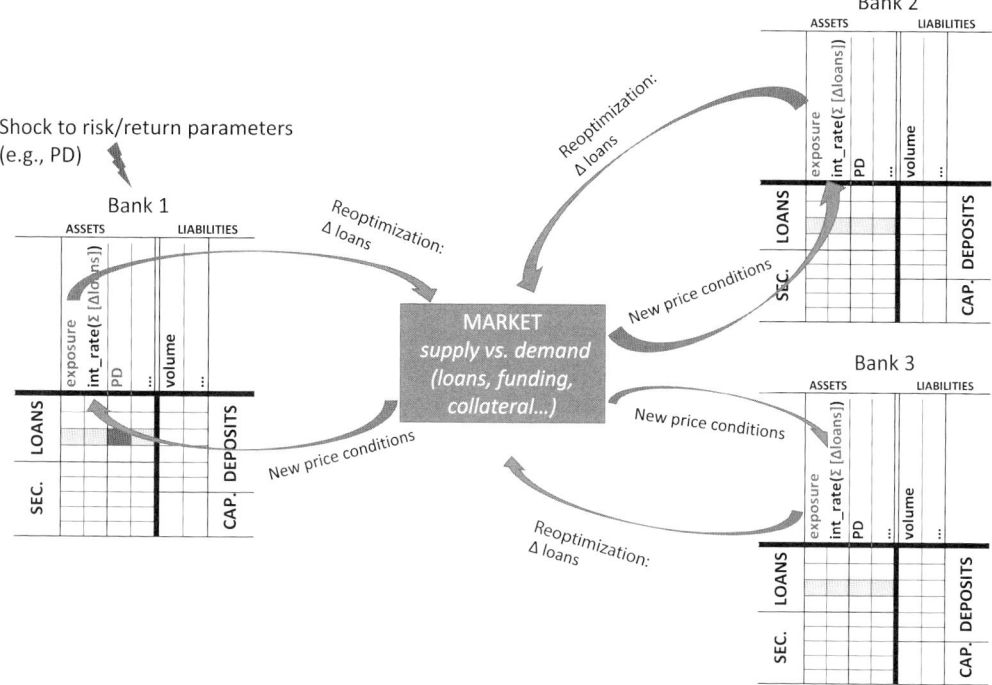

Figure 25.11 System-wide feedback effects affecting management actions
following a shock to a loan portfolio.
Note: The outcome of individual banks' decisions (change in loan supply Δ loans)
is aggregated and affects the pricing conditions considered in the reoptimization problem.

optimization-based time series of balance-sheet structures match the time series of observed
balance sheets? In this way, we capture the impact of the whole sum of financial-market
conditions on the sensitivities of expected returns. What matters for the sensitivities is the
depth of the market and potential actions of other market players (e.g., investment funds),
which does not need to be revealed for the estimation method to work. The relevant actions
are behind the observed changes of market prices and banks' realized returns used in the
calibration of the sensitivities.

The initial shock in the stylized example is an increase of default probability of loans L3.
It translates into higher expected losses affecting banks' capital position and the expected
return on L3, making it relatively less profitable than initially.

Figure 25.12 illustrates how accounting for banks' strategic responses to the hypothetical
shock affects the composition of balance sheets and may amplify the impact of the initial
shock. The asset composition before the shock is denoted by "initial." "No game" presents
the optimal asset structure derived under the assumption that banks rebalance their portfolios
in isolation (i.e., without accounting for the feedback effects of other banks' actions). "Step
n" represents the asset allocation in the *n*th step of the algorithm converging to the Nash
equilibrium. Finally, "Nash Eq." presents the Nash equilibrium assets structure, that is, the
optimal structure when banks account for the strategic response of other banks to shocks
that affect the pricing of loans and securities prices in fire sales. The first observation when

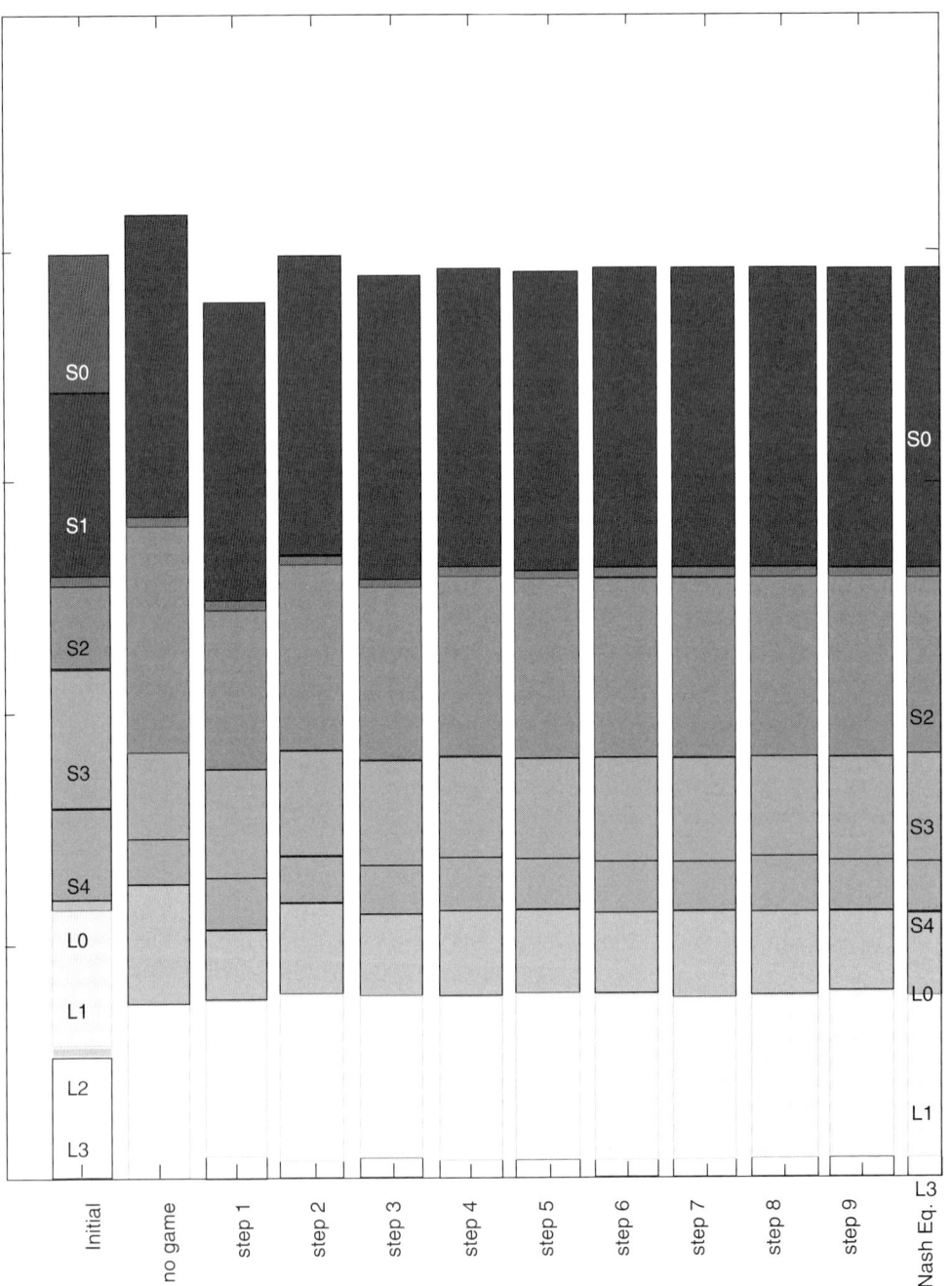

Figure 25.12 Feedback effect in banks' strategic decisions to reoptimize their assets following
a 100-bps shock to the PD of loan portfolio L3.
Source: Hałaj and Priazkhina (2021).
Note: *y*-axis: volumes; "initial," balance sheet structure before the shock; "no game," optimal
structure after the shock, assuming no interactions between banks' balance sheets
(or, equivalently, with no sensitivity of expected returns to changes in volumes of balance sheet
categories); "step *n*," balance-sheet structure after *n* responses of each bank to the other bank's
move in the Nash game; "Nash Eq.," equilibrium balance-sheet structure after the shock.

comparing the individually optimal with the system-wide optimal allocation of assets relates to the size of the balance sheet: the Nash game implies a smaller balance sheet. This is driven in part by the divestment in the loan category affected by the initial credit shock (L3), although the impact of the credit-risk shock is smoothed by the interactions of agents in the Nash equilibrium, as evidenced by the lesser decline in L3 exposures than under the isolated optimization. However, because of the correlation between returns of different balance-sheet categories, other categories change as well: loans L0 and securities S2 are higher under the Nash equilibrium than under the individual optimization, whereas securities S0 and S2 are lower.

The model is still in development, and we are considering two major extensions.

- First, we will consider relaxing the assumption of a constant capital base within the optimization horizon and allowing for equity issuance. We will also replace the current mechanistic rules governing the funding side with an approach where the liquidity characteristics of potential funding sources are factored into the objective function. As a result, the evolution of total assets would be financed with an optimal funding mix.
- Second, we will consider a nonlinear relationship between asset prices and traded volumes to reflect increasing stickiness of prices for large transacted volumes or a market freeze.

Calibrating the model poses significant challenges. Not only is it difficult to specify the theoretical mechanisms of banks' behaviors and their mathematical formulation, but it is also complicated to find and apply data to anchor the key parameters of the model. We use supervisory data to reconstruct the detailed structure of banks' balance sheets, but many parameters of the model related to the overall market functioning are difficult to infer from regulatory data: risk tolerance, elasticities of asset demand and supply, and correlations between profitability of assets and costs of funding. Banks are required to explicitly define and monitor risk tolerance following the principles of an effective risk-appetite framework (Financial Stability Board [FSB], 2013), but the risk tolerance is difficult to quantify, and we treat it as a parameter subject to an expert judgment during the execution of a specific stress test. The elasticities are difficult to estimate, given short time series of granular balance-sheet data for Canadian domestic systemically important banks (D-SIBs) and we must use simple calibration techniques, for instance, as in the example described earlier in this section. Short time series also impede the estimates of correlation. However, even longer time series would be subject to structural changes related to changing regulatory regimes, such as adaptation of Basel III rules. Some of the parameter, that are not bank specific can be estimated based on market data (e.g., correlation between yields of traded bonds). However, the other ones can only be pinned down consistently with a given severe stress scenario, using expert judgment and market intelligence.[38]

We validate our model by studying the responses of the model to some stylized, isolated shocks. We do not have long enough time series of granular data to perform an econometric validation; the data became available in 2015, a period during which there were no major

---

[38] We are exploring the application of potential statistical techniques to estimate some of the parameters based on matching moments of the model-implied variation of the balance sheets in time and the variation of the reported figures.

stress events in the Canadian financial system.[39] Moreover, banks were subject to liquidity regulatory requirements that have been phasing in (completed for LCR and in process for net stable funding ratio [NFSR]). Therefore, our validation relies on responses to some simple stresses; we verify whether the model produces intuitive outcomes (i.e., consistent with banking theory and expert knowledge).

Despite modeling and calibration challenges, attempts to model banks' behaviors under stress is important because these behaviors can contribute to systemic risk.

4 Conclusion

Stress test models constitute a component of the toolkit used at the Bank of Canada to assess financial stability risks: by allowing us to quantify the impact of severe but plausible risk scenarios on the financial system, stress test models inform our assessment of financial system resilience. Given the importance of systemically important banks in the Canadian financial system, we have initially focused on stress test models for the banking sector. Moreover, because our interest lies in systemic risk, given its effects on the real economy, our bank stress test modeling efforts have been geared toward accounting for the amplification and propagation channels that are at the origin of systemic risk.

Although some central banks use results from stress tests to inform prudential policies for banks, the Bank of Canada does not have a supervisory mandate for banks; hence, the results from our stress test models do not feed directly into prudential policies.[40] However, our analysis informs policymakers in different ways. First, stress test results inform our assessment of financial stability risks, which is the primary purpose of doing stress tests for the Bank of Canada.[41] Because our models capture different channels through which systemic risk materializes, we can disentangle and explain the different drivers of systemic risk. Secondly, the results from our stress test models can also provide information on aspects that are of direct interest to the Bank of Canada. For instance, the results from MFRAF can provide information related to the potential for banks to face funding strains and engage in fire sales, which can inform our assessment of liquidity risk, an important consideration, given our lender-of-last resort role. By quantifying how banks could adjust lending to different sectors of the economy under stress, the BDBS model will help inform our assessment of the impact of financial stress on the real economy, thus contributing to integrating financial stability considerations into monetary policy. Because enhancing our toolkit to assess systemic risk remains a priority for the Bank of Canada, we will continue to advance and expand our stress test models, both for the banking system and for other segments of the financial system (e.g., asset managers). These efforts will continue to be driven by the need to capture systemic risk and its drivers to inform our overall assessment of financial stability.

---

[39] Except for the COVID-19 stress period, which has not yet been included in the time series of data.

[40] In contrast, for instance, the Federal Reserve uses Comprehensive Capital Analysis and Review (CCAR) results to set capital requirements, and the Bank of England uses results from its annual cyclical stress test to inform the countercyclical capital buffer (CCyB).

[41] When MFRAF is used in the context of the Bank of Canada–OSFI Macro Stress Test, it serves to quantify the impact of second-round effects on bank capital, ultimately informing the overall analysis.

References

Adam A. (2007), *Handbook of asset and liability management: From models to optimal return strategies*, 1st ed., Wiley.

Ahnert, T., K. Anand, J. Chapman, and P. Gai (2019), "Asset encumbrance, bank funding and financial fragility, *Review of Financial Studies*," 3(6), 2422–2455.

Aikman, D., P. Alessandri, B. Eklund, P. Gai, S. Kapadia, E. Martin, N. Mora, G. Sterne, and M. Willison (2009), "Funding liquidity risk in a quantitative model of systemic stability," Bank of England Working Paper No. 372.

Anand, K., G. Bedard-Page, and V. Traclet (2014, June), "Stress testing the Canadian banking system: A system-wide approach," *Bank of Canada Financial System Review*, 61–68.

Anderson, R., C. Baba, J. Danielsson, U. S. Das, H. Kand, and M. Segoviano (2018, February), "Macroprudential stress tests and policies: Searching for robust and implementable frameworks," Systemic Risk Centre Report.

Ang, A., D. Papanikolaou, and M. Westerfield (2014), "Portfolio choice with illiquid assets," *Management Science*, 60(11), 2381–2617.

Arora, R., G. Bedard-Page, G. Ouellet-Leblanc, and R. Shotlander (2019), "Bond funds and fixed-income market liquidity: A stress-testing approach," Bank of Canada Technical Report No. 115.

Baranova, Y., J. Coen, P. Lowe, J. Noss, and L. Silvestri (2017, July), "Simulating stress across the financial system: The resilience of corporate bond markets and the role of investment funds," Bank of England Financial Stability Paper No. 42.

Barucca, P., M. Bardoscia, F. Caccioli, M. D'Errico, G. Visentin, G. Caldarelli, and S. Battiston (2020), "Network valuation in financial systems," *Mathematical Finance*, 30(4), 1181–1204.

Bassett, W., M. Chosak, J. Driscoll, and E. Zakrajsek (2014), "Changes in bank lending standards and the macroeconomy," *Journal of Monetary Economics*, 62(C), 23–40.

Battiston, S., G. Caldarelli, M. D'Errico, and S. Gurciullo (2016), "Leveraging the network: A stress-test framework based on DebtRank," Working Paper, available at https://arxiv.org/pdf/1503.00621.pdf

Brunnermeir, M., and L. Pedersen (2009), "Market liquidity and funding liquidity," *Review of Financial Studies*, 22(6), 2201–2238.

Burrows, O., D. Learmonth, J. McKeown, and R. Williams (2012), "RAMSI: A top-down stress-testing model developed at the Bank of England," *Quarterly Bulletin 2012 Q3.*

Christensen, I., G. Kumar, C. Meh, and L. Zorn (2015, June), "Assessing vulnerabilities in the Canadian financial system," *Bank of Canada Financial System Review*, 37–46.

Davis, M. H. A., and A. Norman (1990), "Portfolio selection with transaction costs," *Mathematics of Operations Research*, 15(4), 676–713.

Dees, S., and J. Henry (2017), "Stress-test analytics for macroprudential purposes: Introducing STAMP€," in S. Dees, J. Henry, and R. Martin (eds.), *STAMP€: Stress-test analytics for macro-prudential purposes in the euro area*, European Central Bank.

Dees, S., J. Henry, and R. Martin (eds.) (2017), *STAMP€: Stress-test analytics for macroprudential purposes in the euro area,* European Central Bank.

Doyne Farmer, J., A. M. Kleinnijenhuis, P. Nahai-Williamson, and T. Wetzer (2020), "Foundations of system-wide financial stress testing with heterogeneous institutions," Bank of England Staff Working Paper No. 861.

Dufour, A., and R. Engle (2000), "Time and price impact of a trade," *Journal of Finance*, 55, 2467–2498.

Easley, D., and M. O'Hara (1987), "Price, trade size and information in securities markets," *Journal of Financial Economics*, 19, 69–90.

Eisenberg L., and T. Noe (2001), "Systemic risk in financial networks," *Management Science*, 47(2), 236–249.

Fique, J. (2017), "The MacroFinancial Risk Assessment Framework (MFRAF), version 2.0," Bank of Canada Technical Report No. 111.

Financial Stability Board (2013, November 18), "Principles for an effective risk appetite framework," Financial Stability Board, available at www.fsb.org/2013/11/r_131118/.

Gauthier, C., Z. He, and M. Souissi (2010), "Understanding systemic risk: The trade-offs between capital, short-term funding and liquid asset holdings," Bank of Canada Staff Working Paper 2010–29.

Greenwood, R., A. Landier, and D. Thesmar, (2015), "Vulnerable banks," *Journal of Financial Economics*, 115(3), 471–485.

Hałaj, G. (2016), "Dynamic balance sheet model with liquidity risk," *International Journal of Theoretical and Applied Finance*, 19(7), 1–37.

Hałaj, G., and S. Priazkhina (2021), "Stressed but not helpless: Strategic behaviour of banks under adverse market conditions," Bank of Canada Staff Working Papers, forthcoming.

Hilberink, B., and L. C. G. Rogers (2002), "Optimal capital structure and endogenous default," *Finance and Stochastics*, 6(2), 237–263.

International Monetary Fund (2014), "Canada Financial Sector Assessment Program stress testing technical note," IMF Country Report No. 14/69.

Kuhn, D., and D. Luenberger (2010), "Analysis of the rebalancing frequency in log-optimal portfolio selection, *Quantitative Finance*, 10(2), 221–234.

Kusy, M. I., and W. T. Ziemba (1986), "A bank asset and liability management model," *Operations Research*, 34(3), 356–376.

MacDonald, C., and V. Traclet (2018), "The Framework for Risk Identification and Assessment (FRIDA)," Bank of Canada Technical Report No. 113.

Markowitz, Harry (1952), "Portfolio selection," *Journal of Finance*, 7(1), 77–91.

Merton, R. (1969), "Lifetime portfolio selection under uncertainty: The continuous-time case," *The Review of Economics and Statistics*, 51(3), 247–257.

Merton, R. (1971), "Optimal consumption and portfolio rules in continuous-time model," *Journal of Economic Theory*, 3, 373–413.

Morris, S., and H. Shin (2016), "Illiquidity component of credit risk," *International Economic Review*, 57(4), 1135–1148.

Obizhaeva, A., and J. Wang (2013), "Optimal trading strategy and supply/demand dynamics," *Journal of Financial Markets*, 16(1): 1–32.

Office of the Superintendent of Financial Institutions (2018), "Capital adequacy requirements (CAR) guideline," Office of the Superintendent of Financial Institutions, October 2018 Guideline, available at www.osfi-bsif.gc.ca/Eng/fi-if/rg-ro/gdn-ort/gl-ld/Pages/CAR19_chpt1.aspx.

Rochet, J-C., and X. Vives (2004), "Coordination failure and the lender of last resort: Was Bagehot right after all," *Journal of European Economic Association*, 2(6), 1116–1147.

Roncorini, A., S. Battiston, M. D'Errico, G. Hałaj, and C. Kok (2019), "Interconnected banks and systematically important exposures," Bank of Canada Staff Working Paper 2019–44.

Trullo, D. (2016), "Next steps in the evolution of stress testing," Speech at Yale University School of Management Leaders Forum, New Haven, CT, available at www.federalreserve.gov/newsevents/speech/tarullo20160926a.htm.

Stress Testing at the International Monetary Fund

Tobias Adrian, James Morsink, and Liliana Schumacher[*]

1 Introduction

International Monetary Fund (IMF) staff members use macroprudential stress tests to assess systemic risk as part of the IMF's mandate to monitor global financial stability. Stress tests help assess the resilience of financial systems in IMF member countries and underpin policy advice to preserve or restore financial stability. This assessment and advice are mainly provided through the Financial Sector Assessment Program (FSAP). IMF staff members also provide technical assistance in stress testing to a large number of its member countries.

An IMF macroprudential stress test is a methodology to assess financial vulnerabilities that can trigger systemic risk and the need for system-wide mitigating measures. The difference between a macroprudential and a supervisory stress test lies in the nature of the assessment and the consequences of the results:

- A microprudential stress test is a forward-looking supervisory tool that assesses the adequacy of individual banks' capital (or liquidity) conditional on their portfolio risks. Key to the supervisory purpose is the ability of the bank "to pass or not to pass the test," as well as the subsequent supervisory measures that may be needed to beef up cushions when the bank does not pass the test.
- A macroprudential stress test instead focuses on financial vulnerabilities that can trigger systemic risk. Financial vulnerabilities are imbalances and other financial characteristics of the financial environment (e.g., high leverage, mispricing, concentration of risk, liquidity mismanagement, and others) that amplify adverse shocks. Although an important part of IMF stress testing involves assessing the health of individual financial institutions, the final objective is not to determine whether individual banks are adequately capitalized based on a hurdle rate but to assess whether the identified vulnerabilities can compromise financial stability for the whole economy. Results by institution are not published; instead, they are discussed with the authorities and used to support the financial stability assessment and the recommendations that are at the core of the IMF FSAP reports.[1] Although recommendations could include the need to boost capital cushions, they can

[*] This chapter draws on the stress testing research and work by IMF staff in the Financial Sector Assessments and Policies (FS) Division of the IMF Monetary and Capital Markets (MCM) Department. The views expressed in this paper are those of the authors, and do not necessarily represent the views of the International Monetary Fund, its management, or its executive directors.

[1] These reports, called *Financial Sector Stability Assessments* (FSSAs), as well as the technical notes and detailed assessment reports that support their findings, can be found at www.imf.org/en/countries.

also include the adoption of other macroprudential measures, such as measures targeting credit demand (debt-to-income and loan-to-value ratios), surcharges (countercyclical or risk-specific surcharges), or liquidity requirements.

The definition of systemic risk as used by the IMF is relevant to understanding the role of its stress tests as tools for financial surveillance and the IMF's current work program. the Financial Stability Board (FSB) and the Bank for International Settlements (BIS, 2009) defined *systemic risk* at the onset of the Global Financial Crisis (GFC) as the risk of disruptions to the provision of financial services caused by an impairment of all or parts of the financial system that has the potential to cause serious negative consequences for the real economy. Most of the vulnerability analysis conducted by the IMF, until recently, as part of its stress testing exercises has been related to vulnerabilities that could lead to a financial crisis. More recently, however, IMF work on financial stability also includes identifying financial vulnerabilities that may not necessarily lead to a financial crisis but, through the operation of the financial system, could create downside risks to growth. Both systemic financial disruptions and milder reversals of financial vulnerabilities could create downside risks to growth ("growth at risk" [GaR]).[2] Consistently, the goal of the IMF financial surveillance function and stress testing at present is not only to assess the risk of systemic failures of significant financial institutions but also to identify financial vulnerabilities that can create risks for sustainable economic growth, even if these vulnerabilities may not lead to a financial crisis.

Stress tests were first applied to banks and then extended to nonbank financial intermediaries and nonfinancial corporates when relevant for the analysis of systemic risk. Following the identification of specific sources of systemic risk, IMF staff members have also included stress tests of nonbanks, such as insurance and asset management companies and nonfinancial firms, as well as estimates of stress for households, in the FSAPs.[3]

Stress testing at the IMF is adapted to the diversity of its member countries. The IMF membership is diverse, which presents challenges for stress testing. By contrast with national agencies that typically focus on one or a limited number of national financial sectors over time, the IMF uses stress tests as part of financial stability assessments in 12–14 different financial systems each year. (In addition, the IMF helps develop country authorities' capacity in stress testing—through so-called FSSRs and other technical assistance missions—in about 18 different financial systems each year.) Although this schedule helps countries to gain experience in understanding sources of vulnerabilities, it also imposes the need to adapt to different types of threats to financial stability, uneven data availability, and diverse complexity of financial systems. To benefit from local knowledge, stress testing at the IMF usually combines top-down stress tests (conducted by IMF staff, sometimes in collaboration with national supervisors) and bottom-up stress tests (produced by financial institutions), based on agreed-on methodology and scenarios with IMF staff.

The plan of this chapter is as follows: After a brief section on the evolution of stress tests at the IMF, we present the key steps of an IMF staff stress test. They are followed

---

[2] See IMF (2017a). See also Adrian et al. (2019).

[3] Stress tests of nonbanks rely on methodologies and toolboxes different from those used for banks. Their description is not included in this chapter, but more detail can be found in Appendix A and in Broszeit et al. (2014).

by a discussion on how IMF staff members use stress tests results for policy advice. The chapter concludes by identifying remaining challenges and emerging risks to make stress tests more useful for the monitoring of financial stability and an overview of an IMF staff work program in that direction, including in relation to risks arising from climate change.

2 Stress Testing at the IMF: From the Beginnings to Now

Risk management is a relatively recent field within finance. Its origins can be traced to the interplay between the wave of financial innovation triggered by the pathbreaking advances in option pricing in the 1970s and the increasingly volatile financial environment of the 1980s and 1990s, which included shifts in the policy framework of major central banks, failure or almost failure of large investment banks, the 1987 stock market crash, the savings and loans crisis in the United States, and financial crises in major emerging market economies.[4]

It did not take long for *stress testing* to emerge as a more articulated version of risk-management techniques. IMF staff members were among the first to adopt stress testing for banks. The Asian crisis of 1997–1998 was a wake-up call for IMF staff, who had hitherto not placed much emphasis on the macroeconomic impact of a bank's performance. The FSAP, inaugurated by the IMF and the World Bank in 1999, was an effort to respond to the lessons from the Asian crisis (as well as to bring in the World Bank's expertise on financial-sector development in emerging-market and developing economies). An FSAP for a country included a supervisory component, centered on the assessment of supervisory principles, done jointly by the IMF and the World Bank; an IMF-led quantitative component centered on stress tests; and a development component led by the World Bank. Stress tests were adopted as the key tool of assessment because of their forward-looking dimension, as opposed to historical balance-sheet-based indicators used at the time (e.g., capital adequacy, asset quality, management, earnings, liquidity, and sensitivity [CAMELS]; Čihák, 2007).

Initial stress tests were limited to the analysis of individual banks and focused mostly on solvency. These microprudential stress tests focused on the assessment of individual institutions, and the results were communicated in terms of the number of banks that had passed the test (and/or the share of banking assets). The crises—which affected entire asset classes and markets across many types of financial intermediaries, including nonbanks—propagated rapidly in ways that stress tests had not anticipated. Also, losses were much larger than stress tests had estimated in previous years, exposing the limitations of the bank-by-bank analysis.[5]

Since the GFC, IMF staff stress tests have emphasized the need to assess systemic risk, rather than the risk of individual institutions. Clear conceptual and functional separation was achieved between supervisory and macroprudential stress tests. Staff members have also improved their tools for macrofinancial analysis and scenario modeling, extended the stress testing framework to cover forms of risk that had received less attention before (e.g., sovereign, funding, and market-liquidity risks), introduced contagion models to assess negative externalities from interconnectedness, developed stress tests for nonbanks,

---

[4] Adrian (2017) provides a history of the evolution and development of risk management and how it has related to regulations.

[5] See Chapters 18 and 20 in this volume.

and prioritized the development of methodologies that can capture systemic losses from amplification mechanisms, including the interaction between solvency and liquidity risks and that between financial vulnerabilities and the real economy.

Importantly, following the crises, the IMF decided that financial stability assessments—including stress tests—would be mandatory for jurisdictions whose financial systems were determined to be systemically important. From its inception to the time of the crisis, the FSAP program was voluntary. In particular, the United States did not volunteer and was not assessed by the IMF before the crisis. In 2010, the IMF decided that henceforth, financial stability assessments under the FSAP would be mandatory for jurisdictions with systemically important financial sectors. At the time of the 2010 decision, 25 jurisdictions were deemed to have systemically important financial sectors. Following a review of the decision in 2013, the number of jurisdictions subject to mandatory financial stability assessments increased to 29.[6]

The postcrisis era saw the incorporation of stress testing into the supervisory toolkits of major countries around the world. Following a stint at the IMF, Timothy Geithner—first as president of the US Federal Reserve Bank of New York and later as secretary of the US Department of Treasury—proposed and helped operationalize in 2009 a new idea for deploying public funds from the Troubled Asset Relief Program (TARP) program to recapitalize banks.[7] The plan was to use stress tests as a forward-looking exercise to help markets distinguish between viable banks and weak banks in need of capital. European countries subsequently adopted bank stress testing starting in 2010, in the face of widespread banking distress, but initially lacked the elements that made US stress tests successful, namely: a severe scenario, transparency of results,[8] and a plan for recapitalization of weak banks.[9] The introduction of the Transparency Exercise in 2013, with its focus on current asset quality, was a useful complement to enhance the credibility of the European stress tests.

3 Key Steps of an IMF Staff Stress Test

The IMF staff's approach to stress testing broadly consists of the following steps:

- *Initial assessment of vulnerabilities (Section 4):* This is conducted using a range of financial indicators and more recently using the GaR methodology. The preliminary identification of vulnerabilities helps guide the stress test scenario and its severity, as well as the channels of risk amplification that the stress tests need to explore. The first step also involves the definition of the perimeter of stress testing, discussed in Appendices A and B, together with data issues.

---

[6] The 29 jurisdictions are Australia, Austria, Belgium, Brazil, Canada, China, Denmark, Finland, France, Germany, Hong Kong SAR, India, Ireland, Italy, Japan, Korea, Luxembourg, Mexico, Netherlands, Norway, Poland, Russia, Singapore, Spain, Sweden, Switzerland, Turkey, the UK, and the United States.

[7] See Geithner (2014), introduction and page 286. Geithner was the director of the IMF Policy Development and Review Department from 2001 to 2003. On the credibility of stress testing during crisis periods, see Ong and Pazarbasioglu (2013).

[8] On the value of information during periods of stress, see Chapter 12 in this volume.

[9] The Committee of European Banking Supervisors (CEBS), in charge of the European stress tests, clarified at the time that "this is not a stress test to identify individual banks that may need recapitalization, as the assessment of specific institutions' needs for recapitalization remains a responsibility of national authorities."

- *Scenario design (Section 5):* The scenario-design module is responsible for the choice of shocks and the calibration of the macrofinancial variables that will characterize the adverse scenario. Shocks are usually based on an FSAP Risk Assessment Matrix (RAM). The RAM is an organizing framework guiding the FSAP work and connecting potential shocks to vulnerabilities. It helps articulate scenarios by providing a "tail-risk story." Potential shocks can be country specific or of a global nature. The latter are taken from the Global RAM, a core risk assessment framework prepared by the IMF each quarter, assessing risks in the global economy consistently across IMF departments. In addition, the GaR framework is used to provide a consistent metric of scenario severity across countries by enabling the calibration of tail scenarios with a similar probability of occurrence across countries conditional on each country's position in its financial cycle.
- Stress tests of solvency, liquidity, and contagion (Section 6): These comprise multiple "satellite" models to translate adverse scenario variables as well as other features of the country's financial cycle (e.g., corporate and household leverage) into balance-sheet items and profits and losses that affect financial institutions, capital and capital requirements; liquidity inflows and outflows are also stressed to assess the sufficiency of banks' counterbalancing capacity (liquid assets) in the presence of shocks; the contagion analysis assesses, among others, risks from different networks to propagate shocks within and across countries
- Risk-amplification mechanisms (Section 7): The different stress tests are integrated to obtain an overall picture of systemic risk, including by assessing the interaction among types of risks, between solvency and liquidity factors, and among real and financial effects and other forms of risk amplification. Most of this module is still a work in progress, focusing on how to better calibrate the financial system's power to amplify shocks.

4 Assessment of Vulnerabilities

As part of the preparatory work, IMF staff members start by identifying systemic vulnerabilities that can threaten financial stability. Systemic vulnerabilities are imbalances and other characteristics and financial conditions that can magnify shocks. They emerge from market failures that can lead to leverage, excessive maturity transformation, interconnectedness, and complexity. In the context of adverse shocks, they can lead to negative-feedback loops, asset fire-sale dynamics, and reduced credit supply. Their monitoring comes from the recognition that shocks are inherently hard to predict; therefore, financial stability assessments should be based on those vulnerabilities that build over time, creating threats to bank resilience and/or downside risks to gross domestic product (GPD) growth[10]

Staff members use a range of vulnerability indicators, balance-sheet analysis (including of banks, households, and nonfinancial firms), and more recently, GaR tools.

- GaR and stress testing are forward-looking methodologies for the assessment of systemic risk. Both are based on the identification of vulnerabilities that can trigger systemic risk.

---

[10] Adrian et al. (2015) propose a two-part framework for the assessment of financial stabilities: first, a set of metrics for primary vulnerabilities to be monitored is created and reported regularly by the IMF; then these vulnerabilities are mapped into a metric of risks to macroeconomic performance.

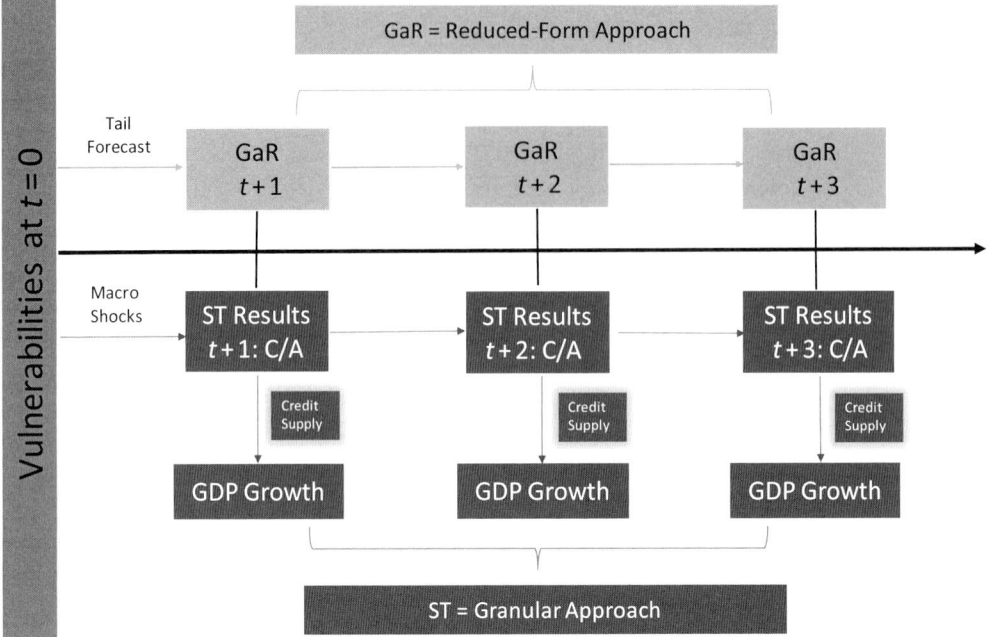

Figure 26.1 The relation between stress testing and GaR.

Vulnerabilities are then mapped into metrics of financial distress (stress testing) and downside risks to growth (GaR). Both measures have implications for the real economy and the financial system, and given that both methodologies measure developments in the tail, these implications are expected to be consistent across the two approaches. Figure 26.1 shows the relation between GaR and stress testing.

- GaR also helps detect the key vulnerabilities that stress tests need to focus on. The GaR framework estimates the downside risk of future GDP growth as a function of financial vulnerabilities, and in this way, it helps teams identify the key factors that need to be explored further, in a more granular setting, through stress testing.
- Finally, the term structure of GaR provides information on whether some financial characteristics and conditions could mitigate macrofinancial risks at some horizons and exacerbate them at others. Adrian et al. (2018) show, using panels of 11 advanced and 10 emerging-market economies, that loose financial conditions mitigate downside risks to growth at short-term horizons but then forecast higher risks at medium-term horizons. That is, the term structure of GaR features an intertemporal trade-off; this knowledge can be used for the choice of the risk horizon of the stress tests.

Staff members distinguish among cyclical, structural, and institutional vulnerabilities:

- *Cyclical vulnerabilities stem from a financial system's position in the financial cycle.* During boom cycles, a general optimism tends to prevail, leading to a low price of risk and correspondingly high asset valuations. Funding constraints are relaxed for individuals and companies, credit expands, and leverage builds up. When the general optimism starts to dissipate, possibly owing to the realization of adverse shocks (e.g., an economic contraction or a change in investors' sentiment), lenders become more risk averse, the price of risk increases, and asset values fall. As pointed out by many authors, no single indicator can provide a perfect guide to the assessment of systemic risks; therefore, IMF staff members look at a range of signals.[11]

- *Leverage is used as a key indicator of cyclical vulnerabilities.*[12] In line with a large literature showing that financial crises tend to be preceded by credit booms, staff members start by looking at credit growth in domestic and foreign currency by sectors and by type of loans; staff members also check other measures, such as the credit-to-GDP gap that was proposed by the Basel Committee on Banking Supervision[13] as an indicator of the financial cycle for the implementation of the countercyclical capital buffer (CCyB); the leverage of nonfinancial firms and households; banks' exposures to maturity transformation, such as the loan-to-deposit ratio; and the prevalence of types of volatile funding (e.g., wholesale funding in foreign currency).

- *The price of risk is also used as a cyclical indicator.* Brunnermeier et al. (2013) offer a rationale for the idea that the financial system is prone to crises even if measured risk is low. Similarly, Gilchrist and Zakrajšek (2012) and Stein (2014) argue that when the risk premium is low, there is a greater probability of a subsequent upward spike in credit spreads. To help address the volatility paradox, the term structure of GaR enables the IMF analyst to approximate the time horizon in which vulnerabilities would bring about a realization of risks, thus enabling vulnerabilities to be addressed with concrete policy recommendations. Staff membres also look at the deviation of real estate and other asset prices from long-term trends and fundamentals, market volatility, and market-based measures of risk.

- *Structural vulnerabilities also amplify shocks.* Some structural vulnerabilities are particularly prevalent in developing economies and emerging markets. For example, international capital inflows may play a stabilizing role in the short term, but they carry the risk of future reversal. High dollarization of financial intermediation, typical of many emerging-market and developing economies, also acts as an amplifier following shocks that require exchange rate adjustments, by increasing the debt service of unhedged borrowers. Other vulnerabilities are present in advanced economies, such as high interconnectedness across firms, which exposes entities to counterparty credit risk or concentration of funding sources, instruments, or products that increase the likelihood that a substantial portion of funding will be withdrawn at the same type. It also increases the correlation between sources of funds and market conditions. Risk concentration on financial institutions' balance sheets define another source of structural vulnerabilities, such as excessive exposures to particular types of assets or markets and a high share of nonperforming loans (NPLs), among others.

- Institutional vulnerabilities are also part of staff analysis. In some cases, their impact may be quantified and included in an adverse scenario when they represent a threat to financial stability (e.g., weak framework for anti–money laundering and countering financing of

---

[11] For a review of indicators, see IMF (2016a) and Adrian et al. (2015).
[12] See chapter 21 in this volume.
[13] See Drehmann et al. (2011) and Drehman and Tsatsaronis (2014).

terrorism [AML-CFT] that could trigger blacklisting, loss of correspondent services, or rating downgrades).

An example of how GaR and stress testing interact in an IMF staff stress test is available in the technical note prepared for the Peru FSSA.[14] The team used a set of macrofinancial variables representing financial conditions such as leverage, the domestic price of risk, and structural factors to forecast the probability distribution of future GDP growth at horizons of up to 3 years through quantile regressions. The separation of a large set of financial indicators into these three predetermined categories was made as a reasonable compromise between maintaining parsimony, allowing various classes of indicators to provide separate signals about risks to growth at different horizons, and being able to provide a more direct economic interpretation to the various subindexes. The analysis concluded that external conditions are the main drivers of Peru's growth over short-term horizons, both on the baseline and also on the tails; their impact is twice as large as that for other variables. This was consistent with the fact that Peru is a small, open, dollarized economy with a strong impact from China and exchange rate movements. The impact of external conditions on growth was also found to be mostly short term (up to 1 year), which is consistent with commodity exporters facing volatile markets. This information was then used to map vulnerabilities into an adverse scenario including the scenario path.

5 Scenario Design

IMF stress tests are based on at least two scenarios. These are a baseline scenario using the *World Economic Outlook* (WEO) projections and at least one adverse scenario. The risk horizon spans 3 to 5 years, with the choice of length and shape based on the nature of vulnerabilities characterizing the country as well as the ability of staff (based on data and type of vulnerabilities) to perform reliable projections beyond the first 3 years.

A scenario describes forward-looking, severe, consistent, and robust trajectories for a comprehensive set of macrofinancial variables that react following the materialization of shocks. A forward-looking approach is necessary because stressful scenarios may respond to new triggers. Scenarios need to be severe enough given current vulnerabilities. Consistency across countries is important for the IMF because of the need for evenhanded treatment of member countries. Robustness poses the question of the ability of models to capture features of tail events and the number of scenarios that may be needed to assess resilience. This section presents the steps followed in scenario design and illustrates how these requirements are met in practice.

Scenario design is divided in three phases:

- Selection of shocks that can exacerbate identified financial vulnerabilities
- Assessment of sufficient severity
- Simulation of the complete set of macrofinancial variables that are consistent with the shock

Scenario design starts with a narrative about how the realization of tail risks could interact with financial vulnerabilities to generate severe but plausible macrofinancial impact.

---

[14] See IMF (2018a).

Country A: Risk Assessment Matrix		
Shocks or Triggers	Likelihood of the shock (low, medium, or high)	Impact on the financial sector, including role of amplification mechanisms (low, medium, or high)
Trigger 1		
Trigger 2		
Trigger 3		

Figure 26.2 Stylized representation of an FSAP RAM.

In modeling terms, this step comprises the choice of one or more shocks. The related domestic and external financial stability episodes that could trigger the shocks are drawn from the RAM, discussed previously. This initial narrative represents the forward-looking aspect of stress testing (Figure 26.2).

The choice of the shocks helps align the scenario with vulnerabilities, as well as changes in institutions' business models and policies and the financial-system structure. A truly consistent scenario cannot be defined independently of financial-sector vulnerabilities and conditions—both initial conditions and behavioral responses. The whole idea of systemic risk (and the use of GaR as a measure of financial-sector impact on the real economy) hinges on the notion that the financial sector is macro relevant. Hence, a properly defined scenario must have an impact on the underlying conditions and behavior of the financial sector.

Scenario severity is typically measured in terms of the fall in the level of real GDP below baseline or, equivalently, the cumulative fall in real GDP growth. Real GDP is the anchor variable of the scenario because a recession typically defines the worst macrofinancial environment for most financial institutions. Consideration is also given to other variables that can be used to anchor severity. In some countries, scenario severity could be better captured by variables other than GDP growth. Examples are the United States, where unemployment is generally considered as the best anchor variable for stress; Gulf region countries, where the evolution of oil prices has been used to proxy severity; and countries with material foreign currency loans, where foreign exchange shocks are key drivers of default risk.

Choosing the right severity is challenging. For years, IMF staff used a rule of thumb developed over time by which shocks to GDP growth should represent a deviation from the IMF baseline projection over the first 2 to 3 years of the scenario of at least two historical standard deviations from the mean; that is, the key benchmark was provided by the unconditional historical distribution of GDP growth. At present the choice is more complex. It starts by using GaR as a minimum severity related to cyclical vulnerabilities. This benchmark is chosen consistently across countries because it is derived conditional on each country's cyclical phase.[15]

---

[15] Adrian et al. (2019) model the full distribution of future US real GDP growth as a function of current financial and economic conditions. They find that the estimated lower quantiles of the distribution of future GDP growth exhibit strong variation as a function of current financial conditions, whereas the upper quantiles are stable over time. They also find that current economic conditions forecast the median of the distribution but do not contain information about the other quantiles of the distribution. Similarly, Brunnermeier and Sannikov (2014) build a continuous-time model to study full-equilibrium dynamics, not just near the steady state. The model shows that the financial system exhibits some inherent instability due to highly nonlinear effects. The effects are asymmetric and only emerge in the downturns.

Consistency is needed across countries. Consistency across countries does not mean an identical decline in GDP growth. The scenario and its calibration should reflect the principle that systemic risk depends, in part, on the phase of the financial cycle in which the financial system is operating. This approach uses the intuitive understanding that the larger the vulnerabilities, the larger could be the amplification mechanisms; therefore, the assessment of vulnerabilities should be used to calibrate the type and size of shock that will drive the scenario. Among other drivers, these could include a fall of asset prices, a shock to interest rates, a reassessment of risk premiums, or a large depreciation to correct an external imbalance. For example, in the Sweden FSAP,[16] against the background of rising household debt and a large stock of interest-only mortgage loans, shocks were represented by a reassessment of global and Sweden-specific risks that raised interest rates by about 250–500 basis points (bps), along with a large fall in real estate prices.

Many countries are also exposed to vulnerabilities that do not have a clear cyclical evolution. Therefore, the final severity may be stronger in countries with significant structural and/or institutional features. The Spain FSAP's[17] core shocks were related to international disruption, given the important role the Spanish banks' international operations play in compensating declining domestic profitability and domestic NPLs. More importantly, legacy issues appeared to still hold the power to impair financial stability, and therefore the FSAP mostly concentrated on the recognition of losses generated during the 2012 crisis. Many developing and small emerging-market economies are characterized by dependence on one or a few export products, low credit-to-GDP ratios, high concentration of banks' loan portfolios due to a concentrated economy, significant lack of access by households and small and medium enterprises to financial services, and significant lending to the government. For these countries, vulnerabilities would mostly be structural, and scenarios will be motivated by capital-flow reversals, terms-of-trade shocks, or other global financial conditions.[18] For example, in Armenia, which is a small, open, highly dollarized economy, the recent FSAP's adverse scenario[19] focused on a terms-of-trade shock and a market reassessment of sovereign and private risks.

Once the severity of the scenario is chosen (in terms of GDP or another variable, as explained previously), a set of consistent macrofinancial variable paths (expressed as deviations from the WEO baseline) is simulated using macrofinancial models and targeting the chosen severity benchmark. For major advanced and emerging-market economies, scenarios are simulated using either the Global Macrofinancial Model (GFM) or the Flexible System of Global Models (FSGM). The GFM is a structural macroeconometric model of the world economy, covering 40 economies at the quarterly frequency, documented in Vitek (2018). This New Keynesian dynamic stochastic general equilibrium (DSGE) model features a range of nominal and real rigidities, extensive macrofinancial linkages with both bank and capital marketbased financial intermediation, and diverse spillover transmission channels. The FSGM is a semistructural macroeconometric model of the world economy, covering 24 economies at the annual frequency, documented in Andrle et al. (2015). For simulating

---

[16] See IMF (2016c, 2017b).

[17] See IMF (2017c).

[18] Capital flows can also be cyclically driven.

[19] See IMF (2018b).

stress test scenarios in which macrofinancial linkages or international financial spillovers are important, the use of the GFM is preferable because the FSGM lacks a banking sector and international financial linkages. In contrast, the use of the FSGM is preferable for simulating FSAP stress test scenarios in which structural shifts are important because the GFM lacks permanent shocks. For economies not currently calibrated by the GFM or the FSGM, IMF staff members generate their own scenarios using structural vector autoregressive models. On occasions, staff members use the authorities' macroeconomic models.

Given computational feasibility considerations, the GFM and the FSGM are linear. This is a disadvantage shared by all main structural macroeconometric models used by major central banks to help inform their policy analysis and is necessitated by the high computational cost of solving, estimating, and simulating nonlinear models, which increases exponentially with their scale. In turn, their advantage is that they can generate consistent macrofinancial variables for a large number of countries, which is particularly useful for international banks operating under different macrofinancial conditions. In addition to recommendations for the FSAP country, this kind of setup can also help identify spillovers and support policies of macroprudential coordination. Nevertheless, some important nonlinearities are captured in FSAP stress test scenarios simulated using linear structural macroeconometric models through the scenario design (Figure 26.3). These include incorporating discrete asset-price adjustments calibrated based on empirical analysis, asymmetric default rate adjustments calibrated based on debt-at-risk indicators, and effective lower-bound restrictions imposed on monetary policy.

Robustness may require the use of several adverse scenarios. This is the case when there are several potential threats or a combination of cyclical and structural vulnerabilities. More than one scenario is sometimes used when banks have different business models or

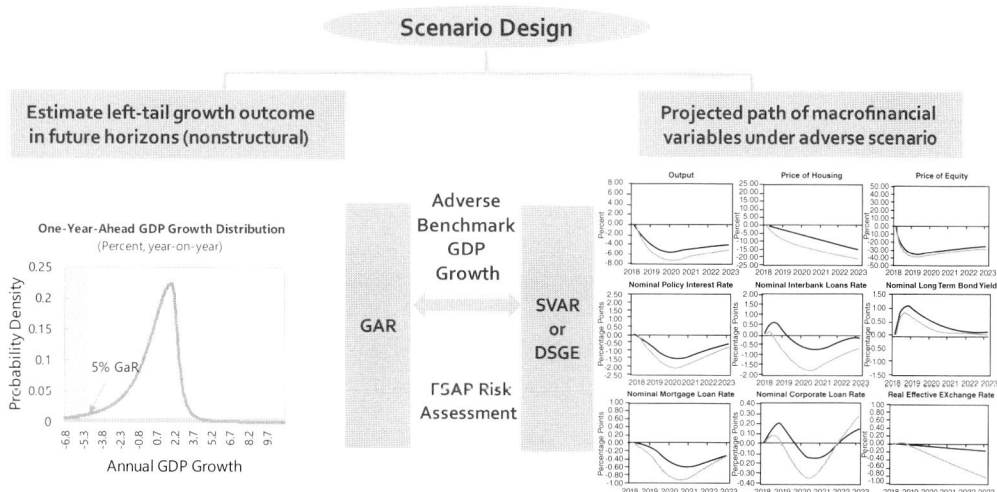

Figure 26.3 From severity to a fully fledged scenario.

geographical coverage of operations (e.g., for Spanish banks because some of them operate mainly internationally, whereas others operate mainly domestically). In other occasions, a layer-of-shocks approach is used whereby all material geographies to individual banks are subject to simultaneous shocks (e.g., 2018 euro-area FSAP). Typically, teams use sensitivity stress tests to complement scenario stress tests to explore resilience to wider shifts of risk factors. Sensitivity stress tests can also include ad hoc tests of failure of large or concentrated exposures.

IMF staff members have recently added to their tools an analytical framework that allows simulating full distributions of financial institutions' capital position conditional on a range of scenarios.[20] The Macro-Financial System Simulator (MASS) allows the ex post (rather than ex ante, as currently done) identification of those scenarios that are associated with downside risks to banks. The semistructural model framework allows for the capture of macrofinancial feedback as well as state-dependent (nonlinear) economic-cycle dynamics. A worst-case scenario search methodology is embedded in the framework, which connects naturally to the framework whose focus lies on distributions, thereby also reflecting scenario uncertainty.

6 Stress Testing Banks: The Basic Framework

We define *basic framework* (Figure 26.4) as the set of modules for solvency, liquidity, and contagion. The advanced framework comprises the gradual integration of all sources of risk (including solvency liquidity and contagion) in one single exercise.

6.1 Solvency Stress Testing

Solvency stress tests measure institutions' resilience to adverse scenarios and identify those vulnerabilities that may be responsible for weaknesses. Resilience is assessed by the adequacy of bank capital under stress. Adequacy depends on a comparison between *actual* capital under stress and *required* capital under stress.

- Actual capital under stress is the sum of the initial capital and net losses that are expected to materialize under the adverse scenario.
- Stress tests also estimate capital requirements under the adverse scenario. For example, for credit risk, the calculation of capital required under stress would reflect the evolution of the portfolio size as well as changes in the credit-risk parameters under the scenario.

Although IMF stress tests focus on losses due to different sources of risk, all elements of the income statement need to be forecasted in the adverse scenario to arrive at a new measure of actual capital under stress. This is composed of the net interest income (including the impact of interest rate risk and funding shocks), provisions for credit risk, trading losses, fee income, operational expenses, and taxes. Staff members estimate "satellite" models to

---

[20] See Gross, Leika et al. (forthcoming).

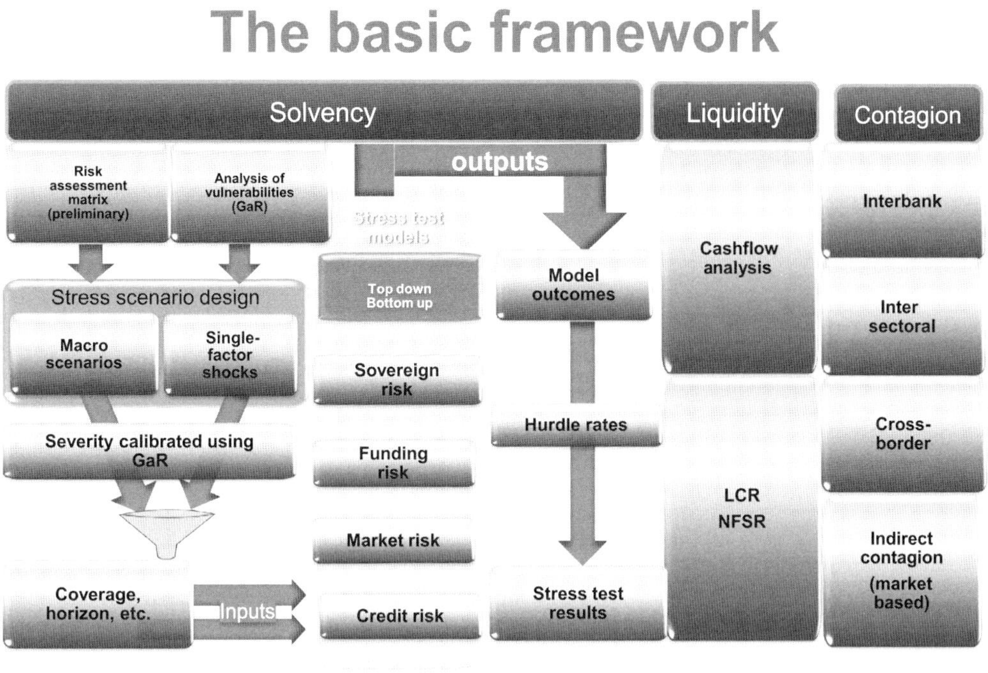

The basic framework

Figure 26.4 The basic stress testing framework.[21]

measure how changes in the macrofinancial variables would affect the different components of the income statement. The final stress test results consist of projections of complete income statements, balance sheets, and capital ratios under the adverse scenario for all years included in the risk horizon.

A hurdle or a passing rate is used as the first step to identify the drivers and not as a goal in itself. These hurdle rates are based on the regulatory approach chosen by the country; that is, they would be based on some measure of the ratio of capital to risk-weighted assets (RWAs). For banks following Basel III, hurdle rates include the minimum capital requirement (Pillar 1) and capital required by the supervisory review to address idiosyncratic risk not included in Pillar 1 (Pillar 2). Under the baseline scenario, typically, the systemically important banks, capital conservation, and the countercyclical buffers are part of the hurdle rate. The inclusion of the CCyB is relaxed under the adverse scenario because banks may be allowed to use this capital to pay for cyclical losses. Leverage ratios are being increasingly used by the IMF as supplementary hurdle rates for global banks. On occasions, FSAPs have used the asset-maintenance ratio for systemically important branches. The asset-maintenance ratio is the amount of local assets that could be available to pay for local deposits under stress.[22]

---

[21] This framework is described in detail by Leika et al. (forthcoming).
[22] See IMF (2013).

The presentation of results intends to maximize the information on potential vulnerabilities while preserving confidentiality. The IMF framework produces stress tests for each individual bank, as shown in Table 26.1. However, this information on individual banks is not released to the public. Instead, for public release, the results are presented in terms of vulnerabilities unveiled by the analysis as well as recommendations to mitigate risks. Table 26.1 shows that for each bank, staff can identify the source of risk that was responsible for the reduced profits or capital losses. Because the losses are related to actual bank exposures and client data, the stress tester can identify drivers and suggest policy solutions to mitigate this potential outcome, as described in Section 6.

Table 26.1. Presentation of results of IMF solvency stress tests

a. [Country] FSAP. Bank 1: Summary Results							
	Dec. -14	Dec. -15	Dec. -16	Dec. -17	Dec. -18	Dec. -19	Dec. -20
SUMMARY RESULTS							
Total capital adequacy ratio (CAR)	9.4%	5.6%	4.2%	4.4%	4.2%	5.8%	5.9%
Tier 1 capital ratio (T1R)	7.1%	4.0%	2.7%	2.9%	2.7%	4.0%	4.2%
Common/Core tier 1 ratio (Core T1R)	4.3%	2.3%	0.8%	1.0%	1.0%	1.6%	1.7%
Did the banks fail the test?		Yes	Yes	Yes	Yes	Yes	Yes
CAR		Yes	Yes	Yes	Yes	Yes	Yes
T1R		Yes	Yes	Yes	Yes	Yes	Yes
Core T1R		Yes	Yes	Yes	Yes	Yes	Yes
Necessary recapitalization (in monetary units)		521.1	809.3	774.3	836.8	515.1	524.8
CAR		521	809	774	837	402	389
T1R		419	704	667	721	366	341
Core T1R		480	785	762	767	515	525
Necessary recapitalization (% of current total capital)							
CAR		12.0%	18.6%	17.8%	19.3%	9.3%	9.0%
T1R		9.6%	16.2%	15.4%	16.6%	8.4%	7.8%
Core T1R		11.0%	18.1%	17.6%	17.7%	11.9%	12.1%
Capital and RWAs (in monetary units)							
Total capital	1,300	1,188	910	953	926	1,043	1,105
Tier 1 capital	975	863	585	628	601	718	780
Common/Core capital	600	482	183	209	225	298	316
Sum of RWAs	13,820	21,358	21,494	21,595	22,034	18,061	18,681
Change of total capital		−112	−277	43	−27	117	62
Change of RWAs		7,538	136	101	439	−3,973	620
Net profit	770	−80	−283	47	119	234	250
Net interest income	1,143	651	619	594	603	602	611
Loss provisions	−373	−549	−770	−469	−453	−338	−330
Net noninterest income	25	16	11	10	10	10	11
Trading income/losses	−21	−197	−142	−82	−27	−14	−14
Others	1	0	0	0	0	0	0
Taxes	−5	0	0	−5	−13	−26	−28

Table 26.1. (cont.)

b. [Country] FSAP. Bank 1: Decomposition of Results						
	Dec. -14	**Dec. -15**	**Dec. -16**	**Dec. -17**	**Dec. -18**	**Dec. -19**	**Dec. -20**

	Dec. -14	Dec. -15	Dec. -16	Dec. -17	Dec. -18	Dec. -19	Dec. -20
IN MONETARY UNITS							
Change in total capital		−111	−276	43	−28	117	62
Net profit		−78	−282	47	119	234	250
Net profit (before losses due to stress)		1,154	1,113	1,087	1,080	1,076	1,096
Loss provisions		−549	−770	−469	−453	−338	−330
Funding risk		−453	−465	−470	−480	−489	−501
Interest rate risk (refinancing and rollover risks)		−34	−18	−19	0	0	0
Spread risk		−15	−10	−5	−1	−1	−1
Repricing risk in the trading and available-for-sale (AFS) books		−27	−14	−14	0	0	0
Foreign exchange rate risk		−148	−114	−63	−27	−14	−14
Commodity risk		−6	−4	0	0	0	0
Losses from derivatives		0	0	0	0	0	0
Other market risks (including equity)		0	0	0	0	0	0
Other comprehensive income		−27	−13	−10	0	0	0
Dividends		0	0	−23	−60	−117	−188
Other equity corrections		0	0	0	0	0	0
Basel III capital adjustments		−6	19	29	−87	0	0
Own-sovereign risk		−31	−16	−12	0	0	0
Foreign-sovereign risk		−6	−3	−4	0	0	0
Sum of RWAs	12,540	21,358	21,494	21,595	22,034	18,061	18,681
Credit-risk RWAs	12,290	20,908	21,044	21,195	21,684	17,716	18,336
Market-risk RWAs	150	200	200	150	100	95	95
Other	100	250	250	250	250	250	250
IN % CONTRIBUTION TO RWAs							
Change in total capital		−0.5%	−1.3%	0.2%	−0.1%	0.6%	0.3%
Net profit		−0.4%	−1.3%	0.2%	0.5%	1.3%	1.3%
Net profit (before losses due to stress)		5.4%	5.2%	5.0%	4.9%	6.0%	5.9%
Loss provisions		−2.6%	−3.6%	−2.2%	−2.1%	−1.9%	−1.8%
Funding risk		−2.1%	−2.2%	−2.2%	−2.2%	−2.7%	−2.7%
Interest rate risk (refinancing and rollover risks)		−0.2%	−0.1%	−0.1%	0.0%	0.0%	0.0%
Spread risk		−0.1%	0.0%	0.0%	0.0%	0.0%	0.0%
Repricing risk in the trading and AFS books		−0.1%	−0.1%	−0.1%	0.0%	0.0%	0.0%
Foreign exchange rate risk		−0.7%	−0.5%	−0.3%	−0.1%	−0.1%	−0.1%
Commodity risk		0.0%	0.0%	0.0%	0.0%	0.0%	0.0%
Losses from derivatives		0.0%	0.0%	0.0%	0.0%	0.0%	0.0%
Other market risks (including equity)		0.0%	0.0%	0.0%	0.0%	0.0%	0.0%
Other comprehensive income		−0.1%	−0.1%	0.0%	0.0%	0.0%	0.0%
Dividends		0.0%	0.0%	−0.1%	−0.3%	−0.6%	−1.0%
Other equity corrections		0.0%	0.0%	0.0%	0.0%	0.0%	0.0%
Basel III capital adjustments		0.0%	0.1%	0.1%	−0.4%	0.0%	0.0%
Own-sovereign risk		−0.1%	−0.1%	−0.1%	0.0%	0.0%	0.0%
Foreign-sovereign risk		0.0%	0.0%	0.0%	0.0%	0.0%	0.0%
Change in RWAs		−4.3%	0.0%	0.0%	−0.1%	0.9%	−0.2%
LEVERAGE RATIOS							
Capital/interest-bearing assets	4.9%	4.4%	3.4%	3.6%	3.4%	3.7%	3.9%
Tier 1 capital/interest-bearing assets	3.7%	3.2%	2.2%	2.3%	2.2%	2.6%	2.7%
Tier 1/total assets	3.1%	2.6%	1.8%	1.9%	1.8%	2.1%	2.2%

Table 26.1. (cont.)

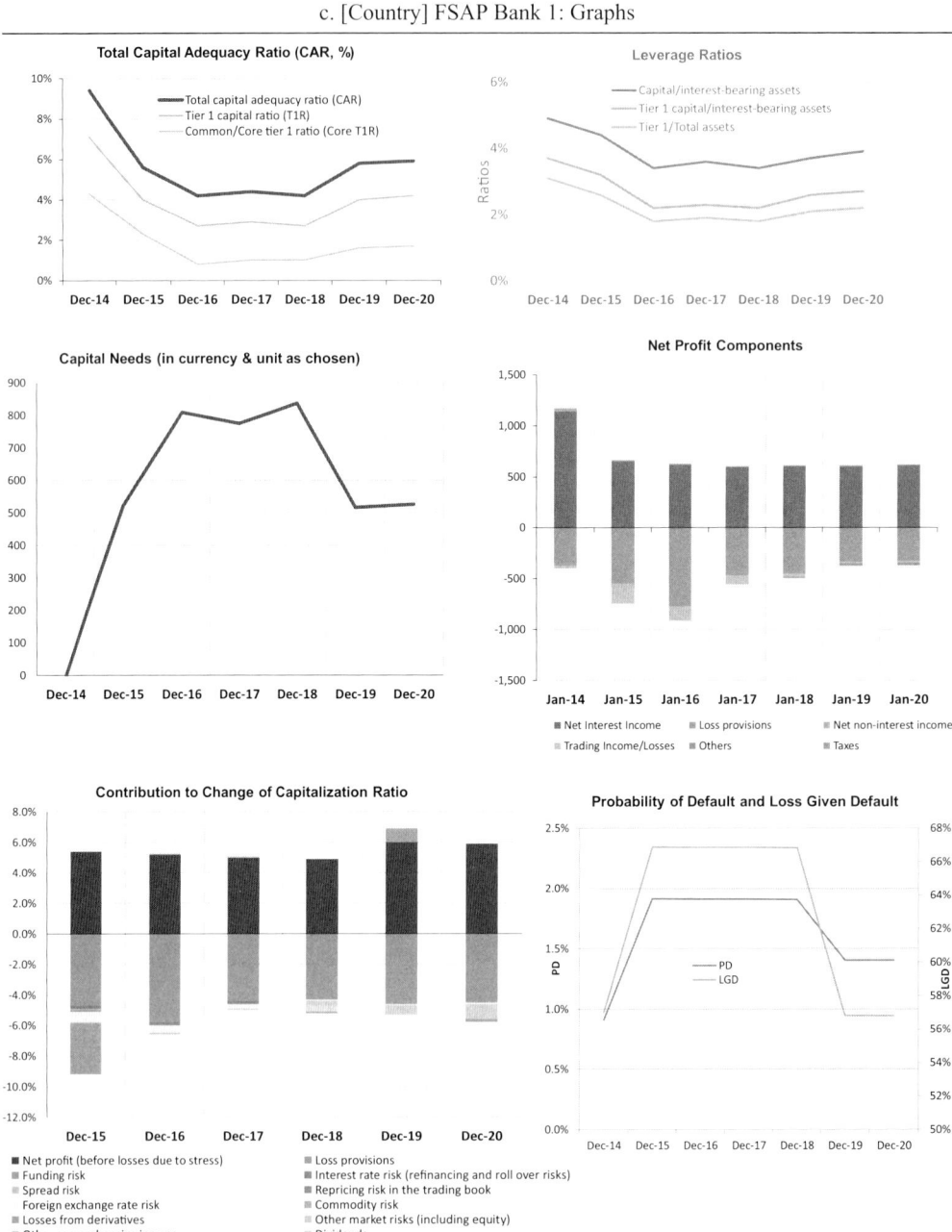

c. [Country] FSAP Bank 1: Graphs

6.2 Liquidity Stress Testing

Liquidity stress tests assess the capacity of individual banks and the banking system to withstand extreme but plausible funding shocks (funding liquidity) and/or a significant decline in the value of banks' assets (market liquidity). The tests are a means to explore weaknesses in banks' management of liquidity risk, assess their preparedness to deal with financial stress, and help the authorities identify priorities for policy actions—including those aimed at reducing specific exposure or building buffers. Although not fully integrated into a single exercise with solvency stress tests, liquidity stress tests benefit from inputs from the solvency stress tests (see Section 7).

Liquidity stress testing is usually composed of an assessment of compliance with current liquidity standards as well as a cash-flow analysis that goes beyond the minimum resilience required by current standards. There are both similarities and significant differences between the IMF staff's cash-flow stress test and the liquidity coverage ratio (LCR). The LCR is designed to ensure that banks have a sufficient stock of high-quality liquid assets (HQLAs) that can be easily converted into cash to meet their liquidity needs in a 30-day risk horizon. Historical experience in emerging-market economies and the global crises demonstrate that illiquidity can last for a period of more than 30 days. Moreover, IMF staff experience is that the management of liquidity risk suffers from cliff effects at the end of the 30 days (as banks mainly try to comply with the regulation) and can also reveal mismatch risk for time horizons shorter than 30 days. This is why the cash-flow test is performed over a wide range of time buckets. For the LCR, the stress scenario is defined by internationally "harmonized" and predetermined parameters, including runoff rates on funding sources, rollover rates on assets, and haircuts that simulate reduced market liquidity. Instead, parameters in the cash-flow analysis are customized to reflect the characteristics of funding markets and include all unencumbered marketable securities (rather than focusing on HQLAs), acknowledging the fact that all marketable assets can be used to raise funding (at appropriate haircuts) to address negative funding gaps. The cash-flow analysis is also more demanding in terms of granular supervisory data to gauge potential liquidity shortages accurately.

The cash-flow stress test simulates the behavior of inflows and outflows in a baseline and in one or more adverse scenarios. In a baseline scenario, there are no funding shocks, asset markets operate with normal depth, and banks may benefit from net behavioral cash inflows. In contrast, an adverse scenario is characterized by funding shocks (proxied by runoff and rollover rates), reduced liquidity in asset markets (proxied by haircuts), and the absence or reduction of net new behavioral cash inflows. The test includes explicit assumptions on the evolution of the exchange rate.

The data for a cash-flow-based stress test are more granular than those for solvency stress tests and need to be collected at higher frequency. The template is consistent with Basel liquidity-monitoring tools for contractual maturity mismatches. It requires contractual and behavioral assumptions for all types of inflows and outflows in all material currencies, as well as for all currencies, and a detailed breakdown of available unencumbered assets. Inflows and outflows are classified in time buckets, the first and second ones comprising all inflows and outflows that are expected to materialize in the next 24 hours and between 2 and 7 days, respectively. Typical time buckets are up to 30 days, 31–60 days, 61–90 days, 91–120 days, 121–150 days, 151–180 days, 181 days–1 year, 1–3 years, and beyond 3 years.

Some of these inflows and outflows are determined by contracts, but others require behavioral assumptions such as demand deposits and callable bonds. There are also contingent flows that depend on changes in prices of financial instruments or on a downgrade in the bank's credit rating; therefore, behavioral assumptions need to be included.

The calibration of parameters in the adverse scenario is challenging. Parameters are comprised of runoff rates, rollover rates, and haircuts. Ideally, the calibration should use the results of the solvency stress tests that could trigger liquidity dry-outs as well as past information related to crises episodes in the FSAP country. However, this information is usually not available. Instead, stress testers use a combination of information on the characteristics of the FSAP country and information available in the literature on the behavior of inflows and outflows, including their behavior during the crisis. To address uncertainty, teams use multiple scenarios with various levels of parameter severity. A good guidance for liquidity scenario design is the literature review in Basel Committee on Banking Supervision (BCBS, 2013) on factors related to liquidity stress, which suggests that core deposits are more stable and that the deposit status (insured vs. noninsured) and the bank–depositor relationship matter during times of stress. It also points to the procyclicality of secured funding with regard to haircuts, valuation, tenors, and counterparty credit-risk limits, highlighting that even for the relatively stable tri-party repo market, asset credit quality is important. Regarding wholesale funding, evidence from the GFC reveals that it did not entirely freeze; instead, tenors shortened, and rates became more sensitive to borrower characteristics. Recent FSAPs where contractual cash flow data were available (Japan,[23] euro area,[24] Switzerland[25]) simulated multiple scenarios to find the level of liquidity risk tolerance (i.e., survival horizon under different stress scenarios) of various banks. The metrics for scenario severity were based on the episodes of the GFC, as well as linked to asset-price shocks derived from solvency scenarios.

A bank fails the test when only emergency liquidity assistance (ELA) from the central bank would allow the bank to continue fulfilling its obligations. Banks can cover negative balances of cash inflows relative to cash outflows by using their counterbalancing capacity to obtain liquidity in secondary markets or through standard central bank facilities. At this point, the bank has depleted all its cash items, its eligible collateral to access the standard central bank facilities is insufficient to offset negative funding gaps, and only access to ELA can keep it open.

6.3 Contagion Analysis

IMF teams assess contagion risks at three levels: interbank, cross-sectoral (i.e., among different types of financial intermediaries), and cross-border.[26] The analysis is conducted using exposure data and market-based indicators. Exposure data are typically used to assess domino effects. To complement the analysis, selected measures of codependence embedded

---

[23] See IMF (2017d).

[24] See IMF (2018c).

[25] See IMF (2019a).

[26] Bricco and Xu (2019) survey current approaches at the IMF for analyzing interconnectedness within the interbank, cross-sector, and cross-border dimensions through an overview and examples of the data and methodologies used in the FSAP.

in asset prices are used by staff to assess potential contagion among banks, among banks and other financial intermediaries, and among banks and sovereigns. The use of market-based methodologies is important because of the limited availability of exposure data and because they capture forms of contagion related to indirect linkages (e.g., exposures to common risk factors) other than domino effects based only on direct claims.[27] Methodologies combining exposure and market-based data have also been used.[28]

The IMF usually relies on the authorities for the assessment of potential domino effects in the interbank market. Authorities use scenarios and parameters that were agreed on with staff. Results (not the underlying data) are shared with IMF teams because the underlying counter-party exposure data are typically confidential. The Espinosa-Vega and Solé (2010) approach is often used for interbank analysis, capturing both credit and funding shocks. The credit shock arises from the failure of a client. The funding shock arises from the failure of a funding source. Whenever relevant,[29] IMF teams request that bank exposures include derivatives that are valued at the prices prevailing in the adverse scenario, and therefore results would capture correlated market, credit, and contagion risks.

Systemic risk resulting from cross-sectoral exposures is composed of the interaction between banks and nonbanks. These can take place because of different channels, including the group's structure,[30] the funding channel[31], the exposure channel, the common exposure channel,[32] and other network relations.[33] The conglomerate structure may affect banks mainly through reputational risks when other intermediaries in the conglomerate are in distress. The funding channel can affect banks' funding and liquidity positions through the response of nonbanks to the risk of banks in which they invest; the exposure channel is mainly credit risk from bank investments in other nonbanks; and finally, the failure or downgrading of common exposures could trigger contagion among bank and nonbanks. The assessment of conglomerate risks is mainly assessed through common adverse scenarios for banks and nonbanks. Staff members also assess regulations to ensure that bank capital is not exposed to risks arising from other parts of the conglomerate. Funding and exposure data are mainly used when available for interconnectedness analysis. The analysis of interaction from common exposures is usually jeopardized by the lack of sufficient data to assess potential contagion.

The analysis is usually more limited for cross-sectoral exposures than for the interbank market because of data constraints. Data on bilateral claims between a bank and a nonbank financial institution are often not collected by countries, including advanced economies. At best, countries may have fragmented data or data that are not easy to use because they are

---

[27] These additional methodologies comprise, among others, those developed by Adrian and Brunnermeier (2016), Diebold and Yilmaz (2008), Segoviano and Goodhart (2009), and Gray and Jobst (2013). See also Chapter 9 in this volume.

[28] See Cortes et al. (2018).

[29] See IMF (2013), page 14.

[30] For example, as revealed by the GFC, banks may be exposed to shadow-banking entities through their common membership in a corporate group, through the provision of explicit or implicit backstops, or indirectly through their common exposures. See BCBS (2017) for a discussion of explicit and implicit backstops.

[31] See IMF (2016d).

[32] See Garcia Luna and Hardy (2019).

[33] See Bricco and Xu (2019).

not specifically reported for this type of analysis; therefore, data preparation may take too long to be useful within the timeframe of the IMF assessment. As a substitute, teams use flow-of-funds data or aggregate data for intersectoral exposures, which give a bird's-eye view of interconnectedness based on exposure and funding data among types of financial intermediaries (i.e., not at individual-institution level). The use of individual data, when available, can provide important information among potential sources of systemic risk. In Sweden, for example, the mission had access to information on holdings of covered bonds issued by some banks and held by other banks or by insurance companies. This led to a fruitful exchange with the authorities on potential financial stability issues associated with cross-sectoral interconnectedness. In the 2017 Japan FSAP,[34] the team obtained granular exposure data from banks, insurers, pension funds, and mutual funds and was therefore able to expand the interbank network to capture risks from nonbank financial institutions; the recent euro-area FSAP (2018)[35] analyzed two complementary networks: one focused exclusively on large institutions, the other on the most material extra-euro-area exposures of each of those institutions.

Cross-border network effects are particularly important for the IMF, given its mandate to monitor global financial stability. Cross-border network effects could emerge from exposure to other financial intermediaries or intragroup exposures that put pressures on the parent bank to support entities abroad. Similarly, subsidiaries or branches can be affected by problems in the parent bank located in another country.

Cross-border interconnectedness is assessed using both exposure and market data. Typically, staff members use the framework developed by Espinosa-Vega (2010) for cross-border exposures, enabling the simulation of both credit and funding shocks. Although data on intragroup exposures are usually not publicly available at the individual-institution level, the IMF performs network analysis of (aggregate) national banking systems using BIS locational or consolidated data. Locational data classify funds by country based on the "resident" criteria, whereas consolidated data are based on country of incorporation. The initial negative credit or funding shock is propagated through the network of bilateral claims across countries. If any banking system incurs losses larger than their total Tier 1 or regulatory capital, the system "fails." This failure can subsequently cause some other banking system to fail, triggering domino effects, where a failure of a banking system in a network transmits to other banking systems.

The exposure-based approach is often complemented by market-based methodologies. These studies usually examine indirect linkages among banks or across countries that may not be properly captured by domino effects—for example, banks or countries exposed to the same type of assets (e.g., bonds issued by a sovereign in distress) or associated with a bank or country in distress by investors because of a lack of specific knowledge and/or asymmetric information. Furthermore, in the recent Spain FSAP, the mission applied the global vector autoregressive (VAR) methodology (Dees et al., 2007) to examine the outward spillover of credit shocks originated in Spain on banking systems in Europe and Latin America.[36]

---

[34] See IMF (2017d).

[35] See IMF (2018c).

[36] More broadly, IMF staff members have also examined interconnectedness among global systemically important banks (GSIBs) and global systemically important insurers (GSIIs; Malik and Xu, 2017).

For the final assessment of whether cross-border banking promotes or threatens financial stability, the quantitative cross-border analysis is complemented by qualitative information. This information includes the core characteristics of cross-border relations, including whether funding flows from or to parent banks, the home country's rules regarding the provision of liquidity, existing (and possible) liquidity requirements and ring-fencing imposed by home and host supervisors (e.g., the application of LCR at group and/or sub/branch levels), and resolution modalities (including whether there is a resolution plan in place that also comprises foreign subsidiaries and single vs. multiple point of entry), among others.

7 Risk Amplification

IMF staff members currently focus on better integrating the different elements of its basic framework through the identification of channels of risk amplification. IMF staff members have been at the forefront of introducing mechanisms in stress tests to capture how the financial system amplifies shocks. A first stylized but integrated approach was presented in the *Global Financial Stability Report* (April 2011),[37] in which systemic risk was simulated through a combination of runs, asset fire sales, and solvency deterioration. Using publicly available information and a forward-looking simulation of integrated market and credit risks applied to the stylized balance sheets of a set of US banks, the paper estimated the probability that multiple banks would fail or experience liquidity runs simultaneously. Liquidity runs were modeled as a response to elevated solvency risk and uncertainties and were shown to increase the probability of correlated bank failures. Under this setting, lower funding liquidity arose from increased uncertainty over counterparty risk and lower asset valuations that induced banks and investors to hoard liquidity, leading to systemic liquidity shortfalls. The simulation featured fire sales of assets as stressed banks sought to meet their cash-flow obligations. Lower asset prices affected asset valuations and margin requirements for all banks in the system, and these in turn affected funding costs and profitability and generated further systemic solvency concerns.

Following this initial mapping, several amplification channels have been analyzed in FSAPs (Figure 26.5 and Table 26.2), including the following:

- Amplification effects from interconnectedness in the interbank market, cross-sectoral, and cross-border. Although some of this work was already in progress before the GFC, it has become a more important feature of the FSAP since then.
- The solvency and liquidity interaction. This channel covers the solvency consequences of liquidity risk as well as liquidity risks that emerge as a result of solvency and asset-quality deterioration. The bank-sovereign nexus is included here.
- The multiple feedback effects between the real economy and the financial sector. This covers the behavioral response of financial intermediaries to stress conditions, including the pass-through of funding costs to borrowing rates, and the reallocation of bank assets, including the impact of stress on the credit supply and other forms of deleveraging.

---

[37] See IMF (2011), page 106, and Barnhill and Schumacher (2011).

Table 26.2. Mechanisms of risk amplification

Mechanism	How it operates	Implementation
	Interconnectedness	
Interbank market and cross-border links	Domino effects through credit and funding channels (e.g., unforeseen withdrawal of interbank funding or counterparty default)	Implemented in most FSAPs; in some cases, supplemented with market-based assessments
	Liquidity/funding factors included in solvency stress tests	
Bank-solvency nexus	Banks' funding costs and liquidity access are affected by the valuation of sovereign bonds in bank portfolios as a result of sovereign distress and excess concentration on own-sovereign assets.	Implemented in most FSAPs
Funding channel	Poor asset quality and/or changes in the market value of equity motivate investors to reassess bank risk, affecting funding costs, profitability, and potentially bank capital; this in turn gives rise to further rounds of higher funding costs, replicating the typical nonlinear effect.	Implemented in the euro-area FSAP and in most FSAPs, with granularity dependent on data availability
Market-liquidity channel	Value of bank assets shrinks as a result of sales forced by lack of access to funding liquidity.	Implemented in FSAPs when data are available
	Solvency factors included in liquidity stress tests	
The probability-of-default (PD) impact	Estimated bank assets' PDs affect bank asset values, including those used for collateral for central bank lending (i.e., haircuts) and those including in the counterbalancing capacity.	Implemented in most FSAPs
The NPL channel	NPLs stop making interest payments, affecting liquidity inflows.	Implemented in most FSAPs

Table 26.2. (Cont.)

Mechanism	How it operates	Implementation
Liquidity dry-out channel	Several versions: troubled banks are shut off from funding over a year for one quarter; all banks are deprived from a significant fraction of their wholesale structure of funding.	Implemented in most FSAPs using ad hoc assumptions based on local banks' structure of funding
Counterparty credit risk and lower derivative inflows	Issuer default risk is related to default events on underlying names to which banks' positions refer. The default of underlying issuers has a direct impact on solvency through an increase in credit-risk losses and an indirect impact on liquidity from a lower inflow rate of maturing inflows.	Implemented in the Switzerland FSAP
Real–financial interaction		
Pass-through rates	Higher funding costs can be passed to bank borrowers depending on market structure. Pass-through rate affects loan demand and GDP growth.	Implemented in most FSAPs and implicitly incorporated in the adverse scenario
Portfolio reallocation I	Optimizing under regulatory and market funding constraints, banks may choose a new asset allocation, including by reducing loan supply to the real economy.	Explored in Switzerland FSAP, based on IMF staff research
Portfolio reallocation II	Banks respond to economic shocks by deleveraging, which affects credit and GDP growth. Consistency between macrofinancial variables and banks' balance sheets is accomplished by embedding a standard stress testing framework based on individual banks' data in a semistructural macroeconomic model.	Used to support macrofinancial analysis in the 2016 and 2017 Brazil Article IV consultation

533

Amplification mechanisms

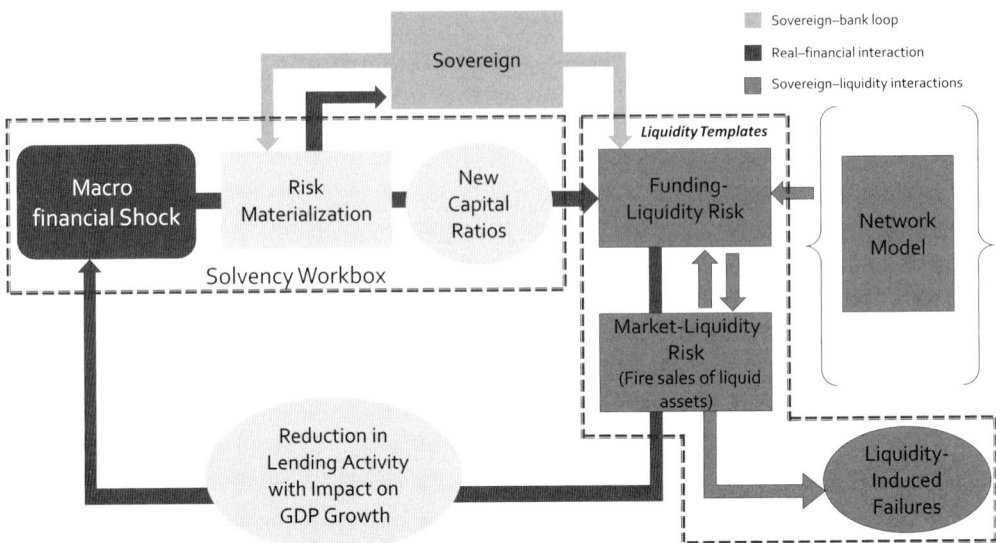

Figure 26.5 Key risk-amplification mechanisms captured in FSAP stress tests.

7.1 Liquidity Factors Included in Solvency Stress Tests

Current efforts are addressed at obtaining robust estimates of the interaction between liquidity and solvency risk. Cont et al. (2019) propose a structural framework for the joint stress testing of solvency and liquidity to derive relations between solvency and liquidity shocks. These relations are then used to model liquidity and solvency risk in a coherent framework, involving external shocks to solvency and endogenous liquidity shocks. This allows for the introduction of solvency–liquidity diagrams to analyze the resilience of banks. This framework was used in the Switzerland FSAP.

Most FSAPs capture the interaction between solvency and liquidity following a simplified approach pioneered by the National Bank of Austria.[38] The simultaneous interdependence of solvency and liquidity risks and cross-bank contagion is difficult to model with sufficient granularity. Therefore, the IMF staff approach follows a shortcut by which solvency and liquidity stress tests are still run independently, but solvency stress tests incorporate the impact of liquidity factors, and liquidity stress tests incorporate the impact of solvency factors. Key liquidity factors that affect the outcome of solvency stress tests are the funding cost–solvency loop, including the impact of the sovereign–bank nexus, and the market-liquidity channel. Key solvency factors that affect the outcome of liquidity stress tests are changes in the risk characteristics of assets used by banks as collateral for central bank lending ("haircuts") and on assets held as counterbalancing capacity, as well as the impact

---

[38] Puhr and Schmitz (2013).

of NPLs on the flow of interest earnings and the shutoff of funding markets as a result of solvency concerns or uncertainty about the quality of bank assets.

7.1.1 Funding Cost and Solvency

Because a bank needs to tap funding markets to fund its balance sheet, funding cost can affect bank solvency. This could lead to additional asset encumbrance, deleveraging with further impact on the economy, and asset liquidation, which could further reduce asset prices—increasing MTM losses, reducing banks' value of counterbalancing capacity (CBC), and impairing capital positions. These negative-feedback loops could significantly amplify the effects of stress scenarios (Adrian and Shin, 2010).

Several recent papers have addressed the empirical problem of linking funding costs with banks' solvency position, especially the endogeneity of funding costs and capital of banks. The recent literature suggests that the interaction between solvency and funding cost is indeed statistically significant, and it might be economically relevant, in particular during period of stress.[39] Aymanns et al. (2016), using publicly available information for the largest US banks, find that a decline in solvency ratios leads to higher unsecured funding costs, as well as a decline in net interest margin. Schmitz et al. (2017) analyze changes in bank funding costs for 54 global banks. Their finding suggests that a 100-bps increase in regulatory capital ratios is associated with a decrease of bank funding costs of about 105 bps. Higher funding costs also reduce the capital ratio (100-bps increase of funding costs reduces CAR by 32 bps on average).

These papers also report empirical evidence of nonlinearities in funding costs. A key finding: during periods of stress, interbank funding cost is more sensitive to solvency than in normal times, and the relationship between funding cost and solvency appears more sensitive at lower levels of solvency.

Recent IMF staff stress tests (Japan,[40] New Zealand,[41] United Kingdom,[42] Poland,[43] euro area,[44] and Switzerland[45]) focused on integrating the solvency and funding feedback loop. This was conducted via a sequential-step procedure by which capital ratios resulting from the first round of solvency test are used to update results for the next (year) round's funding costs, and this iteration continues during the risk horizon. The process is repeated until convergence in funding costs is achieved. Funding costs are projected at the bank, instrument, and contract level. For Japan, the IMF staff stress tests used the ratio of interest expenses to interest-earning liabilities as proxy for funding costs; the final equation of bank funding costs was based on individual bank data and included a bank-specific term that related funding costs to individual capital ratios as well as a systemic component based on the behavior of base rates; the systemic component was controlled by a Markov regime-switching regressor capturing the typical overreaction (nonlinear effect) that affects markets

---

[39] For a review of the literature on the interdependence between funding costs and solvency and country practices, see BCBS (2015).
[40] See IMF (2017d).
[41] See IMF (2017e).
[42] See IMF (2016e).
[43] See IMF (2018e).
[44] See IMF (2018c).
[45] See IMF (2019a).

in times of stress. The team also modeled funding costs in foreign currency using a Markov regime-switching model to analyze the liquidity risk premium in the USD/JPY swap market.

7.1.2 Solvency, Equity Valuation, and Funding Cost

Although the solvency stress test focuses on regulatory capital ratios, the risk of insolvency can be amplified by a market-equity correction that triggers a bank credit-rating down-grade. The euro-area FSAP (IMF, 2018c) estimated the amplification channel from stressed bank equity valuations and solvency risk to funding costs using a Merton-based approach. The analysis was based on Bloomberg's Default Risk (DRSK) module that calculates the implied probability of default (PD), implied rating, and implied bank credit default swap conditional on bank financials and aggregate market variables. The key inputs of the model were drawn from the worst period in the FSAP solvency stress test, which included stressed effective short-term debt, effective long-term debt, loan loss reserves, NPLs, and net income. The market category variables, including bank share price, market capitalization, and share price volatility, were projected using a simple model linking the beta of individual banks' share price and volatility to the path of the market share price and volatility calibrated in the scenario.

7.1.3 The Sovereign–Bank Nexus

The credit quality of government securities can be an important source of uncertainty in bank credit quality that affects the funding constraint. National approaches usually have benign rules for provisions and capital requirements for government bonds that are held to maturity (i.e., will not be traded) under the assumption that losses will not take place because full payment will be made at maturity.[46] However, investors are more interested in the value of bank equity. As such, the impact of sovereign risk on funding costs may be larger for investors than for capital ratios measured by regulators.

For this reason, IMF stress tests may assume, on occasion, losses for sovereign bonds held to maturity when calculating the capital ratio under stress. For stress testing, IMF staff members—while still allowing for low or zero risk weights—typically estimate potential losses due to credit/sovereign risk by means of a market valuation under stress; these losses will reduce the numerator of the capital ratio. The decision to incorporate market-valuation losses depends in part on staff's assessment of the sustainability of sovereign debt.

---

[46] Provisions for government debt held to maturity are typically low for banks under the standardized approach, when loss provisions are based on past default. Similarly, for countries under the IRB approach, provisions are low because the PDs are typically low for most sovereigns that have never defaulted. Regarding capital requirements (the denominator of the capital ratio), IRB banks need to set aside capital according to sovereigns' PDs, but low PDs reduce the risk weights to nearly zero. Under the standardized approach, national authorities are given discretion to give zero risk weights to own-sovereign securities issued in their own currency for both credit and market risks. Even among advanced economies, many banks choose the standardized approach for sovereign securities: the European credit directive on capital requirements allows zero weight on sovereign bonds denominated in euros issued by any euro-area economy (e.g., Italian sovereign bonds in euros held by German banks can be zero risk weighted) when banks choose the standardized approach.

7.1.4 Solvency and Asset Sales

Abrupt and large asset fire sales amplify capital losses. Liquidity shortages may force solvent financial institutions to liquidate assets under fire-sale prices, face losses, and become insolvent. Because of common exposures among banks, the fire-sale channel also captures the contagion effects of reduced market liquidity, affects the value of banks' CBC, and leads to marked-to-market losses.

This channel is also relevant in the presence of central bank support. After the GFC, central banks granted massive liquidity support to banks via quantitative easing programs and refinancing operations with full allotment. This provided banks with an alternative source of funding instead of selling securities in the market; however, selling prices and repo haircuts applied by central banks are still based on market values of securities, thus exposing banks to market-liquidity risks.

The applied literature has suggested different approaches to modeling this channel. The Bank of England approach incorporates changes in market-liquidity premia under stress scenarios. In particular, Baranova et al. (2017) calibrate a partial-equilibrium model of dealer intermediation in bond markets, which is a complex exercise and subject to model uncertainty. Han and Leika (2019) take an empirical approach to estimate the changes in market-liquidity premiums by using Markov regime-switching models for market liquidity.[47] In their approach, asset prices are affected as market liquidity premiums change from one regime to another (nonlinearly), which in turn affects the value of banks' CBC, either through the existing stock of encumbered assets (margin calls), additional asset encumbrance (in case of negative funding gaps), or asset fire sales (in worst-case scenarios), leaving banks' liquidity endogenous to asset prices. As an application to euro-area banks, the paper estimated the market-liquidity premiums of key European sovereign bonds in both normal and stress regimes using transaction-level data. The analysis showed that liquidity premia in a stress regime, compared with that in normal times, could have a much larger impact on banks' capital ratios, and the severity of the impact depends on system-wide funding shortages. Similarly, in the 2017 Japan FSAP,[48] they calculated the Amihud (2002) measure of stock market liquidity—that is, the ratio of price change to the trading volume—to capture the price impact of sales in equities for banks that face a liquidity shortfall.

7.2 Solvency Factors Included in Liquidity Stress Tests

IMF staff members also incorporate solvency factors affecting the outcome of liquidity stress tests. Staff members use the results from the solvency stress tests to calibrate some of the liquidity parameters in the liquidity stress test. For example, solvency factors affect liquidity collateral requirements, PDs estimated for the assessment of credit risk and the scenario variables are used for haircuts set by the central bank and for the revaluation of counterbalancing capacity, and counterparty credit risk plays an important role for secured funding; therefore, solvency deterioration of a bank (implying a downgrade) would lead to an increase in collateral requirements and a potential reduction in funding. Liquidity inflows, from interest income, are reduced by the amount of interest payments on NPLs.

---

[47] This measure was also used in IMF (2015b), page 55.
[48] See IMF (2017d).

7.2.1 Liquidity Dry-Out Due to Solvency Concerns

Concerns about bank insolvency could lead to the inability to raise funding at any cost ("liquidity dry-out"). Gorton and Metrick (2009) characterized the GFC as a system-wide "run" on the securitized banking system—more precisely, a "run on the repo market"—similar to the banking panics of the 19th century. Both episodes, in their view, were triggered by insolvency problems. They find that during 2007–2008, changes in the London Interbank Offer Rate-overnight indexed swap (LIBOR-OIS) spread, a proxy for counterparty risk in the interbank market, was strongly correlated with changes in credit spreads and repo rates for securitized bonds. These changes implied higher uncertainty about bank solvency and lower values for repo collateral. They conclude that the market slowly became aware of the risks associated with the subprime market, which then led to doubts about repo collateral and bank solvency. At some point—August 2007—a critical mass of such fears led to the first run on repo, with lenders no longer willing to provide short-term finance at historical spreads and haircuts.

Building a fully integrated solvency–liquidity dry-out model is difficult. Episodes like the deposit run suffered by Northern Rock in 2007, the failure of Lehman Brothers in 2008, and the evaporation of liquidity in the US securitization markets are rare episodes; data for estimates are needed with high granularity (e.g., by type of liabilities and maturity), as well as with high frequency, because these episodes develop very fast, and circumstances can be very idiosyncratic, making it difficult to find a model that can captures observed regularities. Moreover, liquidity in the banking system is more of a systemic phenomenon that requires modeling the behavior of different types of financial intermediaries at the same time.

Instead, IMF staff liquidity stress tests use several ad hoc scenarios to motivate liquidity dry-out.[49] They are based on projections of banks' cash flows under stress and consist of assumptions on runoff and rollover rates, as well as haircuts in case of collateral use. The type of scenario and parameters depend on system-specific factors, in particular, the banks' prevailing funding structure in each country. Deposits are typically considered the most stable source of funding because of their retail nature.[50] The stress tests take into consideration the currency, the residence of the depositor (resident or nonresident), and the deposit insurance coverage to choose the runoff rates for the scenarios. In spite of the fact that wholesale funding did not entirely dry out during the financial crisis, many stress test scenarios assume a total lack of access to this market. Interest rates on wholesale funding are also typically modeled because they depend on asset quality, although this would typically be part of the additional losses for a solvency stress test; commitments are also an important source of liquidity stress, not so much loan commitments but liquidity support to asset-backed commercial paper (ABCP) programs, whose relevance was clear from the crisis.[51] An important consideration for runoff rates and haircuts is also whether funding is secured or nonsecured. When relevant, stress tests have also assessed the potential impact of withdrawing central bank support. For example, given the large liquidity support from the European Central Bank (ECB), the IMF liquidity stress tests for Spanish banks used

[49] For a review of cases and the literature on how liquidity stress tests can address differences in funding sources, see BCBS (2013).
[50] See Puri and Ufier (2018).
[51] See Covitz et al. (2009).

assumptions on the reduction of central banks' funding to assess the sufficiency of banks' liquidity buffers and the quality of banks' liquidity management. IMF stress tests for Sweden assumed disruptions in the wholesale market, given the reliance of Swedish banks on short-term funding (covered bonds) to finance long-term loans. In the most severe scenario, the euro-area stress test assumed that nonmarketable assets, such as credit claims, would not be included in counterbalancing capacity. There are also some ad hoc versions of this type of scenario, such as assuming that all banks are shut off from funding over a year for one quarter, or all banks are deprived from a significant fraction of their wholesale funding.

7.3 Banks' Reaction to Stress: Feedback Effects between the Financial Sector and the Real Economy

The IMF staff stress testing framework allows bank reactions to stress in a limited manner. Typically, staff members calibrate the impact of funding costs on lending rates, and in turn, these new rates affect credit demand and GDP growth. Lending rates are split into a reference rate and a client spread. The calibration of lending rates is based on the country's market structure as well as historical pass-through rates of higher funding costs. For example, staff used euro-area countries' data on maturing loans and other assets to estimate the pass-through effect on the flow of different types of loans and securities.

Most IMF staff stress tests assume that the banking system maintains the supply of credit in the adverse scenario. In line with other stress test methodologies used by major central banks, such as the Federal Reserve and the Bank of England,[52] this feature is incorporated in IMF staff stress scenarios by specifying paths for aggregate bank lending to the economy, and it has been adopted for two reasons. First, it prevents institutions from achieving higher capital ratios by deleveraging; second, it allows the use of the stress test for additional policy purposes because the test assesses not only the banks' ability to remain resilient but also their capacity to preserve economic growth at the same time by maintaining the needed supply of credit.

The assumption of no change in the credit supply in the adverse scenario is not a realistic assumption. One important feature of a financial crisis is the procyclicality of leverage (Adrian and Shin, 2010, 2014). Leverage is procyclical in part because intermediaries target a fixed leverage; therefore, changes in their asset value lead to an adjustment in leverage. In turn, variation in leverage has a huge impact on the price of assets, contributing to economic bubbles and busts. Moreover, the GFC showed that banks' optimization process is constrained not only by capital requirements imposed by regulators but also by a market-funding constraint.[53] When banks hit minimum regulatory ratios, they are likely to stop funding new operations, leading to a credit freeze; if they lose market access, they are likely to deleverage their positions, leading to a credit contraction and fire sales. This has the potential to unleash self-fulfilling adverse dynamics, whereby banks' reaction under stress leads the system toward a new equilibrium characterized by lower credit growth,

---

[52] See Bank of England (2015), Box 1, and Federal Reserve Board (2014).
[53] See Committee on the Global Financial System (2018).

underpricing of assets, and subdued economic growth. Therefore, the current framework underestimates a risk-amplification element that comes from banks' optimization response to market-funding constraints.

IMF staff members are beginning to incorporate banks' responses to shocks and the associated externalities triggered through macrofinancial feedback effects:

- Valderrama (forthcoming) developed an agent-based dynamic portfolio optimization wherein banks rebalance their portfolios subject to regulatory and market-funding constraints, enabling not only a full integration of solvency and liquidity stresses but also financial–real feedbacks. The provision of credit is endogenous to the model and fluctuates with banks' solvency and the underlying economic environment. The outcome of the model allows for an examination of the aggregate impact of banks' unwinding of portfolios on financial markets through asset-price dynamics, credit-growth developments, banking-system resilience, and the trajectory of key macroeconomic variables under different scenarios.
- Krznar and Troy (2017) developed a modeling framework that facilitates the analysis of both the direct effects of macroeconomic shocks on the solvency of individual banks and feedback effects that allow for the amplification and propagation of shocks that result from bank deleveraging and credit crunches. At the same time, the framework ensures consistency in the key relationships between macroeconomic and financial variables and banks' balance sheets. This is accomplished by embedding a standard stress testing framework based on individual banks' data in a semistructural macroeconomic model. The paper also provides an avenue for many extensions that address the challenges of incorporating other second-round effects important for comprehensive systemic-risk analysis, such as interactions between solvency, liquidity, and contagion risks.
- Gross, Hilberg et al. (forthcoming) developed a macrofinancial, microfounded agent-based model (ABM) for the purpose of conducting macroprudential policy analyses with capital- and borrower-based policy instruments. The model generates endogenous, self-evolving business and financial cycles. The essential features of the model for a policy-analysis purpose pertain to a housing, mortgage, and rental market; to an advanced bank balance-sheet and loan-supply process; and in the latter context, a detailed structural model of bank funding-cost dynamics and the pass-through of funding-cost changes to loan prices. A CCyB-based policy, for example, renders bank defaults less frequent and reduces the extreme positive and negative moves in housing prices while compressing the business- and financial-cycle variance.
- Catalán and Hoffmaister (2020) incorporate a disaggregated banking sector into an otherwise standard macroeconomic structural vector autoregression (SVAR) framework. Following an initial round of bank losses, second-round effects from banks' lending decisions collectively affect macroeconomic developments, leading to subsequent adjustments in banks' profits and capitalization levels. Because banks and their lending responses are heterogeneous, accounting for feedback loops has a differentiated impact across banks. The model can also be used to quantify individual banks' contributions to systemic risk by assessing the dynamic effects of shocks on macroeconomic outcomes under counterfactual assumptions about the initial capitalization of individual banks.

8 Using Stress Test Results for Policy Advice

IMF staff members are increasingly using stress test results to support macroprudential policy advice. Stress tests have long been used to support recommendations of a micro-prudential nature, such as higher provisions or enhanced rules for related-party transactions. Increasingly, recommendations of a macroprudential nature to mitigate systemic risk are also made. They have thus far included calls on additional capital cushions, limits on credit demand (e.g., adoption of stressed debt-to-service ratios), and floors to capital ratios, among others. For example:

- The 2019 Switzerland FSAP[54] advocated the introduction of borrower-based measures to strengthen banks' resilience to losses in the residential property segment (owner occupied and investment led) to tighten lending criteria. The recommendation was supported by the amount of mortgage losses exposed by the stress test exercise.
- In the context of the 2017 Netherlands FSAP[55] and the 2016 Ireland FSAP,[56] staff used micro-level household survey data to assess households' debt-repayment capacity and default rates on mortgage loans under adverse scenarios, which provided the basic frame-work for a macroprudential limit on loan-to-value (LTV) ratios. To examine the feedback effects from the financial sector to the real economy, the Netherland FSAP team used a DSGE model to simulate the effects of house-price shocks on consumption and investment under higher LTV limits through higher mortgage burden, less credit demand, and banks' deleveraging. For the calibration of specific limits on LTV ratios, granular micro-level data on household income, characteristics of households and mortgage loans, and collateral values are needed. Because these data are usually confidential, IMF staff collaborated closely with country authorities to design and conduct the macroprudential stress test framework.
- Based on data from the Household Finance and Consumption survey used for the cali-bration of PDs, the 2016 Finland FSAP[57] found weaknesses in banks' estimates of risk parameters that fed into their risk weights and recommended the introduction of regulatory floors for internal models.
- In the 2016 Sweden FSAP,[58] staff recommended adding a cap on the debt-to-income (DTI) ratio to the macroprudential policy toolkit, as a tool to help contain the risks from high household indebtedness.[59]

IMF staff members have also *calibrated* macroprudential measures using stress tests in Austria and Romania:

- The Austria FSAP (forthcoming) used a semistructural model to project losses on banks' mortgage portfolios and to analyze the potentially mitigating role of different macro-prudential policies, including LTV, debt-service-to-income (DSTI) ratio, combined tools,

---

[54] See IMF (2019a).
[55] See IMF (2017f).
[56] See IMF (2016d)
[57] See IMF (2016b).
[58] See IMF (2016c).
[59] See IMF (2016c).

and speed limits. The model was modified to incorporate Austria-specific characteristics and used to simulate mortgage default rates and losses in the stress test scenarios. The drivers of borrowers' debt-servicing capacity included changes in house prices, income, the unemployment rate, and mortgage interest rates.

- The 2018 Romania FSAP[60] recommended a systemic risk buffer (SRB) to address risks stemming from the sovereign–bank nexus and proposed a calibration based on the stress test results. The Romanian banking system is characterized by exceptionally concentrated and large exposures toward the domestic sovereign. The risk analysis found that these exposures pose significant interest rate risk, contributing to a large part of capital losses in the baseline and adverse scenario as rates increase. As such, the team proposed an SRB calibrated to absorb the losses that a bank would face should interest rates rise by as much as anticipated in the baseline scenario.

- The 2018 Romania FSAP also used a loan-level model to calibrate a limit on the DSTI ratio for household lending. The FSAP identified household borrowing as a growing vulnerability, particularly because loans are extended at variable rates, leaving the borrowers vulnerable to interest rate increases in an environment of accelerating growth. The policy recommendation was thus to build resilience at the borrower level by imposing a maximum *stressed* DSTI limit for households. A loan-level PD model for mortgage and consumer loans was estimated based on data from the Central Credit Register. The results suggest that the PD of a borrower is highly sensitive to any changes in DSTI at DSTI ratios around 50 percent, particularly for mortgage loans. Therefore, it was recommended to set the limit such that loans do not exceed this sensitivity threshold.

Staff members have also developed a general equilibrium model to signal when to loosen or tighten macroprudential policy measures. The model was used in the recent France FSAP.[61] It derives risk measures and assesses fluctuations in default risk from macroeconomic and financial time series. It differentiates between two sources of risk, idiosyncratic and aggregate risk. Idiosyncratic risk is measured by the PD of nonfinancial firms. Aggregate risk is measured by the risk of systemic failure (financial intermediary PD), for example, the risk that the aggregate capital ratio of the banking sector falls below the hurdle rate. The dynamics of the model indicate that bank capital is undershooting during downturns and overshooting during booms, leading to overshooting of risks and risk premiums during downturns and undershooting of risk and risk premia during booms. The model can be used to estimate nonfinancial firm and financial intermediary PDs and quantify the deviation from the optimum. If risks are too high, macroprudential policies should be tightened. On the contrary, if risks are low, macroprudential policies can be loosened.

9 A New Stress Testing Frontier: Climate Risk

Given the likely massive financial stability challenges due to climate change, IMF staff members are prioritizing the assessment of the macrofinancial transmission of climate risk.

---

[60] See IMF (2018f).
[61] See IMF (2019b).

There are two main channels of risk transmission that can affect the financial sector:[62] physical risks arising from damage to property and transition risks arising from changes in policies and technologies. For financial institutions, physical risks can materialize directly, through their exposures to corporations, households, and countries that experience climate shocks, or indirectly, through the effects of climate change on the wider economy and feedback effects within the financial system. One early manifestation of the physical risks is the annual global weather-related insured losses, which increased from about US$10 billion in the 1980s to about US$50 billion in the last decade and to US$138 billion in 2017—the highest since 1980. Exposures also manifest themselves through increased default risk of loan portfolios and lower values of assets.

FSAP stress tests have often captured the physical risks, such as insurance losses and NPLs associated with storms, floods, and droughts. These types of stress tests have become common in FSAPs for small island states (e.g., Samoa, Jamaica, and the Bahamas) and other countries prone to natural disasters. FSAPs for major economies with systemically important financial sectors (e.g., the United States, France, Belgium, Denmark, and Sweden) have also typically covered natural catastrophe risks as part of insurance stress testing. Obtaining comparable data on the financial costs of natural disasters is often a challenge (information is more readily available on numbers of displaced persons and deaths than on financial costs), but IMF staff members have been working with counterparts and experts to improve our understanding of the effects of more frequent and more damaging natural disasters. This has allowed for a broadening of the stress tests for natural disasters from narrow calculations focused on soundness of nonlife insurance to more integrated exercises capturing the macrofinancial effects of more frequent and larger natural disasters. FSAPs for the Bahamas and Jamaica are recent published examples in this regard.

Work is ongoing to examine, on a pilot basis, the financial stability risks associated with the transition to a low-carbon economy. The transition risks are potentially relevant for all countries, with many authorities recognizing that energy transition is unlikely to be smooth and that abrupt changes in policies or technological breakthroughs may change asset valuations. The financial system can exacerbate shocks through leverage and interconnectedness, amplifying instability. Transition risks are challenging to quantify, but recent data show how disruptive changes cause sharp changes in valuations, such as drops in the values of "stranded assets."[63] To deepen our understanding of these risks and get a better sense of both the data needs and analytical limitations, IMF staff members are analyzing financial-system exposure to transition risks in an FSAP for an oil-producing advanced economy.

The transition risks are multifaceted and inherently hard to model. Climate change and the adjustment to a low-carbon economy are subject to fundamental uncertainty. The risks affect a broad range of geographies, sectors, and business models. They are large, nonlinear, and irreversible. They are only partly foreseeable, they depend on short-term actions, and their time horizon is long. A key next step will be to capture second-round effects, in which a decline in asset prices leads to fire sales that further depress asset prices, generating a

---

[62] See the December 2019 issue of the IMF's *Finance and Development.*

[63] An early illustration of the potential disruptions is the market valuation of top US coal producers, which fell by 95 percent between 2010 and 2017.

vicious cycle and an amplifying mechanism for the initial shock. Preliminary attempts at quantifying second-round effects suggest that they can be sizable.

An essential element in the assessment of climate risk is the availability of sufficiently detailed information. The IMF supports public- and private-sector efforts to adopt climate-risk disclosures across markets and jurisdictions, particularly by following the recommendations of the Task Force on Climate-Related Financial Disclosures (2017). A well-defined, internationally comparable taxonomy of green assets, as well as disclosure standards, would help incentivize market participants to reflect climate risks in prices. Unfortunately, disclosures are still uneven across asset classes and jurisdictions.[64] Comprehensive climate stress testing would require improved provision and accessibility of high-quality data.

Going forward, IMF staff members plan to expand and deepen the coverage of climate-related risk in assessments under the FSAP. A better understanding of the macrofinancial transmission of climate risks will allow the IMF staff members to build on their comparative advantages, help country authorities strengthen their policy frameworks, and contribute to the global debate. On physical risks, staff members plan to move toward broader stress tests that explicitly examine the effects of an increased frequency and impact of natural disasters not only on nonlife insurance but also on the rest of the financial system and the economy and are more integrated with the rest of the financial stability assessment. On transition risks, IMF staff will build on the lessons from the ongoing pilot exercise and conduct assessments for a larger number of countries (in collaboration with World Bank staff for jurisdictions where the FSAP is done jointly by the two organizations). IMF staff members are engaging on these topics with experts from central banks and supervisory agencies, think tanks, and academia. The IMF has joined the Central Banks and Supervisors Network for Greening the Financial System (NGFS) as an observer and is collaborating with its members on developing an analytical framework for assessing climate-related risks.

10 IMF Stress Tests: What's Next?

Although stress testing by IMF staff has made significant progress in recent years, important challenges remain. These have to do with improving scenario models to capture nonlinearities, including by using the GaR framework; better integrating stress testing modules and using them to adopt and calibrate macroprudential measures; and upgrading the stress testing framework to assess new risks and use improved inputs.

Work is in progress to define a methodological approach to link GaR to the set of macrofinancial variables needed for a complete stress scenario. GaR captures, at a high level, many of the relationships that stress tests attempt to capture in a granular way. It provides a reduced-form model that helps to pin down the financial stability consequences of imbalances and other frictions in the financial system. GaR also provides a benchmark

---

[64] The IMF's October 2019 *Global Financial Stability Report* (www.imf.org/en/Publications/GFSR/Issues/2019/10/01/global-financial-stability-report-october-2019) examines the issue in depth, pointing out that disclosures remain fragmented and sparse, in part because of associated costs, the often voluntary nature of disclosure, and the lack of standardization. It calls on policymakers to help develop standards, foster disclosure and transparency, and promote the integration of sustainability considerations into investments and business decisions.

for scenario severity so that severity is at the same time consistent across countries (because the same methodology is applied) and consistent with the severity of the country's financial imbalances (given the conditional nature of its forecasts). However, a methodology to link GaR with the broader set of variables required for the definition of a complete adverse scenario is still a challenging work in progress. Relatedly, work is also in progress to improve the ability of approaches (e.g., the DSGE models currently used by staff or others) to capture the behavior of macrofinancial variables in low-probability events to reconcile their use with the evidence that the impact of financial conditions on GDP growth exhibits significant nonlinearities.

IMF staff members see many benefits from further integrating stress testing modules and using them for policy advice. Progress has been made over time by integrating credit and market risk; solvency and liquidity feedback; and interaction among banks, between banks and other financial intermediaries, and between domestic and international financial systems. However, there is still room to make stress testing exercises more realistic by integrating the real economy and regulatory responses and by modeling the reaction of financial institutions to stress. Most importantly, stress tests need to become suitable tools to mitigate systemic risks by using their results to adopt and calibrate macroprudential measures.

Fintech, cybersecurity, and climate change create new challenges and opportunities for stress testing. FinTech may bring new channels of risk transmission, including interactions with banks in terms of credit provisions, to the economy, and interconnectedness via similar trading patterns. Sound supervisory practices to strengthen cybersecurity in the financial sector is a priority for financial stability;[65] stress tests in FSAPs have also incorporated climate change in the stress testing framework (as discussed previously) and cyber-risk events as shocks for adverse scenarios (e.g., bank runs, fall in asset prices, loss of reputation).[66]

Stress testing methodology and models may benefit from big-data analytics. New technologies provide the opportunity for financial institutions and regulatory agencies to collect vast amounts of data about the behavior and financial conditions of firms and individuals. These data in turn allow financial institutions to conduct much more granular scenario and sensitivity analysis of the underlying risks in the loan portfolios. For example, climate-related risks, such as increased frequency and severity of flooding, may be stress tested using data about banks' exposures to vulnerable cities (up to a level of each individual home's location), and changes in energy-consumption patterns may be stress tested by modeling the cash flows of individual firms operating in carbon-intensive industries. Machine learning tools can be used on granular data about each borrower's assets, liabilities, income, and expenses to simulate macrofinancial shocks. Going forward, big-data techniques hold promise, although their practical use thus far in FSAP has been limited, partly as a result of computational challenges and constraints on IMF staff's access to big data compiled by commercial vendors and supervisory agencies.

---

[65] Wilson et al. (2019).
[66] See IMF (2019c).

Appendix A: Perimeter of Stress Tests

Banks are covered at the highest level of consolidation:

- International standards dictate that the home supervisor is responsible for financial consolidation and consolidated supervision of subsidiaries and branches abroad. Therefore, IMF stress tests are applied to all branches and foreign subsidiaries of banks incorporated in the FSAP country on a consolidated basis, including subsidiaries of foreign subsidiaries (i.e., a subgroup operating abroad). Stress tests should also cover banks that represent an economic unit with local banks, even if the home supervisor is not the FSAP country, such as parallel-owned banks operating abroad. Scenarios will need to spell out the set of macrofinancial variables in the countries in which banks operate, and risk-measurement models for these countries will be estimated.
- Subsidiaries whose parent bank is incorporated outside the FSAP country are tested on a stand-alone basis as local banks.[67]
- In the case of conglomerates made of different types of financial intermediaries, banking stress tests would apply only to the consolidated set of banks. This is because prudential requirements, including capital regulations, are set separately for each type of financial intermediary, such as banks, insurance companies, and financial-market infrastructures. However, stress testers will examine initial capital ratios for the conglomerate as well as the regulations to assess the possibility of double or multiple gearing (i.e., whether sectoral regulations have adequate rules in place to sterilize crossed shareholdings between financial sectors).

IMF stress tests can also involve nonbanks:

- Stress tests of insurance companies have been performed in about 15 percent of IMF stress tests; some recent examples include Sweden,[68] Japan,[69] and Belgium.[70] Although traditional insurance business usually does not give rise to systemic risks because insurers do not engage in maturity and liquidity transformation like banks, stress testing the sector might nevertheless be relevant to understanding the consequences of adverse scenarios and insurers' reaction functions when insurance companies provide significant long-term

---

[67] On rare occasions, systemically important branches of foreign banks have been stress tested on a stand-alone basis. The 2012 Singapore FSAP (IMF, 2013a), in addition to the local banks, covered three branches given their *significant* retail presence. The FSAP examined whether branches posed material threats to financial stability under an adverse scenario by assessing their ability to maintain sufficient assets vis-à-vis local liabilities through the asset maintenance ratios (AMRs) imposed by the Monetary Authority of Singapore (MAS). It also assessed potential spillovers from their direct and indirect exposures in the domestic and cross-border interbank markets.

[68] See IMF (2016c, 2017b).

[69] See IMF (2017d).

[70] See IMF (2013b).

funding to banks, the public sector, and the real economy. This becomes even more relevant in countries where insurers are broadening the scope of their investments and venturing more into direct lending, both to corporates and to households. In many countries, the insurance sector is highly interconnected with the banking sector, not only via the funding channel but also via (cross-)shareholdings and joint distribution channels. All these aspects are weighed in to decide whether to stress test them.

- The asset-management sector has been included, usually in liquidity stress tests, when it represents a large source of funding for the banking sector (such as in the United States,[71] Luxembourg,[72] Sweden[73] and Brazil[74] FSAPs). Bank-sponsored off-balance-sheet wealth-management products were included in the solvency stress tests in the China FSAP[75] because of their multiple interlinkages with the banking sector.

IMF stress tests also comprise tests of nonfinancial corporations and estimates of DSTI ratios for households:[76]

- Complete stress tests of nonfinancial corporates have been developed as a tool to assess emerging-market firms with increasing borrowings in foreign currency, owing to easy access to global capital markets, prolonged low interest rates, and good investment opportunities. The tests are based on the share of *corporate debt at risk*, which shows how much of corporate debt is vulnerable to a shock to corporate earnings, interest rates, or exchange rates as a result of the firm's weak debt-servicing capacity.[77] The shocks are calibrated based a combination of historical observations and macrofinancial models that relate corporate earnings to economic growth or global trade.[78] A rapid increase and relatively high share of *debt at risk* is an early warning signal that a country may be more susceptible to corporate distress from macroeconomic and financial shocks.
- When household borrowings represent a vulnerability and data are available, staff members also estimate DSTI ratios. These can be used to assess the impact of rising interest rates and to support policy recommendations for a maximum stressed DSTI limit for households.

---

[71] See IMF (2015a).
[72] See IMF (2017g).
[73] See IMF (2016c, 2017b).
[74] See IMF (2018d).
[75] See IMF (2017h), Box 2, page 25.
[76] See Chow (2015).
[77] The debt at risk for each country is computed as the percentage of the total debt from corporates whose interest coverage ratio (ICR) is less than 1.5. The ICR is defined as the ratio of earnings before interest and depreciation to total interest payments. The lower the ratio, the more the company is burdened by debt expense relative to earnings. An ICR of less than 1 implies that the firm is not generating sufficient revenues to pay interest on its debt without making adjustments, such as reducing operating costs, drawing down its cash reserves, or borrowing more. During the Asian financial crisis, countries whose corporate sector had a median ICR of below 1.5 were more vulnerable.
[78] For details, see Corbacho and Peiris (2018).

Appendix B: Data

Data Challenges

Solvency stress tests conducted by IMF staff are primarily balance-sheet based. Although the final stability assessment is also informed by market-based methodologies, the stress tests themselves depend on very granular prudential data that in many cases are not available to IMF staff unless provided by the authorities. Data challenges can take different forms:

- Data do not exist. This was typical of funding and liquidity data until recently. For example, the ECB has only recently started to collect monthly price data on funding instruments by instrument and by bank since 2014. In the absence of time series of pricing data, proxies are used; for example, credit default swap spreads in advanced economies and the spread of a government's issued securities in advanced economies and developing countries can be used as proxies for funding costs because developments over the GFC showed that investors price the increase in fiscal contingent liabilities arising from a potential bailout of a banking system in distress.

- Data are constrained because markets are not developed. For example, developing countries and emerging-market economies may lack term structures of riskless and risky securities, or they exist but data are missing. Dollarized economies may typically lack term structures of domestic securities issued in US dollars; these countries also suffer from liquid markets that can be used to derive (publicly available) market-based measures of risk in the absence of supervisory data.

- Data exist but are not sufficient. Time series of actual PDs or NPLs may be too short in many developing countries. In this case, teams resort to estimating asset quality using information for countries with similar risk factors or a regional model using information on the region to which the FSAP country belongs. Recently, this approach was used for IMF stress tests in Namibia.

- On occasions, data exist and are reliable, but the authorities are unable (e.g., in the case of legal restrictions) or unwilling to share confidential data with IMF staff. Despite the IMF's mandate to monitor financial stability, Article VIII of the IMF's Articles of Agreement does not include the countries' obligation to provide related data; data are provided to IMF staff on a voluntary basis.

- Data are obtained from different sources. When information is not shared, the FSAP team needs to use publicly available information and, in many cases, combine data from different sources. For example, it is typical to estimate bank clients' PDs by asset class and geography from asset prices, such as stock indexes[79] and sovereign spreads. The projected changes (based on market data) are then used to adjust initial PDs (based on exposure data) when publicly available in annual or Pillar 3 reports. Sometimes, complete information

---

[79] For example, using Moody's Expected Default Frequency data.

on bond portfolios is not available; therefore, bonds need to be repriced using a duration approach applied to bond classes roughly aggregated by their duration. Loss given default for mortgages under stress can be approximated by the scenario's decline in real estate prices when more granular information does not allow a model-based approach. All these limitations are explained in technical notes that are publicly available via the IMF website, imf.org.

Platform

All stress tests tools used by the IMF are based on an Excel-based framework to provide a transparent account of the methodology to the authorities. This is basically an organizing framework that requires estimates of parameters (e.g., the sensitivity of bank asset quality or funding costs to macrofinancial and bank-specific factors) to be calculated outside the framework. The primary focus of the framework is to provide a comprehensive analysis of the banking system and individual banks based on supervisory reporting; hence, the stress tester can easily use results to compare its output with authorities' and banks' results. The granular approach provides an opportunity to identify sources of risks (credit, market, sovereign, funding, etc.) rather than reporting the overall dynamics of the capital adequacy ratio or leverage. The granularity of input data and input parameters allows individual, bank-by-bank customization of feedback loops due to the market, funding liquidity, funding costs, credit growth, deleveraging, and the changing composition of the balance sheets of banks. As the platform becomes more complex, plans have started to translate the Excel sheet into a programming language with an Excel interface to facilitate the work of the user.

Appendix C: Solvency Stress Testing

This appendix provides some detail on how IMF stress tests approach the measurement of key forms of financial risks and their impact on profitability[80] and banks' balance sheets.

Credit Risk

For most banks, credit risk is the largest driver of profits, actual capital, and required capital. Estimates of credit risk under stress affect both the numerator and the denominator of the ratio because new loan-loss provisions representing expected losses will reduce actual capital if they exceed profits, and capital requirements will also increase. Stress tests follow a prudential practice that classifies interest payments on NPLs as nonaccruals. Therefore, in addition to affecting profits through provisions, credit risk will also have an impact on interest income as accrued interests on NPLs are excluded.

The Basel framework provides the basis for estimates of expected and unexpected losses under stress. For exposures under the internal risk-based approach (by which banks use their models to estimate their clients' credit-risk parameters), expected losses are defined as the product of each asset class's default probability, LGD, and exposures at default; therefore, for stress testing purposes, the projection of provisions depends on the projection of these credit-risk parameters: stressed PDs, stressed LGDs, and stressed exposures at default (EADs). Minimum capital is equivalent to the losses that can take place at the 99th percentile of the loss distribution calibrated by the asymptotic single risk factor (ASFM) chosen by the Basel Committee[81] minus the expected losses. Therefore, for stress testing purposes, the same formula is applied with parameters estimated under stress. For exposures under Basel's standardized approach to credit risk, expected losses are replaced by provisions for MPLs, usually defined as 90 days past due, whereas capital requirements are based on a weight, which depends on the type and rating of the exposure. FSAP teams project NPLs and may adjust RWAs under an assumption of downgrading under the adverse scenario.

The credit-risk measurement is undertaken by the FSAP team at the highest possible level of granularity. For example, the euro FSAP[82] estimated bank clients' default probabilities by country and asset class conditional on macrofinancial conditions in all countries in which

---

[80] For a detailed example of how staff members project profitability, see also Xu et al. (2019).
[81] See BCBS (2005), page 10, and BCBS (2017).
[82] See IMF (2018c).

European banks operate. Material geographies included 37 jurisdictions, of which 9 were-core euro area countries, 11 other EU countries, and 17 outside-EU countries. Asset classes comprised seven portfolios (government, corporate, small and medium enterprise [SME], specialized lending, retail, secured by real estate, other); the valuation impact on sovereign exposures as computed for 45 sovereign issuers. This enabled a full integration of market and credit risks at different dimensions that are captured by the "satellite models," such as the impact of changes in interest rates or exchange rates on PDs, LGDs, and EADs, magnifying bank clients' leverage.

Robust econometric frameworks are always implemented using a range of approaches to deal with tail prediction and adjusted to each case. Some approaches that were needed to estimate PDs in the context of the euro FSAP team included the following:

- Newey–West heteroskedasticity and autocorrelation consistent (HAC)-robust standard errors to obtain consistent estimators in the context of heteroskedastic or autocorrelated error terms (or both), once the regressors were proved stationary and ergodic
- A quantile regression approach to address the concern that the drivers of the conditional mean might be different in the higher tail of the credit-risk distribution; the distribution was divided into quartiles (4 segments) as well as deciles (10 segments) to explore cliff-effects.
- A Bayesian model averaging (BMA) technique to address the uncertainty of the drivers of credit-risk dynamics using a normal diffuse prior distribution.[83] The use of a BMA methodology in a stress test context has been emphasized by Gross and Población García (2017), who promote the use of BMA methods for taking explicit control of model uncertainty, which can imply relatively wide error bounds around the scenario-conditional capital position of financial institutions.
- A combination of criteria to inform final projections using the following order: out-of-sample forecast performance, in-sample forecast performance for the overall period, goodness of fit of the regression, sign of coefficients according to theory, and expert judgment applied over the projected paths benchmarked against the 2008 financial crisis and the 2012 European sovereign debt crisis.

When data are available, teams also model LGDs and EAD under stress.

- LGD for collateral loans are usually related to the value of collateral, which can be part of the stress scenario (e.g., in the case of real estate prices). Therefore, haircuts may be applied as necessary.
- The specific features of the legal and judiciary system are also considered to estimate the fraction that banks could expect to recover, including by discounting the expected payment by the time it could take the courts to resolve the issues. World Bank information is sometimes used as a starting approach for LGDs. This initial estimate can be refined as staff members learn more about the country's legal system, including by the presence of staff from the IMF's legal department in many missions.
- When data are available, some refined models can be implemented. For example, the recent IMF stress tests for the UK estimated LGDs for mortgage loans using four key

---

[83] See, for example, IMF (2017f, 2018c,e,f, 2019b, 2020a,b).

parameters: the distribution of original LTV ratios by vintage, the outstanding value of each loan vintage net of amortization, the house-price fall assumed under the scenario, and the forced sales discount on the property's market price under foreclosure.[84]

- In addition to loans, EADs also comprise credit lines and guarantees. The framework allows the stress tester to choose a percentage trigger in each year of the risk horizon based on discussions with banks. The framework assumes that no new credit lines and guarantees are granted during the risk horizon, so no new conversion factors are used. Counterparty credit risk from derivatives is estimated with the aid of discussions with banks.

Net Interest Income

Changes in net interest income are decomposed into two parts: a quantity effect and an interest rate effect. The quantity effect is due to changes in quantities assuming no changes in interest rate; the framework uses the historical effective interest rate for each relevant balance-sheet item; the interest rate effect depends on losses or gains due to changes in the interest rate affecting banking exposures with different repricing times. The interest rate effect also includes the banks' sensitivity to higher funding costs (discussed in a previous section)

Two forms of interest rate risk in the banking book are estimated by significant currency: an earnings-based measure and a valuation approach. Under the earnings approach, interest-sensitive assets and interest-sensitive liabilities are sorted into buckets based on the repricing dates of their cash flows. The analysis is based on 10 time buckets, with an emphasis on repricing within the first year. Given banks' role in liquidity transformation, more assets than liabilities will tend to reprice early in the risk horizon, and therefore, banks would be affected by positive shocks to interest rates unless their exposures are hedged. Although hedging instruments are included, sensitivity stress tests sometimes are undertaken to assess potential losses in the case of the failure of hedging markets. The valuation approach measures equity losses due to valuation effects in the banking book. Valuation losses are estimated using a duration approach.

Market Risk and Trading Losses

Because of its granularity, market risk is measured with limitation. The framework takes into consideration the interest rate risk and credit-spread risk of corporate and sovereign instruments held for trading (HFT), or available for sale (AFS), foreign exchange risk in the banking and trading books, and equity and commodity risks. Losses in the value of HFT and AFS securities due to interest rate and credit-spread risks are assessed through a full revaluation when a detailed description of the portfolio and prices are available; foreign exchange risk and equity and commodity risks are based on the potential losses of open positions in the respective asset when the scenario shocks materialize. For banks with large trading portfolios, staff will rely on estimates made by the banks, and staff will also usually engage in discussions with them to form an opinion on nonlinear risks as well as models used by banks for the calibration of market risk. Similarly, estimates of counterparty risk,

---

[84] See IMF (2016e).

credit-valuation adjustment (CVA) losses, and securitization risks are discussed with banks and rely on a mix of staff and banks' inputs.

The impact of shocks to real estate prices is handled through different features of the stress testing process. The adverse scenario will typically feature a shock to real estate prices in countries where household and corporate leverage is high. The stressed real estate prices will be used to calibrate the decline in banks' collateral (i.e., the increase in the loss given default). Although a complete revaluation would not be typically undertaken, securities that are backed by real estate assets will also suffer haircuts, estimated as a function of the decline in property prices. Moreover, models of PD or NPLs may include real estate prices as a prominent explanatory variable for forecasting the deterioration of asset quality in residential or commercial real estate loans.

References

Adrian, Tobias (2017), "Risk management and regulation," *Journal of Risk*, 21(1), 23–57.

Adrian, Tobias, Nina Boyarchenko, and Domenico Giannone (2019), "Vulnerable growth," *American Economic Review*, 109(4), 1263–1289.

Adrian, Tobias, and Markus K. Brunnermeier (2016), "CoVaR," *American Economic Review*, 106(7), 1705–1741.

Adrian, Tobias, Daniel M. Covitz, and J. Nellie Liang (2015), "Financial stability monitoring," *Annual Review of Financial Economics*, 7, 357–395.

Adrian, Tobias, Federico Grinberg, Nellie Liang, and Sheheryar Malik (2018), "The term structure of growth-at-risk," IMF Working Paper 18/180.

Adrian, Tobias, and Hyun Song Shin (2010), "Liquidity and leverage," *Journal of Financial Intermediation* 19(3), 418–437.

Adrian, Tobias, and Hyun Song Shin (2014), "Procyclical leverage and value-at-risk," *Review of Financial Studies*, 27, 373–403.

Amihud, Yakov (2002), "Illiquidity and stock returns: Cross-section and time series effects," *Journal of Financial Markets*, 5(1), 31–56.

Andrle, M., P. Blagrave, P. Espaillat, K. Honjo, B. Hunt, M. Kortelainen, R. Lalonde et al. (2015), "The flexible system of global models—FSGM," IMF Working Paper 15/64.

Aymanns, Christoph, Carlos Caceres, Christina Daniel, and Liliana Schumacher (2016), "Bank solvency and funding cost," IMF Working Paper 16/64.

Bank of England (2015), *The Bank of England's approach to stress testing the UK banking system*.

Baranova, Yuliya, Zijun Liu, and Tamarah Shakir (2017b), "Dealer intermediation, market liquidity and the impact of regulatory reform," Working Paper 665, Bank of England.

Barnhill, Theodore M., and Liliana B. Schumacher (2011), "Modeling correlated systemic liquidity and solvency risks in a financial environment with incomplete information," IMF Working Paper 11/263.

Basel Committee on Banking Supervision (2005), *An explanatory note on the Basel II IRB risk weight functions*.

Basel Committee on Banking Supervision (2013), "Liquidity stress testing: A survey of theory, empirics and current industry and supervisory practices," BCBS Working Paper No. 24.

Basel Committee on Banking Supervision (2015), "Making supervisory stress tests more macroprudential: Considering liquidity and solvency interactions and systemic risk," BCBS Working Paper 29.

Basel Committee on Banking Supervision (2017a), *Basel III: Finalizing post-crisis reforms*.

Basel Committee on Banking Supervision (2017b), *Identification and management of step-in risks*.

Bricco, Jana, and TengTeng Xu (2019), "Interconnectedness and contagion analysis: A practical framework," IMF Working Paper 19/220.

Broszeit, Timo, Andreas A. Jobst, and Nobuyasu Sugimoto (2014), "Macroprudential solvency stress testing of the insurance sector," IMF Working Paper 14/133.

Brunnermeier, Markus K, Thomas Eisenbach, and Yuliy Sannikov (2013), "Macroeconomics with financial frictions: A survey," in Daron Acemoglu, Manuel Arellano, and Eddie Dekel (eds.), *Advances in economics and econometrics*, Cambridge University Press.

Brunnermeier, Markus K., and Yuliy Sannikov (2014), "A macroeconomic model with a financial sector," *American Economic Review*, 104(2), 379–421.

Catalán, Mario, and Alexander Hoffmaister (2020), "When banks punch back: Macrofinancial feedback loops in stress tests," IMF Working Paper 20/72.

Chow, Julian (2015), "Stress testing corporate balance sheets in emerging economies," IMF Working Paper 15/216.

Čihák, Martin. (2007), "Introduction to applied stress testing," IMF Working Paper 07/59.

Committee on the Global Financial System (2018), "Structural changes in banking after the crisis," Report prepared by a Working Group established by the Committee on the Global Financial System, CGFS Papers No. 60.

Cont, Rama, Artur Kotlicki, and Laura Valderrama (2019), "Liquidity at risk: Joint stress testing of solvency and liquidity," IMF Working Paper.

Corbacho, Ana, and Shanaka J. Peiris (2018), *The ASEAN way: Sustaining growth and stability*, DC: International Monetary Fund.

Cortes, Fabio, Peter Lindner, Sheheryar Malik, and Miguel Segoviano (2018), "A comprehensive multi-sector tool for analysis of systemic risk and interconnectedness (SyRIN)," IMF Working Paper 18/14.

Covitz, Daniel, Nellie Liang, and Gustavo Suarez. (2009), "The evolution of a financial crisis: Panic in the asset-backed commercial paper," Finance and Economics Discussion Series 2009–36, Divisions of Research & Statistics and Monetary Affairs, Federal Reserve Board.

Dees, Stephane, Filippo de Mauro, M. Hashem Pesaran, and Vanessa Smith (2007), "Exploring the international linkages of the euro area: A global VAR analysis," *Journal of Applied Economics*, 22(1), 1–38.

Diebold, Francis X., and Kamil Yilmaz (2008), "Measuring financial asset return and volatility spillovers, with application to global equity markets," NBER Working Paper 13811, National Bureau of Economic Research.

Drehmann, Mathias, Claudio Borio, and Costas Tsatsaronis (2011), "Anchoring countercyclical capital buffers: The role of credit aggregates," Bank for International Settlements Working Paper 355.

Drehmann, Mathias, and Kostas Tsatsaronis (2014, March), "The credit-to-GDP gap and countercyclical capital buffers: Questions and answers," *BIS Quarterly Review*, 55–73.

Espinosa-Vega, Marco A., and Juan Solé (2010), "Cross-border financial surveillance: A network perspective," IMF Working Paper 10/105, International Monetary Fund.

Federal Reserve Board (2014), "Policy statement on the scenario design framework for stress testing," available at www.federalregister.gov/documents/2017/12/15/2017-26858/policy-statement-on-the-scenario-design-framework-for-stress-testing.

Garcia Luna, Pablo, and Bryan Hardy (2019, September), "Nonbank counterparties in international banking," *BIS Quarterly Review*.

Geithner, Timothy (2014), *Stress test. Reflections on financial crises*, Crown Publishers.

Gilchrist, Simon, and Egon Zakrajšek (2012), "Credit spreads and business cycle fluctuations," *American Economic Review*, 102(4), 1692–1720.

Gorton, Gary, and Andrew Metrick (2009), "Securitized banking and the run on repo," *Journal of Financial Economics*, 104, 425–451.

Gray, Dale, and Andreas Jobst (2013), "Systemic contingent claims analysis," IMF Working Paper 13/54, International Monetary Fund.

Gross, Marco, Bjorn Hilberg, Sander Van der Hoog, and Dirk Kohlweye (forthcoming), "The Eurace@IMF+ECB model."

Gross, Marco, and Francisco Javier Población Garcia (2017), "Implications of model uncertainty for bank stress testing," *Journal of Financial Services Research*, 55(1), 31–58.

Gross, Marco, Mindaugas Leika, and Laura Valderrama (forthcoming), "Capital at risk," IMF Working Paper, International Monetary Fund.

Han, Fei, and Mindaugas Leika (2019), "Integrating solvency and liquidity stress tests: The use of Markov regime-switching models," IMF Working Paper 19/250, International Monetary Fund.

International Monetary Fund (2007, October), "Do market risk management techniques amplify systemic risks?" *Global Financial Stability Report.*

International Monetary Fund (2011, April), "How to address the systemic part of liquidity risk," *Global Financial Stability Report.*

International Monetary Fund (2013), *Singapore financial system stability assessment.*

International Monetary Fund (2015a), "United States: Financial sector assessment program–stress testing," Technical Note.

International Monetary Fund (2015b, October), "Market liquidity—resilient or fleeting?" *Global Financial Stability Report.*

International Monetary Fund (2016a), "Staff guidance note on macroprudential policy," Policy Paper Series.

International Monetary Fund (2016b), "Finland: Financial sector assessment program," Technical Note.

International Monetary Fund (2016c), *Sweden: Financial system stability assessment.*

International Monetary Fund (2016d), "Ireland: Financial sector assessment program," Technical Note.

International Monetary Fund (2016e, October), "Monetary policy and the rise of nonbank finance," *Global Financial Stability Report.*

International Monetary Fund (2017a, October), "Is growth at risk?" *Global Financial Stability Report.*

International Monetary Fund (2017b), "Sweden: Financial sector assessment program," Technical Note.

International Monetary Fund (2017c), "Spain: Financial sector assessment program," Technical Note.

International Monetary Fund (2017d), "Japan: Financial sector assessment program," Technical Note.

International Monetary Fund (2017e), "New Zealand: Financial sector assessment program," Technical Note.

International Monetary Fund (2017f), "Netherlands: Financial sector assessment program," Technical Note.

International Monetary Fund (2017g), "Luxembourg: Financial sector assessment program," Technical Note.

International Monetary Fund (2017h), *People's Republic of China: Financial system stability assessment.*

International Monetary Fund (2018a), "Peru: Using GaR for macro-financial analysis and forecasting: Application to Peru," Technical Note.

International Monetary Fund (2018b), "Armenia: Financial sector assessment program," Technical Note.

International Monetary Fund (2018c), "Euro Area: Financial sector assessment program," Technical Note.

International Monetary Fund (2018d), "Brazil: Financial sector assessment program," Technical Note.

International Monetary Fund (2018e), "Poland: Financial sector assessment program," Technical Note.

International Monetary Fund (2018f), "Romania: Financial sector assessment program," Technical Note.

International Monetary Fund (2019a), "Switzerland: Financial sector assessment program," Technical Note.

International Monetary Fund (2019b), "France: Financial sector assessment program," Technical Note.

International Monetary Fund (2019c), "*Singapore: Financial sector assessment program. Financial Stability and Stress Testing.*

International Monetary Fund (forthcoming), "Austria: Financial sector assessment program," Technical Note.

International Monetary Fund, Financial Stability Board, and Bank for International Settlements (2009), "Guidance to assess the systemic importance of financial institutions, markets and instruments: Initial considerations," Report to the G-20 Finance Ministers and Central Bank Governors Washington, DC, and Basel.

International Monetary Fund (2020a), "Austria: Financial sector assessment program. Financial stability analysis, stress testing, and interconnectedness," Technical Note.

International Monetary Fund (2020b), "Korea: Financial sector assessment program. Systemic risk analysis, financial sector stress testing, and an assessment of demographic shift in Korea," Technical Note.

Krznar, Ivo, and Troy Matheson (2017), "Towards macroprudential stress testing: Incorporating macro-feedback effects," IMF Working Paper 17/149.

Leika, Mindaugas, Liliana Schumacher, and Laura Valderrama (forthcoming), "Workbox 2020: The IMF solvency stress testing tool," IMF Working Paper.

Malik, Sheheryar, and TengTeng Xu (2017), "Interconnectedness of global systemically-important banks and insurers," IMF Working Paper 17/210, International Monetary Fund.

Ong Li Lian, and Ceyla Pazarbasioglu (2013), "Credibility and crisis stress testing," IMF Working Paper 13/178, International Monetary Fund.

Puhr, Claus, and Stephan W. Schmitz (2013), "A view from the top—The interaction between solvency and liquidity stress," *Journal of Risk Management in Financial Institutions*, 7(1), 38–51.

Puri, Manju, and Alexander Ufier (2018), "Deposit inflows and outflows in failing banks: The role of deposit insurance," FDIC CFR Working Paper 2018–02.

Schmieder, Christian, Maher Hasan, and Claus Puhr (2011), "Next generation balance sheet stress testing," IMF Working Paper 11/83, International Monetary Fund.

Schmitz, Stefan, Michael Sigmund, and Laura Valderrama (2017), Bank solvency and funding cost: New data and new results," IMF Working Paper 17/116, International Monetary Fund.

Segoviano, Miguel, and Charles Goodhart (2009), "Banking stability measures," IMF Working Paper No 09/04, International Monetary Fund.

Stein, Jeremy (2014), "Incorporating financial stability considerations into a monetary policy framework," Remarks at the International Research Forum on Monetary Policy, Board of Governors of the Federal Reserve System, Washington, DC.

Task Force on Climate-Related Financial Disclosures (2017), "Final report: Recommendations of the Task Force on Climate-Related Financial Disclosures," Financial Stability Board.

Valderrama, Laura (forthcoming), "An agent-based model for stress testing," IMF Working Paper.

Vitek, Francis (2018), "The global macrofinancial model," IMF Working Paper 18/81, International Monetary Fund.

Wilson Christian, Tamas Gardosch, Frank Adelmann, and Anastassia Morozova (2019), "Cybersecurity risk supervision," IMF Working Paper 19/15, International Monetary Fund.

Xu, TengTeng, Kun Hu, and Udaibir S. Da (2019), "Bank profitability and financial stability," IMF Working Paper 19/5, International Monetary Fund.

A Comprehensive Approach to Macroprudential Stress Testing

Anthony Bousquet, Jérôme Henry, and Dawid Żochowski[*]

1 From Microprudential to Macroprudential Stress Testing

Following the Global Financial Crisis (GFC), system-wide stress tests became a standard tool. They were initially set up with a view to recapitalizing banking systems after the crisis, thereby reestablishing confidence as well as protecting banks against a range of possible future shocks. In more recent years, departing from a crisis-management purpose, stress tests have evolved, with an emerging differentiation between those conducted primarily for supervisory purposes and others that have more macroprudential objectives. The latter include calibrating macroprudential measures or systemic-risk assessment (see Baudino et al. [2018] for an elaborated discussion on the relations between the design of stress tests and their specific policy objectives).

In particular at the European Central Bank (ECB), stress testing and related inputs from ECB staff took a variety of forms and have substantially developed over time. Initially, work concentrated on scenario design, in cooperation with the Committee of European Banking Supervisors and thereafter the European Banking Authority (EBA; process documented in Henry [2015]; see also Chapter 5 in this handbook), which were both in charge of leading EU-wide banking sector stress tests. Then ECB staff also contributed to the development of the methodology for banks to follow in these stress tests and eventually put together a quality-assurance framework to review banks' results, inter alia, in the light of alternative models (Mirza and Żochowski, 2017). During the euro-area crisis, regular work was conducted in relation to EU country programs along similar lines, especially for the calibration of banking-sector recapitalization measures (International Monetary Fund [IMF], 2019). ECB staff also contributed to the Single Supervisory Mechanism (SSM) Comprehensive Assessment for the stress test leg and subsequently to the SSM exercises,

[*] The views expressed in this chapter are those of the authors and do not necessarily reflect those of the European Central Bank. The current version of this chapter benefited from helpful comments and suggestions by the editors and a referee. The text also reflects discussions with participants in the International Monetary Fund FSAP@20 conference, as well as in seminars at the Bank of Greece, Central Bank of Lithuania, Central Bank of Luxembourg, National Bank of Poland, Asian Bureau of Finance and Economic Research, Center for Monetary and Financial Studies, Center for Latin American Monetary Studies, South East Asian Central Banks, Columbia University, London School of Economics Systemic Risk Centre, Nova Lisboa, University of Maastricht, and University of Montreal. Reflections presented have been also fueled by exchanges held over the years with colleagues at the Bank of England, Banque de France, Bank for International Settlements, European Central Bank, European Systemic Risk Board, International Monetary Fund, and US Federal Reserve.

in particular for benchmarking purposes (ECB, 2014). While continuing with supervisory-oriented stress tests (i.e., EBA and SSM exercises), stress testing at the ECB has also taken a macroprudential angle and plays a specific and important role in the broader context of the ECB macroprudential framework (see Constâncio, 2019). Inspired by this ECB staff experience, we intend in this chapter to identify what ingredients could be needed and gathered, with a view to substantiating the macroprudential function of stress testing in the euro area.

Beyond supervisory stress tests, there is a specific value of conducting stress tests that pursue differing, more "macro" objectives, in particular by emphasizing the system-wide perspective instead of that of the single entity (Constâncio, 2015). To achieve this, there is a pressing need to account for banks' reactions, implying a need to model dynamics, feedbacks, and spillovers or contagion effects. Banks' reactions create externalities, via price and quantity channels, both within the banking sector and in other sectors. Supervisory stress tests, in turn, usually simulate the impact of selected adverse scenarios on single banks in isolation, moreover, for the EU and euro-area-wide exercises, without any change in banks' balance-sheet size or composition, over the stress test horizon. Based on such a counterfactual assumption, these stress tests do not consider real–financial feedback loops, nor do they account for interconnectedness. See Anderson (2016) for a comprehensive review of issues and options to address them, for example, as done in the UK or by the IMF, the latter having been a forerunner in such ventures.

Another critical dimension of the approach taken in various stress tests is the distinction of top-down versus bottom-up testing, that is, respectively, a centralized stress test conducted by an authority running its own models (as done, e.g., by the US Federal Reserve) versus an exercise where banks produce their own results, also using their own tools. Supervisory stress tests such as the EBA's are mostly of the bottom-up kind, albeit using a scenario and methodology defined by the responsible authority (Baudino et al., 2018). For macroprudential purposes, the need to incorporate system-wide interactions within the banking sector as well as cross-feedbacks with other sectors, especially the real sphere, calls for a top-down approach instead, so as to capture all desirable transmission channels of stress. Banks themselves are hardly in a position to account for each other's results and reactions in a fully consistent manner, hence the need for a top-down approach to better capture macro impacts.

In practical terms, to inform policymakers, simulations are carried out by authorities' staff, possibly with a range of models, conditional on scenarios that include relevant risks or changes in policy measures. Models may also have to cover actors beyond banks, especially those on the real side, via the credit channel as a core component, but also other financial sectors, such as through valuation and asset cross-holding linkages. All of these require building a specific toolkit to address systemic issues, for which it is nonetheless key to adopt a bank-level approach. Systemic events involve various banks to a differing extent, so their respective specific balance-sheet structures and reactions to shocks may result in a possibly complex and nonlinear overall system response.

From 2008 onward, ECB staff gathered expertise, constructed data sets, and estimated models that were eventually published as an occasional paper (Henry and Kok, 2013), based on a consistent bank-by-bank, risk-by-risk modeling approach. The number of banks in scope ranged between 25 and 130 (for the so-called "large and complex banking groups" [LCBGs] and the SSM Comprehensive Assessment banks, respectively). The data initially employed were those publicly available in the pre-SSM universe. Indeed, before the creation of the single supervisor for the euro area, supervisory data were not readily accessible outside irregular, but comprehensive and more granular, releases of EU-wide stress test data. To best inform the needed risk-by-risk "macro" impact assessment, data must be granular enough to capture the specific response of a number of balance-sheet items to a variety of shocks, without, however, requiring the extensive granularity needed for microprudential purposes.

Although ECB staff regularly conducted internal "macro" stress tests, with some reported in the ECB *Financial Stability Review*, the complete use of its top-down framework to derive a systemic and macroprudential exercise from euro-area supervisory stress test results was first published in the ECB *Macroprudential Bulletin* of October 2016 (Dees et al., 2016). The comprehensive suite of models (Dees et al., 2017) that had been employed for this macroprudential stress test analysis was, shortly afterward, fully documented in an ECB e-book, with a foreword by its then vice president (Constâncio, 2017). The framework can cater to balance-sheet analyses, real economy feedback, systemic liquidity assessment, interconnectedness quantification, and policy measure-impacts. It is centered on banks and therefore does not cover all financial-sector agents in an equally detailed fashion, a possible limitation with macroprudential objectives in mind.

Largely building on this work conducted and published by ECB staff over the last decade, the next section of this chapter presents top-down models that were developed for banks specifically, focusing on the so-called "first-round effects." The latter we define as the impact stress has on individual banks' balance sheets in isolation, hence not considering externalities that result in additional, so-called "second-round effects." The following section describes how a similar approach can be envisaged for insurers, providing a proxied impact of adverse conditions on this sector overall, also using simplified balance sheets. The subsequent section presents preliminary thoughts on how the impact of stress on other financial sectors could be captured, in which case the analysis requires, to be relevant from a "macro" perspective, the inclusion of cross-sectoral interactions in the first place. Finally, methods to capture second-round effects are presented, covering spillovers from stressed banks and insurers, leading to a discussion of financial-system-wide stress testing.

2 Bank Balance Sheets under Stress—the Direct Impact of Macrofinancial Shocks

Globally, sector-wide stress tests for banks are by now a well-established and commonly employed tool. Banks are, from a systemic-risk perspective, the most relevant financial sector, be it due to their asset size or as a result of their dual interactions with both the real economy and the financial markets and all of its actors. In the euro area, for instance, banks hold about half of total financial sector's assets and contribute heavily to the financing of the economy, their systemic role being enhanced by a strong degree of interconnection within the sector itself, via interbank loans and security holdings (ECB, 2017a).

Following the GFC, most authorities developed a banking-system-wide stress test framework (see Baudino et al. [2018][1] for a survey of such sector-wide stress tests and their

---

[1] Annexed to this document is a tabular presentation of the ECB stress testing framework, covering both micro- and macroprudential aims.

objectives and design). This was for essentially supervisory purposes, such as calibrating bank-specific regulatory capital requirements, as one tool to weather the crisis. At the same time, some authorities also attempted to develop a setup dedicated to macro-oriented purposes. Exercises with such dominantly macroprudential objectives could, for example, seek to evaluate system-wide capital needs per se to assess the resilience of the sector as a whole, or to calibrate a recapitalization backstop. They could also aim to get a wider picture of the banking sector, for example, to discuss its profitability or reaction to specific monetary or macroprudential policy measures—even in the absence of marked stress. Results obtained on the impact of stress on banks' balance sheets and profit and loss (P&L) are also a first step toward estimating further consequences for the economy—to the real side via the credit channel but also within the broader banking sector and beyond within the financial system, thereby identifying and quantifying the first- as opposed to second-round effects of a given scenario.

2.1 STAMP€: A Comprehensive Framework, Nesting First- and Second-Round Effects

Although specific to the euro area, STAMP€ can be considered as a representative and illustrative framework for conducting top-down macroprudential stress tests. It accounts for the transmission of stress via a number of channels, both at the microeconomic and macroeconomic levels. It also exploits a variety of data sets of differing degrees of granularity, up to survey and transaction data. It is a rather unique product, in the sense of providing an integrated and detailed documentation on the various building blocks comprised in a macroprudential stress test infrastructure. In particular, it includes a thorough modeling of the first-round impact of stressed conditions on individual banks' balance sheets, which we see as a prerequisite before modeling further rounds of impacts of any sector-wide stress—in this case covering all euro-area countries. The toolkit was nonetheless developed from the outset as a macro-impact assessment framework for the euro area, albeit generating bank-level results, that is, not with a view to employing it for microprudential purposes, which would require a more granular modeling and extensive use of corresponding bank-level data. Such data are, in any case, not necessarily readily available to nonsupervisory modelers.

This framework is modular and rests on four pillars (see Figure 27.1; Bank of England [Dent et al., 2016], Banque de France [Bennani et al., 2017], and IMF [2014] documents show similar flowcharts; see also Chapter 26 in this volume). The process starts with the scenario design, the cornerstone of any stress test, by definition a conditional exercise. It then involves so-called "satellite models" to derive bank-specific risk parameters, computed conditional on any given scenario. After this, a quasi-accounting block delivers the full balance sheet, P&L, and capital for all banks in scope. Finally a feedback block covers real–financial interactions and a variety of other spillovers within the banking sector and beyond. Running the latter block would then imply updating the scenario and reiterating the full sequence on this new basis, stopping whenever the additional rounds would yield only marginal changes to the overall "macro" picture.[2]

---

[2] The micro nature of the framework, as well its complexity (e.g., multistep, nonlinearities, no closed-form solution) and the size of the databases involved, would call for computations via an iterative process anyhow.

At all stages (i.e., for each module), a variety of empirical approaches can be taken, exploring possibilities to exploit available data—ranging from macro-modeling at the euro-area or country level to individual-bank-level behavior calibration. Modeling is generally subject to restrictions due to confidentiality pertaining to supervisory or stress test data. Although ECB staff can generally access such data, their use remains limited by the lack of timely, granular, or regular collections, especially at the entity or, even more, transaction level. In the euro area, the coverage of the regular supervisory data (such as in Finrep-Corep templates) can, for example, be less detailed than what is occasionally provided by banks during stress test exercises (e.g., for interest income) and backdata sparse. As a result, macroeconomic or macrosectoral data are used to estimate banking-sector-specific parameters that are meant to proxy banks' behavior. In some cases, panel data are instead sufficiently rich to be exploited and thereby generate parameters that are institution specific. Associating sectoral elasticities to bank-specific portfolio information eventually provides a micro-specific flavor to the results, without a need to account for bank-specific elasticities or pricing behavior. Such an approach may eventually strike the right balance, reaching a degree of detail sufficient to broadly capture balance-sheet impacts, including for individual banks, without going as far as exploiting and having to rely on very granular and possibly confidential supervisory data.

This section focuses on the modeling elements that feature within the gray box in Figure 27.1, which, by and large, replicates the supervisory approach to system-wide stress testing in the EU and the euro area. The strategy to estimate the first-round impact of macrofinancial variables on banks is indeed inspired, albeit less granular, by that adopted for the EU-wide microprudential solvency stress test (EBA, 2018). The STAMP€ top-down modeling framework accordingly considers, for each bank, the following risk areas: credit risk, net interest income, and market risk (these three categories being generally the main risk and loss drivers in stress testing), as well as models for fees and commission income and operational risk.

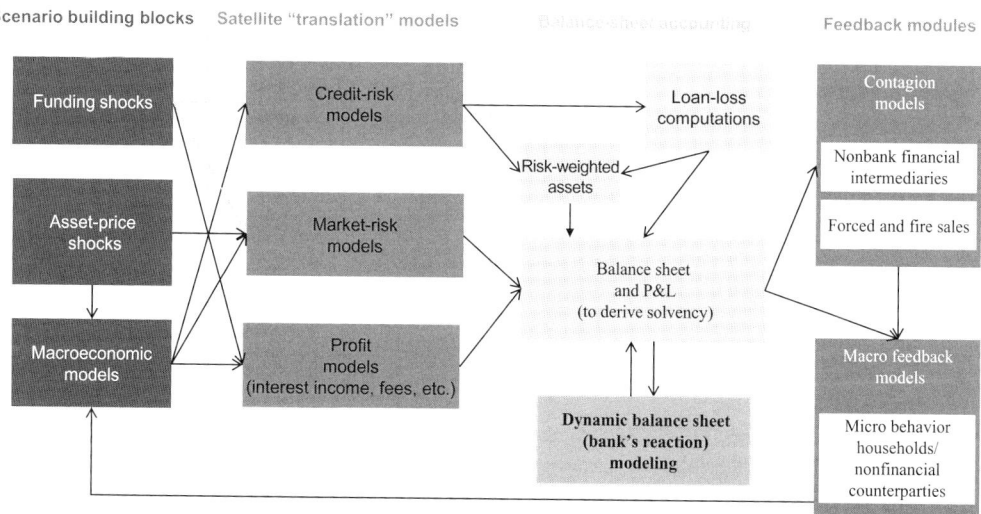

Figure 27.1 A standardized process for macroprudential stress testing of banks.
Source: Dees et al. (2017).

2.2 *The Main First-Round Drivers: Credit, Interest Rate, and Market Risks*

Credit risk is modeled at the banking-sector level for 28 EU Member States and 20 areas for the rest of the world (see Gross et al., 2017a). For each of these, it is estimated for various segments, across loan purposes (consumption, mortgages, investment, commercial real estate) and borrowers (households, large nonfinancial corporations, small and medium-sized enterprises [SMEs]). The resulting set of granular risk-parameter equations relates, for example, the probability of default on mortgages in a given country to corresponding activity and interest rate variables. Over the horizon, bank-level parameters are generated for a given bank by linking the average sector developments in a nonlinear fashion to the banks' last reported own parameters.

Models have been estimated with the Bayesian model averaging (BMA) method, combining a range of possible explanatory variables to select and average across a large set of relevant and significant models (see Gross and Poblacion [2015], for details). Because the method provides a distribution for all estimated elasticities, instead of only point estimates, the stress impact can be enhanced by using parameters stronger than the estimated medians—to capture, for example, nonlinearities or simply to increase the exercise's degree of conservatism. Such equations have regularly been used to compute EBA benchmarks, also accounting for International Financial Accounting Standards (IFRS) 9 in more recent exercises.

All relevant risk parameters are computed under the baseline and the adverse scenarios (dark- and light-gray dots, respectively, in Figure 27.2). In all cases, probabilities of default (PDs) are, as expected, substantially higher under the adverse scenario, with gross domestic product (GDP) and unemployment as major drivers. Losses given default (LGDs) are also

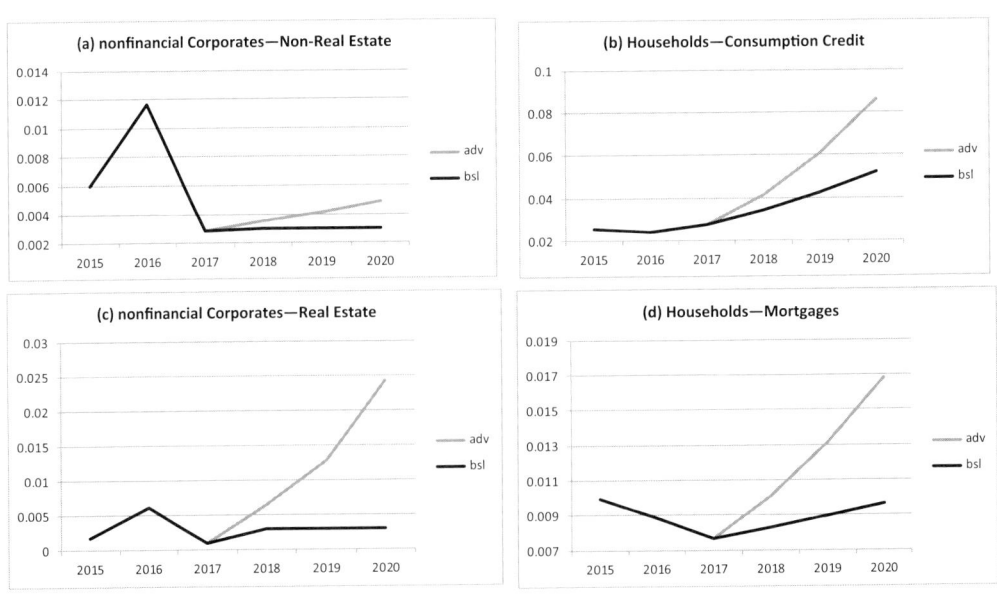

Figure 27.2 Sectoral probability of default, nonfinancial counterparties (NFCs) and households, adverse ("adv") and baseline ("bsl"), %.
Source: Authors' computations.

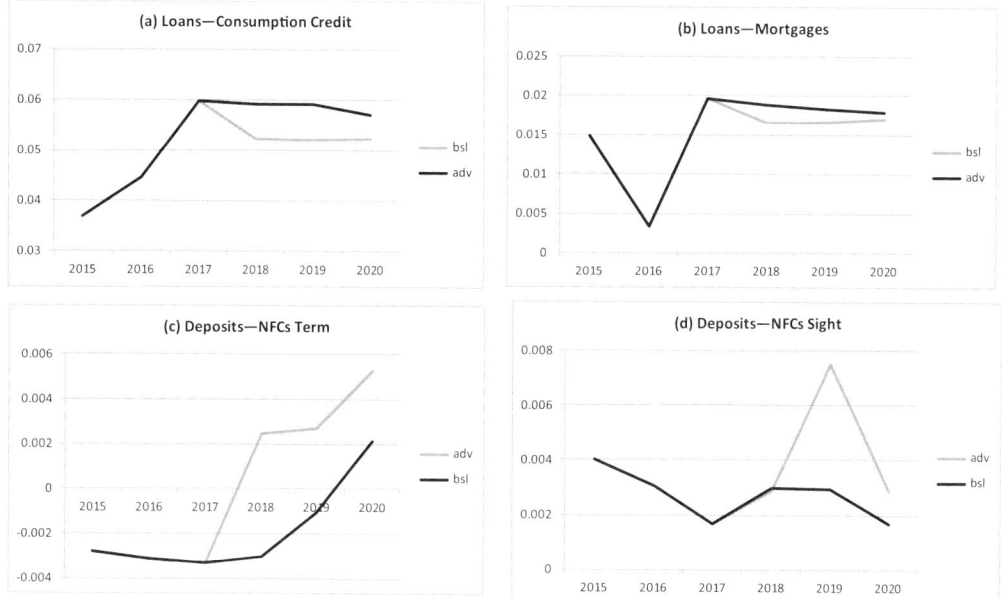

Figure 27.3 Country-specific household loan and corporate deposit rates, adverse ("adv") and baseline ("bsl"), %.
Source: Authors' computations.

modeled, but in a simpler, quasi-accounting fashion—LGDs depend on loan-to-value ratios, which themselves are indexed on the scenario assumptions for real estate prices.

In STAMP€ **net interest income** (NII) is split into two components, income and expenditure, with a focus on interest rates, themselves broken down into reference rates and product-specific spreads, all of which are functions of macrofinancial variables, estimated via BMA (see Gross et al., 2017b). Rollover and repricing assumptions also affect results. For each country or area, with the same set of geographies as for credit risk, lending rates are modeled for different sectors (SMEs, large corporates, households for consumption and mortgages). Similarly, funding costs are estimated for each aggregated banking sector for different types of instruments (sight and term deposits, wholesale, bonds). Alternative, panel bank-level estimates are also available, delivering each bank's net interest margins directly. For any given bank, funding costs can also be set to decline with its solvency ratio—a relevant loss-amplification channel under stress.

Figure 27.3 illustrates results for both the baseline and adverse scenarios, taking the example of specific lending rates (loans to households) and funding costs (corporate deposits). Lending and borrowing rates are both higher under the adverse scenario, reflecting a steeper yield curve along with dampened activity. In some adverse scenarios, the net income impact could even be positive for some banks, depending on their asset and liability mix. This standard paradox in model-based stress testing may be tackled in supervisory exercises, such as EBA ones, by imposing caps and floors on income and expenditure, respectively, as well as pass-through constraints (EBA, 2018). Macroprudential exercises may deviate from this supervisory approach and still use estimated actual parameters.

Market-risk parameters (see Laliotis and Mehta, 2017) are modeled for mark-to-market valuation, credit-valuation adjustment, liquidity reserve, or counterparty risk (i.e., defaults of selected counterparties). Mark-to-market computations simply require the transformation of scenario yield shocks into equivalent price movements. In turn, other market-risk parameters are derived from more complex models, which are informed by portfolio information and supervisory stress test results (regressing the latter on risk factors and synthetic portfolios for each bank). Simulations must also account for hedging, on which data are partially available, again from previous stress test results.[3]

Sovereign holdings treatment has to account for both credit and market risks. These assets are treated in differing manners, depending on whether they are on the banking or trading book (i.e., reported at amortized costs or at fair value, for held-to-maturity vs. available-for-sale assets, respectively). The latter is subject to mark-to-market in line with the scenario assumption on the corresponding bond rates, whereas the former is associated with an estimated credit risk, whereby the respective PD is linked to scenario macrofinancial variables (see Henry [2015] as an example of such relations).

Miscellaneous risk items also need to be modeled to complete the P&L picture. Net fees and commission income (NFCI) is derived from panel estimations involving macrofinancial scenario variables and bank-specific indicators (Mirza et al., 2017). Finally, for operational risk (Bousquet and Dubiel-Teleszynski, 2017), records of past loss events help fit probability distributions for conduct or nonconduct risk losses, clustered by amounts. As a result, this item is based on a purely statistical approach, that is, with no role for macrofinancial variables (see also Wei et al. [2017], for a survey about such data collections in a variety of institutions).

2.3 The First-Round Impact of a Given Scenario

All of the previously described blocks can be employed to simulate the impact of specific scenarios on the euro-area banking sector, as regularly reported (e.g., in the ECB *Financial Stability Review*). Informed by staff risk identification, simulations conducted for alternative scenarios deliver a quantified and documented analysis, thereby helping rank risks according to their impact on the system. The employed top-down approach, moreover, facilitates the use of more than a single adverse scenario and allows the modeler to combine risks that trigger one another (e.g., low growth leading to a degradation in domestic debt situations). In addition, the results inform on the relative weight of the various risk drivers of capital depletion (see Figure 27.4), hence suggesting specific mitigation measures and guiding further monitoring tasks, both at the macro- and microprudential levels.

Even without banks' reactions or second-round effects, such exercises indeed provide relevant information to policymakers (and possibly the public). Simulations carried out on a regular basis generate an updated overview of how the banking system would be directly affected by a degraded macrofinancial environment—barring, by construction, further potentially significant second-round effects of various kinds. Such assessments can also

---

[3] This area or, similarly, net traded income, poses special challenges to the (macro)modeler, to the extent that underlying determinants can be extremely granular—thousands of risk factors per bank—and, moreover, very volatile because banks may adjust their market portfolio with much higher frequency than the reporting one.

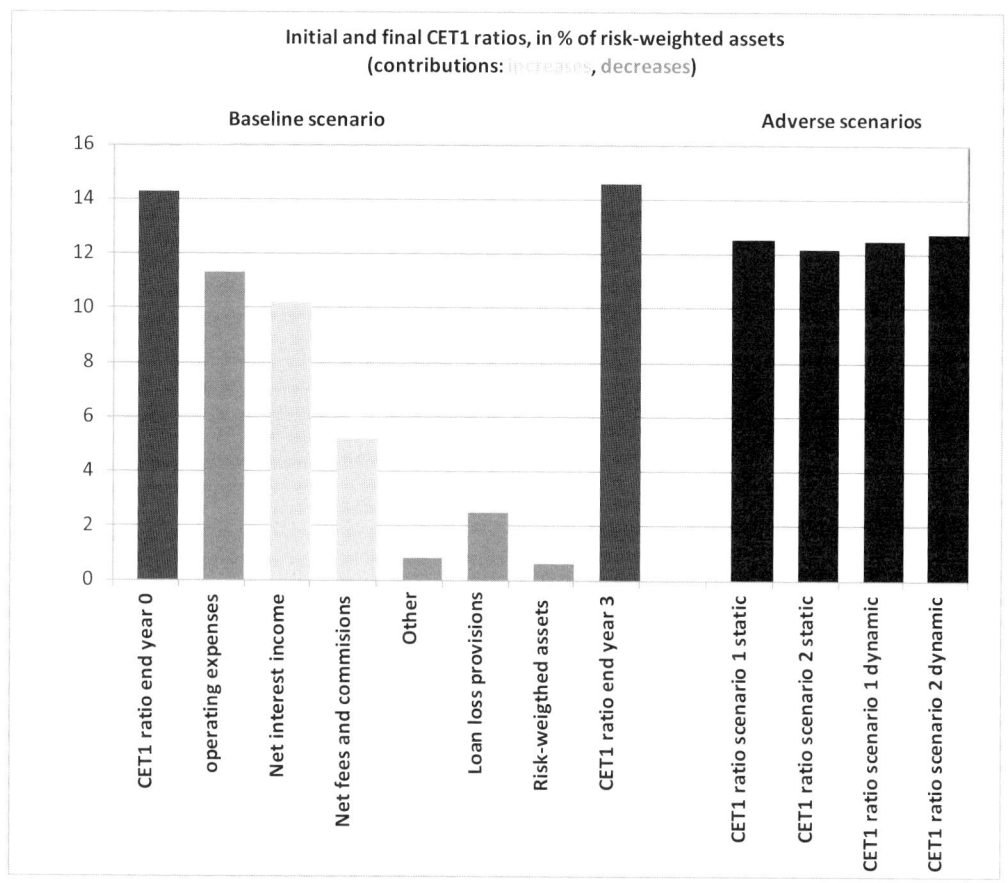

Figure 27.4 Banking system capital ratios (common equity Tier 1 [CET1])
under alternative scenarios and main drivers.
Source: ECB (2018a).

be conducted for a baseline or mildly adverse scenario, in which case sizeable second-round effects may be less likely, anyhow, to arise with significant magnitude. In any event, a sufficiently granular and accurate computation of first-round impacts is a necessary step before engaging in second-round computations.

With respect to simulations that simply replicate the supervisory approach, macroprudential interpretations of the results still require an adjustment, namely, to credit supply. A critical assumption underlying the EBA stress test is that of a static balance sheet for all banks, whereby both the asset and liability sides remain constant in terms of volume, composition, and risk profile. Correspondingly, scenario macrofinancial assumptions do not include any change in credit variables. Not only does this imply that credit, a key element in the macrofinancial landscape, remains constant under stress (a clearly counterfactual hypothesis), but also interrelations across risk parameters are exceedingly simplified.

To consider a more realistic setup than fully fixed balance sheets, a first step can be to run credit flow/stock macromodels (e.g., in Gross and Venditti, 2017). These deliver aggregate credit consistent with other macrofinancial variables covered by the EBA scenario. Under

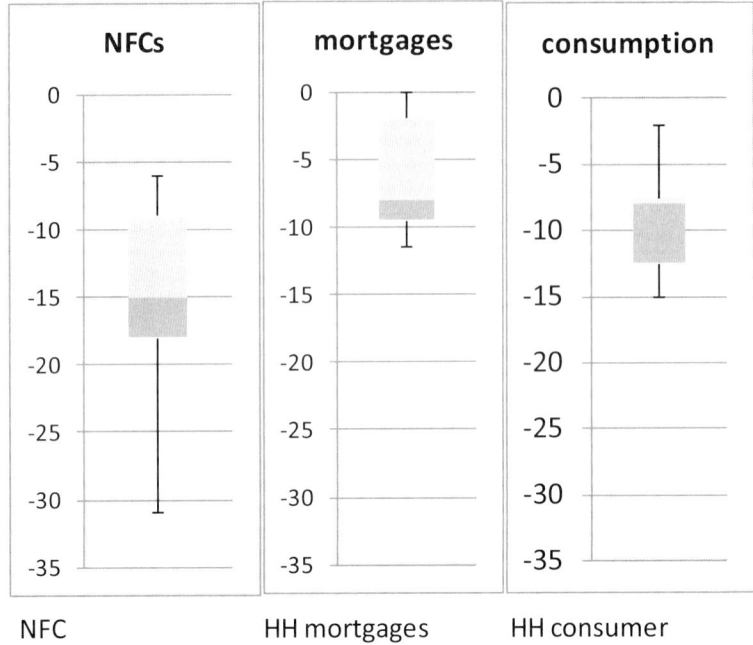

Figure 27.5 Deviation of credit from baseline under the adverse scenario,
distribution across countries, ppts.
Source: ECB (2016).

stress, credit should indeed be negatively affected and would not remain constant even in a baseline configuration. This alignment of credit then simply complements, in a consistent fashion, the scenario typically underlying the EBA exercise, without providing, however, an amplification mechanism per se on the macroeconomic side because all other variables in the scenario remain the same.

Figure 27.5 recalls the result of such an alignment exercise based on the EBA 2016 stress-test. The chart reports the distribution of estimated credit shocks across euro-area banking sectors, consistent with an adverse-scenario GDP loss of some 8 percentage points (ppts) over the horizon. In sum, credit to households declines in line with GDP about one-to-one, whereas the average elasticity for loans to NFCs is of the order of 2. Under the assumption of a (credit) constant market share across banks, for a given country, credit granted by each bank then experiences a proportional reduction. In addition, similar to what is assumed under a static balance sheet, banks' asset and liability mix can be kept unchanged, along with a residual adjustment to funding needs consistent with the now-lower loan books.

As mechanical and neutral to the scenario as it might be, such a consistency step has nonetheless an overall impact on capital that is not straightforward. For instance, a reduction in the loan book affects NII and also possibly credit risk, depending on whether nonperforming loans (NPLs) are disposed of or kept on the bank's balance sheet. Asset deleveraging more generally would crucially affect market-risk parameters and the capital needed to cover for them (because portfolio composition would change). Lower credit implies lower risk-weighted assets (RWAs) and hence stronger capital ratios, but there are counteracting factors, such as NPLs, that may remain on the balance sheet or be sold at a distressed price, or NII may be negatively affected by the shrinkage of the loan book. The overall impact, by construction, would be an aggregation of bank-specific responses, themselves heavily dependent on their respective balance-sheet structures.

As documented in Chapter 3, of STAMP€ elaborating on Dees et al. (2016), only about half of the banks in the reviewed sample improved their capital position by reducing the size of their loan books. The differences between the two populations' end capital was significant on average, at approximately 0.5 ppt. Figure 27.4, moreover, shows that the impact of completing the scenario by adding credit paths can also be scenario dependent. For the sample of banks used in this *Financial Stability Review* exercise, the CET1 ratio for a given scenario can improve or stay unchanged when moving from static to dynamic balance sheets.

Although such simulations that build on the microprudential approach and data sets are relevant per se to inform macroprudential policy analyses, further add-ons are required to increase their macroprudential flavor. This next step would involve modeling further sectors, such as insurance and other financial institutions (see following Sections 3 and 4), as well as a variety of spillovers within the banking system and beyond—how to possibly capture such second-round effects via credit and financial markets is discussed in Section 5.

3 Stress Testing Insurers

When developing a top-down stress testing toolkit for the insurance sector, due account has to be taken of the specific insurer's business model and its own sensitivity to the broader macroeconomic environment. Insurers can rely on more stable cash flows than banks and are less affected by a deterioration of the liquidity conditions because premia are paid ahead of the realization of the related risk (*inverted production cycle*). In addition, the core activity of insurers does not imply maturity transformation, contrary to the banking sector, itself with a structural-maturity mismatch between assets and liabilities.

Business models of insurers consist mainly of two lines of products: life insurance and property and casualty (P&C) insurance. The risk underlying these products can be either supported directly by the insurer engaged with customers (*primary insurance*) or passed on to another insurer (*reinsurance*). The nature of the risk differs substantially across business types. For life insurers, the unit- or index-linked business bears a limited risk because policyholders have their return linked to the return of the underlying investments. The investment risk is then borne by policyholders, whereas in traditional insurance policies, the return is guaranteed by the insurer; hence, the latter bears the risk. The range of risks covered by the P&C business is broad, ranging from natural perils, theft, and neglect to damages or injuries to both property and persons.[4]

These different dimensions call for a tailored approach to insurance top-down stress testing—aligned with relevant risk factors and their transmission channels. In particular, the specific nature of insurers' business models calls for a dedicated scenario-design process. In some cases, a scenario targeted at stressing the banking system may even have limited

---

[4] See Chapter 15 in this volume for more details on the business models coexisting in the insurance market and the risks faced.

impact, if at all, on the insurance companies. The scenarios used by the European Insurance and Occupational Pensions Authority (EIOPA), for example, for its own EU-wide exercise may differ from the ones used by the EBA in terms of specific variable coverage and time horizon (see also Chapter 5 in this handbook). This section accordingly draws on the specificities of the insurance sector, in the case at hand, the euro-area one, to sketch what a top-down stress test methodology should cover in order to appropriately address and assess the risks potentially affecting this financial sector (see Jobst et al. [2014] for the IMF approach).

3.1 Euro-Area Insurers' Structural Features—a Stylized Balance Sheet

In the euro area, insurers represent a much smaller, albeit still systemic, sector than banks, with approximately 13 percent of the total assets of the region's financial sectors. Moreover, insurance firms have very strong interconnections within their own sector, as well as important holdings of banks securities and fund shares, reflecting on the asset side their typical business model (see ECB, 2017).

The total volume of insurers' assets and liabilities shows, at the same time, a substantial heterogeneity across euro-area countries, for example, as a ratio to GDP in Figure 27.6. If Luxembourg stands out with an insurer's total balance sheet representing more than three times its GDP, France and Ireland rank, respectively, second and third, both with insurance assets above 100 percent of GDP. These estimates allocate assets based on the insurer's headquarters country and therefore deviate from a pure locational perspective because groups have cross-border activities.

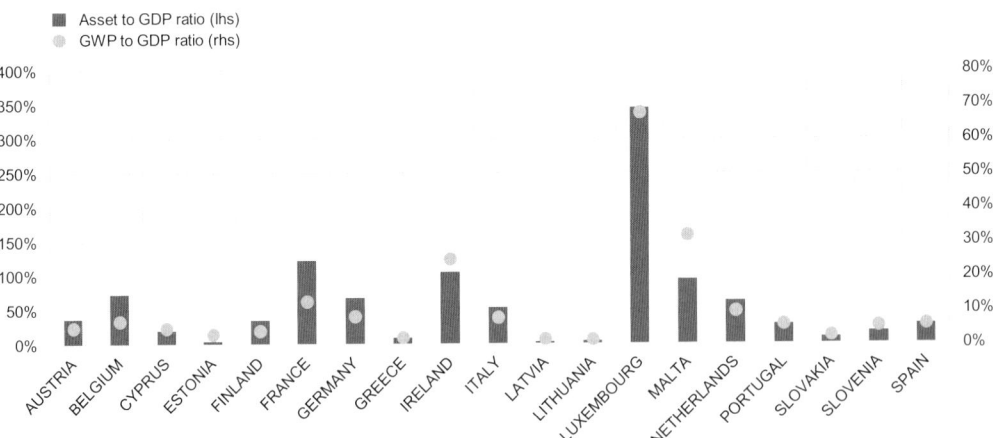

Figure 27.6 Insurers in the euro area: Assets and GWP as of December 2017.
Note: The asset-to-GDP ratio is calculated on a consolidated basis; namely, all assets held by a given insurer are attributed to the jurisdiction where the group is headquartered. The GWP-to-GDP ratio instead is locational, that is, computed on a solo basis, reflecting the jurisdiction where the business is being generated. All countries with a dot above the bar can be seen as net receivers of insurance services, whereas the countries with the dot below the bar could be considered net providers. Source: ECB and EIOPA data.

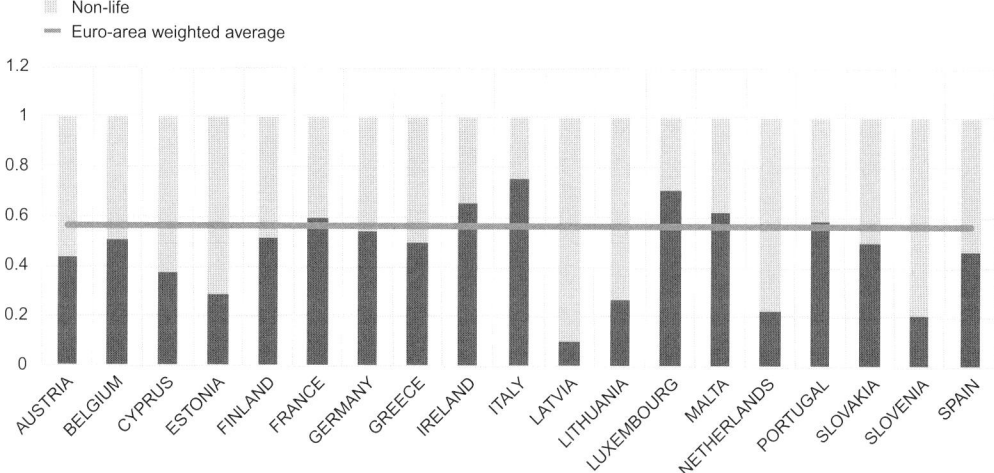

Figure 27.7 Life versus non-life business: Share of business GWP per country as of December 2017.
Source: EIOPA data.

Heterogeneity across countries is less impressive, albeit still present, when based on the insurer's penetration rate, measured by the ratio gross written premium (GWP)[5] over GDP. Beyond the relative heterogeneity of the insurance sector across countries, Figure 27.6 also shows that cross-borders flows of insurance services are significant, by comparing solo and consolidated figures, so that a shock in a given country may well spill over to others via an intragroup channel.

Another key structural feature of the insurance sector is the split between life and non–life insurance activities (see Figure 27.7). Christophersen and Jakubik (2014) found a greater and quicker response of life industry GWP to GDP than for non-life. Respective weights also differ quite substantially across countries. From both perspectives, any area-wide stress test modeling would therefore require risk- and country-specific information as well as a corresponding model specification.

Similar to the approach taken for banks, modelers need to pin down an appropriate, albeit simplified, balance sheet for insurers. To that aim, Figure 27.8 shows the distribution of assets held by euro-area insurers across countries, with bonds of various kinds and the euro area taking the lion's share. The direct holdings of equity represent a meagre 11 percent of total assets. Equities can also be acquired indirectly via collective investment undertakings (CIUs; i.e., funds). On the liability side (see Figure 27.9), insurers are subject to constraints on their technical provisions, estimated on the basis of the likely claims to arise. Their amount for P&C remains below that of technical provisions linked to the life business.

---

[5] The GWP is the total premium to be collected by an insurer during a specific period, before the deduction of reinsurance premiums or ceding commissions.

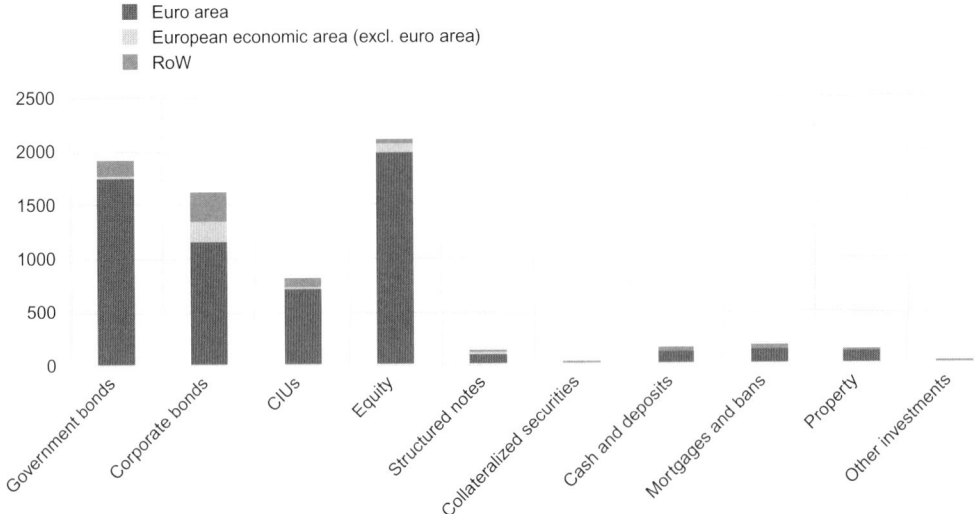

Figure 27.8 Euro-area insurers: Assets per category and counterpart location, Q3 2018, in billions of euros.
Source: EIOPA data.

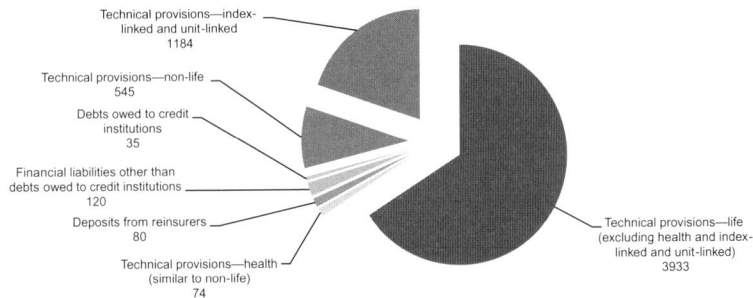

Figure 27.9 Euro-area insurers: Main components of liabilities as of December 2017, in billions of euros.
Source: EIOPA data.

In view of this, the EU insurers' balance-sheet structure can be simplified as reported in Figure 27.10. This stylized approach, based on data largely available for larger insurance groups, would still capture the main channels through which stress affects insurers' solvency. Accordingly, these summary balance-sheet items need to be mapped with the core risk components to be modeled for a macro-oriented top-down stress test exercise to be properly carried out.[6]

---

[6] The modeling approach suggested in this section builds on work done by G. Sher when at the ECB.

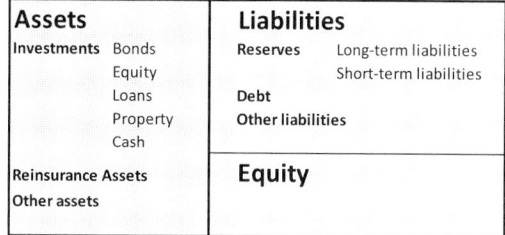

Figure 27.10 Stylized balance sheet: Insurance company.

3.2 Generic and Specific Risks Affecting Insurers

Even though insurers have specific business models, they are nonetheless affected by risks that are also faced across the financial sector by most types of entities, namely:

Credit risk: Even though most of the proceedings are invested in bonds, loans are also subject to credit-risk coverage. The nature and maturity of the investment are somewhat conditioned by those of the related liability as a result of, for example, liquidity constraints stemming from the insurance contract.

Interest rate risk: This risk has an impact mostly on net asset valuation, rather than on interest income, given the dominant share of bond holdings. It affects both asset and liability sides, via the change in present value of the cash-flow stream on mark-to-market portfolios. The change in the shape of the yield curve therefore has strong implications for an insurer profitability and business model.

Market risk: Although the equity holdings of euro-area insurers are limited, market shocks may reinforce a preexisting state of weakness via mark-to-market valuation. Changes in the valuation of property holdings, a standard significant investment for such firms, also affect insurers' balance sheet.

Although these three types of risk are common to most financial entities, insurance companies—by design, investors with long-run planning horizons—are especially sensitive to market risk and yield curve shifts. Being risk specialists, they additionally face risks specific to their business model, namely:

Reinsurance risk: An additional channel whereby credit risks affect insurers specifically is the reinsurance channel. The default of a reinsurer would affect all connected insurers, with possible systemic implications, depending on the relations between the given reinsurer and other firms.

Lapse risk: Lapse risk is the risk that some of the life insurance policies active at the beginning of a period are not renewed at the end of the period. This is measured via the lapse rate, which shows a correlation with the macroeconomic environment, including unemployment. This phenomenon is mitigated by surrender payments from the policy-holder to the insurer, although approximately 50 percent of the EU life insurance business is not attached to penalty payments, with another 40 percent triggering penalty payments that are nonetheless lower than 15 percent of the policy value.[7]

---

[7] Statistics based on EU large life insurers, as per ESRB (2015).

Underwriting risk: Probably the most emblematic of the insurance risks, the underwriting risk refers to the future number, timing, and amount of the claims to arise. By nature, it is linked to the type of risk covered and can be affected by demographic shocks (for life business) but also environmental conditions (for non-life business mainly).

The underwriting mechanisms and impacts are specific to each type of shock. A *demographic shock*, essentially an increase in morbidity and mortality, creates the conditions for early withdrawal of the amounts invested. The induced unexpected sales can happen in a distressed environment where the proceedings of the sales do not match the volume of claims, having an impact for traditional policies (not unit linked). *Natural catastrophes* may have a similar impact on insurers (increase in unexpected sales), especially in a context where climate change tends to increase the frequency of extreme events, possibly beyond that underlying the initially envisaged risk coverage.

3.3 Modeling the Main Transmission Channels of Stress to Insurers

Prior to engaging in risk modeling, publicly available data from the main euro-area insurers (i.e., a dozen players) has to be collected and exploited. Some key behavioral ratios can be computed; for instance, the reported dividend distribution can be scaled against a chosen metric, such as net income. The modeler also can exploit information from the "embedded value report" published by insurers, which features sensitivities to insurance-specific risks (e.g., demographic shock), not computable otherwise. This can also provide reference points and benchmarks for estimated parameters.[8]

Reflecting the aforementioned risk typology, models are required for four main types of risk events that affect insurers' balance sheet, namely: (1) lower credit quality of assets; (2) changes in securities valuation stemming from sovereign yields, equity prices fluctuations or yield curve moves; (3) unexpected redemptions on the life business and implied sales of assets; and (4) increase in claims via the property and casualty lines.

Credit risk can be modeled in a fashion similar to what is commonly done for banks, that is, by applying estimated parameters for a given segment (country–instrument–borrower) to a given insurer's portfolio structure, with parameters then depending on macrofinancial scenario assumptions.

Valuation of assets *at amortized cost* is affected directly by the respective projected rate of default, using a forward path estimated by country and type of counterparty, also covering risks reinsured via the probability of default of the reinsurers (inferred from the reported allocation of reinsurers to ratings). The information required mimics that needed to evaluate credit risk.

Valuation of assets and liabilities accounted for *at fair value* is affected by the changes in market prices, inter alia by changes in interest rates. On the asset side, the modeling can borrow from what is typically done for banks' market-risk parameters. Regarding liabilities, insurer-specific discount factors are to be used, reflecting asset prices assumptions, and the given insurer's liability mix.

---

[8] These reports are however not available on a regular basis from all main insurers. In addition stress test insurer-specific data are not released to the same extent as done for banks.

Redemption impacts (i.e., from customers withdrawing from life insurance) can be modeled by linking the corresponding observed "lapse rate" to changes in GDP and unemployment.

Underwriting effects can be estimated using published data on past claims (i.e., by fitting a distribution to their observed volume and volatility). Similar to what has been presented to assess operational risks for banks, potential losses for property and casualty business can be calibrated at any given percentile, including for catastrophe events.

An example of such a model is further explained in ECB (2017b), where the results of simulations are reported to quantitatively assess the insurance sector's resilience to specific scenarios. Figure 27.11 presents a summary overview of the specific modeling strategy adopted for the risk categories that could be reasonably covered. The modeling strategy again attempted to strike the right balance, given the (limited) available data and the need to remain commensurate with the macro-oriented purpose of this impact assessment exercise for the insurance sector (clearly much less detailed, therefore, than a typical microprudential bottom-up exercise, such as that conducted by EIOPA).

3.4 Sensitivity of Insurers' Balance Sheets: A Variety of Impacts

Stress on insurers can originate in macroeconomic (e.g., unemployment), financial (e.g., as a reshaping of the yield curve), or nonfinancial risk factors (among which, demographic shocks or natural catastrophes). These factors would then be those required to design a relevant scenario, the impact of which would be captured via models developed for each of the risks suggested previously.

Figure 27.12 illustrates how alternative scenarios reflecting specific risk drivers can have contrasted effects on insurers. In the case at hand, the "twin-shock" scenario combined adverse shocks on asset prices with low short-term interest rates, the "flight-to-safety" one mostly consisted of higher risk bond premia, and the third one associated a catastrophe add-on with a negative impact of the nonbank sector on liquidity conditions. Overall, credit risk—relating to adverse economic conditions across scenarios—has a systematically negative impact, whereas the scenario impact via interest rate risk appears strong but can change sign because differing assumptions on the relevant yield curves, level, and slope could equally improve or deteriorate insurers' balance sheets.

Especially for insurers engaged in the pension business, as a result of the long horizon of claims and associated premia, the average maturity of liabilities can exceed that of the asset side. A steepening of the yield curve would then increase net asset value because the value of liabilities would fall more than that of assets. Instead, a flattening of the yield curve draws down insurers' net asset value. Under a market-valuation shock, life business would also be affected because of the guaranteed minimal rate they may have committed to deliver. Insurers may therefore adopt riskier investment policies (i.e., "search for yield") when rates and returns are oriented downward. On the demand side, moreover, savers may shift portfolios toward other investments.

Figure 27.13 shows the relative estimated impact of scenarios over the period 2015–2017 on specific selected risk categories. Although reported weights depend by construction on the employed scenarios, the latter as largely fueled by the risk identification exercises conducted in parallel by ECB staff. As a result, the chart summarizes how prevailing

Risk drivers	Channels of transmission	Technical assumptions
Credit risk	Changes in the credit quality of loan portfolios	Credit-risk assessment carried out using (i) breakdowns by rating or region, depending on data availability, and (ii) loss rate starting levels, which are stressed using the same methodology as that applied for assessing the resilience of euro-area banks.
Interest rate risk transmission	Valuation effects on financial securities and liabilities	Sensitivities to interest rate changes computed for each interest rate–sensitive asset and liability exposure. Relevant yield curves used to project asset and liability cash-flow streams, to calculate internal rates of return, and to discount the cash flows using yield curve shocks.
Market valuations of securities	Valuation effects on financial securities and liabilities	Haircuts for debt securities derived from changes in the value of representative securities implied by the increase in interest rates under each shock and uniformly applied across the sample of large euro-area insurers. Valuation haircuts applied to government bond portfolios estimated on the basis of representative euro-area sovereign bonds across maturities. Haircuts for corporate bonds derived from a widening of credit spreads. Stock prices estimated using a representative euro-area benchmark.
Lapse risk	Sales of assets due to unforeseen redemptions resulting from increased lapse rates	Lapse risk quantified by projecting insurers' cash flows over a 2-year horizon, assuming a static composition of contracts and the reinvestment of maturing assets without a change in the asset allocation. Lapse rates linked to macroeconomic variables. Unexpected component of lapses leads to surrender payments. In the case of negative cash flows from surrender payments, the insurer is obliged to use cash reserves or sell assets to meet obligations. Lapse risk equals the cash or other assets needed to cover surrender payments.
Catastrophe risk	Variations in the projected claims	Catastrophe risk estimated by fitting a log-normal distribution to historical loss payments and then drawn via Monte Carlo simulations to estimate the annual loss distribution. The percentile is given by the scenario.
Other assumptions specific to the sensitivity of investment income		Investment income earned from reinvested assets shocked on the basis of investment income earned at the beginning of the simulation horizon. All other assets assumed to earn the initial investment income throughout the simulation horizon. Maturing fixed-income assets reinvested, retaining the initial asset composition. Underwriting business component of operating profit assumed to remain constant throughout the simulation horizon. No distribution of dividends assumed.

Figure 27.11. A bird's-eye view of a top-down impact assessment model for euro-area insurers. Source: ECB (2017b).

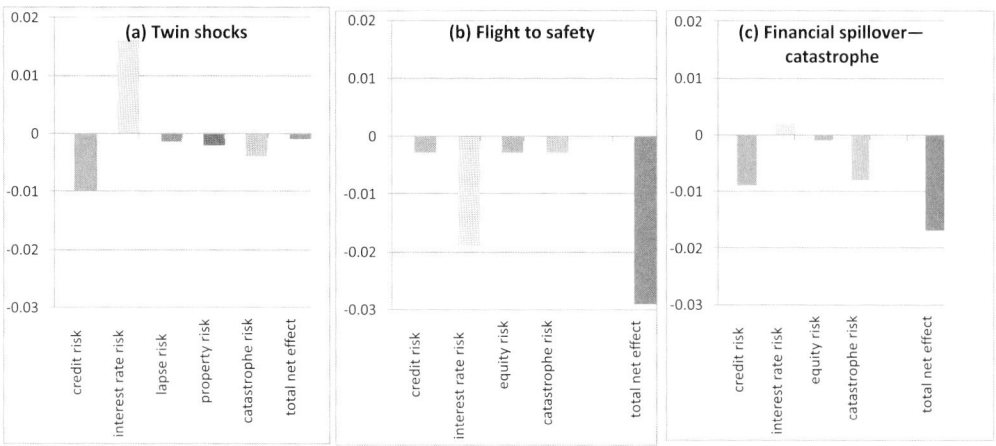

Figure 27.12 Impact on insurers of specific risk scenarios. Source: ECB (2017a).

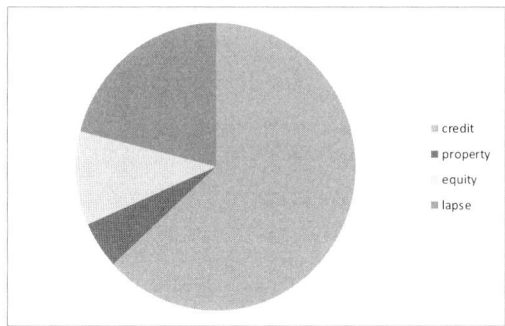

Figure 27.13 Relative impacts of macrofinancial factors on insurers' risks, excluding underwriting and interest rate risks.
Note: These relative weights are derived from stress test exercises performed on euro-area insurers from 2015 to 2017. They are clearly dependent on the scenario design, their representation here in aggregate being solely meant to give a simple representation of the output of the model presented in this chapter, throughout these 2 years. Two risks are not reported, the underwriting risk as defined outside of the macroeconomic scenario and the interest rate risk, for its sensitivity to the yield curve and the related lack of stability of its impact.
Source: Authors' computations.

macrofinancial risks would have affected the insurance sector over the period. Credit risk had been one of the primary risks, not so volatile across adverse scenarios but continuously driving the overall estimated impact. Lapse risk, equity risk, and property risk contributed to a lesser extent.

The effects of interest rate risks proved highly dependent on the scenario design, as driven by the shape of the yield curve. Given its highly varying impact and its scenario-dependent sign, it has been excluded, but it can outweigh all other risks under specific scenarios that embed, for example, a flattening of the yield curve. Underwriting risk has been excluded as well, as defined outside the macroeconomic scenario. It does, however, play a significant role, too, highly (nonlinearly) dependent on the quantile selected to pin down the catastrophe stress severity.

The proposed framework or a similar model could be used for macroprudential stress testing purposes because, although simplified, it duly accounts for insurers' specific features and behavior. The simulations should also be adequately designed; that is, the scenario should address risks relevant to insurers (possibly with add-ons, such as for catastrophe risks) and cover an appropriate—possibly very long by banks' standards—horizon. Data availability may, however, act as a constraint and remain a continuous challenge for the insurance stress tester to overcome—publications for insurers stress test similar to that repeatedly undertaken for banking exercises would, in this respect, greatly facilitate the development of a relevant top-down macroprudential approach to the sector.

4 Stress Testing the Broader Financial System

Next to the systemic sectors of banks and insurers, other financial institutions (OFIs) are subject to stress testing, as defined by supervisors, that can also be conducted sector wide, such as those coordinated by the ESMA for the EU for various groups of entities (European Securities and Markets Authority, 2019a,b,c). These microprudential exercises all require specific scenarios, horizons, and methodologies. The OFI sector indeed comprises a myriad of agents with very diverse business models and behavior. It is, however, deeply nested in the rest of the financial sector, with large exposures to banks and also insurers.[9]

To the extent that OFIs may not be systemic per se, a macro-oriented stress test needs primarily to account for their interaction with other players because, for example, OFI illiquidity could spread out to other larger players, and OFIs are largely dependent on other sectors' ability to fund or back them. Similarly, for central clearing counterparties (CCPs), interactions with other intermediaries are key by design. CCPs are systemic or even super-systemic as a sector, representing, along with their clearing members (CM), a network of networks. This is precisely why a macroprudential stress test should go beyond assessing CCPs individually, for example, computing their own daily liquidity balance under stress without getting to the systemic view. Instead, a macroprudential-oriented exercise should also investigate spillovers both ways with CMs (for credit- or market-risk factors as well) and possible interactions across CCPs. These can arise also indirectly, via common connections with some especially larger CMs, which can be on both the lending or borrowing side for a given CCP.

In what follows, we propose avenues to explore, along which fit-for-purpose models could be developed for OFIs and CCPs, bearing in mind that work is ongoing on such issues and clearly not as mature as for systemic sectors such as banks and insurers.

4.1 Structure of the OFI Sector and Main Transmission Channels at Play

The assets of OFIs, unlike the assets of banks, grew almost without interruption throughout the GFC and the euro-area sovereign crisis, both globally and in the euro-area. This underscores the role of nonconjunctural factors behind this structural shift. In fact, the theory

---

[9] See details in ECB (2020).

suggests that the shadow-bank-like institutions have a natural tendency to grow until the point where they would become systemically important for the entire financial system by endangering the stability of the banking sector (Ari et al., 2017). Stress testing of shadow banks could unveil vulnerabilities in this sector and help assess the potential for spillover of the stress in that sector to the rest of the financial sector.

After decades of strong growth, OFIs have reached, overall for the euro area, a size comparable to that of the banking sector. In March 2017, according to the ECB (2018),[10] OFI assets, including money market funds (MMFs), represented approximately 43 percent of the €76 trillion total assets held by all types of financial corporations in the euro area (see Figure 27.14). Statistics collected by the FSB (2018) confirm that the euro-area OFI sector now has an importance similar to that of banks (see Figure 27.15). At the end of 2016, the total assets of euro-area OFIs represented 32 percent of assets of global OFI assets, the largest share of all reporting countries—considering the euro area as a single jurisdiction.

The term *OFI* covers a large variety of financial institutions: MMFs, hedge funds, investment funds, real estate investment trusts (REITs) and real estate funds, trust companies, finance companies, broker-dealers, structured finance vehicles, central counterparties, captive financial institutions, and money lenders. In the euro area, OFIs are mostly composed of MMFs, fixed-income funds, mixed funds, credit hedge funds and real estate funds, which together account for more than 80 percent of the euro-area OFI assets. Then securitization vehicles, structured finance vehicles, and asset-backed securities together amount to approximately 10 percent of total OFI assets.[11]

With the increasing size of the nonbank financial institutions, the inclusion of OFIs in a comprehensive framework for macroprudential stress testing becomes particularly relevant. This is because a financial sector characterized by a large share of unregulated OFIs poses externalities to the entire financial sector, including the regulated banking sector. In particular, open-ended funds, such as open-ended fixed-income funds, credit hedge funds, and MMFs, may be subject to runs because of the intrinsic maturity mismatch between their long-term assets and short-term liabilities. The latter is not covered by an explicit or implicit guarantee in this sector, contrary to what is the case for bank deposits, so funds may face abrupt redemptions.

A potential materialization of a run on investment or money market funds could, moreover, adversely affect the rest of the banking sector, mostly via two contagion channels:

Direct contagion channel: quantity effects, via borrowing and lending links between OFIs and banks via the repo, credit, or funding markets—including defaults.

Indirect contagion channel: price effects via the valuation of assets held by banks or OFIs as a result of the implicit (bank's banking books) or explicit (OFIs' assets, banks' trading and investment books) impact on equity. In extreme scenarios, self-reinforcing fire sales can occur.[12]

---

[10] The latter also shows (pp. 13–14) that OFIs are both very much interconnected within their own sector and, via securities and equities holdings, exposed to banks and insurers.

[11] ECB and FSB data provide broadly similar results—differences relate to classification issues as well as timing of valuation.

[12] See Duarte and Eisenbach (2013). Subsequent versions of this seminal paper provide further insights into this issue.

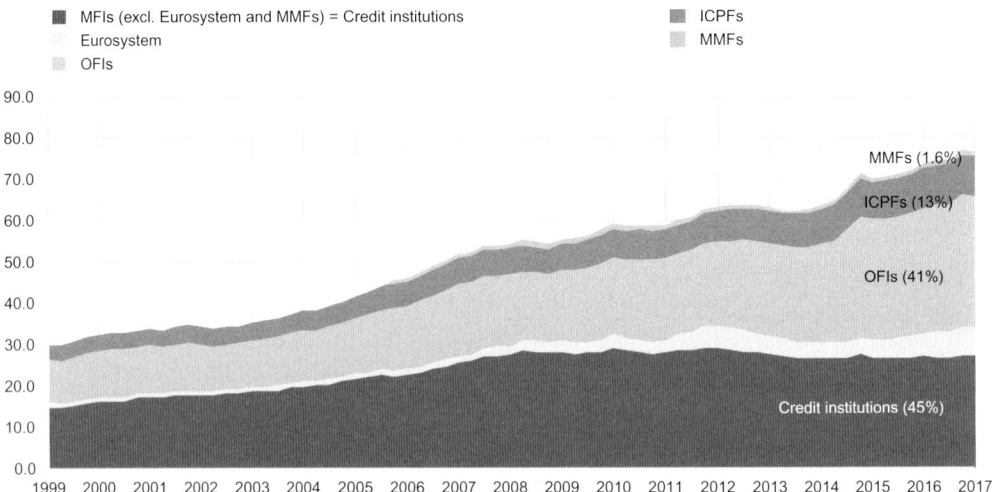

Figure 27.14 Assets of financial corporations by type in the euro area.
Source: ECB (2018b).

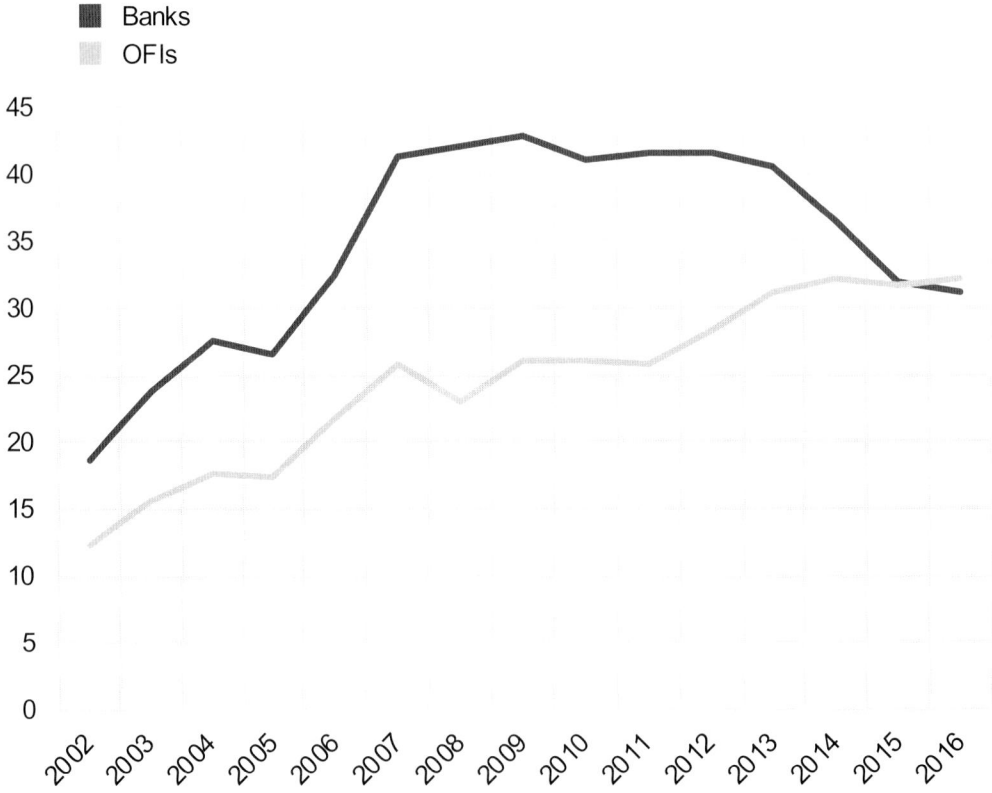

Figure 27.15 Assets of euro area banks and OFIs. Source: FSB (2018).

4.2 OFIs: Theoretical Agent-Based Models for Shadow Banking and Banks

To assess how a shock to OFIs can propagate into the financial system, the modeling framework needs to feature both the direct and indirect contagion channels. Econometric models, estimated on historical data, may not be able to capture the possible nonlinearities that characterize stress events. General-equilibrium models (GEMs), using independent representative agents, cannot capture the key feature of the contemporary financial sector that is at the root of contagion—that is, its high level of interconnectedness. Furthermore, by design, GEMs do not allow for temporary market failures, such as the dry-up of the interbank funding market after the collapse of Lehman Brothers. This in turn may trigger adjustments in the behavior of economic agents under adverse conditions. Even if agents continuously optimize, they may not internalize the impact of their collective behavior on the entire system. In other words, an optimal behavior at the level of economic agents may be suboptimal at the system level, with a resulting "bad" Nash equilibrium emerging.

It seems instead that agent-based models are better suited for modeling interactions among financial intermediaries and related externalities. They allow for a sufficient level of granularity and heterogeneity as regards the composition of the assets and liabilities and, as such, are flexible enough to model various types of financial institutions and a variety of possible business models. Importantly, they can be used to model the key feature of the financial sector—interconnectedness—whereby an asset held by one entity is a liability of another. Also, indirect connection via asset-holdings commonality needs to be captured in such a setup for it to be fully relevant.

Such an illustrative agent-based framework involving banks and shadow banks can be found in Calimani et al. (2019), embedding both *direct* and *indirect* contagion, as defined previously. In this model, fire sales can emerge as a feature of market equilibrium. Financial institutions, although optimizing agents, do not internalize their actions' impact on the system. This is critical when agents operate in an environment of liquidity and capital constraints that may force them to deleverage. Such a framework, albeit with simplified mechanisms, can account for externalities arising from a large shadow-bank sector, as now seen now in the euro-area financial system.

The model involves two types of financial intermediaries: banks and asset managers. Banks are funded by equity, deposits, and interbank borrowings, and they can invest in illiquid loans, a liquid security, interbank lending, and cash. Asset managers are fully funded by redeemable participations and invest in liquid securities and cash (see Figure 27.16). To reflect the heterogeneity of business models in the financial sector, the market is assumed to comprise a number of bank and shadow-bank agents that can, moreover, structure their balance sheets in varied ways. Interbank borrowing and interbank lending costs depend on the counterparty's risk premium, which in turn is a function of its respective leverage. In addition, the yield on the security reflects the differing prices at which individual agents bought the security in the secondary market.

Banks are *directly* interconnected with each other in the interbank market, whereas all institutions are *indirectly* linked with each other via price effects—by holding the security and selling or buying it. Although, for the sake of simplicity, the model does not include an explicit wholesale funding market, in itself an important shock-propagation channel, the liquidity and price of interbank borrowing are determined endogenously, which can then

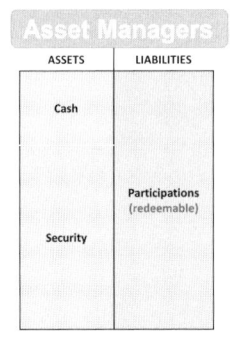

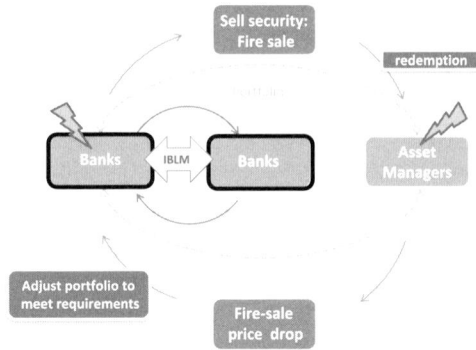

Figure 27.16 Stylized balance sheet of banks and asset managers in an agent-based framework.
Source: Calimani et al. (2019).

Figure 27.17 Direct and indirect linkages between banks and shadow banks in an agent-based framework.

mimic the behavior of stressed funding markets. The role of the funding channel is explored in more in detail in Arnould et al. (2021).

All agents maximize their profits, subject to a minimum capital requirement for banks only and to a minimum liquidity requirement for both banks and asset managers. At the inception of the system, agents are endowed with equity and a random internal return on the security, and they optimally allocate the remaining liquidity into available assets to maximize profits. At the beginning of the simulations, one bank or one asset manager at a time is subject to an idiosyncratic liquidity shock, which triggers a specific balance-sheet adjustment. Banks redeem interbank liquidity or sell the security, whereas asset managers sell the security to meet the capital and liquidity constraints. This, in turn, may trigger further adjustments to the balance sheets of other agents.

The model embeds a clearing mechanism with an endogenous formation of asset prices. In such a setup, an initial exogenous liquidity shock may lead to a fire-sale spiral. Banks, which are subject to liquidity requirements, may be forced to sell a security, which affects its endogenously determined market price. As the price of the security then decreases, all agents update their equity and adjust their balance sheets by making decisions on whether to sell or buy the security. This endogenous process may trigger a cascade of sales, creating a fire-sale configuration (see Figure 27.17).

Simulations suggest that mixed-portfolio banks (i.e. those banks that both invest in the security and are present in the interbank market) act as plague-spreaders in a context of financial turmoil. This is because they can transmit the shocks between the two types of agents. In addition, tighter capital requirements may accelerate contagion because stricter capital requirements incentivize banks to hold similar assets (see Figures 27.18 and 27.19). Based on such results, it appears that macroprudential policy actions, if any, should instead focus on funds, namely, the origin of the problem in the case at hand.

Building on such a setup, also to cater to more shocks and transmission channels within the broader financial system, additional various types of financial agents could be incorporated. They would be characterized by differing balance-sheet structures, and there would as well be further nonlinearities emerging from their interactions in various markets. Such a framework would help assess the impact of capital, liquidity, and margin measures on

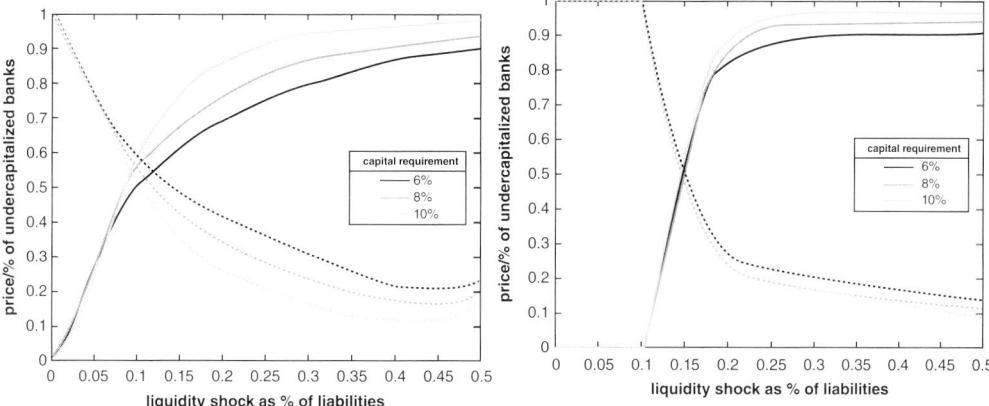

Figure 27.18 Simulation of a fire sale in a system of banks and asset managers as a result of a shock originating in the banking sector. Source: Calimani et al. (2019).

Figure 27.19 Simulation of a fire sale in a system of banks and asset managers as a result of a shock originating in the asset-management sector. Source: Calimani et al. (2019).

systemic stability in a multisector environment. An illustrative agent-based model as outlined in this section can be a first step toward building such an ambitious framework (see Chapter 30 in this handbook on related attempts).

4.3 Preliminary Thoughts on Modeling CCP Interactions with Clearing Members

CCPs are an important element of the postcrisis financial ecosystem. They emerged as part of G20 reforms triggered by the collapse of Lehman Brothers. Given that this investment bank was involved in a very large number of contracts in the over-the-counter (OTC) market, its failure brought the derivative market close to a collapse.[13] CCPs were set up to reduce the systemic risk resulting from bilateral counterparty connections in the OTC market. The latter creates a network that is highly sensitive to spillovers of idiosyncratic shocks, possibly resulting in a cascade of defaults of highly interconnected counterparties.

A CCP can provide clearing members with access to the complete network of all counterparties. In principle, netting of exposures, centralized collateral, and settlement reduce the operational risk for dealers. In addition, collateral costs are materially lower thanks to efficient portfolio cross-margining. Nevertheless, because a CCP is counterparty to all its CMs, it is by design "super-systemic," and its default could therefore endanger the entire financial system. Furthermore, the procyclicality of margin requirements may aggravate liquidity stress when the price of collateral falls.

Under systemic stress, CCPs can then act as shock absorbers. Duffie and Zhu (2011) illustrate the specific contribution of CCPs to the reduction of counterparty risk for certain classes of derivatives. At the same time, CCPs can also act as potential triggers for a complex set of waterfall reactions that should also be modeled (e.g., margin, default funds, liquidity

---

[13] See G20 Leaders' Statement following the 2009 Pittsburgh Summit at www.fsb.org/wp-content/uploads/g20_leaders_declaration_pittsburgh_2009.pdf.

tools, see, e.g., Cont, 2015), including in terms of the impact they would have on the rest of the system.

To assess how CCPs contribute to the stability of the financial system, Krug and Żochowski (2022) have put forward a partial-equilibrium model of a CCP and its CMs, again using a stylized balance-sheet approach (see Figure 27.20). CMs are profit-optimizing agents exposed to market risk. The balance sheet and leverage of the CCP and CMs are assumed to change in tandem with market-price movements. They compare a system based on clearing collateral exposures via a CCP versus a system of bilateral counterparty exposures by analyzing the impact of a stochastic market-price shock on the stability of the system. As shown in Figure 27.21, the CCP-based one mitigates the effect of stress.

Such a model can be one building block in a macroprudential stress testing framework covering OTC and collateral markets, with, for example, large broker-dealers that act as market-makers or liquidity providers. Ultimately, modeling efforts could focus on endogenous feedback effects between CMs' defaults and market liquidity and on contagion to nondefaulting CMs, as well as from service functions provided by CMs to CCPs, such as liquidity provision or collateral and investment services. The framework would then feature CCPs as potential absorbers or transmitters of adverse shocks, including to CMs' solvency, also in interaction with market liquidity. A recent Bank for International Settlements (BIS, 2019) report has indeed proposed to develop a multi-CCP, macroprudential-oriented setup to cover credit and liquidity risks; the main challenges mentioned there being to link market-risk factors and default assumptions, as well as dealing with the widely different horizons of liquidity and solvency stress.

5 Stress Testing the Whole Financial System

System-wide stress testing has been conducted by microprudential authorities for subparts of the whole system, namely, for banks, insurers, and OFIs—for example, for the EU by the EBA, EIOPA, and the ESMA, respectively (see previously made references). These tests, however, did not consider much interaction within a given sector, let alone with players outside it. From a macroprudential perspective, although they provide useful information, these (bottom-up) exercises come short of what is eventually desirable. In addition, even

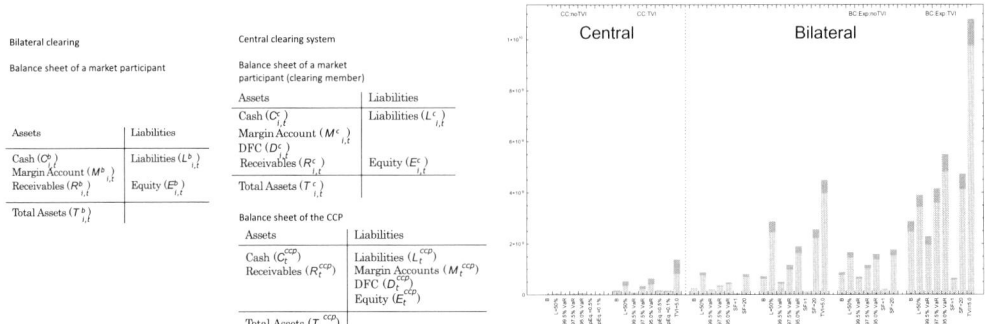

Figure 27.20 Stylized balance sheet of CCP and market participants (CMs).
Source: Krug and Żochowski (2022).

Figure 27.21 Simulated losses in the system—central clearing system (left panel) versus bilateral system (right panel).
Source: Krug and Żochowski (2022).

though top-down stress-testing models have been developed for a variety of financial sectors and players, much of the attention appeared to be put on the banking sector, as more directly systemic but also with easier access to sufficiently comprehensive data with respect to other financial players, especially those that are less regulated.

There is therefore a need to go both beyond banking-sector stress per se and adopt a system rather than an isolated firm-by-firm (first-round-only) approach, with a view toward incorporating spillovers and amplification channels (second-round impacts). The latter involves real as well as financial aspects. First of all, from a macroeconomic real-side viewpoint, the credit channel via bank lending and banks' balance sheets operates as a core transmission mechanism, possibly adversely affecting the real economy when dysfunctional. This effect can be aggravated by the impact of disruptions in insurance services. In addition, on the purely financial side, markets and interconnections among agents can further amplify shocks that initially affected only banks or insurers—via waves of contagion similar to those outlined previously. In both cases, shock propagation creates ex post additional stress beyond that envisaged in the initial scenario because the latter did not capture ex ante all these real or financial spillovers, thereby leading to second-round effects that need to be assessed in their own right and fully accounted for in the impact analysis.

5.1 Spillovers from Banks' Stress

Banks provide credit to the economy, are very much interconnected with each other and the rest of the financial system, and also play a key role in a number of financial markets. All of these characteristics create potential amplification channels when stress affects the banking sector.

Credit channel—deleveraging: In response to regulatory or market-driven unanticipated capital requirements (i.e., not considered ex ante in the macrofinancial scenario), banks would have to react. Banks' responses would be nonstandard historically speaking and would not be captured in the estimated relations underpinning the scenario, the latter built with the assumption that regulatory requirements are unchanged.[14] Although banks may raise capital, this could prove unfeasible in times of crisis, in which case they may instead act on their RWAs and engage in deleveraging. If concentrated on loans, this step would initiate a real–financial feedback, to the extent that the implied decline in credit supply would come on top of those consistent with the initial adverse scenario and thereby lead to even lower activity.

The implied additional negative credit-supply shock can be estimated via reduced/structural form models (e.g., Rancoita and Hilberg, 2017; Gross and Żochowski, 2017, resp.). These macromodels differ from the previously mentioned loan-flow ones because they need to relate credit and activity conditional on banks' capital or leverage ratios—now set at new levels. The loan path can be set consistent with the banks' historical average responses to capital requirements (mixing loan deleveraging with profits or right issues) or alternatively under the extreme assumption that the capital target is reached solely via loan-

---

[14] Against the background of upward-trending capital needs, corresponding requirements can be related to activity, interest rates, and credit, but there are instead only a few available cases that could be the basis for calibrating banks' responses to unexpected requirement changes (e.g., the EBA capital exercise of 2011).

book reductions.[15] Such models deliver an updated GDP path consistent with the new risk premia (investment or consumption specific) required to get to the desired loan-book decline. Additional effects on GDP can be substantial, as expected, strongest when loans only contribute to reaching the ex post set capital target; they also increase with the hurdle rate to define capital shortfalls. In addition, estimated additional GDP losses are larger as well as when cross-country spillovers are explicitly accounted for (see, e.g., ECB, 2016).

A number of additional avenues could be explored to get a better handle on these effects, such as using bank-level vector autoregression (VAR) modeling,[16] a wider range of actions (e.g., deleveraging assets other than loans), considering NPL-related actions (including sales), basing the selection of assets to be deleveraged on a pecking order (be it loans or other assets, with, e.g., foreign assets being shed first), or even a risk–return model to optimally reshape banks' portfolios and achieve the newly required capital.

Exposure channel—direct contagion: A first immediate channel is that of interbank lending, via corresponding exposures. The resulting contagion can again be direct first, that is, via "cascades of defaults" (as in Eisenberg and Noe, 2001). Banks that have the lowest post-stress capital ratios are assumed to be pushed into default, leading to losses for other banks, which in turn can also enter the default zone. This can be seen as a quantity mechanism spreading stress.

Asset-sale channel—indirect contagion: Next to "cascade of defaults," a price mechanism can also enhance system-wide stress via "fire sales." Even before reaching the default point, that is, when in distress (exhibiting low but still solvent capital ratios), banks would engage in generalized deleveraging, selling financial assets as well. In turn, such asset disposal by weaker banks would lead to declines in the asset valuation of all banks, albeit sound, that hold assets being disposed of.

In practical terms, unsecured interbank lending became too limited after the crisis in the euro area to generate very significant effects via a pure default contagion—even including fire sales. Illustrative experiments show that accounting for herding behavior and solvency–liquidity links is critical to obtain sizeable effects via this channel (documented by Hałaj and Henry, 2017).

5.2 Spillovers from Insurers' Stress

From a historical perspective, the systemic risk attached to insurance companies does not compare in magnitude with that from the banking sector. An insurance default does not necessarily imply a customer's loss on the ongoing contracts. Insurers hold long-maturity assets against the payment of claims, to be settled in a usually long resolution process. At the same time, if insurers do not generally engage in maturity transformation, therefore with a significantly lower potential liquidity mismatch than banks, systemic risk may develop via reinsurance or guarantees and their possible concentration in entities that are systemic within the insurance sector. At the same time, insurers may also act as a shock absorber because the sector is driven by policyholders' saving behavior, which is less prone to bank-run types of reactions in the case of crisis.

---

[15] See also Gross et al. (2018) for a similar approach but involving nonlinear effects.
[16] In the FAVAR spirit of, for example, Buch et al. (2014).

There are several channels via which the insurance sector may amplify and transmit a shock, whether exogenous or stemming from the insurance sector itself.[17] Similar to banks, they can affect primarily the real economy or, instead, primarily the financial sphere:

Disruption in insurance services: The economy depends on the provision of cost-efficient insurance services. The disruption in the provisioning of these services, due to an insurer failing or withdrawal from a business line, may create and propagate a shock to the real economy as a result of other actors not immediately substituting for the vanished service. This risk especially concerns economic sectors that rely heavily on insurance services (e.g., the airline industry). Insurance failure on mortgages would also affect the LGDs to be recorded by banks, in some countries to a very sizeable extent.

Housing channel: Insurers, in their potential quality of loan distributors and covered bond holders, contribute to the depth of the housing market. Insurers' vulnerabilities may then in turn result in an overall hampered capacity to extend such loans, with a negative impact on GDP and real estate prices—the latter affecting both banks' and insurers' asset valuation.

Exposure channel: Financial and nonfinancial entities are exposed to insurance companies. As such, a shock to one or more insurers may spill over to the rest of the system. The nature of the contagion effect is dependent on the legal structure of the entity—especially in the specific case of financial holding companies providing both banking and insurance services (*bancassurance*).

Asset-sale channel: Typically, procyclicality can be fed by insurers' investment behavior because it is sensitive to credit-rating downgrades and propagate further to the financial system via asset liquidations. Especially, in a low-for-long environment, insurers would tend to invest in riskier assets than usual, which would prove costlier to liquidate at times of stress. Also, contractual commitments can lead insurers to engage in fire sales, with the expected impact on prices and valuations.

5.3 Cross-System Spillovers via Equity Valuation

Banks' or, equally, insurers' stress can spill over to other sectors particularly via equity valuation—to the extent that agents holding equities of those financial firms that are hit by the stress would also see their own asset valuation reduced (Candelon and Sy, 2015). In the European case, this can be calibrated using the flow-of-fund matrices from the quarterly financial accounts, which, for each country and institutional sector, show how much of their assets are held in the form of financial instruments issued by another sector (the "who-to-whom" principle). Using these matrices, an iterative process can be set in motion, whereby a stress test leading to a given aggregate capital depletion is reflected one-to-one in the equity valuation of firms subject to the stress. This then affects the asset valuation of all sectors holding such firms' equities. Further to which, sectors also issuing equities, such as OFIs or NFCs, transmit the drop in value of their own mark-to-market equities to the sectors holding their shares. The time dimension of such iterations can be assumed as immediate because market participants' expectations underlying equity prices are formed on a continuous basis. This approach can be implemented based on, for example, results from the EIOPA or the

---

[17] Building on the categorization from the ESRB (2018).

EBA exercises that provide a calibration for the initial equity-valuation loss for each sector and country affected.

An illustration of such a chain reaction has been published by the ECB (2016) (reproduced in Figure 27.22), taking as an input the banking-sector supervisory stress test conducted in that year. Based on a stress test leading to some 400 basis points (bps) of capital depletion for banks, the completed iterative process has substantial second-round effects on other sectors, by construction, on those holding bank equities in a sizeable manner. This would be the case especially for non-MMFs, pension funds, and OFIs, which can lose close to 10 percent of their own asset value if the stress test results were to be immediately translated into banks' (and other sectors') market valuation. This simple but powerful approach could readily be extended to debt securities holdings, the valuation of which also suffers from capital declines (linking, e.g., credit default swap premia/pricing to solvency or net asset valuation).[18]

5.4 System-Wide Stress Test and Beyond: Work in Progress

Beyond equity-valuation effects, a more ambitious avenue, not yet fully explored, is to consider a system-wide stress test covering not only banks and insurers but all market participants, such as funds, asset managers, insurers, and so forth, and all possible channels of propagation, including the macrofinancial feedback loop via the credit channel, the funding and rating channels, and the contagion-based channels (see Constâncio, 2016). This could be done by extending the available banking liquidity and solvency stress test toolkits, adding interactions between various sectors and a CCP network. Beyond bank or insurer events, sources of shocks should also involve funds' contract redemptions, CCP liquidity, and credit risk (after market-risk shocks and possibly default of counterparties). Both the frequency of

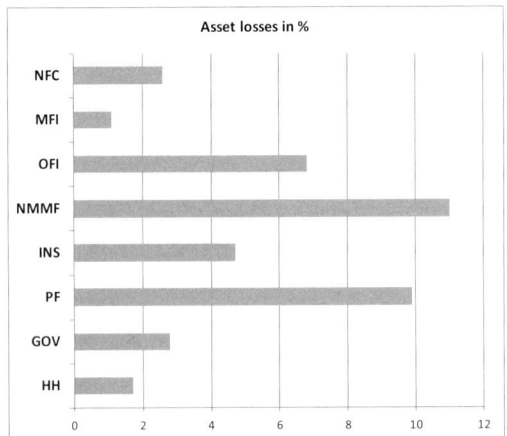

Figure 27.22 Impact on institutional sectors of banks, equity-valuation
loss after stress (asset percent).
Source: ECB (2016).

---

[18] For a comprehensive modeling using data on individual banks' large exposures, see Covi et al. (2019).

shocks and the horizons of agents' reactions should then end up much shorter than those typically considered for banks' or insurers' solvency stress testing.

Expanding on the banking-sector liquidity stress test apparatus (focused on counterbalancing capacity)[19] would allow modelers to integrate elements that are key to market-channeled interactions. These comprise haircuts on assets held and funding collateral; market dry-out (i.e., no funding sources for weaker players); collateral pool limits; herding behavior with, for example, banks facing restrictions similar to those put on their weaker peers; and access to a lender of last resort before failing, including emergency lending facilities, among others. Beyond these, the two-way interaction between liquidity and solvency also needs to enter the modeling setup, as recommended by the Basel Committee on Banking Supervision (BCBS, 2015) and explored specifically by Coen et al. (2019).[20] Moreover solvency ratios below a certain (distress) threshold, possibly well above the default one, will also need to have a negative impact on banks' values (e.g., Battiston et al., 2015) and their funding costs. This relation in turn would reinforce the likelihood of liquidity distress (see Aikman et al., 2009) as well as the likelihood of funding dry-out episodes.

With a view toward modeling the whole financial system, although the work on banks is already quite advanced and comprehensive and somewhat developed for insurers, further work on funds and CCPs needs to bring theoretical models to the data. The latter remains largely to be collected or accessed, whenever available. Connecting these various blocks, as is certainly necessary, would also require additional work on, for example, multilayer networks (as in Kok and Montagna, 2016), that is, capturing channels other than equity holdings or lending, both for direct and indirect contagion. Aikman et al. (2019) propose a promising first step in that direction: modeling solvency and liquidity jointly, also considering interactions with asset, funding, repo, and derivative markets. They show that nonbanks crucially matter in shaping market developments, and this is precisely the area for which data and, as a result, behavior analyses are scarce. Hopefully, incoming transaction data sets for the euro area, such as EMIR or SFTDS, collected via trade repositories and in line with ESMA regulations, will help.

At the same time, such extensions would not fully substitute banking-solvency stress tests conducted for macroprudential purposes, themselves needed in their own right, especially for banks and insurers. System-wide exercises apparently aim to replicate market crisis events, which are very concentrated over time, from within a day to up to a quarter, involving differing and higher-frequency mechanisms with respect to long-lasting solvency crises and their real-side consequences, spreading possibly over decades. In a nutshell, system-wide stress test as being developed seem to cover (close-to) collapse events, whereas macroprudential exercises, especially if focusing on credit, would rather investigate lasting slump situations or a phase of risk buildup. Last but not least, a trade-off between complexity and tractability is likely to arise when interpreting, possibly deciphering, such intricate simulation outcomes before deriving and justifying policy moves. This issue would be even

---

[19] See the frameworks for systemic liquidity stress testing in the seminal Schmieder et al. (2003) or the more recent Hałaj and Henry (2017).

[20] See also Cont and Kotlicki (2019).

more pressing in the event where such system-wide exercises would be deemed eventually relevant only if conducted globally, given the prevailing market structure.[21]

Further down the road, stress tests could be extended to nonfinancial sectors, such as the household and corporate sectors. Using individual survey data, the Bank of England and the ECB have published studies constituting one step in that direction for the household sector in relation to housing markets. Such models allow for estimating stress test elasticities in response to an interest rate shock, an income shock, and a house-price shock at a household level, thus allowing for more precise estimates of the credit and house-price macro-feedback loops to the bank balance sheets by taking into account households' specific liquidity and collateral buffers. For the same reasons, similar tools would be valuable at the individual corporate level.[22]

Although still in the making, such an approach appears highly desirable, especially for policy purposes, such as for evaluating borrower-based instruments, in particular with a cost–benefit analysis.[23] This is, similar to financial-sector exercises, done by running scenario-based simulations that account for household's characteristics (income, debt, assets, housing, and unemployment status). Unemployment then appears as the key bridge variable between macro and micro results, in other words, the pivotal variable through which the scenario transmits to individual households. Imposing LTV/loan-to-income binding ratios mechanically reduces credit supply and thereby activity, which is the cost of the measure. At the same time, fewer household defaults are found when risks eventually materialize. Banks and the economy thereby end up better off after to a severe downturn if such macroprudential measures were taken ahead of it, which is the benefit of the measure. The policy decision would then subsequently have to be made based on the respective weights of estimated costs and benefits (e.g., based on a specified policymaker's loss function).

6 Conclusions

- In this chapter, we have shown how a top-down modeling framework can be developed for macroprudential stress testing of the euro-area financial sectors, building on individual-entity data and modeling. This can be done on the basis of a simplified balance-sheet approach across sectors, which can be specifically detailed for banks and insurers, where data and experience have been gathered after the GFC. The approach has more of a work-in-progress status for OFIs and CCPs, with recent developments in this area, especially on the academic side.
- We have covered a number of modeling options that are available to supplement the direct impact of shocks embedded in a given scenario on individual financial entities in isolation—as typically done by supervisors—with real–financial and financial–financial feedbacks that eventually contribute to building the system-level view, as required for macroprudential purposes.

---

[21] This could be termed the "everything is in everything and conversely" risk (after P. Dac, a French philosopher).

[22] See, for example, Anand et al. (2013) (Dees et al., 2017) for a three-layer network (firms, domestic and foreign banks) with default and price effects.

[23] See Baptista et al. (2016) or Chapter 15 of STAMP€ for agent-based models with household- and borrower-based macroprudential measures.

- Such an infrastructure has already been employed to generate top-down results for a number of macroprudential purposes—relevant illustrations are provided in the chapter, based, inter alia, on a series of ECB publications. There is further room for additional use of such toolkits outside stress testing to, for example, analyze banks' profitability prospects or insurers' situation in a low-for-long interest rate environment. In the current context, impact analyses could also be conducted for measures targeting NPLs or those put together as mitigation for the spread of COVID-19.
- Most importantly for macroprudential policy matters, comparing baseline and stress results under differing assumptions for capital buffers or second-round effects (essentially via dynamic balance sheets and the credit channel) can inform a cost and benefit analysis supporting, for example, countercyclical buffer calibrations. Borrower-based measures may, moreover, call for specific household modeling, also available in some form. Forthcoming credit-register data for the EU (Anacredit) should boost the analysis of empirics around loans, including on the borrower side.
- Modelers may likely also wish to deepen the analysis of banks' reactions to stress in terms of asset shedding and credit deleveraging, possibly integrating the first and second rounds in one single model. The latter setup would, however, run the risk of oversimplification as well as resulting in too-specific (and thus prone to model risk) design choices. A modular approach that instead involves a suite of, possibly competing, models may still appear relevant, in particular to mitigate model risk but also to facilitate the communication, interpretation, and replicability of results.
- On the modeling of entities beyond banks, that of insurers is already advanced but not yet up to challenging firm-level results, contrary to what has been already carried out for banks, whereas OFI models remain in infancy. Developments for insurers or, for that matter, OFIs more generally would hinge on getting more firm-specific data, including from supervisory stress tests. Looking ahead, a first opportunity to expand and refine the setup would be to fill data gaps on nonbanks' balance sheets or agents' transactions (EU-wide data collections such as EMIR/SFTDS for derivatives and repo operations, respectively, appear promising in this respect).
- The availability of extended data would accordingly enrich the scope of contagion studies, as already done with the IMF and ECB "CoMap" project exploiting large-exposure data and combining this with a set of specified reactions to stress for banks. This would also support the development of further system-wide liquidity/market stress testing—as initiated by the Bank of England.
- For any further (and welcome) attempt to enrich and enhance such frameworks, modelers should, however, bear in mind the usual trade-off that arises between the complexity or granularity of a given model, or set of models, and its policy relevance, to the extent that the latter crucially depends on the extent to which results can be readily presented and understood.

References

Aikman, D., P. Alessandri, B. Eklund, P. Gai, S. Kapadia, E. Martin, N. Mora, G. Sterne, and M. Willison (2009), "Funding liquidity risk in a quantitative model of systemic stability," Bank of England Working Paper 372.

Aikman, D., P. Chichkanov, G. Douglas, Y. Georgiev, J. Howat, and B. King (2019), "System-wise stress test simulation," Bank of England Working Paper 809.

Ampudia, M., H. van Vlokhoven, and D. Żochowsk (2016), "Financial fragility of euro area households," *Journal of Financial Stability*, 27(C), 250–261.

Anand, K., S. Brennan, P. Gai, and S. Kapadia (2013), "A network model of financial system resilience," *Journal of Economic Behaviour and Organisation*, 85(1), 219–235.

Anderson R. (ed.) (2016), *Stress testing and macroprudential regulation: A transatlantic assessment*, CEPR Press.

Ari, A., M. Darracq-Paries, C. Kok, and D. Żochowski (2017), "Shadow banking and market discipline on traditional banks," IMF Working Paper 17/285.

Arnould, G., G. Avignone, C. Pancaro, and D. Żochowski (2021), "Bank funding costs and solvency," *The European Journal of Finance*, DOI: 10.1080/1351847X.2021.1939753.

Bank for International Settlements (2019), "Framework for supervisory stress-tests of CCPs," CPMI-IOSCO Report.

Baptista, R., J. Doyne Farmer, M. Hinterschweiger, K. Low, D. Tang, and A. Uluk (2016), "Macroprudential policy in an agent-based model of the UK housing market," Bank of England Working Paper 619.

Battiston, S., G. di Iasio, L. Infante, and F. Pierobonet (2015), "Capital and contagion in financial networks," in Bank for International Settlements, *Indicators to support monetary and financial stability analysis: Data sources and statistical methodologies*, vol. 39.

Baudino, P., R. Goetschmann, J Henry, K Taniguchi, and W Zhu (2018), "Stress testing frameworks: A comparative assessment," FSI Insight #12, Bank for Internatiuonal Settlements.

Basel Committee on Banking Supervision (2015), "Making supervisory stress tests more macroprudential: Considering liquidity and solvency interactions and systemic risk," Working Paper 29.

Bennani, T., C. Couaillier, A. Devulder, S. Gabrieli, J. Idier, P. Lopez, T. Piquard, and V. Scalone (2017), "An analytical framework to calibrate macroprudential policy," Banque de France Working Paper 648.

Bousquet, A., and T. Dubiel-Teleszynski (2017), "Operational risk module of the top-down stress test framework," in S. Dees, J. Henry, and R. Martin, *STAMP€: Stress-test analytics for macroprudential purposes in the euro area*, European Central Bank.

Buch, C., S. Eickmeier, and E. Prieto (2014), "Macroeconomic factors and microlevel bank behavior," *Journal of Money Credit and Banking*, 46(4), 715–751.

Calimani, S., G. Hałaj, and D. Żochowski (2019), "Simulating fire sales in a system of banks and asset managers," *Journal of Banking & Finance*, 105707.

Candelon, B., and A. Sy (2015), "How did markets react to stress-tests?" IMF Working Paper 15/75.

Coen, J., C. Lepore, and E. Schaanning (2019), "Taking regulation seriously: Fire sales under solvency and liquidity constraints," Bank of England Working Paper 793.

Constâncio, V. (2015), "The role of stress-testing in supervision and macroprudential policy," Keynote address at the LSE Conference "Stress Testing and Macroprudential Regulation: A Trans-Atlantic Assessment."

Constâncio, V. (2016), "Principles of macroprudential policy," Speech at the ECB-IMF Conference on Macroprudential Policy, Frankfurt, Germany.

Constâncio, V. (2017), "A new framework for macroprudential stress tests," Conference Paper, available at www.researchgate.net/publication/342610483_Chapter__A_comprehensive_approach_to_macroprudential_stress-testing.

Constâncio, V. (ed.) (2019), "Macroprudential policy at the ECB," ECB Occasional Paper 227.

Cont, R. (2015), "The end of the waterfall: Default resources of central counterparties," *Journal of Risk Management in Financial Institutions*, 8(4), 365–389.

Cont, R., and A. Kotlicki (2019), "Liquidity at risk: joint stress testing of liquidity and solvency," available at www.bostonfed.org/-/media/Documents/events/2020/FR-STRC/presenter-slides/Liquidity-at-Risk-Cont-Kotlicki-Valderrama.pdf.

Covi, G., M. Grope, and C. Kok (2019), "CoMap: Mapping contagion in the euro area banking sector," IMF Working Paper 19/102.

Dees, S., G. Gaiduchevici, M. Grodzicki, M. Gross, B. Hilberg, K. Maliszewski, E. Rancoita, R. Silva, S. Testi, F. Venditti, and M. Volk (2016), "Macroprudential effects of systemic bank stress,"

Macroprudential Bulletin, Issue 2, Chapter 1.

Dees, S., J. Henry, and R. Martin (eds.) (2017), *STAMP€: Stress-test analytics for macroprudential purposes in the euro area*, European Central Bank.

Dent, K., B. Westwood, and S. Segoviano (2016), "Stress testing of banks: An introduction," *Bank of England Quarterly Bulletin*, Q3.

Duarte, F., and T. Eisenbach (2013), "Fire-sale spillovers and systemic risk," FRB New York Staff Report 645.

Duffie, D., and H. Zhu (2011), "Does a central clearing counterparty reduce counterparty risk?" *Review of Asset Pricing Studies*, 1(1), 74–95.

Eisenberg, L., and T. Noe (2001), "Systemic risk in financial systems," *Management Science*, 47(2), 236–249.

European Banking Authority (2018), "EU-wide stress-test 2018," Methodological Note.

European Central Bank (2014, August), "Comprehensive assessment stress test manual," available at www.ecb.europa.eu/pub/pdf/other/castmanual201408en.pdf.

European Central Bank (2016, October), *Macroprudential Bulletin*.

European Central Bank (2017a), "Report on financial structures," available at www.ecb.europa.eu/pub/pdf/other/reportonfinancialstructures201710.en.pdf.

European Central Bank (2017b, November), *Financial Stability Review*.

European Central Bank (2018a, May), *Financial Stability Review*.

European Central Bank (2018b), *Report on financial structure*.

European Central Bank (2020), "Financial integration and structure in the euro area," available at www.ecb.europa.eu/pub/fie/html/ecb.fie202003$\sim$197074785e.en.html.

European Insurance and Occupational Pensions Authority (2014, May), "Insurance and the macroeconomic environment," *Financial Stability Report*.

European Insurance and Occupational Pensions Authority (2017), "Systemic risk and macroprudential policy in insurance," available at www.eiopa.europa.eu/sites/default/files/publications/pdfs/003systemic_risk_and_macroprudential_policy_in_insurance.pdf.

European Securities and Markets Authority (2019a), "Guidelines for money market fund stress test (scenario, redemption, market risk)."

European Securities and Markets Authority (2019b), "Stress simulations for investment funds."

European Securities and Markets Authority (2019c), "Framework for the 2019 CCP stress-test exercise."

European Systemic Risk Board (2015), "Report on systemic risks in the EU insurance sector," available at www.esrb.europa.eu/pub/pdf/other/2015-12-16-esrb_report_systemic_risks_EU_insurance_sector.en.pdf.

European Systemic Risk Board (2018), "Macroprudential provisions, measures and instruments for insurance," available at www.esrb.europa.eu/pub/pdf/reports/esrb.report181126_macroprudential_provisions_measures_and_instruments_for_insurance.en.pdf.

Financial Stability Board (2018), *Global shadow banking monitoring report 2017*.

Gross, M., O. M. Georgescu, and B. Hilberg (2017a), "Credit risk satellite models," in S. Dees, J. Henry, and R. Martin (eds.), *STAMP€: Stress-test analytics for macroprudential purposes in the euro area*, European Central Bank.

Gross, M., J. Henry, and W. Semmler (2018), "Destabilizing effects of bank overleveraging on real activity: An analysis based on a threshold MCS-GVAR," *Macroeconomic Dynamics*, 22(7), 1750–1768.

Gross, M., B. Hilberg, and C. Pancaro (2017b), "Satellite models for bank interest rates and net interest margins," in S. Dees, J. Henry, and R. Martin (eds.), *STAMP€: Stress-test analytics for macroprudential purposes in the euro area*, European Central Bank.

Gross, M., and J. Poblacion (2015), "A false sense of security in applying handpicked equations for stress test purposes," ECB Working Paper No. 1845.

Gross, M., and F. Venditti (2017), "Loan flow satellite models," in S. Dees, J. Henry, and R. Martin (eds.), *STAMP€: Stress-test analytics for macroprudential purposes in the euro area*, European Central Bank.

Gross, M., and D. Żochowski (2017), "Assessing second-round effects using a mixed-cross-section GVAR model," in S. Dees, J. Henry, and R. Martin (eds.), *STAMP€: Stress-test analytics for macroprudential purposes in the euro area*, European Central Bank.

Hałaj, G., and J. Henry (2017), "Sketching a roadmap for systemic liquidity stress tests (SLST)," *Journal of Risk Management in Financial Institutions*, 10(4), 319–340.

Henry, J. (2015), "Macrofinancial scenarios for system-wide stress tests: Process and challenges," in M. Quagliarello (ed.), *Europe's new supervisory toolkit*, Risk.

Henry, J., and C. Kok (eds.) (2013), "A macro stress testing framework for assessing systemic risks in the banking sector," ECB Occasional Paper 152.

International Monetary Fund (2014), "A guide to IMF stress-testing," available at www.elibrary.imf.org/view/books/071/20952-9781484368589-en/20952-9781484368589-en-book.xml.

International Monetary Fund (2019), "Greece—selected issues," IMF Country Report No. 19/341.

Jobst, A., N. Sugimoto, and T. Broszeit (2014), "Macroprudential solvency stress testing of the insurance sector," IMF Working Paper 14/133.

Kok, C., and M. Montagna (2016), "Multi-layered interbank model for assessing systemic risk," ECB Working Paper 1944.

Krug, S., and D. Żochowski (2022), "An agent-based stress testing framework of central clearing," ECB Working Paper Series.

Laliotis, D., and W. Mehta (2017), "Top-down modelling for market risk," in S. Dees, J. Henry, and R. Martin (eds.), *STAMP€: Stress-test analytics for macroprudential purposes in the euro area*, European Central Bank.

Mirza, H., D. Moccero, and C. Pancaro (2017), "Satellite model for top-down projections of banks' fee and commission income," in S. Dees, J. Henry, and R. Martin (eds.), *STAMP€: Stress-test analytics for macroprudential purposes in the euro area*, European Central Bank.

Mirza, H., and D. Żochowski (2017), "Stress-test quality assurance from a top-down perspective," *ECB Macroprudential Bulletin*, No. 3.

Rancoita, E., and B. Hilberg (2017), "Estimating the macroeconomic feedback effects of macroprudential measures—DSGE," in S. Dees, J. Henry, and R. Martin (eds.), *STAMP€: Stress-test analytics for macroprudential purposes in the euro area*, European Central Bank.

Schmieder, C., H. Hesse, B. Neudorfer, C. Puhr, and S. W. Schmitz (2003), "Next generation system-wide liquidity stress testing," IMF Working Paper 12/3.

Wei, L., J. Li, and X. Zhu (2017), "Operational data collection: A literature review," *Annals of Data Science*, 5(2).

A Complex Systems Perspective on Macroprudential Regulation

Stefan Thurner

1 Introduction

One central role of the financial system is to distribute the risks that emerge from the financial demands of the real economy (Schweitzer et al., 2009). Providing funding, liquidity, and other types of investment creates networks of financial contracts (asset-liability or exposure networks). These networks provide an efficient service by sharing and diversifying the burden of economic risks. However, financial networks create additional risk by themselves—*systemic risk*, which is the risk that significant portions of the networks stop functioning and collapse, sometimes in cascades of failing agents that can be triggered by the default of a single institution or bank. The very structure of financial networks, termed *network topology* (see Appendix A for network terminology), plays a crucial role for how cascading failure and financial contagion unfold and lead to financial and maybe even economic crises.

Financial networks largely emerge and organize randomly; they are neither designed nor optimized for stability or resilience in ways such that the possibility for cascading events would be minimized. The central question we address in this contribution is to what extent is it possible to improve the stability and resilience of financial networks under the constraint that the efficiency of financial markets is not affected. In this sense, our objectives are aligned with those of macroprudential regulation: to monitor the levels of systemic risk, to identify the weak spots in the system, to reduce systemic risk, and to ensure that systemic cascading events are minimized in case of individual defaults—and all of this at a minimum loss of market efficiency. In particular, we shall focus on the question of what complexity science can add to the toolkit of traditional stress testing.

The main contribution of complexity science—the science of self-organized, networked, dynamical systems (Thurner et al., 2018)—is that it provides a systematic understanding of network-based systemic risk and the resilience of financial networks. Whereas traditional approaches to stress testing estimate the performance of financial agents under distress in computer-generated scenarios—and thus monitor balance-sheet information; capital buffers; and exposure to interest rate, foreign exchange (FX), and other types of shock (independent of counterparties)—complexity science additionally focuses on the role of exposure networks. It covers three main areas. First, how do contract-based interconnections between financial agents create systemic risk, and how can we systematically estimate the size and likelihood of *systemic* events, such as cascading bank failures and system-wide collapse? Answers to these questions become accessible by a new era of specifically designed network measures in combination with *agent-based models* (ABMs). These enable us to index institutions and financial transactions with respect to their systemic-risk contribution to the financial system, which opens a number of new possibilities for policy interventions. Second, complexity science explores ways to compute (hypothetical) network structures that carry a minimum of systemic risk but that are otherwise equally efficient as the current suboptimal networks. In this way, optimal benchmark levels of systemic risk can be derived and compared with observed risk levels. This allows us to monitor and map systemic risk on a regional and global scale and to relate its changes to economic developments or policy interventions. Third, complexity science studies how the individual behavior of agents leads to aggregate systemic properties, such as efficiency, stability, and resilience. In the context of financial networks, this can be used to devise incentive schemes for agents' behavior that lead toward the formation of systemically optimal network structures.

We argue that the current complexity of networked dynamical financial markets requires a more sophisticated apparatus to design robust policy instruments to monitor, control, and manage systemic risk than traditional stress testing can currently offer. Complex systems science provides natural extensions to current stress testing techniques (Aymanns et al., 2018; see also Chapter 30 in this handbook). In combination with ABMs calibrated to actual data, these extensions offer a significantly more realistic assessment of systemic risk. They will become increasingly more important for macroprudential regulation as more time-resolved data on financial networks become available to central banks and regulators. The policy implications of the novel possibilities that emerge with these technologies are discussed at the end of the chapter.

1.1 Complexity Science and Financial Networks

The science of complex systems provides a conceptual framework for studying the systemic properties of evolving, networked systems, in particular their efficiency, resilience, and proneness to collapse. Complex systems are composed of many elements (e.g., banks) that interact with each other (e.g., through financial contracts). Nodes are characterized by "states" (e.g., balance-sheet information). Interactions (contracts) are represented by links in a time-evolving network (see Appendix A). In the financial system, this could be a network of assets and liabilities in a credit market. Neither states nor interactions are static but evolve in a tightly related manner. States and interactions "update" each other—they co-evolve (Thurner et al., 2018). The understanding of co-evolutionary dynamics in networked systems is crucial to understand the origin of their *systemic properties*, systemic risk in particular. Systemic risk arises whenever the underlying networks can no longer perform their "function." A loss of function is associated with a massive change in the way that links are created between nodes. For example, the fear of cascading events following the default of a big financial player might reduce the willingness of banks to issue loans, and the function of the financial system—to provide credit to the economy—might be lost; a credit freeze results. In more technical language, a "trust evaporation" in the nodes (banks) is followed by "credit-link evaporation" (no more loans are issued). With the combined knowledge of states and networks in a system, as well as how they mutually influence each other, it becomes possible to quantify systemic risk in networks. The concept of network-based systemic risk has been studied in various contexts, including ecosystems (May, 1972), epidemiology (Gross et al., 2006; Pastor-Satorras and Vespignani, 2001), international trade (Klimek et al., 2015), and banking systems (Haldane and May, 2011). The availability of empirical data on banking networks and recent developments in systemic-risk quantification in financial markets (Thurner and Poledna, 2013; Battiston et al., 2012) have triggered a wave of recent empirical research. It has become clear by now that financial systemic risk is tightly related to the detailed network structure of financial exposures. We shall argue that the management of systemic risk is essentially a matter of restructuring financial networks from data. By knowing these, the probability of cascading failure can be computed, and subsequently reduced or even eliminated.

1.2 Economic Risk, Credit Default Risk, Systemic Risk

Economic risk originates from the uncertainty of investments in the real economy, which might not pay off or might fail. The financial system diversifies and absorbs these risks and can be seen as a "service" to share the burden of economic risk. This service should, however, not produce additional risks, such as cascading or systemic risk. As long as systemic risk is endogenously created by financial networks, they are systemically suboptimal. We shall see how systemic risk endogenously emerges in financial networks in Section 2.3, and how they can be optimized with respect to overall systemic risk.

Credit risk is the risk that a borrower fails to fulfill prespecified repayments. It is a risk that emerges between two counterparties: the lender is the sole bearer of credit risk and accounts for it by demanding a risk premium. Lenders charge higher interest rates to borrowers that are more likely to default. Credit risk can be mitigated in various ways (Duffie and Singleton, 2012); the Basel accords provide a corresponding framework (Balen, 2008; Bank for International Settlements [BIS], 1988, 2006). If two counterparties are part of a larger financial system, for example, as nodes in a financial-exposure network, their transactions may affect the financial system as a whole. Lenders are no longer the sole bearers of credit risk, nor does credit risk depend on the financial conditions of the borrower alone. The impact of the default of a borrower is no longer limited to the lender but may affect the creditors of the lender, which in turn may affect their creditors, and so on. Lenders become vulnerable not only to the defaults of their borrowers but also to the defaults of all debtors of their borrower, as well as their debtors. In financial networks, credit risk is no longer limited to two counterparties but becomes *systemic*.

Systemic risk is the risk that the financial system as a whole, or a large part of it, can no longer perform its function as a financial service provider and collapses. It is a result of the interconnected nature of exposures that is created through financial contracts, claims, and liabilities. Systemic risk is created by the possibility of cascading defaults on contracts, triggered by defaults of individual counterparties (Eisenberg and Noe, 2001). These cascades can be potentially large and may be amplified by, for example, deleveraging cycles. Lenders have an incentive to mitigate credit risk but are usually not concerned about systemic risk. Institutions manage their own risks but do not (cannot) consider their impact on the system as a whole (Acharya et al., 2009). Unless institutions are required to internalize the costs of systemic risk, institutions have no incentive to minimize the risks that are borne by the public (Acharya et al., 2012). The management of systemic risk is in the public interest. Various possibilities for the taxation of systemic risk have been discussed (Acharya et al.,

2012, 2013; Adrian and Brunnermeier, 2011; Cooley et al., 2009; Markose et al., 2012; Masciandaro and Passarelli, 2013; Poledna and Thurner, 2016; Thurner and Poledna, 2013; Zlatić et al., 2014).

1.3 Systemic Risk as a Structural Property of Networks

Systemic risk is tightly related to the structure of the underlying exposure (liability) networks (Battiston et al., 2012; Boss et al., 2004, 2005; Roukny et al., 2013; Thurner et al., 2003; Thurner and Poledna, 2013). Empirical data indicate an abundance of scale-free connectivity patterns in interbank liability networks (Bech and Atalay, 2010; Boss et al., 2005; Cajueiro, 2009; Martínez-Jaramillo et al., 2014; Soramäki et al., 2007; Upper and Worms, 2002), see Appendix A. Such patterns imply the existence of hubs. Scale-free degree distributions are also found in overnight markets (Iori et al., 2008), financial flows (Kyriakopoulos et al., 2009), and mutual cross-holdings (Huang et al., 2013).

Time-resolved data on financial network topology are now becoming available to many central banks and regulators. Such data pose a number challenges (see Chapter 10 in this handbook). Financial exposure networks exist not only for interbank credit but also for exposures created through contracts in derivatives, FX, securities markets, and so on. These different types of exposure lead to a collection of networks between the same set of financial agents; they are conveniently integrated into so-called *multi-layer networks*. We will discuss systemic risks in multilayer networks in Section 3.

Systemic risk, when interpreted as a network property, can be quantified by the use of network centrality measures that take contract networks and balance-sheet information (e.g., capitalization) into account (Bardoscia et al., 2015; Battiston et al., 2012; Thurner and Poledna, 2013). In particular, systemic risk can be assigned to nodes (institution-specific systemic risk; Battiston et al., 2012), to individual exposures between counterparties (liability-specific systemic risk; Thurner and Poledna, 2013), and even to individual transactions (contract-specific systemic risk; Poledna and Thurner, 2016). The total systemic risk of a financial multilayer network can be assigned to an entire financial economy and can be monitored over time (Poledna et al., 2015). We discuss details in Section 2.

These new possibilities to assign systemic risks to individual financial agents and even to transactions based on objective data allow us to devise incentive schemes that will lead to rearrangements of exposure networks such that systemic risk on the level of institutions and transactions, as well as globally, can be drastically reduced. That this is indeed possible without increasing economic costs is shown in Section 4.3, where we prove the existence of systemically risk-efficient equilibria in a mathematical theorem. These equilibria operate at the same efficiency levels as the systemically risky ones. We discuss why systemically optimal networks cannot be expected to emerge without explicitly incentivizing systemic risk, for example, through a regulator.

Different financial network topologies are associated with different collapse probabilities and sizes of cascading events. In this context, the question arises as to what is an optimal network topology with respect to systemic risk. We address this question in Section 5. In this spirit, the management of systemic risk reduces to a technical problem of reshaping the topology of financial networks, such that a minimum of systemic risk is obtained while maintaining a maximum of efficiency in financial markets in terms of credit provision, transaction volumes, insurance, and costs. In Section 6 we show how exposure networks can be optimized explicitly (Diem et al., 2020; Pichler et al., 2018). Such optimal networks can be used as benchmarks for monitoring systemic risk levels in actual financial markets.

In this contribution, I focus on recent work done together with O. Bochmann, C. Diem, M. V. Leduc, S. Martínez-Jaramillo, A. Pichler, and S. Poledna.

2 Quantification of Systemic Risk in Financial Networks

Financial systemic risk—the risk that significant parts of the system collapse—is a notion of contagion, or "impact," that starts with the failure of an institution (or set of institutions) and percolates through the financial system and potentially to the real economy (BIS, 2010; De Bandt and Hartmann, 2000). Systemic risk generally emerges and propagates by two mechanisms, either by synchronization of agents' behavior (fire sales, margin calls, herding) or by the interconnectedness of agents through contracts (network-based systemic risk). For a description of basic properties of financial networks, see Chapter 23 in this handbook.

2.1 Defining Systemic Risk without Networks

Information about exposure networks is generally not publicly available but is restricted to central banks. Attempts have been made to quantify systemic risk without taking them into account. Systemic-risk measures were proposed that focus on the statistics of losses in combination with a potential shortfall during periods of synchronized behavior. We mention four important statistical measures: (1) The conditional value-at-risk (CoVaR) is defined as the value at risk (VaR), conditional on institutions being in distress. An institution's contribution to systemic risk is the difference between CoVaR, conditional on the institution being in distress, and CoVaR in its median state (Adrian and Brunnermeier, 2011). It is computed with quantile regressions. (2) The systemic expected shortfall (SES) quantifies the propensity to be undercapitalized, given that the system as a whole is undercapitalized (Acharya and Pedersen, 2012). SES is tightly related to leverage. (3) Systemic risk (SRISK) is closely related to SES, and as such, it depends on the size of an institution, its leverage, and its marginal expected shortfall (Brownlees and Engle, 2012). Finally, (4) the distressed insurance premium (DIP) quantifies the price of insurance against systemic financial distress in a banking system. It is closely related to the expected shortfall (Huang et al., 2012). The advantage of these measures is that they can be computed with openly available data.

If network data are available, however, one can do substantially better in quantifying systemic risk. In this case one is no longer restricted to statistical statements but can relate systemic risk "mechanistically" to individual financial institutions and transactions. The knowledge of the current linking structure in the financial system allows us to become predictive in the sense that the costs of future cascading events can be anticipated by computing what would occur under specific scenarios.

2.2 Financial Networks and Systemic Risk

By now, the regulatory framework of Basel III fully recognizes the importance of networks (BIS, 2010; Georg, 2011). For an overview of network-based financial regulation, see

Enriques et al. (2019). Network-based systemic risk is potentially extremely harmful because of the possibility of cascading failure, where the default of a financial agent may trigger the defaults of others (Boss et al., 2004; Eisenberg and Noe, 2001). Secondary defaults might cause avalanches of defaults percolating through the network that can potentially bring the financial system to a halt through an amplifying deleveraging cascade (Adrian and Shin, 2008; Aymanns and Farmer, 2015; Brunnermeier and Pedersen, 2009; Caccioli et al., 2012; Fostel and Geanakoplos, 2008; Geanakoplos, 2010; Minsky, 1992; Poledna et al., 2014; Thurner et al., 2012). The fear of cascading failure is one of the reasons why distressed institutions are often bailed out (or bailed in) at tremendous public costs. The understanding of the importance of financial networks in the context of systemic risk inspired much recent empirical research on financial networks (Bech and Atalay, 2010; Boss et al., 2004, 2005; Cajueiro et al., 2009; Fricke and Lux, 2014; Greenwood et al., 2015; Iori et al., 2008, 2015; Markose, 2012; Soramäki et al., 2007; Upper and Worms, 2002). It was shown that network topology can be associated with probabilities for systemic collapse (Haldane and May, 2011; Roukny et al., 2013) and a number of network centrality measures (see Appendix A), were suggested to quantify systemic risk (Billio et al., 2012; Boss et al., 2004; Caballero, 2012; Markose, 2012; Minoiu et al., 2013; Puhr et al., 2012). A disadvantage of some of these measures is that the corresponding systemic-risk values for financial institutions cannot be interpreted as the expected losses generated by cascading failure events that would follow the default of the institution. A network measure that improves this problem is the so-called *DebtRank* (Battiston et al., 2012), which inspired recent work on systemic financial risk, including (Aoyama et al., 2013; Bardoscia et al., 2015; Leduc and Thurner, 2017; Poledna et al., 2015; Poledna and Thurner, 2016; Thurner and Poledna, 2013). DebtRank allows us to quantify not only institutional systemic risk (Thurner and Poledna, 2013) but also the total systemic risk in a country (Poledna et al., 2015), as well as the transaction-specific systemic risk of individual contracts (Poledna and Thurner, 2016).

2.3 Systemic Risk of Agents—DebtRank

DebtRank is a quantity that measures the fraction of the total economic value, V, in the network that is potentially affected by the default and distress of a node or a set of nodes in a network (see 7.3). It relates the network topology and the equity of institutions to their systemic-risk contribution. To compute it, we denote the liability (exposure) network at a given time t by $L_{ij}(t)$. For example, think of credit relations in the following. L_{ij} is the liability bank i has toward bank j. For credit networks, $L_{ij} = \sum_k l_{ijk}$ is the sum of all loans, l_{ijk}, that bank j currently extends to bank i. C_i is the equity capital of bank i. If bank i defaults and cannot repay its loans, bank j loses the loans in L_{ij} (j has exposure to i). If j does not have enough capital available to cover the loss, j also defaults. Given the liability matrix, L, and the capital vector, C, that contains the capital of all banks, C_i, the DebtRank for institution i, $R_i(L,C)$, can be computed as a (recursive) function of the network and the capital (see 7.3). More generally, the DebtRank can be computed for sets of institutions, S, that are under distress; we denote that by $R_S(L,C)$.

In Figure 28.1 (top) we show a snapshot of the Austrian interbank network at the end of the third quarter of 2006 (see 7.3). The nodes are the banks in the Austrian banking system; links show interbank lending (credit exposure). Node shading represents the systemic impact

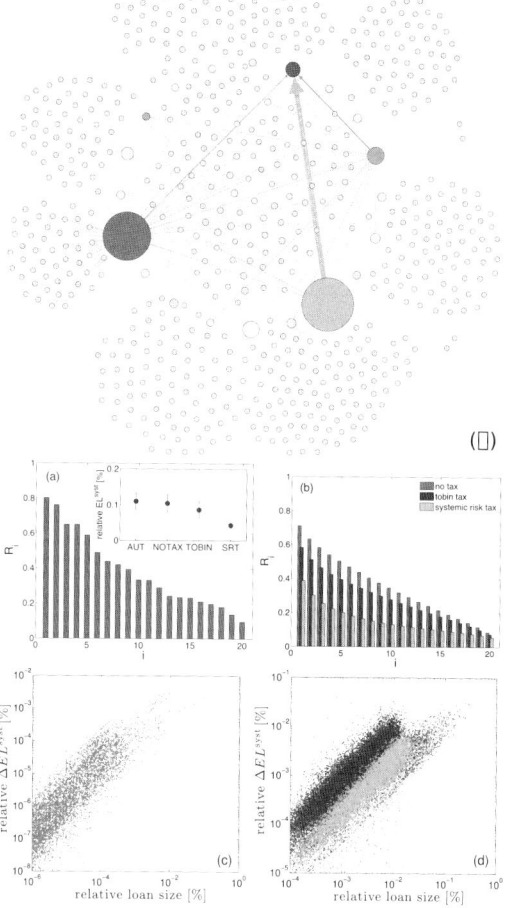

Figure 28.1 (Top) Representation of the Austrian banking sector in 2006. Nodes are banks; links show interbank lending (credit exposure), L_{ij}. Shading represents DebtRank. Systemically important banks are dark gray; systemically irrelevant ones are light gray. Node size is the capitalization of banks. (a) DebtRank, R_i, of the 20 largest banks, the most risky being to the very left. (b) ABM results for R_i: without tax (dark gray), with Tobin tax (gray), and with systemic risk tax (SRT; light gray) SRT drastically reduces the systemic risk contributions of individual banks by about half and more. The situation without SRT resembles the empirical distribution. Inset: Expected systemic loss from all banks for the data, along with the three model outcomes. (c) Marginal effects on expected systemic loss, $\Delta \mathrm{ESL}^{kl}$, of individual interbank liabilities L_{kl} versus the relative size of interbank loans in logarithmic scale. Every data point represents an interbank liability, L_{kl}^{data}. The loan size captures the credit risk for lenders, whereas $\Delta \mathrm{ESL}^{kl}$ is the systemic risk of the liability. (d) Marginal effects for the ABM simulations. SRT reduces transaction-specific systemic risk by at least an order of magnitude. Source: From Poledna and Thurner (2016), reprinted by permission of the publisher Taylor & Francis Ltd., www.tandfonline.com.

defined by DebtRank (including the direct loss of i): systemically important banks are dark gray; systemically irrelevant ones are light gray. Node size represents the capitalization of banks; the widths of links correspond to exposure sizes, L_{ij}. We define the *SR profile* of a country at time t as the rank-ordered DebtRank values for all financial institutions in a country. The SR profile is the distribution of systemic impact across the institutions in a country. Institutions with a high SR level are shown to the left. Panel (a) of Figure 28.1 shows the SR profile for the 20 largest (with respect to capitalization) banks of the Austrian banking sector at the end of the third quarter of 2006.

2.4 Expected Systemic Loss—Total Systemic Risk

The meaning of DebtRank as the fraction of the total economic value, V, that is potentially affected as a consequence of distress in the network allows us to define the *expected systemic loss*, $ESL(t)$, for an entire economy at a given time. Expected systemic loss estimates the expected size of losses from systemic events in the entire network in the case of no interventions, such as bailouts or other resolution schemes (Poledna and Thurner, 2016). To compute the expected systemic loss, first consider the situation where only one institution i defaults, and the other $N - 1$ survive. The expected loss is ESL_i (one default) $= V \cdot p_i \cdot (1 - p_1) \cdots (1 - p_{i-1}) \cdot (1 - p_{i+1}) \cdots (1 - p_b) \cdot R_i$, where p_i is the default probability for institution i, and $(1 - p_j)$ is the probability of survival for j. The general case occurs when we consider joint defaults, meaning that a set of institutions, S, is in distress. Taking into account *all* possible combinations of initially defaulting institutions, the expression for the expected systemic loss for an economy with N institutions is obtained:

$$ESL = V \sum_{S \in \mathcal{P}(B)} \prod_{i \in S} p_i \prod_{j \in B \setminus S} (1 - p_j) R_S. \tag{28.1}$$

Here, $\mathcal{P}(B)$ is the power set of the set of financial institutions, B, and R_S is the DebtRank of the set of nodes, S, that are initially in distress. Note that this is an approximation and assumes independent defaults. If the joint-default probabilities of multiple banks are known, these should be used in a similar expression. However, reasonable joint probabilities are even harder to obtain than individual-default probabilities, p_i. Equation (28.1) is computationally feasible only for relatively small numbers of financial institutions and is impossible to compute for large system sizes. Poledna et al. (2015) suggest a practical approximation for Equation (28.1), where we assume $R_S \sim \sum_{i \in S} R_i$, and obtain

$$ESL \sim V \sum_{i=1}^{N} p_i R_i. \tag{28.2}$$

It is valid if the default probabilities are small ($p_i \ll 1$) and interconnectedness is low. The approximation works extremely well for realistic scenarios (Poledna et al., 2015). A central idea here is to separate the individual default risk, p_i, from contagion risk. Contagion risk is the risk that a default by one institution leads to defaults of other institutions; it is captured by DebtRank.

   Note the difference between expected systemic loss, ESL, and the expected loss (or credit risk), EL_i, for a single institution i, which is

$$EL_i(t) = \sum_{j=1}^{N} p_j \, LGD_j \, L_{ji}(t), \tag{28.3}$$

where LGD_j is the loss given default for j, and L_{ji} is the exposure at default of i to j.

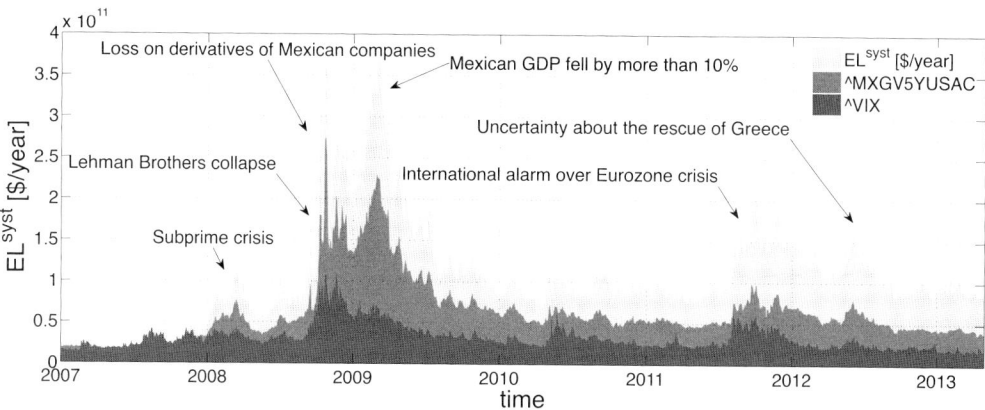

Figure 28.2 Total systemic risk in Mexico from 2007 to 2013. Expected systemic loss (ESL) (in pesos [Mex$] per year; light gray). In 2013 ESL is about four times higher than before the crisis. For comparison, two market-based risk indices, the Volatility Index VIX (black) and the credit default swap (CDS) spreads of 5-year Mexican government bonds in USD (MXGV5YUSAC; dark gray) are shown. All indicators are scaled to the same value at the first time point. Market-based indices relax to precrisis levels, whereas ESL does not. ESL is determined to a large extent by network topology, which cannot be observed and priced by markets. In this sense markets may consistently underestimate systemic risk. Source: From Poledna et al. (2015), with permission from Elsevier.

Figure 28.2 shows the time series of the ESL for Mexico for 2007–2013 from the combined exposures (see Section 3 and 7.3). In Mexico there are no quotes for CDS spreads for banks, and ratings are only obtainable for specific securities. This makes it difficult to estimate the default probabilities for individual banks, p_i. Here we approximate them with sovereign-default probabilities; for details, see Poledna et al. (2015). For comparison, we show two market-based risk indicators, the VIX and CDS spreads. We indicate important historical events. Even though VIX and ESL are correlated, the VIX and spreads return quickly to precrisis levels, whereas ESL does not. Note that ESL in 2013 is approximately a factor of four higher than before the crisis. The fact that network effects are not visible to market participants indicates that markets might drastically underestimate systemic risk.

2.5 Systemic Risk of Individual Transactions—Marginal Systemic Risk

The systemic-risk contribution of individual exposures to total systemic risk, the *marginal systemic risk*, ΔESL, was defined by Poledna and Thurner (2016). To compute it, we define an individual exposure, X_{kl}, as a matrix that contains precisely one nonzero element: the exposure size between banks k and l. The marginal systemic-risk contribution of exposure X_{kl} to the ESL is the difference of total expected systemic loss:

$$\Delta\text{ESL}^{kl} = \sum_{i=1}^{N} p_i \left[V(L + X_{kl}) R_i(L + X_{kl}, C) - V(L) R_i(L, C) \right], \tag{28.4}$$

where $R_i(L + X_{kl}, C)$ and $V(L_{ij} + X_{kl})$ are the DebtRank and the total economic value of the liability network including the specific exposure, X_{kl}, respectively. A positive marginal systemic risk, ΔESL^{kl}, means that exposure X_{kl} increases total systemic risk. Similarly, this method allows us to compute the systemic-risk contribution of individual contracts that build up the exposure X_{kl}. In this way the marginal systemic-risk contribution of every existing exposure, as well as every hypothetical transaction, can be computed.

Using the same notation, the marginal contribution to credit risk of an individual exposure, X_{kl}, is the increase in the credit risk of the bank with the additional exposure (risk taken by lender):

$$\Delta \text{EL}^{\text{credit}} = \sum_{i=1}^{N} \text{EL}_i^{\text{credit}}(L + X_{kl}) - \text{EL}_i^{\text{credit}}(L). \qquad (28.5)$$

Here $\text{EL}_i^{\text{credit}}(.)$ means that $\text{EL}_i^{\text{credit}}$ is computed with equation (28.3) with the network in the argument.

In panel (c) of Figure 28.1 the marginal systemic risk from equation (28.4) is presented for all Austrian individual interbank liabilities, L_{kl}^{data}, as a function of the relative size of the interbank loans. Every data point is an individual interbank liability, L_{kl}^{data}, from bank k to l in the first quarter of 2006. The loan size represents the maximum credit risk for lenders. ΔESL^{kl} is the marginal systemic risk of the liability. See also panel (b) of Figure 28.4 for Mexican interbank data. The situation is very similar.

2.6 Systemic Risk Is an Externality

If the marginal systemic risk of an individual contract is less than or equal to its associated credit risk, $\Delta \text{ESL} \leq \Delta \text{EL}^{\text{credit}}$, a default of the exposure will affect the involved parties (for credit that would be the lender) but would not involve third parties. For transactions, where $\Delta \text{ESL} > \Delta \text{EL}^{\text{credit}}$, third parties also might be affected by the default because the loss is larger than the value of the asset. The obvious abundance of transactions where $\Delta \text{ESL} > \Delta \text{EL}^{\text{credit}}$ in panel (b) of Figure 28.4 is a clear indicator of the existence of an incentive problem: systemic risk generated by bilateral exposures appears to be an externality. This means that those counterparties that create systemic risk might not be able to pay for the damage created in a systemic event. These costs have to be borne by third parties—generally the public.

3 Financial Markets Are Multilayer Networks

Typically, financial institutions engage in more than one type of financial contract. Banks not only operate in the credit market but also in securities, derivatives, or foreign exchange. They are also invested in stocks, bonds, and other financial assets. Each of these contracts and assets might create bilateral exposures for the involved institutions. We have seen in Section 2 that these can be represented as links in an exposure matrix, L_{ij}. For example, institution j issues securities that are bought by institution i. By holding these securities, i exposes itself to the possibility that institution j will fail to fulfill the contract. Similarly, foreign exchange transactions may generate large exposures between banks. Here exposures might, for example, emerge through settlement risks. Issuing and trading financial derivatives can create yet other types of exposure. All exposures can contribute substantially to the total systemic risk of an institution, as we shall see later in the chapter. To arrive at an assessment of total systemic risk, it is practical to represent the different types of exposure in different layers of a multilayer network (see Appendix A). In this framework, the total systemic risk of an institution can be decomposed into systemic-risk contributions that emerge in the different markets. Note that exposures from different layers are not necessarily independent, and correlations may exist. In that case, one cannot expect that the total systemic risk is a linear superposition of the systemic risk of the different layers. This fact is discussed by Poledna et al. (2015). For notation, we label the different exposure types by α. The exposure size of type α between institution i and j at time t is denoted by $L_{ij}^{\alpha}(t)$. As before, we use the convention that L_{ij}^{α} represents the "liabilities" i has toward j. The transpose of matrix L gives the exposures. Superscript $\alpha = 1, 2, 3, \ldots$ labels the exposure layers: "credit," "derivatives," "securities," "foreign exchange," "deposits and loans," "overlapping portfolio risk," "overnight liquidity," and so on. Note that in many countries, the transaction data of financial institutions are available to central banks. From these, bilateral exposures can be obtained and represented in a multilayer network, L_{ij}^{α}.

3.1 Systemic Risk in Multilayer Networks

Although exposures in different layers arise from very different types of financial risk, exposures in all layers should have the same meaning as the total loss that might arise for an institution as the consequence of the default of another. If this is ensured, DebtRank values can be computed for each layer of a multilayer network, L_{ij}^{α}. The economic value in each layer is given by $v_i^{\alpha} = L_i^{\alpha} / \sum_j L_j^{\alpha}$, where $L_i^{\alpha} = \sum_j L_{ji}^{\alpha}$. DebtRank can also be calculated for the *combined liability network*, $L_{ij}^c = \sum_{\alpha} L_{ij}^{\alpha}$; we denote it by R_i^{comb}. The corresponding total economic value, $V^c = \sum_i L_i^c$, is given by total interbank assets in all layers combined. To compare R_i^{α} between different layers, it should be defined as a percentage of V^c. The appropriately normalized DebtRank for layers α is thus

$$\hat{R}_i^{\alpha} = \frac{V^{\alpha}}{V^c} R_i^{\alpha}, \tag{28.6}$$

where $V^{\alpha} = \sum_i L_i^{\alpha}$ is the total economic value of the interbank assets in layer α.

3.2 Different Types of Direct Exposure

We discussed the credit layer in Section 2. Here we present additional layers that we studied empirically for Mexican interbank markets on a daily basis (Poledna et al., 2015) (see 7.3).

Deposits and loans. Exposures arise from interbank deposits and loans and from credit lines extended for settlement purposes. The definition of exposure is straightforward. $L_{ij}^{\text{dep}}(t)$ is obtained by adding all deposits and loans between bank i and j. Usually the gross exposure (not netted) is the quantity of interest.

Securities cross-holdings. Exposures emerge from cross-holding of securities between banks, securities lending, securities used for collateral, and the trading of securities. Cross-holding means that j holds securities issued by i. Typically, one uses gross exposures because security contracts must be honored, even when the counterparty defaults. Cross-holdings gross exposures, $L_{ij}^{\text{sec}}(t)$, are obtained by adding all securities cross-holdings between bank i and j.

Derivatives. Exposures arise from the valuation of different types of derivative transaction, such as swaps, forwards, options, or repo transactions. For each type, the contract is valued, and the resulting net exposure (at the contract level) is assigned to the corresponding bank. It is reasonable that derivative exposures are netted by each type of contract. For instance, options with the same underlying asset are all added on each side. The exposure is then assigned to the counterparty with the positive net position. This is done for each type of derivative with the same underlying, and the resulting net exposures are added to obtain $L_{ij}^{der}(t)$.

Foreign exchange. FX transactions create exposures associated with settlement risk—the risk that a counterparty does not pay its obligations at the time of maturity. If banks settle FX transactions between themselves by using a clearance service, there might be no exposure. Otherwise, the exposure, $L_{ij}^{fx}(t)$, includes both foreign currency receivable and foreign currency payable between bank i and bank j.

Figure 28.3 shows the exposure layers of the Mexican financial interbank network on a day in September 2013. Exposures from derivatives (light gray), securities cross holdings

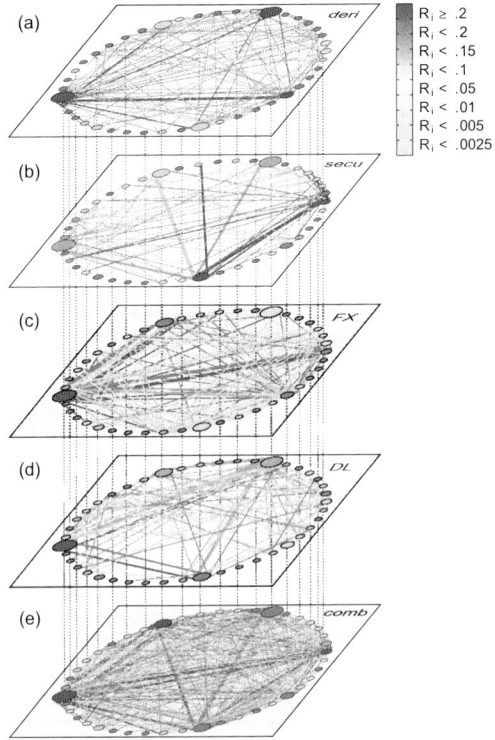

Figure 28.3 Layers of the multilayer network of Mexican financial markets on a specific day in September 2013. (a) Exposure network from derivatives, (b) securities cross-holdings, (c) FX exposures, (d) deposits and loans, and (e) the combined network, $L_{ij}^{c}(t)$. Nodes (banks) are shaded according to the systemic risk, R_i^{α}, in the respective layer: systemically important banks are dark gray; systemically safe banks are light gray. Node size represents total assets, link width exposure size. Source: From Poledna et al. (2015), with permission from Elsevier.

(lightest gray), foreign exchange (medium gray), and deposits and loans market (dark gray) are shown. Nodes represent banks; node size is proportional to their total assets. Nodes i are shaded according to their systemic risk, R_i^α. Systemically risky banks are represented by dots in dark gray, systemically irrelevant ones are light gray. Link width represents exposure size. The total exposures in layers α equal 2 to 4 is $\sum_{i,j} L_{ij}^\alpha(t) \sim 5\text{E}10$ Mex\$; they are similar in size. The total exposure in the derivatives layer is smaller, $\sum_{i,j} L_{ij}^{\text{der}}(t) \sim 1\text{E}10$ Mex\$. Derivative exposures also contain exposures from repo transactions. The respective amounts are relatively small (< 2 percent) because repo transactions involve collateral. In panel (e) of Figure 28.3, the combined exposures, $L_{ij}^c(t)$, are shown.

Panel (a) of Figure 28.4 shows the SR profile for the different layers as shaded bars, $\hat{R}_i^\alpha$, and the combined exposures, R_i^c (black line), for a day in September 2013. Different systemic-risk contributions from the different layers reflect the different trading strategies of banks. Some smaller banks have a systemic impact in the securities market only. FX and securities markets dominate the contributions. The systemic impact of the combined layers (line) is always larger than the sum of the layers separately, $R_i^c > \sum_\alpha \hat{R}_i^\alpha$, for all banks. In panel (b) of Figure 28.4, we compare the marginal systemic risk, ΔESL, for individual exposures (y-axis) to credit risk ΔEL (x axis). Every exposure between Mexican banks between January 2, 2007, and May 30, 2013, is represented as a data point. Layers are distinguished by shading. $\Delta\text{EL}^{\text{syst}} > \Delta\text{EL}^{\text{credit}}$ for the vast majority of transactions. We verified that this cannot be explained by exposure size relative to equity capital or by capital ratios. This demonstrates that transaction-specific systemic-risk contributions depend not only on the two involved parties but also on the conditions of other nodes in the network. Note that small and medium-sized liabilities can have contributions that vary by three orders of magnitude. Compare to the Austrian interbank credit data in panel (c) of Figure 28.1.

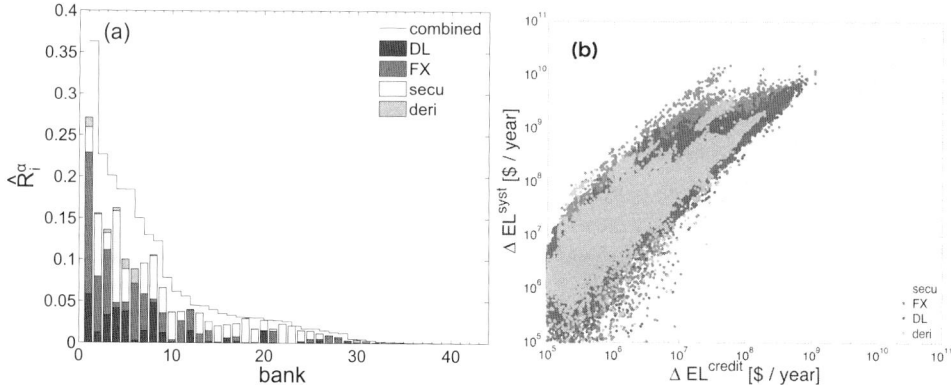

Figure 28.4 (a) Systemic-risk profile for different exposure layers in the multilayer network. Normalized DebtRank values, $\hat{R}_i^\alpha$, are stacked for each bank. (b) Marginal systemic risk, ΔESL, versus credit risk, $\Delta\text{EL}^{\text{credit}}$, for individual exposures. Points represent individual interbank liabilities, L_{ij}^α, on a given layer (indicated by shading). Data are aggregated from Mexican banks over all days from January 2, 2007, to May 30, 2013. Note that $\Delta\text{ESL} > \Delta\text{EL}^{\text{credit}}$ for almost all contracts. This means that defaults would not affect the "lending party" alone, but also third parties: Systemic risk is an externality. Source: From Poledna et al. (2015), with permission from Elsevier.

3.3 Indirect Exposures—Overlapping Portfolios

Although the network layers discussed in the previous section represent direct financial exposures, one important layer is one created by the effects of holding similar portfolios—overlapping portfolio risk. Financial stress on this layer may appear when an institution rapidly sells parts of its asset holdings, which then devaluate as a result of market impact. If the same assets are held by other institutions, their portfolios also suffer losses. This may trigger further selling and cause a fire-sale cascade that devaluates portfolios even further (Cont and Schaanning, 2017; Thurner et al., 2012).

Following Pichler et al. (2018), to compute indirect exposures from overlapping portfolios, consider a bipartite asset-holder network, A, with two types of node. One represents N institutions (e.g., banks), labeled by $i = 1, \ldots, N$, the other, K different assets, labeled by $k = 1, \ldots, K$. If bank i holds asset k, a link exists between i and k, with the weight, A_{ki}, being the value of the investment in monetary units. Even though institutions may not be linked directly, i can experience an exposure toward j if they both hold the same (not-perfectly-liquid) asset. We assume a linear price impact as in Bouchaud (2010) and Kyle (1985), where price change, $\Delta p_k(z) = z/M_k$, is linear in the net signed trading volume, z in monetary units. Here M_k is a parameter that estimates the liquidity of a security. It is defined such that selling (buying) the value $M_k/100$ of security k, moves the price down (up) by 1 percent. Following the approach of Braverman and Minca (2014), Guo et al. (2016), and Cont and Schaanning (2017), it can be estimated by

$$M_k = c \ \mathrm{ADV}_k/\sigma_k, \tag{28.7}$$

where $c > 0$ is a scaling parameter: ADV_k is the average traded daily volume (in monetary units); and σ_k is the empirical (not the implied) volatility of a particular security, measured as the standard deviation of daily log-returns. We set $c = 0.4$, as suggested by Cont and Schaanning (2017). The value of asset k in the portfolio of bank i is $A_{ki} = \beta_{ki} p_k$, where β_{ki} is the number of shares of k, and p_k is the corresponding price. Consider institution j that holds the same asset k. The maximum loss that j can experience from i's sales of k is $A_{kj} \frac{A_{ki}}{M_k}$. The overall exposure of j to i is

$$w_{ij} = \sum_{k=1}^{K} A_{kj} A_{ki} / M_k. \tag{28.8}$$

The systemic risk of overlapping portfolios is now obtained by using w_{ij} in the impact matrix that is used in the definition of DebtRank (see 7.3; Pichler et al., 2018):

$$\tilde{W}_{ij} = \min\left(1, \frac{w_{ij}}{C_j}\right). \tag{28.9}$$

The relative economic value in an overlapping portfolio setting is given by

$$\tilde{v}_i = A_i / \sum_{k,j} A_{kj}, \tag{28.10}$$

where $A_i = \sum_k \beta_{ki}\, p_k$ is the portfolio value of bank i. With these quantities, DebtRank is computed. Poledna et al. (2018) find that systemic risk from overlapping portfolios in the Mexican securities markets accounts for a substantial portion of the total systemic risk in the markets. Focusing only on direct exposures underestimates total systemic risk levels by up to 50 percent.

4 Managing Systemic Risk in Financial Networks

An effective way to reduce network-based systemic risk is to rearrange the topology of the underlying exposure networks (Thurner and Poledna, 2013; Poledna and Thurner, 2016). We shall now discuss this in more detail. First, we show that there indeed exist optimal networks with respect to systemic risk. Then we demonstrate how optimal networks would be structured and how much systemic risk can be eliminated in the best of all worlds. We finally propose a way to incentivize institutions such that systemically optimal network structures would emerge in a self-organized and robust way.

We have seen in Section 2 that network-based systemic risk is an externality. Losses associated with individual transactions that would occur in the case of a systemic event may exceed the capability to cover them by those who created them. Then third parties need to cover these excess losses. We have shown that conventional bilateral contracting mechanisms fail to internalize the systemic risk externality that it generates (Leduc and Thurner, 2017; Thurner and Poledna, 2013). In a bilateral loan contract, the lender only considers the default risk of the borrower, whereas a borrower is only concerned with the interest rate. Neither party has incentives to internalize the systemic-risk externality created by the contract.

To incentivize systemic-risk awareness, the basic idea is to make systemically risky transactions more expensive. We have seen in Figures 28.1 (panel [c]) and 28.4 (panel [b]) that for a given contract size, there exist transactions that carry systemic risk that is approximately three orders of magnitude higher than others. If transactions with large marginal systemic loss were avoided, it is reasonable to assume that the total systemic risk can be drastically reduced. In the following, we shall see that this is indeed the case. To avoid transactions with high marginal systemic risk, we propose to tax them with a *systemic risk tax* (SRT). Such a tax would be computed by a regulator with the means to compute the ΔESL of all requested transactions. Before a transaction is made—say bank i wants to take a loan from bank j—the regulator computes ΔESL^{ij} for that transaction and charges a tax that is proportional to its marginal systemic risk. This incentive scheme leads to a reduction of systemic risk in a self-organized manner: market participants looking for credit will try to avoid the tax by looking for credit from other institutions that do not increase systemic and are thus tax-free. We shall see that as a result, the network rearranges toward a topology that, in combination with the financial conditions of individual institutions, will lead to a de facto elimination of systemic risk (Poledna and Thurner, 2016). This is due to the fact that the likelihood of cascading failure is suppressed in the newly emerging topology of the exposure networks. Note that the tax should not be paid; it just serves as an incentive to avoid systemically risky transactions and thus rearrange exposure networks toward systemically optimal configurations. In the following, we first show that under an SRT, there exists an equilibrium that is indeed optimal with respect to systemic risk. We then use an ABM to estimate the fraction of systemic risk that can be eliminated.

4.1 Matching Markets and Expected Systemic Risk

The idea of the following proof is that under the scheme of an SRT, there exists a systemically risk-efficient equilibrium matching of lenders and borrowers in financial markets. This matching then constitutes an optimal network. The corresponding theorem is phrased in the context of credit between borrowers and lenders; however, the result is more general and also applies to exposures of different kinds. We follow Leduc and Thurner (2017) and assume a slightly simplified definition of DebtRank, which, however, does not change the generality of the theorem.

Let us assume that at any time, $t \in \{0, 1, 2, \ldots\}$, a regulator with information about the current interbank credit exposure matrix, L_t, would like to control the formation of the interbank network by influencing the matching of credit between the sets (groups) of potential lenders $\mathcal{L}_t$ and borrowers $\mathcal{B}_t$ to achieve a desired level of systemic risk. To define a *matching*, we denote by **P** the set of preferences:

$$\mathbf{P} = \{P_\beta^a, P_\beta^b, P_\beta^c, \ldots, P_\lambda^d, P_\lambda^e, P_\lambda^f, \ldots\}. \tag{28.11}$$

$P_\beta^j(a) \succ P_\beta^j(b)$ means that bank j prefers to borrow from a rather than from b. Similarly, $P_\lambda^i(d) \sim P_\lambda^i(e)$ means that i is indifferent between lending to d or to e. Subscripts β and λ indicate borrowing and lending, respectively.

The interbank market for liquidity at time t is denoted by the triple $(\mathcal{B}_t, \mathcal{L}_t, \mathbf{P})$. The outcome of the interbank market at time t is a bipartite network representing a set of matches between potential lenders and borrowers. All banks have a reservation rate, $\bar{r}_j$, at which they prefer not to borrow from a bank i that offers a rate (to j), $r_{ij} > \bar{r}_j$, that is too high. In this case, banks will not trade. A matching, μ_t, at time t is a one-to-one map from the set of banks onto itself, such that for any borrower $b \in \mathcal{B}_t$, if $\mu_t(b) \neq b$, then $\mu_t(b) \in \mathcal{L}_t$, and for any $l \in \mathcal{L}_t$, if $\mu_t(l) \neq l$, then $\mu_t(l) \in \mathcal{B}_t$. For example, let the set of banks be $\{1, 2, \ldots, 9\}$, $\mathcal{L}_t = \{1, 2, 3, 4\}$ and $\mathcal{B}_t = \{5, 6, 7, 8, 9\}$. Then a matching could be

$$\mu_t = \begin{matrix} 4 & 1 & 2 & 3 & (5) \\ 6 & 7 & 8 & 9 & 5 \end{matrix}, \tag{28.12}$$

so that $\mu_t(4) = 6$ and $\mu_t(6) = 4$, and thus bank 4 lends to bank 6, $\mu_t(1) = 7$ and $\mu_t(7) = 1$, and bank 1 lends to bank 7, and so on. Note that $\mu_t(5) = 5$, and no one lends to bank 5, which remains unmatched. For an example of how different matchings influence the expected systemic loss, see Figure 28.5.

4.2 Transaction-Specific Tax

How can the regulator now incentivize banks such that they form a low-ESL equilibrium matching, $\hat{\mu}_t^*$? This is achievable by means of a transaction-specific tax, which has the effect of reordering the *borrowers'* preferences for the lenders.

Assume $\mathcal{T} = \{\tau_{ij}\}$, where $i \in \mathcal{L}_t$ and $j \in \mathcal{B}_t$. $\mathcal{T}$ is a $|\mathcal{L}_t| \times |\mathcal{B}_t|$ matrix of transaction-specific taxes, with $\tau_{ij} \geq 0$. τ_{ij} is the mark-up that is applied to the interest rate paid by bank j when it borrows from bank i. The borrowing bank j pays $r_{ij}^{\mathcal{T}} = r_i + h_{ij} + \tau_{ij}$ instead of $r_{ij} = r_i + h_{ij}$. Here r_i is the interest rate at which bank i can lend excess liquidity on

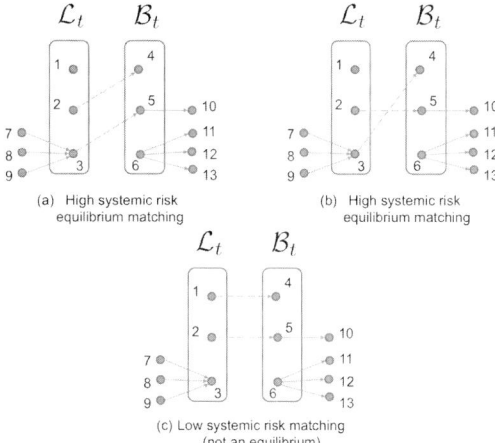

Figure 28.5 Example for systemic risk under different matchings. Assume that banks default exogenously during time span $[t, t + S]$ with the same independent probability, $p_i(S)$, and $C_t^i = \$50$ million for all banks, and each link represents a \$60-million loan. The exogenous failure of any bank will trigger the failures of all banks down its path. Then the *total* failure probabilities follow $\bar{p}_6(t, S) > \bar{p}_5(t, S) > \bar{p}_4(t, S)$. Further, assume $r_3 < r_2 < r_1$, so that borrowers prefer lending from bank 3 over bank 2 over bank 1, and consider $r_{3,4} < r_{2,4} < r_{1,4} < \bar{r}_4$ and $r_{3,5} < r_{2,5} < r_{1,5} < \bar{r}_5$, so that borrowing banks 4 and 5 are willing to borrow from any lending bank, whereas $\bar{r}_6 < r_{3,6} < r_{2,6} < r_{1,6}$, so that bank 6's higher default risk makes borrowing too expensive. (a) and (b) show two possible equilibria with high ESL. The ESL in panel a is obviously higher than that in panel b. (c) A low-ESL matching, achieving the same transaction volume. Source: From Leduc and Thurner (2017), with permission from Elsevier.

the interbank market, and h_{ij} is a risk premium that reflects bank i's view on the probability that bank j will default on the loan. For details, see Leduc and Thurner (2017). Under the tax $\mathcal{T}$, a borrower's expected payoff is

$$\Pi^j_{\beta, \mathcal{T}}(i) = 1 - \frac{1}{(1 + r_j)^S}(1 + r_i + h_{ij} + \tau_{ij})^S. \tag{28.13}$$

S is the number of time steps to maturity (counted from t). The lender's expected payoff is left unchanged. Tax $\mathcal{T}$ effectively reorders the preferences of each borrower over the set of lenders. This allows the regulator to create heterogeneous preferences; that is, each borrower can now have a different preference list, P^j_β. The tax, τ_{ij}, does not affect the lenders' but only the borrowers' behavior.

Assume that the borrowers' default probabilities, $p_j(t, S)$, for $[t, t + S]$, and the lenders' lending rates, r_i, are known to the regulator. From this, the risk premia, h_{ij}, can be computed and also the banks' payoffs. Now, by setting $\mathcal{T}$, the regulator can reorder each borrower j's preference list, P^j_β, such that *any* desired matching, $\hat{\mu}_t$, is obtainable as a unique equilibrium. Because this tax allows the regulator to require a systemic-risk-efficient equilibrium, we call this tax a *systemic risk tax* (SRT) for details, see Leduc and Thurner (2017).

The special case where $\mathcal{T}_{ij} = \kappa$ is a constant for all i and j reduces to a Tobin-like tax, which does not allow a regulator to reorder the preference lists of borrowers. Borrowers have homogenous preferences, but some transactions now become too expensive, which might result in a reduction of transaction volume and loss of efficiency. For a proof of this statement, see Leduc and Thurner (2017).

4.3 Fundamental Theorem on Network-Based Systemic-Risk Management

We now discuss a theorem stating that a regulator can always choose a transaction-specific tax, $\mathcal{T}$, to achieve lower systemic risk *without sacrificing* transaction volume (Leduc and Thurner, 2017). A Tobin-like tax, on the other hand, indiscriminately taxes every transaction equally. This has the effect of reducing the set of lenders with which a borrower is willing to trade, *without* reordering preferences. It does reduce transaction volume by reducing the number of possible matchings.

Theorem. Systemic risk under the SRT. Let $(\mathcal{B}_t, \mathcal{L}_t, \mathbf{P})$ be a market for liquidity at time t. Given a net exposure matrix, L_{t-1}, at time $t-1$, let $L_t^{*,\mathcal{T}}$, $L_t^{*,\kappa}$, and L_t^* be the net exposure matrices at time t with an SRT, $\mathcal{T}$, with a Tobin-like tax, κ, and without tax by the equilibrium matchings, $\mu_t^{*,\mathcal{T}}$, $\mu_t^{*,\kappa}$, and μ_t^*, respectively. Let $\mathcal{E}_t$ denote the set of all possible stable equilibrium matchings at time t; then:

- For any $\mu_t^* \in \mathcal{E}_t$ such that the trading volume $\mathrm{Vol}(\mu_t^*) = v$, there exists a $\mathcal{T}$, such that the expected systemic loss $\mathrm{ESL}(L_t^{*,\mathcal{T}}, C_t) \leq \mathrm{ESL}(L_t^*, C_t)$, and $\mathrm{Vol}(\mu_t^{*,\mathcal{T}}) \geq \mathrm{Vol}(\mu_t^*)$. In particular, there exists a $\mathcal{T}$ such that $\mu_t^{*,\mathcal{T}}$ is systemic-risk efficient.
- For any $\mu_t^{*,\kappa} \in \mathcal{E}_t^\kappa$ such that $\mathrm{Vol}(\mu_t^{*,\kappa}) = v$, there exists a $\mathcal{T}$ such that $\mathrm{ESL}(L_t^{*,\mathcal{T}}, C_t) \leq \mathrm{ESL}(L_t^{*,k}, C_t)$, and $\mathrm{Vol}(\mu_t^{*,\mathcal{T}}) \geq \mathrm{Vol}(\mu_t^{*,k})$.

For the proof, see Leduc and Thurner (2017). The theorem implies that for any outcome of a market for liquidity under a bilateral contracting mechanism, one can design an SRT that achieves lower systemic risk at potentially even higher transaction volume, whereas a Tobin tax sacrifices transaction volume without having any impact on network topology. For an example, see Figure 28.6.

To address the question of how to set the tax at a particular time t, the regulator first specifies a desired transaction volume, v, and then computes $\hat{\mathcal{T}}$ by solving the optimization problem

$$\hat{\mathcal{T}} \in \underset{\mathcal{T}:\mathrm{Vol}(\mu_t^{*,\mathcal{T}})=v}{\mathrm{argmin}} \quad \mathrm{ESL}(L_t^{*,\mathcal{T}}, C_t), \tag{28.14}$$

where C is the capitalization vector, and $L_t^{*,\mathcal{T}}$ is the net exposure matrix formed as a result of the equilibrium matching, $\mu_t^{*,\mathcal{T}}$. $\mathrm{ESL}(\cdot)$ is the one-period-ahead expected systemic loss. Note that there can be multiple $\mathcal{T}$ values that all lead to the same matching, $\mu_t^{*,\mathcal{T}} = \hat{\mu}_t$. A meaningful way is to tax deviations from the desired equilibrium matching, $\hat{\mu}_t$, proportional to the amount of systemic risk that it generates, that is, the marginal systemic risk (Poledna and Thurner, 2016). The desired equilibrium remains untaxed. This means that $\forall j \in \mathcal{B}_t$, the regulator sets $\mathcal{T}_{\hat{\mu}_t(j),j} = 0$, and

$$\mathcal{T}_{ij} = r_{\hat{\mu}_t(j),j} - r_{ij} + \epsilon + \zeta \max(0, \Delta\mathrm{ESL}^{ij}), \tag{28.15}$$

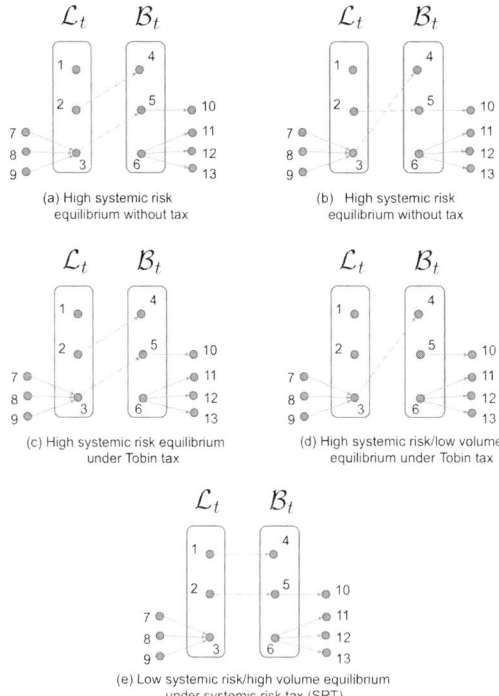

Figure 28.6 Example of how an SRT leads to a systemic-risk-efficient equilibrium. Panels (a) and (b) show two possible equilibria without tax, as in panels (a) and (b) of Figure 28.5. Panels (c) and (d) show the situation for a Tobin-like tax, κ, that causes $r^K_{15} > r^K_{25} > \bar{r}_5$, and $r^K_{35} < \bar{r}_5$, so that the transaction between lender 2 and borrower 5 becomes too expensive. The Tobin-like tax has $r^K_{24} < \bar{r}_4$ because of bank 4's lower total default probability; it leaves the first equilibrium unchanged, panel (c) versus (a), whereas it reduces the volume of the second equilibrium, panel (d) versus (b). Panel (e) shows an equilibrium matching that can be achieved with an SRT, $\mathcal{T}$. It is systemic-risk efficient for a transaction volume of two unit loans. A choice of $\mathcal{T}$ is, for example, $\tau_{14} = 0$, $\tau_{15} = \tau_{16} > 0$, whereas $\tau_{25} = 0$, $\tau_{24} = \tau_{26} > 0$, and $\tau_{36} > 0$. The desired matches are untaxed; the undesired ones are taxed. This guarantees that the desired lenders are on top of each borrower's preference lists and allows this systemic-risk-efficient matching to be sustained as a unique equilibrium, without a reduction of credit volume.
Source: From Leduc and Thurner (2017), with permission from Elsevier.

where ϵ and ζ are some positive constants. This has the effect of reordering a borrower's preferences in decreasing order of their contribution to systemic risk. $r^{\mathcal{T}}_{ij}$ now reorders the preferences of the borrowers with the desired match on top and taxes the other matches proportional to marginal systemic loss, that is, the systemic risk they create.

5 How Much Systemic Risk Can Be Eliminated?

Although the theorem in the previous section tells us that systemic-risk-efficient networks (matchings) exist, it says nothing about to what extent systemic risk can be reduced in realistic scenarios, or about the structure of these optimal networks. To estimate the potential of systemic-risk reduction in financial networks, we employ an ABM that is calibrated to an actual financial economy.

5.1 Demonstration of Systemic-Risk Elimination in an Agent-Based Model

5.1.1 The Agent-Based Model

To test the economic and financial implications of the SRT, we employ the CRISIS ABM) (Klimek et al., 2015; Poledna and Thurner, 2016). This is an economic simulator that combines a well-studied macroeconomic ABM (Delli Gatti et al., 2008, 2011; Gaffeo et al., 2008) with an ABM of financial markets. We use a modified version of the model in Delli Gatti et al. (2011) that includes an interbank market and that is a *closed* economic system in the sense that allows no in- or out-flows of cash. For a comprehensive description of the model, see Delli Gatti et al. (2008) and Gualdi et al. (2015); for its modifications, see Poledna and Thurner (2016). The model features three types of agent: households, banks, and firms. They interact in the four markets shown in Figure 28.7: (1) Firms and banks interact in the credit market. (2) Banks interact with banks in the interbank market. (3) Households and firms interact in the job market. (4) Households and firms interact in the consumption goods market.

Households. There are H households of two types: firm owners and workers. Each worker j has a bank account, $BA_{j,b}(t)$, at one of the B banks, b. Accounts are randomly assigned to banks. Workers apply for jobs at F different firms. Once hired, they receive a fixed wage, w, at every time step and provide a fixed labor productivity, α^l. Firm owners receive dividends from their firm's profits. At every time step, households spend a fixed percentage, c, of their accounts on consumption. They are price sensitive and compare prices of goods from z firms and buy at the most economical.

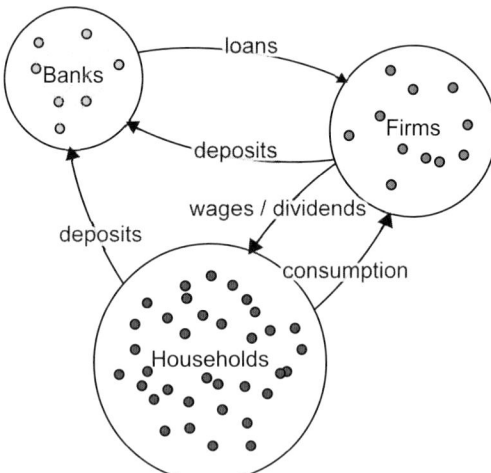

Figure 28.7 Schematic overview of the ABM, showing the three agent types (banks, firms, and households) and their interactions. Firms pay dividends to their owners and wages to their workers (financed through sales and loans). Households consume goods produced by the firms. Households and firms deposit money in banks; banks grant loans to the firms. If a firm defaults, a corresponding lending bank writes off the loss by reducing its C_i accordingly. Over time, repeated losses may drive banks into default, which in turn may cause a systemic event. Source: From Poledna and Thurner (2016), reprinted by permission of the publisher Taylor & Francis Ltd., www.tandfonline.com.

Firms. F firms produce substitutable goods. At every time step, firms compute an expected demand, $d_i^g(t)$, and an estimated price, $p_i^g(t)$ (subscript labels the firm), based on a rule that takes into account both excess demand/supply and the deviation of the price, $p_i^g(t-1)$, from the average price at the previous time point (Delli Gatti et al., 2011). Each firm computes the number of required workers to supply the expected demand. If wages for the respective workforce exceed the firm's current liquidity, firms look for credit at n randomly chosen banks and take the loan with the lowest rate. If it exceeds a threshold rate, $r^{\max}$, the firm only asks for ϕ percent of the originally desired loan volume. Based on the outcome of this credit request, firms reevaluate the needed workforce and hire or fire workers. Firms sell their goods on the consumption market. Firms go bankrupt if they have negative liquidity. Each of the bankrupted firm's debtors (banks) incurs a capital loss in proportion to their investment in the company. Firm owners of bankrupted firms are personally liable; their accounts are divided by the debtors pro rata. They immediately start a new company, with initially zero equity. Their initial estimates for $d_i^g(t)$ and $p_i^g(t)$ are equal to the respective averages in the population.

Banks. B banks offer loans to firms at rates that factor in an individual specificity of banks (modeled by a uniformly distributed random variable) and the firms' creditworthiness. Firms pay a risk premium that reflects their creditworthiness. It is modeled by a monotonically increasing function of their financial "fragility" (Delli Gatti et al., 2011). Banks grant firm loans only if they have enough liquidity. If they do not have enough they try to get it on the interbank market. If a bank does not have enough cash and cannot raise the necessary amount for a requested firm loan on the interbank market, it does not pay out the loan. Interbank and firm loans have the same duration. At each time step, firms and banks repay τ percent of their outstanding debt (principal plus interest). If banks have excess liquidity, they offer it on the interbank market for a nominal interest rate. The interbank relation network is modeled as a fully connected network (see Appendix A). Banks choose the interbank offer with the lowest rate. Interbank rates, r_{ij}, offered by bank i to bank j take into account the specificity of bank i and the creditworthiness of bank j. If a firm goes bankrupt, the respective creditor bank writes off the outstanding loans. If the bank does not have enough equity capital to cover these losses, it defaults. Following a bank default, a default cascade may unfold for interbank creditors. For simplicity, we assume no recovery for interbank loans. This assumption is reasonable for short-term liquidity Cont et al. (2011). A cascade of bankruptcies happens within one time step. After the last bankruptcy is taken care of, the simulation stops.

Regulator. A regulator knows all interbank loans and is able to compute the DebtRank for the individual agents as well as the marginal systemic risk for all exposures and contracts. In particular, the regulator can compute the marginal systemic risk for requested interbank transactions. The regulator operates in three modes: (1) It does not impose any tax. (2) It computes an SRT, τ_{ij}^k, for every intended interbank transaction, k, between i and j, and communicates its value to the prospective borrower. (3) It imposes a Tobin tax that taxes all transactions, irrespective of their systemic-risk contribution.

5.1.2 Implementation of the Systemic-Risk Tax

A *systemic-risk premium*, in the form of an SRT, can be imposed by the regulator on interbank transactions. Let us index by k one particular transaction (loan) of size l_{ijk} between

i and j, so that $L_{ij}(t) = \sum_k l_{ijk}(t)$. In this case, before entering a desired loan contract, $l_{ijk}(t)$, the borrowing bank i obtains quotes for the SRT, $\bar{\tau}_{ij}^{(+k)}(t)$, from the regulator, maybe for various lending banks, j. The borrower takes the offer with the lowest total rate, say, from bank j. The total rate becomes $r_{ij}^{\text{effective}}(t) = r_{ij}(t) + \bar{\tau}_{ij}^{(+k)}(t)$. This effective interest rate reflects both the creditworthiness of the borrowing counterparty (in r_{ij}) and the marginal systemic risk associated with the transaction (in $\bar{\tau}_{ij}^{(+k)}$). The SRT is collected into a "bailout fund." The SRT (in monetary units) for a transaction $l_{ijk}(t)$ between two banks i and j is calculated as

$$\tau_{ij}^{(+k)}(t) = \zeta \max \left[0, \sum_{i=1}^{B} p_i \left(V^{(+k)} R_i^{(+k)} - V(t) R_i \right) \right], \qquad (28.16)$$

where ζ is some constant, $V(t)$ is the economic value, is $R_i(t)$ DebtRank, and $p_i(t)$ represents the independent default probabilities. To convert it to a rate, $\bar{\tau}_{ij}^{(+k)} = \tau_{ij}^{(+k)}/l_{jik}$. ζ specifies how strongly marginal systemic risk is taxed. $\zeta = 1$ means that 100 percent of the marginal systemic risk will be charged. If ζ is too large, however, this might result in an overall reduction of credit volume. We show that ζ can be chosen such that the efficiency (trading volume) is kept practically the same as in the untaxed world. For details, see Poledna and Thurner (2016).

For comparison, a Tobin-like financial transaction tax (Tobin, 1978) for interbank loans is implemented by imposing a constant rate of 0.2 percent of the transaction. The Tobin-like tax also makes lending less attractive for firms that borrow from banks that need liquidity on the interbank market because refinancing costs remain with the firms.

5.1.3 How Much Systemic Risk Can Be Eliminated? Model Results

We implement the model for $B = 20$ banks, $F = 100$ firms, and $H = 1,300$ households. The model is run in three independent modes: without any tax, with SRT, and with a Tobin-like financial transaction tax. Results are shown as averages over $10,000$ independent, identical simulations across 500 time steps. We set $p_i = 0.01$, $V = \sum_{i=1}^{B} \sum_{j=1}^{B} L_{ij}$, and $\zeta = 0.02$. We calibrate the model to historical, anonymized, and linearly transformed interbank liability data provided by the Austrian Central Bank (OeNB; see Poledna and Thurner (2016) and 7.3). The ABM results for R_i are presented in panel (b) of Figure 28.1: without tax (gray), with a Tobin tax (dark gray), and with SRT (light gray). Distributions are from $10,000$ independent simulation runs at time step $t = 100$. The SRT drastically reduces the systemic risk of the individual banks by more than a factor of 2. The situation without tax approximately resembles the empirical distribution from panel (a) of Fig. 28.1. Panel (d) of Figure 28.1 shows the marginal systemic risk for the ABM simulations of the three modes. SRT reduces the marginal systemic loss of liabilities by about an order of magnitude (note the log scale) and leaves contract sizes practically unchanged.

The effects of the SRT and the Tobin-like tax on the loss distribution of banks that occur as a consequence of bank defaults and secondary cascading events are shown in panel (a) of Figure 28.8. The mode without tax (medium gray) produces well-known fat tails in the loss distributions in the banking sector. The Tobin tax slightly reduces losses (dark gray), and the SRT gets completely rid of large losses in the system (light gray). The remaining losses are mainly due to firm defaults (i.e., economic risk; see Section 1.2). In the SRT scenario, effects

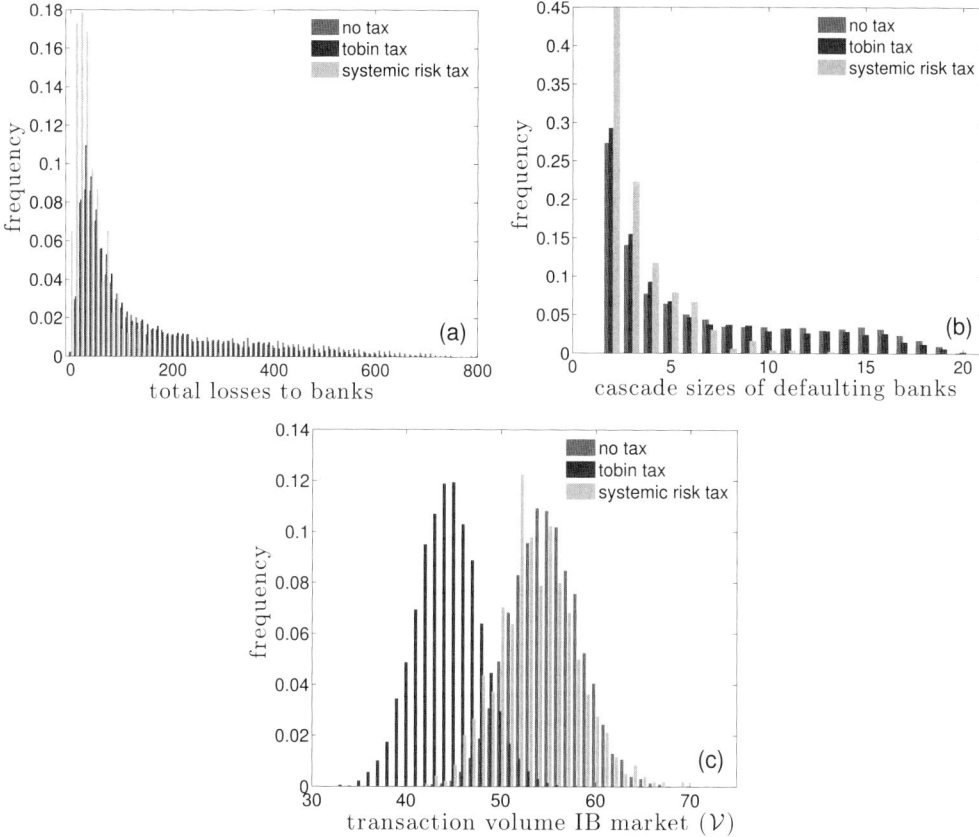

Figure 28.8 Comparison of no transaction tax gray on interbank loans, SRT (light gray), and Tobin tax (dark gray). (a) Distribution of total losses to banks, (b) distribution of cascade sizes of secondary defaulting banks, and (c) distribution of total transaction volume v in the interbank market. Based on 10,000 independent, identical simulations, each with 500 time steps, 20 banks. Source: After Poledna and Thurner (2016), reprinted by permission of the publisher Taylor & Francis Ltd., www.tandfonline.com.

from suboptimal network structures are practically eliminated. This elimination of losses is due to the fact that under the SRT, the possibility for cascading defaults is suppressed. This is captured in panel (b) of Figure 28.8 showing the distributions of cascade sizes (number of banks that default in one systemic event triggered by one defaulting bank) for the three modes. The untaxed scenario produces considerable cascade sizes involving up to all 20 banks; the maximum cascade sizes under the SRT is only about 10. The Tobin tax is comparable with the untaxed case. Interbank loan sizes are practically unchanged under the SRT. This is also seen in the total transaction volume v in the interbank market in panel (c) of Figure 28.8, where the distribution of transaction volume at time step 100 is shown. As expected, the situation for the SRT (light gray) is almost identical to the untaxed case (medium gray), whereas the Tobin tax drastically reduces transaction volume (dark gray).

We observe that the potential to reduce network-based systemic risk is large. On the basis of individual transactions, the SRT reduces about an order of magnitude of systemic risk; on the level of banks, it reduces about a factor of 2 (and more) in the DebtRank.

5.2 Why Basel III Does Not Reduce Network-Based Systemic Risk

We now study the capacity for systemic-risk reduction with an ABM implementation of the Basel III regulation with capital surcharges for globally systemically important banks (GSIBs). We first estimate the systemic-risk-reduction potential of the Basel III indicator-based measurement approach. Then we study the effects of different weight assignments in the Basel III approach. There we are interested in whether it is more effective to place additional capital surcharges on "super-spreaders" or on "super-vulnerable" financial institutions. For full details, see Poledna et al. (2017).

5.2.1 Basel III Indicator-Based Measurement Approach

The Basel III indicator-based approach has five broad categories to which it assigns equal weights: size, interconnectedness, lack of readily available substitutes or financial institution infrastructure, global (cross-jurisdictional) activity, and "complexity." Categories may contain subcategories. The score, S_j, of the Basel III indicator-based measurement approach for bank j and each indicator, D^i, for example, cross-jurisdictional claims, is calculated as the fraction of the individual banks with respect to all B banks and is then weighted by the indicator, β_i. The score, given in basis points (bps), is

$$S_j = \sum_{i \in I} \beta_i \frac{D_j^i}{\sum_j^B D_j^i} 10,000. \qquad (28.17)$$

Here I is the set of indicators, D^i, and β_i represents the weights from Table 28.1. Based on this score, banks are divided into four equally sized "buckets" of systemic importance; see Table 28.2. The cutoff score and bucket thresholds have been calibrated by the Basel Committee on Banking Supervision (BCBS) such that the magnitude of the higher loss-absorbency requirements for the highest-populated bucket is 2.5 percent of risk-weighted assets (RWAs), with an initially empty bucket of 3.5 percent of RWAs. The loss-absorbency

Table 28.1. Basel III indicators and weights for the indicator-based measurement approach

Category (all 20%)	Indicator	Weight
Cross-jurisdictional activity	Cross-jurisdictional claims	10%
	Cross-jurisdictional liabilities	10%
Size	Total exposures as defined for use in the Basel III leverage ratio	20%
Interconnectedness	Intrafinancial system assets	6.67%
	Intrafinancial system liabilities	6.67%
	Securities outstanding	6.67%
Substitutability/financial institution infrastructure	Assets under custody	6.67%
	Payments activity	6.67%
	Underwritten transactions in debt and equity markets	6.67%
Complexity	Notional amount of over-the-counter (OTC) derivatives	6.67%
	Level 3 assets	6.67%
	Trading and available-for-sale securities	6.67%

Table 28.2. Categories of systemic importance in the BCBS "bucketing approach"

Bucket	Score range	Bucket thresholds	Loss absorbency requirement (common equity % of RWA)
5	D–E	530–629	3.50%
4	C–D	430–529	2.50%
3	B–C	330–429	2.00%
2	A–B	230–329	1.50%
1	Cutoff A	130–229	1.00%

requirement for the lowest bucket is 1 percent of RWAs. It must be met with common equity (BIS, 2010). Bucket 5 is empty initially. As soon as the bucket becomes populated, a new bucket is added in such a way that it is equal in size (scores) to each of the other populated buckets, and the minimum higher-loss-absorbency requirement is increased by 1 percent of RWAs.

5.2.2 Basel III Capital Surcharges for GSIBs

In the ABM, we implement the size indicator by calculating the total exposures including all assets, that is, loans to firms and loans to other banks and excluding cash. Interconnectedness is measured by interbank assets (loans) and interbank liabilities (deposits). As a proxy for substitutability, we use the payment activity of banks that is estimated by the sum of all payments. This includes wages paid by firms to households, dividend payments, and payments from interactions on the goods market. In the model, we do not have cross-jurisdictional activity, and banks do not trade complex financial products such as derivatives or level 3 assets. The weights for cross-jurisdictional activity and complexity are set to zero, and the others are adjusted correspondingly. When banks in the model observe the capital requirements according to Basel III, they are required to hold 4.5 percent of common equity (up from 2 percent in Basel II) of RWAs. These are calculated following a standardized approach, that is, with fixed weights for all asset classes. For fixed weights, we use 100 percent for interbank loans and commercial loans. We define the equity capital of banks in the model as common equity. Banks are assigned to the buckets in Table 28.2, based on their scores from equation (28.17). They have to meet the loss-absorbency requirements shown in Table 28.2. The score is calculated at every time step, and capital requirements must be observed before providing new loans. As the baseline model, we implement a regulation similar to Basel II like, where we require banks to hold 2 percent of common equity of their RWAs. For more details, see Poledna et al. (2017).

5.2.3 Systemic-Risk Reduction in the Basel III Approach

The results from simulations are summarized in Figure 28.9. As before, the SRT drastically suppresses large losses to the banking system as a consequence of a reduction of cascade sizes, see Poledna et al. (2017). Panel (a) of Figure 28.9 shows the SR profiles (the highest systemic risk is to the left). The baseline model (Basel II capital requirements) is shown in medium gray, the Basel III implementation in dark gray, and the SRT in light gray. As before, the SRT drastically reduces DebtRank by about a factor of 2 and leads to a more

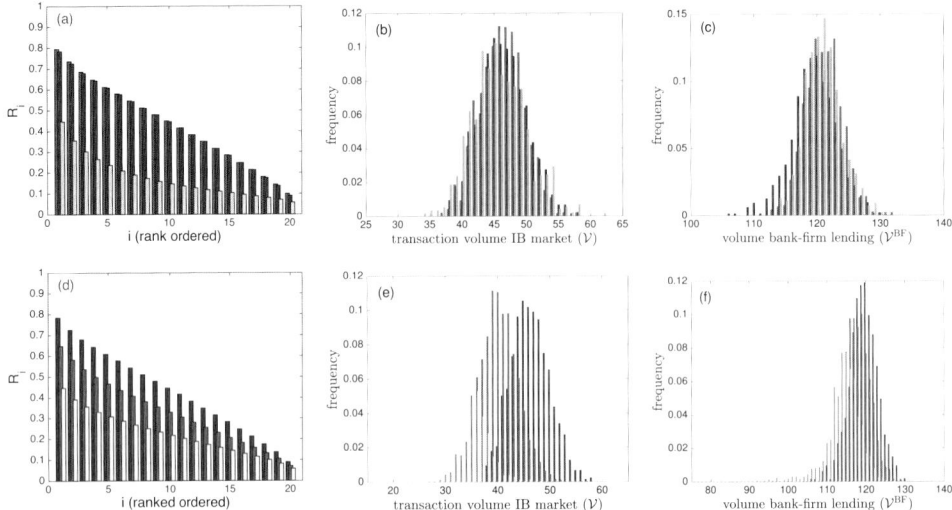

Figure 28.9 Comparison of different regulation modes of systemic risk. (a) SR profile given by DebtRank, R_i, for the baseline model with Basel II capital requirements (medium gray), the Basel III implementation (dark gray), and the SRT (light gray). Basel III does *not* appear to reduce systemic risk. (b) Distribution of interbank transaction volume, $\mathcal{V}$ (same shading scheme), and (c) bank–firm lending volume, $\mathcal{V}^{BF}$ (same shading scheme). Credit volume is not affected much by the different regulation policies. (d) SR profile for the model that represents the Basel III requirements (dark gray), the Basel III implementation with twice the surcharges (medium gray), and three times the surcharges (light gray). (e) Interbank transaction volume for the normal, double, and triple surcharges and (f) bank–firm lending volume for the various surcharge levels. Source: After Poledna et al. (2017), with permission from Elsevier.

homogeneous distribution of systemic risk across banks. Basel III with capital surcharges for GSIBs (dark gray) shows practically no reduction of systemic importance. Panels (b) and (c) of Figure 28.9 show that the total interbank transaction volume, v, and the total bank–firm lending volume, v^{BF}, are practically unaffected by the different regulation policies. Just Basel III (dark gray) slightly reduces total bank–firm lending volume v^{BF}.

If Basel III regulation with the planned capital surcharges for GSIBs does not considerably reduce network-based systemic risk, will it do so with higher surcharges? To understand the trade-off between reducing systemic risk and losses of efficiency under increased capital requirements, we consider three settings: (1) one with capital surcharges for GSIBs as specified in Basel III as before, (2) a scenario where we double capital surcharges for GSIBs, and (3) one where we triple them. In panel (d) of Figure 28.9, we show the systemic-risk profiles for the planned surcharge in dark gray, the doubled surcharge ($\times 2$) in medium gray, and the tripled ones ($\times 3$) in light gray. With larger capital surcharges, Basel III now has a visible effect on systemic-risk reduction. However, the reduced systemic losses are paid for with a tremendous loss of efficiency: note the 40 percent loss in the total transaction volume in the interbank market, v, in panel (e) of Figure 28.9, and the significant losses in the bank–firm lending volume, v^{BF} shown in panel (f).

It might be argued that under certain circumstances, the reduction of credit volume might be desirable; see, for example, D'Errico and Roukny (2019). However, this is not what we address here. We are interested in the following question: Given an exogenous demand for credit, how much does credit volume reduce under the different systemic-risk-reduction schemes? In the context of comparing different methods for systemic-risk reduction, such as the SRT and Basel III, it is necessary to compare them with respect to the same economic benchmarks, such as credit volume. Obviously and trivially, it is always possible to reduce systemic risk by reducing credit volume.

6 Systemically Optimal Financial Network Topologies

In the theorem of Section 4.3, we proved that systemically optimal financial networks do exist. In the previous section, we demonstrated that a simple incentive scheme such as the SRT can reduce the systemic risk of institutions by approximately a factor of 2. We have shown that the SRT as implemented in the ABM indeed drives networks toward optimal configurations where the likelihood for cascading events is suppressed. We now study how much systemic risk can be avoided in a (hypothetical) best-of-all-worlds scenario, under the condition that economic efficiency is not reduced. We discuss one scenario for direct exposures and another for overlapping portfolio risk. In both cases, the idea is to minimize the systemic risk by finding an optimal network topology.

6.1 Optimal Systemic-Risk Levels for Direct Exposure Networks

Following Diem et al. (2020), given an interbank exposure network, L_{ij}, we approximate its systemic risk not by DebtRank but by a modified quantity that can be used as an objective function in a standard optimization problem. Because DebtRank cannot be formulated in closed form (as a formula), it would be computationally awkward to use it as an objective function. We therefore define the *direct impact* of an institution (bank) i on its neighbors as

$$I_i = \frac{1}{V} \sum_{j=1}^{N} \min\left(\frac{L_{ij}}{C_j}, 1\right) a_j, \qquad (28.18)$$

where L_{ij} is the liability matrix, C_i is the equity of i, $a_i = \sum_{j=1}^{N} L_{ji}$ is the assets of i, and $l_j = \sum_{j=1}^{N} L_{ij}$ is the liabilities; the total volume is $V = \sum_{i=1}^{N} a_i = \sum_{i=1}^{N} l_i$. The sum of all direct impacts (of all N banks), $I = \sum_{i=1}^{N} I_i$, is a simple approximation of DebtRank that can be minimized by a mixed-integer linear programming (MILP) algorithm. The optimization procedure yields the optimal network, L^*. Optimization should keep the total assets, a_i; the liabilities, l_i; and the total network volume, V, unchanged. This is realized by the constraints in the optimization problem:

$$\min_{L \in \{M: \, M \in \mathbb{R}_+^{N \times N}, \, M_{ii}=0\}} \sum_{i=1}^{N} \sum_{j=1}^{N} \min\left(\frac{L_{ij}}{C_j}, 1\right) a_j$$

$$\text{subject to} \quad l_i = \sum_{j=1}^{N} L_{ij}, \quad \forall i \quad [\text{liabilities}]$$

$$a_i = \sum_{j=1}^{N} L_{ji}, \quad \forall i \quad [\text{assets}]. \qquad (28.19)$$

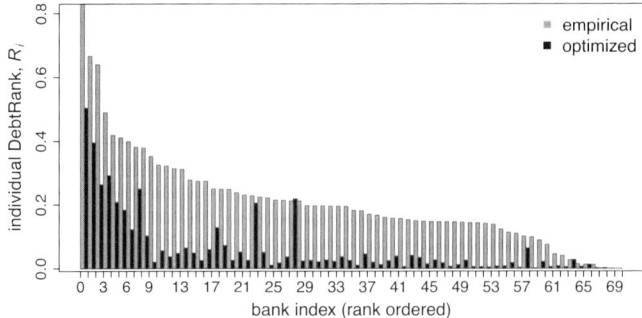

Figure 28.10 Systemic risk profile (DebtRank, R_i) of Austrian banks for the empirical (light gray) and the minimized (dark gray) liability networks in the first quarter of 2006. Systemic risk is drastically reduced for practically all banks, with only one exception. For the 10 most systemically relevant banks, there appears a reduction of DebtRank by approximately a factor of 2. From Diem et al. (2020), with permission from Elsevier.

Values for C, l, a, and V can be obtained from balance-sheet data. Note that because only row and column sums of L are needed for the optimization, detailed knowledge of the asset-liability network, L_{ij}, is not necessary. A global optimum must exist; however, it is not necessarily unique. For details, see Diem et al. (2020). Figure 28.10 shows the resulting SR profile (DebtRank) for the Austrian interbank liability network (in the first quarter of 2006) before and after optimization. DebtRank is approximately a factor of 2 lower after optimization for the 10 systemically most relevant banks, and it is even more (approximately a factor of 5) for the others. Note that DebtRank is shown in the figure, whereas the objective function used for the optimization was the direct impact from equation (28.19). Also note that one bank slightly increased its systemic risk after optimization.

6.2 Optimal Systemic Risk from Indirect Exposures—Systemic-Risk-Efficient Portfolio Allocation

It is possible to optimize the network topology of indirect exposures that arise from overlapping portfolios (Pichler et al., 2018) by minimizing systemic risk by optimally rearranging overlapping portfolio networks. The idea is that agents (hypothetically) suffle their assets, such that the resulting new portfolios are equivalent in value but are composed such that the overlapping portfolio network is systemically optimal. The optimization is performed under the constraints that the portfolio values of the individual banks must not decrease. We characterize the values of the banks' portfolios within the classical mean-variance framework of (Markowitz, 1952). This means that the expected return of any portfolio should be equal or higher after rewiring; the variance should be equal or less. Under these constraints, we minimize the average DebtRank of the overlapping portfolio layer. Again, DebtRank is not directly suitable as an objective function; however, it can be approximated by the direct impacts, $\tilde{W}\tilde{v}$, as defined in equations (28.9) and (28.10). A quadratic optimization problem can now be formulated: Let σ_{kl}^2 be the covariance of bond k and l, and let r_k be the expected return of bond k. The expected return and variance of portfolio i are $\tilde{r}_i = \sum_k A_{ki} r_k$, and $\tilde{\sigma}_i^2 = \sum_k \sum_l A_{ki} A_{li} \sigma_{kl}^2$, respectively. The total value of bond k is S_k. We now have the nonconvex, quadratically constrained, quadratic programming problem:

$$\min_{x_{ki} \geq 0 \; \forall k, i} \quad \sum_i \sum_j \frac{\tilde{v}_j}{C_j} \sum_k x_{ki} x_{kj} \frac{1}{M_k}$$

$$\text{subject to} \quad V_i = \sum_k x_{ki}, \quad \forall i,$$

$$S_k = \sum_i x_{ki}, \quad \forall k, \tag{28.20}$$

$$\tilde{r}_i \leq \sum_k x_{ki} r_k, \quad \forall i,$$

$$\tilde{\sigma}_i^2 \geq \sum_k \sum_l x_{ki} x_{li} \sigma_{kl}^2, \quad \forall i,$$

where x_{ki} denotes the investments that can be reallocated in A. $\tilde{v}_j$ is the relative economic value, defined in equation (28.10). The quadratic programming problem in equation (28.20) minimizes the total direct impact if defaults occur without any deterioration in the banks' risk profiles. Total portfolio volumes and the total outstanding volumes are kept constant; that is, the network is only rewired.

We demonstrate the approach on the overlapping portfolio network of sovereign exposure between major European banks, using data from the European Banking Authority stress test of 2016 (European Banking Authority, 2016). To obtain an estimate for the bipartite asset-holder network $A_{ki}(t)$, we use data from investments of European banks in government bonds. This network connects institutions i with assets k; A_{ki} is the amount in assets k held by institution i at time t. We include 49 major European banks that are invested in 36 different sovereign bonds. The associated total market volume is EUR 2,617.39 billion, or roughly 10 percent of these banks' total assets. To estimate the market depth of the bonds, M_k from equation (28.7), we pool market-price data with reported data on trading activity and outstanding volume. For details, see Pichler et al. (2018).

Solutions to equation (28.20) are NP-hard in general (Anstreicher, 2009), however, for the small network size with 49 banks, standard optimization solvers can handle the problem. To estimate the average total systemic overlapping portfolio risk, we compute the DebtRank for every bank two times, once before the optimization, denoted by $\bar{R}^{\text{orig}} = \sum_i R_i^{\text{orig}}$ (original systemic risk), and once after the optimization, $\bar{R}^{\text{opt}} = \sum_i R_i^{\text{opt}}$ (optimized systemic risk). For the European government bond market, we find that $\bar{R}_{\text{orig}} = 6.66$ percent before optimization, and $\bar{R}_{\text{opt}} = 2.89$ percent after optimizing the portfolio holdings in equation (28.20). Systemic risk on the market level is reduced by more than a half, and the maximum DebtRank in the network decreases from 0.22 to 0.09. Banks with originally high systemic-risk levels lose systemic relevance, whereas some of the least systemic banks increase slightly. Overall, a significantly more systemic risk-efficient allocation is achieved. Consistent with the findings in the previous sections, systemic risk from indirect exposures from overlapping portfolios can be reduced by more than a factor of 2, without any detrimental effects for the individual banks. We also find that diversification in the portfolios is practically not affected by the optimization. For more details, see Pichler et al. (2018).

7 Summary and Policy Implications

In this contribution, we demonstrated that methodology from complexity and network science allows us to quantify systemic risk in ways that are readily interpretable in terms of contagion risk, size of cascading events, expected costs of financial crises, and resilience in general. The essence of these methods is that they are based on the detailed information of financial interactions between financial agents and therefore allow for more realistic estimates of systemic risk than conventional stress testing. These methods can be used in combination with ABMs that are calibrated to actual data to anticipate and quantify the response and resilience to specific endogenous or exogenous shocks under specific economic conditions. Here we showed how to use network-based measures to quantify the systemic risk of institutions, regions, and individual financial contracts and how to use these measures to proactively manage systemic risk. We further showed how to estimate the amount systemic risk that can be eliminated without additional economic costs or losses of efficiency. Our findings can be summarized in more detail as follows.

7.1 Quantifying and Monitoring Systemic Risk

7.1.1 Systemic Risk of Institutions

We use the network centrality measure DebtRank to quantify the systemic-risk contributions of individual institutions on the basis of exposure networks and basic balance-sheet information. We introduce a systemic-risk profile that displays the systemic-risk contributions of all financial institutions in a country.

DebtRank in combination with estimates of default probabilities of institutions allows us to define the *expected systemic-loss index*, ESL, for a financial economy. It quantifies the expected losses of hypothetical cascading events at any point in time, given that no interventions, such as bailouts or bail-ins, occur. This makes it possible to monitor expected costs from systemic events in monetary units per year.

When ESL is compared to expected costs for bailouts or other resolution schemes, decisions for resolution strategies can be based on quantitative, transparent, and rational grounds (Poledna et al., 2015). Klimek et al. (2015) found indications that the economic context of an economy determines the optimal resolution scheme. In the present context, it would be a fascinating question how the systemic-risk measures discussed here would change under different resolution schemes. In particular, if the presented measures would have to be improved if it was known beforehand which resolution scheme would be applied for a given default. Questions like these come within reach with the availability of detailed data on bail-in scenarios in combination with multilayer simulation methods (de Souza et al., 2019; Hüser et al., 2019).

Finally, ESL can be used to compare temporal trends of systemic risk between economies and to monitor the systemic impact of policy interventions. We find that conventional financial risk measures tend to systematically underestimate systemic-risk levels. Comparing the ESL index of Mexico across the recent financial crisis with the VIX and the CDS spreads of government bonds, the VIX quickly returned to precrisis levels, and spreads have doubled since the crisis. In the same period, the ESL index has quadrupled. This means that the potential costs for a cascading failure would have been four times higher in 2013 than before the crisis. We think that the ESL should be implemented as a monitoring tool for financial economies.

7.1.2 Financial Multilayer Networks

We showed that to obtain a comprehensive picture of systemic risk, it is necessary to consider systemic risk from its various sources; in other words, the multilayered nature of financial exposure networks must be taken into account. We explicitly demonstrated the importance of the multilayer approach to systemic risk, including direct and indirect exposures. In analyzing systemic-risk contributions from various direct-exposure layers of the Mexican interbank network, we find that by relying on a single layer, one may drastically underestimate the total systemic risk in the system, missing up to 90 percent of it (Poledna et al., 2015). We found that systemic risk from the indirect-exposure layer of overlapping portfolio risk in Mexican securities markets accounts for a substantial fraction of the total systemic risk. Focusing on direct exposures only can also underestimate total systemic-risk levels by up to 50 percent.

7.1.3 Systemic Risk of Transactions

We introduce the concept of *marginal systemic risk* that allows us to unambiguously quantify the systemic risk of individual exposures and transactions. This opens two new possibilities: First, to demonstrate that systemic risk is indeed an *externality* whose costs (created in a systemic event) might not be payed for by those agents who created it. Second, it allows us to design a market-based incentive scheme to mitigate it. When the marginal systemic risk of individual liabilities is plotted against their associated credit risk, one observes that for a given level of credit risk, there exists a huge variation (3 orders of magnitude) in the marginal systemic-risk contributions. The fact that marginal systemic risk generally exceeds the associated credit risk for individual liabilities is a direct indicator that systemic risk is indeed an *externality*.

7.2 Managing Systemic Risk

In a mathematical theorem, we prove that systemic-risk-efficient equilibria do exist under an incentive scheme that appropriately prices the externality of systemic risk. It is essential to reduce systemic risk without reducing the efficiency of financial markets. In particular, we demonstrate rigorously that a regulator (e.g., a central bank) with information about the exposure network can design a transaction-specific tax that incentivizes institutions to form contract networks (matchings) that are more resilient to default cascades. This SRT allows the regulator to select equilibrium network configurations that minimize systemic risk. Without SRT, networks arise in different equilibria that are generally systemic-risk inefficient. We prove that a standard financial transaction tax such as a Tobin-like tax reduces economic efficiency while only having marginal power on reducing systemic risk. A Tobin-like tax does not account for the fact that different liabilities have different impacts on systemic risk. Although the theorem states the existence of optimal networks, it says nothing about how much systemic risk can be eliminated in actual financial markets without a loss of efficiency.

7.2.1 Agent-Based Models for Stress Testing

How much systemic risk can be eliminated? To clarify the issue, we test the SRT within the framework of a macrofinancial ABM. The model is able to produce systemic-risk profiles that are very similar to actually observed interbank liability data. Implementing the SRT in the ABM drastically reduces the probability for cascading failure as a result of restructured liability networks. We show that the SRT is able to restructure financial-liability networks without loss of transaction volume. The average degrees and clustering coefficients of the systemically improved liability networks become somewhat smaller under the SRT. In contrast, a Tobin-like tax does not restructure networks and only reduces systemic risk by reducing transaction volume (and thus efficiency) since credit becomes more expensive. We find that an SRT can eliminate about half the systemic risk of institutions without introducing inefficiencies.

7.2.2 Implementing an SRT

We believe that the proposed incentive scheme to internalize the externalities created by systemic risk, or similar ones, can be used for actual regulation. Technically, the requirements for its implementation include an electronic market that is accessible to the regulator. This market would work similar to airline ticketing systems by guaranteeing a quote for the SRT for a limited time. The computational requirements to compute the SRT even for thousands of banks is a technical triviality. Central banks would have to record transactions in real time, which is already routinely done in many countries. The risk that SRT-quotes of institutions could reveal critical information about market participants is limited. The free-riding problem that generally applies to financial trans-action taxes—if they are not implemented on a global scale—however, also applies for the SRT. The implementation of the SRT would lead to some additional costs for institutions (e.g., for establishing relations with new counterparties). For banks to become systemically risk-aware will require experts and teams for strategic systemic-risk aversion so as not to jeopardize lending opportunities because of high SRTs. However, when compared to the levels of reduction of systemic risk under the SRT scheme, these costs are marginal.

7.2.3 Why Basel III Might Not Reduce Systemic Risk

As an alternative to mitigate systemic risk, it has been proposed to tax systemically impor-tant financial institutions (SIFIs; Acharya et al., 2012, 2013; Adrian and Brunnermeier, 2011; Cooley et al., 2009; Markose et al., 2012; Zlatić et al., 2014) or—similar in spirit—to increase capital requirements, such as proposed in Basel III. We studied the ability of the Basel III framework for network-based systemic-risk reduction with an ABM. Results indicate that to obtain any reasonable reduction, capital surcharges for SIFIs must be sub-stantially larger (by a factor of 3) than currently envisioned in Basel III, which is, of course, completely unrealistic. Strategies that simply tax institutions based on their systemic-risk levels might have limited macroprudential efficiency, simply because they do not restructure the underlying networks in appropriate ways.

7.2.4 Systemically Optimal Networks

Finally, we showed that it is possible to estimate the theoretical optimum for systemic risk in a given network. This is obtained by finding systemic risk-optimal networks in a straightforward optimization problem. We find that the maximum systemic-risk reduction that is theoretically possible is about a factor of 1/2, showing that the SRT scheme already leads to almost systemically optimal networks. The knowledge of the optimal network topology could further serve as a benchmark to monitor if empirical markets are approaching the optimum or are diverging from it.

7.3 Policy Implications

The management of network-based systemic risk is to a large extent associated with the restructuring of networks toward optimal structures that suppress the likelihood of cascading failure. Complexity science endows us with a set of quantitative tools that allows us to quantify and monitor aspects of systemic risk that complement and extend current stress testing techniques. In particular, these extensions from complexity science are threefold: First, it systematically links those aspects of systemic risk that are associated with contagion and cascading failure with network structures that are observable in balance sheets and central bank data. Second, it allows us to compute those network structures that are optimal with respect to systemic risk. Third, complexity science systematically studies how the individual behavior of agents leads to the macroscopic properties of a system. This allows us to derive policy interventions that affect the behavior of individual institutions, which in turn lead toward optimally resilient networks. These new possibilities lead to a number of policy implications.

1. **Monitoring systemic-risk levels of countries and regions.** The newly introduced *expected systemic-loss index*, ESL Poledna et al. (2015), monitors variations in regional systemic-risk levels. This in turn can be used to objectivize policy interventions, in particular, to see if interventions lead to a decrease or increase of regional systemic stability. The ESL index quantifies the network-based systemic risk that is invisible to market-based risk measures, such as, for example the VIX or CDS spreads spreads.
2. **Identifying systemic relevance of institutions.** The so-called *DebtRank*, R_i (Battiston et al., 2012), quantifies the network-based systemic risk of financial institutions in a given exposure network. It can be interpreted as the fraction of the total market volume that is affected by the hypothetical default of institution i. To obtain realistic values, the multilayer nature of exposure networks must be taken into account appropriately (Poledna et al., 2015). DebtRank can also be used to quantify the systemic risk of indirect exposures that originate, for example, from overlapping portfolios of institutions (Poledna et al., 2018).
3. **Estimating costs of financial crisis.** The *expected systemic loss index* monitors the expected cost of hypothetical systemic events at any point in time, assuming that there are no interventions, such as bail-ins or bailouts. These estimates can be used, in combination with agent-based simulations, such as in Klimek et al. (2015), for optimal decision-making on banking resolution schemes.
4. **Identifying systemically risky transactions.** The index of *marginal systemic risk*, ΔESL, (Poledna and Thurner, 2016), quantifies the systemic-risk contribution of individual transactions between institutions. It can be used to identify especially dangerous transactions that create large levels of systemic risk. The *marginal systemic risk* effectively prices the systemic risk and the associated externality costs, that is, the expected damage of systemic events that cannot be paid for by those who produce it.

5. **Incentivizing financial agents to manage systemic risk.** In a regulatory framework, the *marginal systemic-risk* index can be used to incentivize institutions to avoid and thereby manage and reduce systemic risk. By feeding information back to institutions about which transactions are particularly dangerous, it allows them to look for alternative, however, equivalent transactions with other transaction partners—and thereby restructure the liability network. The SRT (Poledna and Thurner, 2016) is one market-based possibility to incentivize this restructuring process; there exist other alternatives (Thurner and Poledna, 2013). A brute-force approach would be to forbid systemically risky transactions above a certain threshold.

6. **Benchmarking observed systemic risk to optimal levels.** The knowledge of the minimal systemic-risk levels possible within a given economic environment, both for direct (Diem et al., 2020) and indirect exposures (Pichler et al., 2018), allows one to benchmark observed systemic-risk levels. The ratio of actual systemic risk to the optimal level is a measure that monitors *systemic-risk effectiveness*. This can be done at the country level or for individual agents. For current interbank liability networks, this ratio is approximately 2; at the optimum, it would be 1. The change of this ratio over time indicates the systemic-risk effectiveness of policy interventions.

7. **Reconsidering of the Basel III with respect to systemic risk.** Agent-based simulations indicate that Basel III as provisioned now will practically not reduce network-based systemic risk (Poledna et al., 2017). This is not surprising because it does not rewire networks toward optimal structures. Unlike the SRT, Basel III introduces economic costs, makes credit more expensive and the financial system less efficient.

Up to this day, exposure network topologies that arise randomly as a consequence of, say, exogenous demand for credit and other services are typically suboptimal with respect to systemic risk. The central message of this contribution is that restructuring these networks— such that economic benchmarks such as cost for credit or credit volume are not reduced—is an extremely effective approach to macroprudential regulation. About half of the systemic risk that is currently present in financial networks can be eliminated by restructuring financial networks toward systemically optimal topologies.

Acknowledgments

This work was supported by FFG Projects 857136 and 857136, and the Austrian National Bank Project P17795.

Appendix A Mini-Glossary on Networks

- **Network.** In mathematical terms, an undirected network is defined as a pair of sets, $\mathcal{G} = (\mathcal{N}, \mathcal{L})$, with the node set, $\mathcal{N}$, containing all nodes i and the link set, $\mathcal{L}$, containing unordered pairs $L_{ij} = \{i, j\}$, the undirected links. A directed network has a link set $\mathcal{L}$ that contains ordered pairs $L_{ij} = (i, j)$ or directed links going from i to j.
- **Degree and degree distribution.** In undirected networks, the degree k_i of a node i is the number of links connecting to it. All k_i nodes that are directly attached to i are called *(nearest) neighbors* of i. The average degree of all nodes in a network is denoted by $\bar{k}$. In directed networks, the in-degree k_i^{in} of node i is the number of its incoming links; the out-degree k_i^{out} is the number of outgoing links. The degree distribution, $p(k)$, is the probability that a randomly chosen node in the network has a degree of k. The degree distribution is characteristic to networks of different nature.
- **Clustering coefficient.** The clustering coefficient, C_i, for node i in an undirected graph is the ratio between the number of links, y_i, between its k_i neighbors and the number of all possible links $k_i(k_i - 1)/2$ between them, $C_i \equiv \frac{2y_i}{k_i(k_i-1)}$. The clustering coefficient of the entire network, C, is the average over all clustering coefficients, C_i.
- **Network topology/centrality measures**. The network topology refers to the detailed structure of a network. The overall structure is described by so-called "network measures" that include the degree distribution, the clustering coefficient, the eigenvalue spectrum of L_{ij}, and many more. To describe the structure of networks in a meaningful way, multiple measures are necessary. Network centrality measures assign quantities, such as degree, clustering coefficient, eigenvector component, betweenness, and so forth, to specific nodes and links in a network.
- **Multilayer network.** A multilayer network is a network where a set of nodes is connected by a set of links of a different type. It is usually defined as a set of nodes and a set of link-sets, $\mathcal{G}^n = (\mathcal{N}, \mathcal{L}^1, \mathcal{L}^2, \cdots, \mathcal{L}^n)$. Each $\mathcal{L}^\alpha$ contains ordered, pairs $L_{ij}^\alpha = (i, j)$, or unordered pairs, $L_{ij}^\alpha = \{i, j\}$. In real-world scenarios, often links in one layer are not independent of links in another.
- **Random graph.** The term *random graph* is often used for networks where every pair of nodes is linked with a fixed probability, ρ. These networks are characterized by a Poissonian degree distribution. Random graphs do not have hubs. The clustering coefficient of a random graph is $C_r = \bar{k}/N$. Networks in realistic settings often have a larger clustering coefficient than random graphs.

- **Scale-free network.** A network is called *scale-free* if its degree distribution follows an asymptotic power law, $p(k) \sim k^{-\gamma}$, where γ is the *scaling exponent*. As a consequence, scale-free networks have relatively many nodes with a very high degree—so-called "hubs." Most networks observed in realistic situations are approximately scale-free. Many financial networks, including interbank liability networks, are scale-free (Boss et al., 2005).

Appendix B DebtRank

DebtRank is a recursive method suggested by Battiston et al. (2012) for determining the systemic relevance of nodes in financial networks. It is the fraction of the total economic value in the network that is potentially affected by a defaulting node or set of nodes. L_{ij} denotes the liability network at a given moment (e.g., loans of j to i), and C_i is the capital of institution i. If i defaults and cannot repay its loans, bank j loses the loans L_{ij}. If j does not have enough capital available, j also defaults. The impact of institution i on j (in the case of the default of i) is

$$W_{ij} = \min\left[1, \frac{L_{ij}}{C_j}\right].$$
(B.1)

Given the total outstanding loans of bank i, $L_i = \sum_j L_{ji}$, its *economic value* is defined as $v_i = L_i / \sum_j L_j$. The value of the impact of bank i on its neighbors is $I_i = \sum_j W_{ij} v_j$. To take the impact of nodes at distance 2 and higher into account, it has to be computed recursively:

$$I_i = \sum_j W_{ij} v_j + \beta \sum_j W_{ij} I_j,$$
(B.2)

where β is a damping factor. If the network W_{ij} contains cycles, the impact can obviously exceed 1. To avoid this problem, an alternative was suggested (Battiston et al., 2012), where two state variables, $h_i(t)$ and $s_i(t)$, are assigned to each node. h_i is a continuous variable between 0 and 1; s_i is a discrete state variable for three possible states: undistressed, distressed, and inactive, $s_i \in \{U, D, I\}$. The initial conditions are $h_i(1) = \Psi, \forall i \in S_f$; $h_i(1) = 0, \forall i \notin S_f$, and $s_i(1) = D, \forall i \in S_f$; $s_i(1) = U, \forall i \notin S_f$ (parameter Ψ quantifies the initial level of distress: $\Psi \in [0,1]$, where $\Psi = 1$ means default). The dynamics of h_i are then specified by

$$h_i(t) = \min\left[1, h_i(t-1) + \sum_{j \mid s_j(t-1)=D} W_{ji} h_j(t-1)\right].$$
(B.3)

The sum extends over indices j, for which $s_j(t-1) = D$,

$$s_i(t) = \begin{cases} D & \text{if } h_i(t) > 0; s_i(t-1) \neq I, \\ I & \text{if } s_i(t-1) = D, \\ s_i(t-1) & \text{otherwise.} \end{cases}$$
(B.4)

The DebtRank of set S_f (set of nodes in distress at time 1), is $\hat{R} = \sum_j h_j(T) v_j - \sum_j h_j(1) v_j$, and it measures the distress in the system, excluding the initial distress.

If S_f is a single institution, DebtRank measures its systemic impact on the network. The DebtRank of S_f containing only a single institution i is

$$\hat{R}_i = \sum_j h_j(T)v_j - h_i(1)v_i. \tag{B.5}$$

DebtRank has the meaning of economic loss, measured, for example, in dollars, caused by the distress or default of a node (Battiston et al., 2012). DebtRank in Equation (B.5) excludes the loss generated directly by the default of the node itself and measures only the impact on the rest of the system through default contagion. For some purposes, however, it can be useful to include the direct loss of a default of i as well. The total loss of a default of i on the whole system, including the loss caused directly by i, is $R_i = \sum_j h_j(T)v_j$.

Appendix C Data

C.1 Austrian Interbank Data

Data are provided by the Austrian Central Bank (OeNB) and contain fully anonymized and linearly transformed interbank liabilities/exposures $L_{ij}^{data}(t)$ from the entire Austrian banking system, composed of approximately 800 banks over 12 consecutive quarters from 2006–2008. The data set additionally includes the total assets, total liabilities due from banks and due to banks, and liquid assets (without due from banks) for all banks, again in anonymized form.

C.2 Mexican Interbank Data

The data set is collected and owned by the Banco de México and contains various types of daily exposure between the major Mexican financial intermediaries (banks) over the period 2007–2013. Banks interact simultaneously in four different markets, generating four different types of exposures: (unsecured) interbank credit, securities, FX, and derivative markets. Hence, institutions are connected by contracts of four different types. The various exposure networks can be represented as layers in a financial multilayer network. The data further contain the capitalization of banks for every month.

References

Acharya, V., L. Pedersen, T. Philippon, and M. Richardson (2009), "Regulating systemic risk," in V. Acharya and M. P. Richardson (eds.), *Restoring financial stability: How to repair a failed system*, Wiley.

Acharya, V., L. Pedersen, T. Philippon, and M. Richardson (2012), "Measuring systemic risk," Technical Report, CEPR Discussion Papers, available at http://ssrn.com/abstract=1573171.

Acharya, V., L. Pedersen, T. Philippon, and M. Richardson (2013), "Taxing systemic risk," in J. Fouque and J. Langsam (eds.), *Handbook on systemic risk*, Cambridge University Press.

Adrian, T., and M. Brunnermeier (2011), "Covar," Technical Report, National Bureau of Economic Research.

Adrian, T., and H. S. Shin (2008), "Liquidity and leverage," Technical Report 328, Federal Reserve Bank of New York.

Anstreicher, K. M. (2009), "Semidefinite programming versus the reformulation-linearization technique for nonconvex quadratically constrained quadratic programming," *Journal of Global Optimization*, 43(2), 471–484.

Aoyama, H., S. Battiston, and Y. Fujiwara (2013), "DebtRank Analysis of the Japanese Credit Network," Discussion Paper 13087, Research Institute of Economy, Trade and Industry (RIETI).

Aymanns, C., and D. Farmer (2015), "The dynamics of the leverage cycle," *Journal of Economic Dynamics and Control*, 50, 155–179.

Aymanns, C., J. D. Farmer, A. M. Kleinnijenhuis, and T. Wetzer (2018), "Models of financial stability and their application in stress tests," in C. Hommes and B. LeBaron (eds.), *Handbook of Computational Economics*, Vol. 4. Elsevier.

Balin, B. J. (2008), "Basel I, Basel II, and emerging markets: A nontechnical analysis," Working Paper, available at http://ssrn.com/abstract=1477712.

Bank for International Settlements (1988), *International convergence of capital measurement and capital standards*.

Bank for International Settlements (2006), *International Convergence of Capital Measurement and Capital Standards: A Revised Framework Comprehensive Version*.

Bank for International Settlements (2010), *Basel III: A global regulatory framework for more resilient banks and banking systems*.

Bardoscia, M., S. Battiston, F. Caccioli, and G. Caldarelli (2015), "DebtRank: A microscopic foundation for shock propagation," *PloS One*, 10(6), 1–13.

Battiston, S., M. Puliga, R. Kaushik, P. Tasca, and G. Caldarelli (2012), "Debtrank: Too central to fail? Financial networks, the FED and systemic risk," *Scientific Reports*, 2(541).

Bech, M. L. and E. Atalay (2010), "The topology of the federal funds market," *Physica A: Statistical Mechanics and its Applications*, 389(22), 5223–5246.

Billio, M., M. Getmansky, A. W. Lo, and L. Pelizzon (2012), "Econometric measures of connectedness and systemic risk in the finance and insurance sectors," *Journal of Financial Economics*, 104(3), 535–559.

Boss, M., H. Elsinger, M. Summer, and S. Thurner (2005), "The network topology of the interbank market," *Quantitative Finance*, 4, 677–684.

Boss, M., M. Summer, and S. Thurner (2004), "Contagion flow through banking networks," *Lecture Notes in Computer Science*, 3038, 1070–1077.

Bouchaud, J.-P. (2010), Price impact, in R. Cont (ed.), *Encyclopedia of Quantitative Finance*, John Wiley and Sons.

Braverman, A., and A. Minca (2014), "Networks of Common Asset Holdings: Aggregation and Measures of Vulnerability," Working Paper, available at https://papers.ssrn.com/sol3/papers.cfm?abstract_id=2379669.

Brownlees, C. T., and R. F. Engle (2012), "Volatility, correlation and tails for systemic risk measurement," Working Paper, available at https://faculty.washington.edu/ezivot/econ589/VolatilityBrownlees.pdf.

Brunnermeier, M., and L. Pedersen (2009), "Market liquidity and funding liquidity," *Review of Financial Studies*, 22(6), 2201–2238.

Caballero, J. (2012), "Banking crises and financial integration," IDB Working Paper Series No. IDB-WP-364.

Caccioli, F., J.-P. Bouchaud, and J. D. Farmer (2012), "Impact-adjusted valuation and the criticality of leverage," *Risk*, 74–77.

Cajueiro, D. O., B. M. Tabak, and R. F. Andrade (2009), "Fluctuations in interbank network dynamics," *Physical Review E*, 79(3).

Cont, R., A. Moussa, and E. Santos (2011), "Network structure and systemic risk in banking systems," at http://ssrn.com/abstract=1733528.

Cont, R., and E. Schaanning (2017), "Fire sales, indirect contagion and systemic stress-testing," Norges Bank Working Paper 02/17.

Cooley, T., T. Philippon, V. Acharya, L. Pedersen, and M. Richardson (2009), "Regulating systemic risk," in V. Acharya, and M. P. Richardson (eds.), *Restoring Financial Stability: How to Repair a Failed System*, John Wiley & Sons.

De Bandt, O., and P. Hartmann (2000), "Systemic risk: A survey," Technical Report, CEPR Discussion Papers.

de Souza, S. R. S., T. C. Silva, and C. E. de Almeida (2019), "Bailing in banks: Costs and benefits," *Journal of Financial Stability*, 45, 100705.

Delli Gatti, D., S. Desiderio, E. Gaffeo, P. Cirillo, and M. Gallegati (2011), *Macroeconomics from the Bottom-up*, Springer Milan.

Delli Gatti, D., E. Gaffeo, M. Gallegati, G. Giulioni, and A. Palestrini (2008), *Emergent Macroeconomics: An Agent-Based Approach to Business Fluctuations*, Springer.

D'Errico, M., and T. Roukny (2019), "Compressing over-the-counter markets," ESRB Working Paper No. 44, available at www.esrb.europa.eu/pub/pdf/wp/esrbwp44.en.pdf.

Diem, C., A. Pichler, and S. Thurner (2020), "What is the minimal systemic risk in financial exposure networks?" *Journal of Economic Dynamics and Control*, 116, e103900.

Duffie, D., and K. Singleton (2012), "*Credit Risk: Pricing, Measurement, and Management,*" Princeton Series in Finance, Princeton University Press.

Eisenberg, L., and T. H. Noe (2001), "Systemic risk in financial systems," *Management Science*, 47(2), 236–249.

Enriques, L., A. Romano, and T. Wetzer (2019), "Network-sensitive financial regulation," *Journal of Corporation Law*, forthcoming.

European Banking Authority (2016), "Eu-wide stress testing 2016," available at www.eba.europa.eu/risk-analysis-and-data/eu-wide-stress-testing/2016.

Fostel, A., and J. Geanakoplos (2008), "Leverage cycles and the anxious economy," *American Economic Review*, 98(4), 1211–44.

Fricke, D., and T. Lux (2014), "Core–periphery structure in the overnight money market: Evidence from the e-mid trading platform," *Computational Economics*, 1–37.

Gaffeo, E., D. Delli Gatti, S. Desiderio, and M. Gallegati (2008), "Adaptive microfoundations for emergent macroeconomics," *Eastern Economic Journal*, 34(4), 441–463.

Geanakoplos, J. (2010), "The leverage cycle," in D. Acemoglu, K. Rogoff, and M. Woodford (eds.), *NBER Macro-economics Annual 2009*, Vol. 24, University of Chicago Press.

Georg, C.-P. (2011), "Basel III and systemic risk regulation—what way forward?" Technical Report 17, Working Papers on Global Financial Markets.

Greenwood, R., A. Landier, and D. Thesmar (2015), "Vulnerable banks," *Journal of Financial Economics*, 115(3), 471–485.

Gross, T., C. J. D. D'Lima, and B. Blasius (2006), "Epidemic dynamics on an adaptive network," *Physical Review Letters*, 96, 208701.

Gualdi, S., M. Tarzia, F. Zamponi, and J.-P. Bouchaud (2015), "Tipping points in macroeconomic agent-based models," *Journal of Economic Dynamics and Control*, 50, 29–61.

Guo, W., A. Minca, and L. Wang (2016), "The topology of overlapping portfolio networks," *Statistics & Risk Modeling*, 33(3–4), 139–155.

Haldane, A. G., and R. M. May (2011), "Systemic risk in banking ecosystems," *Nature*, 469(7330), 351–355.

Huang, X., I. Vodenska, S. Havlin, and H. E. Stanley (2013), "Cascading failures in bi-partite graphs: Model for systemic risk propagation," *Scientific Reports*, 3(1219).

Huang, X., H. Zhou, and H. Zhu (2012), "Systemic risk contributions," *Journal of Financial Services Research*, 42(1–2), 55–83.

Hüser, A.-C., G. Hałaj, C. Kok, C. Perales, and A. van der Kraaij (2019), "The systemic implications of bail-in: A multi-layered network approach," ECB Working Paper No. 2010.

Iori, G., G. De Masi, O. V. Precup, G. Gabbi, and G. Caldarelli (2008), "A network analysis of the italian overnight money market," *Journal of Economic Dynamics and Control*, 32(1), 259–278.

Iori, G., R. N. Mantegna, L. Marotta, S. Micciche, J. Porter, and M. Tumminello (2015), "Networked relationships in the e-mid interbank market: A trading model with memory," *Journal of Economic Dynamics and Control*, 50, 98–116.

Klimek, P., M. Obersteiner, and S. Thurner (2015), "Systemic trade risk of critical resources," *Science Advances*, 1(10).

Klimek, P., S. Poledna, J. Farmer, and S. Thurner (2015), "To bail-out or to bail-in? Answers from an agent-based model," *Journal of Economic Dynamics and Control*, 50, 144–154.

Kyle, A. S. (1985), "Continuous auctions and insider trading," *Econometrica: Journal of the Econometric Society*, 53(6), 1315–1335.

Kyriakopoulos, F., S. Thurner, C. Puhr, and S. W. Schmitz (2009), "Network and eigenvalue analysis of financial transaction networks," *The European Physical Journal B: Condensed Matter and Complex Systems*, 71(4), 523–531.

Leduc, M. V., and S. Thurner (2017), "Incentivizing resilience in financial networks," *Journal of Economic Dynamics and Control*, 82, 44–66.

Markose, S. (2012), "Systemic risk from global financial derivatives: A network analysis of contagion and its mitigation with super-spreader tax," IMF Working Paper WP/12/282.

Markose, S., S. Giansante, and A. R. Shaghaghi (2012), "'Too interconnected to fail' financial network of US CDS market: Topological fragility and systemic risk," *Journal of Economic Behavior and Organization*, 83(3), 627–646.

Markowitz, H. (1952), "Portfolio selection," *Journal of Finance*, 7(1), 77–91.

Martínez-Jaramillo, S., B. Alexandrova-Kabadjova, B. Bravo-Benitez, and J. P. Solórzano-Margain (2014), "An empirical study of the mexican banking system's network and its implications for systemic risk," *Journal of Economic Dynamics and Control*, 40, 242–265.

Masciandaro, D., and F. Passarelli (2013), "Financial systemic risk: Taxation or regulation?" *Journal of Banking and Finance*, 37(2), 587–596.

May, R. M.(1072), "Will a large complex system be stable?" *Nature*, 238(1), 413–414.

Minoiu, C., C. Kang, V. S. Subrahmanian, and A. Berea (2013, December), "Does financial connectedness predict crises?" Technical Report 13/267, International Monetary Fund.

Minsky, H. P. (1992), "The financial instability hypothesis," The Jerome Levy Economics Institute Working Paper.

Pastor-Satorras, R., and A. Vespignani (2001), "Epidemic spreading in scale-free networks," *Physical Review Letters*, 86(14), 3200.

Pichler, A., S. Poledna, and S. Thurner (2021), "Systemic-risk-efficient asset allocation: Minimization of systemic risk as a network optimization problem," *Journal of Financial Stability*, 52, 100809.

Poledna, S., O. Bochmann, and S. Thurner (2017), "Basel III capital surcharges for g-sibs are far less effective in managing systemic risk in comparison to network-based, systemic risk-dependent financial transaction taxes," *Journal of Economic Dynamics and Control*, 77, 230.

Poledna, S., S. Martinez-Jaramillo, F. Caccioli, and S. Thurner (2021), "Quantification of systemic risk from overlapping portfolios in the financial system," *Journal of Financial Stability*, 52, 100808.

Poledna, S., J. L. Molina-Borboa, S. Martínez-Jaramillo, M. van der Leij, and S. Thurner (2015), "The multi-layer network nature of systemic risk and its implications for the costs of financial crises," *Journal of Financial Stability*, 20, 70–81.

Poledna, S., and S. Thurner (2016), "Elimination of systemic risk in financial networks by means of a systemic risk transaction tax," *Quantitative Finance*, 16(10), 1599–1613.

Poledna, S., S. Thurner, J. D. Farmer, and J. Geanakoplos (2014), "Leverage-induced systemic risk under Basel II and other credit risk policies," *Journal of Banking and Finance*, 42, 199–212.

Puhr, C., R. Seliger, and M. Sigmund (2012), "Contagiousness and vulnerability in the Austrian interbank market," *OeNBs Financial Stability Report*, 24.

Roukny, T., H. Bersini, H. Pirotte, G. Caldarelli, and S. Battiston (2013), "Default cascades in complex networks: Topology and systemic risk," *Scientific Reports*, 3(2759).

Schweitzer, F., G. Fagiolo, D. Sornette, F. Vega-Redondo, A. Vespignani, and D. R. White (2009), "Economic networks: The new challenges," *Science*, 325(5939), 422–425.

Soramäki, K., M. L. Bech, J. Arnold, R. J. Glass, and W. E. Beyeler (2007), "The topology of interbank payment flows," *Physica A: Statistical Mechanics and its Applications*, 379(1), 317–333.

Thurner, S., J. Farmer, and J. Geanakoplos (2012), "Leverage causes fat tails and clustered volatility," *Quantitative Finance*, 12(5), 695–707.

Thurner, S., R. Hanel, and S. Pichler (2003), "Risk trading, network topology and banking regulation," *Quantitative Finance*, 3(4), 306–319.

Thurner, S., P. Klimek, and R. Hanel (2018), "*Introduction to the Theory of Complex Systems*," Oxford University Press.

Thurner, S., and S. Poledna (2013), "Debtrank-transparency: Controlling systemic risk in financial networks," *Scientific Reports*, 3(1888).

Tobin, J. (1978), "A proposal for international monetary reform," Technical Report 506, Cowles Foundation for Research in Economics, Yale University.

Upper, C., and A. Worms (2002), "Estimating bilateral exposures in the german interbank market: Is there a danger of contagion?" Technical Report 9, Deutsche Bundesbank Research Centre.

Zlatić, V., G. Gabbi, and H. Abraham (2014), "Reduction of systemic risk by means of Pigouvian taxation," *PLoS ONE* 10(7), e0114928.

Stress Testing a Central Bank's Own Balance Sheet

Nathanaël Benjamin, Abigail Haddow, and David Jacobs*

Introduction

The 2008 financial crisis marked a shift in the way central banks use their balance sheets. The Bank of England (BoE) was among a number of central banks that took unprecedented action to stabilize national economies and financial systems. This environment drove a large expansion in the scale and complexity of financial-market operations that the BoE conducts and, as a consequence, in its actual and potential exposures. Those operations were effective in supporting both financial stability and the real economy, which ultimately was also consistent with protecting the BoE's balance sheet. But as a result of these interventions, the demand and need for effective risk management of the BoE's own financial exposures have become more important than ever. Stress testing is an integral part of that, and it leverages the broad range of mechanisms available to central banks—beyond just financial instruments. The BoE now uses stress testing to (1) measure the adequacy of its financial resources, (2) articulate its risk tolerance, and (3) test its readiness for various scenarios, bringing together the full spectrum of central banking across market operations and supervisory tools.

This chapter explores why and how central banks can use forward-looking stress testing as a tool for risk management, focusing on the institutional setup at the BoE, and how it stress tests its own balance sheet in practice.

- Section 1 considers how financial risk management differs in central versus commercial banks, which frames the key issues around risk management and the role of stress testing.
- Section 2 explores the changing landscape for central bank balance sheets since the crisis.
- This sets the scene for Section 3, which presents the strengthened risk management framework that has been developed at the BoE, with stress testing as one of its core pillars.
- Section 4 sets out the different potential applications of stress testing for a central bank.
- Finally, Section 5 examines how central bank stress tests work in practice.

Like the rest of this book, this chapter was written before the outbreak of the COVID-19 pandemic. For sure, the stress testing approach described here contributed to informing the Bank of England's response to the COVID-19 crisis. If anything, this underscored even further its value as a central bank risk-management tool.

---

* The authors want to acknowledge the substantial guidance and input over several years from Andrew Hauser, then executive director for Banking, Payments and Financial Resilience at the Bank of England, during the period over which the approach presented here was developed. Several aspects of this discussion were presented in Hauser (2017). We are also grateful for the assistance of Nasreen Hussain and Mathew Sim in producing this chapter.

1 How Does Financial Risk Management Differ in Central versus Commercial Banks?

The starting point when considering central bank risk management, and the role that can be played by forward-looking stress tests, has to be the nature of the balance sheets. These differ from commercial banks' balance sheets in a number of important ways.

1.1 Nature of the Balance Sheet

The shape of a central bank's balance sheet is typically a direct consequence of its mission to maintain monetary and financial stability.[1] Central bank balance sheets have a number of features in common. That said, their precise structure can also vary widely depending on the features of their financial system, their institutional setup, and which monetary policy tools are employed.

The liability side of central banks' balance sheets primarily comprises money, in the form of bank reserves and banknotes. For the BoE, the liability side is composed principally of reserves from commercial banks remunerated at the policy interest rate (bank rate, thereby transmitting that base rate to the economy), banknotes, deposits from institutional customers, and some bond issuance in foreign currency.

Central banks also have capital. The majority of central banks are publicly owned, including the BoE, which is wholly owned by the UK's Treasury (Her Majesty's Treasury [HMT]). That said, other central banks remain partly or wholly owned by private-sector shareholders. For example, the US Federal Reserve is wholly owned by private banks, whereas the Bank of Japan is partly owned by private, nonbank shareholders (and shares are even traded on the stock exchange).[2]

The asset side of the central bank balance sheet mainly reflects the operations that have been used to implement policy, as well as holdings of foreign assets ("foreign reserves"). In the decade following the 2008 crisis, the BoE's assets have been dominated by quantitative easing (QE). In particular, QE shows up as a loan to the asset-purchase facility—a subsidiary of the BoE that is indemnified by the sovereign—through which QE is conducted.[3] For central banks that have not engaged in QE, assets tend to be dominated by short-term secured lending in money markets. Indeed, the BoE's assets also include loans and repos to financial institutions through the BoE's sterling monetary framework (SMF[4]). Central banks may also extend loans as a lender of last resort, and the BoE can do so as part of its financial stability toolkit. Finally, in broad terms, the asset side is completed by some foreign currency reserves and the investment of capital in sterling bonds (which match to corresponding liabilities). Foreign reserves comprise a larger portion of assets for some other central banks; the extent of holdings depends on factors such as the extent to which central banks intervene in their exchange rate and whether these foreign reserves are owned by the central bank or the finance ministry.

---

[1] See Rule (2015).

[2] See Bholat and Martinez Gutierrez (2019) and Archer and Moser-Boehm (2013).

[3] That loan, in turn, constitutes the entire liability side of the purchase facility, a vehicle that is indemnified directly by HMT (to whom all net profits or losses from QE are therefore transferred).

[4] See www.bankofengland.co.uk/markets/bank-of-england-market-operations-guide/our-objectives.

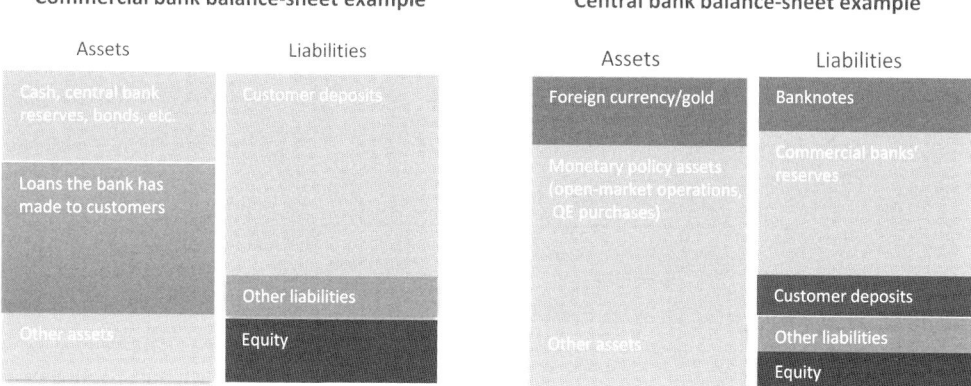

Figure 29.1 Stylized comparison between commercial and central banks' balance sheets.

Such a balance-sheet profile, and the way in which it differs from commercial banks', is illustrated in Figure 29.1. Commercial banks' balance sheets are a consequence of their business model, traditionally based on maturity transformation: customer deposits often dominate the liability side, and these, together with any wholesale funding, are usually reinvested in customer loans and other assets with varying degrees of liquidity.

1.2 Objectives and Risk Tolerance

As illustrated previously, central banks use their balance sheets, and capital, to deliver the policy goals of monetary and financial stability. This is a key difference from commercial banks' objectives, which are primarily to maximize economic returns for their shareholders. This is important because it affects both the nature and severity of the financial risks that the BoE is required to take onto its balance sheet to meet its policy objectives.

The BoE offers liquidity facilities to provide assistance to individual firms or financial markets going through a period of stressed conditions, for example, via its SMF, in order to contain financial instability that could otherwise generate wider contagion. The SMF includes all market-wide sterling liquidity facilities, such as indexed long-term repo operations, shorter-term open market operations, intra-day liquidity provision, or the discount window. In stressed circumstances, the BoE has a role to play as lender of last resort to participants of the SMF: stepping in to provide temporary liquidity to firms or the market when private liquidity providers are either unable or unwilling to do so given their risk appetite—and then stepping out.

Clearly, the BoE takes a number of measures that substantially reduce the risks inherent in such circumstances. For example, it lends only to institutions assessed as solvent and viable, takes high-quality collateral with conservative haircuts, charges penal interest rates, and typically lends only for limited periods of time. But because of its policy objectives, by design, the BoE is more likely to take on risk as the degree of systemic stress increases. So the BoE's risk tolerance increases as the degree of systemic risk increases, which is the opposite of the pattern for private firms (Figure 29.2).

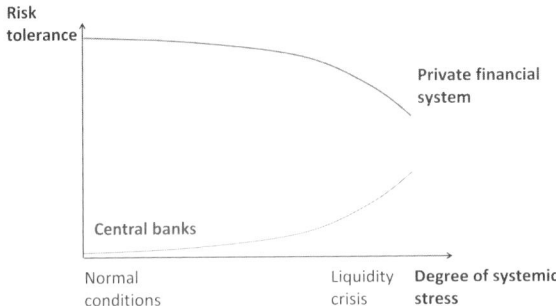

Figure 29.2 Central bank versus private financial system risk tolerance.

This all has important implications for risk management and the need for stress tests. First, it means that risks need to be examined in a forward-looking fashion because the current balance sheet may provide only a very limited picture of risks. That is because most of the BoE's exposure to risk stems from contingent balance-sheet expansions: because of its role as the backstop, the central bank, by design, has to expose itself to wrong-way risk. Second, stress tests are important because they spell out concretely the severe scenarios for which the BoE has to stand ready.

1.3 Endogeneity of Balance-Sheet Risks

As is apparent from this discussion, the central bank balance sheet cannot be thought about in isolation precisely because of its policy role. Central bank policies affect commercial banks' risks, whereas commercial banks' actions also affect central bank balance sheets when risks crystallize. That "endogeneity" with the wider financial system is a crucial difference when stress testing a central bank.

One implication is that, in some circumstances, the extension of credit by the central bank and/or lending on easier terms may also be consistent with *protecting* its balance sheet (even if this is not the direct objective). By taking risks onto its balance sheet, the central bank can reduce risks in the financial sector and real economy and thus its own balance sheet. That is particularly pertinent during crises. Bagehot (1873)[5] articulated this point, describing the "only safe plan" for the BoE as the "brave plan." Bindseil and Jabłecki (2013)[6] formalize this feedback aspect of central bank risk management using a general equilibrium model. They show that in stressed market conditions, central bank losses can sometimes *increase* with larger haircuts (and thus a less supportive liquidity policy), even though larger haircuts would, in isolation, be expected to provide *more* protection.

That said, there are a number of counterpoints and qualifications to Bindseil and Jabłecki (2013) argument. They acknowledge that there are drawbacks to a liquidity regime that is too supportive, some of which could increase risks in some circumstances (and their model captures some such trade-offs). These include moral hazard, which might affect risk-taking

---

[5] See http://public-library.uk/ebooks/58/60.pdf; Bindseil and Jabłecki (2013) provide a further discussion of the history of this argument during the nineteenth century.

[6] See www.ecb.europa.eu/pub/pdf/scpwps/ecbwp1542.pdf?1deaeaa1be0d40e3512ef9bab21bc4f8.

behavior ex ante. Excess support might also see inefficiencies arise in the economy. Their findings could also imply that the more risk a central bank takes onto its balance sheet, the more risk it should keep adding as a mitigant, and so forth.

These feedback effects of central bank actions do have some parallels with private-sector risk management. Banks can help to support customers through a liquidity shock, so long as the customer remains solvent. But for a central bank, these effects are amplified considerably by the fact that borrowers can be systemic, and so their failure could trigger wider contagion.

It is important to build stress testing with the general-equilibrium nature of the central bank balance sheet in mind. For example, as will be discussed later, more severe scenarios than those applied to commercial banks are required to test the BoE's readiness and the resources required to support its financial operations during periods of heightened financial stress. In addition, another implication of this endogeneity is that the more that commercial banks are able to self-insure against risks, then the further out the tail central bank support would be expected to kick in—that might influence the nature of scenarios that a central bank would have to run. Accordingly, one potential response to a central bank stress test might be to have commercial banks modify their risk-taking behavior. However, as we discuss later in the chapter, bringing these insights into a practical risk management framework is difficult, and more work remains to be done.

1.4 The Role of Capital

Central banks also have different solvency constraints than private firms. Central banks can create money, which means they can always meet liabilities denominated in their own currency. So in theory, central banks can operate with negative capital. That could put in question the need for stress testing.

However, central banks need to avoid a situation described as "policy insolvency."[7]. That is, they need capital (or a powerful ability to recover from losses, e.g., through income) to convince financial markets that they retain the unconstrained ability to pursue their broader policy objectives without resorting to fiscal resources or currency debasement. Otherwise, absorbing losses could well conflict with monetary policy objectives. Insufficient financial buffers and/or frequent recourse to the finance ministry to carry out operations within its core mandate might also erode confidence in the central bank's operational independence.

Ensuring financial preparedness for a range of plausible events is consistent with central banks acting as responsible stewards of public funds. Stress testing is a powerful tool for the BoE to prepare for forward-looking risks to its balance sheet and to articulate the level of risk it is willing to bear—and seeks to have sufficient capital for—*on its own*. The finance ministry is the ultimate bearer of risk. So it is in the interest of the central bank to develop, with the ministry, a common understanding ex ante about which institution directly stands behind which part of the risk tail. For example, in the UK, both organizations have developed such a structure as part of a new Memorandum of Understanding (MoU), discussed later in this chapter. It sets out publicly the principles for which operations should be backed by the central bank's capital versus government indemnities.

---

[7] cf. Stella (2005).

Having touched on what risk management and financial resilience mean for central banks, let us turn to the environment in which their balance sheets have evolved in recent years.

2 The Changing Landscape for Central Bank Balance Sheets

The balance sheets of many central banks around the world have changed substantially. In the decades leading up to the financial crisis, the size of the BoE's balance sheet represented less than 5 percent of the UK's GDP. And its main exposures were small and short term. The assets on the balance sheet were mainly high-quality, low-duration assets. But this changed with the financial crisis. As for many central banks, the BoE's balance sheet expanded dramatically. The BoE now holds assets worth about a quarter of the UK's annual GDP (Figure 29.3).

The changing nature of the balance sheet has driven a need for risk management to evolve. There are three driving forces:

- Firstly, the increase in the size and scope of balance sheets has seen a corresponding increase in direct financial risk taking for the central bank—although that arguably reduced risk in the banking system as a whole, thereby indirectly making the central bank's balance sheet safer. Much of the increase in the early financial crisis was driven by large-scale liquidity assistance to the banking sector. For example, the BoE lent $61.5 billion to the Royal Bank of Scotland (RBS) and HBoS in 2008–2009, and the Special Liquidity Scheme peaked at $185 billion in 2009.[8] Meanwhile, in October 2008, the

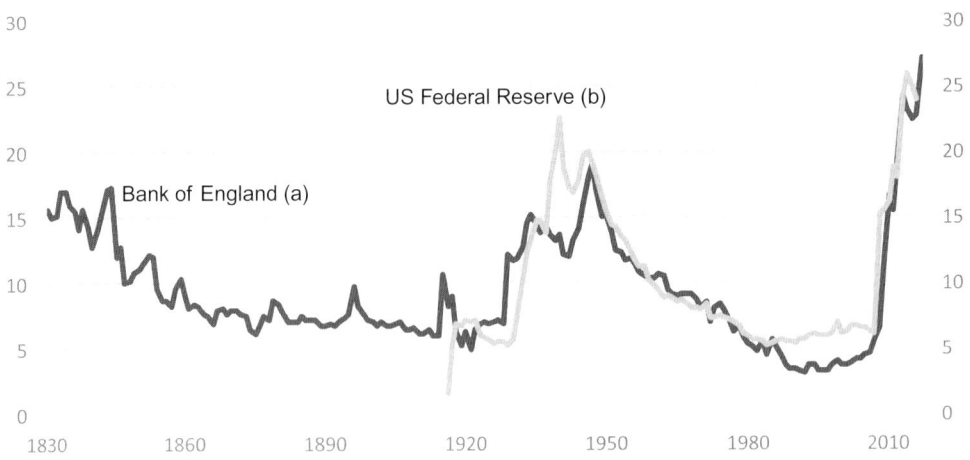

Figure 29.3 Bank of England and Federal Reserve balance sheets compared, as percentages of the countries' annual nominal gross domestic product (GDP).
Note: (a) MeasuringWorth data as of February. Final data point shows the latest published balance-sheet data, on February 29, 2016, plus the increase in the loan to the asset-purchase facility since this time. Excludes TARGET balances held at the BoE by European central banks in 1999–2001. (b) Consolidated assets of all Federal Reserve Banks. Data as at end December. Source: BoE et al. (2017).

[8] cf. Plenderleith (2012).

UK's Treasury injected a total of $37 billion into the RBS, HBoS, and Lloyds TSB; as mentioned earlier, the BoE acts as the liquidity backstop only for *solvent* banks. The introduction of long-term lending schemes and outright purchases designed to deliver monetary stability led to further balance-sheet exposure. In many cases, the transmission of the monetary stimulus from these schemes stemmed directly from the transfer of risk from private to public balance sheets.

- Secondly, looking ahead, these demands are expected to become a more permanent feature of central bank balance sheets. That implies a rise in contingent risks. After crisis, the BoE was given a broader range of policy responsibilities by the UK's parliament, and its liquidity insurance toolkit has become more comprehensive. The BoE now has responsibilities for banking supervision, macroprudential policy, and resolution. To some extent, these developments underscore long-standing risks that could come to bear on the BoE's balance sheet. But in many cases, these new responsibilities also imply potential new calls on its balance sheet. Operational reform and the BoE's "open-for-business"[9] stance have also widened the range of banks and nonbanks it may need to lend to, as well as the range of collateral it might need to lend against. This open-for-business stance is nevertheless accompanied by robust risk mitigants, beyond supervising the solvency of eligible institutions. Importantly, these mitigants also include applying conservative collateral haircuts to central bank lending. These haircuts provide a high degree of public fund protection, designed to be effective in the severe circumstances in which these funds could be at risk. Another powerful risk mitigant is the recent structural reform of the UK banking system, which ring-fences deposit-taking activities away from trading books. Other than liquidity-assistance operations, the future steady-state framework for controlling interest rates[10] could also be an important factor in the overall profile of the central bank's balance sheet.
- Thirdly, and as a direct result of these developments, the demand for greater accountability and transparency has risen since the financial crisis.

Given these fundamental changes to central bank operations, the BoE has pioneered a new financial risk framework. It includes an expanded range of stress testing tools as well as structures to monitor and challenge the financial risk it takes onto its own balance sheet. Much of the framework is based on private-sector risk management practices. But there are key differences in the way that central banks are exposed to financial risks that have shaped the BoE's internal stress testing framework. This is an important context for the approach the BoE has taken to stress testing its own balance sheet.

3 Strengthening the Management of the Bank of England's Balance-Sheet Risks

In addition to the increasing scale and complexity of its balance-sheet risks, it also became clear that the BoE needed to do more to strengthen its risk framework. Its balance-sheet risk management expectations of regulated firms had also started to become relevant for the central bank itself. So the BoE has strengthened its financial risk framework in recent

---

[9] See Mark Carney's 2013 speech, available at www.bankofengland.co.uk/-/media/boe/files/speech/2013/the-uk-at-the-heart-of-a-renewed-globalisation.pdf?la=en&hash=75D7639033A4653557DAE972D22649D8C 374BF2E.

[10] See www.bankofengland.co.uk/paper/2018/boe-future-balance-sheet-and-framework-for-controlling-interest-rates.

years. Stress testing fits into a wider set of institutional and governance changes that mark a stronger set of arrangements to guard against financial risk.

The BoE has long had a first-line risk division embedded in its Markets directorate. That division reviews pricing, applies haircuts to the wide range of collateral prepositioned at the central bank, monitors the specific risks taken in live operations on a day-to-day basis, and evaluates the creditworthiness of its counterparties. In doing so, it liaises with supervisory colleagues in the BoE's Prudential Regulation Authority (PRA).

Until recently, however, the BoE had not had a dedicated second-line financial-risk function, reflecting a perception that central banking operations did not give rise to tensions of the sort seen between profit-making and risk control in private firms. The amount of residual risk to which the BoE was actually exposed, after taking account of large collateral haircuts and government indemnities, was also felt to be relatively small. But more recently, the risks had become more complex. Contingent exposures have become larger and more difficult to measure, aggregate across the balance sheet, and compare against resources. Also, to hold itself accountable, the BoE needed more effective internal challenge mechanisms.

So in June 2015, the BoE set up a new second-line financial-risk function,[11] staffed with some of the risk experts who previously worked in its supervisory arm to challenge commercial banks—in the spirit of doing unto itself as it does to others. In 2018 the BoE's risk governance was further strengthened by the creation of a fully fledged and independent Risk Directorate, of which the financial-risk function was a founding constituent—reporting directly to the governor.

Much of this will feel familiar to risk professionals in commercial banks. And in some aspects, it is only central banks catching up. But in other aspects, the BoE has had to pioneer an uncharted path, reflecting the unique and unusual nature of some aspects of risk management at a central bank.

3.1 A New Toolkit

The objective of the BoE's new financial-risk framework has been, all along, to provide forward-looking assessments and challenges of financial risks to the balance sheet across all its operations. This is so that the BoE can achieve monetary and financial stability without exposing public funds to greater risk than necessary. The keystone of that system has been a risk-tolerance statement, which embodies governors' and the Court of Directors' stance that

1. the risk implications of policy proposals should be thoroughly evaluated before decisions are made; and
2. such decisions should be implemented in a way that takes an amount of risk that is justified by the policy's objective, but not more than that.

Quite a different way of articulating risk tolerance from the way commercial firms do. It reflects the fact that central banks may have to take a lot of risk when everyone else is trying to take less. To achieve that, the BoE has deployed a suite of second-line risk management tools. They fall into the following four categories of activities:

---

[11] See www.bankofengland.co.uk/speech/2017/watching-the-watchers-forward-looking-assessment-and-challenge-of-a-central-banks-own-financial-risk.

1. **Detecting vulnerabilities** of the balance sheet and asset-purchase facility to potential financial loss under a range of scenarios due to overall credit, market, and liquidity risks arising from the combination of the BoE's operations. The tools used for that are stress tests and targeted risk reviews of particular exposures.
2. **Evaluating exposures** across operations. That implies developing and using methodologies to measure the BoE's total current and forward-looking exposures to financial risks across all combined market operations, including implications for capital needs.
3. **Articulating tolerance**. The second line owns the framework to articulate the BoE's financial risk tolerance, translates it practically into usual risk levels, and maintains risk standards accordingly. It also monitors and reports periodically to governors and the Court of Directors on the level of financial risk present on the balance sheet.
4. **Challenging** new policy proposals from the perspective of their risk implications and challenging risk and pricing tools built (e.g., haircuts), or risk assessments made, by the first line of defense.

Stress testing has represented a fundamental ingredient sprinkled across all these activities. The next section describes the richness of what a central bank can use it for. Although the BoE is used as an example here, the use of stress tests and the way they work depend on the specific priorities of the central bank. That is because across jurisdictions, the nature of central banks' balance sheets and other tools can vary markedly, as does the nature of risk sharing with the sovereign.

4 Use of Stress Testing in a Central Bank

Broadly, the BoE has started to stress test its balance sheet for two main purposes:

- **To articulate risk tolerance, including implications for capital needs.** The BoE has determined that it should be able to withstand severe but plausible scenarios without exhausting its independent financial resources; for example, "the BoE needs to have sufficient capital to cover several scenarios, including one in which house prices fall x% (worse than what banks are asked by the PRA to hold capital for), several banks experience liquidity outflows as bad as the crisis and Y counterparties fail." The scenarios that the BoE uses to articulate its own risk tolerance are informed by those set by the PRA and the Financial Policy Committee (FPC) for the resilience of the banking sector, which set out how much self-insurance commercial banks need to have.
- **To check that it is ready, from a risk standpoint, should a given scenario occur.** That means helping the BoE identify what it might need to do in advance need to have be prepared. This is an area where the BoE's own stress tests are beginning to inform the supervisory stress tests, as discussed later in this chapter.

4.1 A Tool to Measure the Adequacy of Central Bank Financial Resources

Central bank capital is different from that of a commercial bank; as mentioned earlier, central banks need to meet a test of "policy solvency" rather than a "financial solvency." It also sits at the base of a particularly steep waterfall. That is because any loss in a lending operations would be expected to be met first by a solvent firm's own resources, second

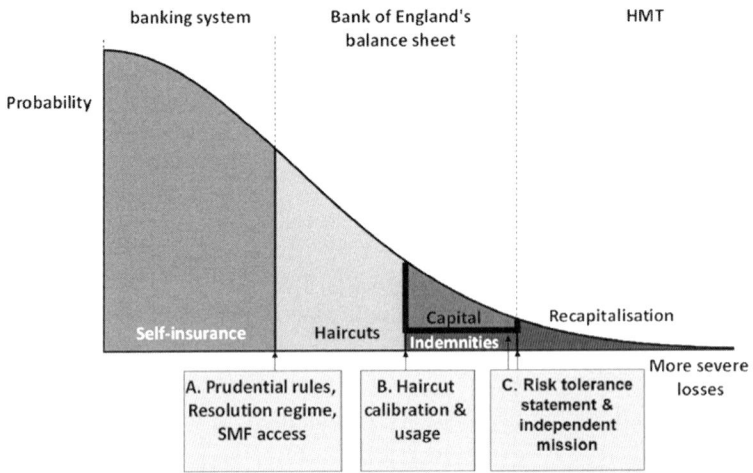

Figure 29.4 The central bank capital waterfall.

by the central bank's collateral haircuts, and only then by its capital (unless the particular operation is indemnified directly by the government). Any loss beyond that implicitly falls to the government, as the ultimate owner of the central bank (Figure 29.4).

Central banks across the globe differ in where they sit in this spectrum. In the UK, up until recently, the BoE has operated with a relatively low level of capital, with few means of rapidly replenishing that capital if it is depleted (Figure 29.5). As a result, the BoE has historically been much more reliant on government indemnities than other central banks.

Of particular importance is the fact that the BoE does not have access to seigniorage income. Seigniorage is the income from the issuance of banknotes and has historically constituted the main financial resource of many central banks. In an accounting sense, it is the flow of net income that the central bank earns as a result of holding income-generating assets against its (non-interest-bearing) banknote liabilities, less the cost of producing banknotes.[12] In the UK, unusually, the entirety of that income is always passed to the UK's Treasury, as it has since 1844 by law.[13]

In contrast, other central banks in major developed economies typically do retain certain rights to seigniorage income. For example, that is the case for the US Federal Reserve, the European Central Bank (ECB), the Bank of Canada, and many other central banks.[14] In those cases, seigniorage income is earned by the central bank (rather than another accounting entity) and can effectively be retained when needed if equity capital has been depleted by a large shock (rather than automatically distributed to shareholders). More generally, central banks have a range of risk-sharing arrangements that determine the extent to which

---

[12] Although there are different concepts of seigniorage in practice, this is the relevant definition for our discussion.

[13] The Bank Charter Act 1844 (see www.legislation.gov.uk/ukpga/Vict/7-8/32). Seigniorage income is separated out via a distinct accounting entity, the Issue Department.

[14] See Archer and Moser-Boehm (2013).

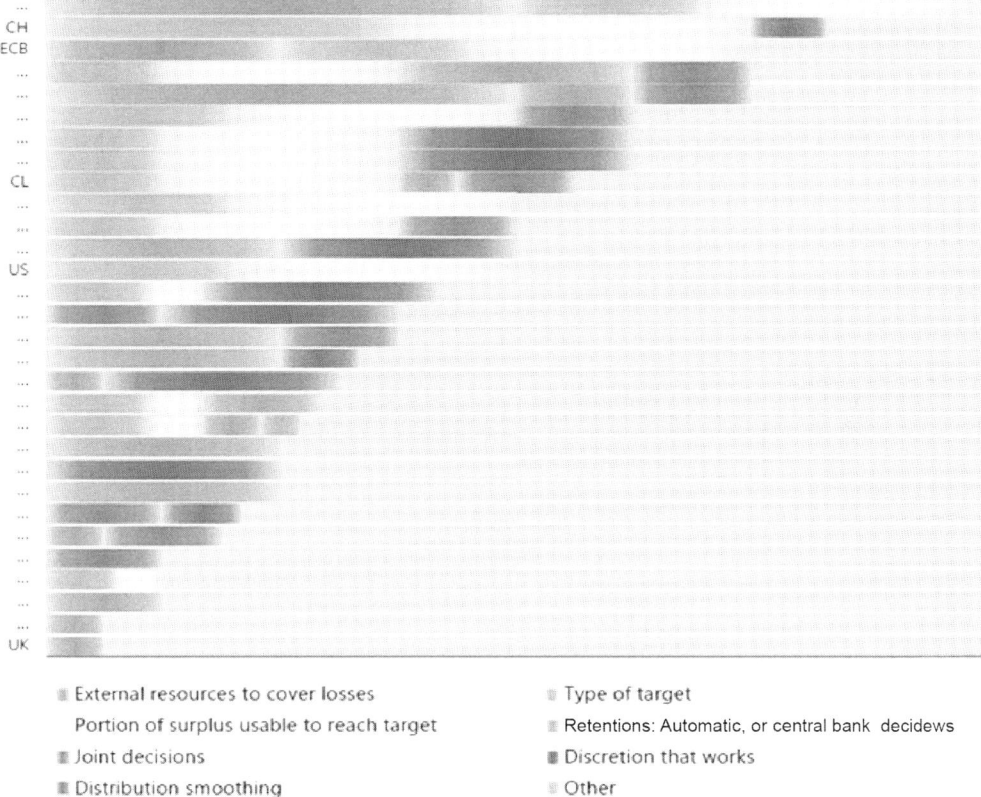

■ External resources to cover losses
 Portion of surplus usable to reach target
■ Joint decisions
■ Distribution smoothing

 Type of target
■ Retentions: Automatic, or central bank decidews
■ Discretion that works
 Other

Figure 29.5 Central banks' ability to recover from a loss. Note: The authors of this chart made the bars intentionally fuzzy because their evaluation is necessarily subjective; note that most central banks in the sample have been anonymized by the authors.
A higher index indicates greater financial strength, based on components of distribution arrangements that aid in the retention of financial resources.[15]
Source: Archer and Moser-Boehm (2013). Full publications are available at www.bis.org/.

distributions to shareholders can be adjusted over time (potentially extending to injections of fresh capital).

Accordingly, as was announced in the 2018 Mansion house speeches[16] of the governor and the chancellor of the Exchequer, the BoE and the UK's Treasury recently agreed on a new capital framework for the BoE.

4.1.1 The Bank of England's New Capital Framework

That new framework was codified in a public MoU.[17]. At its heart is a risk-based capital target, reflecting forward-looking risks to the balance sheet over a 5-year horizon. The BoE's capital is capped by a ceiling, above which all net profits are transferred to the

---

[15] This assessment was conducted prior to the BoE's new capital framework outlined later in the chapter.
[16] See Carney (2018).
[17] See www.bankofengland.co.uk/letter/2018/banks-financial-framework-june-2018.

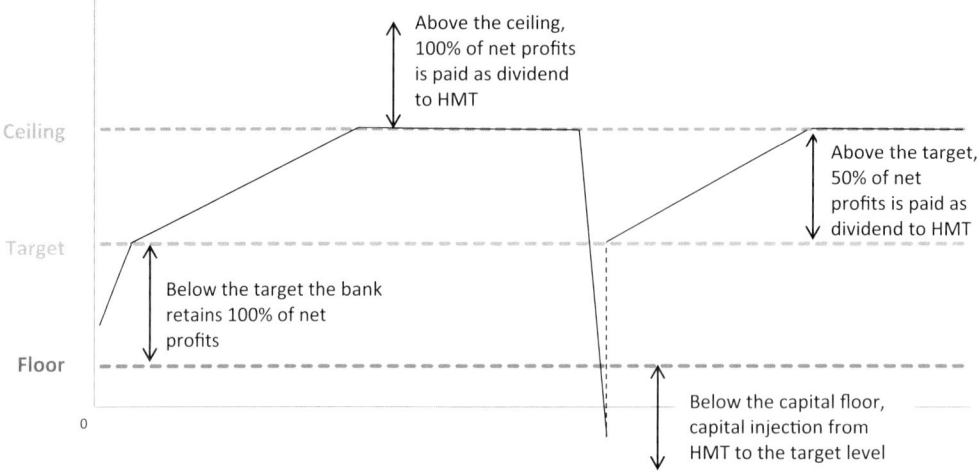

Figure 29.6 Mechanism of the BoE's new capital framework.

HMT as dividends (Figure 29.6). The framework also ensures that capital remains above a floor, below which a rapid recapitalization to the target is triggered. The floor is set as the level below which the credibility of the BoE's unconstrained ability to deliver its mission autonomously would be in sufficient jeopardy to warrant timely action. In addition, a variable income-retention mechanism was introduced, whereby no dividend is paid to the UK's Treasury when capital is below target, and 50 percent of net profits is paid as dividend when capital is between the target and the ceiling.

The risk-based target is at the heart of this new capital framework. Its level is determined by stress tests of the BoE's own balance sheet, evaluating the loss impact of severe but plausible stress scenarios, which are reviewed annually. As will be covered in the next section, the key building blocks of these central bank stress tests are the liquidity stress tests of banks that may require liquidity assistance from the central bank in some scenarios.

The initial values of the capital parameters have been $0.5 billion for the floor, $3.5 billion for the target, and $5.5 billion for the ceiling. In addition, the UK's Treasury injected $1.2 billion into the BoE in March 2019, which brought the BoE's loss-absorbing capital from its previous level of $2.3 billion to the $3.5 billion target. The $3.5 billion capital target is risk based and an outcome of the BoE's own stress tests. These parameters are to be formally reviewed by the BoE and the UK's Treasury every 5 years. However, both organizations have agreed that in circumstances where the risk environment faced by the BoE changes fundamentally, an intermediate review may be warranted.

The capital injection of $1.2 billion enabled the BoE to take the Term Funding Scheme (TFS; a market-wide liquidity facility) on balance sheet without an indemnity from the Treasury.

In his letter to the chancellor on the day of the 2018 Mansion House Dinner, the governor wrote that the new financial framework "provides a robust and transparent system that ensures the credibility of the BoE's policy actions in even the most stressed environment. The framework reflects the BoE's much wider range of responsibilities including banking supervision, macro-prudential policy and resolution. These responsibilities create a range of potential new calls on the BoE's balance sheet. The new framework also reflects the new ways in which the BoE provides liquidity. It is capable of lending against a wider range of collateral, to a wider range of counterparties, for longer terms, and at lower prices than ever before."

Having more capital enables the BoE to do more without indemnities. However, the new framework does not mean that the BoE would take on any level of financial risk without recourse to the Treasury. Particularly where that risk might be so large that holding capital against it would not be the best use of public funds—or would involve potentially sensitive actions—the BoE would expect to consult with HMT and, where necessary, seek indemnities. The BoE and HMT have agreed on a set of principles, codified in the public MoU, for which risks should be backed by the central bank's capital versus government indemnities.

The governor said in his Mansion House speech: "The additional capital will significantly increase the amount of liquidity the BoE can provide through collateralised, market-wide facilities without needing an indemnity from HM Treasury to more than half a trillion pounds." The link between the capital stress test of the central bank and the liquidity needs of commercial banks under stress is the key underpinning here.

4.2 A Tool to Articulate Risk Tolerance and Manage Risk

The BoE is using stress tests to articulate its own risk tolerance: it needs sufficient financial resources to deploy its balance sheet and withstand stresses without recourse to the Treasury, and it is willing to take an amount of risk that is appropriate given expected policy benefits, but not more than that. Indeed, the role of the second line is to make sure the organization challenges itself that the risks are worth it. Yet such a concept of risk return is more difficult to measure for a central bank than for a commercial bank. The former seeks policy benefits; the latter seeks profits.

The BoE now uses stress tests, and stressed exposures, as a framework to help with such trade-offs of risk versus policy benefit. Importantly, the stress tests of risk exposures do not prevent risk taking; remember, a central bank has to take more risks in times of stress. Rather, the BoE uses them to delineate the boundaries of the "usual risk levels": the levels of stressed contingent exposure beyond which an active decision is needed that the extra risk is justified by the expected policy benefit.

Another use of the BoE's own stress testing framework as a tool is to articulate the maximum amount of lending of any given type (collateral, currency, counterparty concentration) that the BoE can afford without government indemnity.

4.3 A Tool to Test Readiness

Other than in a capital context, the BoE has been using stress tests in business planning. Understanding the risk implications of stress scenarios has been a key tool for the BoE to challenge itself that it is prepared should some of the scenarios occur. The typical questions that the BoE has been using such exploratory stress tests to answer are as follows:

- *Sensitivity analysis:* Which scenario assumptions matter most, in that they make the greatest difference to the losses that would be experienced under stress? Examining this question can help ensure a focus of risk management on the aspects of defenses that are economically of greatest importance.
- Does the central bank have the right type of facilities to address the nature of liquidity stress under the circumstances being examined?
- Do firms have enough collateral available to secure the amount of liquidity assistance they would require?
- Do some firms need to be asked in advance to preposition at the central bank more collateral of a given type?
- Without being procyclical, are the collateral haircuts appropriate for particular types of secured lending under the scenarios considered?
- Should the central bank explore with some firms whether they are able to substitute the prepositioned collateral at the central bank so that it is either of a different type or in a different currency?
- Should some of the risks identified be borne by the private sector rather than the public sector? Is it preferable for firms to extinguish potential liquidity needs at their source so that, should the scenario occur, the central bank would no longer need to step in and thereby put public funds at risk? If so, the central bank, as supervisory authority, can have that conversation in advance with the firms concerned.
- Is there a need to get in touch with fellow central banks or supervisors in relation to firms with international footprints in order to agree in advance on a preferred course of action should some of the liquidity needs materialize?
- Under which circumstances would resolution be a better central bank tool than the use of its balance sheet?

Stress tests have proved to be particularly useful tools for the BoE to answer these types of questions. They help determine in advance what its preferred course of action would be, should the scenarios occur. And they help the BoE think broadly across the full central banking toolkit at its disposal, rather than only the traditional financial tools (e.g., haircuts). Some of these actions can be taken ex ante in normal times—to be as prepared as practicable—whereas others would only be taken if the scenarios crystallized.

For the BoE, the wider power of its own stress tests was demonstrated in 2016 when we ran an exercise to evaluate the potential implications of different possible outcomes of the UK referendum on membership of the European Union. The results of that work helped to inform judgments on many of the issues outlined earlier in the chapter, as well as engagement with fellow central banks and our own supervisory actions. Importantly, it confirmed that the BoE had sufficient lending capacity in place to meet an extremely severe liquidity shock. That allowed the BoE to make reassuring statements about its readiness to act. More recently, its own stress tests also played a core role in the BoE's preparations for a range of possible Brexit scenarios. For example, they contributed to the creation of its new Liquidity Facility in Euros.

Now that we have described how useful stress tests can be for a central bank, let us turn to how they work in practice.

5 How to Stress Test a Central Bank Balance Sheet in Practice

How does one go about stress testing a central bank? In summary:

- A natural starting point is to build on financial stability assessments and supervisory stress tests, as part of "scanning the horizon" for key stress events.
- But it is also important to recognize that central bank balance sheets are unique, and tests need to be tailored to their specific vulnerabilities.
- Calibrating the tests involves decisions around three core issues (the three "Ss"): scope, scenarios, and severity.
- The BoE's tests are inherently forward-looking. They build on supervisory information (especially liquidity) to gauge the potential need to expand its balance sheet in a stress.
- As well as measuring potential losses, tests can be used as a vehicle for assessing a wide managerial toolkit to prepare for and respond to stress events.
- And finally, central banks are themselves a part of the financial ecosystem. As far as possible, the stress tests of central banks versus commercial banks should strive to recognize this endogeneity/general equilibrium.

5.1 Scanning the Horizon

When it comes to exploring stress events, central banks are not starting from scratch. A core function of modern central banks is to assess risks to the stability of the financial system. For example, the BoE's June 2018 Financial Stability Report highlights several key risks to UK financial stability: the provision of financial services around Brexit, global debt-market conditions, UK external financing, and household indebtedness.[18] As prudential supervisor, the BoE also designs a set of common stress tests for the financial institutions that it supervises.[19]

These same risks and scenarios may be relevant to stress testing of the central bank balance sheet. Should they crystallize, there would likely be large movements in asset prices and exchange rates that would directly affect any market positions through outright asset holdings. The BoE might ultimately expand its balance sheet, either to ease monetary conditions or to support financial stability. Some counterparties may draw upon central bank facilities at the same time as their creditworthiness is deteriorating.

However, stress tests also need to be tailored to the unique circumstances of a central bank. Taking supervisory tests "off the shelf," or focusing only on the risks identified in financial stability reviews, may not sufficiently capture the shocks that would be most damaging. For example:

- Some types of stress events in parts of the financial system (including in banks) might have little direct consequence for the central bank balance sheet if there would be little need for, or likelihood of, liquidity support.

---

[18] See www.bankofengland.co.uk/financial-stability-report/2018/june-2018.
[19] See www.bankofengland.co.uk/-/media/boe/files/stress-testing/2017/stress-testing-the-uk-banking-system-2017-results.pdf?la=en&hash=ACE1E2FB54482F5DC3412864C6907928B622044A.

- Conversely, some events that might not be systemic might have a large financial impact on central banks. For policy reasons, some have large open-market positions in exchange rates or interest rates, against which private-sector firms would tend to be well protected.

Indeed, the risks that matter most might differ drastically from one central bank to another. The key vulnerabilities for a given institution will depend on how monetary policy is implemented, the nature of liquidity facilities in place, accounting approaches, and the extent to which some risks are absorbed directly by the government.[20] These factors all come into play when setting up the appropriate stress testing framework. The reality is that from a central bank stress test perspective, the relevant lens for whether a scenario is "severe" or not is whether it constitutes a material channel of loss for its balance sheet.

5.2 Designing a Stress Testing Framework

What are the specific steps in building a suite of concrete stress tests? There are three questions that need to be answered—the three Ss:

1. What is the *scope* of the stress test?
2. Which *scenarios* push a central bank's financial vulnerabilities?
3. What is the appropriate *severity* of the scenarios?

5.2.1 What Is the Scope of the Stress Test?

The first decision is which aspects of a central bank's *operations* are within the scope of the stress tests. This question has several aspects:

- *Which operations is capital held against?* For the BoE, certain operations are out of scope for capital planning because those financial risks are absorbed directly by the government. That is true for much of the BoE's unconventional monetary-policy operations (QE), and the portion of the balance sheet that relates to the issue of banknotes.[21] The new capital framework MoU has a section that sets out the agreement between the BoE and the UK Treasury on the rules for which operations should be backed by the capital of the central bank versus by government indemnities.
- *What is the scope for the balance sheet to change over time?* In addition to the current balance sheet, as already discussed, many of the financial risks to a central bank are contingent—they would arise through new operations during a period of financial or economic instability. This is true of lender-of-last-resort operations and potentially also monetary-policy operations. The BoE's stress tests are very much forward-looking in nature and allow for the possible expansion of the balance sheet within the existing operational framework as a lender of last resort.
- *Precisely which new operations might be undertaken in the event of a stress?* Should this be limited to market-wide facilities and operations with published terms, or should it also include firm-specific emergency liquidity assistance or lending in resolution? Should it include lending only in local currency or also foreign currency, potentially facilitated by central bank swap lines? Might new counterparties seek eligibility for central bank

---

[20] See Rule (2015).
[21] See annual report.

operations?[22] These aspects of operations have changed dramatically for many central banks since the crisis, and for a central bank with a broad mandate, the future is difficult to judge.

At the BoE, for capital-setting purposes, the scope of stress testing, at least initially, has focused on its existing unindemnified balance sheet, as well as potential expansions as per its role as lender of last resort.[23] The BoE's approach has been to focus on the public commitments to provide market-wide and firm-specific liquidity insurance through the BoE's current set of facilities and operations. That is embodied in the key principles that have been agreed on between the BoE and HMT, which state that the following types of operations should be backed by capital:

- Secured lending in line with the BoE's published frameworks, including against eligible collateral.
- Asset-purchase operations to support conventional monetary-policy implementation, the BoE's official customer business, or the funding of the BoE.

5.2.2 Which Scenarios Probe the Central Bank's Financial Vulnerabilities?

Within the scope of the stress test, scenarios need to be selected that will probe the vulnerabilities of the balance sheet. Any number of scenarios could be examined. The key is to design them to "push where it hurts"—the same idea as reverse stress tests.

For outright market exposures, the stress will need to involve a shock to the combination of market prices to which the central bank is most exposed. For market risk, the BoE uses a stressed loss metric to measure the potential mark-to-market losses from shocks to asset prices.

For secured lending (including as lender of last resort), there are multiple layers of strong protection. Financial losses arise only if a counterparty defaults *and* the amount recovered from the collateral is insufficient to reimburse the BoE's loan. The scenarios of most interest are therefore those where a number of things go wrong in a way that diminishes the multiple layers of protection—that is, they give rise to "wrong-way risk." Several aspects of wrong-way risk have been of particular interest in the BoE's scenario design:

- The counterparties that draw upon liquidity facilities in the largest volume also experience a decline in creditworthiness.
- A shock affects both a particular counterparty and the collateral that it has pledged. For example, a housing shock might affect both the capital position (and hence the creditworthiness) of a bank and the value of the mortgage collateral that it has pledged to the central bank.
- Lending and the collateral pledged are in different currencies, and the default of a major institution is associated with a large movement in the exchange rate.

---

[22] A stress test could allow for the possibility that the number of eligible counterparties grows during a crisis. For example, that might occur under Section 13 of the US Federal Reserve Act or related provisions for other central banks.

[23] The asset-purchase facility is also subject to stress tests, but these are separate from considerations of capital capacity.

For the stress scenarios applied to tease out vulnerabilities in secured lending operations, the BoE uses a stressed exposure-at-default (EAD) metric to measure the potential financial loss that could be incurred in the event of counterparty default, net of collateral held, where that collateral is also stressed.

5.2.3 What Is the Appropriate Severity of the Scenarios?

Finally, the severity of the scenarios needs to be decided. How much do asset prices fall? How large is the demand for borrowing from the central bank? How many counterparties fail, of what size, and how badly impaired is their collateral? Choosing such severity, either for capital planning or to test readiness, constitutes an articulation of the BoE's risk tolerance.

For capital planning purposes, the scenarios would generally need to be very severe, given the robust buffers in place against losses. In particular:

- They need be at least as severe as supervisory stress tests (including both capital and liquidity stress tests). This is consistent with the notion that the supervisory tests describe events that firms should be able to withstand using their own resources. It is when firms' liquidity needs go beyond the level that could reasonably be covered by their own resources that the central bank might be called upon to act in size.
- They also need to consider events that may be more severe than assumed by haircuts on collateral. The BoE's haircuts are conservative, but they cannot eliminate financial risk— nor should they aim to. For example, the stress might consider scenarios in which it takes longer to liquidate collateral than is typically assumed by haircuts. That might be the case if the central bank ends up having to recover value from a large concentration of collateral assets of the same type, possibly from different counterparties, and possibly representing a material portion of the overall market for such assets. It is not the role of haircuts to protect against such concentration risk.

These are clearly events deep in the tail. However, the scenarios chosen should not be so extreme that they are implausible. If the scenarios are too extreme, it would be an inefficient use of public funds for the central bank to hold capital against them. That is, they should be "severe but plausible." Although that terminology is also used to describe the severity of supervisory stress tests, as described earlier, for the BoE this refers to states of the world that are typically more in the tail. This reflects the fact that because the central bank has naturally stronger defenses, it takes more for an outcome to feel "severe."

As noted previously, the actions of the central bank are designed to reduce the financial stability impact—and hence the severity—of the shock, restoring confidence in banks and mitigating fire sales of assets. That is, severity itself is endogenous. However, building this endogeneity into the design of stress tests is challenging in practice. Firstly, the actions of the central bank can mitigate the liquidity aspects of a shock, but not necessarily its solvency aspects. Secondly, incorporating the feedback effects of central bank actions requires an understanding of the effects of policy interventions through the broader system of banks, nonbanks, and markets that compose the financial system. These effects are difficult to calibrate. Bindseil and Jabłecki's (2013) model assumes that there is full information on the structure of the system. Currently, these parts of the financial system are either stress tested at an individual-institution level (and then separately for solvency and liquidity shocks) or not at all, meaning that many of the interactions are not well understood. In the absence of a fully developed model of financial system behavior under stress, it is prudent from a risk management perspective to ensure the central bank's defenses are at least prepared for the worst.

For exploratory (rather than capital-setting) purposes, less severe scenarios might be of more interest. Given the strong protections in place, the central bank may want to know whether some lower-severity events might generate an unexpectedly large (but avoidable) loss or otherwise test certain constraints of the operational framework. There is a lot central banks can do before scenarios reach extreme severities, and indeed, that in itself might well help make such extreme outcomes even less likely. So solely focusing on the extreme may not be the most productive thing to do.

5.3 Linking the Scenario to the Balance Sheet: Liquidity Demand

Many of the issues outlined previously have parallels in other forms of stress testing, and thus their practice is well established. But one area that is unique is assessing the potential demand for central bank liquidity assistance in a stress.

At some level, this is similar to a commercial bank's customers drawing down on precommitted lines of credit during a period of economic stress. But in the case of a central bank, (1) the size of potential drawdowns can be particularly large relative to its business-as-usual balance sheet; (2) the size of the commitment is not clearly capped, except perhaps by the available collateral; and (3) the customers in this case include systemic financial institutions, which can otherwise raise liquidity in the market before turning to the central bank. They have a complex array of options and behave as a broad system.

As a starting point, the size of the domestic financial system can have a bearing on the scale of a potential liquidity event. In the case of the UK, the banking system is large relative to the size of the economy, owing to certain advantages in the provision of these services, agglomeration benefits, and history (Figure 29.7).[24] Indeed, one potential test might be to

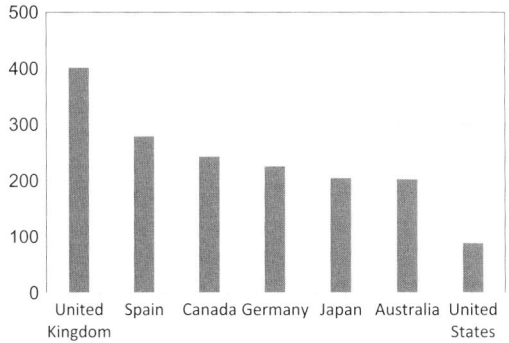

Figure 29.7 Selected banking-system assets as a share of GDP.
Sources: US Federal Reserve, IMF, Bank of Japan data.

---

[24] See Bush et al. (2014).

presume that the system borrows in a stress up to the value of all currently eligible collateral (in the spirit of a reverse stress test).

History can provide some guidance for the size of potential liquidity events relative to the scale of the financial system. The International Monetary Fund (IMF) has collected data on central bank lending in 151 systemic banking crises across economies since 1970. These data indicate that, for those crises where there was material liquidity support, this amounted to 20 percent of the value of deposits in the system on *average* (Figure 29.8). That is, around one-fifth of broad money was converted into high-powered money on average.[25]

This approach provides a useful benchmark. But it is lacking in two important respects:

- The size of liquidity insurance has varied significantly from one episode to another and across economies, depending on the size of the shock and resilience of the financial system. For example, the demand for central bank liquidity in a stress will depend (for a given size of the financial system) on the extent to which institutions are leveraged and take liquidity risk, the amount of self-insurance via holdings of liquid assets, access to central bank facilities, the degree of stigma around using these facilities, and other institutional features/policy measures (e.g., the presence of deposit insurance).[26]
- This historical approach sheds little light on which parts of the system are most at risk, the nature of the lending that the central bank might need to extend (e.g., which currency), and the likely type of or sufficiency of collateral. That is particularly problematic when it comes to examining event risks, such as the impact of the UK referendum on Brexit.

A better alternative is to use liquidity stress testing data for individual banks that might draw upon central bank facilities. This approach has the advantage of capturing the nature of the liquidity risks in the financial system as well as how these change over time. Liquidity stress tests set out circumstances in which banks (or nonbanks) are unable to roll over certain funding or experience withdrawals, resulting in a need to raise liquidity. That is, they identify "stress outflows." However, supervisors often run different variants of such stresses of differing severity, for example, a stress event that lasts for several months rather than just one month.

How can stress outflows be mapped into demand for central bank liquidity? This is a difficult exercise that requires judgment around a range of factors. Figure 29.9 provides a stylized illustration. The mapping requires assumptions around the following:

1. *What portion of the system is subject to outflows?* In the short term, the banking system cannot destroy liquidity; while some firms experience outflows, others will experience corresponding inflows.
2. *To what extent are liquidity outflows met through the market rather than the central bank?* Banks hold liquid assets; indeed, the purpose of supervisory stress tests is to ensure that these holdings are adequate. Accordingly, their first resource is to sell/repo these assets in the market, effectively ensuring that the system recycles liquidity back to the firms that require it. However, banks may still have the *option* to resort to central bank facilities. They might do so because this provides a liquidity upgrade, converting less

---

[25] Note that these data capture any liquidity extended by treasury authorities as well as central banks.
[26] See discussion in Bush et al. (2014).

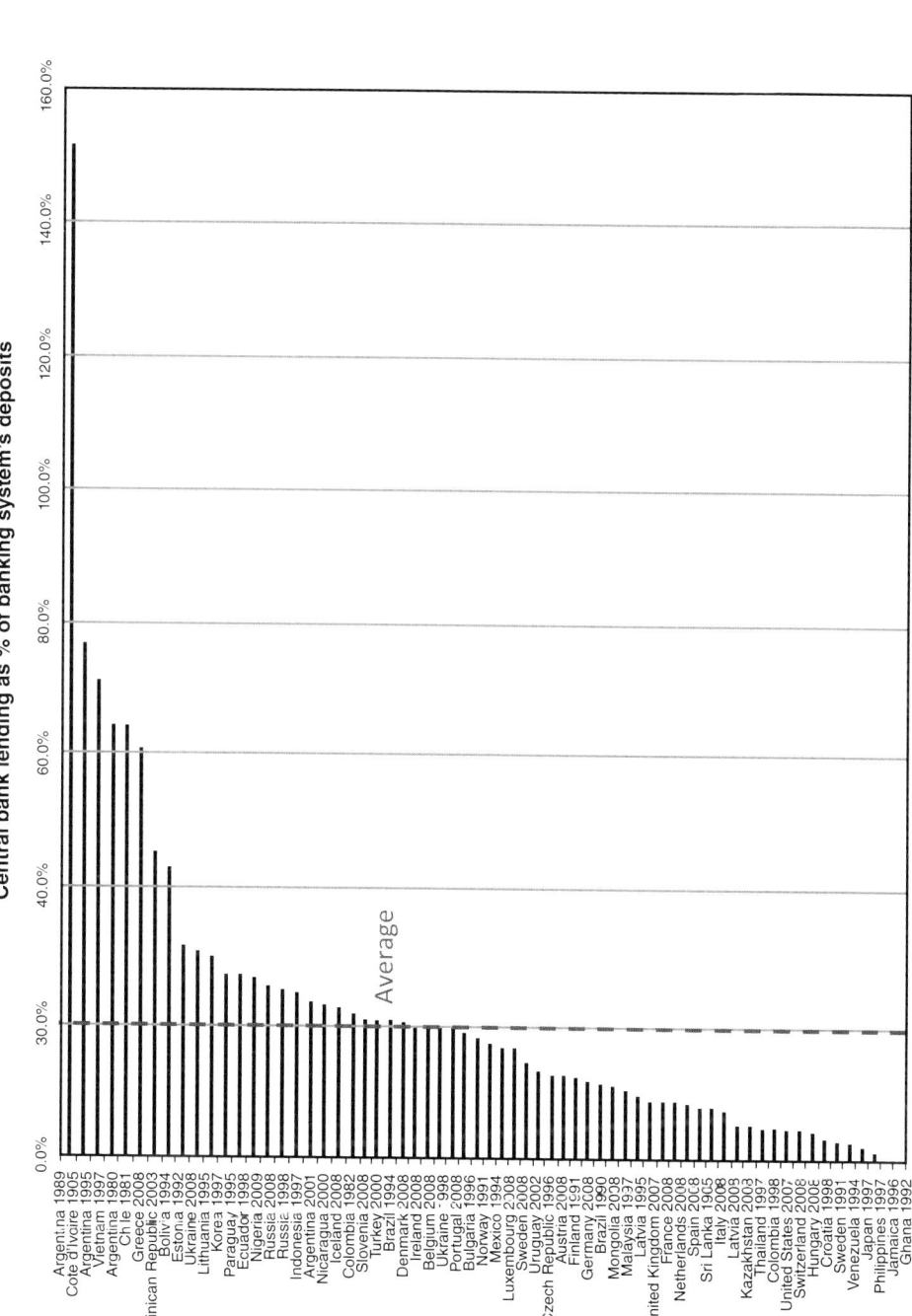

Figure 29.8 Central bank lending during systemic banking crises as a share of banking-system deposits (1970–2017).
Source: Laeven and Valencia (2018).

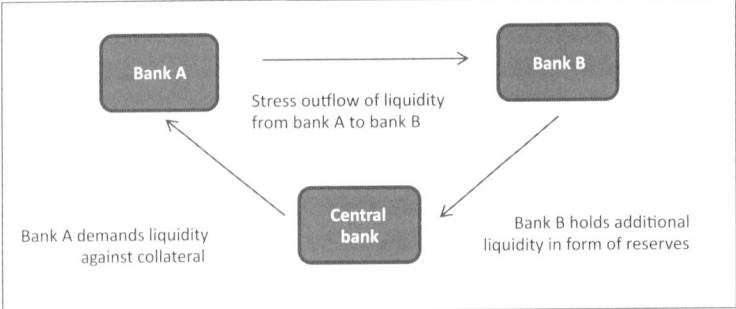

Figure 29.9 Stylized link between stressed liquidity outflows and the demand for central bank liquidity.

liquid assets (e.g., mortgages) to liquid assets, ensuring that they maintain a war chest of liquid assets. Banks may choose to hoard liquidity, increasing the demand for reserves and thus pushing up its cost in the market. At some point, the cost of raising liquidity in the market may make central bank borrowing a more attractive option. Finally, the degree of perceived stigma around central bank facilities will be an important factor. These are all behavioral variables, which will in part depend on the expectations of bank supervisors.

This section described the mechanics at play in central bank balance-sheet stress tests. A natural question is then how that knowledge can inform the supervisory stress tests to which the prudential authorities subject the banking system—and thereby close a key loop on macroprudential stress tests, reflecting the role of central banks within the financial system. The following topics are relevant to that feedback loop:

- Scenarios: states of the world where firms might struggle to provide enough (or the right type of) collateral to receive the liquidity assistance they need from the central bank.
- Market concentrations: many firms in isolation relying on the same asset class against which to obtain central bank funding.

With its upcoming 2019 Biennial Exploratory Scenario (BES),[27] the BoE is taking further steps toward exploring the endogeneity pertaining to its own role within the financial system. Indeed, the theme of the stress test is the implication of a broad-based liquidity shock. And two of the test's four objectives relate directly to the role of the BoE within the system:

- To improve public understanding about the role of the BoE in mitigating liquidity risk to the UK financial system. The BES will help to raise awareness of how the BoE's liquidity facilities—which include the ability to lend to banks in all major currencies—operate in a liquidity stress and how they interact with the PRA's regulatory framework.
- To understand banks' possible demands on the BoE's liquidity insurance facilities in stress, as well as the risks the BoE itself would be exposed to through providing that

---

[27] cf. July 2019 Financial Stability Report (www.bankofengland.co.uk/-/media/boe/files/financial-stability-report/2019/july-2019.pdf). This exercise took a more internal form given COVID-19 developments.

liquidity. The BES will enhance understanding of how banks will use its facilities and thereby continue to inform the design of the liquidity-assistance framework.

The interplay between commercial and central bank stress tests is firmly in focus.

6 Conclusion

Hopefully this chapter will have given the reader a flavor of the broad range of issues involved in central bank stress testing. It is altogether a new discipline, at the vanguard of modern central bank risk management, which is definitely characterized by a number of "twists" away from normal stress testing practice in the private sector. Shaping up this new discipline, and putting it to immediate and powerful use, has been a real focus of initiative at the BoE—and we are proud of it. The BoE is used as an example here, and naturally, the role stress testing plays as a central bank risk management tool will vary across jurisdictions. But much of what we have learned applies universally, and the material here will be valuable for anyone considering the risks a central bank faces.

Many questions merit greater exploration. For example, the true depth of the feedback loop between central and commercial bank stress tests is a key issue. By understanding the impact of some macroeconomic scenarios on both commercial firms and central banks that come to the rescue, as well as the effects of policy intervention itself, can we better measure the size and nature of the risks to the central bank? And ultimately, can that make the financial system—and incidentally, the central bank's balance sheet—safer? The depths of these general- versus partial-equilibrium questions, thinking about the central bank's own role within the financial system, are yet to be fully explored. But they represent a key difference between what a central versus a commercial bank can and should consider when it stresses itself.

Yet some wider lessons are already clear. The more regulated firms improve the quality of their internal liquidity stress tests, the more effectively central banks will be able to provide liquidity assistance during crises. Firms can put greater focus on how they would deal with situations where they are shut out from funding and/or currency markets. They can earmark in advance the collateral (beyond what is routinely prepositioned) they would pledge at the central bank to borrow liquidity against in times of stress. These are all steps firms can take to help the central bank help them without putting public funds unduly at risk.

References

Archer, David, and Paul Moser-Boehm (2013), "Central bank finances," BIS Papers No. 71, available at www.bis.org/publ/bppdf/bispap71.pdf.

Bagehot, Walter (1873), "Lombard Street: A description of the money market," available at http://public-library.uk/ebooks/58/60.pdf.

Bank of England, Federal Reserve, US Bureau of Labour Statistics, ONS, and Samuel H. Williamson (2017), "What was the U.S. GDP then?" available at www.measuringworth.com/datasets/usgdp/.

Bholat, David, and Karla Martinez Gutierrez (2019, October 18), "The ownership of central banks," *Bank Underground*, Bank of England.

Bindseil, Ulrich, and Juliusz Jabłecki (2013, May), "Central bank liquidity provision, risk-taking and economic efficiency," ECB Working Paper No. 1542, available at www.ecb.europa.eu/pub/pdf/scpwps/ecbwp1542.pdf?1deaeaa1be0d40e3512ef9bab21bc4f8.

Bush, Oliver, Samuel Knott, and Chris Peacock (2014), "Why is the UK banking system so big and is that a problem?" *Bank of England Quarterly Bulletin*, 53(4), 385–395, available at www.bankofengland .co.uk/quarterly-bulletin/2014/q4/why-is-the-uk-banking-system-so-big-and-is-that-a-problem.

Carney, Mark (2018, June), "New economy, new finance, new bank," Speech given at the Mansion House Bankers' and Merchants' Dinner, London, available at www.bankofengland.co.uk/-/media/boe/files/ speech/2018/new-economy-new-finance-new-bank-speech-by-mark-carney.pdf?la=en&hash=F362C661 810257FE779E3D5BCA17144EC65E8C63.

Hauser, Andrew (2017, March), "Watching the watchers," Speech given at the GARP 18th Annual Risk Management Convention, New York, available at www.bankofengland.co.uk/speech/2017/watching-the-watchers-forward-looking-assessment-and-challenge-of-a-central-banks-own-financial-risk.

Laeven, Luc, and Fabian Valencia (2018, September), "Systemic banking crises revisited," IMF Working Paper No. 18/206, available at www.imf.org/en/Publications/WP/Issues/2018/09/14/Systemic-Banking-Crises-Revisited-46232.

Plenderleith, Ian (2012, October), "Review of the Bank of England's provision of emergency liquidity assistance in 2008–09," Report presented to the Court of the Bank of England, available at www .bankofengland.co.uk/-/media/boe/files/news/2012/november/the-provision-of-emergency-liquidity-ass istance-in-2008-9.

Rule, Garreth (2015), "Understanding the central bank balance sheet," *Centre for Central Banking Studies Handbook No. 32*, Bank of England.

Stella, Peter (2005), "Central bank financial strength, transparency, and policy credibility," IMF Staff Papers, 52(2), available at www.imf.org/External/Pubs/FT/staffp/2005/02/pdf/stella2.pdf.

Part VI

Concluding Thoughts: The Road Ahead

30

Stress Testing the Financial Macrocosm

J. Doyne Farmer, Alissa M. Kleinnijenhuis, and Thom Wetzer[*]

What kind of models do we need to guide us through the next crisis? If past crises are any indication, we need to explore new approaches. During the Global Financial Crisis (GFC), the models that existed at the time were of little value because they focused on firm-level interactions and did not capture the system-wide dynamics that fueled the crisis. Conventional equilibrium models have been of more help in understanding the COVID crisis, which was in several senses much simpler from an economic point of view. However, it is worth noting that some of the best-detailed predictions were made by dynamic, granular simulation models operating out of equilibrium (e.g., Pichler et al., 2021). The Pichler et al. model is formulated at the sector level, which makes it able to deal with the heterogeneous effects of the pandemic on the economy. The fact that it is a dynamic model that does not assume equilibrium makes it better equipped for taking into account the extremely rapid impact of the pandemic. This underscores once again that there is value in breaking away from traditional equilibrium models.

In this chapter, we sketch a vision for a new approach to understanding and mitigating financial and economic crises. Current data-collection efforts run far short of what is possible. Just imagine, for a moment, that we had up-to-date and granular knowledge of the balance sheets of all the major institutions that comprise the global financial system. Such dynamic balance-sheet-level data could be used to develop and calibrate detailed models that would simulate the financial system from the bottom up. By doing so, these models could predict the response of each institution to shocks, simulate the interactions of institutions with each other through multiple channels of contagion, and provide a platform for accurate system-wide stress testing. This approach is already being investigated by a range of leading institutions, including the European Central Bank (ECB), the Bank of England, and the Bank of Canada, with promising results.[1]

These models, and the data they are based on, should cover as comprehensive a view of the financial macrocosm as possible. One of the most important lessons of the GFC is that it is not possible to separate the financial system from the real economy. Suppose that in addition to having comprehensive data about the financial system, we also had good data on supply chains at the level of individual products, consumer demography, and the geographic diversity of occupational labor. This would allow us to build a similar model that could simulate the economy at the level of individual firms and fine-grained demographic groups

[*] We thank Maarten Scholl, Til Schuermann, and Garbrand Wiersema for their valuable comments and suggestions. The usual disclaimers apply.

[1] See, for example, Chapters 4, 5, and 25 in this handbook.

of households. Finally, suppose we were able to track the investment flows that connect the financial system to the real economy. Then we could couple the models of the financial system and the real economy together to simulate the macrofinancial dynamics of the global economy.

Such models would allow us to investigate counterfactual "what-if" scenarios and provide us with a laboratory to explore alternative futures corresponding to different policy choices. They could help us spot crises as they are building on the horizon and help us formulate policies to avoid them or navigate them once they occur. Regulators and supervisors, supported by real-time data capture translated into actionable insights and flexible models, would have a far better understanding of systemic risk in the financial system and its interactions with the real economy.

If we manage to build and implement this vision and embed it into policymaking, we would be better able to deal with a host of different financial regulatory challenges. This would give us the ability to measure and model the systemic risks of the financial system, anticipate when we are about to get into trouble again, and test new policy ideas to improve financial stability outcomes. In a rapidly evolving financial system that is constantly facing a host of new challenges that might spur financial crises, from endogenous risks to pandemics and climate change, such a toolkit would mark a sea change in regulatory capacity and put central bankers back in the driver's seat.

Although such a vision might sound ambitious, we believe that if we commit the necessary resources and solve some of the administrative bottlenecks, it could be achieved within the next decade. The technical challenges, although substantial, are not insurmountable. System-wide stress tests are feasible using current or proximate technology and modeling techniques. There are already several examples where the success of simulation models in this vein, but with more limited scope, provides a proof of principle. The purpose of this chapter is to make this vision plausible and to articulate what it will take to pull it into reality. We outline why doing so is necessary (Section 1), introduce the modeling techniques (Section 2) and conceptual framework (Section 3) that such an approach would be based on, outline the modeling innovations (Section 4) and the enabling conditions (Section 5) that are needed to make it real, and conclude with a brief overview of the improvements to financial regulatory policy such stress tests would enable (Section 6).

1 The Challenge: Finance Moves Faster Than Regulators

A financial system is a classic example of a complex system.[2] It consists of a diverse set of actors, from banks and insurers to hedge funds and broker-dealers. All these actors interact with each other, the real economy (itself a complex system), and the broader environment (also a complex system). In such a highly interconnected system, seemingly localized shocks can propagate and amplify to take on systemic importance. Furthermore, in some cases, complex dynamics can emerge endogenously, without any external shocks (Aymanns et al., 2016). The financial crisis of 2007–2009, that is, the GFC, provides a good example of an emergent phenomenon, which is the hallmark of a complex system.

---

[2] See, for example, May and Arinaminpathy (2010).

The GFC has made everyone aware of the complex nature of the interactions and feedback loops in the economy and has driven an explosive amount of research attempting to better understand the financial system from a systemic point of view. It has also underlined the policy relevance of the complex-systems approach. Systemic risk occurs when the decisions of individuals, which might be prudent if considered in isolation, combine to create risks at the level of the whole system that may be qualitatively different from the simple combination of their individual risks. By its very nature, systemic risk is an emergent phenomenon that comes about as a result of the (frequently nonlinear) interaction of individual agents. To understand systemic risk, we therefore need to understand the collective dynamics of the system that gives rise to it. Looking at individual institutions with the hope that aggregating these results will suffice would be misguided—"the system is not the sum of its parts (Brazier, 2017). This realization has motivated a push to develop macroprudential policies[3] and system-wide stress tests. The challenge for regulators is to understand the diverse financial institutions involved, their interconnections, and their interactions under stressed financial conditions.

That challenge is substantial. Modern finance is not only a "complex system" in the technical sense but also exceedingly complicated in the more basic sense (Awrey and Judge, 2020). Financial instruments and the financing structures they create, whether through derivatives or securitization vehicles, are increasingly arcane.[4] At the firm level, corporate structures are bourgeoning, sometimes with thousands of legal entities operating across business lines and jurisdictions and being supervised by a plethora of different regulatory bodies.[5] Finally, the rapid rise of nonbank financial intermediation ("shadow banking") vastly complicates the financial network through a lengthening of intermediation chains and obscures the location of risk (Pozsar et al., 2010). In this constellation, keeping track of what is happening, let alone attempting to understand how the financial system functions and evolves, is no easy feat. It is a major challenge simply to draw the map of the "plumbing" of the financial system, which changes all the time.

This complex financial ecosystem is highly dynamic: it is a complex *adaptive* system. The rapid rate of change means that any understanding of a specific state of the financial system is quickly outdated. Such dynamism, illustrated by the emergence of mortgage-backed securities (MBSs) at the start of the millennium, is the product of rapid innovation, driven by the introduction of new technologies and the development of novel financial products, services, and markets. Critically, regulatory change is itself also a significant driver of the dynamism of financial markets; the introduction of new rules incentivizes market participants to find new ways to arbitrage them.[6] Financial-market participants constantly adapt, adjusting their strategies and moving into different niches to seek a higher level of fitness, as measured by (for instance) profitability.[7] This means that the dynamism of financial markets is inevitable, all-encompassing, and at least partly endogenous. Tomorrow's financial system will not look like today's. Likewise, understanding today's risks to financial stability will almost certainly

---

[3] See, for example, Borio (2003), Claessens (2015), and Hanson et al. (2011).
[4] See, for example, Awrey (2012), Célérier and Vallée (2013), and Judge (2012).
[5] See, for example, Carmassi and Herring (2016), Herring and Santomero (1990), and Lumsdaine et al. (2021).
[6] See, for example, Goodhart (2008).
[7] See, for example, Farmer (2002).

be insufficient to understand tomorrow's. Regulators need to be able to grapple with that change.

Unfortunately, there is a mismatch between the dynamism and complexity of modern finance and regulators' ability to understand and govern it.[8] Finance moves faster than financial regulators can, in part because of administrative inertia and rigidly defined authorities in the public sector and in part because of substantial resource differentials between the regulator and the regulated (Armour et al., 2016). The legislative and regulatory process that generates financial regulation, which tends to be a slow-moving and crisis-driven process, may also be to blame (Coffee et al., 2011; Romano, 2004). As a result, regulators face pervasive information gaps as well as uncertainty, and they may not be able to anticipate risks or to respond appropriately as these risks arise. This is particularly true when the risks themselves are new, as is the case for pandemics, climate change, and cyber-risk. The need to understand the complex interactions that generate systemic risk, coupled with the complexity and dynamism of modern finance, represents a profound regulatory and supervisory challenge (Awrey and Judge, 2020; Enriques et al., 2019). Drawing on such observations, Levin and Lo (2015) have suggested that "the financial system has crossed a threshold of complexity where the system is evolving faster than regulators and regulations can keep pace."

Regulators have responded to this challenge. They complemented their microprudential thinking with a macroprudential perspective, developed new regulatory tools in line with that thinking (Claessens, 2015; Enriques et al., 2019), and reshuffled the regulatory architecture to create oversight bodies that take a system-wide view (e.g., the Financial Stability Oversight Council [FSOC] in the United States). In theory, financial stress testing also fits with this thinking because it shows an appreciation for the dynamism of risk and the need to adjust capital standards accordingly. Stress tests are inherently forward-looking and can—at their best—help prepare regulators for emerging risks and the financial stability implications of changes in the financial system. These innovations reflect, at the very least, an implicit understanding that the core regulatory challenge is to understand how interactions within a vastly complex, rapidly evolving, and incompletely understood financial system can generate systemic risk and to evaluate and assess the efficacy of policies designed to mitigate that risk.

But practice remains miles away from theory. System-wide thinking, as well as the tools and processes needed to stay on top of rapidly escalating complexity, is in its infancy. The state of stress testing, particularly system-wide stress testing, reflects this diagnosis. This is not to say regulators have not made progress—the richness of the stress testing models described in this book shows how much has been achieved over the last 10 years. But stress tests, like the Dodd–Frank Act Stress Tests (DFAST) and the Comprehensive Capital Analysis and Review (CCAR) in the United States, remain focused on banks, and system-wide stress tests—although under development—are generally not (yet) being used as part of regulatory stress test exercises. That means these exercises miss out on the interconnections and feedback loops that characterize finance and its risks, leaving regulators ill-prepared to deal with them. In sum, despite commitments to macroprudential regulation, regulatory and supervisory action is still based too much on antiquated views of the financial system. That

---

[8] For a more comprehensive argument along these lines, see Awrey and Judge (2020).

translates into a focus on form over function, leading to an excessively atomistic and siloed view of finance and its regulation.[9] We have only done a small fraction of what could be done to make the financial system more stable and secure.

Even though there are substantial hurdles to overcome, we are optimistic about the potential for financial regulation to better measure up to the challenges presented by a complex and dynamic financial system. We do not contest the seriousness of the fundamental challenges facing regulators or suggest that these problems will ever be fully resolved. There is no silver bullet. Instead, our optimism is rooted in the observation that we can leverage new modeling techniques and analytical perspectives by drawing on sustained advances in data availability and computing power—and thereby go a long way toward advancing the ability of regulators to measure and model the systemic risk of the financial system, anticipate scenarios that might get us into trouble again, and test new policy ideas (Farmer and Foley, 2009). But to get there, we need to innovate—in our financial models, our mental models, and our approach to studying financial systems. We need to change the regulatory architecture and how these powerful new methods feed into policymaking. In what follows, we set out what needs to happen.

2 Economic Simulation Models

Roughly speaking, there are two approaches for modeling systemic risk.[10]

The traditional approach, which we will call *equilibrium modeling*, relies on highly stylized models that are analytically tractable. It has focused on situations where it is possible to compute an equilibrium. This typically requires making very strong simplifications, such as studying only a few actors and interactions at a time. Generally speaking, this approach begins by assigning each agent a utility function and assuming that all agents maximize their utility subject to their model of beliefs. In the canonical macroeconomic approach, the beliefs are rational, meaning that expectations, formed through observation of historical data points, are on average equal to real-world outcomes. In more modern work, this is modified by imposing frictions, and with the increasing influence of behavioral economics, alternative models for beliefs are beginning to be explored that allow for "imperfections (Gennaioli and Shleifer, 2018; Gabaix, 2020).

Equilibrium models place great emphasis on the incentives and information structure of agents in a financial market. Given those, agents behave strategically, taking into account their beliefs about the state of the world and other agents' beliefs. The objects of interest are then the game-theoretic equilibria of this interaction. This allows for studying the effects of properties such as asymmetric information, uncertainty, or moral hazard on the stability of the financial system. Such efforts have greatly improved our understanding of a wide range of phenomena related to systemic risk, ranging from bank runs (Diamond and Dybvig, 1983; Morris and Shin, 2001) and credit cycles (Kiyotaki and Moore, 1997; Brunnermeier and Sannikov, 2014) to financial-contagion channels.[11, 12]

---

[9] See Chapter 20 in this handbook.

[10] This section is based on Aymanns et al. (2018).

[11] On information contagion, see, for example, Acharya and Yorulmazer (2008). On balance-sheet contagion, see, for example, Allen and Gale (2000).

[12] See, for example, Brunnermeier (2009).

However, there are two problems with the equilibrium approach, even in its new, more behaviorally relevant forms. The first is empirical realism. Although this is hotly debated, many have argued that utility maximization is not well supported by empirical evidence. The second problem is that it rapidly becomes too complicated to apply this approach in real-world situations where there are many different actors with nonlinear interactions. To maximize utility, each agent has to take all the other agents into account, and as the situation becomes complicated, this becomes intractable. As a result, equilibrium models are forced to incorporate strong assumptions to remain simple enough to be solvable. When the economic setting becomes complicated, this often forces the omission of important structural features of the economy. Particularly during a crisis, the assumptions of rationality and equilibrium may be too strong. Equilibrium models can be very useful in stylized contexts when one seeks to understand basic principles. However, they are typically very difficult to build for complicated features of the economy, such as institutional and market structure or agent heterogeneity.

The second modeling approach, and the one that we will focus on here, uses economic simulation models.[13] In economics, one class of simulation models (and the one we will focus on) is agent-based models (ABMs). Such models have a very different conceptual framework than standard equilibrium models in economics and have complementary strengths and weaknesses.[14]

In contrast to equilibrium models, agent-based simulation models are built from the bottom up in a way that makes them scalable to complicated situations. Unlike equilibrium models, which are built using a tightly prescribed standard template, agent-based simulation models are better characterized as a flexible toolkit for solving complex problems involving heterogeneous agents. Designing an agent-based simulation model does not start with axioms but rather with a clear-eyed look at reality: the system, its agents, and their relationships. Each agent has an algorithm that implements the agent's decisions, and these algorithms can differ across agents to reflect their heterogeneity. The decision rules can be static, or they can evolve over time under a learning algorithm. The computer simulates the decisions of each agent and changes the state of the world accordingly, for example, by adjusting prices and quantities, and then simulates the new decisions that agents make in response, and so on. Like equilibrium models, agent-based simulations can be used to illustrate conceptual points in a qualitative manner, but they can also mimic the world in a more realistic manner.

An important feature of agent-based simulation models is the approach for incorporating the decisions of economic actors. Rather than assuming utility maximization, simulations model agent behaviors directly. There is a great deal of flexibility in how this can be done, but the key is that the agents use behavioral rules that are based on plausible knowledge using feasible computation that could plausibly be implemented by a real human being. The rules can be simple heuristics, such as "buy undervalued assets," or they may emerge from learning algorithms, for example, if each agent acts like an econometrician. The rules can be backward-looking—for example, following a trend—but they can also be forward-looking.

---

[13] Because simulation models are not very familiar to many economists, we will begin by giving a brief conceptual summary. For a more comprehensive set of examples, see Aymanns et al. (2018).

[14] See, for example, Haldane and Turrell (2018).

A good example of plausible forward-looking behavior is "level K reasoning," in which each agent reasons about the next K steps of the other agents and the environment and develops a best response. The rules often represent goal-oriented behavior, even if they are not optimal and are not even explicitly derived in order to achieve the goal.

The key point is that when direct behavioral models are used, the computational burden on each agent is small, keeping the overall model simple enough that it becomes feasible to run in complicated settings with very large numbers of agents. This is why, using agent-based simulation modeling, it is possible to incorporate the detailed structure of an economy more realistically. This not only makes simulation models tractable. In fact, when needed, they can closely mimic real-world processes, a property that we refer to as *verisimilitude* (Farmer, 2019). In comparison to equilibrium models, agent-based simulation models therefore place less emphasis on complicated strategic interactions and incentives in favor of (relatively) simple, empirically motivated behavioral rules and learning algorithms and more realistic institutional and market structure. From this perspective, the key drivers of systemic risk that the simulation models are well suited for studying are the amplification of dynamic instabilities and contagion processes in financial markets. Agent-based simulation models show that even when agents act on relatively simple behavioral rules, their behavior can lead to complex aggregate phenomena, such as financial crises. In such contexts, these models can show how outcomes are shaped by the structure of this interaction and the heterogeneity of the agents. If appropriately calibrated, they can also yield quantitative insights.

Equilibrium and agent-based simulation models are complements rather than substitutes. The most appropriate approach depends on the context and the goals of the modeling exercise. Analytic models have the benefit of the relative ease with which they can be used to understand the concepts driving structural cause-and-effect relationships. But many aspects of the economic world are not simple, and in an increasingly complex and interconnected financial system characterized by complex emergent properties, agent-based simulation models may be better suited to characterizing system-wide dynamics.[15] The strengths of the simulation approach stand out for use cases such as system-wide stress tests, where there are many interacting heterogeneous agents and where we need scalable models with a high degree of verisimilitude in order to get reliable answers.

Network dynamics play a key role in determining financial stability. Diverse channels of contagion—including counterparty risk,[16] funding risk,[17] and overlapping portfolio[18] contagion—cause (nonlinear) interactions that can create positive-feedback loops that amplify external shocks or even generate purely endogenous dynamics, such as booms and busts.[19] Positive feedback loops can also be amplified by bounded rationality (often in the context of incomplete information and learning) and by behavioral and institutional constraints. Often, such constraints are implied by contractual terms between private parties or imposed by regulators with a view toward increasing financial stability. In many cases,

---

[15] Similarly, see Bookstaber (2019) Bookstaber highlights four broad phenomena that are endemic to financial crises—emergent phenomena, non-ergodicity, radical uncertainty, and computational irreducibility—and argues that in such situations, simulation models are superior to equilibrium models.

[16] See, for example, Cont and Moussa (2010).

[17] See, for example, Gai et al. (2011).

[18] See, for example, Caccioli et al. (2014), Cont and Schaanning (2017), and Greenwood et al. (2015).

[19] See, for example, Aymanns and Farmer (2015).

however, these constraints are designed to increase the resilience of an individual financial institution to idiosyncratic shocks rather than the resilience of the system as a whole—and simulation models are well positioned to tease out when the two come into conflict and to investigate possible solutions when this happens.

Leverage constraints provide a good example of the conflict between microprudential and macroprudential regulation.[20] If a financial institution has high leverage, a small shock may be enough to push it into insolvency. Hence, from a regulatory perspective, a cap on leverage seems like a good idea. However, a leverage constraint may have the adverse side effect that it forces distressed institutions to sell into falling asset markets, causing prices to fall further and amplifying a crisis.[21] Of course, leverage constraints are needed, but the point is that their effects can go far beyond the failure of individual institutions, and the way in which they are enforced can make a big difference in the resulting level of systemic risk. Simulation models using simple behavioral rules, grounded in empirical evidence of bank behavior (Adrian and Shin, 2010), can be used to study such remarkable and unexpected dynamics and propose policy changes. For example, a simple model of leverage targeting by investment banks leads to an endogenously generated bubble, followed by a crash that closely resembles the behavior of prices, leverage, and volatility in the decade leading up to the GFC (Aymanns and Farmer, 2015; Aymanns et al., 2016).

Bounded rationality often plays an important role in generating endogenous dynamics. Although equilibrium models can generate endogenous dynamics, bounded rationality means that expectations do not match outcomes, which can lead to cyclical or chaotic behavior. This is particularly true in complicated settings where agents are competing with one another (Pangallo et al., 2019). Simulation models can naturally capture bounded rationality and make it possible to study instabilities that are suppressed by equilibrium models based on rational expectations.

Another advantage of agent-based simulation models over equilibrium models is that they can be implemented modularly. It is thus possible to construct subcomponents of a model independently and couple them together without having to substantially rework each subcomponent. Consequently, it also becomes easier to update or refine the subcomponents without modifying all the other subcomponents. This means that it is much easier to begin with a simple model and build it up and refine it to systematically incorporate realism. When designing the models of a system as complex and dynamic as the financial system, modularity provides the flexibility regulators need to rapidly respond to changing circumstances.

Of course, agent-based simulations can present serious challenges. There is a danger that models can become too complicated, making it difficult to estimate parameters, perform model validation, and disentangle cause and effect. A poorly designed ABM can easily become a black box whose predictions are difficult to test. However, in recent years, great strides have been made, and there are now good examples in many fields where simulations have overcome these problems and are the state of the art. This is particularly true in fields

---

[20] See Chapter 21 in this handbook. More generally, see Aymanns et al. (2018).

[21] This basic mechanism has been discussed by many authors, including Gennotte and Leland (1990), Geanakoplos (2010), Shleifer and Vishny (1997), and Thurner et al. (2012).

outside of economics, such as epidemiology or traffic modeling, but there are now good examples in economics as well (Platt, 2020).

Cutting-edge macroprudential stress tests increasingly rely on agent-based simulation models. Pioneering work includes that of Hałaj (2018) and that described by Bousquet et al. in this handbook,[22] as well as research carried out at the Bank of England that focuses on particular markets and the interactions among representative sectors (Aikman et al., 2019). Hałaj (2018) builds a system-wide stress test tool to assess how a funding shock to banks and nonbanks, in particular asset managers, can propagate across the financial system. He assesses how such shocks may be amplified via agents' behaviors and their interconnectedness. Bousquet et al. have started to develop a top-down macroprudential stress test of the euro-area financial sectors, including banks and insurers, as well as (to a more limited extent) other financial institutions and central clearing parties.

Although these models represent significant progress, there is much left to do. Truly system-wide stress tests, which include not only the interaction between banks but also model nonbanks, are still in their infancy (Anderson et al., 2018; Aymanns et al., 2018). For these models to really take off, simulation models would have to comprehensively capture the amplification of solvency and liquidity shocks and fully take account of the heterogeneity of institutions and their responses to these shocks, given the constraints they face. Such simulations need to be scalable to ensure a high degree of verisimilitude, should be readily adjustable in response to the dynamism of the financial sector, and must allow for calibration and verification (Farmer et al., 2020).

Farmer et al. (2020) provide a general system-wide stress testing simulation toolkit that starts to address these challenges. The paper outlines a structural framework for the development of system-wide financial stress tests with multiple interacting contagion and amplification channels, as well as heterogeneous financial institutions. This framework conceptualizes financial systems through the lens of five building blocks: financial institutions, contracts, markets, constraints, and behavior. These blocks can be flexibly implemented to form a dynamic multiplex network using the accompanying software engine and library (the "Economic Simulation Library" [ESL]).[23] This modular design allows users to build simulation models that closely track data about the financial system. Depending on the needs of regulators and researchers and the data they have access to, this framework (and the software that implements it) supports both stylized stress testing models and large-scale, data-driven models that map out the financial system with a high degree of verisimilitude.

As a proof of principle, the ESL has been used to implement a system-wide stress test model for the European financial system. It provides a stress testing framework that captures solvency and liquidity channels and incorporates four interacting amplification channels: default contagion, price-mediated contagion via asset sales, funding contagion, and liquidity stress via margin calls. It takes account of the heterogeneity of the financial institutions by modeling three classes of financial institutions (i.e., banks, investment funds, and hedge funds) and allowing for heterogeneity within these classes. The system-wide model is

---

[22] See Chapter 27 in this handbook.

[23] This software package is freely accessible at https://github.com/topics/economic-simulation-library. The code for this paper, as well as accompanying documentation, sample implementations, and robustness checks, is freely accessible at https://github.com/ox-inet-resilience/resilience.

implemented as a macroprudential overlay on top of the regulator microprudential European Banking Authority (EBA) stress test from 2018 so that the micro- and macroprudential results can be easily compared.

A key finding of Farmer et al. (2020) is that microprudential stress tests can dramatically underestimate the true size of financial risk and therefore lead to false confidence. Systemic risk is often highly nonlinear and thus effects are invisible below a threshold and then dramatically appear at large size above that threshold. The analysis shows that the same shock may be amplified to a completely different extent depending on factors such as the size of capital buffers. Hence, microprudential stress tests must be complemented with macroprudential stress tests in order to be reliable. Missing that perspective can lead to undesirable policy outcomes. The failure to capture shock amplifications in stress exercises led the Federal Reserve to underestimate the risk of a global financial crisis in the run-up to the 2007–2008 GFCs. As Battiston (2015) wrote, the Fed calculated that "even if subprime mortgages defaulted at extraordinarily high rates, [...], the resulting financial losses would be smaller than those from a single bad day in global stockmarkets. We came to realize that the ultimate economic costs of the panic far outweighed the magnitude of the trigger," as occurred in the 1907 Panic.

For all the reasons just given, agent-based simulation models are the appropriate tool for system-wide stress tests. It is the only method that can capture the complicated architecture of the financial system and render its nonlinear interactions with enough verisimilitude to provide the quantitative guidance that we need to mitigate financial crises. Although there is a great deal of work that remains to be done (Section 4), the challenges are tractable. The key hurdle that needs to be surmounted is to acquire the resources needed to implement this vision (Section 5).

3 Finance through the Lens of Ecology

The use of agent-based simulation models allows regulators to study important problems that current stress tests fail to elucidate. In order to leverage that potential, however, a different conceptual framework is required. Market ecology theory provides the perspective required.

The theory of market ecology provides a conceptual framework that facilitates interpretation. Market ecology (Farmer, 2022; Lo, 2004) borrows concepts and methods from biology and applies them to financial markets. Financial trading strategies are analogous to biological species. In the same way that living organisms are specialists that evolve to fill niches that provide food, financial trading strategies evolve to exploit market inefficiencies. Capital invested in each strategy is the equivalent of the abundance of a species. Market conditions change as the capital invested in each strategy changes, and the market evolves as old strategies fail while new strategies appear. The conceptual framework that market ecology provides is useful for interpreting what happens in the financial system and, in particular, for understanding why it malfunctions.

Market ecology can be viewed as a step beyond the theory of market efficiency. The market-efficiency hypothesis postulates that prices fully incorporate all available information, so that it is not possible to consistently make excess profits (or to develop a strategy that makes excess profits in expectation; Malkiel and Fama, 1970). This means that markets function perfectly, and prices always reflect fundamental values. Prices only change when new information enters the market. By definition, this implies that any problems with price setting are caused by factors outside the market. This provides a useful conceptual foundation for some problems, such as pricing options (Black and Scholes, 2019), but it is not useful for understanding systemic risk. This is because systemic risk is caused by market failures, and by their very nature, market failures involve a breakdown of market efficiency (Lo, 2005). Thus, to understand market failures, we need to understand why markets are not efficient and understand the forces that keep them from being so. Market ecology provides a useful conceptual framework for conceptualizing, designing, and interpreting the output of simulation models and for getting insight into the causes of market failures.[24]

Markets are necessarily inefficient because perfect market efficiency is inherently self-contradictory (Grossman and Stiglitz, 1980). Market efficiency depends on arbitrageurs, who buy and sell assets and thereby incorporate information into prices. But if markets are perfectly efficient, then there are no profits for arbitrageurs, and therefore there is no reason for them to participate in the market. And if arbitrageurs are not present in the market, there is no mechanism to make markets efficient. Ergo: while markets may be efficient in some approximate sense, they cannot be perfectly efficient.

In contrast to the theory of market efficiency, the theory of market ecology begins with the assumption that markets may be close to efficiency but necessarily deviate from it. It asks what drives these inefficiencies, and it categorizes and studies the profit-making strategies that exploit them. The word *ecology* enters by analogy with biology: just as plants and animals are specialists that have evolved to fill niches that provide food, trading strategies are specialized information processors that buy and sell assets to exploit inefficiencies in order to make profits. There are many different specialized strategies interacting with each other in the marketplace, and these interactions may affect each other and cause market inefficiencies to change over time. Similarly, changes in the regulatory system may change the ecosystem, affecting profit-making opportunities and triggering adaptation, which may in turn change the nature of the interactions. To understand how the market functions and why it breaks down, we need to understand the key strategies that are present in the market and their interactions with each other. These questions are quite different from those addressed by the theory of market efficiency, making the two approaches complementary.

Attempts to understand the origin of clustered volatility illustrate the conceptual difference between market efficiency and market ecology. If markets are efficient, then by definition, prices fully reflect all available information, and changes in prices reflect the arrival of new information. This means that the sole cause of clustered volatility in prices is intermittency in the rate of information arrival. If markets are efficient, then changes in the rate of information arrival are the only possible cause of changes in price volatility. Empirical studies indicate that this view is wrong: the lion's share of market volatility is not due to information arrival. Cutler et al. (1989) showed that only about a third of the largest daily changes in the Standard & Poor's (S&P) index are caused by news—the other two-thirds of the large moves occurs on days when there is no discernible new information.

The theory of market ecology offers an alternative explanation for clustered volatility that is more consistent with the data. In a market ecosystem, there are a variety of different kinds

---

[24] This part of the chapter builds on Scholl et al. (2021).

of strategies active in the market, but their level of activity fluctuates in time. For example, two strategies typically studied in market ecology research are value investing (whose practitioners are also called "fundamentalists") and trend following (whose practitioners are also called "momentum traders"). When value investors are more abundant, they cause prices to revert to fundamental values, but when trend followers are more abundant, they cause persistence in price movement that can drive prices away from fundamental values. Trend following is inherently destabilizing, and as a result, periods of high trend following cause high volatility. This does not mean that information arrival does not cause volatility as well—value investors trade based on mispricings, which can occur either because new information has arrived or because prices have changed for other reasons. Market ecology predicts that there will be endogenous price movements driven by fluctuations in the wealth invested in the different strategies in the market while also acknowledging the role of information arrival.

Using a simple stylized financial ecosystem consisting of value investors, trend followers, and noise traders, Scholl et al. (2021) have demonstrated how these ideas can be used to understand market malfunctions. The three strategies are each endowed with an initial wealth and begin trading with each other. The profits and losses of the strategies redistribute their wealth under the assumption of proportional reinvestment. Although the system tends to evolve toward an efficient equilibrium, statistical fluctuations in the performance of the strategies cause large deviations in their wealth, which can be substantially different from their equilibrium values. By tracking these deviations, it is possible to make a good prediction of market volatility and mispricing (relative to fundamentals).

These ideas could be used by regulators to understand the stability of real market ecosystems. Regulators are uniquely positioned to do this because of their access to data on the balance sheets of market participants and transaction data with counterparty identifiers. Such data could be collected in real time so that regulators could understand the state of the financial ecosystem at any point in time. Simulations of this ecosystem could give a useful understanding of its stability. Farmer et al. (2020), as discussed previously, provide an illustration of how this could be done.

The introduction of MBSs in the run-up to the GFC provides an illustration of how market failures can occur when the market ecosystem is disrupted by financial innovation. MBS began to be an important factor in markets in the late 1980s, but it took time (and the GFC) for them to be well understood. Prior to the crisis, pricing formulas for MBSs made unwarranted independence assumptions (Hellwig, 2009).[25] (The fact that this happened is in and of itself proof that market participants are not rational and markets are not efficient.) There was a neglect of systemic risk, which can strongly correlate with the price movements of many different assets, including assets that may not appear to be closely related. The resulting dramatic underestimation of risk meant that MBSs were purchased with far too much leverage. The vast amounts of borrowing and lending that took place as a result strongly coupled MBSs to the rest of the global financial market. When housing markets entered a downturn this set off a chain reaction, sending shocks through the entire financial network, creating systemic risk and correlating asset price movements (Brunnermeier, 2009). The resulting nonlinear feedback effects made the problems with MBSs worse but also affected a wide variety of other global assets (Mishkin, 2011).

---

[25] For a slightly different take, see Ashcraft and Schuermann (2008).

The GFC illustrates that global financial markets form a tightly coupled complex system that cannot be fully understood based on simple stylized models that use assumptions of efficiency, rationality, and equilibrium. During the financial crisis, outcomes did not match expectations, and the economy was not in equilibrium. Simulation models allow for such deviations to occur (Scholl et al., 2021). They make it possible to investigate the specialized behavior of individual strategies and how the introduction of new strategies, potentially in response to regulatory change, can knock an existing ecosystem out of balance. Market ecology provides an intellectual framework for understanding out-of-equilibrium behavior, including market failures. This perspective can be invaluable to regulators trying to grapple with the dynamism of financial markets and struggling to understand, let alone anticipate, their role in triggering such change.

Current state-of-the-art systemic risk models that capture multiple types of institutions in a heterogeneous setting represent a first step toward understanding market ecology.[26] We say this because such models represent the financial system out of equilibrium, and we study how specialized institutions such as banks or hedge funds interact with one another— in this case it is the type of institution that is analogous to a "species." These early models differentiate behavior only based on differences in balance sheets and differences in binding constraints. More complete market ecology models in the future may also study the sources of profits for these institutions and model their behavior even during normal times.

Financial strategies evolve through time in response to each other and in response to regulatory changes or changes in the real economy. In the future, it may be possible to simulate not only the ecological dynamics of markets but also the evolution of strategies through time. This provides one way to respond to the Lucas critique within a framework of bounded rationality and to better grapple with the dynamism of financial markets.

4 Pushing the Modeling Frontier

Agent-based simulation models open up opportunities for a new generation of system-wide stress testing models. Over the past decade, progress in model design has already been rapid. But as much as has been achieved, there are a number of significant outstanding modeling challenges that the next generation of simulation models should address to truly be able to capture the system-wide dynamics of the economic macrocosm.

4.1 Model Innovations Relating to the Financial System

A first challenge is to better exploit the ability of simulation models to capture the heterogeneity that characterizes financial systems. The vast complexity of the financial system is evident on multiple levels, which would ideally be reflected in simulation models that have a claim to verisimilitude.

The diversity of institutions in the financial system is staggering.[27] There are large and small institutions, with different objectives, balance sheets, strategies, and regulatory requirements. Capturing this heterogeneity of actors is central to understanding the dynamics of the financial system under distress. This is true for the heterogeneity of institutions of the same type, with banks providing an example of dizzying complexity within their

---

[26] See, for example, Farmer et al. (2020).
[27] See Chapters 10, 19, and 20 in this handbook.

corporate structures that can conceivably affect their resilience.[28] Similarly, there are, of course, many different types of financial institutions.[29] Early stress tests that looked at groups of institutions focused on banks alone,[30] and failed to capture the endogenous shock amplification that arises from their interconnection with other institutions.[31] Nonbanks are starting to play an increasingly important role in most financial systems[32] and cumulatively are already equal in size to the banking system in many jurisdictions, which makes the exclusive focus on banks in stress tests problematic. Credit-intermediation channels, for example, increasingly run through both the banking and nonbanking sectors, which are strongly interconnected (Brazier, 2017). As Goodhart notes in his contribution to this volume, the more "such nonbank financial intermediation partakes of, and takes over, credit expansion and the provision of means of payment, the more that such new channels need to be brought within the ambit of stress testing procedures."[33]

Stress tests that take a more diverse range of financial institutions into account are still at an early stage. Pioneering work has focused on particular markets and the interactions among representative sectors,[34] which represents an important but incomplete advance in understanding system-wide dynamics. But nonbanks are a heterogeneous lot, which include pension funds, insurers, hedge funds, broker-dealers, mutual funds, central clearing parties, and many more.[35] Moreover, the heterogeneity and interactions within each "sector" may matter for financial stability outcomes, too.[36] In their chapter, Adrian et al. note that systemic risk arising from cross-sectoral exposures can take place through a variety of channels (including the financial group's corporate structure, the funding channel, the exposure channel, the common exposure channel, and other network connections).[37] Abstracting away from that detail by using representative sectors is therefore risky.

In recent years, macroprudential stress tests have become more system-wide and started to capture a greater variety of nonbank financial institutions. Calimani et al. (2019) capture both banks and asset managers in their model (Calimani et al., 2019),[38] whereas Aikman et al. (2019) feature seven representative agents (pension fund, insurance company, hedge funds, money market fund, commercial bank, investment agent, broker-dealer), each representing a part of the UK financial sector.[39] However, this captures only the heterogeneity of the sectors themselves but not the heterogeneity within sectors. The implementation of the European stress test in Farmer et al. (2020) includes a generic hedge fund and investment fund. However, the structural modeling framework it outlines is well suited to capturing heterogeneity between and within sectors and could provide a basis for further advances in system-wide stress testing.

---

[28] See, for example, Chapter 10 in this handbook; see also Wetzer (2019, 2021).

[29] See, for example, Chapter 29 in this handbook.

[30] See, for example, Bookstaber et al. (2014).

[31] See Chapter 19 in this handbook.

[32] See Chapters 4 and 27 in this handbook; see also Aldasoro et al. (2020).

[33] See Chapter 20 in this handbook.

[34] See, for example, Aikman et al. (2019). Farmer et al. (2020).

[35] See Chapter 27 in this handbook.

[36] See Chapter 19 in this handbook.; see also Cont and Schaanning (2019).

[37] See Chapter 26 in this handbook.

[38] See Calimani et al. (2019) and Chapter 27 in this handbook.

[39] See Chapter 4 in this handbook.

The need to capture granularity and heterogeneity also applies to the constraints institutions face. Participants in financial markets are subject to a number of constraints that help shape their behavior. Constraints can originate from many sources. These include regulatory constraints (e.g., risk-weighted capital), contractual constraints (e.g., margin requirements), market-based constraints (e.g., losing access to funding), internal risk limits (e.g., maximum value at risk in a given portfolio), reputation (e.g., cannot be seen to cut a client that is performing poorly). At any point, multiple constraints may bind. The effects of multiple binding constraints at the same time can be destabilizing in ways that are difficult to foresee, which is why models might help. It is important not only to model multiple constraints but also to do so at a high level of granularity (e.g., contractual constraints modeled at the level of individual contracts).

Sometimes such constraints, although designed to protect an individual firm or its counterparties, generate destabilizing externalities.[40] One example, which we discussed previously, is the leverage cycle that can be endogenously generated when participants are subjected to the Basel capital framework.[41] The first generation of contagion models typically only modeled one constraint. Early models of overlapping portfolio contagion, for example, only explain how the leverage requirement may drive booms and busts in asset prices.[42] More recent system-wide stress tests start to capture the interaction of multiple (types of) constraints. Recent models study how contagious spillovers may be driven by institutions that have to meet not only a leverage ratio and contractual obligations but also other capital requirements, such as the risk-weighted capital requirement, as well as liquidity requirements, such as the liquidity leverage ratio.[43] That trend should continue, and not only because the interaction of multiple constraints can lead to unexpected outcomes. As Goodhart points out in his contribution to this volume,[44] when modeling multiple constraints it may become obvious that there is redundancy in the regulatory framework so that there is room to simplify it without sacrificing efficacy.[45]

In addition to capturing greater heterogeneity of market participants and the constraints they face, their behavior should be more realistically modeled. Many system-wide stress test models tend to assume that constraints drive behavior in times of distress because in times of distress constraints more often bind—but that is clearly only part of the story. In these models, the binding nature of constraints reduces the set of possible actions an institution can take, and within that smaller subset of actions heuristics reflecting behavioral assumptions govern the choices of the institution. When a particular institution is forced to delever to avoid breaching a constraint, an assumption might be that it will do so by liquidating assets proportionally to initial holdings.

Some heterogeneous multiagent simulation models already use myopic optimization, and in the future, the line between the two approaches may become increasingly blurred. As methods such as computational game theory or multiagent reinforcement learning mature, it may become possible to introduce strategic interactions into computational heterogeneous simulation models. To better capture system-wide dynamics, the next generation of stress

---

[40] See also Chapter 25 in this handbook.
[41] See Section 6.
[42] For a review, see Aymanns et al. (2018).
[43] See, for example, Adrian and Shin (2010). See Chapters 4, 25, 26, and 27 in this handbook.
[44] See, for example, Chapter 20 in this handbook.
[45] More generally, see Greenwood et al. (2017).

tests should capture a richer set of responses to shocks in order to meet some objective, such as maximizing profits, the Sharpe ratio, or the mean-variance return.

The Bank of Canada is pushing the frontier in this regard, with its latest models not only relying on constraint-driven behavior but also including profit-seeking objectives. As Hałaj and Traclet note in their contribution to this volume, the institutions in their model evaluate the "cost–benefit trade-offs of different management actions to select those that balance risk and return subject to multiple regulatory requirements."[46] Similarly, the Bank of England is making rapid advances, with Haldane and others stressing that modeling how institutions behave when they are unconstrained and have the potential to support market functioning is critical.[47] Finally, Adrian et al. explain that the models used as part of the IMF's Financial Sector Assessment Program (FSAP) used to model behavior "in a limited manner," but recent models incorporate behavioral responses more explicitly.[48]

With more granular modeling of a greater variety of (heterogeneous) institutions and constraints and more realistic and potentially even adaptive behavior, it becomes possible to study the systemic implications of multiple interacting contagion channels. Whereas early contagion models typically only captured one contagion mechanism, such as exposure-loss contagion or overlapping-portfolio contagion, more and more system-wide stress tests now incorporate a host of interacting contagion channels.[49] Some recent models capture different types of channels of contagion on different layers of the network, thus creating a multiplex network.[50] The interaction of different contagion channels then results from the connectivity and interactions between these network layers. Alternatively, the different layers in a multiplex network of the financial system could represent various functions (e.g., asset layer, funding layer, collateral layer[51]), which provides a different basis to arrive at a better understanding of the dependency, connectivity, and potential for shock transmission and amplification of different parts of the financial system (Bookstaber, 2019).

It is clear that interacting contagion channels can dramatically amplify each other. Wiersema et al. (2019) consider four contagion channels (funding contagion, overlapping-portfolio contagion, counterparty risk, and leverage-targeting contagion) and show how the interaction of solvency (valuation) and liquidity shocks can lead to instability that far exceeds the instability that results from the sum of individual channels acting alone. Other recent models of the solvency–liquidity nexus have produced similar results,[52] highlighting the importance of system-wide stress tests taking account of both dimensions. This solvency–liquidity nexus manifests itself in various ways. It could be that shocks to solvency can (endogenously) raise banks' funding costs, thus placing additional pressure on liquidity and solvency conditions. Similarly, a better-capitalized bank may be less likely to face liquidity runs in the first place (Diamond and Dybvig, 1983). There is a trend toward incorporating such dynamics in stress test models. Adrian et al., for example, model liquidity risk as a response to elevated solvency risk and uncertainty.[53] In that setting, lower

---

[46] See Chapter 25 in this handbook.

[47] See Chapter 4 in this handbook.

[48] See, for example, Chapter 26 in this handbook.

[49] Research at the frontier of this development in presented in this volume; see, for example, Chapters 4, 24, 25, 26, 27, and 28 in this handbook. See also Farmer et al. (2020).

[50] See, for example, Chapter 28 in this handbook; see also Montagna and Kok (2016).

[51] See, for example, Bookstaber and Kenett (2016).

[52] See, for example, Cont et al. (2020) and Farmer et al. (2020).

[53] See Chapter 26 in this handbook.

funding liquidity arises from increased uncertainty over counterparty risk and lower asset valuations that induce banks and investors to hoard liquidity, leading to systemic liquidity shortfalls. As stressed banks try to meet their cash-flow obligations, they engage in fire sales, thus depressing asset prices and margin requirements for all banks in the system and generating further systemic solvency concerns.

When modeling interacting contagion channels, there are at least two novel challenges. First, financial and economic shocks materialize over different time horizons (e.g., liquidity shocks are felt almost instantaneously, whereas losses on real-economy loans may materialize more slowly), and market participants act at different speeds (e.g., with high-frequency traders acting in splits seconds and pension funds moving more slowly). By contrast, system-wide stress test models, including agent-based simulation models, tend to assume that all shocks and institutions move at the same speed. That may obscure some critical financial stability dynamics. For example, even if pension funds following a fundamental value strategy would be interested in buying securities during a fire sale, they may act too slowly to create a floor underpinning asset prices. This problem is particularly salient when studying interacting contagion channels because liquidity and solvency shocks are traditionally modeled using different timescales (and with liquidity stress being modeled using a higher frequency of interactions).[54] Similarly, the wider the scope of the system-wide stress test, the more pronounced the different speeds at which interactions act may be (e.g., when a model includes central clearing parties, insurance and pension funds, trend-following hedge funds, and money market mutual funds).[55] The next generation of system-wide stress testing models should account for the differences between the tortoises and the hares of financial markets—with some characterizing this as "a critical priority."[56]

A second challenge in relation to modeling interacting contagion channels is to account for hedges and collateral. Hedges may limit the degree of shock spillovers in some cases[57] while increasing connectivity and thereby contagion risk in others.[58] Because of that uncertainty, it is important that stress tests can be relied upon to study their impact. Unfortunately, current system-wide stress tests tend to either model derivatives markets[59] or other asset classes—but not the two together.

4.2 *The Financial System and the Real Economy*

One of the most important lessons of the GFC is that it is not possible to consider the financial system and the real economy in isolation from one another. What began as a global financial crisis rapidly became a global economic crisis. This should not be surprising: the financial system plays an essential function in allocating credit, which controls activity in the real economy.

The ongoing COVID-19 crisis demonstrates how causality can also flow from the real economy to the financial system. As infections spread and governments imposed lockdowns, the scope and scale of global economic disruption became apparent in mid- to late March 2020. Short-term funding markets came under acute stress as market participants responded

---

[54] See, for example, Chapter 5 in this handbook.
[55] See, for example, Chapter 18 in this handbook.
[56] See Chapter 4 in this handbook.
[57] See, for example, Chapter 25 in this handbook.
[58] See, for example, Chapter 6 in this handbook.
[59] See, for example, Paddrik and Young (2017) and Bardoscia et al. (2019).

to the advent of this low-probability adverse scenario. Financial market participants repriced and repositioned their portfolios, asset prices dropped, and financial institutions largely pulled back from the market. In the resulting dash for cash, volatility spiked and spreads in dollar funding markets widened sharply. Financial authorities around the world intervened at an unprecedented scale to stabilize markets and restore orderly market functioning.[60] What had started as a pandemic quickly morphed into a global economic and financial crisis.

The COVID-19 crisis left no part of the economy untouched. Risks to lives yielded lockdowns and other restrictions, which threatened livelihoods. Understanding these effects jumpstarted research that linked up various parts of the economy, ranging from production networks to labor markets. The COVID-19 crisis also showed how quickly the macroeconomy can respond to an external shock, demonstrating the need for out-of-equilibrium models. For example, del Rio-Chanona et al. (2020) made a prediction of the shocks by using characteristics of occupations to predict which occupations could work from home and how this would affect output for each sector, and combining this with previous studies on how pandemics affect the demand for different industries. These shock predictions were then used by Pichler et al. (2021) to make accurate predictions about supply-chain bottlenecks in the UK economy and how these would affect output on quarterly timescales (Pichler et al., 2021). A key to the success of this model was that it could operate out of equilibrium and incorporated features such as inventory dynamics. Future work could connect such macroeconomic models with financial models for credit and investment.[61]

The lesson is that to truly appreciate the risks facing the financial system, we need to understand how it relates to the real economy. This implies that system-wide stress tests should, ideally, capture that relationship too. Looking ahead, the next big challenge (and potential financial risk) may well be the climate crisis. This is not a place for a comprehensive review of climate risk, but two points are worth emphasizing because they illustrate both the necessity of expanding the scope of system-wide stress tests even further and the value of agent-based simulation models in doing so.

First, both the climate and the financial system are examples of complex systems and share several qualities, including significant feedbacks, thresholds, tipping points and nonlinearities, fat-tailed distributions of outcomes, nonequilibrium system dynamics (often chaotic), and emergent properties that are sensitive to initial conditions (Hepburn and Farmer, 2020). Making sense out of this will require complex systems thinking and agent-based simulation models.

Second, the climate and financial system interact.[62] The physical impacts of climate change—storms, floods, draughts, heatwaves, and so forth—damage assets, change asset values, and affect economic productivity.[63] Such physical risks will quickly find their way into the financial system through the insurance and banking systems, as well other channels.

---

[60] See a speech by Lael Brainard, Governor of the US Federal Reserve (March 1, 2021), at www.federalreserve.gov/newsevents/speech/brainard20210301a.htm.

[61] As noted elsewhere: "A thorough attempt to understand the whole economy through agent-based [simulation] modelling will require integrating models of financial interactions with those of industrial production, real estate, government spending, taxes, business investment, foreign trade and investment, and with consumer behaviour. The resulting simulations could be used to evaluate the effectiveness of different approaches to economic stimulus, such as tax reductions versus public spending." See Farmer and Foley (2009).

[62] For an outstanding review, see Bolton et al. (2020).

[63] See, for example, Burke et al. (2015).

This can happen, for example, because credit risk on a mortgage portfolio or a long-term loan to a factory rises.[64] Notwithstanding the losses that are being incurred in the insurance and reinsurance industry, such physical climate risks are likely to be comparatively less extreme in the short term. However, if we remain on track for an increase of 3–4 degrees Celsius in global average temperatures, it is likely that dramatic (and costly) shifts in climate and habitats for humans[65] and other species will result, with large economic costs at the country level (Dibley et al., 2021).

In the short(er) run (within the next decade), action taken by governments, citizens, and businesses to reduce the worst impacts of climate change will change the returns to asset classes. To a large extent, emission-heavy industries have been economically feasible because of the implicit subsidy that allowed firms to dump carbon pollution into the common atmosphere for free. Once that subsidy is removed, because emissions are either priced (e.g., a carbon tax) or capped, cost curves will adapt and economic activity will shift. More broadly, meeting the goals outlined in the Paris Agreement will require nothing less than a fundamental rewiring (and decarbonization) of the global economic system. That transition will create winners and losers, which will be reflected in financial valuations. There are already more fossil-fuel reserves on the books of companies than can be permitted to be emitted into the atmosphere (Hepburn et al., 2014), and there is a surplus of assets (refineries, power plants) that use them (Pfeiffer et al., 2016). It is likely that these assets have to be stranded.[66] As Mark Carney (former governor of the Bank of England) observed, the scale of the potential revaluations implies that concentrating them in a single climate "Minsky moment" rather than having a smooth adjustment pathway could have significant financial stability consequences.[67]

Climate stress tests could by well suited to measuring the physical and transition risks emerging from (and conditional on) climate change and mitigation policies. A climate stress test poses a hypothetical but plausible scenario of the increase in temperature and the set of policies in place to limit climate change. Each financial institution then assesses what losses from physical risk and transition risk it may face given its current balance sheet and business model. Once the risks are measured and disclosed, investors can price the risk. Moreover, the stress-tested institution can use these insights to adapt its business model to limit its risks and seize the opportunities presented by the transition. In this way, climate stress tests help to adapt to climate change and accelerate the transition.

The COVID-19 crisis and climate risk are instructive of a broader challenge: many emerging risks require an understanding of the linkages between the financial system, the real economy, and the broader environment in which society operates. Accordingly, system-wide stress tests need to be extended so that rather than considering the financial system in isolation, they consider the real economy as well as the environment. This will require a much more ambitious conception of what a "system-wide stress test" looks like. We need to understand, and these models should capture, the interlinkages among climate, economic, financial, and financial stability risks. Different economies, geographies, sectors, and even

---

[64] See, for example, Ouazad and Kahn (2019).

[65] Generally, see Allen et al. (2019). See also, for example, Xu et al. (2020).

[66] See, for example, Mercure et al. (2018) and Kruitwagen et al. (2021).

[67] See www.bankofengland.co.uk/-/media/boe/files/speech/2015/breaking-the-tragedy-of-the-horizon-climate-change-and-financial-stability.pdf.

firms are likely to face distinct risks from climate change, with each taking actions that affect the others (Brunetti et al., 2021).

Luckily, this is a challenge to which agent-based simulation models are well suited. Their capacity for modular approaches is particularly useful because (as noted) this means each of the modules can be developed and fit to the data separately while providing flexibility, scalability, and the ability to do justice to heterogeneity. Such models can be extended to operate at the scale of individual products and firms. The result is not a single monolithic model but, rather, a library of modules with standardized interconnections (see, e.g., the ESL referred to in Section 2). Box 30.1 illustrates how adding modules containing models of real-economy sectors and the climate system to models of the financial system could work.

Box 30.1 A nonequilibrium model of the green energy transition

Figure 30.1 shows a schematic diagram of a model under development at the University of Oxford to understand the green energy transition. It consists of six modules: (1) the global energy system (Pichler, Lafond et al., 2020), (2) input–output and growth (McNerney et al., 2018), (3) competitiveness and the adjacent possible (del Rio-Chanona et al., 2021), (4) innovation and technology (Pichler, Pangallo et al., 2021), (5) occupational labor (Way et al., 2020), and (6) investment. Each of these modules has been tested on its own using historical data, but they

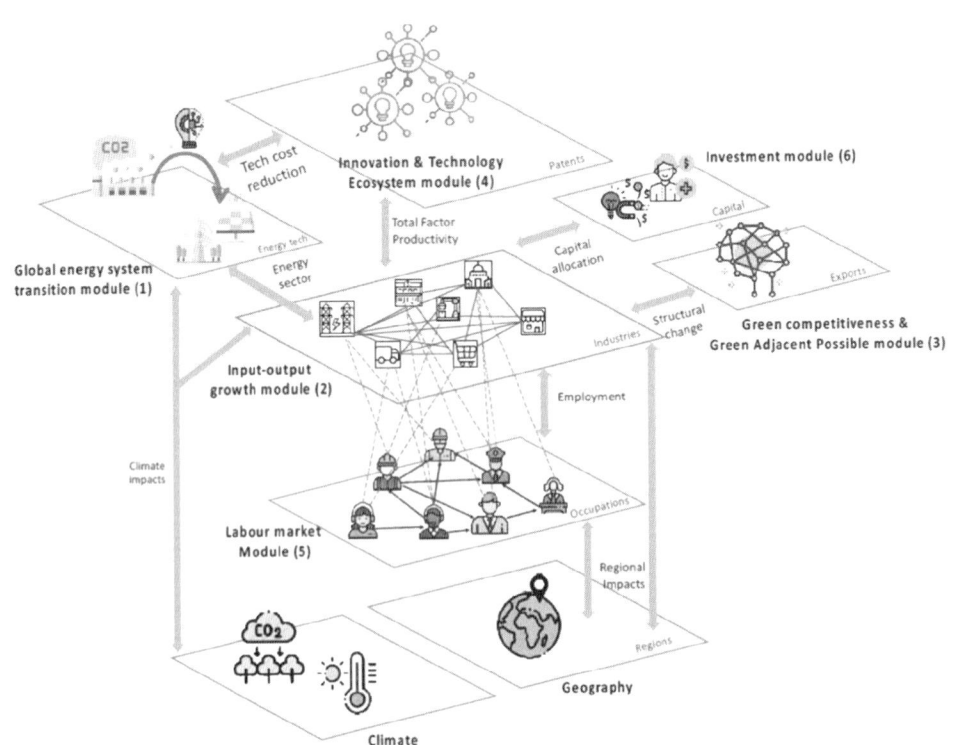

Figure 30.1 A schematic diagram of a model for the green energy transition.

can also be linked to provide a more comprehensive model. For example, the link between occupational labor (5) and input–output (2) was an essential ingredient for making accurate forecasts of the economic effects of the COVID-19 pandemic (see the previous discussion). The investment module (6), which provides the link to the financial system, is still under development.

4.3 Sense-Checking Model Outputs

Finally, and critically, we also think about how to make sure these models are sensible and reliable. Up until recently, the use of simulation models as quantitative tools in social science has been limited by the problems of parameter estimation and model validation. Simulation models can become complicated, and they are typically not differentiable and in many cases are not simply expressed in terms of equations. Recent progress in the development of Bayesian methods and the importation of techniques such as data assimilation from meteorology provide a proof of principle that these problems can be solved (Platt, 2020). We anticipate a new wave of quantitative simulation models that can be used to understand the dynamics of financial networks and their interaction with the real economy. These models will make full use of the rapidly increasing availability of data and computational resources.

5 Creating Conditions for Successful System-Wide Stress Testing

To get the models right and implement them successfully will take a revamping of the institutional structure into which system-wide stress tests are embedded. In this section, we briefly reflect on four enabling conditions that are necessary to realize and effectively execute our vision for system-wide stress testing: better data gathering, access, and usage (Section 5.1); a macroprudential apex regulator (Section 5.3); a high-tech culture at macroprudential authorities (Section 5.3); and a healthy dose of intellectual humility in interpreting and relying on model outcomes (Section 5.4).

5.1 Data Gathering, Access, and Usage

Data sits at the heart of the ongoing challenge financial regulators face in their attempts to understand the financial system and its evolution. Information gaps are a significant source of regulatory failure and, more broadly, stand in the way of the effective design and implementation of policies aimed at enhancing resilience. Hence, it is no surprise that the Financial Stability Board and the International Monetary Fund observed that "[the global financial crisis of 2007–2009] has reaffirmed an old lesson—good data and good analysis are the lifeblood of effective surveillance and policy response at both the national and international levels (IMF Staff and FSB Secretariat, 2009). Solid data gathering is also the lifeblood for system-wide stress tests. After all, to implement system-wide stress tests, regulators and supervisors first need to have a clear view of the financial system. That, in turn, requires detailed information on financial institutions and the precise linkages between them. You cannot stress test what is not measured. Then vice-chair of the Federal Reserve

Janet Yellen has argued that "[w]ithout such comprehensive and detailed data, it is simply not possible to understand how stress in one part of the network may spread and affect the entire system" (Yellen, 2013).

To make it work, we need en masse data capture at many scales and with high frequency. This is a common element in trying to understand complex systems. In the climate system, for example, time-series paleoclimatology data from ice cores, tree rings, corals, and sediments "have been vital in understanding the sorts of states that are semi-stable and the types of transitions that are possible." (Hepburn and Farmer, 2020). Such data should cover many scales because it is impossible to understand a complex system with aggregate data alone. In the context of the financial system, that means gathering granular, contract-level data that allow for the pricing of the individual contracts (in the context of derivatives, for example, this should enable a determination of the "Greeks") in order to study linkages and the potential for shock propagation. Finally, the data should have a sufficiently high frequency. Given the dynamism of the financial system, the baseline risk that regulators should avoid is that they are effectively looking to the past to the point where the data lose relevance. More ambitiously, regulators would ideally have frequent access to the balance sheets of all financial-market participants so that they can track the evolving ecosystem of the markets they regulate (Scholl et al., 2021).

Increasing digitization and technological advancements in data gathering and analysis hold out the promise that data collection can indeed be implemented at a large scale (Cristano et al., 2020). to facilitate up-to-date and truly system-wide stress tests.[68] Leveraging that opportunity may well be the only realistic prospect regulators and supervisors have to start to come to grips with the complexity and dynamism of the financial system.[69] This has implications for regulatory reporting requirements, which should become more extensive. Luckily, the last 10 years have seen real progress on this front.[70] In over-the-counter derivatives markets, Dodd–Frank and European Market Infrastructure Regulation (EMIR) provide US and EU authorities, respectively, with detailed contract-level data that are reported to large-scale data repositories. Beyond derivatives, the granularity (and scope) of reporting has expanded—for example, through the Securities Holdings Statistics in Europe that collects quarterly data on the holdings of banking groups.

The next 10 years should see current progress accelerate. In particular, the scope of granular, contract-level reporting should increase to encompass more asset classes. Data on various repo and securities lending markets remain substantially incomplete,[71] for example. Moreover, the evolution of reporting requirements, and in particular their perimeter, has not kept pace with the dynamism of the financial system; nonbanks increasingly play a role in shaping market developments, but this is precisely the area for which data remain poor.[72] Although the data quality is improving, the data that *are* currently being gathered are frequently patchy, and their quality is not typically mandated and enforced, which compromises the data's usefulness in calibrating regulatory stress tests.

---

[68] See Chapters 10 and 26 in this handbook.

[69] See Section 2 of this chapter.

[70] For a more extensive overview, see Chapter 10 in this handbook.

[71] Chapter 19 in this handbook.

[72] Chapter 4 in this handbook.

Too frequently, the data that are gathered by different regulators and across different parts of the financial system are also incompatible, making it more difficult to use such data in system-wide models.[73] This suggests that data standardization, a topic that may appear banal, deserves much more attention (Judge and Berner, 2019). Some progress has been made with the introduction of the Legal Entity Identifier (LEI) initiative, a joint effort by the Group of 20, the Financial Stability Board, and regulators around the world. An LEI is a 20-digit, alphanumeric, standardized code that connects to an increasing amount of key reference information about legal entities participating in financial transactions, answering the questions "who is who" and "who owns whom."[74] Although the number of LEIs is growing rapidly, full adoption remains elusive in many jurisdictions (including the United States). The next step would be to create unique product identifiers (UPIs) for each type of financial claim, which could work in tandem with the LEIs to create the building blocks of a more comprehensive and accurate map of the financial system (Armour et al., 2016).[75]

Because the financial system is not siloed along jurisdictional lines, data should not be either. Finance spans borders; system-wide models should too. In much the same way that the data-reporting perimeter should be expanded *within* jurisdictions to capture the full range of market participants, so too should the perimeter encompass interactions between jurisdictions. That will require a degree of international data sharing that, so far, has been absent.

Regulatory data will never be perfect. It may remain slow to recognize losses and vulnerabilities, thus complicating regulatory and supervisory efforts to intervene in a timely manner. In addition, banks may have significant discretion in recognizing asset impairments, which means that they can essentially avoid fair-value accounting on their loan books.[76] Making matters worse, banks can easily arbitrage these regulatory reporting requirements.[77] These observations suggest that regulatory data should be augmented by market data, which have complementary strengths (and weaknesses).[78] Financial institutions should face stress tests that draw on both data sets, with supervisors insisting that they pass both. In case of agreement in results, that boosts confidence. Where there are discrepancies in stress test outcomes, it provides an opportunity to dig deeper into the sources of these differences, which could either uncover latent risks or help improve models.[79]

Finally, when system-wide models expand beyond the financial system, data-gathering efforts need to do the same. Models that capture the interaction of the financial system and the real economy will inevitably require data types that are not traditionally used in financial stress tests. The problem of climate risk is again illustrative for this data challenge. At some level, quantifying climate-related risks should be similar to quantifying any other financial stability risk. In practice, it involves several unique challenges to measurement that require investment to address. For physical risk modeling, regulators will need a wide range of

---

[73] Chapter 10 in this handbook.

[74] See www.gleif.org/en/about-lei/introducing-the-legal-entity-identifier-lei. See also Awrey and Judge (2020).

[75] More ambitiously, regulators could require portfolio identifiers too—something that would help advance market ecology modeling by making it easier to study the evolution of trading strategies.

[76] See Chapter 9 in this handbook.

[77] See Chapter 9 in this handbook.

[78] See Chapter 9 in this handbook; see also Vickers (2019).

[79] Chapters 8 and 9 in this handbook.

(geospatial and temporal) sector-, institution-, and even asset-level data, covering not only the real economy but also climatological (and meteorological) factors (Fiedler et al., 2021). Transition risk modeling also requires a wide-ranging set of data that is novel in this context, from global input–output data[80] to experience curves to assessments of sectoral innovation trends.[81] This creates a challenge for regulators and supervisors, who will have to collect, understand, and learn to work with a much wider range of data (sources). In some cases, the data might not yet exist or might be uneven across sectors and regions, making comparative assessments difficult. In others, the data might exist but be spread out across the agencies and jurisdictions.

5.2 A Macroprudential Apex Regulator

Regulatory fragmentation, with different agencies taking responsibility for different parts of the financial system, is problematic from the perspective of systemic-risk regulation and oversight. In the United States, for example, the institutional structure is highly fragmented. A plethora of regulatory agencies, at different levels of government (e.g., state, federal), divides the responsibility for regulating the financial system. This creates coordination problems in the case of overlapping authorities but also lacunas where no agency at all is responsible for parts of the financial system. Regulatory fragmentation also affects the flow of information, creating data siloes that make it cumbersome if not impossible to form a picture of the whole system and study system-level dynamics. Finally, different areas of expertise and resources become siloed too, so that it becomes much more difficult to develop an integrated regulatory strategy for financial stability regulation that cuts across different agencies. In sum, regulatory fragmentation exacerbates all of the regulatory challenges discussed in Section 1 of this chapter.[82]

Implementation of macroprudential policies, including system-wide stress tests, requires an authority tasked with identifying systemic risk and invoking or coordinating the use of macroprudential tools (Armour et al., 2016). The objective of such a macroprudential authority (MPA) should be to preserve financial stability, an objective that should be given primacy over other regulatory objectives in case they conflict. In that sense, the MPA would be the apex authority within the financial regulatory system (Armour et al., 2016) with the ability to take a holistic approach to financial stability.[83] MPAs should have access to as rich a set of information about the financial system as possible and should be well equipped to process and analyze the data. Given these requirements, central banks appear generally well placed to take up this role.

Since the previous financial crisis, MPAs have been created in a number of jurisdictions. In the United States, the Dodd–Frank Act established the FSOC. This interagency "apex committee" coordinates among all significant financial regulators and policymakers in the United States and has the broad duty to identify and respond to threats to US financial stability. In carrying out this task, it is assisted by the Office of Financial Research (OFR),

---

[80] See, for example, Kruitwagen et al. (2021).

[81] See, for example, Way et al. (2020).

[82] See Section 1.

[83] See Chapter 20 in this handbook.

which provides the FSOC and its member agencies with aggregate financial data, standards, and analysis relating to the financial system as a whole. In the European Union, the European Systemic Risk Board (ESRB) is responsible for the macroprudential oversight of the financial system and the prevention and mitigation of systemic risk. In practice, the ECB fulfills an important role behind the scenes. Not only does the ECB president chair the ESRB, but the ECB also provides structural and institutional support and delivers much of the data and analysis on which the ESRB relies.

These reforms to the financial regulatory architecture are a good start but need to be revived to live up to their promise. For starters, neither the FSOC nor the the ESRB is an "action agency." Their role is primarily to provide information and coordinate policy across the broader financial regulatory system—they lack the incisive powers to do much more. And although both authorities can request specific information from financial institutions, their data mandates remain limited. They have the ability to gather data from other agencies, but these agencies may themselves have limited data-gathering mandates. They also cannot compel data standardization across regulatory agencies, a basic but valuable function that would lower the costs of cleaning, processing, and merging the data in order to use them in a system-wide model. Both in terms of resources and mandate, neither institution is equipped to be a full-fledged MPA. That impairs their ability to develop and run truly system-wide stress tests.

The handling of emerging financial stability risks of climate change in the US system illustrates this point. Such a task would naturally fall within the remit of an MPA, in this case the FSOC and OFR, but that has not been the case. Instead, the Federal Reserve Board is establishing a new Financial Stability Climate Committee (FSCC) to identify, assess, and address climate-related risks to financial stability. The FSCC is a system-wide committee that will approach this task from a macroprudential perspective—that is, one that considers the potential for complex interactions across the financial system. Lael Brainard, member of the Board of Governors of the US Federal Reserve System, noted that the FCSS's objectives are "to promote the resilience of the financial system to climate-related financial risks, to ensure coordination with the Financial Stability Oversight Council (FSOC) and its member agencies, and to increase the Federal Reserve's international engagement and influence on this issue" (Brainard, 2013). Although this new initiative shows that an awareness of the importance of an MPA is there, it is an ad hoc institutional solution that further fragments the already fragmented US regulatory system.

5.3 A High-Tech Culture at Macroprudential Authorities

To build and deploy the next generation of system-wide stress tests, we need a high-tech culture at the MPAs. Simulation models, and the data and computational infrastructure required to run them, require a different skill set and infrastructure than that traditionally on offer. Complementing economists and lawyers, central banks will need expertise in computer and data science, in artificial intelligence and agent-based models, and in a range of adjacent fields (e.g., climate change). This means that teams become multidisciplinary and will be required to innovate. This is needed to ensure that data are gathered, evaluated, cleaned, and aggregated using automated processes so that regulators and supervisors can run real-time system-wide stress tests of the financial system to continuously monitor resilience against a range of scenarios.[84] Over time, it may even become possible to automate stress test development and analysis.[85] "[D]ata scraping tools, big data analysis techniques, and artificial intelligence or machine learning algorithms, [...] together allow for the aggregation and effective use of amounts of information that flesh-and-blood compliance officers and supervisors could never muster and master" (Enrique, 2017).

The skill sets and culture at central banks should enable it to leverage the advances in computational resources and large volumes of data. Technological advances in hardware and software open up the potential of simulation models. Whether that potential is used depends on the ability and willingness of central bank teams to embrace it.

Central banks are starting to become more tech-savvy and innovative. At the ECB and the Bank of England, for example, multidisciplinary teams are increasingly common, which is reflected in the diversity of the methodologies used in their research and policymaking. The Bank for International Settlements (BIS) has launched a global Innovation Hub, an effort to spearhead central banks' response to digital innovation (including regulatory and supervisory technology).[86] At the Organisation for Economic Co-operation and Development (OECD), the New Approaches to Economic Challenges (NAEC) Innovation Lab allows researchers across the OECD to work together to experiment with new analytical tools and techniques to encourage their adoption throughout the wider organization.[87]

But challenges remain. For starters, the competition to hire (and retain) the best data scientists may be one that is challenging for regulators to win. This challenge is a familiar one, with regulators and the regulated being engaged in a constant cat-and-mouse game for talented lawyers and economists. However, the problem may be even more acute in the context of data scientists.[88] Their outside options, both in the financial sector and in the tech industry, are extremely compelling—and the advantage of having experience working at a regulator may be less meaningful for them than it might be for economists and lawyers (Enriques, 2017). We are more optimistic, partly because of the unique data that regulators could gather and because of the attraction that comes with serving the public good, but we recognize this as a serious challenge going forward.[89] Central banks agree. In a recent survey, they ranked recruiting and retaining top technology talent as their most pressing tech challenge going forward—ahead of keeping up with FinTech products and strained resources.[90]

On top of that, the highly sensitive nature of regulatory data may impede fluent collaboration—for example, because the data cannot leave the (dated) systems of the central bank. The same problem also complicates collaboration with academia and other outside experts, who could otherwise contribute to regulatory innovation. Frontier-pushing research to develop the next generation of system-wide stress tests requires data access,

---

[84] See Chapter 26 in this handbook.

[85] See Chapter 3 in this handbook.

[86] See www.bis.org/about/bisih/about.htm; see also www.thebanker.com/Comment-Profiles/Viewpoint/Central-banks-must-be-at-cutting-edge-of-technology.

[87] See www.oecd.org/naec/projects/naecinnovationlab/.

[88] See www.law.ox.ac.uk/business-law-blog/blog/2017/07/hr-challenge-fintech-financial-regulators.

[89] See also Chapter 10 in this handbook.

[90] See www.centralbanking.com/benchmarking/7816641/cyber-security-and-retaining-top-talent-are-biggest-tech-challenges.

both to calibrate the models and to study their outputs. But if the data cannot leave the central banks' servers and cannot be ported to the cloud-based infrastructure that supports large-scale simulation models, or worse, if external researchers cannot get direct access at all, that type of collaboration becomes infeasible. As Cumming notes elsewhere in this volume, "the challenge for the custodians of this highly sensitive stress testing data is to identify avenues to expand access to these data in a controlled manner that preserves the confidentiality of individual banking institutions. Access to these data could be made available to scholars in much the same highly controlled manner as US census data are made available for research."[91]

5.4 *A Healthy Dose of Intellectual Humility*

Ambition to improve stress tests models is necessary, but we need intellectual humility too. Excessive reliance on canonical theories of financial economics—including modern portfolio theory, the capital asset pricing model, and the efficient-market hypothesis—without questioning the validity of their assumptions created regulatory blindspots that paved the way for the financial crisis of 2007–2009 (Awrey, 2012). That should remain a teachable moment: overconfidence in models is dangerous because it affects regulatory outcomes. Models are going to be imperfect, and data are going to be partial and frequently misleading (sometimes in a way that is endogenous, for example, as a consequence of regulatory arbitrage). Models are always abstractions of reality, which means that the choice between models is essentially a choice between what is captured and what is omitted by the model. Aspiring to verisimilitude does not mean capturing every detail or every dimension of a situation. We are not trying to create *The Matrix*, and that would not be a realistic goal either. The challenge—especially when studying a system that is as complex and dynamic as the financial system—is to differentiate between what is known and what is not, a distinction that experts have not always appreciated.

Less precision in reporting model results is likely to mean more truth (Hepburn and Farmer, 2020). Overconfidence is pervasive, and there is no reason why estimates relating to financial systems should be immune.[92] Regulators should acknowledge that uncertainty and report stress test results as a distribution around some mean, with confidence intervals.[93] Current stress tests often present stress tests results with point-precise estimates, downplaying the underlying uncertainties. The underlying problem in the financial system is that it is rife with nonlinear positive-feedback loops that can enormously amplify modest shocks. Modeling everything precisely is therefore impossible. It is also less important than getting an order-of-magnitude idea of the key risks and a sense of the narrative as to how these risks might unfold (Bookstaber, 2019). Moreover, some data will not be known (because they are too costly to gather or analyze) and other data cannot be known (because of Knightean uncertainty [Knight, 1921]). That distinction should instill a sense of modesty,

---

[91] See Chapter 18 in this handbook.

[92] *Accuracy* is defined as the proximity of a measurement to its true value. *Precision* is defined as the degree to which repeated measurements show the same results and is commonly expressed in statistics as the reciprocal of the variance. See Hepburn and Farmer (2020).

[93] See Chapter 2 in this handbook.

which is particularly important when reviewing stress test results that are conditional on a particular scenario. Other, unforeseen scenarios might lead to wildly different results, implying that passing a stress test should never be considered a "clean bill of health."[94] Parameter selection is itself an art as much as a science, which will be driven by regulators' (potentially biased) perceptions of what sort of situations might cause problems. We should avoid only "looking for keys under the lantern."

In sum, even with the best possible models and data, there will inevitably remain a high degree of residual judgment and analysis when interpreting stress test results. That places additional value on the ability to understand the outcomes produced by system-wide stress tests and calls for caution when relying on stress test results to formulate policy positions.

6 Stress Testing to Regulate and Supervise the Financial Macrocosm

In this chapter, we have sketched how simulation models—enabled by highly granular data—can enable truly system-wide stress testing. We close with observations about the financial regulatory policy innovations that such novel stress tests would enable.

Truly system-wide stress tests informed by granular and up-to-date data enable a far more dynamic approach to (macro)prudential policy. Given the dynamism of the financial system, financial regulation must continuously evolve and adapt too, in order to promote financial stability. That adaptive process should be informed by up-to-date insights on the state of the financial system, feeding into dynamic macroprudential policy that is forward-looking (Gordon, 2018). System-wide stress tests, calibrated using up-to-date data, can help make that real.

In practice, this opportunity has not yet been grasped. Not only do most "system-wide" stress tests only capture a subset of the financial system (as noted previously), but they are also not commonly used to calibrate prudential policy. That is inconsistent with the wider shift toward a macroprudential perspective in systemic risk regulation, and—rather predictably—means that some risks are left unseen (Haldane and May, 2011). If capital requirements are to be designed to strengthen the financial system, rather than only reduce idiosyncratic risks to the balance sheet of an individual institution, this is problematic. Similar arguments apply to the regulation of liquidity requirements, such as the liquidity coverage ratio and the net stable funding ratio. Liquidity stress can propagate to take on systemic proportions, too, so it is important that the binding liquidity targets in one part of the system do not amplify stresses elsewhere. System-wide stress tests can help elucidate these risks and also shed light on the interactions between solvency and liquidity risks at the level of individual institutions or the system as a whole (Cont et al., 2020; Wiersema et al., 2019). That would represent a major advancement in regulators' ability to set macroprudential policy.

Recent studies have confirmed that interacting contagion channels produce significantly higher rates of bank failure than suggested by the sum of contagion channels when they act in isolation (Farmer et al., 2020). Bank capital requirements that are calibrated using microprudential stress tests are therefore likely to provide false comfort, leading to capital levels that are too low. The same study also highlighted how various regulatory requirements

---

[94] See www.ft.com/content/c55e9466-9da7-11e5-8ce1-f6219b685d74.

might interact to generate instability, in line with the concept of holistic regulation suggested by Goodhart,[95] which should inform a redesign of capital requirements. The study specifically showed that the degree to which bank buffers are genuinely "usable," in the sense that they are shock-absorbing in an ongoing-concern scenario, has significant implications for the resilience of the financial system. Higher usable buffers limit predefault contagion, an effect that is not captured by microprudential stress tests.[96]

Beyond capital and liquidity requirements, system-wide stress testing could be used to inform a range of other macroprudential interventions. Several studies have highlighted that the network topology of the financial system has implications for the state and dynamics of systemic risk (Haldane and May, 2011). This has led to a host of regulatory initiatives designed to alter the structure of the financial system, ranging from large-exposure regulation to mandatory central clearing (Duffie and Zhu, 2011; Enriques et al., 2019). In this volume, Thurner suggests introducing a systemic risk tax, levied to optimize the network structure to minimize systemic risk (Pichler et al., 2021).[97] A similar logic has been applied to study the financial stability implications of creating greater modularity in the financial network (Haldane and May, 2011). One set of initiatives in this vein attempts to ring-fence the "critical" functions of banking groups (e.g., deposit-taking) from the "more risky" parts of the banking group (e.g., proprietary trading)—with notable examples being the Volcker rule in the United States and retail bank ring-fencing in the United Kingdom.[98]

Macroprudential calibration using system-wide stress tests could be used more creatively to make a much larger part of the financial regulatory toolkit more dynamic and forward-looking. The list of options is long. Stricter governance requirements could be applied to more systemically important financial institutions (Enriques et al., 2019) recovery planning could be informed by a range of scenario analyses that study (for example) the systemic impact of bail-in design (Hüser et al., 2018) or calibrate total loss-absorbing capacity (TLAC) requirements[99], and the implications of bank mergers could be evaluated ex ante. Regulatory requirements to standardized risk models may lead to increased homogeneity in assets and liabilities, which may breed fragility[100]—stress tests can be used to assess such risks, too. In each of these cases, it is important to note (in the spirit of the intellectual humility we advocated previously) that stress tests can help inform policy design but may not account for the inevitable higher-order effects that manifest when market participants adapt to the new rules (Awrey and Judge, 2020).[101]

Doing this well requires that macroprudential stress tests become truly system-wide and are indeed used for prudential purposes across the system. Current practice remains to limit regulatory stress testing to a small subset of the financial system (in most cases, banks and insurers). However, as noted earlier (Sections 1 and 4), nonbank financial intermediation is growing. That underscores the need for microprudential stress tests for nonbanks that are

---

[95] See Chapter 20 in this handbook.
[96] See also Kleinnijenhuis et al. (2020).
[97] Discussed in See Chapter 28 in this handbook.
[98] See, for example, Vickers (2011) and Wetzer (2019).
[99] See www.fsb.org/2021/03/evaluation-of-the-effects-of-too-big-to-fail-reforms-final-report/.
[100] See, for example, Heinrich et al. (2021).
[101] However, scenario-based stress testing informed by the perspective of market ecology may help explore what these higher-order dynamics *could* look like.

coordinated with those carried out at banks[102], but also the need for more system-wide stress tests that capture their interaction with the banking system.

In addition to expanding the coverage of stress tests models, there may be scope for expanding the use of (system-wide) stress tests as a policy tool beyond the remit of the financial system. Several events, ranging from the fallout during the COVID-19 crisis to the fallout in global trade after the blockage of the Suez Canal by the *Ever Given* (a giant container ship) in 2021, have illustrated the concern that the resilience of the real economy has been sacrificed in favor of efficiency. Similarly, there is now an increasing concern that non–financial-sector companies can generate systemic risks.[103] This set of observations has led several commentators to call for some form of (macro)prudential regulation for the real economy.[104] An in-depth assessment of what such real-economy macroprudentialism would look like is beyond the scope of this chapter, but it is worth noting that simulation models of the type discussed in this chapter would, in principle, be able to facilitate such real-economy analyses.

In addition to expanding the scope of the sectoral coverage of stress tests, the scope of the types of stress scenarios should be expanded. Specifically, stress tests could also be used for contingency planning in the face of emerging risks. Such horizon scanning, and systemic risk modeling more broadly, is, in principle, a pure information-gathering exercise that is not automatically accompanied by regulatory consequences like raising more capital.[105] The Federal Reserve's financial stability monitoring framework is making inroads along these lines (Brunetti et al., 2021), as is the Bank of England, which recently introduced exploratory biennial stress scenarios (Bank of England, 2019).

Central bankers and regulators at MPAs should explore a wide range of potential scenarios, something that flexible, system-wide simulation models can facilitate. To date, regulators have focused on macroeconomic and financial shocks, but shocks that are nonfinancial in origin can be devasting to financial stability as well. Think, for example, about a cyberattack that disrupts financial services,[106] a pandemic that lays low entire economies,[107] or a climate crisis that affects asset values in a synchronized fashion (Battison et al., 2016, 2017). What are the impacts of such a shock on the financial system? What would be appropriate (macro)prudential preventative actions, and what would be appropriate responses should such an event occur (Hepburn and Farmer, 2020)? The value of such an exercise, which is necessarily speculative, lies not in the point estimate that the stress test produces[108] but, rather, in the learning process that takes place as regulators run through these scenarios with regulated institutions. Ideally, regulators test for multiple scenarios in each stress testing exercise to obtain a more holistic view of the resilience of the institution or system. Currently, stress test exercises typically do not go down that route but instead focus on one hypothetical stress scenario.

---

[102] See Chapter 2 in this handbook.

[103] See, for example, Wu et al.'s (2021) study on systemic risk across global energy companies.

[104] See www.ft.com/content/ed64386e-ef9e-4fa0-b8e0-b9311beb9a88.

[105] See Chapter 2 in this handbook.

[106] See, for example, Mee and Schuermann (2018).

[107] See, for example, Potter and Schuermann (2020).

[108] Especially because the outcome of stress tests is highly conditional on the parameters chosen in each scenario, which is as much art as it is science. See, for example, Borio et al. (2014).

As noted previously (Section 4), climate risk is one of the most salient (see, e.g., Battiston et al., 2017; Dietz et al., 2016). Academics and central bankers have already started laying the groundwork for models that capture the physical and transition risks that climate change presents,[109] and regulators around the world are preparing to feature these risks in horizon-scanning exercises.[110] Some regulators already contemplate going further, using climate risk stress tests to calibrate institution-specific capital requirements.[111] To do so in a meaningful way would require a step change in the development of system-wide stress test models, data gathering, and institutional governance along the lines advocated in this chapter. In fact, such stress tests highlight the broader point that this chapter has advocated: truly system-wide stress tests that capture not only the full financial sector but also its interactions with the real economy, and that are informed by granular and real-time data, can help regulators to be more responsive to emerging risks and make prudential regulation more dynamic and versatile.

The past 10 years have seen rapid innovation in the sophistication and use of stress tests. Let the next 10 years mark the move toward system-wide stress testing of the whole financial macrocosm.

References

Acharya, V. V., and T. Yorulmazer (2008), "Information contagion and bank herding," *Journal of Money, Credit and Banking*, 40(1), 215–231.

Adrian, T., and H. S. Shin (2010), "Liquidity and leverage," *Journal of Financial Intermediation*, 19(3), 418–437.

Aikman, D., P. Chichkanov, G. Douglas, Y. Georgiev, J. Howat, and B. King (2019), "System-wide stress simulation," Bank of England Staff Working Paper No. 809.

Aldasoro, I., W. Huang, and E. Kemp (2020), "The global picture of non-bank financial intermediation from FSB data," Working Paper, Bank for International Settlements.

Allen, F., and D. Gale (2000), "Financial contagion," *Journal of Political Economy*, 108(1), 1–33.

Allen, M., P. Antwi-Agyei, F. Aragon-Durand, M. Babiker, P. Bertoldi, M. Bind, S. Brown, M. Buckeridge, I., Camilloni, and A. Cartwright (2019), "Technical summary: Global warming of 1.5° C," An IPCC Special Report, available at www.ipcc.ch/site/assets/uploads/sites/2/2019/05/SR15_Citation.pdf.

Anderson, R., J. Danielsson, C. Baba, M. U. S. Das, M. H. Kang, and M. A. S. Basurto (2018), *Macroprudential stress tests and policies: Searching for robust and implementable frameworks,"* International Monetary Fund.

Armour, J., D. Awrey, P. L. Davies, L. Enriques, J. N. Gordon, C. P. Mayer, and J. Payne (2016), *Principles of financial regulation*, Oxford University Press.

Ashcraft, A. B., and T. Schuermann (2008), "Understanding the securitization of subprime mortgage credit," *Foundations and Trends in Finance*, 2(3), 191–309.

Awrey, D. (2012), "Complexity, innovation, and the regulation of modern financial markets," *Harvard Business Law Review*, 2, 235.

Awrey, D., and K. Judge (2020), "Why financial regulation keeps falling short," Columbia Law and Economics Working Paper 617.

---

[109] See, for example, Brunetti et al. (2021).

[110] See also Chapter 2 in this handbook. For examples, see Bank of England (2019), European Systemic Risk Board (2020), Dutch Central Bank (2018), and Bank of France (2020).

[111] Frank Elderson speaking to Morgan Stanley, as quoted in the *Financial Times* (www.ft.com/content/7b734848-1287-4106-b866-7d07bc9d7eb8?shareType=nongift).

Aymanns, C., F. Caccioli, J. D. Farmer, and V. W. Tan (2016), "Taming the Basel leverage cycle," *Journal of Financial Stability*, 27, 263–277.

Aymanns, C., and J. D. Farmer (2015), "The dynamics of the leverage cycle," *Journal of Economic Dynamics and Control*, 50, 155–179.

Aymanns, C., J. D. Farmer, A. M. Kleinnijenhuis, and T. Wetzer (2018), "Models of financial stability and their application in stress tests," In C. Hommes and B. LeBaron (eds.), *Handbook of computational economics*, Vol. 4, Elsevier.

Bank of England (2019), "The 2021 biennial exploratory scenario on the financial risks from climate change," Discussion Paper, available at www.bankofengland.co.uk/paper/2019/biennial-exploratory-scenario-climate-change-discussion-paper.

Bank of France (2020), "Climate-related scenarios for financial stability assessment: An application to France," Working Paper Series No. 744, available at https://publications.banque-france.fr/en/climate-related-scenarios-financial-stability-assessment-application-france.

Bardoscia, M., G. Bianconi, and G. Ferrara (2019), "Multiplex network analysis of the UK over-the-counter derivatives market," *International Journal of Finance and Economics*, 24(4), 1520–1544.

Battiston, S., J. D. Farmer, A. Flache, D. Garlaschelli, A. G. Haldane, H. Heesterbeek, C. Hommes, C. Jaeger, R., May, and M. Scheffer (2016), "Complexity theory and financial regulation," *Science*, 351(6275), 818–819.

Battiston, S., A. Mandel, I. Monasterolo, F. Schütze, and G. Visentin (2017), "A climate stress-test of the financial system," *Nature Climate Change*, 7(4), 283–288.

Bernanke, B. S. (2015), The courage to act: A memoir of a crisis and its aftermath, W. W. Norton & Company.

Black, F., and M. Scholes (2019), "The pricing of options and corporate liabilities," In M. Crouhy, D. Galai, Z. Wiener (eds.), *World Scientific reference on contingent claims analysis in corporate finance: Volume 1: Foundations of CCA and equity valuation*, World Scientific.

Bolton, P., M. Despres, L. A. P. Da Silva, F. Samama, and R. Svartzman (2020), *The green swan*, BIS Books.

Bookstaber, R. (2019), *The end of theory: Financial crises, the failure of economics, and the sweep of human interaction*, Princeton University Press.

Bookstaber, R., J. Cetina, G. Feldberg, M. Flood, and P. Glasserman (2014), "Stress tests to promote financial stability: Assessing progress and looking to the future," *Journal of Risk Management in Financial Institutions*, 7(1), 3–21.

Bookstaber, R., and D. Y. Kenett (2016), "Looking deeper, seeing more: A multilayer map of the financial system," *OFR Brief*, 16(06).

Borio, C. (2003), "Towards a macroprudential framework for financial supervision and regulation?" *CESifo Economic Studies*, 49(2), 181–215.

Borio, C., M. Drehmann, and K. Tsatsaronis, K. (2014), "Stress-testing macro stress testing: Does it live up to expectations?" *Journal of Financial Stability*, 12, 3–15.

Brazier, A. (2017), "How to: MACROPRU. 5 principles for macroprudential policy," Speech given at the London School of Economics Financial Regulation Seminar.

Brunetti, C., B. Dennis, D. Gates, D. Hancock, D. Ignell, E. K. Kiser, G. Kotta, A. Kovner, R. J. Rosen, and N. K. Tabor (2021, March 19), "Climate change and financial stability," FEDS Notes, available at www.federalreserve.gov/econres/notes/feds-notes/climate-change-and-financial-stability-20210319.htm.

Brunnermeier, M. K. (2009), "Deciphering the liquidity and credit crunch 2007–2008," *Journal of Economic Perspectives*, 23(1), 77–100.

Brunnermeier, M. K., and Y. Sannikov (2014), "A macroeconomic model with a financial sector," *American Economic Review*, 104(2), 379–421.

Burke, M., S. M. Hsiang, and E. Miguel (2015), "Global non-linear effect of temperature on economic production," *Nature*, 527(7577), 235–239.

Caccioli, F., M. Shrestha, C., Moore, and J. D. Farmer (2014), "Stability analysis of financial contagion due to overlapping portfolios," *Journal of Banking and Finance*, 46, 233 245.

Calimani, S., G. Hałaj, and D. Żochowski (2019), "Simulating fire sales in a system of banks and asset managers," *Journal of Banking and Finance*, Article 105707.

Carmassi, J., and R. Herring (2016), "The corporate complexity of global systemically important banks," *Journal of Financial Services Research*, 49(2), 175–201.

Célérier, C., and B. Vallée, B. (2013), July 1, "What drives financial complexity? A look into the retail market for structured products," Paris Conference Paper, December 2012 Finance Meeting EUROFIDAI-AFFI.

Claessens, S. (2015), "An overview of macroprudential policy tools," IMF Working Paper WP/14/214.

Coffee, J. C., Jr. (2011), "Political economy of Dodd-Frank: Why financial reform tends to be frustrated and systemic risk perpetuated," *Cornell Law Review*, 97, 1019.

Cont, R., A. Kotlicki, and L. Valderrama (2020), "Liquidity at risk: Joint stress testing of solvency and liquidity," *Journal of Banking and Finance*, 118, 105871.

Cont, R., and A. Moussa (2010), "Network structure and systemic risk in banking systems," Working Paper, available at https://papers.ssrn.com/sol3/papers.cfm?abstract_id=1733528.

Cont, R., and E. Schaanning (2017), "Fire sales, indirect contagion and systemic stress testing," Norges Bank Working Paper No. 2/2017.

Cont, R., and E. Schaanning (2019), "Monitoring indirect contagion," *Journal of Banking and Finance*, 104, 85–102.

Crisanto, J. C., K. Kienecker, J. Prenio, and E. Tan (2020, December 16), "From data reporting to data-sharing: How far can suptech and other innovations challenge the status quo of regulatory reporting?," FSI Insights No. 29, available at www.bis.org/fsi/publ/insights29.htm.

Cutler, D. M., J. M. Poterba, and L. H. Summers (1989), "What moves stock prices?" *Journal of Portfolio Management*, 15, 4–12.

del Rio Chanona, R. M., P. Mealy, M. Beguerisse-Diaz, F. Lafond, and J. D. Farmer (2021), "Occupational mobility and automation: A data-driven network model," *Journal of the Royal Society Interface*, 18(1074), available at https://doi.org/10.1098/rsif.2020.0898.

del Rio-Chanona, R. M., P. Mealy, A. Pichler, F. Lafond, and J. D. Farmer, (2020), "Supply and demand shocks in the COVID-19 pandemic: An industry and occupation perspective," *Oxford Review of Economic Policy*, 36(Suppl. 1), S94–S137.

Diamond, D. W., and P. H. Dybvig (1983), "Bank runs, deposit insurance, and liquidity," *Journal of Political Economy*, 91(3), 401–419.

Dibley, A., T. Wetzer, and C. Hepburn (2021), "National COVID debts: Climate change imperils countries' ability to repay," *Nature*, 592(7853), 184–187.

Dietz, S., A. Bowen, C. Dixon, and P. Gradwell (2016), "'Climate value at risk' of global financial assets," *Nature Climate Change*, 6(7), 676–679.

Duffie, D., and H. Zhu (2011), "Does a central clearing counterparty reduce counterparty risk?" *Review of Asset Pricing Studies*, 1(1), 74–95.

Dutch Central Bank (2018), "An energy transition risk stress test for the financial system of the Netherlands," available at www.dnb.nl/en/publications/research-publications/occasional-studies/nr-7-2018-an-energy-transition-risk-stress-test-for-the-financial-system-of-the-netherlands/.

Enriques, L. (2017), "Financial supervisors and regtech: Four roles and four challenges," *Revue Trimestrielle de Droit Financier*, 53.

Enriques, L., A. Romano, and T. Wetzer (2019), "Network-sensitive financial regulation," *Journal of Corporation Law*, 45, 351.

European Systemic Risk Board (2020, June), "Positively green: Measuring climate change risks to financial stability," available at www.esrb.europa.eu/pub/pdf/reports/esrb.report200608_on_Positively_green_-_Measuring_climate_change_risks_to_financial_stability~d903a83690.en.pdf.

Farmer, J. D. (2002), "Market force, ecology and evolution," *Industrial and Corporate Change*, 11(5), 895–953.

Farmer, J. D. (2019), "The future of complexity economics: Better solutions to the world's problems," in W. B. Arthur, E. D. Beinhocker, and A. Stanger (eds.), *Complexity economics: Dialogues of the Applied Complexity Network*, SFI Press.

Farmer, J. D., and D. Foley (2009), "The economy needs agent-based modelling," *Nature*, 460(7256), 685–686.

Farmer, J. D., A. M. Kleinnijenhuis, P. Nahai-Williamson, and T. Wetzer (2020), "Foundations of system-wide financial stress testing with heterogeneous institutions," Bank of England Working Paper Series.

Fiedler, T., A. J. Pitman, K. Mackenzie, N. Wood, C. Jakob, and S. E. Perkins-Kirkpatrick (2021), "Business risk and the emergence of climate analytics," *Nature Climate Change*, 11(2), 87–94.

Gabaix, X. (2020), "A behavioral New Keynesian model," *American Economic Review*, 110(8), 2271–2327.

Gai, P., A. Haldane, and S. Kapadia, (2011), "Complexity, concentration and contagion," *Journal of Monetary Economics*, 58(5), 453–470.

Geanakoplos, J. (2010), "The leverage cycle," *NBER Macroeconomics Annual*, 24(1), 1–66.

Gennaioli, N., and A. Shleifer (2018), *A crisis of beliefs*, Princeton University Press.

Gennotte, G., and H. Leland (1990), "Market liquidity, hedging, and crashes," *American Economic Review*, 999–1021.

Goodhart, C. (2008), "The boundary problem in financial regulation," *National Institute Economic Review*, 206(1), 48–55.

Gordon, J. N. (2018), "'Dynamic precaution' in maintaining financial stability: The importance of FSOC," S. O'Halloran and T. Groll (eds.), *Ten years after the crash*, Columbia University Press.

Greenwood, R., A. Landier, and D. Thesmar (2015), "Vulnerable banks," *Journal of Financial Economics*, 115(3), 471–485.

Greenwood, R., J. C. Stein, S. G. Hanson, and A. Sunderam, (2017), "Strengthening and streamlining bank capital regulation," *Brookings Papers on Economic Activity*, 2017(2), 479–565.

Grossman, S. J., and J. E. Stiglitz, (1980), "On the impossibility of informationally efficient markets," *American Economic Review*, 70(3), 393–408.

Hałaj, G. (2018), "Agent-based model of system-wide implications of funding risk," European Central Bank Working Paper Series No. 2121.

Haldane, A. G., and R. M. May (2011), "Systemic risk in banking ecosystems," *Nature*, 469(7330), 351–355.

Haldane, A. G., and A. E. Turrell (2018), "An interdisciplinary model for macroeconomics," *Oxford Review of Economic Policy*, 34(1–2), 219–251.

Hanson, S. G., A. K. Kashyap, and J. C. Stein (2011), "A macroprudential approach to financial regulation," *Journal of Economic Perspectives*, 25(1), 3–28.

Heinrich, T., J. Sabuco, and J. D. Farmer (2021), "A simulation of the insurance industry: The problem of risk model homogeneity," *Journal of Economic Interaction and Coordination*, advance online publication, available at https://link.springer.com/article/10.1007/s11403-021-00319-4.

Hellwig, M. F. (2009), "Systemic risk in the financial sector: An analysis of the subprime-mortgage financial crisis," *De Economist*, 157(2), 129–207.

Hepburn, C., E. Beinhocker, J. D. Farmer, and A. Teytelboym (2014), "Resilient and inclusive prosperity within planetary boundaries," *China & World Economy*, 22(5), 76–92.

Hepburn, C., and J. D. Farmer (2020), "Less precision, more truth: Uncertainty in climate economics and macroprudential policy," in G. Chichilnisky and A. Rezai (eds.), *Handbook on the economics of climate change*, Edward Elgar.

Herring, R. J., and A. M. Santomero (1990), "The corporate structure of financial conglomerates," *Journal of Financial Services Research*, 4(4), 471–497.

Hüser, A.-C., G. Haüaj, C. Kok, C. Perales, and A. van der Kraaij (2018), "The systemic implications of bail-in: a multi-layered network approach," *Journal of Financial Stability*, 38, 81–97.

IMF Staff and FSB Secretariat (2009), "The financial crisis and information gaps," Report to the G-20 Finance Ministers and Central Bank Governors, available at www.imf.org/external/np/g20/pdf/102909.pdf.

Judge, K. (2012), "Fragmentation nodes: A study in financial innovation, complexity, and systemic risk," *Stanford Law Review*, 64, 657.

Judge, K., and R. Berner (2019), "The data standardization challenge." In E. A. Douglas W. Arner, D. Busch, and S. L. Schwarcz (eds.), *Systemic risk in the financial sector: Ten years after the great crash*, CIGI Press.

Kiyotaki, N., and J. Moore (1997), "Credit cycles," *Journal of Political Economy*, 105(2), 211–248.

Kleinnijenhuis, A., L. Kodres, and T. Wetzer (2020). Usable bank capital," available at https://voxeu.org/article/usable-bank-capital.

Knight, F. H. (1921), *Risk, uncertainty and profit*, Vol. 31, Houghton Mifflin.

Kruitwagen, L., J. Klaas, A. Baghaei Lakeh, and J. Fan (2021), "Asset-level transition risk in the global coal, oil, and gas supply chains," Working Paper, available at https://papers.ssrn.com/sol3/papers.cfm?abstract_id=3783412.

Levin, S. A., and A. W. Lo (2015), "Opinion: A new approach to financial regulation," *Proceedings of the National Academy of Sciences*, 112(41), 12543–12544.

Lo, A. W. (2004), "The adaptive markets hypothesis," *Journal of Portfolio Management*, 30(5), 15–29.

Lo, A. W. (2005), "Reconciling efficient markets with behavioral finance: The adaptive markets hypothesis," *Journal of Investment Consulting*, 7(2), 21–44.

Lumsdaine, R. L., D. N. Rockmore, N. J. Foti, G. Leibon, and J. D. Farmer (2021), "The intrafirm complexity of systemically important financial institutions," *Journal of Financial Stability*, 52, 100804.

Malkiel, B. G., and E. F. Fama (1970), "Efficient capital markets: A review of theory and empirical work," *Journal of Finance*, 25(2), 383–417.

May, R. M., and N. Arinaminpathy (2010), "Systemic risk: The dynamics of model banking systems," *Journal of the Royal Society Interface*, 7(46), 823–838.

McNerney, J., C. Savoie, F. Caravelli, and J. Doyne Farmer (2018), "How production networks amplify economic growth," available at https://arxiv.org/abs/1810.07774.

Mee, P., and T. Schuermann (2018), "How a cyber attack could cause the next financial crisis," *Harvard Business Review*, available at https://hbr.org/2018/09/how-a-cyber-attack-could-cause-the-next-financial-crisis.

Mercure, J.-F., H. Pollitt, J. E. Viñuales, N. R. Edwards, P. B. Holden, U. Chewpreecha, P. Salas, I. Sognnaes, A. Lam, and F. Knobloch (2018), "Macroeconomic impact of stranded fossil fuel assets," *Nature Climate Change*, 8(7), 588–593.

Mishkin, F. S. (2011), "Over the cliff: From the subprime to the global financial crisis," *Journal of Economic Perspectives*, 25(1), 49–70.

Montagna, M., and C. Kok (2016), "Multi-layered interbank model for assessing systemic risk," ECB Working Paper No. 1944.

Morris, S., and H. S. Shin (2001), "Global games: Theory and applications," Cowles Foundation Discussion Paper No. 1275R.

Ouazad, A., and M. Kahn (2019), "Mortgage finance in the face of rising climate risk," NBER Working Paper w26322.

Paddrik, M. E., and P. Young (2017), "How safe are central counterparties in derivatives markets?" Working Paper, available at www.financialresearch.gov/working-papers/2017/11/02/how-safe-are-ccps/.

Pangallo, M., T. Heinrich, and J. D. Farmer (2019), "Best reply structure and equilibrium convergence in generic games," *Science Advances*, 5(2), eaat1328.

Pfeiffer, A., R. Millar, C. Hepburn, and E. Beinhocker (2016), "The '2 C capital stock' for electricity generation: Committed cumulative carbon emissions from the electricity generation sector and the transition to a green economy," *Applied Energy*, 179, 1395–1408.

Pichler, A., F. Lafond, and J. D. Farmer (2020, February), "Technological interdependencies predict innovation dynamics," available at https://arxiv.org/abs/2003.00580.

Pichler, A., M. Pangallo, R. M. del Rio-Chanona, F. Lafond, and J. D. Farmer (2020, May), "Production networks and epidemic spreading: How to restart the UK economy?" INET Oxford Working Paper No. 2020-12, available at www.inet.ox.ac.uk/publications/no-2020-12-production-networks-and-epidemic-spreading-how-torestart-the-uk-economy/.

Pichler, A., M. Pangallo, R. M. del Rio-Chanona, F. Lafond, and J. D. Farmer (2021), "In and out of lockdown: Propagation of supply and demand shocks in a dynamic input-output model," available at https://arxiv.org/abs/2102.09608.

Pichler, A., S. Poledna, and S. Thurner (2021), "Systemic risk-efficient asset allocations: Minimization of systemic risk as a network optimization problem," *Journal of Financial Stability*, 52, 100809.

Platt, D. (2020), "A comparison of economic agent-based model calibration methods," *Journal of Economic Dynamics and Control*, 113, 103859.

Potter, S., and T. Schuermann (2020), "Stressing the Fed stress tests against COVID-19," Working Paper, available at https://papers.ssrn.com/sol3/papers.cfm?abstract_id=3635187.

Pozsar, Z., T. Adrian, A. Ashcraft, and H. Boesky (2010), "Shadow banking," Federal Reserve Bank of New York Staff Report 458.

Romano, R. (2004), "The Sarbanes-Oxley Act and the making of quack corporate governance," *Yale Law Journal*, 114, 1521.

Scholl, M. P., A. Calinescu, and J. D. Farmer (2021), "How market ecology explains market malfunction," *PNAS*, 118(26) e2015574118.

Shleifer, A., and R. W. Vishny (1997), "The limits of arbitrage," *Journal of Finance*, 52(1), 35–55.

Thurner, S., J. D. Farmer, and J. Geanakoplos (2012), "Leverage causes fat tails and clustered volatility," *Quantitative Finance*, 12(5), 695–707.

Vickers, J. (2019), "The case for market-based stress tests," *Journal of Financial Regulation*, 5(2), 239–248.

Vickers, J. S. (2011), "*Independent Commission on Banking final report: Recommendations*," The Stationery Office.

Way, R., P. Mealy, and J. D. Farmer (2020), "Estimating the costs of energy transition scenarios using probabilistic forecasting methods," INET Oxford Working Paper No. 2021–01.

Wetzer, T. (2019), "In two minds: the governance of ring-fenced banks," *Journal of Corporate Law Studies*, 19(1), 197–249.

Wetzer, T. (2021), "Segmented banks," Working Paper, forthcoming.

Wiersema, G., A. M. Kleinnijenhuis, T. Wetzer, and J. D. Farmer (2019), "Scenario-free analysis of financial stability with interacting contagion channels," INET Working Paper Series.

Wu, F., D. Zhang, and Q. Ji (2021), "Systemic risk and financial contagion across top global energy companies," *Energy Economics*, 97, 105221.

Xu, C., T. A. Kohler, T. M. Lenton, J.-C. Svenning, and M. Scheffer (2020), "Future of the human climate niche," *Proceedings of the National Academy of Sciences*, 117(21), 11350–11355.

Yellen, J. L. (2013, January 4), "Interconnectedness and systemic risk: Lessons from the financial crisis and policy implications," Speech at the American Economic Association/American Finance Association Joint Luncheon, San Diego, CA, available at www.federalreserve.gov/newsevents/speech/yellen20130104a.htm.

Index